Springer Proceedings in Physics

Volume 162

More information about this series at http://www.springer.com/series/361

Kaoru Yamanouchi · Steven Cundiff
Regina de Vivie-Riedle
Makoto Kuwata-Gonokami
Louis DiMauro
Editors

Ultrafast Phenomena XIX

Proceedings of the 19th International
Conference, Okinawa Convention Center,
Okinawa, Japan, July 7–11, 2014

 Springer

Editors
Kaoru Yamanouchi
Department of Chemistry
The University of Tokyo
Tokyo
Japan

Makoto Kuwata-Gonokami
Department of Physics
The University of Tokyo
Tokyo
Japan

Steven Cundiff
JILA
University of Colorado
Boulder, CO
USA

Louis DiMauro
Department of Physics
The Ohio State University
Columbus, OH
USA

Regina de Vivie-Riedle
Department of Chemistry
Ludwig-Maximilians-University
Munich
Germany

ISSN 0930-8989   ISSN 1867-4941   (electronic)
Springer Proceedings in Physics
ISBN 978-3-319-13241-9   ISBN 978-3-319-13242-6   (eBook)
DOI 10.1007/978-3-319-13242-6

Library of Congress Control Number: 2014957404

Springer Cham Heidelberg New York Dordrecht London

Printed on acid-free paper

Springer International Publishing AG Switzerland is part of Springer Science+Business Media (www.springer.com)

Preface

This volume is a compilation of research papers presented at the Nineteenth International Conference on Ultrafast Phenomena held at the Okinawa Convention Center, Okinawa, from July 7 to 11, 2014. The Ultrafast Phenomena conferences are held every two years and are the premier international forum for discussion of the latest and most important results in ultrafast science. These meetings bring together researchers from a variety of research fields in laser science and engineering to deliberate the latest advances in ultrafast optics and their applications. The conferences and associated published proceedings effectively disseminate the most recent scientific advances using ultrashort coherent light pulses. More than 327 papers were presented at *Ultrafast Phenomena XIX*.

Significant progress in attosecond pulses and high-order harmonic generation and their applications were reported. The response of atoms and molecules to intense ultrashort-pulsed laser pulses was discussed. Interfacial phenomena, photovoltaic, and light-harvesting systems were popular topics. Advances in time-resolved electron and x-ray diffraction and spectroscopy were presented, providing detailed information on atomic and electronic dynamics in molecular systems and solids. New light sources in THz and x-ray regions were described, opening new research frontiers of ultrafast phenomena. These examples are but a small subset of the research summaries gathered in this volume, which provides a valuable synopsis of the recent advances and impact of ultrafast technology in illuminating fundamental processes being investigated in physics, chemistry, and biology.

There were more than 370 attendees at the meeting, and among them more than 100 were graduate students. The presence of the young attendees energized the discussion during the conference, and the discussion was further enhanced by the beautiful ocean side setting in Okinawa. This year, unpredictably, the sessions scheduled to be held on Tuesday were postponed by the typhoon Neoguri, but thanks to the cooperation of all the attendees, all the scheduled sessions and events were rearranged and went through smoothly.

Many people, organizations, and sponsor companies made invaluable contributions to the success of the conference. The international program committee reviewed over 400 submissions, and organized the scientific program. The conference was co-organized by Japan Intense Light Field Science Society, Center for Ultrafast Intense Laser Science, the University of Tokyo. The technical program was supported by the Optical Society of America, and the conference was supported by the European Physical Society and Innovative Center for Coherent Photon Technology, the University of Tokyo.

We are particularly grateful to Ms. Chie Sakuta and Ms. Ayane Maezawa for coordinating the submissions of the proceedings manuscripts and for their help in compiling them in the form of this book.

General Co-chairs

Tokyo
Boulder
Munich

Kaoru Yamanouchi

Steven Cundiff

Regina de Vivie-Riedle

Program Co-chairs

Tokyo
Columbus

Makoto Kuwata-Gonokami

Louis DiMauro

Contents

Part V Chemistry—Liquid Phase

Contributors

K. Abe Department of Physics, Graduate School of Science, Tohoku University, Aoba-Ku, Sendai, Japan

Shunsuke Adachi Department of Chemistry, Graduate School of Science, Kyoto University, Kyoto, Japan; RIKEN Center for Advanced Photonics, Wako, Japan

M. Ahmed Lawrence Berkeley National Laboratory, Chemical Sciences Division, University of California, Berkeley, CA, USA

Kyoko Aikawa Graduate School of Science, Kobe University, Kobe, Japan

S. Akbar Ali Physics Department, Indian Institute of Technology Madras, Chennai, India

I.A. Akimov Experimentelle Physik 2, Technische Universität Dortmund, Dortmund, Germany

Skirmantas Ališauskas Photonics Institute, Vienna University of Technology, Vienna, Austria

A.E. Almand-Hunter JILA, National Institute of Standards and Technology, University of Colorado, Boulder, CO, USA

A.S. Alnaser Max-Planck-Institut für Quantenoptik, Garching, Germany; Department of Physics and Astronomy, King-Saud University, Riyadh, Saudi Arabia

Benjamín Alonso Universidad de Salamanca, Salamanca, Spain

Jan Alster Department of Chemical Physics, Lund University, Lund, Sweden

A. Amani Eilanlou RIKEN Center for Advanced Photonics, Saitama, Japan

Michele Amato Institut d'Électronique Fondamentale, UMR8622, CNRS, Universitè Paris-Sud, Orsay, France

Yorai Amit The Institute of Chemistry and the Center for Nanoscience and Nanotechnology, Hebrew University, Jerusalem, Israel

Erik Anderson Lawrence Berkeley National Laboratory, Center for X-Ray Optics, Berkeley, CA, USA

G. Andriukaitis Photonics Institute, Vienna University of Technology, Vienna, Austria

S. Anumula Department of Physics, Politecnico di Milano, Milan, Italy

Kotaro Araki Department of Physics, Faculty of Engineering, Yokohama National University, Yokohama, Japan

M. Arbeiter Institute of Physics, University of Rostock, Rostock, Germany

F. Ardana-Lamas SwissFEL, Paul Scherrer Institute, Villigen, Switzerland

Masaaki Ashida Department of Materials Engineering Science, Osaka University, Osaka, Japan

A. Auger Kapteyn-Murnane Labs Inc, Boulder, CO, USA

D. Ayuso Departamento de Química, Modulo 13, Universidad Autonoma de Madrid, Madrid, Spain

Julia Bahrenburg Institute of Physical Chemistry, Christian-Albrechts-University Kiel, Kiel, Germany

Huib J. Bakker FOM Institute AMOLF, Amsterdam, The Netherlands

Artem A. Bakulin FOM Institute AMOLF, Amsterdam, The Netherlands

T. Balčiūnas Photonics Institute, Vienna University of Technology, Vienna, Austria

Andrius Baltuška Photonics Institute, Vienna University of Technology, Vienna, Austria

André D. Bandrauk Laboratoire de Chimie Théorique, Faculté des Sciences, Université de Sherbrooke, Sherbrooke, QC, Canada

Uri Banin The Institute of Chemistry and the Center for Nanoscience and Nanotechnology, Hebrew University, Jerusalem, Israel

Dmitry Baranov Department of Chemistry and Biochemistry, University of Colorado, Boulder, CO, USA

M. Barthelemy Institut de Physique et Chimie Des Matériaux de Strasbourg, Unité Mixte 7504 CNRS, Université de Strasbourg, Strasbourg, France

P. Bartz Fakultät für Physik, Universität Bielefeld, Bielefeld, Germany

G. Batignani Physics Department, University "Sapienza", Rome, Italy

Victor S. Batista Department of Chemistry, Yale University, New Haven, CT, USA

P. Baum Ludwig-Maximilians-Universität München, Garching, Germany; Max-Planck-Institute of Quantum Optics, Garching, Germany

M. Bayer Experimentelle Physik 2, Technische Universität Dortmund, Dortmund, Germany

P. Beaud Swiss Light Source, Paul Scherrer Institut, Villigen PSI, Switzerland

Samuel Beaulieu Centre ÉMT, Institut National de la Recherche Scientifique, Quebec, QC, Canada

A. Jaron-Becker JILA and Department of Physics, University of Colorado at Boulder, Boulder, CO, USA

L. Belshaw School of Maths and Physics, Centre for Plasma Physics, Queen's University Belfast, Belfast, UK

F. Benabid GPPMM Group, Xlim Research Institute, CNRS UMR 7252, University of Limoges, Limoges, France

I. Ben-Itzhak James R. Macdonald Laboratory, Physics Department, Kansas State University, Manhattan, KS, USA

K.K. Berggren Research Laboratory of Electronics, Department of EECS, Massachusetts Institute of Technology, Cambridge, MA, USA

B. Bergues Max-Planck-Institut für Quantenoptik, Garching, Germany

Paolo Biagioni Dipartimento di Fisica, Politecnico di Milano, Milan, Italy

Jason D. Biggs Department of Chemistry, University of California, Irvine, CA, USA

J.-Y. Bigot Institut de Physique et Chimie Des Matériaux de Strasbourg, Unité Mixte 7504 CNRS, Université de Strasbourg, Strasbourg, France

Thomas Binhammer VENTEON Laser Technologies GmbH, Hannover, Germany

Prem B. Bisht Physics Department, Indian Institute of Technology Madras, Chennai, India

Éric Bisson Centre ÉMT, Institut National de la Recherche Scientifique, Quebec, QC, Canada

H. Bluhm Lawrence Berkeley National Laboratory, Chemical Sciences Division, University of California, Berkeley, CA, USA

R. Boge Physics Department, ETH Zurich, Zurich, Switzerland

Mischa Bonn Max Planck Institute for Polymer Research, Mainz, Germany

Andreas Borgschulte Swiss Federal Laboratories for Materials Testing and Research, Laboratory for Hydrogen and Energy, EMPA, Dübendorf, Switzerland

Rocìo Borrego-Varillas Instituto de Nuevas Tecnologías de la Imagen (INIT), Castellón, Spain; Departamento de Física Aplicada, Universidad de Salamanca, Salamanca, Spain; IFN-CNR, Dipartimento Di Fisica, Politecnico Di Milano, Milan, Italy; Universitat Jaume I, Castelló de la Plana, Spain

Maximilian Bradler LS für BioMolekulare Optik, Ludwig-Maximilians-Universität München, Munich, Germany

M. Braun Institute of Physical and Theoretical Chemistry, Johann Wolfgang Goethe-University, Frankfurt, Germany

Jens Bredenbeck Institute of Biophysics, University of Frankfurt, Frankfurt, Germany

Marshall T. Bremer Department of Physics and Astronomy, Michigan State University, East Lansing, MI, USA

Christian Bressler European XFEL, Hamburg, Germany

D. Brida Department of Physics and Center for Applied Photonics, University of Konstanz, Konstanz, Germany; IFN-CNR, Dipartimento di Fisica, Politecnico di Milano, Milan, Italy

Tobias Brixner Universität Würzburg, Würzburg, Germany

B.D. Bruner Department of Physics of Complex Systems, Weizmann Institute of Science, Rehovot, Israel

D.B. Bucher BioMolecular Optics, Department of Chemistry, Center for Integrated Protein Science, Ludwig-Maximilians-Universität München, Munich, Germany

Philip H. Bucksbaum SLAC National Accelerator Laboratory, Stanford PULSE Institute, Menlo Park, CA, USA

T. Buckup Physikalisch-Chemisches Institut, Ruprecht-Karls-Universität Heidelberg, Heidelberg, Germany

Nediljko Budisa Department of Chemistry, Technical University Berlin, Biocatalysis Group, Berlin, Germany

David Cahen Weizmann Institute of Science, Rehovot, Israel

James F. Cahoon Department of Chemistry, University of North Carolina, Chapel Hill, NC, USA

F. Calegari IFN-CNR, Milan, Italy

Anne-Laure Calendron Department of Physics, Center for Free-Electron Laser Science, Deutsches Elektronen Synchrotron, University of Hamburg, Hamburg,

Germany; The Hamburg Centre for Ultrafast Imaging, Universität Hamburg, Hamburg, Germany

F. Campi Department of Physics, Lund University, Lund, Sweden

Alessia Candeo IFN-CNR, Dipartimento Di Fisica, Politecnico Di Milano, Milan, Italy

Hüseyin Çankaya Department of Physics, Center for Free-Electron Laser Science, Deutsches Elektronen Synchrotron, University of Hamburg, Hamburg, Germany; The Hamburg Centre for Ultrafast Imaging, Universität Hamburg, Hamburg, Germany

W. Cao James R. Macdonald Laboratory, Physics Department, Kansas State University, Manhattan, KS, USA

Benjamin W. Caplins Department of Chemistry, University of California, Berkeley, CA, USA; Chemical Sciences Division, Lawrence Berkeley National Laboratory, Berkeley, CA, USA

T. Carell Department of Chemistry, Center for Integrated Protein Science, Ludwig-Maximilians-Universität München, Munich, Germany

Anne-Marie Carey Institute for Molecular, Cell and Systems Biology, University of Glasgow, Glasgow, UK

Luca Castiglioni Department of Physics, University of Zurich, Zurich, Switzerland

Pedro J. Castro Departement de Química Física I Inorgànica, Tarragona, Spain

Emma E.M. Cating Department of Chemistry, University of North Carolina, Chapel Hill, NC, USA

Andrea Cattoni Laboratoire de Photonique et de Nanostructures, Route de Nozay, Marcoussis, France

A. Cavalleri Max Planck Institute for the Structure and Dynamics of Matter, Hamburg, Germany; Clarendon Laboratory, Department of Physics, Oxford University, Oxford, UK

A. Caviezel Swiss Light Source, Paul Scherrer Institute, Villigen, Switzerland

Giulio Cerullo Dipartimento Di Fisica, Politecnico Di Milano, Milan, Italy; IFN-CNR, Dipartimento di Fisica, Politecnico di Milano, Milan, Italy

Deying Chen National Key Laboratory of Science and Technology on Tunable Laser, Harbin Institute of Technology, Harbin, China

M.-C. Chen JILA and Department of Physics, University of Colorado at Boulder, Boulder, CO, USA; Institute of Photonics Technologies, National Tsing Hua University, Hsinchu, Taiwan

S. Chen Department of Physics and Astronomy, Louisiana State University, Baton Rouge, LA, USA

Zhuoying Chen ESPCI/CNRS/UPMC UMR 8213, Paris, France

Ya Cheng State Key Laboratory of High Field Laser Physics, Shanghai Institute of Optics and Fine Mechanics, Chinese Academy of Sciences, Shanghai, China

K. Chevalier Department of Physics, TU Kaiserslautern, Kaiserslautern, Germany

Shih-Hsuan Chia Center for Free-Electron Laser Science, Deutsches Elektronen-Synchrotron DESY, Hamburg, Germany; The Hamburg Center for Ultrafast Imaging, Hamburg, Germany

D. Chiappe Laboratorio MDM, IMM-CNR, Agrate Brianza, Italy

Alex Chin Cavendish Laboratory, University of Cambridge, Cambridge, UK

L. Chipperfield Max Born Institute, Berlin, Germany

Hana Cho Chemical Sciences Division, Lawrence Berkeley National Laboratory, Berkeley, CA, USA

Byungmoon Cho Department of Chemistry and Biochemistry, University of Colorado, Boulder, CO, USA

Joseph D. Christesen Department of Chemistry, University of North Carolina, Chapel Hill, NC, USA

Wei Chu State Key Laboratory on Integrated Optoelectronics, College of Electronic Science and Engineering, Jilin University, Changchun, China; State Key Laboratory of High Field Laser Physics, Shanghai Institute of Optics and Fine Mechanics, Chinese Academy of Sciences, Shanghai, China

E. Cinquanta Laboratorio MDM, IMM-CNR, Agrate Brianza, Italy

C. Cirelli Physics Department, ETH Zurich, Zurich, Switzerland

Giovanni Cirmi Center for Free-Electron Laser Science, Deutsches Elektronen-Synchrotron DESY, Hamburg, Germany; The Hamburg Center for Ultrafast Imaging, Hamburg, Germany

C.L. Cocke James R. Macdonald Laboratory, Physics Department, Kansas State University, Manhattan, KS, USA

Richard J. Cogdell Institute for Molecular, Cell and Systems Biology, University of Glasgow, Glasgow, UK

Oren Cohen Solid State Institute and Physics Department, Technion, Haifa, Israel

Eric Collet Institut de Physique de Rennes, Campus de Beaulieu, University Rennes 1, Rennes, France

S.D. Conte Istituto Di Fotonica E Nanotecnologie IFN-CNR, Milan, Italy

Amy Cordones-Hahn Chemical Sciences Division, Lawrence Berkeley National Laboratory, Berkeley, CA, USA

Paul Corkum Joint Laboratory for Attosecond Science of the National Research Council, University of Ottawa, Ottawa, Canada

Jérôme Cornil Service de Chimie des Materiaux Nouveaux, Université de Mons, Mons, Belgium

Rene Costard Max Born Institut für Nichtlineare Optik und Kurzzeitspektroskopie, Berlin, Germany

Trevor L. Courtney Department of Chemistry, University of Washington, Seattle, Washington, USA; Department of Chemistry and Biochemistry, University of Colorado, Boulder, CO, USA

Olivier Crégut Institut de Physique et Chimie Des Matériaux de Strasbourg, Strasbourg, France

S.T. Cundiff JILA, National Institute of Standards and Technology, University of Colorado, Boulder, CO, USA

Ana V. Cunha Zernike Institute for Advanced Materials, University of Groningen, Groningen, The Netherlands

Paul M.G. Curmi School of Physics and Centre for Applied Medical Research, St. Vincents Hospital, The University of New South Wales, Sydney, NSW, Australia

S. Dal Conte IFN-CNR, Milan, Italy; Dipartimento di Fisica, Politecnico di Milano, Milan, Italy

Marcos Dantus Department of Chemistry, Michigan State University, East Lansing, MI, USA; Biophotonic Solutions Inc, East Lansing, MI, USA; Department of Physics and Astronomy, Michigan State University, East Lansing, MI, USA

L.V. Dao Centre for Quantum and Optical Science, Swinburne University of Technology, Melbourne, Australia

Thang Duy Dao International Center for Materials Nanoarchitectonics, National Institute for Materials Science, Tsukuba, Japan; CREST, Japan Science and Technology Agency, Kawaguchi, Japan

Jeffrey A. Davis Centre for Quantum and Optical Science, Swinburne University of Technology, Melbourne, Australia

K.B. Dinh Centre for Quantum and Optical Science, Swinburne University of Technology, Melbourne, Australia

J.R. Vázquez de Aldana Instituto de Nuevas Tecnologías de la Imagen (INIT), Castellón, Spain

S. De Camillis School of Maths and Physics, Centre for Plasma Physics, Queen's University Belfast, Belfast, UK

Sandro De Silvestri IFN-CNR, Dipartimento Di Fisica, Politecnico Di Milano, Milan, Italy

Antonietta De Sio Center of Interface Science, Institut für Physik, Carl von Ossietzky Universität, Oldenburg, Germany

Regina de Vivie-Riedle Department of Chemistry, Ludwig-Maximilians-Universität München, Munich, Germany

Tushar Debnath Radiation and Photochemistry Division, Bhabha Atomic Research Centre, Mumbai, India

P. Decleva Dipartimento di Scienze Chimiche e Farmaceutiche, CNR-IOM, Università di Trieste, Trieste, Italy

J. Demsar Physics Department, Universitaet Konstanz, Konstanz, Germany; Institute of Physics, Ilmenau University of Technology, Ilmenau, Germany

Yunpei Deng Fritz-Haber Institut der Max Planck Gesellschaft, Berlin, Germany

P.M. Derlet Paul Scherrer Institute, Condensed Matter Theory Group, Villigen, Switzerland

Detlef Diesing Fakultät für Chemie, Universität Duisburg-Essen, Essen, Germany

M. Devetta Instituto di Fotonica e Nanotecnologie-CNR, Milan, Italy

Dominik Differt Fakultät für Physik, Universität Bielefeld, Bielefeld, Germany

R. Diller Department of Physics, TU Kaiserslautern, Kaiserslautern, Germany

Chengyuan Ding JILA and Department of Physics, University of Colorado, Boulder, CO, USA

Katharina Doblhoff-Dier Institute for Physical Chemistry, Friedrich-Schiller University Jena, Jena, Germany

Franklin Dollar JILA and Department of Physics, University of Colorado, Boulder, CO, USA

Péter Dombi Max-Planck-Institut für Quantenoptik, Garching, Germany; MTA "Lendület" Ultrafast Nanooptics Group, Wigner Research Centre for Physics, Budapest, Hungary

Pablo Nahuel Dominguez BioMolekulare Optik and Center of Integrated Protein Science, CIPSM Ludwig-Maximilians-Universität München, Munich, Germany

Shuo Dong School of Physical and Mathematical Sciences, Nanyang Technological University, Nanyang, Singapore

Zhiwei Dong National Key Laboratory of Science and Technology on Tunable Laser, Harbin Institute of Technology, Harbin, China

Konstantin E. Dorfman Department of Chemistry, University of California, Irvine, CA, USA

C. Dornes Institute for Quantum Electronics, Physics Department, ETH Zurich, Zurich, Switzerland

H. Dube Department of Chemistry, Ludwig-Maximilians-University, Munich, Germany

N. Dudovich Department of Physics of Complex Systems, Weizmann Institute of Science, Rehovot, Israel

Patrick Durkin Department of Chemistry, Technical University Berlin, Biocatalysis Group, Berlin, Germany

Hermann A. Dürr Stanford Institute for Materials and Energy Sciences, SLAC National Accelerator Laboratory, Menlo Park, CA, USA

R.J. Dwayne Miller Departments of Chemistry and Physics, University of Toronto, Toronto, ON, Canada; Max Planck Institute for the Structure and Dynamics of Matter, Hamburg, Germany

K.E. Echternkamp Physical Institute, University of Göttingen, Göttingen, Germany

Dassia Egorova Institut für Physikalische Chemie, Christian-Albrechts-Universität zu Kiel, Kiel, Germany

Tamar Eliash Department of Organic Chemistry, Weizmann Institute Rehovot, Rehovot, Israel

K. Ellen Keister JILA and Department of Physics, University of Colorado, Boulder, CO, USA

Jennifer L. Ellis JILA and Department of Physics, University of Colorado, Boulder, CO, USA

Thomas Elsaesser Max-Born-Institut Für Nichtlineare Optik Und Kurzzeitspektroskopie, Berlin, Germany

T. Endo Department of Chemistry, Nagoya University, Nagoya, Aichi, Japan

J.W. Engels Institute of Organic Chemistry and Chemical Biology, Johann Wolfgang Goethe-University, Frankfurt, Germany

Sonia Erattupuzha Photonics Institute, Vienna University of Technology, Vienna, Austria

R. Ernstorfer Fritz-Haber-Institut der MPG, Berlin, Germany

Martin Essig Institute of Biophysics, University of Frankfurt, Frankfurt, Germany

D. Facialá Dipartimento di Fisica, Politecnico di Milano, Milan, Italy

Sarah M. Falke Institut für Physik, Carl von Ossietzky Universität, Oldenburg, Germany; Center of Interface Science, Carl von Ossietzky Universität, Oldenburg, Germany

Arya Fallahi Center for Free-Electron Laser Science, Deutsches Elektronen-Synchrotron DESY, Hamburg, Germany; The Hamburg Center for Ultrafast Imaging, University of Hamburg, Hamburg, Germany

G. Fan Photonics Institute, Vienna University of Technology, Vienna, Austria

M. Fanciulli Laboratorio MDM, IMM-CNR, Agrate Brianza, Italy; Dipartimento Di Scienza Dei Materiali, Università Degli Studi Di Milano-Bicocca, Milan, Italy

Shaobo Fang Center for Free-Electron Laser Science, Deutsches Elektronen-Synchrotron DESY, Hamburg, Germany; The Hamburg Center for Ultrafast Imaging, Hamburg, Germany

Paolo Farinello IFN-CNR, Dipartimento di Fisica, Politecnico di Milano, Milan, Italy

Adam Faust The Institute of Chemistry and the Center for Nanoscience and Nanotechnology, Hebrew University, Jerusalem, Israel

Elena Fedulova Max-Planck-Institut für Quantenoptik, Garching, Germany

Xinliang Feng Max Planck Institute for Polymer Research, Mainz, Germany

Franziska Fennel Institute of Physics, University of Rostock, Rostock, Germany

T. Fennel Institute of Physics, University of Rostock, Rostock, Germany

C. Ferrante Physics Department, University "Sapienza", Rome, Italy

A. Ferrer Institute for Quantum Electronics, Physics Department, ETH Zurich, Zurich, Switzerland; Swiss Light Source, Paul Scherrer Institut, Villigen PSI, Switzerland

Henk Fidder Max Born Institut für Nichtlineare Optik und Kurzzeitspektroskopie, Berlin, Germany

S. Fiechter Institut für Solare Brennstoffe, Helmholtz Zentrum Berlin, Hahn-Meitner-Platz 1, Berlin, Germany

R.L. Field Departments of Chemistry and Physics, University of Toronto, Toronto, ON, Canada; Centre for Free Electron Laser Science, Max Planck Institute for the Structure and Dynamics of Matter, Hamburg, Germany

Benjamin P. Fingerhut Max-Born-Institut für Nichtlineare Optik und Kurzzeit-spektroskopie, Berlin, Germany

J. Fischer Department of Physics and Center for Applied Photonics, University of Konstanz, Konstanz, Germany

Avner Fleischer Solid State Institute and Physics Department, Technion, Haifa, Israel; Department of Physics and Optical Engineering, Ort Braude College, Karmiel, Israel

Graham R. Fleming Department of Chemistry, University of California, Berkeley, CA, USA; Physical Biosciences Division, Lawrence Berkeley National Laboratory, Berkeley, CA, USA

Christos Flytzanis Laboratoire Pierre Aigrain, École Normale Supérieure, Paris, France

Paul Fons Nanoelectronics Research Institute, National Institute of Advanced Industrial Science and Technology, Tsukuba Central 4, Tsukuba, Japan

Mark E. Foord Physics Division, Physical and Life Sciences, Lawrence Livermore National Laboratory, Livermore, CA, USA

Andrew Forbes Council for Scientific and Industrial Research, Pretoria, South Africa

C. Fourcade-Dutin GPPMM Group, Xlim Research Institute, CNRS UMR 7252, University of Limoges, Limoges, France

Emmanuel P. Fowe Laboratoire de Chimie Théorique, Faculté des Sciences, Université de Sherbrooke, Sherbrooke, QC, Canada

Zachary W. Fox Department of Chemistry, University of Washington, Seattle, Washington, USA

F. Frassetto IFN-CNR, Padua, Italy

Frank Friedriszik Institute of Physics, University of Rostock, Rostock, Germany

Richard H. Friend University of Cambridge, Cambridge, UK

Takao Fuji Institute for Molecular Science, Okazaki, Japan

Naoki Fujii Department of Chemistry, Graduate School of Science, Osaka University, Toyonaka, Osaka, Japan

R. Fujii The Osaka City University Advanced Research Institute for Natural Science and Technology (OCARINA), Osaka, Japan; JST/PRESTO, Kawaguchi, Saitama, Japan

Yuichi Fujimura Department of Chemistry, Graduate School of Science, Tohoku University, Sendai, Japan; Department of Applied Chemistry, Institute of Molecular Science and Center for Interdisciplinary Molecular Science, National Chiao-Tung University, Hsin-Chu 300, Taiwan

Tomotsumi Fujisawa Molecular Spectroscopy Lab, RIKEN, Wako, Saitama, Japan

Takehisa Fujiwara Department of Applied Physics, Hokkaido University, Sapporo, Kita-ku, Japan

K. Fukumoto Tokyo Institute of Technology, Meguro-Ku, Tokyo, Japan; JST-CREST, Kawaguchi, Saitama, Japan

Y. Furukawa RIKEN Ceneter for Advanced Photonics, Wako, Saitama, Japan

M. Fushitani RIKEN, Sayo, Hyogo, Japan; Department of Chemistry, Nagoya University, Nagoya, Aichi, Japan

S. Fusi Dipartimento Di Chimica, Università Degli Studi Di Siena, Siena, Italy

M.B. Gaarde Department of Physics and Astronomy, Louisiana State University, Baton Rouge, LA, USA

Michelle M. Gabriel Department of Chemistry, University of North Carolina at Chapel Hill, Chapel Hill, NC, USA

Jim A. Gaffney Physics Division, Physical and Life Sciences, Lawrence Livermore National Laboratory, Livermore, CA, USA

Isabel Gallardo-González Centro de Láseres Pulsados, Salamanca, Spain

Lukas Gallmann Department of Physics, Institute for Quantum Electronics, ETH Zurich, Zurich, Switzerland; Institute of Applied Physics, University of Bern, Bern, Switzerland

B. Galloway JILA and Department of Physics, University of Colorado at Boulder, Boulder, CO, USA

Meng Gao Departments of Chemistry and Physics, University of Toronto, Toronto, ON, Canada; Max Planck Institute for the Structure and Dynamics of Matter, Hamburg, Germany

Alastair Gardiner Institute of Molecular Cell and Systems Biology, University of Glasgow, Glasgow, Scotland

Fabien Gatti CTMM, Institut Charles Gerhardt Montpellier, Montpellier, France

W. Gawelda European XFEL, Hamburg, Germany

Itay Gdor Institute of Chemistry, Hebrew University, Jerusalem, Israel

Henning Geiseler The Institute for Solid State Physics, The University of Tokyo, Kashiwa, Chiba, Japan

Simon Gelinas University of Cambridge, Cambridge, UK

F. Gérôme GPPMM Group, Xlim Research Institute, CNRS UMR 7252, University of Limoges, Limoges, France

O. Gessner Ultrafast X-Ray Science Laboratory, University of California, Berkeley, CA, USA; Chemical Sciences Division, Lawrence Berkeley National Laboratory, University of California, Berkeley, CA, USA

Hirendra Nath Ghosh Radiation and Photochemistry Division, Bhabha Atomic Research Centre, Mumbai, India

I. Gierz Max Planck Institute for the Structure and Dynamics of Matter, Hamburg, Germany

Mathieu Giguére Centre ÉMT, Institut National de la Recherche Scientifique, Quebec, QC, Canada

E. Gindensperger Laboratoire de Chimie Quantique, Institut de Chimie, CNRS - Université de Strasbourg, Strasbourg, France

P.-A. Glans Advanced Light Source, University of California, Berkeley, CA, USA

A. Gliserin Ludwig-Maximilians-Universität München, Garching, Germany; Max-Planck-Institute of Quantum Optics, Garching, Germany

Adelheid Godt Fakultät für Chemie and Center for Molecular Materials, Universität Bielefeld, Bielefeld, Germany

D. Golde Department of Physics, University of Marburg, Marburg, Germany

Leticia González's Institute of Theoretical Chemistry, University of Vienna, Vienna, Austria

Kazuki Goto Department of Physics, Tohoku University, Sendai, Japan

Stefanie Gräfe Institute for Physical Chemistry, Friedrich-Schiller University Jena, Jena, Germany

F.F. Graupner Faculty of Physics, Center for Integrative Protein Science, Ludwig-Maximilians-University Munich, Munich, Germany

Alexander X. Gray Stanford Institute for Materials and Energy Sciences, SLAC National Accelerator Laboratory, Menlo Park, CA, USA; Department of Physics, Temple University, Philadelphia, PA, USA

C. Grazianetti Laboratorio MDM, IMM-CNR, Agrate Brianza, Italy

J. Greenwood Centre for Plasma Physics, School of Maths and Physics, Queen's University Belfast, Belfast, UK

Michael Greif Department of Physics, University of Zurich, Zurich, Switzerland

Christian Greve Max Born Institut für Nichtlineare Optik und Kurzzeitspektroskopie, Berlin, Germany

Jakob Grilj Laboratory de Spectroscopie Ultrarapide, Ecole Polytechnique Federale de Lausanne, Lausanne, Switzerland; Stanford PULSE Institute, SLAC National Accelerator Laboratory, Menlo Park, CA, USA

Petra Groß Institut für Physik, Carl von Ossietzky Universität, Oldenburg, Germany; Center of Interface Science, Carl von Ossietzky Universität, Oldenburg, Germany

S. Gruebel Swiss Light Source, Paul Scherrer Institut, Villigen PSI, Switzerland

Erik M. Grumstrup Department of Chemistry, University of North Carolina at Chapel Hill, Chapel Hill, NC, USA

C. Grünewald Institute of Organic Chemistry and Chemical Biology, Johann Wolfgang Goethe-University, Frankfurt a. M., Germany

A. Grupp Department of Physics and Center for Applied Photonics, University of Konstanz, Konstanz, Germany

Patrik Grychtol Department of Physics and JILA, University of Colorado, Boulder, CO, USA

Xiaokun Gu Department of Mechanical Engineering, University of Colorado, Boulder, CO, USA

M. Gueye Institut de Physique et Chimie Des Matériaux de Strasbourg & Labex NIE, CNRS - Université de Strasbourg, Strasbourg, France

M. Gühr Stanford PULSE Institute, SLAC National Accelerator Laboratory, Menlo Park, CA, USA

M. Gulde 4. Physical Institute, University of Göttingen, Göttingen, Germany

J.-H. Guo Advanced Light Source, University of California, Berkeley, CA, USA

Lanjun Guo Institute of Modern Optics, Nankai University, Key Laboratory of Optical Information Science and Technology, Ministry of Education, Tianjin, China

S. Haacke Institut de Physique et Chimie des Matériaux de Strasbourg, Strasbourg University—CNRS, Strasbourg, France

S. Haessler Photonics Institute, Vienna University of Technology, Vienna, Austria

R.F. Haglund Jr. Department of Physics and Astronomy, Vanderbilt University, Nashville, TN, USA

Christopher R. Hall Centre for Quantum and Optical Science, Swinburne University of Technology, Melbourne, Australia

Norio Hamada Science & Technology Entrepreneurship Laboratory, Osaka University, Suita, Japan

P. Hannaford Centre for Quantum and Optical Science, Swinburne University of Technology, Melbourne, Australia

Yu Harabuchi Department of Chemistry, Hokkaido University, Sapporo, Kita-Ku, Japan

Charles B. Harris Department of Chemistry, University of California, Berkeley, CA, USA; Chemical Sciences Division, Lawrence Berkeley National Laboratory, Berkeley, CA, USA

Muneaki Hase Nanoelectronics Research Institute, National Institute of Advanced Industrial Science and Technology, Tsukuba Central 4, Tsukuba, Japan; Institute of Applied Physics, University of Tsukuba, Tsukuba, Japan

Hirokazu Hasegawa Department of Integrated Sciences, Graduate School of Arts and Sciences, The University of Tokyo, Meguro-ku, Tokyo, Japan

H. Hashimoto The Osaka City University Advanced Research Institute for Natural Science and Technology (OCARINA), Osaka, Japan; Department of Physics, Graduate School of Science, Osaka City University, Osaka, Japan

C.P. Hauri SwissFEL, Paul Scherrer Institute, Villigen-PSI, Switzerland; Ecole Polytechnique Federale de Lausanne, Lausanne, Switzerland

Ping He College of Foundation Science, Harbin University of Commerce, Harbin, China

Christoph T. Hebeisen National Research Council of Canada, Ottawa, ON, Canada; Department of Physics, University of Ottawa, Ottawa, ON, Canada

Ingo Heesemann Fakultät für Chemie and Center for Molecular Materials, Universität Bielefeld, Bielefeld, Germany

Timo Hefner Institute for Physical and Theoretical Chemistry, Department of Chemistry and Pharmacy, University of Wuerzburg, Würzburg, Germany

U. Heinzmann Fakultät für Physik, Universität Bielefeld, Bielefeld, Germany

Ismael A. Heisler Max-Born-Institut f. Nichtlineare Optik Und Kurzzeitspektroskopie, Berlin, Germany

Jan Helbing Department of Chemistry, University of Zurich, Zurich, Switzerland

Matthias Hengsberger Department of Physics, University of Zurich, Zurich, Switzerland

Sarah Henry Institute for Molecular, Cell and Systems Biology, University of Glasgow, Glasgow, UK

Matthias Hensen Fakultät für Physik, Universität Bielefeld, Bielefeld, Germany

G. Herink 4. Physical Institute, University of Göttingen, Göttingen, Germany

Jorge N. Hernandez-Charpak JILA and Department of Physics, University of Colorado, Boulder, CO, USA

C. Hernández-García JILA and Department of Physics, University of Colorado at Boulder, Boulder, CO, USA; Grupo de Investigación en Óptica Extrema, Universidad de Salamanca, Salamanca, Spain

J. Herrmann Department of Physics, Institute for Quantum Electronics, ETH Zurich, Zurich, Switzerland

Tobias Hertel Institute for Physical and Theoretical Chemistry, Department of Chemistry and Pharmacy, University of Wuerzburg, Würzburg, Germany

J. Herz Physikalisch-Chemisches Institut, Ruprecht-Karls-Universität Heidelberg, Heidelberg, Germany

S. Heuser Physics Department, ETH Zurich, Zurich, Switzerland

E. Heyer Laboratoire de Chimie Organique et Spectroscopies Avances (ICPEES-LCOSA), Strasbourg University—CNRS, Strasbourg, France

Daniel D. Hickstein JILA and Department of Physics, University of Colorado, Boulder, CO, USA

Takuya Higuchi The University of Tokyo, Bunkyo-Ku, Tokyo, Japan

Y. Hikosaka RIKEN, Sayo, Hyogo, Japan; Department of Environmental Science, Niigata University, Niigata, Niigata, Japan

Matthias Himmelstoss BioMolekulare Optik and Center of Integrated Protein Science, CIPSM Ludwig-Maximilians-Universität München, Munich, Germany

Sho Hiraoka Graduate of Science, Kobe University, Kobe, Japan

H. Hirori Institute for Integrated Cell-Material Sciences (WPI-ICeMS), Kyoto University, Kyoto, Sakyo-Ku, Japan; CREST, Japan Science and Technology Agency, Saitama, Kawaguchi, Japan

Kenichi Hirosawa Department of Electronics and Electrical Engineering, Keio University, Kohoku-Ku, Yokohama, Kanagawa, Japan

A. Hishikawa Department of Chemistry, Graduate School of Science, Nagoya University, Nagoya, Aichi, Japan; RIKEN, Sayo, Hyogo, Japan

R.G. Hobbs Department of EECS and Research Laboratory of Electronics, Massachusetts Institute of Technology, Cambridge, MA, USA

C. Hofmann Physics Department, ETH Zurich, Zurich, Switzerland

M. Hohenleutner Department of Physics, University of Regensburg, Regensburg, Germany

Marcel Holtz Max-Born-Institut Für Nichtlineare Optik Und Kurzzeitspektroskopie, Berlin, Germany

Asami Honda Department of Applied Physics, Hokkaido University, Sapporo, Japan

Kathleen Hoogeboom-Pot JILA and Department of Physics, University of Colorado, Boulder, CO, USA

Takuya Horio Department of Chemistry, Graduate School of Science, Kyoto University, Kitashirakawa-Oiwakecho, Sakyo-Ku Kyoto, Japan; RIKEN Center for Advanced Photonics, RIKEN, Wako, Japan; CREST, Japan Science and Technology Agency, Chiyoda-Ku, Tokyo, Japan

Zoltán L. Horváth Department of Optics and Quantum Electronics, University of Szeged, Szeged, Hungary

Aruto Hosaka Department of Electronics and Electrical Engineering, Keio University, Yokohama, Japan

W. Hu Max Planck Institute for the Structure and Dynamics of Matter, Hamburg, Germany

Weijie Hua Department of Chemistry, University of California, Irvine, CA, USA

P.-C. Huang Institute of Photonics Technologies, National Tsing Hua University, Hsinchu, Taiwan

L. Huber Institute for Quantum Electronics, Physics Department, ETH Zurich, Zurich, Switzerland

R. Huber Department of Physics, University of Regensburg, Regensburg, Germany

T. Huber Institute for Quantum Electronics, Physics Department, ETH Zurich, Zurich, Switzerland

C.R. Hunt Max Planck Institute for the Structure and Dynamics of Matter, Hamburg, Germany

U. Huttner Department of Physics, University of Marburg, Marburg, Germany

Heide Ibrahim Centre ÉMT, Institut National de la Recherche Scientifique, Quebec, QC, Canada

Masayoshi Ichimiya Department of Electronic Systems Engineering, The University of Shiga Prefecture, Shiga, Japan; Department of Materials Engineering Science, Osaka University, Osaka, Japan

Junichi Ichimura Department of Physics, Tohoku University, Sendai, Japan

Ryo Iikubo Department of Applied Physics, Hokkaido University, Sapporo, Kitaku, Japan

Takushi Iimori Institute for Solid State Physics, The University of Tokyo, Chiba, Japan

N. Ikeda Department of Physics, Okayama University, Okayama, Japan

Yuki Ikeda Department of Chemistry, School of Science, the University of Tokyo, Bunkyo-ku, Tokyo, Japan

D. Imanbaew Department of Chemistry, Forschungszentrum OPTIMAS, TU Kaiserslautern, Kaiserslautern, Germany

K. Imura School of Advanced Science and Engineering, Waseda University, Shinjuku, Tokyo, Japan

Akiko Inagaki PRESTO, Japan Science and Technology Agency (JST), Kawaguchi, Saitama, Japan; Graduate School of Science and Engineering, Tokyo Metropolitan University, Hachioji, Tokyo, Japan

G. Ingold Swiss Light Source, Paul Scherrer Institut, Villigen PSI, Switzerland

Ken-ichi Inoue Molecular Spectroscopy Laboratory, Riken, Wako, Japan

Hajime Ishihara Department of Physics and Electronics, Osaka Prefecture University, Osaka, Japan

S. Ishihara Department of Physics, Tohoku University, Sendai, Japan

Nobuhisa Ishii Institute for Solid State Physics, The University of Tokyo, Kashiwa, Chiba, Japan

Haruto Ishikawa Department of Chemistry, Graduate School of Science, Osaka University, Toyonaka, Osaka, Japan

Manabu Ishikawa Research Center for Low Temperature and Materials Sciences, Kyoto University, Kyoto, Japan

T. Ishikawa Department of Physics, Tohoku University, Sendai, Japan; RIKEN, Sayo, Hyogo, Japan

Akihito Ishizaki Institute for Molecular Science, National Institutes of Natural Sciences, Okazaki, Japan

Jiro Itatani Institute for Solid State Physics, University of Tokyo, Kashiwa, Chiba, Japan

Yuta Ito Department of Chemistry, Niigata University, Niigata, Japan

H. Itoh Deaprtment of Physics, Tohoku University, Sendai, Japan; JST-CREST, Sendai, Japan

K. Itoh Deaprtment of Physics, Tohoku University, Sendai, Japan

Ivan Ivanov Max Planck Institute for Polymer Research, Mainz, Germany

S. Iwai Department of Physics, Tohoku University, Sendai, Japan; JST-CREST, Sendai, Japan

Munetaka Iwamura Graduate School of Science and Engineering, University of Toyama, Toyama, Japan

Atsushi Iwasaki Department of Chemistry, School of Science, the University of Tokyo, Bunkyo-ku, Tokyo, Japan

Clemens Jakubeit Max-Planck-Institut für Quantenoptik, Garching, Germany

Thomas L.C. Jansen Zernike Institute for Advanced Materials, University of Groningen, Groningen, The Netherlands

A. Jaron-Becker JILA and Department of Physics, University of Colorado at Boulder, Boulder, CO, USA

Søren Jensen Max Planck Institute for Polymer Research, Mainz, Germany

Jaewoo Jeong IBM Almaden Research Center, San Jose, CA, USA

Y. Jiang Department of Chemistry and Physics, Centre for Free Electron Laser Science, Max Planck Institute for the Structure and Dynamics of Matter, University of Hamburg, Hamburg, Germany

Jose L. Jimenez Department of Chemistry, University of Colorado, and CIRES, Boulder, CO, USA

Chengrui Jing State Key Laboratory of High Field Laser Physics, Shanghai Institute of Optics and Fine Mechanics, Chinese Academy of Sciences, Shanghai, China

Hogyun Jinn Graduate School of Science, Kobe University, Kobe, Japan

J.A. Johnson Swiss Light Source, Paul Scherrer Institut, Villigen PSI, Switzerland

S.L. Johnson Institute for Quantum Electronics, Physics Department, ETH Zurich, Zurich, Switzerland

David M. Jonas Department of Chemistry and Biochemistry, University of Colorado, Boulder, CO, USA

T. Joo Department of Chemistry, Pohang University of Science and Technology (POSTECH), Pohang, Korea

Vincent Juvé Max-Born-Institut Für Nichtlineare Optik Und Kurzzeitspektroskopie, Berlin, Germany

H. Kageyama Institute for Integrated Cell-Material Sciences (WPI-ICeMS), Kyoto University, Kyoto, Sakyo-Ku, Japan; Department of Energy and Hydrocarbon Chemistry, Graduate School of Engineering, Kyoto University, Kyoto, Nishikyo-Ku, Japan

S. Kaiser Max Planck Institute for the Structure and Dynamics of Matter, Hamburg, Germany

T. Kajikawa Department of Chemistry, School of Science and Technology, Kwansei Gakuin University, Sanda Hyogo, Japan

T. Kambe Department of Physics, Okayama University, Okayama, Japan

Hironari Kamikubo Graduate School of Materials Science, Nara Institute of Science and Technology, Ikoma, Japan

Teruto Kanai The Institute for Solid State Physics, The University of Tokyo, Kashiwa, Chiba, Japan

Keisuke Kaneshima The Institute for Solid State Physics, The University of Tokyo, Kashiwa, Chiba, Japan

Fumihiko Kannari Department of Electronics and Electrical Engineering, Keio University, Kohoku-Ku, Yokohama, Kanagawa, Japan

Manabu Kanno Department of Chemistry, Graduate School of Science, Tohoku University, Sendai, Japan

J. Kano Department of Physics, Okayama University, Okayama, Japan

Reika Kanya Department of Chemistry, School of Science, The University of Tokyo, Bunkyo-ku, Tokyo, Japan

C.-H. Kao Department of Electrophysics, National Chiao-Tung University, Hsinchu, Taiwan

Henry C. Kapteyn JILA and Department of Physics, University of Colorado, Boulder, CO, USA

Nicholas Karpowicz Max-Planck-Institut für Quantenoptik, Garching, Germany

Daniil Kartashov Photonics Institute, Vienna University of Technology, Vienna, Austria

Franz X. Kärtner Center for Free-Electron Laser Science, Deutsches Elektronen-Synchrotron DESY, Hamburg, Germany; Department of Physics, University of Hamburg, Hamburg, Germany; The Hamburg Center for Ultrafast Imaging, University of Hamburg, Hamburg, Germany; Department of Electrical Engineering and Computer Science and Research Laboratory of Electronics, Massachusetts Institute of Technology, Cambridge, MA, USA

L. Kasmi Department of Physics, Institute for Quantum Electronics, ETH Zurich, Zurich, Switzerland

Mikio Kataoka Graduate School of Materials Science, Nara Institute of Science and Technology, Ikoma, Japan

Ikufumi Katayama Department of Physics, Graduate School of Engineering, Yokohama National University, Yokohama, Japan

Takeo Kato Institute for Solid State Physics, The University of Tokyo, Kashiwa, Chiba, Japan

S. Katsumura Department of Chemistry, School of Science and Technology, Kwansei Gakuin University, Sanda Hyogo, Japan

Nina Kausch-Busies Conductive Polymers Division, Heraeus Precious Metals GmbH & Co. KG, Leverkusen, Germany

Y. Kawakami Department of Physics, Tohoku University, Sendai, Japan

Tomohiro Kawasaki Institute for Solid State Physics, The University of Tokyo, Chiba, Japan

Sakae Kawato Graduate School of Engineering, University of Fukui, Fukui, Japan

Phillip D. Keathley Department of Electrical Engineering and Computer Science and Research Laboratory of Electronics, Massachusetts Institute of Technology, Cambridge, MA, USA

Sabine Keiber Max-Planck-Institut für Quantenoptik, Garching, Germany; Fakultät für Physik, Ludwig-Maximilians-Universität, Garching, Germany

B. Keimer Max Planck Institute for Solid State Research, Stuttgart, Germany

Sharon Keinan Department of Chemistry, Ben-Gurion University of the Negev, Be'er Sheva, Israel

U. Keller Physics Department, ETH Zurich, Zurich, Switzerland

C. Kerner Department of Chemistry, Forschungszentrum OPTIMAS, TU Kaiserslautern, Kaiserslautern, Germany

Ofer Kfir Solid State Institute and Physics Department, Technion, Haifa, Israel

Munira Khalil Department of Chemistry, University of Washington, Seattle, WA, USA

Anatoli Kheifets Research School of Physical Sciences, The Australian National University, Canberra, Australia

Jean-Claude Kieffer Centre ÉMT, Institut National de la Recherche Scientifique, Quebec, QC, Canada

Jens S. Kienitz Center for Free-Electron Laser Science, Deutsches Elektronen-Synchrotron DESY, Hamburg, Germany; The Hamburg Center for Ultrafast Imaging, University of Hamburg, Hamburg, Germany

S. Kim Department of Chemistry, Korea Advanced Institute of Science and Technology (KAIST), Daejeon, Korea

Takashi Kinoshita Department of Physics and Electronics, Osaka Prefecture University, Osaka, Japan

M. Kira Department of Physics, University of Marburg, Marburg, Germany

F.O. Kirchner Ludwig-Maximilians-Universität München, Garching, Germany; Max-Planck-Institute of Quantum Optics, Garching, Germany

Justin R. Kirschbrown Department of Chemistry, University of North Carolina at Chapel Hill, Chapel Hill, NC, USA

Hideo Kishida Department of Applied Physics, Nagoya University, Nagoya, Japan; JST, CREST, Sendai, Japan

Masahiro Kitajima Department of Physics, Graduate School of Engineering, Yokohama National University, Yokohama, Japan; International Center for Materials Nanoarchitectonics, National Institute for Materials Science, Tsukuba, Japan; CREST, Japan Science and Technology Agency, Kawaguchi, Japan; LxRay Co., Ltd, Nishinomiya, Japan; Department of Applied Physics, National Defense Academy, Yokosuka, Japan

Tsuyoshi Kitamura Corporate R&D Headquarters, CANON Inc., Utsunomiya, Japan

Kenta Kitano The Institute for Solid State Physics, The University of Tokyo, Kashiwa, Chiba, Japan

Markus Kitzler Photonics Institute, Vienna University of Technology, Vienna, Austria

M. Kling Department für Physik, LMU München, Garching, Germany; Max-Planck-Institut für Quantenoptik, Garching, Germany

Ronny Knut Department of Physics and JILA, University of Colorado, Boulder, CO, USA; Electromagnetics Division, National Institute of Standards and Technology, Boulder, CO, USA

T. Kobayashi Department of Electrophysics, National Chiao-Tung University, Hsinchu, Taiwan; Core Research for Evolutional Science and Technology, Japan Science and Technology Agency, Chiyoda-ku, Tokyo, Japan; Department of Applied Physics and Chemistry and Institute for Laser Science, University of Electro-Communications, Chofu, Tokyo, Japan; Institute of Laser Engineering, Osaka University, Suita, Osaka, Japan

Yohei Kobayashi Institute for Solid State Physics, University of Tokyo, Kashiwa, Chiba, Japan

M. Koch Stanford PULSE Institute and Institute of Experimental Physics, Graz University of Technology, Graz, Austria, EU

S.W. Koch Department of Physics, University of Marburg, Marburg, Germany

Benjamin Koeppe Max Born Institut für Nichtlineare Optik und Kurzzeitspektroskopie, Berlin, Germany

Toshiro Kohmoto Graduate School of Science, Kobe University, Kobe, Japan

Alexander V. Kolobov Nanoelectronics Research Institute, National Institute of Advanced Industrial Science and Technology, Tsukuba Central 4, Tsukuba, Japan

Fumio Komori Institute for Solid State Physics, The University of Tokyo, Chiba, Japan

Arkaprabha Konar Department of Chemistry, Michigan State University, East Lansing, MI, USA

Hirohiko Kono Department of Chemistry, Graduate School of Science, Tohoku University, Sendai, Japan

Frank H.L. Koppens The Institute of Photonic Sciences, Mediterranean Technology Park, Castelldefels, Barcelona, Spain

Shiro Koseki Department of Chemistry, Graduate School of Science and Research Institute for Molecular Electronic Devices (RIMED), Osaka Prefecture University, Sakai, Japan

S. Koshihara Tokyo Institute of Technology, Meguro-Ku, Tokyo, Japan; JST-CREST, Kawaguchi, Saitama, Japan

D. Kosumi The Osaka City University Advanced Research Institute for Natural Science and Technology (OCARINA), Osaka, Japan

D. Kosumi The Osaka City University Advanced Research Institute for Natural Science and Technology (OCARINA), Osaka, Japan

Attila P. Kovács Department of Optics and Quantum Electronics, University of Szeged, Szeged, Hungary

Oleg V. Kozlov Zernike Institute for Advanced Materials, University of Groningen, Groningen, The Netherlands; International Laser Center and Faculty of Physics, Moscow State University, Moscow, Russian Federation

F. Krausz Fakultät für Physik, Max-Planck-Institut für Quantenoptik, Ludwig-Maximilians-Universität München, Garching, Germany; Max-Planck-Institute of Quantum Optics, Garching, Germany

Peter R. Krogen Department of Electrical Engineering and Computer Science and Research Laboratory of Electronics, Massachusetts Institute of Technology, Cambridge, MA, USA

Shian Ku Max-Born-Institut Für Nichtlineare Optik Und Kurzzeitspektroskopie, Berlin, Germany

T. Kubacka Institute for Quantum Electronics, Physics Department, ETH Zurich, Zurich, Switzerland

M. Kübel Department für Physik, LMU München, Garching, Germany

T.S. Kuhlman Department of Chemistry, Technical University of Denmark, Kongens Lyngby, Denmark

Oliver Kühn Institute of Physics, University of Rostock, Rostock, Germany

Benjamin E. Van Kuiken Department of Chemistry, University of Washington, Seattle, WA, USA

Jochen Küpper Center for Free-Electron Laser Science, Deutsches Elektronen-Synchrotron DESY, Hamburg, Germany; Department of Physics, University of Hamburg, Hamburg, Germany; The Hamburg Center for Ultrafast Imaging, University of Hamburg, Hamburg, Germany

Hikaru Kuramochi Molecular Spectroscopy Laboratory, RIKEN, Wako, Japan

Yoshiyuki Kuramoto Corporate R&D Headquarters, CANON Inc., Utsunomiya, Japan

Takayuki Kurihara Institute for Solid State Physics, The University of Tokyo, Kashiwa, Chiba, Japan

Miyuki Kusaba Department of Electronics and Electrical Engineering, Keio University, Kohoku-ku, Yokohama, Kanagawa, Japan

Makoto Kuwata-Gonokami The University of Tokyo, Bunkyo-Ku, Tokyo, Japan

S. Lahme Ludwig-Maximilians-Universität München, Garching, Germany; Max-Planck-Institute of Quantum Optics, Garching, Germany

Pengfei Lan RIKEN Center for Advanced Photonics, Wako, Saitama, Japan

Jesús Lancis Universitat Jaume I, Castelló de la Plana, Spain

J. Láncis Instituto de Nuevas Tecnologías de la Imagen (INIT), Castellón, Spain

A.S. Landsman Physics Department, ETH Zurich, Zurich, Switzerland

B. Langdon Kapteyn-Murnane Labs Inc, Boulder, CO, USA

C. Lange Department of Physics, University of Regensburg, Regensburg, Germany

F. Langer Department of Physics, University of Regensburg, Regensburg, Germany

Guglielmo Lanzani Center for Nano Science and Technology@PoliMi, Istituto Italiano Di Tecnologia, Milan, Italy

Benjamin Lasorne CTMM, Institut Charles Gerhardt Montpellier, Montpellier, France

C. Laulhe Synchrotron SOLEIL, L'Orme Des Merisiers, Saint-Aubin, Gif-Sur-Yvette Cedex, France; Université Paris-Sud, Orsay Cedex, France

G. Laurent Research Laboratory of Electronics, Massachusetts Institute of Technology, Cambridge, MA, USA

David Lauvergnat Laboratoire de Chimie Physique, Université Paris-Sud and CNRS, Orsay, France

M. Le Tacon Max Planck Institute for Solid State Research, Stuttgart, Germany

Taegon Lee Department of Chemistry and Chemistry Institute for Functional Materials, Pusan National University, Busan, Korea

François Légaré Centre ÉMT, Institut National de la Recherche Scientifique, Quebec, QC, Canada

Florian Lehner BioMolekulare Optik and Center of Integrated Protein Science, CIPSM Ludwig-Maximilians-Universität München, Munich, Germany

Alfred Leitenstorfer Department of Physics and Center for Applied Photonics, University of Konstanz, Konstanz, Germany

J. Léonard Institut de Physique et Chimie Des Matériaux de Strasbourg and Labex NIE, CNRS - Université de Strasbourg, Strasbourg, France

Nicholas H.C. Lewis Department of Chemistry, University of California, Berkeley, CA, USA; Physical Biosciences Division, Lawrence Berkeley National Laboratory, Berkeley, CA, USA

Guihua Li State Key Laboratory of High Field Laser Physics, Shanghai Institute of Optics and Fine Mechanics, Chinese Academy of Sciences, Shanghai, China

H. Li JILA, National Institute of Standards and Technology, University of Colorado, Boulder, CO, USA

Helong Li State Key Laboratory on Integrated Optoelectronics, College of Electronic Science and Engineering, Jilin University, Changchun, China

Jialin Li School of Physical and Mathematical Sciences, Nanyang Technological University, Nanyang, Singapore

Stephen B. Libby Physics Division, Physical and Life Sciences, Lawrence Livermore National Laboratory, Livermore, CA, USA

Chelsea E. Liekhus-Schmaltz PULSE Institute, Stanford University, Stanford, CA, USA

Christoph Lienau Institut für Physik, Carl von Ossietzky Universität, Oldenburg, Germany; Center of Interface Science, Carl von Ossietzky Universität, Oldenburg, Germany

Manho Lim Department of Chemistry and Chemistry Institute for Functional Materials, Pusan National University, Busan, Korea

L. Liu Institut de Physique et Chimie des Matériaux de Strasbourg, Strasbourg University—CNRS, Strasbourg, France

Lai Chung Liu Departments of Chemistry and Physics, University of Toronto, Toronto, ON, Canada; Centre for Free Electron Laser Science, Max Planck Institute for the Structure and Dynamics of Matter, Hamburg, Germany

W. Liu DESY and Department of Physics, Center for Free-Electron Laser Science, University of Hamburg, Hamburg, Germany

Weiwei Liu Institute of Modern Optics, Nankai University, Key Laboratory of Optical Information Science and Technology, Ministry of Education, Tianjin, China

XiaoJun Liu State Key Laboratory of Magnetic Resonance and Atomic and Molecular Physics, Wuhan Institute of Physics and Mathematics, Chinese Academy of Sciences, Wuhan, China

Y.-S. Liu Advanced Light Source, University of California, Berkeley, CA, USA

Stefan Lochbrunner Institute of Physics, University of Rostock, Rostock, Germany

Reto Locher Department of Physics, Institute for Quantum Electronics, Zurich, Switzerland

T. Loew Max Planck Institute for Solid State Research, Stuttgart, Germany

Zhi-Heng Loh School of Physical and Mathematical Sciences, Nanyang Technological University, Nanyang, Singapore

Stefano Longhi Dipartimento di Fisica, Politecnico di Milano, Milan, Italy

Paul H.M. van Loosdrecht Zernike Institute for Advanced Materials, University of Groningen, Groningen, AG, The Netherlands

Erik Lötstedt Laser Technology Laboratory, RIKEN, Wako, Saitama, Japan; Department of Chemistry, School of Science, the University of Tokyo, Tokyo, Japan

Robert Lovrincic Weizmann Institute of Science, Rehovot, Israel; Innovationlab, Braunschweig University, Heidelberg, Germany

Vadim V. Lozovoy Department of Chemistry, Michigan State University, East Lansing, MI, USA

Cheng Lu Departments of Chemistry and Physics, University of Toronto, Toronto, ON, Canada

Peixiang Lu Wuhan National Laboratory for Optoelectronics, School of Physics, Huazhong University of Science and Technology, Wuhan, China; Key Laboratory of Fundamental Physical Quantities Measurement of Ministry of Education, Wuhan, China

Matteo Lucchini Department of Physics, Institute for Quantum Electronics, ETH Zurich, Zurich, Switzerland

A. Ludwig Department of Physics, Institute for Quantum Electronics, ETH Zurich, Zurich, Switzerland

S. Ludwigs Institute of Polymer Chemistry, University of Stuttgart, Stuttgart, Germany

A. Luebcke Swiss Light Source, Paul Scherrer Institut, Villigen PSI, Switzerland; Laboratoire de Spectroscopie Ultrarapide, EPF Lausanne, Lausanne, Switzerland

Larry Lüer Department of Nanoscience, Madrid Institute for Advanced Studies, Cantoblanco, Spain

J. Luning Université Pierre et Marie Curie, LCPMR, UMR CNRS 7614, Paris, France

Chih-Wei Luo Department of Electrophysics, National Chiao Tung University, Hsinchu, Taiwan, Republic of China

Yuriy N. Luponosov Institute of Synthetic Polymeric Materials of the Russian Academy of Science, Moscow, Russian Federation

Matthias Lütgens Institute of Physics, University of Rostock, Rostock, Germany

Michael S. Lynch Department of Chemistry, University of Washington, Seattle, Washington, USA

Keisuke Maekawa Department of Physics, Graduate School of Engineering, Yokohama National University, Yokohama, Japan

B. Maerz Chair for Biomolecular Optics, Department of Physics, Ludwig-Maximilians-University, Munich, Germany

Sebastian Mai Institute of Theoretical Chemistry, University of Vienna, Vienna, Austria

Partha Maity Radiation and Photochemistry Division, Bhabha Atomic Research Centre, Mumbai, India

M. Maiuri IFN-CNR, Dipartimento di Fisica, Politecnico di Milano, Milan, Italy

Margherita Maiuri IFN-CNR, Dipartimento di Fisica, Politecnico di Milano, Milan, Italy

Balázs Major Department of Optics and Quantum Electronics, University of Szeged, Szeged, Hungary

A. Makida Department of Applied Physics, Hokkaido University, Sapporo, Kita-Ku, Japan

K. Makino Nanoelectronics Research Institute, National Institute of Advanced Industrial Science and Technology, Tsukuba Central 4, Tsukuba, Japan

C. Mancuso JILA and Department of Physics, University of Colorado at Boulder, Boulder, CO, USA

Samansa Maneshi Max Planck Institute for Structure and Dynamics of Matter, Hamburg, Germany

Cristian Manzoni IFN-CNR Dipartimento di Fisica, Politecnico di Milano, Milan, Italy

J.P. Marangos Blackett Laboratory, Imperial College, London, UK

M. Marek Physikalisch-Chemisches Institut, Ruprecht-Karls-Universität Heidelberg, Heidelberg, Germany

S.O. Mariager Swiss Light Source, Paul Scherrer Institut, Villigen PSI, Switzerland

A. Marini ISM-CNR, Monterotondo Stazione, Italy

Philipp Marquetand Institute of Theoretical Chemistry, University of Vienna, Vienna, Austria

F. Martín Departamento de Química, Modulo 13, Universidad Autonoma de Madrid, Madrid, Spain; Instituto Madrileno de Estudios Avanzados en Nanociencia, Madrid, Spain

T.J. Martínez PULSE Institute and Department of Chemistry, Stanford University, Stanford, USA

S. Maruta Department of Physics, Graduate School of Science, Osaka City University, Osaka, Japan

R.E. Marvel Department of Physics and Astronomy, Vanderbilt University, Nashville, TN, USA

Yuta Masaki Department of Electronics and Electrical Engineering, Keio University, Kohoku-Ku, Yokohama, Kanagawa, Japan

Akitaka Matsuda Department of Chemistry, Graduate School of Science, Nagoya University, Nagoya, Aichi, Japan

T. Matsuki Tokyo Institute of Technology, Meguro-Ku, Tokyo, Japan

Yoshiyasu Matsumoto Department of Chemistry, Graduate School of Science, Kyoto University, Kyoto, Japan

Jan Matyschok VENTEON Laser Technologies GmbH, Garbsen, Germany; Institute of Quantum Optics, Hannover, Germany

B. Mayer Department of Physics and Center for Applied Photonics, University of Konstanz, Konstanz, Germany

T. Meier Department of Physics, University of Paderborn, Paderborn, Germany

Omel Mendoza-Yero Instituto de Nuevas Tecnologías de la Imagen (INIT), Universitat Jaume I, Castelló de la Plana, Spain

C. Merschjann Institute of Physics, University of Rostock, Rostock, Germany

F. Merschjohann Fakultät für Physik, Universität Bielefeld, Bielefeld, Germany

Andreas T. Messmer Institute for Biophysics, Goethe University, Frankfurt, Germany

Hans-Dieter Meyer Theoretische Chemie, Physikalisch-Chemisches Institut, Ruprecht-Karls Universität, Heidelberg, Germany

Jeff Michelmann BioMolekulare Optik and Center of Integrated Protein Science, CIPSM Ludwig-Maximilians-Universität München, Munich, Germany

Zoltán Mics Max Planck Institute for Polymer Research, Mainz, Germany

K. Midorikawa RIKEN Ceneter for Advanced Photonics, Wako, Saitama, Japan

Katsumi Midorikawa Laser Technology Laboratory, RIKEN Center for Advanced Photonics, Wako, Saitama, Japan

T. Miki Physikalisch-Chemisches Institut, Ruprecht-Karls-Universität Heidelberg, Heidelberg, Germany

R.J.D. Miller Departments of Chemistry and Physics, University of Toronto, Toronto, ON, Canada; Centre for Free Electron Laser Science, Max Planck Institute for the Structure and Dynamics of Matter, Hamburg, Germany

Yasuo Minami Department of Physics, Graduate School of Engineering, Yokohama National University, Yokohama, Japan; Department of Physics, Faculty of Engineering, Yokohama National University, Yokohama, Japan

Gladys Mínguez-Vega Instituto de Nuevas Tecnologías de la Imagen (INIT), Universitat Jaume I, Castelló de la Plana, Spain

Kazuhiko Misawa Tokyo University of Agriculture and Technology, Koganei, Japan

Roland Mitric Institut Für Physikalishce Und Theoretische Chemie, Am Hubland, Universität Würzburg, Würzburg, Germany

Kirill Mitrofanov Nanoelectronics Research Institute, National Institute of Advanced Industrial Science and Technology, Tsukuba Central 4, Tsukuba, Japan

Mitsuhiro Miyamoto Department of Chemistry, Graduate School of Science, Osaka University, Toyonaka, Osaka, Japan

Hiroshi Miyasaka Graduate School of Engineering Science, Osaka University, Osaka, Japan; Center for Quantum Science and Technology Under Extreme Conditions, Osaka University, Osaka, Japan

Kiyoshi Miyata Department of Chemistry, Graduate School of Science, Kyoto University, Kyoto, Japan

K. Mizoguchi Department of Physical Science, Osaka Prefecture University, Osaka, Sakai, Japan

R. Mizokuchi Tokyo Institute of Technology, Meguro-Ku, Tokyo, Japan

Misao Mizuno Department of Chemistry, Graduate School of Science, Osaka University, Toyonaka, Osaka, Japan

Yasuhisa Mizutani Department of Chemistry, Graduate School of Science, Osaka University, Toyonaka, Osaka, Japan

Elisa Molinari Istituto Nanoscienze—CNR, Modena, Italy; Dipartimento di Scienze Fisiche, Matematiche e Informatiche, Università di Modena e Reggio Emilia, via Campi 213a, Modena, Italy

A. Molle Laboratorio MDM, IMM-CNR, Agrate Brianza, Italy

K.B. Møller Department of Chemistry, Technical University of Denmark, Kongens Lyngby, Denmark

Daniele M. Monahan Department of Chemistry, University of California, Berkeley, USA; Physical Biosciences Division, Lawrence Berkeley National Laboratory, Berkeley, USA

B. Monoszlai SwissFEL, Paul Scherrer Institute, Villigen-PSI, Switzerland

M. Mootz Department of Physics, Philipps-University Marburg, Marburg, Germany

Sarah E. Morgan Cavendish Laboratory, University of Cambridge, Cambridge, UK

Uwe Morgner Institute of Quantum Optics, Hannover, Germany

Yuya Morimoto Department of Chemistry, School of Science, The University of Tokyo, Bunkyo-ku, Tokyo, Japan

Ryuji Morita Department of Applied Physics, Hokkaido University, Sapporo, Japan

Takeshi Moriyasu Graduate School of Science, Kobe University, Kobe, Japan

Jeffrey Moses Department of Electrical Engineering and Computer Science and Research Laboratory of Electronics, Massachusetts Institute of Technology, Cambridge, MA, USA

Marcus Motzkus Physikalisch-Chemisches Institut, Ruprecht-Karls-Universität Heidelberg, Heidelberg, Germany

Oliver D. Mücke Center for Free-Electron Laser Science, Deutsches Elektronen-Synchrotron DESY, Hamburg, Germany; The Hamburg Center for Ultrafast Imaging, Hamburg, Germany

Y. Mukai Department of Physics, Graduate School of Science, Kyoto University, Kyoto, Sakyo-Ku, Japan

Shaul Mukamel Department of Chemistry, University of California, Irvine, CA, USA

Tatsuhiko Mukuta Graduate School of Science and Engineering, Tokyo Institute of Technology, Midori-ku, Yokohama, Japan

Klaus Müllen Max Planck Institute for Polymer Research, Mainz, Germany

M. Müller Fritz-Haber-Institut der MPG, Berlin, Germany

N. Müller Fakultät für Physik, Universität Bielefeld, Bielefeld, Germany

Henrike M. Müller-Werkmeister Institute of Biophysics, University of Frankfurt, Frankfurt, Germany; Departments of Chemistry and Physics, University of Toronto, Toronto, ON, Canada

Hiroyuki Murata Department of Materials Engineering Science, Osaka University, Osaka, Japan

Kei Murata Chemical Resources Laboratory, Tokyo Institute of Technology, Midori-ku, Yokohama, Japan

Margaret M. Murnane JILA and Department of Physics, University of Colorado, Boulder, CO, USA

Margaret Murnane JILA and Department of Physics, University of Colorado, Boulder, CO, USA

Tadaaki Nagao International Center for Materials Nanoarchitectonics, National Institute for Materials Science, Tsukuba, Japan; CREST, Japan Science and Technology Agency, Kawaguchi, Japan

Yutaka Nagasawa Graduate School of Engineering Science, Osaka University, Osaka, Japan; Center for Quantum Science and Technology Under Extreme Conditions, Osaka University, Osaka, Japan; PRESTO, Japan Science and Technology Agency (JST), Saitama, Japan

M. Nagasono RIKEN, Sayo, Hyogo, Japan

T. Nagata Department of Physics, Okayama University, Okayama, Japan

Yota Naito Department of Physics, Tohoku University, Sendai, Japan

Katsunori Nakai Department of Chemistry, School of Science, The University of Tokyo, Tokyo, Japan

Makoto Nakajima Institute of Laser Engineering, Osaka University, Suita, Osaka, Japan

Ryosuke Nakamura Science & Technology Entrepreneurship Laboratory, Osaka University, Suita, Japan

T. Nakano Nanoelectronics Research Institute, National Institute of Advanced Industrial Science and Technology, Tsukuba Central 4, Tsukuba, Japan

Shohei Nambu Graduate School of Engineering Science, Osaka University, Osaka, Japan

M. Nango OCARINA, Osaka City University, Osaka, Japan

Damiano Nardi JILA and Department of Physics, University of Colorado, Boulder, CO, USA

T. Narushima Institute for Molecular Science and The Graduate University for Advanced Studies, Okazaki, Japan

Adi Natan Stanford PULSE Institute, SLAC National Accelerator Laboratory, Menlo Park, California, USA

S. Neb Fakultät für Physik, Universität Bielefeld, Bielefeld, Germany

M. Negro Istituto di Fotonica e Nanotecnologie—IFN-CNR, Piazza L., Milan, Italy

Hans Nembach Electromagnetics Division, National Institute of Standards and Technology, Boulder, CO, USA

A. Nenov Department of Chemistry, Ludwig-Maximilians-University, Munich, Germany

S. Neppl Ultrafast X-Ray Science Laboratory, University of California, Berkeley, CA, USA; Chemical Sciences Division, Lawrence Berkeley National Laboratory, University of California, Berkeley, CA, USA

Jielei Ni State Key Laboratory of High Field Laser Physics, Shanghai Institute of Optics and Fine Mechanics, Chinese Academy of Sciences, Shanghai, China

Erik T.J. Nibbering Max Born Institut für Nichtlineare Optik und Kurzzeitspektroskopie, Berlin, Germany

Daniele Nicoletti Max Planck Institute for the Structure and Dynamics of Matter, Hamburg, Germany

Bai Nie Department of Physics and Astronomy, Michigan State University, East Lansing, MI, USA

Zhaogang Nie School of Physical and Mathematical Sciences, Nanyang Technological University, Nanyang, Singapore

Satoshi Nihonyanagi Molecular Spectroscopy Laboratory, Riken, Wako, Japan; Ultrafast Spectroscopy Research Team, RIKEN Center for Advanced Photonics (RAP), Wako, Japan

Julien Nillon Institut de Physique et Chimie Des Matériaux de Strasbourg, Strasbourg, France; European XFEL, Hamburg, Germany

T. Nishimoto Graduate School of Science, Kobe University, Kobe, Japan

Masatoshi Nishio Graduate School of Engineering, University of Fukui, Fukui, Japan

Y. Nishiyama Institute for Molecular Science and The Graduate University for Advanced Studies, Okazaki, Japan

M. Nisoli IFN-CNR, Milan, Italy; Department of Physics, Politecnico di Milano, Milan, Italy

T. Noguchi Tokyo Institute of Technology, Meguro-Ku, Tokyo, Japan

Yutaka Nomura Institute for Molecular Science, Okazaki, Japan

Y. Nosenko Department of Chemistry, Forschungszentrum OPTIMAS, TU Kaiserslautern, Kaiserslautern, Germany

Patrick Nuernberger Ruhr-Universität Bochum, Bochum, Germany

W. Oba Graduate School of Engineering, Yokohama National University, Yokohama, Japan

S. Oda Tokyo Institute of Technology, Meguro-Ku, Tokyo, Japan

J. Oelmann Department of Physics and Center for Applied Photonics, University of Konstanz, Konstanz, Germany

S. Oesterling Department of Chemistry, Ludwig-Maximilians-University, Munich, Germany

Kaoru Ohta Molecular Photoscience Research Center, Kobe University, Kobe, Japan

H. Okamoto Institute for Molecular Science and The Graduate University for Advanced Studies, Okazaki, Japan

T. Okino RIKEN Center for Advanced Photonics, Saitama, Japan

S. Okumura Department of Chemistry, School of Science and Technology, Kwansei Gakuin University, Sanda Hyogo, Japan

Thomas A.A. Oliver Department of Chemistry, University of California, Berkeley, CA, USA; Physical Biosciences Division, Lawrence Berkeley National Laboratory, Berkeley, CA, USA

Yoann Olivier Service de Chimie des Materiaux Nouveaux, Université de Mons, Mons, Belgium

M. Olivucci Dipartimento Di Chimica, Università Degli Studi Di Siena, Siena, Italy; Chemistry Department, Bowling Green State University, Bowling Green, USA

Ken Onda Graduate School of Science and Engineering, Tokyo Institute of Technology, Midori-ku, Yokohama, Japan; PRESTO, Japan Science and Technology Agency (JST), Kawaguchi, Saitama, Japan

G. Oohata Department of Physical Science, Osaka Prefecture University, Osaka, Sakai, Japan

A. Oriana IFN-CNR, Dipartimento di Fisica, Politecnico di Milano, Milano, Italy

Jürg Osterwalder Department of Physics, University of Zurich, Zurich, Switzerland

A. Paarmann Fritz-Haber-Institut der MPG, Berlin, Germany

Tim Paasch-Colberg Max-Planck-Institut für Quantenoptik, Garching, Germany

S. Pabst Center for Free-Electron Laser Science, DESY, Hamburg, Germany

A. Palacios Departamento de Química, Modulo 13, Universidad Autonoma de Madrid, Madrid, Spain

Brett B. Palm Department of Chemistry, University of Colorado, and CIRES, Boulder, CO, USA

John M. Papanikolas Department of Chemistry, University of North Carolina at Chapel Hill, Chapel Hill, NC, USA

Dmitry Yu. Paraschuk International Laser Center and Faculty of Physics, Moscow State University, Moscow, Russian Federation

Doo Jae Park Institut für Physik, Carl von Ossietzky Universität, Oldenburg, Germany; Center of Interface Science, Carl von Ossietzky Universität, Oldenburg, Germany

Samuel D. Park Department of Chemistry and Biochemistry, University of Colorado, Boulder, CO, USA

Seongchul Park Department of Chemistry and Chemistry Institute for Functional Materials, Pusan National University, Busan, Korea

Stuart S.P. Parkin IBM Almaden Research Center, San Jose, CA, USA

Khaled Parvez Max Planck Institute for Polymer Research, Mainz, Germany

A. Pashkin Department of Physics and Center for Applied Photonics, University of Konstanz, Konstanz, Germany

Gerhard G. Paulus Institute of Optics and Quantum Electronics, Friedrich-Schiller University Jena, Jena, Germany; Helmholtz Institute Jena, Jena, Germany

G.G. Paulus Institute of Optics and Quantum Electronics, Jena, Germany

Jorge Pérez-Vizcaíno Instituto de Nuevas Tecnologías de la Imagen (INIT), Universitat Jaume I, Castelló de la Plana, Spain

Aurelie Perveaux Laboratoire de Chimie Physique, Université Paris-Sud and CNRS, Orsay, France; CTMM, Institut Charles Gerhardt Montpellier, Montpellier, France

George M. Petrov Plasma Physics Division, Naval Research Lab, Washington, DC, USA

Vladimir S. Petrovic PULSE Institute, Stanford University, Stanford, CA, USA

Walter Pfeiffer Fakultät für Physik, Universität Bielefeld, Bielefeld, Germany

Alessandra Picchiotti Max Planck Institute for Structure and Dynamics of Matter, Hamburg, Germany

Björn Piglosiewicz Institut für Physik, Carl von Ossietzky Universität, Oldenburg, Germany; Center of Interface Science, Carl von Ossietzky Universität, Oldenburg, Germany

B.M. Pilles BioMolecular Optics and Center for Integrated Protein Science, Ludwig-Maximilians-Universität München, Munich, Germany

Christopher W. Pinion Department of Chemistry, University of North Carolina at Chapel Hill, Chapel Hill, NC, USA

L. Plaja Grupo de Investigación en Óptica Extrema, Universidad de Salamanca, Salamanca, Spain

L. Poletto IFN-CNR, Padua, Italy

Dario Polli IFN-CNR, Dipartimento di Fisica, Politecnico di Milano, Milan, Italy

Sergei A. Ponomarenko Institute of Synthetic Polymeric Materials of the Russian Academy of Science, Moscow, Russian Federation

Dimitar Popmintchev Department of Physics and JILA, University of Colorado, Boulder, CO, USA

Tenio Popmintchev Department of Physics and JILA, University of Colorado, Boulder, CO, USA

Miguel A. Porras Departamento de Física Aplicada a los Recursos Naturales, Grupo de Sistemas Complejos, Universidad Politécnica de Madrid, Madrid, Spain

Mirabelle Prémont-Schwartz Max Born Institut für Nichtlineare Optik und Kurzzeitspektroskopie, Berlin, Germany

Oliver Prochnow VENTEON Laser Technologies GmbH, Garbsen, Germany

Valentyn I. Prokhorenko Max Planck Institute for the Structure and Dynamics of Matter, Centre for Free Electron Laser Science, Hamburg, Germany

Brian T. Psciuk Department of Chemistry, Yale University, New Haven, CT, USA

Maxim S. Pshenichnikov Zernike Institute for Advanced Materials, University of Groningen, Groningen, The Netherlands

Audrius Pugžlys Photonics Institute, Vienna University of Technology, Vienna, Austria

Michele Puppin Fritz-Haber Institut der Max Planck Gesellschaft, Berlin, Germany

W.P. Putnam Department of EECS and Research Laboratory of Electronics, Massachusetts Institute of Technology, Cambridge, MA, USA

Akshay Rao Cavendish Laboratory, University of Cambridge, Cambridge, UK

Tim Rathje Institute of Optics and Quantum Electronics, Friedrich-Schiller University Jena, Jena, Germany; Helmholtz Institute Jena, Jena, Germany

S. Ravy Synchrotron SOLEIL, L'Orme Des Merisiers, Saint-Aubin, Gif-Sur-Yvette Cedex, France

Olga Razskazovskaya Max-Planck-Institut für Quantenoptik, Garching, Germany

Mar Reguero Departement de Química Física I Inorgànica, Tarragona, Spain

J. Réhault IFN-CNR, Dipartimento di Fisica, Politecnico di Milano, Milan, Italy

Klaus Reimann Max-Born-Institut, Berlin, Germany

A.J. Reuss Institute of Physical and Theoretical Chemistry, Johann Wolfgang Goethe-University, Frankfurt a. M., Germany

D. Reuter Lehrstuhl fuer Angewandte Festkoerperphysik, Ruhr-Universitaet Bochum, Bochum, Germany

Gethin H. Richards Centre for Quantum and Optical Science, Swinnburne University of Technology, Melbourne, Australia

Martin Richter Institute of Theoretical Chemistry, University of Vienna, Vienna, Austria

Eberhard Riedle LS für BioMolekulare Optik, Ludwig-Maximilians-Universität München, Munich, Germany

C. Riehn Department of Chemistry, Forschungszentrum OPTIMAS, TU Kaiserslautern, Kaiserslautern, Germany

Jörg Robin Institut für Physik, Carl von Ossietzky Universität, Oldenburg, Germany; Center of Interface Science, Carl von Ossietzky Universität, Oldenburg, Germany

Michał F. Rode Institute of Physics Polish Academy of Science, Warsaw, Poland

Sebastian Roeding Universität Würzburg, Würzburg, Germany

Stefan Roither Photonics Institute, Vienna University of Technology, Vienna, Austria

T. Roland Institut de Physique et Chimie des Matériaux de Strasbourg, Strasbourg University—CNRS, Strasbourg, France

C. Ropers 4. Physical Institute, University of Göttingen, Göttingen, Germany

Matthew Ross Department of Chemistry, University of Washington, Seattle, WA, USA

Giulio M. Rossi Center for Free-Electron Laser Science, Deutsches Elektronen-Synchrotron DESY, Hamburg, Germany; Department of Physics, University of Hamburg, Hamburg, Germany

N. Rothe Institute of Physics, University of Rostock, Rostock, Germany

Katharina Röttger Institute of Physical Chemistry, Christian-Albrechts-University Kiel, Kiel, Germany

A. Rouzée Max-Born-Institut, Berlin, Germany

Carlo A. Rozzi Istituto Nanoscienze—CNR, Centro S3, via Campi 213a, Modena, Italy

Angel Rubio Nano-Bio Spectroscopy Group and ETSF Scientific Development Centre, Dpto. Física de Materiales, Universidad del País Vasco, Centro de Física de Materiales CSIC-UPV/EHU-MPC and DIPC, San Sebastián, Spain; Fritz-Haber-Institut der Max-Planck-Gesellschaft, Berlin, Germany

A. Ruff Institute of Polymer Chemistry, University of Stuttgart, Stuttgart, Germany

Sanford Ruhman Institute of Chemistry, Hebrew University, Jerusalem, Israel

F. Rupp Department of Physics, TU Kaiserslautern, Kaiserslautern, Germany

M. Sabbar Physics Department, ETH Zurich, Zurich, Switzerland

Y. Sagae Department of Physics, Tohoku University, Sendai, Japan

Özge Sağlam Physik-Department, Technische Universitat München, Garching, Germany

T. Saiki Graduate School of Science and Technology, Keio University, Yokohama, Japan

Y. Saito Nanoelectronics Research Institute, National Institute of Advanced Industrial Science and Technology, Tsukuba Central 4, Tsukuba, Japan

K. Sakaguchi Department of Chemistry, Graduate School of Science, Osaka City University, Osaka, Japan

S. Sakai Department of Life and Materials Engineering, Nagoya Institute of Technology, Nagoya, Japan

M. Salmeron Materials Sciences Division, Lawrence Berkeley National Laboratory, University of California, Berkeley, CA, USA

Mahesh G. Samant IBM Almaden Research Center, San Jose, CA, USA

E. Samoylova Chair for Biomolecular Optics, Department of Physics, Ludwig-Maximilians-University, Munich, Germany

M. Sanches Piaia Institut de Physique et Chimie Des Matériaux de Strasbourg, Unité Mixte 7504 CNRS, Université de Strasbourg, Strasbourg Cedex, France

Joseph Sanderson Department of Physics and Astronomy, University of Waterloo, Waterloo, ON, Canada

D. Sangalli ISM-CNR, Monterotondo Stazione, Italy

R. Santra Center for Free-Electron Laser Science, DESY, Hamburg, Germany; Department of Physics, University of Hamburg, Hamburg, Germany

T. Sasaki JST-CREST, Sendai, Japan; Institute for Material Research, Tohoku University, Sendai, Japan

Kenji Sato Department of Physics, Graduate School of Engineering, Yokohama National University, Yokohama, Japan

Motoki Sato Department of Chemistry, Graduate School of Science, Kyoto University, Kyoto, Japan

T. Sato RIKEN, Sayo, Hyogo, Japan

Takahiro Sato Department of Chemistry, School of Science, the University of Tokyo, Bunkyo-ku, Tokyo, Japan

Masaaki Sato Tokyo University of Agriculture and Technology, Koganei, Japan

Shota Sawai Department of Electronics and Electrical Engineering, Keio University, Yokohama, Japan

Ilyas Saytashev Department of Chemistry, Michigan State University, East Lansing, MI, USA

H. Schaefer Physics Department, Universitaet Konstanz, Konstanz, Germany

K.J. Schafer Department of Physics and Astronomy, Louisiana State University, Baton Rouge, LA, USA

O. Schalk Stockholm University, Stockholm, Sweden; National Research Council, Ottawa, Canada

D. Schimpf DESY and Department of Physics, Center for Free-Electron Laser Science, University of Hamburg, Hamburg, Germany

Frank Schlawin Department of Chemistry, University of California, Irvine, CA, USA; Institute of Physics Albert-Ludwigs University of Freiburg, Freiburg, Germany

R. Schmeissner Laboratoire Kastler Brossel, Université Pierre et Marie Curie, CNRS, ENS, Paris Cedex, France

Bruno E. Schmidt Centre ÉMT, Institut National de la Recherche Scientifique, Quebec, QC, Canada

C. Schmidt Department of Physics and Center for Applied Photonics, University of Konstanz, Konstanz, Germany

Slawa Schmidt Institut für Physik, Carl von Ossietzky Universität, Oldenburg, Germany; Center of Interface Science, Carl von Ossietzky Universität, Oldenburg, Germany

Robert W. Schoenlein Chemical Sciences Division, Lawrence Berkeley National Laboratory, Berkeley, CA, USA

Markus Schöffler Photonics Institute, Vienna University of Technology, Vienna, Austria

Marco Schröter Institute of Physics, University of Rostock, Rostock, Germany

O. Schubert Department of Physics, University of Regensburg, Regensburg, Germany

J. Schulte DESY and Department of Physics, Center for Free-Electron Laser Science, University of Hamburg, Hamburg, Germany

C. Schuster Institute of Physics, University of Rostock, Rostock, Germany

B. Schütte Max-Born-Institut, Berlin, Germany

Michael S. Schuurman National Research Council of Canada, Ottawa, ON, Canada

Alexander Schwarz Max-Planck-Institut für Quantenoptik, Garching, Germany; Fakultät für Physik, Ludwig-Maximilians-Universität, Garching, Germany

T. Scopigno Physics Department, University "Sapienza", Rome, Italy

Francesco Scotognella Dipartimento Di Fisica, Politecnico Di Milano, Milan, Italy

A. Scrinzi Ludwig Maximilians Universität, Munich, Germany

Shawn Sederberg Max-Planck-Institut für Quantenoptik, Garching, Germany

Takashi Seki Corporate R&D Headquarters, CANON Inc., Utsunomiya, Japan

Taro Sekikawa Department of Applied Physics, Hokkaido University, Sapporo, Kita-ku, Japan

D.V. Seletskiy Department of Physics and Center for Applied Photonics, University of Konstanz, Konstanz, Germany

Oleg Selig FOM Institute AMOLF, Amsterdam, The Netherlands

Almis Serbenta Zernike Institute for Advanced Materials, University of Groningen, Groningen, AG, The Netherlands

Carles Serrat UPC—Universitat Politècnica de Catalunya, Terrassa, Barcelona, Spain

Dror Shafir Department of Physics of Complex Systems, Weizmann Institute of Science, Rehovot, Israel

A. Shavorskiy Chemical Sciences Division, Lawrence Berkeley National Laboratory, University of California, Berkeley, CA, USA

Justin Shaw Electromagnetics Division, National Institute of Standards and Technology, Boulder, CO, USA

Alex J. Shearer Department of Chemistry, University of California, Berkeley, CA, USA; Chemical Sciences Division, Lawrence Berkeley National Laboratory, Berkeley, CA, USA

Mordechai Sheves Department of Organic Chemistry, Weizmann Institute Rehovot, Rehovot, Israel

E. Shigemasa RIKEN, Sayo, Hyogo, Japan; Institute for Molecular Science, Okazaki, Aichi, Japan

Noriyuki Shimakura Department of Chemistry, Niigata University, Niigata, Japan

Keisuke Shinokita Zernike Institute for Advanced Materials, University of Groningen, Groningen, The Netherlands

Hideto Shirai Institute for Molecular Science, Okazaki, Japan

R. Siemering Department für Chemie, LMU München, Munich, Germany

Prashant Chandra Singh Molecular Spectroscopy Laboratory, Riken, Wako, Japan

R. Singh JILA, University of Colorado & National Institute of Standards and Technology, Boulder, CO, USA

Emily Sistrunk Lawrence Livermore National Laboratory, Livermore, CA, USA; Stanford PULSE Institute, SLAC National Accelerator Laboratory, Menlo Park, CA, USA

E. Siva Subramaniam Iyer Institute of Chemistry, Hebrew University, Jerusalem, Israel

D.S. Slaughter Chemical Sciences Division, Lawrence Berkeley National Laboratory, University of California, Berkeley, CA, USA

Karla M. Slenkamp Department of Chemistry, University of Washington, Seattle, Washington, USA

Giancarlo Soavi Dipartimento Di Fisica, Politecnico Di Milano, Milan, Italy

Andrzej L. Sobolewski Institute of Physics Polish Academy of Science, Warsaw, Poland

K. Sobue Department of Physics, Graduate School of Science, Tohoku University, Aoba-Ku, Sendai, Japan

H. Soifer Department of Physics of Complex Systems, Weizmann Institute of Science, Rehovot, Israel

H. Soifer Department of Physics of Complex Systems, Weizmann Institute of Science, Rehovot, Israel

Íñigo J. Sola Universidad de Salamanca, Salamanca, Spain

D.R. Solli 4. Physical Institute, University of Göttingen, Göttingen, Germany

Carmine Somma Max-Born-Institut, Berlin, Germany

Ephraim Sommer Institut für Physik, Carl von Ossietzky Universität, Oldenburg, Germany; Center of Interface Science, Carl von Ossietzky Universität, Oldenburg, Germany

Kotaro Sonoda Department of Integrated Sciences, Graduate School of Arts and Sciences, The University of Tokyo, Meguro-ku, Tokyo, Japan

Michael Spanner National Research Council of Canada, Ottawa, ON, Canada

R. Squibb Blackett Laboratory, Imperial College, London, UK

S. Stagira Dipartimento Di Fisica, Politecnico Di Milano, Milan, Italy

André Staudte Joint Laboratory for Attosecond Science of the National Research Council and the University of Ottawa, Ottawa, Canada

Gregory J. Stein Department of Electrical Engineering and Computer Science and Research Laboratory of Electronics, Massachusetts Institute of Technology, Cambridge, MA, USA

Andreas Steinbacher Universität Würzburg, Würzburg, Germany

A. Stolow National Research Council, Ottawa, Canada; Department of Chemistry and Department of Physics, University of Ottawa, Ottawa, Canada

Mathew L. Strader SLAC National Accelerator Laboratory, Menlo Park, CA, USA

Christian Strüber Fakultät für Physik, Universität Bielefeld, Bielefeld, Germany

Mayra C. Stuhldreier Institute of Physical Chemistry, Christian-Albrechts-University Kiel, Kiel, Germany

Haim Suchowski NSF Nanoscale Science and Engineering Center, University of California, Berkeley, CA, USA

Akira Suda Department of Physics, Tokyo University of Science, Noda, Chiba, Japan

Tohru Suemoto Institute for Solid State Physics, The University of Tokyo, Kashiwa, Chiba, Japan

Toshiki Sugimoto Department of Chemistry, Graduate School of Science, Kyoto University, Kyoto, Japan

M. Sugisaki Department of Physics, Graduate School of Science, Osaka City University, Osaka, Japan

David E. Suich Department of Chemistry, University of California, Berkeley, CA, USA; Chemical Sciences Division, Lawrence Berkeley National Laboratory, Berkeley, CA, USA

Takashi Sukegawa Corporate R&D Headquarters, CANON Inc., Utsunomiya, Japan

Toshinori Suzuki Department of Chemistry, Graduate School of Science, Kyoto University, Kyoto, Japan; RIKEN Center for Advanced Photonics, RIKEN, Wako, Japan

Yoshi-Ichi Suzuki Department of Chemistry, Graduate School of Science, Kyoto University, Kyoto, Japan; RIKEN Center for Advanced Photonics, RIKEN, Wako, Japan

Michael E. Swanwick Microsystems Technology Laboratories, Massachusetts Institute of Technology, Cambridge, MA, USA

Tahei Tahara Molecular Spectroscopy Laboratory, Riken, Wako, Japan; Ultrafast Spectroscopy Research Team, RIKEN Center for Advanced Photonics (RAP), Wako, Japan

Hiroshi Takahashi Department of Physics, Tokyo University of Science, Noda, Chiba, Japan

Eiji J. Takahashi Attosecond Science Research Team, RIKEN Center for Advanced Photonics, Wako, Saitama, Japan

S. Takaichi Department of Biology, Nippon Medical School, Nakahara, Kawa-saki, Japan

Jun Takeda Department of Physics, Graduate School of Engineering, Yokohama National University, Yokohama, Japan

Tetsuya Taketsugu Department of Chemistry, Hokkaido University, Sapporo, Kita-ku, Japan

Eisuke Takeuchi Graduate School of Engineering Science, Osaka University, Osaka, Japan

Satoshi Takeuchi Molecular Spectroscopy Laboratory, RIKEN, Wako, Japan; Ultrafast Spectroscopy Research Team, RIKEN Center for Advanced Photonics (RAP), Wako, Japan

Jun Takeya Department of Frontier Sciences, The University of Tokyo, Kashiwa, Chiba, Japan

K. Tanaka Department of Physics, Graduate School of Science, Kyoto University, Kyoto, Sakyo-Ku, Japan; Institute for Integrated Cell-Material Sciences (WPI-IC-eMS), Kyoto University, Kyoto, Sakyo-Ku, Japan; CREST, Japan Science and Technology Agency, Saitama, Kawaguchi, Japan

Sei'ichi Tanaka Graduate School of Science and Engineering, Tokyo Institute of Technology, Midori-ku, Yokohama, Japan

Shunsuke Tanaka Department of Chemistry, Graduate School of Science, Kyoto University, Kyoto, Japan

Friedrich Temps Institute of Physical Chemistry, Christian-Albrechts-University Kiel, Kiel, Germany

Ian Tenney PULSE Institute, Stanford University, Stanford, CA, USA

Sebastian Thallmair Department Chemie, Ludwig-Maximilians-Universität München, Munich, Germany

W.R. Thiel Department of Chemistry, Forschungszentrum OPTIMAS, TU Kais-erslautern, Kaiserslautern, Germany

Nicolas Thiré Centre ÉMT, Institut National de la Recherche Scientifique, Quebec, QC, Canada

Klaas-Jan Tielrooij The Institute of Photonic Sciences, Mediterranean Technology Park, Castelldefels, Barcelona, Spain

J.W.G. Tisch Blackett Laboratory, Imperial College, London, UK

Keisuke Toda Department of Physics, Tokyo University of Science, Noda, Chiba, Japan

Yasunori Toda Department of Applied Physics, Hokkaido University, Sapporo, Japan

T. Togashi Japan Synchrotron Radiation Research Institute, Sayo, Hyogo, Japan

Jonathan O. Tollerud Centre for Quantum and Optical Science, Swinburne University of Technology, Melbourne, Australia

Kazunori Toma Department of Electronics and Electrical Engineering, Keio University, Kohoku-Ku, Yokohama, Kanagawa, Japan

Gaia Tomasello Institut Für Physikalishce Und Theoretische Chemie, Am Hubland, Universität Würzburg, Würzburg, Germany

Junji Tominaga Nanoelectronics Research Institute, National Institute of Advanced Industrial Science and Technology, Tsukuba Central 4, Tsukuba, Japan

Keisuke Tominaga Molecular Photoscience Research Center, Kobe University, Kobe, Japan; Graduate School of Science, Kobe University, Kobe, Japan

A. Trabattoni Department of Physics, Politecnico di Milano, Milan, Italy

A.S. Tremsin Space Sciences Laboratory, University of California, Berkeley, CA, USA

N. Treps Laboratoire Kastler Brossel, Université Pierre et Marie Curie, CNRS, ENS, Paris Cedex, France

Sebastian Trippel Center for Free-Electron Laser Science, Deutsches Elektronen-Synchrotron DESY, Hamburg, Germany

P. Trojanowski Institute of Physical and Theoretical Chemistry, Johann Wolfgang Goethe-University, Frankfurt a. M., Germany

T. Troy Chemical Sciences Division, Lawrence Berkeley National Laboratory, University of California, Berkeley, CA, USA

B. Tudu Université Pierre et Marie Curie, LCPMR, UMR CNRS 7614, Paris, France

Dmitry Turchinovich Max Planck Institute for Polymer Research, Mainz, Germany

Emrah Turgut Department of Physics and JILA, University of Colorado, Boulder, CO, USA

Takafumi Uemura Department of Frontier Sciences, The University of Tokyo, Kashiwa, Chiba, Japan

Takayuki Umakoshi Department of Materials Engineering Science, Osaka University, Osaka, Japan

A.-N. Unterreiner Institute of Physical Chemistry, Karlsruhe Institute of Technology, Karlsruhe, Germany

B. Urbanek Department of Physics, University of Regensburg, Regensburg, Germany

Giuseppe Della Valle Dipartimento di Fisica, Politecnico di Milano, Milan, Italy

Luis F. Velásquez-García Microsystems Technology Laboratories, Massachusetts Institute of Technology, Cambridge, MA, USA

C. Vicario SwissFEL, Paul Scherrer Institute, Villigen-PSI, Switzerland

Daniele Viola Dipartimento Di Fisica, Politecnico Di Milano, Milan, Italy

Jan Vogelsang Institut für Physik, Carl von Ossietzky Universität, Oldenburg, Germany; Center of Interface Science, Carl von Ossietzky Universität, Oldenburg, Germany

M. Vomir Institut de Physique et Chimie Des Matériaux de Strasbourg, Unité Mixte 7504 CNRS, Université de Strasbourg, Strasbourg Cedex, France

H. Vonesh Institut de Physique et Chimie Des Matériaux de Strasbourg, Unité Mixte 7504 CNRS, Université de Strasbourg, Strasbourg Cedex, France

A.A. Voronin M.V. Lomonosov Moscow State University, Moscow, Russia

C. Vozzi Instituto di Fotonica e Nanotecnologie-CNR, Milan, Italy

M.J.J. Vrakking Max-Born-Institut, Berlin, Germany

J. Wachtveitl Institute of Physical and Theoretical Chemistry, Johann Wolfgang Goethe-University, Frankfurt a. M., Germany

Suguru Wakabayashi Graduate School of Science, Kobe University, Kobe, Japan

M. Walbran Ludwig-Maximilians-Universität München, Garching, Germany; Max-Planck-Institute of Quantum Optics, Garching, Germany

Benji Wales Department of Physics and Astronomy, University of Waterloo, Waterloo, ON, Canada

B.C. Walker University of Delaware, Newark, DE, USA

ChuanLiang Wang State Key Laboratory of Magnetic Resonance and Atomic and Molecular Physics, Wuhan Institute of Physics and Mathematics, Chinese Academy of Sciences, Wuhan, China

YanLan Wang State Key Laboratory of Magnetic Resonance and Atomic and Molecular Physics, Wuhan Institute of Physics and Mathematics, Chinese Academy of Sciences, Wuhan, China

Yu-Ting Wang National Chiao Tung University, Hsinchu, Taiwan, Republic of China

Vincent Wanie Centre ÉMT, Institut National de la Recherche Scientifique, Quebec, QC, Canada

Matthew R. Ware Stanford PULSE Institute, SLAC National Accelerator Laboratory, Menlo Park, California, USA

Hiroshi Watanabe Institute for Solid State Physics, The University of Tokyo, Chiba, Japan

Kazuya Watanabe Department of Chemistry, Graduate School of Science, Kyoto University, Kyoto, Japan

Shuntaro Watanabe Research Institute for Science and Technology, Tokyo University of Science, Noda, Chiba, Japan; CREST, Japan Science and Technology Agency (JST), Chiyoda, Tokyo, Japan

Jannick Weisshaupt Max-Born-Institut Für Nichtlineare Optik Und Kurzzeitspektroskopie, Berlin, Germany

Lukas V. Whaley-Mayda Department of Chemistry, University of California, Berkeley, USA; Physical Biosciences Division, Lawrence Berkeley National Laboratory, Berkeley, USA

A.D. Wieck Lehrstuhl fuer Angewandte Festkoerperphysik, Ruhr-Universitaet Bochum, Bochum, Germany

S. Wiedbrauk Department of Chemistry, Ludwig-Maximilians-University, Munich, Germany

Roland Wilcken LS für BioMolekulare Optik, Ludwig-Maximilians-Universität München, Munich, Germany

Luuk J.G.W. van Wilderen Institute for Biophysics, Goethe University, Frankfurt, Germany

Krystyna E. Wilk School of Physics and Centre for Applied Medical Research, St Vincents Hospital, The University of New South Wales, Sydney, NSW, Australia

L. Wimmer 4. Physical Institute, University of Göttingen, Göttingen, Germany

T. Witting Blackett Laboratory, Imperial College, London, UK

Emanuel Wittmann LS für BioMolekulare Optik, Ludwig-Maximilians-Universität München, Munich, Germany

Michael Woerner Max-Born-Institut Für Nichtlineare Optik Und Kurzzeitspektroskopie, Berlin, Germany

Martin Wolf Fritz-Haber Institut der Max Planck Gesellschaft, Berlin, Germany

T.J.A. Wolf Stanford PULSE Institute, SLAC National Accelerator Laboratory, Menlo Park, CA, USA; Institute of Physical Chemistry, Karlsruhe Institute of Technology, Karlsruhe, Germany

Steffen Wolter Institute of Physics, University of Rostock, Rostock, Germany

C.-H. Wu Materials Sciences Division, Lawrence Berkeley National Laboratory, University of California, Berkeley, CA, USA; Department of Chemistry, University of California, Berkeley, CA, USA

M. Wu Department of Physics and Astronomy, Louisiana State University, Baton Rouge, LA, USA

Frank Würthner Institut für Organische Chemie and Center for Nanosystems Chemistry, Universität Würzburg, Würzburg, Germany

Yu Xi Weizmann Institute of Science, Rehovot, Israel

Yuanqin Xia National Key Laboratory of Science and Technology on Tunable Laser, Harbin Institute of Technology, Harbin, China

Rui Xian Max Planck Institute for the Structure and Dynamics of Matter, Centre for Free Electron Laser Science, Hamburg, Germany

Dequan Xiao Department of Chemistry, Yale University, New Haven, CT, USA; Department of Chemistry and Chemical Engineering, University of New Haven, West Haven, CT, USA

Hongqiang Xie State Key Laboratory of High Field Laser Physics, Shanghai Institute of Optics and Fine Mechanics, Chinese Academy of Sciences, Shanghai, China

Xinhua Xie Photonics Institute, Vienna University of Technology, Vienna, Austria

Wei Xiong JILA and Department of Physics, University of Colorado, Boulder, CO, USA

Bingwei Xu Biophotonic Solutions Inc, East Lansing, MI, USA

Huailiang Xu State Key Laboratory on Integrated Optoelectronics, Jilin University, Changchun, China

M. Yabashi RIKEN, Sayo, Hyogo, Japan; Japan Synchrotron Radiation Research Institute, Sayo, Hyogo, Japan

Atsushi Yabushita Department of Electrophysics, National Chiao-Tung University, Hsinchu, Taiwan; Core Research for Evolutional Science and Technology, Japan Science and Technology Agency, Chiyoda-ku, Tokyo, Japan

Kyuya Yakushi Toyota Physical and Chemical Research Institute, Nagakute, Japan

S.V. Yalunin 4. Physical Institute, University of Göttingen, Göttingen, Germany

K. Yamada Deaprtment of Physics, Tohoku University, Sendai, Japan

Y. Yamada Tokyo Institute of Technology, Meguro-Ku, Tokyo, Japan

Keita Yamaguchi Institute for Solid State Physics, The University of Tokyo, Kashiwa, Chiba, Japan; Yokohama Research Laboratory, Hitachi Ltd., Yokohama, Japan

Shoichi Yamaguchi Molecular Spectroscopy Laboratory, Riken, Wako, Japan; Ultrafast Spectroscopy Research Team, RIKEN Center for Advanced Photonics (RAP), Wako, Japan

Kaoru Yamamoto Department of Applied Physics, Okayama University of Science, Okayama, Japan

T. Yamamoto Department of Energy and Hydrocarbon Chemistry, Graduate School of Engineering, Kyoto University, Kyoto, Nishikyo-Ku, Japan

Yo-Ichi Yamamoto Department of Chemistry, Graduate School of Science, Kyoto University, Kitashirakawa-Oiwakecho, Sakyo-Ku Kyoto, Japan

Yusuke Yamanaka Department of Electronics and Electrical Engineering, Keio University, Yokohama, Japan

Keisaku Yamane Department of Applied Physics, Hokkaido University, Sapporo, Japan; CREST, JST, Chiyoda-ku, Japan

Kaoru Yamanouchi Department of Chemistry, School of Science, The University of Tokyo, Bunkyo-ku, Tokyo, Japan

Hideki Yamochi Research Center for Low Temperature and Materials Sciences, Kyoto University, Kyoto, Japan

Kazuhiro Yanagi Department of Physics, Tokyo Metropolitan University, Hachioji, Japan

Chunfan Yang The Institute of Chemistry, Hebrew University, Jerusalem, Israel

Ronggui Yang Department of Mechanical Engineering, University of Colorado, Boulder, CO, USA

K. Yano Department of Chemistry, School of Science and Technology, Kwansei Gakuin University, Sanda Hyogo, Japan

Y. Yang Department of EECS and Research Laboratory of Electronics, Massachusetts Institute of Technology, Cambridge, MA, USA

Jinping Yao State Key Laboratory of High Field Laser Physics, Shanghai Institute of Optics and Fine Mechanics, Chinese Academy of Sciences, Shanghai, China

Hong Ye Center for Free-Electron Laser Science, Deutsches Elektronen-Synchrotron DESY, Hamburg, Germany; Department of Physics, University of Hamburg, Hamburg, Germany

Tien-Tien Yeh Department of Electrophysics, National Chiao Tung University, Hsinchu, Taiwan

Yusuke Yoneda Graduate School of Engineering Science, Osaka University, Osaka, Japan

K. Yonemitsu Department of Physics, Chuo University, Tokyo, Japan

Kento Yonezawa Graduate School of Materials Science, Nara Institute of Science and Technology, Ikoma, Japan

S. Yoshino Department of Physical Science, Osaka Prefecture University, Osaka, Sakai, Japan

M. Yoshizawa Department of Physics, Graduate School of Science, Tohoku University, Aoba-Ku, Sendai, Japan

A. Zaïr Blackett Laboratory, Imperial College, London, UK

Flavio Zamponi Max-Born-Institut für Nichtlineare Optik und Kurzzeitspektroskopie, Berlin, Germany

Julius Zauleck Department Chemie, Ludwig-Maximilians-Universität München, Munich, Germany

I. Zegkinoglou Chemical Sciences Division, Lawrence Berkeley National Laboratory, University of California, Berkeley, CA, USA

Bin Zeng State Key Laboratory of High Field Laser Physics, Shanghai Institute of Optics and Fine Mechanics, Chinese Academy of Sciences, Shanghai, China

Dongfang Zhang Centre for Free Electron Laser Science, Department of Chemistry and Physics, Centre for Free Electron Laser Science, Max Planck Institute for the Structure and Dynamics of Matter, University of Hamburg, Hamburg, Germany

Haisu Zhang State Key Laboratory of High Field Laser Physics, Shanghai Institute of Optics and Fine Mechanics, Chinese Academy of Sciences, Shanghai, China

Li Zhang Photonics Institute, Vienna University of Technology, Vienna, Austria

Sheng Zhang Department of Physics, Harbin Institute of Technology, Harbin, China

Yu Zhang Department of Chemistry, University of California, Irvine, CA, USA

Zhonghua Zhang National Key Laboratory of Science and Technology on Tunable Laser, Harbin Institute of Technology, Harbin, China

Jiayu Zhao Institute of Modern Optics, Nankai University, Key Laboratory of Optical Information Science and Technology, Ministry of Education, Tianjin, China

Yang Zhao National Key Laboratory of Science and Technology on Tunable Laser, Harbin Institute of Technology, Harbin, China

A.M. Zheltikov M.V. Lomonosov Moscow State University, Moscow, Russia; Department of Physics and Astronomy, Texas A&M University, Texas, USA

Yi Ying Zheng School of Physical and Mathematical Sciences, Nanyang Technological University, Nanyang, Singapore

Chun Zhou Research Institute for Science and Technology, Tokyo University of Science, Noda, Chiba, Japan; CREST, Japan Science and Technology Agency (JST), Chiyoda, Tokyo, Japan

Yueming Zhou School of Physics, Huazhong University of Science and Technology, and Wuhan National Laboratory for Optoelectronics, Wuhan, China; Key Laboratory of Fundamental Physical Quantities Measurement of Ministry of Education, Wuhan, China

A. Zielinski Ludwig Maximilians Universität, Munich, Germany

R. Ziessel Laboratoire de Chimie Organique et Spectroscopies Avances (ICPEES-LCOSA), Strasbourg University—CNRS, Strasbourg, France

David F. Zigler Department of Chemistry, University of North Carolina at Chapel Hill, Chapel Hill, NC, USA

Donatas Zigmantas Department of Chemical Physics, Lund University, Lund, Sweden

Wolfgang Zinth BioMolekulare Optik and Center of Integrated Protein Science, CIPSM Ludwig-Maximilians-Universität München, Munich, Germany

Dmitriy Zusin Department of Physics and JILA, University of Colorado, Boulder, CO, USA

Part I
Attosecond and High-Order Harmonic Generation and Applications

Attosecond Control of Electron Emission in Two-Color Ionization of Atoms

G. Laurent, W. Cao, I. Ben-Itzhak and C.L. Cocke

Abstract We demonstrate that the electron emission from atoms generated by an attosecond pulse train made of both odd and even harmonics in the presence of a weak IR field can be temporally controlled on the attosecond time scale. In addition, we show that such electron emission carries information about the relative phase between consecutive odd and even harmonics in the attosecond pulse train.

1 Introduction

Recently, the advent of extreme-ultraviolet (XUV) light pulses in the attosecond time scale has opened up new avenues for experimentalists to observe the electron dynamics with unprecedented precision [1, 2]. Both isolated attosecond pulses and attosecond pulse trains (APTs) have already shown to be very promising tools for the temporal characterization of various atomic processes, such as the time delays in Auger decay, tunneling, and photoionization [3–5]. In this work, we demonstrate that an asymmetric electron emission from atoms can be generated and controlled on the attosecond time scale, by combining an APT composed of both odd and even harmonics and a weak IR field (10^{11} W/cm^2) [6]. In addition, we show that such

G. Laurent (✉)
Research Laboratory of Electronics, Massachusetts Institute of Technology,
Cambridge, MA 02139, USA
e-mail: glaurent@mit.edu

W. Cao · I. Ben-Itzhak · C.L. Cocke
James R. Macdonald Laboratory, Physics Department, Kansas State University,
Manhattan, KS 66506, USA
e-mail: caowei@phys.ksu.edu

I. Ben-Itzhak
e-mail: ibi@phys.ksu.edu

C.L. Cocke
e-mail: cocke@phys.ksu.edu

© Springer International Publishing Switzerland 2015
K. Yamanouchi et al. (eds.), *Ultrafast Phenomena XIX*,
Springer Proceedings in Physics 162, DOI 10.1007/978-3-319-13242-6_1

asymmetric emission is also related to the properties (amplitude and phase) of the APT. The temporal analysis of the modulated electron emission, based on an accurate description of the atomic physics of the two-color ionization process, then provides a way to measure the temporal profile of the attosecond pulse [7].

2 Results and Discussion

The principle of our experiment is presented in Fig. 1. Electron wave-packets are formed by ionizing argon with the APT in the presence of the IR field. Consequently, a mix of energy-degenerate even (s, d) and odd (p, f) parity states is fed into the continuum by one- and two-photon transitions. These interfere, leading to an asymmetric electron emission along the polarization vector. At some appropriate time delays between the APT and IR fields, the even and odd angular wave function resulting from one- and two-photon transitions, respectively, add constructively on one side (up) of the polarization vector and destructively on the other side (down), thus creating a strong up-down asymmetry in the angular emission of the photoelectrons, as shown in Fig. 2. This asymmetry oscillates as the time delay is varied.

The checkerboard pattern observed in Fig. 2 indicates that, at a given time delay, the asymmetric emission of photoelectrons associated with odd harmonics points in a direction along the polarization vector opposite to that associated with the even

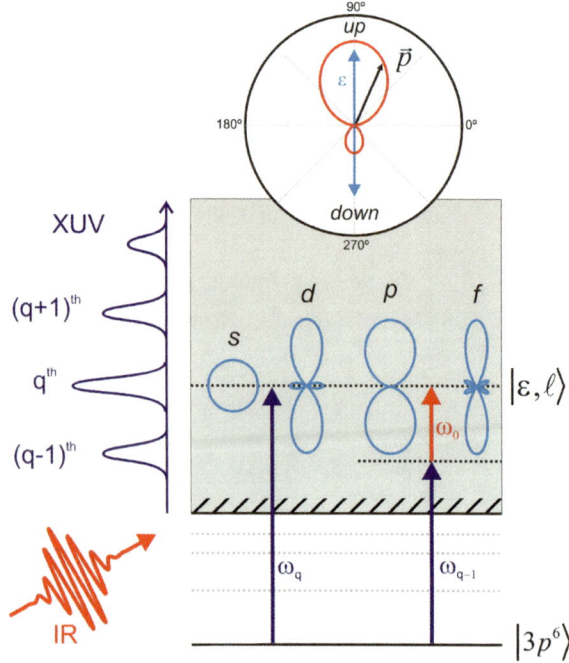

Fig. 1 Principle of the experiment: two-color photoionization of argon

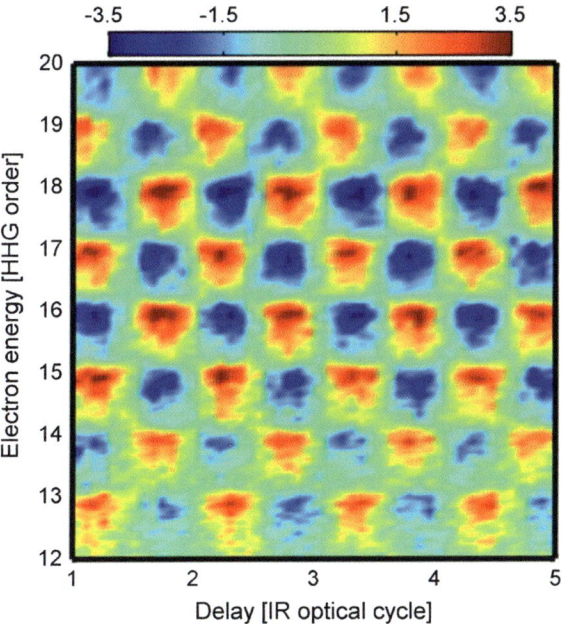

Fig. 2 Asymmetry plot of the electron emission

harmonics. The origin of this effect lies in the intrinsic phases of the harmonics [6]. Our measurement then allows us to determine the relative phase shift between consecutive odd and even harmonic in the comb. To that purpose, we have used a novel retrieval procedure, named iPROOF (improved Phase Retrieval by Omega Oscillation Filtering), which allows for the unique determination of the phase making up the APT by accurately taking into account the atomic physics of the two-color ionization process [7]. We observe a large phase shift between consecutive odd and even harmonics near $\pi/2$. As shown in Fig. 3, the resulting APT has a complex

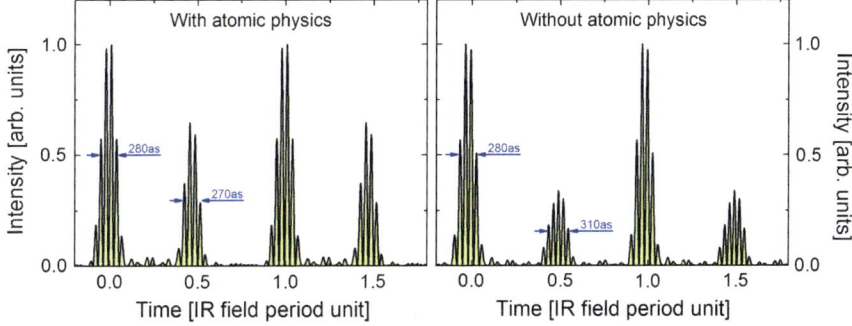

Fig. 3 Temporal profile of the attosecond pulse train retrieved by including and neglecting the contribution of the atomic physics of the photoionization process in the analysis

structure not resembling a single attosecond pulse once per IR period which is the case for zero phase. Finally, to emphasize the importance of an accurate treatment of the two-color ionization process in the retrieval procedure, the temporal profile of the APT retrieved without taking into account the atomic physics of the photoionization process is also shown. Even though the two analysis, with and without including the atomic physics, lead to a similar duration for the pulses (~ 300 as), the temporal envelopes of the reconstructed trains are distinct.

3 Conclusions

We have explored an interference process leading to an asymmetric electron emission from atoms along the laser polarization. The asymmetry can be controlled by varying the time delay between an APT made of odd and even harmonics and an IR pulse. We have shown that such asymmetric emission allows the measurement of the relative phase between consecutive odd and even harmonics in the comb.

Acknowledgments This work was supported by the U.S. Army Research Office, the National Science Foundation, and the Office of Basic Energy Sciences of the U.S. Department of Energy.

References

1. M. Hentschel *et al.*, "Attosecond metrology," Nature **414**, 509 (2001).
2. P. M. Paul *et al.*, "Observation of a Train of Attosecond Pulses from High Harmonic Generation," Science **292**, 1689 (2001).
3. M. Drescher *et al.*, "Time-resolved atomic inner-shell spectroscopy," Nature **419**, 803 (2002).
4. P. Eckle *et al.*, "Attosecond ionization and tunneling delay time measurements in helium," Science **322**, 1525 (2008).
5. K. Klünder *et al.*, "Probing single-photon ionization on the attosecond time scale," Phys. Rev. Lett. **106**, 143002 (2001).
6. G. Laurent *et al.*, "Attosecond Control of Orbital Parity Mix Interferences and the Relative Phase of Even and Odd Harmonics in an Attosecond Pulse Train," Phys. Rev. Lett. **109**, 083001 (2012).
7. G. Laurent *et al.*, "Attosecond pulse characterization," Opt. Express **21**, 16914 (2013).

Probing Xenon Electronic Structure by Two-Color Driven High-Order Harmonic Generation

M. Negro, D. Faccialà, B.D. Bruner, M. Devetta, S. De Silvestri,
N. Dudovich, S. Pabst, R. Santra, H. Soifer, S. Stagira and C. Vozzi

Abstract We studied the two-color HHG emission from xenon in the giant resonance spectral region. We found a substantial departure from the behavior expected for the single-active-electron picture which could be ascribed to electron correlation effects.

1 Introduction

High-order harmonic generation (HHG) is a sensitive probe of atomic and molecular structures. Recently this research field greatly benefited from the exploitation of mid-IR driving pulses that allowed the extension of the harmonic emission to higher photon energies, giving access to several phenomena previously unexplored, such as the giant resonance in xenon [1]. This enhancement in the harmonic generation yield around 100 eV has been interpreted in terms of the electronic structure of xenon, suggesting the key role of single [2] and multi-electron [1] contribution to the

M. Negro · M. Devetta · C. Vozzi (✉)
Istituto di Fotonica e Nanotecnologie—IFN-CNR,
Piazza L. Da Vinci 32, Milan, Italy
e-mail: caterina.vozzi@cnr.it

D. Faccialà · S. De Silvestri · S. Stagira
Dipartimento di Fisica, Politecnico di Milano,
Piazza L. Da Vinci 32, Milan, Italy

B.D. Bruner · N. Dudovich · H. Soifer
Department of Physics of Complex Systems,
Weizmann Institute of Science, Rehovot, Israel

S. Pabst · R. Santra
Center for Free-Electron Laser Science, DESY Notkestrasse 85,
Hamburg, Germany

R. Santra
Department of Physics, University of Hamburg,
Jungiusstrasse 9, Hamburg, Germany

© Springer International Publishing Switzerland 2015
K. Yamanouchi et al. (eds.), *Ultrafast Phenomena XIX*,
Springer Proceedings in Physics 162, DOI 10.1007/978-3-319-13242-6_2

harmonic generation process. The presence of phase-matching effects could play an important role as well, making the interpretation even more complicated [3].

In order to provide further insight into this interesting phenomenon, we exploited HHG by a two-color field, combining this powerful experimental approach with a mid-IR driving source. Harmonic spectroscopy based on two-color driving pulses has already been successfully applied to the study of electronic structure and attosecond dynamics in atoms and molecules [4, 5]. In this work we provide the evidence of a deviation of the xenon response with respect to the expected atomic behavior which could be attributed to electron correlation effects as introduced by Shiner et al. [1] and recently theoretically confirmed by Pabst and Santra [6].

2 The Giant Resonance in Xenon

The giant resonance in xenon [1, 6] involves electron correlation effects. In particular, in the first step of the HHG process (see Fig. 1), the electron is mainly ionized by tunneling from the valence 5p orbital. Under the influence of the external laser field, the electron is driven back to the parent ion. In the recombination step, the electron may either recombine with the 5p hole in the parent ion, or it could exchange energy with the ion promoting an inner shell electron from the 4d shell to the valence shell via Coulomb interaction (dotted arrow in Fig. 1). In this latter case the electron will then recombine with the 4d hole. This inter-channel coupling gives rise to the enhancement in the harmonic yield at high photon energy.

3 Experimental Results

We exploited an optical parametric amplifier (OPA) pumped by an amplified Ti: sapphire laser system (60 fs, 20 mJ, 800 nm). The OPA is based on difference frequency generation and provides driving pulses with 1,500 nm central wavelength, pulse duration of 20 fs and pulse energy 1.2 mJ. We used a BBO crystal followed by

Fig. 1 Sketch of the ionization and recombination steps of HHG (spin states are not considered)

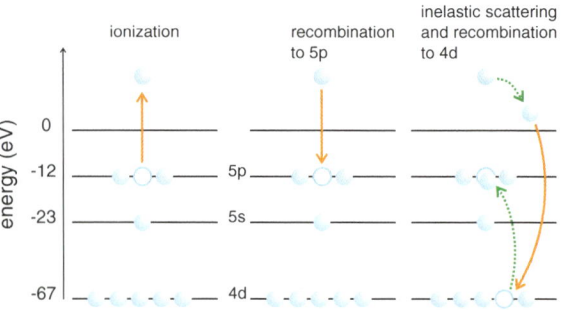

a calcite plate for generating a two color driving field with two components with perpendicular polarization. We changed the relative phase of the two driving field components with a pair of wedges. High order harmonics were generated by focusing the two-color driving field on a pulsed gas jet. Harmonics were detected by means of a flat field XUV spectrometer coupled to an MCP and a CCD detector.

Figure 2a shows a sequence of harmonic spectra generated in xenon for different values of the relative phase between the two components of the driving field as a 2D colormap. By changing this relative phase, the harmonic yield oscillates giving rise to the spectral modulation clearly visible in the figure. The spectra at high photon energy show for some value of the relative phase a feature which can be related to the giant resonance [1]. It is worth noting that the measurement reported in Fig. 2a has been performed with a very low xenon pressure in order to avoid clusterization and to minimize the role of phase-matching in the HHG process [3]. The phase of the spectral oscillation has been retrieved by Fourier transforming the sequence of harmonic spectra for each photon energy and by selecting the component at the oscillation frequency. The retrieved phase is shown as a solid black line on top of the 2D colormap in Fig. 2a. For the sake of comparison we show the same results for a measurement performed in krypton in Fig. 2b. In atoms, in the framework of the single-active-electron picture, one expects to see a smooth monotonic increase in the phase up to the cutoff [5], as the one we retrieved in krypton. In the framework of the single-active electron picture this behavior is intimately connected to the evolution of the electron trajectories related to the harmonic emission. In xenon we observed a very different behavior: a clear non-monotonic change in the slope of the phase corresponding to the spectral region of the giant resonance. Such finding is compatible with the occurrence of multielectron dynamics in HHG. Indeed, two-color HHG has been suggested as a sensitive approach for probing collective effects on an attosecond time scale [5].

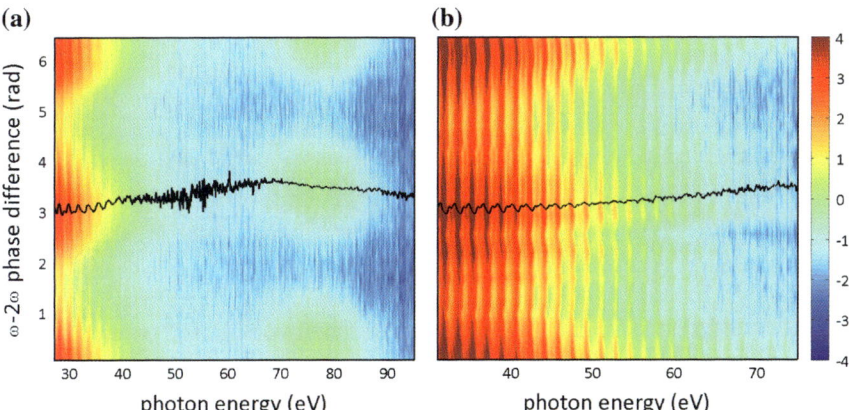

Fig. 2 Scan of harmonic spectra acquired in xenon (**a**) and krypton (**b**) as a function of photon energy and two-color phase difference (log scale). The *black line* corresponds to the phase of the spectral oscillation retrieved by Fourier transforming the sequence of harmonic spectra for each photon energy and by selecting the component at the oscillation frequency

4 Conclusions

In conclusion we investigated the xenon response by two-color HHG spectroscopy. We found a substantial departure from the expected behavior in atoms at high photon energy. This finding might support the evidence of correlation effects occurring in xenon and paves the way to the study of these dynamics on the attosecond time-scale.

References

1. A. D. Shiner et al., Nature Phys. **7**, 464–467 (2011)
2. Jingtao Zhang and D.-S. Guo, Phys. Rev. Lett. **110**, 063002 (2013)
3. C. Vozzi et al., New J. Phys. **13** 073003 (2011)
4. D. Shafir et al., Nature Phys. **5**, 412–416 (2009)
5. D. Shafir et al., Nature **485**, 343–346 (2012)
6. S. Pabst and R. Santra, Phys. Rev. Lett. **111**, 233005 (2013)

Strong Field Applications of Gigawatt Self-compressed Pulses from a Kagome Fiber

T. Balciunas, G. Fan, S. Haessler, C. Fourcade-Dutin, T. Witting,
A.A. Voronin, A.M. Zheltikov, F. Gérôme, G.G. Paulus, A. Baltuska
and F. Benabid

Abstract Nonlinear self-compression of 1.7-μm pulses in a gas-filled Kagome fiber down to a single cycle duration and pulse energies up to 100 μJ provides a uniquely simple driver source for high-harmonic generation and above-threshold ionization experiments.

OCIS codes 320.0320 Ultrafast optics · 020.2649 Strong field laser physics · 060.4005 Microstructured fibers

We demonstrate efficient self-compression of 1.7-μm infrared pulses to quasi-single cycle duration with sustained 100-μJ level pulse energies in an ultra-broadband Kagome-lattice hollow-core photonic crystal fibre (HC-PCF) [1]. This offers a simple solution to the seemingly deadlocked problem of efficient nonlinear pulse compression acting on the whole beam . A scalable (in energy and wavelength) self-compression scheme [2] is based on a newly identified regime of multiple soliton fusion inside a specialty photonic crystal fibre with a large hollow core filled with a

T. Balciunas (✉) · G. Fan · S. Haessler · A. Baltuska
Photonics Institute, Vienna University of Technology,
Gusshausstraße 27/387, 1040 Vienna, Austria
e-mail: tadas.balciunas@tuwien.ac.at

C. Fourcade-Dutin · F. Gérôme · F. Benabid
GPPMM Group, Xlim Research Institute, CNRS UMR 7252,
University of Limoges, Limoges, France

T. Witting
Blackett Laboratory, Imperial College London, SW7 2AZ London, UK

A.A. Voronin · A.M. Zheltikov
M.V. Lomonosov Moscow State University, Moscow, Russia

A.M. Zheltikov
Department of Physics and Astronomy, Texas A&M University,
College Station, Texas, USA

G.G. Paulus
Institute of Optics and Quantum Electronics, Jena, Germany

© Springer International Publishing Switzerland 2015   11
K. Yamanouchi et al. (eds.), *Ultrafast Phenomena XIX*,
Springer Proceedings in Physics 162, DOI 10.1007/978-3-319-13242-6_3

noble gas as a nonlinear optical medium. We show both experimentally and theoretically, that, in a single step, ultra-short infrared pulses with ~ 100 µJ energies undergo a 20-fold nonlinear self-compression to reach a sub-cycle duration and a gigawatt peak power at the fibre exit. We use these self-compressed pulses for ATI electron spectrometry measurements and driving high-order harmonic generation.

Solid-core photonic crystal fibres are particularly well suited for accurate dispersion shaping and controlled solitonic self-compression [3], but they are not energy-scalable as the correct dispersion profile generally corresponds to tiny core diameters. Gas-filled capillaries permit high energy throughput and are routinely used for mJ level few-cycle pulse generation [4], but the negative dispersion is weak and is usually overtaken by the positive dispersion of the gas thus preventing self-compression. The advent of HC-PCF with inhibited coupling between guiding channels created a new paradigm in optical guidance whereby core guided modes could cohabitate with cladding modes with no significant interaction between the two types of modes [1]. Unlike the earlier photonic bandgap guided HC-PCF [5], the ones with inhibited coupling stand out with a significantly larger bandwidth, lower dispersion and higher power handling, opening the opportunity to reach very high intensities capable of ionizing the gas in the fibre core [2].

Finally, by refocusing the self-compressed 10^{14} W/cm^2 beam from the fibre, the scheme can serve as a source for strong field applications. We employ above-threshold-ionization (ATI) of Xe atoms in a stereo time-of-flight spectrometer [6] to act, firstly, as a practical demonstrator that the output pulses lays in the strong-field regime, and as a full electric field characterization tool of the generated pulses, which allowed a preliminary verification of the carrier envelope phase stability of the solitonically self-compressed pulses.

Figure 1 illustrates the uniquely simple Kagome-fibre-based pulse self-compression scheme. Input 80-fs pulses centred at 1.8 µm are launched into a negative dispersion waveguide where they undergo self-compression. Complete temporal, spectral and spatial characterization of the input and output laser waveforms was performed using the technique of spatially encoded spectral phase interferometry for direct electric-field reconstruction (SEA SPIDER). These experimental results are in good agreement with theoretical simulations (not shown here) that allow getting more insight into the physics behind the self-compression process. This lets us identify a new regime of optical pulse compression, in which an optical shock wave, arising as a part of the highly nonlinear guided-wave evolution of an ultra-short laser pulse, enhances solitonic pulse compression. Remarkably, this dynamics takes place in a wave-guided manner, in the single-cycle self-compression limit and a single mode fashion.

Key results of the SEA SPIDER and stereo ATI measurements of the self-compressed pulses are presented in Fig. 2. The experimentally obtained spatio-temporal pulse intensity distribution shown in Fig. 2a proves that the self-compression mechanism is indeed active across the entire beam cross-section and is not limited to its central portion. Figure 2b, c shows ATI spectra obtained for different CEP settings (with a phase difference of π) of the 60-µJ output pulses obtained when the cell is filled with 1 bar of Ar. The dramatic asymmetry in the flux of the

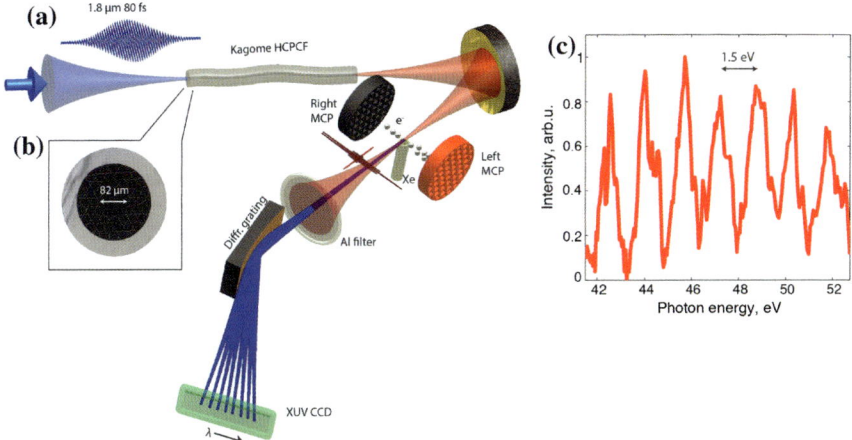

Fig. 1 a Pulse self-compression scheme based on Inhibited Coupling guiding HC-PCF, the field measurement using stereo ATI photoelectron spectrometry and high-order harmonic generation driven by self-compressed pulses. The experiment consists of launching infrared 80-fs pulses with energies of up to 120 μJ into 0.2 m long hypocycloid-core Kagome HC-PC with an *inner-circle* of 82 μm (corresponding to a mode-field diameter of ∼64 μm). The pulses were generated in an optical parametric amplifier, and which wavelengths were tuneable in the 1.4–1.9 μm range. The fibre shown in the *inset* (**b**) was optimized for this spectral range, exhibiting a linear loss of ∼70 dB/km and an S-shaped dispersion curve over 1,000–2,000 nm optical bandwidth and featuring anomalous dispersion ranging from 0 to 0.5 ps/km/nm within the spectral range of 1,150–2,000 nm. The solitonic self-compression in the negative dispersion regime is achieved by filling the fibre with Xenon to a pressure of 4 bar so to provide the necessary optical nonlinearity. **c** Measured spectrum of high-order harmonics generated using self-compressed pulses

photo-ionized electron outgoing in the direction of the highest-intensity half-cycle of the laser pulse serves as a direct proof that the electric field of the pulse carries but a single dominant half-cycle. Note that for the "cosine - pulses" where the peak of the envelope coincides with the peak of the electric field (see Fig. 2b, c), the instantaneous intensity of the adjacent electric field peaks is 50 % less than the main peak. For extreme nonlinear processes like strong-field ionization or high-order harmonic generation this confines the interaction to the single main peak of 1.4 fs duration.

The experimental measurement results are in good agreement with theoretical simulations that allow getting more insight into the physics behind the self-compression process. An experimentally measured as well as a numerically modelled evolution of output pulse temporal and spectral profiles after the propagation in the HC-PCF is shown in Fig. 3 as a function of input pulse energy. The dynamics reveal pulse self-compression in a negative dispersion fibre down to sub-cycle duration. FROG and SEA-SPIDER techniques were used to characterize the initial stages of the self-compression down to few-cycle regime to independently confirm

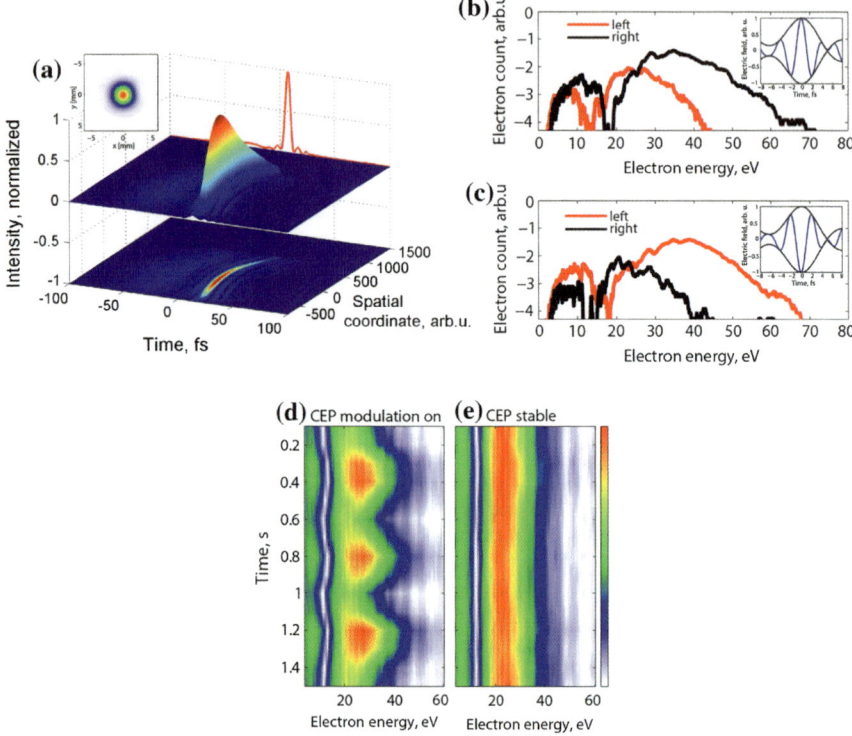

Fig. 2 The spatial uniformity and whole-beam compression is illustrated in the spatio-temporal reconstruction of the pulse depicted in panel **a** and measured beam profile shown in the *inset*. The panel **b** and **c** shows the photoelectron spectrum of ionized Xe atoms in two directions in Stereo-ATI spectrometer using self-compressed pulses for +cos and −cos electric fields. The panels **d** and **e** show that the self-compressed pulses retain the CEP phase of the much longer input pulses despite the very high compression ratio. The CEP of the input pulses was locked and the dependence of the electron spectrum was measured when the CEP of the laser is (**d**) modulated with a liner phase ramp scanned with the speed of 12 rad/s and **e** kept stabilized

the same result. However, with the spectra ultimately spreading into the multi-octave range, only the SEA-SPIDER technique remained adequate pulse characterization in the single-cycle regime.

The measured stereo ATI spectra allow us to calibrate the CEP values and the actual peak intensity, 5×10^{13} W/cm^2 in the ATI apparatus. By engaging an active CEP lock on the laser pumping the IR OPA that supplies 1.8-μm input pulses we were also able to verify that CEP stability is preserved in the self-compressed output pulses as confirmed by an ATI measurement shown in Fig. 2d, e. The generated high order harmonics serve as an example of the application of the compact pulse self-compression scheme for the generation of isolated attosecond pulses.

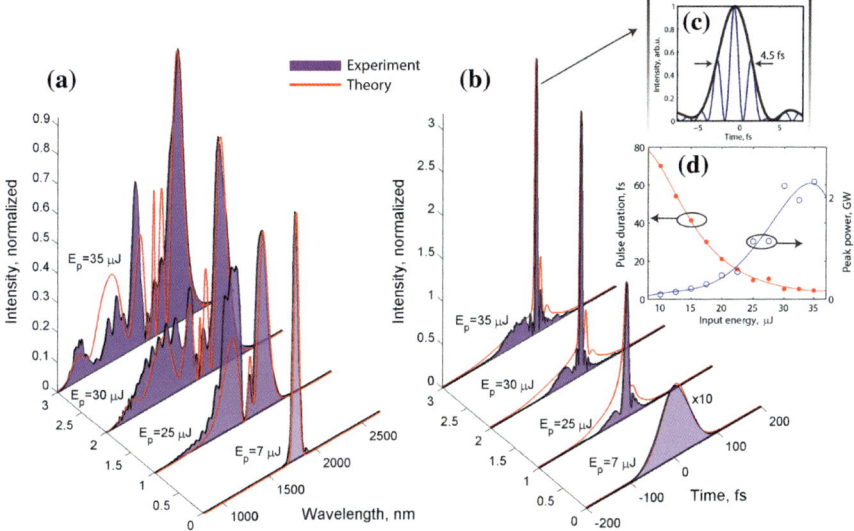

Fig. 3 Measured pulse self-compression at different energy levels. **a** Experimentally measured spectra after the waveguide. **b** Output pulse profiles measured with SEA-SPIDER technique. **c** Intensity (*red*) and instantaneous field intensity (*blue*) profiles of the shortest self-compressed pulse. **d** Dependence of the self-compressed pulse duration and peak power on the input pulse energy

References

1. F. Benabid, J. C. Knight, G. Antonopoulos, P. St. J. Russell, Science **298**, 399 (2002); F. Couny, F. Benabid, P. J. Roberts, P. S. Light, and M. G. Raymer, Science **318**, 1118 (2007); Y. Y. Wang, N. V. Wheeler, F. Couny, P. J. Roberts, and F. Benabid, Opt. Lett. **36**, 669 (2011).
2. P. Hölzer, W. Chang, J. C. Travers, A. Nazarkin, J. Nold, N. Y. Joly, M. F. Saleh, F. Biancalana, and P. St. J. Russell, Phys. Rev. Lett. **107**, 203901 (2011); John C. Travers, Wonkeun Chang, Johannes Nold, Nicolas Y. Joly, and Philip St. J. Russell, J. Opt. Soc. Am. B **28**, A11 (2011).
3. A. A. Amorim, M. V. Tognetti, P. Oliveira, J. L. Silva, L. M. Bernardo, F. X. Kärtner, and H. M. Crespo, Opt. Lett. **34**, 3851 (2009).
4. M. Nisoli, S. De Silvestri, and O. Svelto, Appl. Phys. Lett. 68, 2793 (1996).
5. D. G. Ouzounov, F. R. Ahmad, D. Muller, N. Venkataraman, M. T. Gallagher, M. G. Thomas, J. Silcox, K. W. Koch, and A. L. Gaeta, Science **301**, 1702 (2003).
6. T. Wittmann, B. Horvath, W. Helml, M. G. Schatzel, X. Gu, A. L. Cavalieri, G. G. Paulus, R. Kienberger, Nature Physics **5**, 357 (2009).

Intense Attosecond Pulses for Probing Ultrafast Molecular Dynamics

E.J. Takahashi, P. Lan, T. Okino, Y. Furukawa, Y. Nabekawa, K. Yamanouchi and K. Midorikawa

Abstract Recent developments of high energy high-harmonic beam line pumped by 100 Hz 12 fs Ti:sapphire laser for attosecond nonlinear Fourier transformation spectroscopy and multi-gigawatt isolated attosecond pulses by infrared two-color laser field synthesis are reported.

1 Introduction

With the progress of femtosecond laser technologies, it becomes possible to generate attosecond pulses by utilizing the coherent wavelength conversion process known as high-order harmonic generation. As a straightforward extension of femtochemistry, in which nuclear wavepacket motion is tracked in real-time, attosecond pulses enable us to capture the electron motion in atoms and molecules and the ultrafast nuclear-electron coupled motion in molecules. One of the ultimate goals of attosecond science is to perform attosecond pump and attosecond probe measurement to understand chemical reactions such as ultrafast charge migration and electron correlation with unprecedented temporal resolution of the atomic unit.

To seriously tackle the above interesting research topics, one of the most important issues is the development of high-power attosecond pulses (IAPs) and/or attosecond pulse trains (APTs). IAPs and APTs are produced using high-order hamonic generation in gases. At RIKEN, high-power APTs have successfully been generated owing to research on high harmonic energy scaling using a loose-focusing geometry [1]. Using the intense ATPs, we have developed a new attosecond spectroscopic method called as nonlinear Fourier transformation spectroscopy (NFTS),

E.J. Takahashi · P. Lan · T. Okino · Y. Furukawa · Y. Nabekawa · K. Midorikawa (✉)
RIKEN Ceneter for Advanced Photonics, 2-1 Hirosawa, Wako, Saitama 351-0198, Japan
e-mail: kmidori@riken.jp

K. Yamanouchi
Department of Chemistry, School of Science, The University of Tokyo, 7-3-1 Hongo, Bunkyo-Ku, Tokyo 113-0033, Japan

© Springer International Publishing Switzerland 2015
K. Yamanouchi et al. (eds.), *Ultrafast Phenomena XIX*,
Springer Proceedings in Physics 162, DOI 10.1007/978-3-319-13242-6_4

in which the nonlinear responses of the molecules are encoded in the energy-resolved interferometric autocorrelation traces [2, 3]. More recently, we have proposed and demonstrated a robust generation method of intense isolated attosecond pulses by employing two-color field synthesis and the energy-scaling method of high-order harmonic generation [4]. Here we report these two research topics of attosecond science at RIKEN.

2 Nonlinear Fourier Transformation Spectroscopy with Intense Attosecond Pulse Train

Observation and control of nuclear wavepacket and electron wavepacket are crucial to understanding ultrafast molecular dynamics. In particular, sub-10 fs high harmonics composing APT is one of the effective light sources to launch nuclear and electron wavepackets. Temporal profile of APT and the nonlinear response of molecules can be investigated by NFTS. Recently, we extended our NFTS method by introducing 100 Hz 12 fs 50 mJ Ti:sapphire laser as a pump source and velocity map imaging spectrometer with which the angular distribution of fragment ions and photoelectrons can be measured. The anisotropy of the ejection direction of the fragment ion carries the information of the symmetry of the electronic states involved in the decomposition processes.

Experimentally, intense APT is generated by loosely focusing the output of the femtosecond laser system to a xenon gas cell, and the resultant APT is propagated to a highly stable interferometer. The pair of APT is focused to a molecular beam of target molecules with a SiC concave mirror. The generated fragment ions are extracted by parallel electrodes in the velocity map imaging configuration. The ion signals are detected by a detector composed of chevron type microchannel plates and a phosphor screen. The signal of the fragment ion was selectively detected by imposing the pulse gating voltage to the detector system. Figure 1 is the temporal shape of ATP measured by interferometric autocorrelation using two-photon dissociative ionization of N_2 molecules.

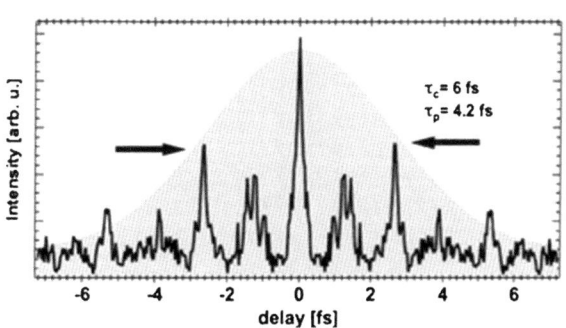

Fig. 1 Temporal profile of ATP. Pulse envelop of 4.2 fs and individual pulse duration of 400 as were obtained

3 Generation of Multi-gigawatt Isolated Attosecond Pulses

We have demonstrated the generation of microjoule IAPs which are intense enough to induce the nonlinear phenomena in atoms and molecules, and performed direct characterization by autocorrelation measurement. Our generation scheme, which is based on the infrared two-color laser field synthesis [5] and the energy-scaling method [1], is robust and straightforward for scaling up the IAP energy. By carefully designing the generation configurations, we obtained IAPs with energy up to 1.3 µJ/pulse around 30 eV region, thus showing its energy enhancement from 100- to 1,000-fold compared with the previous report so far. The conversion efficiency attained was improved to be 1.1×10^{-4} thanks to the phase matching technique. From a 500-as pulse duration determined by autocorrelation method, the peak power of this IAP was evaluated to be 2.6 GW. Furthermore, the output energy of a mid-plateau region (14–29 eV) having a quasi continuum spectrum by the two-color high harmonic generation attained to 10 µJ level. By utilizing this quasi-continuum spectrum, we also generated a quasi-IAP with a pulse duration of 375 as.

We are going to expand this two-color scheme into much shorter wavelength region. Since our method has the advantage that the high harmonic output yield can be linearly scaled up by increasing the high harmonic emission volume, we can estimate exactly a scaled-up configuration in the soft-X-ray region to obtain even shorter IAP durations. In the soft-X-ray region around 100 eV, we have already demonstrated a ~50 nJ energy per harmonic order by one-color excitation scheme in Ne gas. By straightforwardly upgrading the two-color scheme to a main pump energy of 50 mJ and adopting a focusing length of 5 m with a 5-cm-long Ne medium, we expect to achieve an isolated attosecond pulse energies grater than 0.1 µJ at 95–110 eV, which is almost 1,000-fold higher than the energies previously reported. Our simulations indicate that IAPs can be generated with sub-300-as duration. In addition, when we switch the wavelengths of the main and supplementary fields, we can expect to further shorten the wavelength of the generated attosecond pulses to the "water window" region [6].

References

1. E. Takahashi, Y. Nabekawa, T. Otsuka, M. Obara and K. Midorikawa, "Generation of highly coherent sub-mJ soft X-ray by high-order harmonics", Phys. Rev. A 66, 021802 (2002).
2. Y. Nabekawa, T. Shimizu, T. Okino, K. Furusawa, H. Hasegawa, K. Yamanouchi, and K. Midorikawa, "Interferometric autocorrelation of an attosecond pulse train in the single-cycle regime", Phys. Rev. Lett. 97, 153904 (2006).
3. T. Okino, K. Yamanouchi, T. Shimizu, R. Ma, Y, Nabekawa, and K. Midorikawa, "Attosecond nonlinear Fourier transformation spectroscopy of CO_2 in extreme ultraviolet wavelength region," J. Chem. Phys. 129, 161103 (2008).
4. E. J. Takahashi, P. F. Lan, O. D. Mücke, Y. Nabekawa, and K. Midorikawa, "Attosecond nonlinear optics using gigawatt-scale isolated attosecond pulses," Nat. Commun. 4, 2691 (2013).

5. E. J. Takahashi, P. Lan, O. D. Mücke, Y. Nabekawa, and K. Midorikawa, "Infrared two-color multicycle laser field synthesis for generating intense attosecond pulse", Phys. Rev. Lett. 104, 233901 (2010).
6. P. Lan, E. J. Takahashi, and K. Midorikawa, "Wavelength scaling of efficient high-order harmonic generation by two-color infrared laser fields," Phys. Rev. A 81, 061802(R) (2010).

Disentangling Structural and Dynamical Effects via Multidimensional High Harmonic Spectroscopy

B.D. Bruner, H. Soifer, M. Negro, M. Devetta, D. Faccialá, C. Vozzi, S. Stagira, S. de Silvestri and N. Dudovich

Abstract Extending the dimensionality of high harmonic generation (HHG) measurements has the potential to reconstruct structural features in molecules and resolve multielectron dynamics on attosecond time scales. We demonstrate that structural and dynamical effects in molecules can be unambiguously distinguished using multidimensional HHG techniques.

1 Introduction

HHG spectroscopic techniques provide a general tool for obtaining Angstrom-level structural information about atoms and molecules and for resolving dynamics on the attosecond time scale. However one of the major limitations is the intrinsic coupling between spatial and temporal domains that limits the complete reconstruction of the investigated processes. By increasing the dimensionality of the measurement, in close analogy with the NMR and optical regimes [1], it is possible to unravel structural and dynamical features that are often hidden in HHG spectroscopy.

Past measurements have focused on extracting either structural information, such as tomographic reconstructions [2], or time resolved attosecond dynamics. However, even for a relatively simple molecule such as CO_2, there is an ongoing debate regarding the interplay between structure and dynamics in the HHG emission. When generating harmonics at 800 nm, there is a multi-orbital contribution to the

B.D. Bruner (✉) · H. Soifer · N. Dudovich
Department of Physics of Complex Systems, Weizmann Institute of Science,
76100 Rehovot, Israel
e-mail: barry.bruner@weizmann.ac.il

D. Faccialá · S. Stagira · S. de Silvestri
Dipartimento di Fisica, Politecnico di Milano, 20133 Milan, Italy

M. Negro · M. Devetta · C. Vozzi
Instituto di Fotonica e Nanotecnologie-CNR, 20133 Milan, Italy

© Springer International Publishing Switzerland 2015
K. Yamanouchi et al. (eds.), *Ultrafast Phenomena XIX*,
Springer Proceedings in Physics 162, DOI 10.1007/978-3-319-13242-6_5

signal, leading to hole dynamics that are visible in the so-called 'spectral dynamical minimum' [3, 4]. However, with mid-IR lasers, the dominant spectral feature is a fixed minimum in the spectrum that was attributed to a structural signature of the CO_2 HOMO [2]. Here, we increase the dimensionality of the measurement by generating harmonics with a two-color mid-IR field. We observed a clear multi-orbital contribution of to the harmonic signal, nearby to- but easily distinguished from the structural minimum.

When generating HHG with a two-color field, composed of the strong generating field and a weak second-harmonic field polarized in the perpendicular direction, the harmonic intensity oscillates as a function of the phase delay between the two components. The delay at which the signal of a given harmonic is maximal is determined by the temporal evolution of the corresponding electron trajectory [3]. When applying this scheme to a signal in which two orbital contributions interfere, the two electron trajectories have slightly different temporal evolution, and the harmonic intensity is peaked at slightly different delays. When the two contributions interfere destructively, the small delay between them can lead to a large phase jump in the position of the maximum of the total signal, as was observed for 800 nm [3].

2 Experimental Setup and Results

Powerful mid-IR femtosecond pulses (1.2 mJ, 20 fs, 1.5 μm central wavelength) were generated by an optical parametric amplifier (OPA) pumped by an amplified Ti:S laser system (60 fs, 20 mJ, 800 nm). The molecules were impulsively aligned using a portion of the 800 nm beam. The relative phase delay was controlled with a pair of glass wedges.

The harmonic intensity as a function of the two colour phase delay ϕ was measured for CO_2 molecules aligned at 0° relative to the fundamental field polarization, and for two intensities of the fundamental field. A 2D colour plot of the low intensity measurement is shown in Fig. 1a and a clear shift of the maximum HHG signal with respect to ϕ is seen at 70 eV. For the high intensity (2D plot not shown), the shift of the maximum is centered at 75 eV. This is more easily observed by comparing the delay ϕ where the signal maximizes to the calculated delay for a single-orbital contribution (Fig. 1b). A large deviation from the single-orbital case is easily observed, and the maximum difference is shifted by about 5 eV between the two intensities.

Theory predicts that the delay shift should occur at the HHG energies where the relative accumulated phase between electrons ionized from the channels is close to π [3]. The calculated delay shifts are in good agreement with the experiment. The magnitude of the shift depends on the relative signal contributions and contrasts of the two channels, which depends on the laser intensity. The striking difference between the two intensities (both in sign and magnitude) is another strong indication of two-orbital contributions.

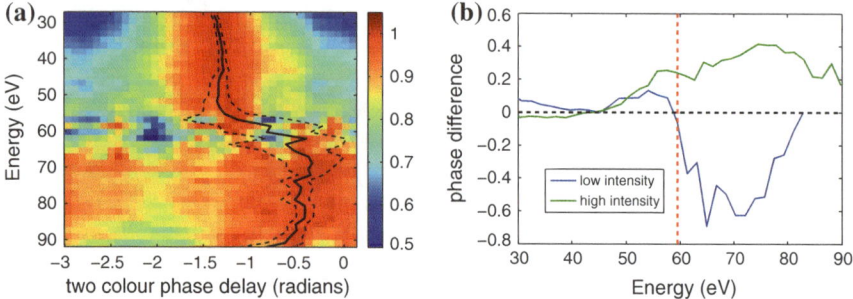

Fig. 1 **a** Normalized 2D colour plot of HHG signal versus phase delay ϕ between the two fields. *Bold line* HHG maximum with respect to phase. *Dashed lines* error in extracting the maximal phase. **b** Maximum two colour phase delay as a function of HHG energy. The phase differences are shown relative to calculated values for a reference atom (Kr). The intensities in the measurement were 1.4×10^{14} W/cm^2 (high) and 1.2×10^{14} W/cm^2 (low)

We note that the structural minimum in CO_2 is centred at 60 eV and selectively suppresses signal from the HOMO channel over a range of about 15 eV. This structurally-based reduction of the signal from the HOMO relative to the HOMO-2 leads to a more comparable contribution from the two channels and allows the delay shift(s) to be resolvable in a region of the HHG spectrum where the relative contributions of multiple channels is typically obscured [5].

In future work, we are aiming to extend these multidimensional schemes to a wider range of molecular systems. These measurements were also performed in aligned ethylene (C_2H_4) molecules (Fig. 2). The harmonic maximum shows a very complex behaviour with respect to ϕ. In CO_2, the HOMO is a nonbonding molecular orbital and therefore removal of a HOMO electron via tunnel ionization will not affect the bonding structure of the molecule. In contrast, the HOMO in C_2H_4 is a bonding orbital, and removal of an electron should lead to rapid conformations of the carbon–carbon bond, which is then probed by the recombining electron at later times.

Fig. 2 2D plot for C_2H_4 aligned at 0°. The signal shows a complex behaviour with respect to ϕ

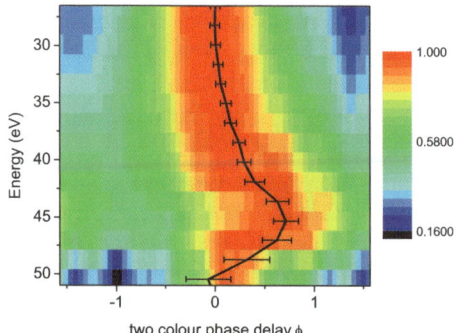

References

1. R. R. Ernst *et. al.*, *Principles of Nuclear Magnetic Resonance in One and Two Dimensions* (Oxford Univ. Press, New York, 1990).
2. C. Vozzi *et. al.*, Nat. Physics **7**, 822–826 (2011).
3. D. Shafir *et. al.*, Nature **485**, 343–346 (2012).
4. O. Smirnova *et. al.*, Nature **460**, 972–977 (2009).
5. A. Rupenyan *et. al.*, Phys. Rev. A **87**, 033409 (2013).

Attosecond Frequency Resolved Momentum Imaging of Two-Photon Dissociative Ionization Dynamics of Nitrogen Molecule

T. Okino, Y. Furukawa, A. Amani Eilanlou, Y. Nabekawa,
E.J. Takahashi, K. Yamanouchi and K. Midorikawa

Abstract Two-photon dissociative ionization processes of nitrogen molecule are investigated with attosecond nonlinear Fourier transformation spectroscopy. Nonlinear absorption spectrum of nitrogen molecule are extracted by measuring interferometric autocorrelation of a-few-pulse attosecond pulse train using non-sequential two-photon processes of nitrogen molecule. The frequency resolved momentum images (FRMI) are reconstructed from the delay dependent momentum images of fragment ion N^+ and the resultant FRMIs show the attosecond nonlinear response of nitrogen molecule.

1 Introduction

When the sub-femtosecond temporal profile in short wavelength region is characterized with autocorrelation method, the efficient nonlinear media such as atoms and molecules should be utilized. The nonlinear response of medium should be known in detail for determining the temporal profile covering several electronic states of atoms and molecules (several eV to dozens of eV). The frequency domain spectra obtained by Fourier transformation of the observed interferometric autocorrelation trace (IAC) of attosecond pulse train (APT) depends on the atomic species and the

T. Okino (✉) · Y. Furukawa · A. Amani Eilanlou · Y. Nabekawa · E.J. Takahashi ·
K. Midorikawa
RIKEN Center for Advanced Photonics, 2-1 Hirosawa, Wako,
Saitama 351-0198, Japan
e-mail: tomoya.okino@riken.jp

K. Yamanouchi
Department of Chemistry, School of Science, The University of Tokyo,
7-3-1 Hongo, Bunkyo-ku, Tokyo 113-0033, Japan

© Springer International Publishing Switzerland 2015
K. Yamanouchi et al. (eds.), *Ultrafast Phenomena XIX*,
Springer Proceedings in Physics 162, DOI 10.1007/978-3-319-13242-6_6

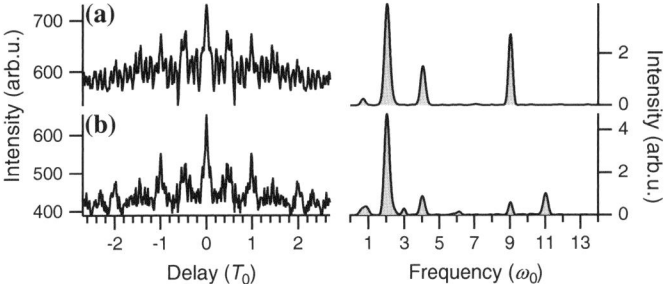

Fig. 1 Fragment ion intensity as a function of delay (*left*) and frequency spectra (*right*). **a** E = 3.0 eV, **b** E = 4.0 eV. T_0 and ω_0 are the optical period and the angular frequency of fundamental light field, respectively

fragment ion species and the kinetic energy of fragment ions (Fig. 1) [1, 2]. In other words, the observed frequency spectrogram is regarded as a convolution of the intrinsic electric waveform of APT and the nonlinear response of atoms and molecules to APT.

2 Experimental Setup

The intense a-few-pulse APT was generated by focusing the output of femtosecond laser system (800 nm, 12 fs, 100 Hz) onto a xenon static gas cell. The generated APT was spatially divided into two with two Si plane mirrors and the reflected APT pulses are introduced into a single-shot capable velocity map imaging spectrometer (VMIS). The APT beams were focused onto a molecular beam of N_2 introduced into a vacuum chamber using a piezo valve integrated into the repeller electrode in the VMIS. The temporal delay Δt was scanned every 35.6 as from −7 to +7 fs by moving one of the two Si plates on a single axis high precision piezo stage. For securing the delay stability, the temperature of piezo stage was actively controlled by a Peltier unit within 0.001 °C.

In each delay, the momentum image of fragment ion N^+ originated from the two-photon dissociative ionization processes $\left(N_2^+ \rightarrow N^+ + N\right)$ was recorded by temporally gating a MCP/phosphor screen assembly with 100 ns duration.

3 Results and Discussion

The fragment ions can be generated via one-photon processes and two-photon processes. While one-photon signal depends on the delay of APT pulses, two-photon signals are composed of delay-dependent signals and delay-independent signals. The delay-independent two-photon signals are originated from one arm of interferometer. The total signal, depending on the kinetic energy of fragment ion E and the ejection angle θ measured from the laser polarization direction and the delay of APT pulses τ, can be expressed as

$$I(E, \theta, \tau) = I_1(E, \theta) + I_{2,\text{incoh}}(E, \theta) + I_{2,\text{coh}}(E, \theta, \tau), \tag{1}$$

where $I_1(E, \theta)$, $I_{2,\text{incoh}}(E, \theta)$, and $I_{2,\text{coh}}(E, \theta, \tau)$, stand for one-photon signals, incoherent two-photon signals, and coherent two-photon signals. Since the correlated signals are originated from APT pulses composed by multiple harmonics, the signals can be expressed explicitly as

$$I_{2,\text{coh}}(E, \theta, \tau) = \sum_j a_j(E, \theta) \cos(\omega_j \tau) \tag{2}$$

where $a_j(E, \theta)$ shows the fraction of ω_j component for the kinetic energy E and the ejection angle θ. Frequency resolved momentum image can be constructed by Fourier transformation of the correlated two-photon signals $I_{2,\text{coh}}(E, \theta, \tau)$ as follows.

$$I_{2,\text{coh}}(E, \theta, \omega_j) = a_j(E, \theta) \tag{3}$$

The odd order frequencies of fundamental light field are originated from the interference fringes of the harmonic field. On the other hand, the even order frequencies of fundamental light field are explained as the difference frequency between the adjacent and the next adjacent harmonic field modes to form the envelope of APT. The FRMIs can be constructed from the correlated two-photon signals $I_{2,\text{coh}}(E, \theta, \omega_j)$.

The recorded momentum image of fragment ions N^+ on the detector was decomposed into the FRMIs as shown in Fig. 2. These momentum images can clearly visualize the angular distribution of fragment ions and the correlations between harmonics leading to the non-sequential two-photon processes. For example, the outer two ring features of 2ω around 4–5 eV shown in Fig. 2b correlate 9ω and 11ω. Since the anisotropy of angular distribution for 9ω is larger than that for 11ω around 4–5 eV, the outermost anisotropic feature is correlated to 9ω and the second outermost anisotropic feature is correlated to 11ω, respectively (Fig. 2).

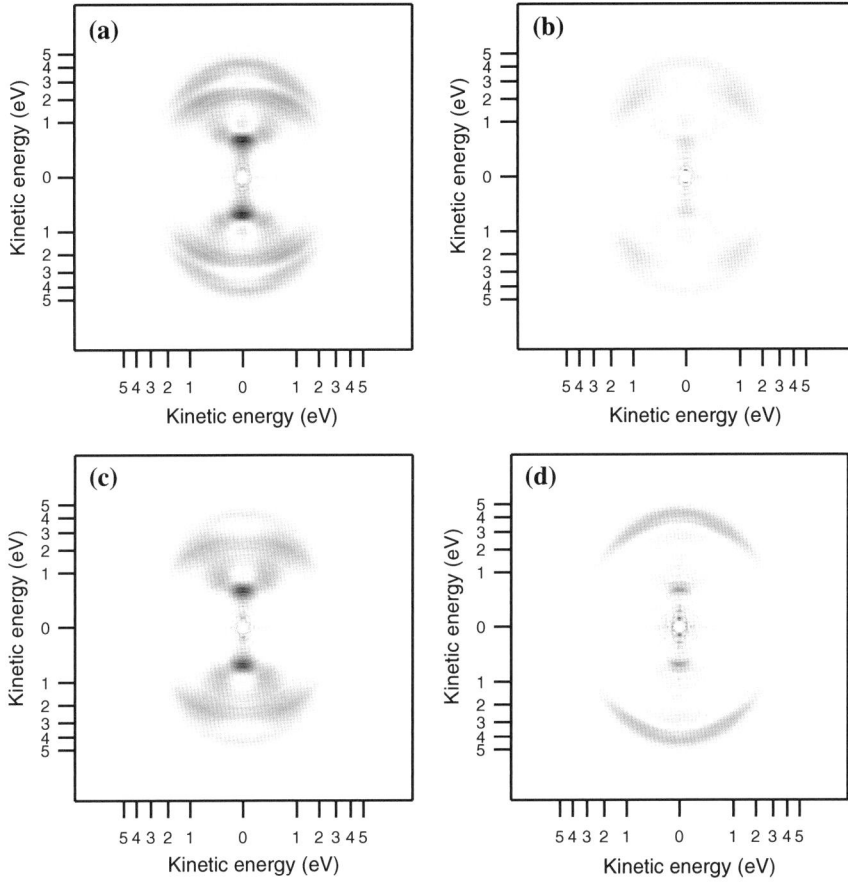

Fig. 2 Frequency resolved momentum images of fragment ion N^+. **a** 2ω, **b** 4ω, **c** 9ω, **d** 11ω

Acknowledgments This research was partially supported by the Photon Frontier Network Program of Ministry of Education, Culture, Sports, Science and Technology, Japan.

References

1. Y. Nabekawa, T. Shimizu, T. Okino, K. Furusawa, H. Hasegawa, K. Yamanouchi, K. Midorikawa, Phys. Rev. Lett. **97**, 153904 (2006).
2. T. Okino, Y. Furukawa, T. Shimizu, Y. Nabekawa, K. Yamanouchi, K. Midorikawa, J. Phys. B: At. Mol. Opt. Phys. **47**, 124007 (2014).

High-Order Harmonics Fourier Transform Spectroscopy of Two-Photon Dissociative Ionization of Hydrogen Molecules

Yusuke Furukawa, T. Okino, Y. Nabekawa, A. Amani Eilanlou,
E.J. Takahashi, K. Yamanouchi and K. Midorikawa

Abstract High-order harmonic generation provides us a coherent femtosecond pulse in a vacuum ultraviolet wavelength region. We have investigated the two-photon dissociative ionization of deuterium molecules by Fourier transform spectroscopy with a high-order harmonic generation pulse. We measured the delay-dependent kinetic energy distribution of the deuteron fragments resulting from the dissociative ionization by scanning the delay between the two replicas of the high-order harmonic generation pulse. From the Fourier transform analysis, we successfully distinguished the harmonic order, which induced the photodissociation and provided the specific kinetic energy to the fragments via the repulsive electronic excited state.

1 Introduction

The high-order harmonic generation (HHG) pulse is one of the powerful light sources to investigate ultrafast phenomena in an extreme-ultraviolet region. However, it is not simple to understand the molecular dynamics induced by the HHG pulse without additional experimental results or theoretical supports, because the HHG pulse is generally composed by a sequence of odd-order harmonics. By measuring the interferometric autocorrelation (IAC) traces of HHG pulses with the molecular responses, one can distinguish which harmonic order is involved in the

Y. Furukawa (✉) · T. Okino · Y. Nabekawa · A. Amani Eilanlou · E.J. Takahashi · K. Midorikawa
RIKEN Center for Advanced Photonics, 2-1 Hirosawa, Wako,
Saitama 351-0198, Japan
e-mail: furukawa-y@riken.jp

K. Yamanouchi
Department of Chemistry, School of Science, The University of Tokyo,
7-3-1 Hongo, Bunkyo-ku, Tokyo 113-0033, Japan

© Springer International Publishing Switzerland 2015
K. Yamanouchi et al. (eds.), *Ultrafast Phenomena XIX*,
Springer Proceedings in Physics 162, DOI 10.1007/978-3-319-13242-6_7

28

Fig. 1 Pump and probe scheme of the sequential two photon dissociative ionization of D_2 with the HH pulse. The pump process is the ionization of D_2 and produce D_2^+ in the $1s\sigma_g$ electronic ground state. The probe process is the dissociation through the electronic transition from the $2p\sigma_u$ state to the $2p\sigma_u$ repulsive state

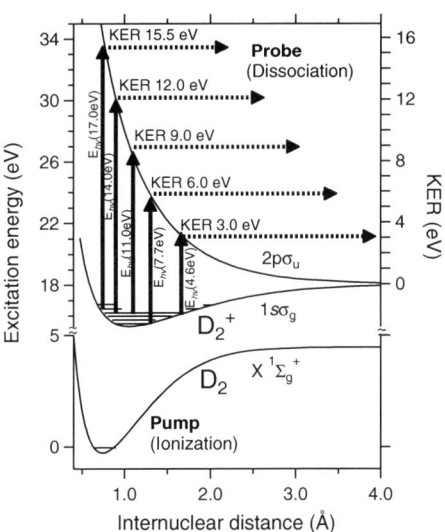

multi-photon process resulting in the molecular response [1, 2]. When the IAC of the HHG pulse is measured with deuterium molecules (D_2), the sequential two-photon dissociative ionization, in which the first D_2 is ionized to the $1s\sigma_g$ state of D_2^+ and then the D_2^+ in $1s\sigma_g$ is excited to the $2p\sigma_u$ repulsive potential, has been observed [3]. The low-order harmonics, whose frequencies appear in the IAC traces of the D_2^+ kinetic energy distribution, tends to excite D_2^+ at the long internuclear distance region. When higher energy photon induces the transition between $1s\sigma_g$ and $2p\sigma_u$, the D_2^+ is excited at a shorter internuclear distance and dissociate with larger kinetic energy release (KER).

In the present work, we investigated the two-photon dissociative ionization of D_2 using High-order harmonics Fourier transform spectroscopy to detect the second excitation process of D_2^+ by the high-order harmonics. The pump and probe scheme of the sequential two-photon dissociative ionization of D_2 in the present work is shown in Fig. 1.

2 Experimental

The HHG pulse was generated in a xenon gas cell by focusing of the fundamental driving laser pulses with a sub-15-fs pulse duration and a center wavelength of approximately 800 nm [4]. The HHG pulse was spatially divided into two replicas by the beam splitter which was made with a pair of silicon mirrors. The resultant HHG pulse pair was focused onto the D_2 molecular beam by a SiC concave mirror with a focal length of 100 mm. The D^+ fragments were produced through the dissociative ionization process induced by the HHG pulses. By scanning the delay between the two HHG pulses, we measured the delay-dependent kinetic energy distribution of the D^+ fragments using velocity map imaging ion mass spectrometer. We executed signal counting analysis for image data in order to detect low probability events.

3 Results and Discussion

The KER distribution of the D_2^+ dissociation parallel to the laser polarization exhibits the multi peaks, the five of which are illustrated in Fig. 1. The 11th-order or higher order harmonic is required to produce the KER component that is observed at 15.5 eV. To investigate how the HHG pulse interacts with D_2^+, we performed the Fourier transform spectroscopy. The delay-dependent KER distribution is obtained by arranging the kinetic energy spectra obtained at each delay in order of increasing the temporal delay in Fig. 2a. By performing the Fourier transformation of the delay-dependent KER distribution, the KER-resolved frequency spectrogram is obtained, as shown in Fig. 2b. In the KER-resolved frequency spectrogram, the five observed frequencies coincide with the optical frequencies of the harmonics up to the 11th-order.

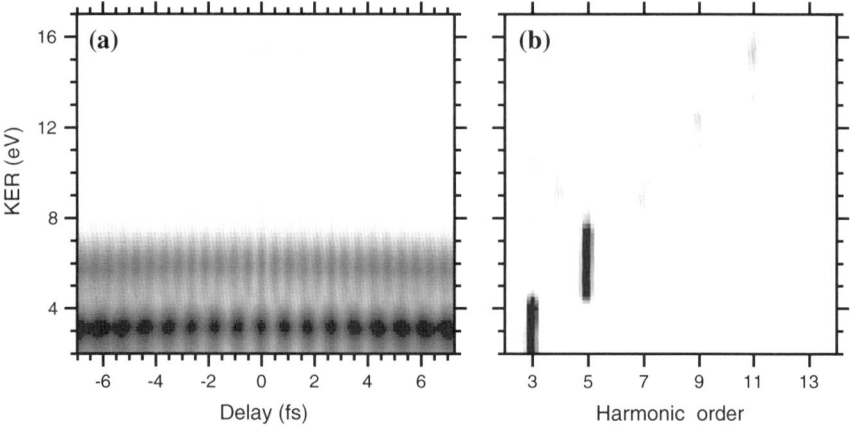

Fig. 2 a The delay-dependent kinetic energy release distribution of the D^+. **b** Kinetic-energy-resolved frequency spectrogram obtained by carrying out Fourier transform of the spectrogram in (**a**)

Acknowledgments This research was supported by the Photon Frontier Network Program of the MEXT, Japan.

References

1. Y. Nabekawa, T. Shimizu, T. Okino et al., Interferometric autocorrelation of an attosecond pulse train in the single-cycle regime. Phys. Rev. Lett. **97**, 153904 (2006).
2. T. Okino, K. Yamanouchi, T. Shimizu et al., Attosecond nonlinear Fourier transformation spectroscopy of CO_2 in extreme ultraviolet wavelength region. Phys. Rev. A. **129**, 161103 (2008).
3. Y. Furukawa, Y. Nabekawa, T. Okino et al., Nonlinear Fourier-transform spectroscopy of D_2 using high-order harmonic radiation. J. Chem. Phys. **82**, 013421 (2010).
4. Y. Furukawa, Y. Nabekawa, T. Okino et al., Resolving vibrational wave-packet dynamics of D_2^+ using multicolor probe pulses. Opt. Lett. **37**, 2922-2924 (2012).

Simultaneous Observation of Vibrational Wavepackets of Nitrogen Molecule in Neutral and Singly-Charged Manifolds

T. Okino, Y. Furukawa, A. Amani Eilanlou, Y. Nabekawa, E.J. Takahashi, K. Yamanouchi and K. Midorikawa

Abstract Vibrational wavepackets of nitrogen molecule were created by attosecond pulse train and their time-evolutions were also probed by attosecond pulse train. Vibrational wavepackets are launched both in neutral electronic states and singly-charged electronic states of nitrogen molecule. The created vibrational wavepackets were excited to the dissociative states and detected as a momentum image of fragment ion using a velocity map momentum imaging ion spectrometer. The fast vibrational wavepackets $(T_{vib} = 15 - 20\,\text{fs})$ were ascribed to the vibrational wavepackets generated in the singly-charged electronic states. The slow vibrational wavepackets $T_{vib} = 50 - 80\,\text{fs}$ were ascribed to the vibrational wavepackets generated in the weakly-bounded neutral electronic states. The co-existence of several vibrational wavepackets indicates the possibility to launch the electron wavepacket between these electronic states.

1 Introduction

When molecules are exposed to a short wavelength light with short pulse duration, vibrational wavepacket (VWP) in the electronic excited states of neutral molecules and/or the ionic electronic states of molecules can be initiated [1]. The manipulation of VWPs can be considered as one of the essential approaches for controlling chemical reactions in diatomic or polyatomic molecules. In most of cases, VWPs are prepared in the ionic electronic states of molecules by a pump pulse and a probe pulse is used for tracking the evolution of the VWPs by detecting the resultant

T. Okino (✉) · Y. Furukawa · A. Amani Eilanlou · Y. Nabekawa · E.J. Takahashi · K. Midorikawa
RIKEN Center for Advanced Photonics, 2-1 Hirosawa, Wako, Saitama 351-0198, Japan
e-mail: tomoya.okino@riken.jp

K. Yamanouchi
Department of Chemistry, School of Science, The University of Tokyo, 7-3-1 Hongo, Bunkyo-ku, Tokyo 113-0033, Japan

© Springer International Publishing Switzerland 2015
K. Yamanouchi et al. (eds.), *Ultrafast Phenomena XIX*,
Springer Proceedings in Physics 162, DOI 10.1007/978-3-319-13242-6_8

fragment ions. In this experimental configuration, it is difficult to launch VWPs in the electronic excited states of neutral molecules because of shortage of DUV and/ or VUV light sources with sub-10 fs pulse duration. In the present work, we observed the VWPs of nitrogen molecule (N_2) launched by sub-5 fs XUV pulses generated with a sub-15 fs, 100 Hz laser system [2] and their evolution are monitored by sub-5 fs XUV pulses composing of attosecond pulse train (APT) by detecting momentum image of fragment ion N^+ originated from sequential two-photon ionization processes with velocity map imaging ion spectrometer.

2 Experimental Setup

The APT was generated by focusing the output of Ti:S femtosecond laser system (800 nm, 12 fs, 100 Hz) loosely onto a Xe gas cell by a concave mirror. The generated APT was spatially divided into two with two Si plates and propagated into a velocity map imaging (VMI) ion spectrometer. The replica of APT pulses was focused by a SiC concave mirror (f = 100 mm) onto a molecular beam of N_2 injected by a piezo valve integrated into the repeller electrode in the VMI spectrometer. By moving one of the two Si plates on a single axis piezo stage every 1,120 nm, the temporal delay Δt between the two split APT was varied with a step of 2 fs from −10 fs to +400 fs. In each delay, we recorded 2D momentum images of N^+ originated mainly from sequential two-photon dissociative ionization processes $\left(N_2^+ \rightarrow N^+ + N\right)$ by temporally gating a MCP/Phosphor detector with 100 ns duration.

3 Results and Discussion

The fragment ion signals and its anisotropy parameter $\langle \cos^2 \theta \rangle$ are plotted as a function of the delay Δt and the kinetic energy of fragment ion E in Fig. 1a, b, respectively. The modulation patterns shown in Fig. 1a, b correspond to the time evolution of vibrational wavepackets created either in the electronic excited states of N_2 or in the electronic excited states of N_2^+. The periodic modulation patterns shown in Fig. 1a, b are quantitatively analyzed by performing a Fourier transformation and the resultant frequency spectrograms are shown in Fig. 1c, d. These modulation periods are ascribed to the periods of vibrational wavepackets as follows. The relevant potential energy curves of N_2 and N_2^+ are shown in Fig. 2. The modulation period $T_{vib} \sim 55$ fs is assigned to the VWP launched at the electronic excited states $b'^1 \sum_u^+$ of neutral nitrogen molecule by 9th order harmonic.

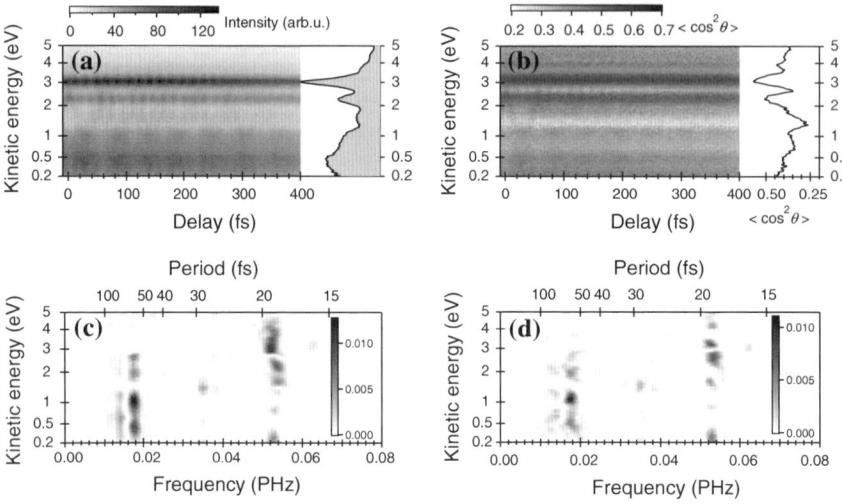

Fig. 1 Time evolution of vibrational wavepackets of N_2 using **a** intensity and **b** anisotropy. The *right panels* in **a** and **b** show the delay averaged kinetic energy distribution of N^+ and the delay averaged anisotropy $\langle \cos^2 \theta \rangle$ of N^+ where the angle θ is measured from the laser polarization direction. **c** Frequency spectrogram of (**a**). **d** Frequency spectrogram of (**b**)

Fig. 2 Potential energy curves of N_2/N_2^+

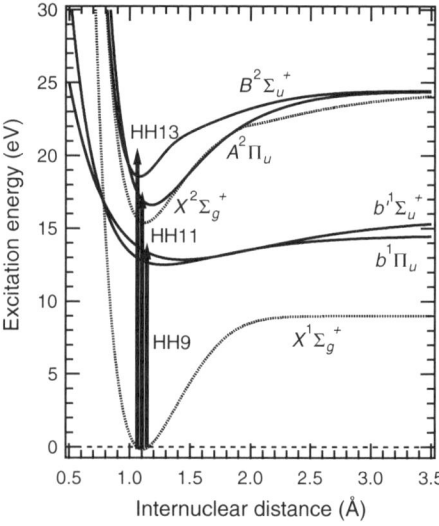

On the other hand, the modulation period $T_{vib} \sim 15-20$ fs is assigned to the VWP launched at the electronic excited states $\left(X^2 \sum_g^+, A^2 \Pi_u \right)$ of singly charged nitrogen molecule by 11th order harmonic.

In both cases, multiple order harmonics would work as a probe pulse to generate the fragment ion N^+ via dissociative ionization $\left(N_2^+ \rightarrow N^+ + N\right)$.

The modulation period $T_{vib} \sim 28$ fs is ascribed to the overtone beat frequency of VWP in the electronic state $b'^1 \sum_u^+$.

The comparison between the frequency spectrogram of intensity shown in Fig. 1c and the frequency spectrogram of anisotropy shown in Fig. 1d leads to find the symmetry of transition process from the VWP launched electronic state to final dissociative state.

Acknowledgments This research was partially supported by the Photon Frontier Network Program of Ministry of Education, Culture, Sports, Science and Technology, Japan.

References

1. M. Magrakvelidze *et al.*, Phys. Rev. A **86**, 013415 (2012).
2. Y. Nabekawa, A.A. Eilanlou, Y. Furukawa, K. L. Ishikawa, H. Takahashi, and K. Midorikawa, Appl. Phys. B **101**, 523-534 (2010).

Photo Ionization Time Delay in Molecular Hydrogen

S. Heuser, M. Sabbar, R. Boge, M. Lucchini, L. Gallmann, C. Cirelli and U. Keller

Abstract Here, we present the first experiments addressing the single-photon ionization time delay for randomly oriented molecular hydrogen (H_2), the simplest non-charged molecule. We measure the difference $\Delta\tau_m^{Ar,H2}$ in time delays between electrons emitted from the $3p^6$ shell of argon (Ar) and the highest occupied molecular orbital of H_2 by means of two distinct methods, employing attosecond streaking and RABBITT.

1 Introduction

Here, we present the first experiments addressing the single-photon ionization time delay for randomly oriented molecular hydrogen (H_2), the simplest non-charged molecule. We measure the difference $\Delta\tau_m^{Ar,H2}$ in time delays between electrons emitted from the $3p^6$ shell of argon (Ar) and the highest occupied molecular orbital of H_2 by means of two distinct methods, employing attosecond streaking and RABBITT. In contrast to previous experiments [1, 2], here, we compare photo ionization time delays for electrons with approximately the same kinetic energy emitted from two different target systems. According to theory, in both methods one part of the observed delay is caused by the influence of the infrared (IR) probe pulse calling for a measurement induced delay τ_{IR}, which depends on the kinetic energy of the emitted electrons [3, 4]. Subtracting this contribution from the measured time delay τ_m, direct access to the target related photo ionization time delay, also called Wigner-Smith time τ_{WS}, is feasible [5]. Since τ_{IR} only depends on the electron

S. Heuser (✉) · M. Sabbar · R. Boge · M. Lucchini · C. Cirelli · U. Keller
Physics Department, ETH Zurich, 8093 Zurich, Switzerland
e-mail: sheuser@phys.ethz.ch

L. Gallmann
Institute of Applied Physics, University of Bern, 3012 Bern, Switzerland

© Springer International Publishing Switzerland 2015
K. Yamanouchi et al. (eds.), *Ultrafast Phenomena XIX*,
Springer Proceedings in Physics 162, DOI 10.1007/978-3-319-13242-6_9

kinetic energy, it is valid to assume that τ_{IR}^{H2} is equal to τ_{IR}^{Ar} and therefore vanishes once the difference is computed [5]: $\Delta\tau_m^{Ar,H2} \approx \Delta\tau_{WS}^{Ar,H2} = \tau_{WS}^{Ar} - \tau_{WS}^{H2}$. Thus, if τ_{WS}^{Ar} is known [6], τ_{WS}^{H2} can be computed.

2 Setup

We generate Attosecond Pulse Trains (APTs) with a cut-off energy of about 40 eV by focusing commercial 800 nm IR laser pulses of approximately 30 fs duration, with a repetition rate of 10 kHz into an Ar gas target. Just after its generation, a 300 nm thick aluminum filter is used to remove the fundamental co-propagating IR and compensate the intrinsic chirp of the attosecond pulses. Single Attosecond Pulses (SAPs) are obtained with driving IR pulses compressed to about 5 fs using a neon filled hollow core fiber and employing the polarization gating technique. With this generation scheme, SAPs centered at about 30 eV, with 10 eV bandwidth and durations of less than 200 as, can be produced.

The XUV-pump beam is then recombined with the delayed IR-probe through a holey mirror and both beams are collinearly focused by a toroidal mirror into a supersonic gas target of a reaction microscope detector [7]. Here, ions and electrons are separated by the electric field of its spectrometer and guided towards space and time sensitive detectors. This allows to retrieve the 3D momentum vector of each individual particle at the moment of ionization and thus enables to measure ions and electrons in coincidence allowing to distinguish between electrons resulting from the photo ionization of Ar or H_2, although they energetically overlap. The gas target contains a proper mixture of Ar and H_2 such that in every measurement the experimental conditions for both species are the same thus minimizing potential sources of errors and ensuring the possibility of directly comparing both RABBITT traces and both attosecond streaking traces, respectively, with each other.

3 Results

In a first experiment, the attosecond streaking method was employed. Figure 1 shows the results achieved with this technique. For every delay step an energy average is calculated resulting in the data points shown in Fig. 1b. The photo ionization time delay between the two species can be extracted from the phase differences obtained by the fits of the data. This procedure is repeated for 52 independent measurements and the distribution of the results is shown in Fig. 1c as a function of the streaking field intensity, which is estimated from the amplitude of the energy modulation. The obtained photo ionization time delay difference $\Delta\tau_{WS}^{Ar,H2}$ is equal to (1.5 ± 15) as at an average photon energy of about 30 eV with its uncertainty representing the width of the distribution in Fig. 1c.

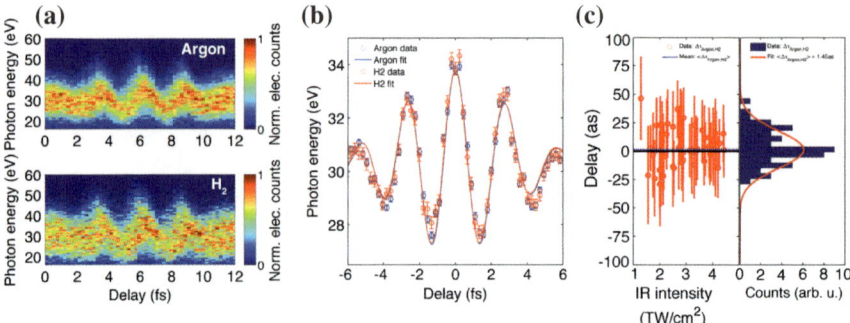

Fig. 1 **a** Attosecond streaking spectra; **b** for every delay step, average energy (*data points*) and corresponding fit (*solid line*); **c** histogram of the time delay between the two species and its distribution as a function of the IR streaking field intensity. Data points are extracted from 52 independent measurements

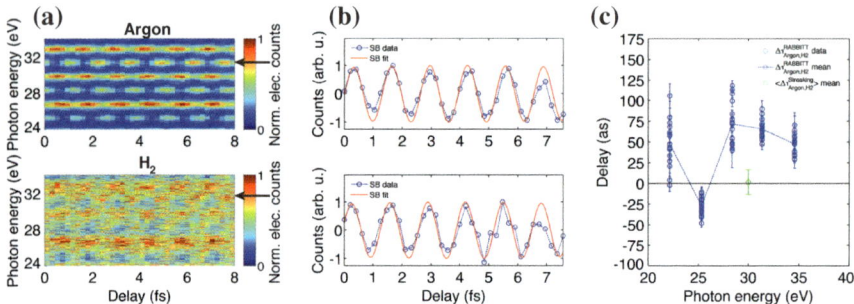

Fig. 2 **a** RABBITT traces of Ar and H_2, respectively; **b** SB oscillations for e.g. SB 18 at about 31 eV, marked by the *black arrows* in (**a**) (*data points*) and corresponding fits (*solid lines*); **c** distribution of the individual time delays obtained by 16 individual measurements (*blue circles*) as well as their mean value (*blue dashed line*) as a function of the SB energy. The *green circle* represents the mean value obtained by the attosecond streaking measurements shown in Fig. 1c

In a second experiment, the RABBITT method was applied. Its results are shown in Fig. 2. The sideband (SB) oscillations of Ar and H_2 are clearly visible (Fig. 2a) and as an example the oscillation of SB 18 at an energy of about 31 eV is illustrated (Fig. 2b). The time delay difference $\Delta\tau_m^{Ar,H2}$ between Ar and H_2 for each SB is obtained by subtracting the phases resulting from the fits of the data for each individual measurement. The time delay difference of SB 18 $\Delta\tau_m^{Ar,H2}$ results to be (75 ± 3) as. The uncertainty is the error of the mean value. This value disagrees by about 74 as with the result extracted from the attosecond streaking measurements, which have been performed at a similar photon energy.

In conclusion, both techniques give complementary access to the photo ionization time delay between Ar and H_2. However, for a photon energy of about 30 eV, which is accessible using both methods, the extracted results differ by about 74 as, calling for a different theoretical interpretation of the measured time delays.

References

1. M. Schultze *et al.*, Science **328**, 1658 (2010).
2. K. Klünder *et al.*, Phys. Rev. Lett. **106**, 143002 (2011).
3. J. M. Dahlström, A. L'Huillier and A. Maquet, J. Phys. B **45**, 183001 (2012).
4. M. Ivanov and O. Smirnova, Phys. Rev. Lett. **107**, 213605 (2011).
5. A. S. Kheifets, Phys. Rev. A. **87**, 063404 (2013).
6. A. S. Kheifets and I. A. Ivanov, Phys. Rev. A. **87**, 063404 (2013).
7. J. Ullrich *et al.*, Rep. Prog. Phys. **66**, 1463 (2003).

Ultrafast Relaxation and Photodissociation Dynamics of 1,3-Butadiene Studied by Probing Molecular Orbitals

A. Makida, T. Fujiwara, Yu Harabuchi, T. Taketsugu and T. Sekikawa

Abstract Femtosecond relaxation and picosecond photodissociation dynamics of 1,3-butadiene were investigated by time-resolved photoelectron spectroscopy with high harmonics pulses, probing the deeper molecular orbitals which are sensitive to the molecular structure.

High harmonic pulses are promising for the investigation of ultrafast phenomena from the femtosecond to attosecond regions. In particular, the extreme ultraviolet (XUV) lights can access to deeper molecular orbitals (MOs) of molecules, characterizing the chemical bonds and the structures. Therefore, time-resolved photoelectron spectroscopy (TRPES) by high harmonics pulses provides us unique opportunities to observe the transient MOs produced by photoexcitation and we can gain insight into the electron dynamics during photo-physical and -chemical reactions.

We investigated the simplest conjugated diene, 1,3-butadiene, which is the fundamental unit of the polyene structure [1]. It has been investigated both experimentally and theoretically to understand energy relaxation processes in polyene chromophores in photosynthesis reactions. In this work, we report the ultrafast recovery of valence electrons, the ultrafast energy redistribution, and structural changes in 1,3-butadiene upon photoexcitation revealed by TRPES.

We employed a time-delay compensated monochromator (TDCM) for TRPES to select a single harmonic order with the temporal duration preserved [2]. The spectral widths of attosecond pulses are too broad to probe the MOs of polyatomic molecules, where the energy separations are less than a few eV. Hence, it is necessary to limit the spectral width of high harmonics to observe several MOs separately. The sample in gas phase was excited by the second harmonic of a Ti:

A. Makida · T. Fujiwara · T. Sekikawa (✉)
Department of Applied Physics, Hokkaido University, Kita 13 Nishi 8,
Sapporo, Kita-Ku 060-8628, Japan
e-mail: sekikawa@eng.hokudai.ac.jp

Y. Harabuchi · T. Taketsugu
Department of Chemistry, Hokkaido University,
Kita 10 Nishi 8, Sapporo, Kita-Ku 060-0810, Japan

© Springer International Publishing Switzerland 2015
K. Yamanouchi et al. (eds.), *Ultrafast Phenomena XIX*,
Springer Proceedings in Physics 162, DOI 10.1007/978-3-319-13242-6_10

sapphire laser (=400 nm) in the two-photon absorption to the 2^1A_g state. The 19th harmonic (=42 nm) was selected for the probe pulses by the TDCM [1]. The time-resolved photoelectron spectra were recorded by a magnetic bottle photoelectron spectrometer by changing the optical delay between the pump and probe pulses.

Figure 1a, b shows the photoelectron spectrogram and the photoelectron spectra at delay times of −270, 40, and 427 fs, respectively. The negative delay means that the pump pulse comes after the probe pulse. In Fig. 1a, the transient MO was observed around 7 eV, of which lifetime was 47 fs. In Fig. 1b, at 40 fs, the number of the photoelectrons was reduced and then was almost recovered at 427 fs. The recovery time of the photoelectron number of the highest occupied MO (HOMO) was 53 fs.

To understand the experimental results and to elucidate the relaxation processes in 1,3-butadiene, the steepest decent path from the Franck-Condon region to the planar minimum in the 2^1A_g state, and the interpolated path from the planar minimum to the minimum energy CI (MECI) point between 2^1A_g and 1^1A_g states, were determined by the multi-state complete-active-space second-order perturbation theory (MS-CASPT2) with cc-pVDZ basis sets, using MOLPRO2012.

Figure 2 shows the predicted potential energy curve; In the Franck-Condon structure, the doubly-excited $\pi^2\pi^{*2}$ state (2^1A_g) is located slightly higher than the

Fig. 1 **a** Photoelectron spectrogram of 1,3-butadiene. **b** Time-resolved photoelectron spectra of 1,3-butadiene at delay times of −293 (*solid line*), 40 (*dotted line*), and 427 fs (*dashed-dotted line*) and the assignment of the peaks

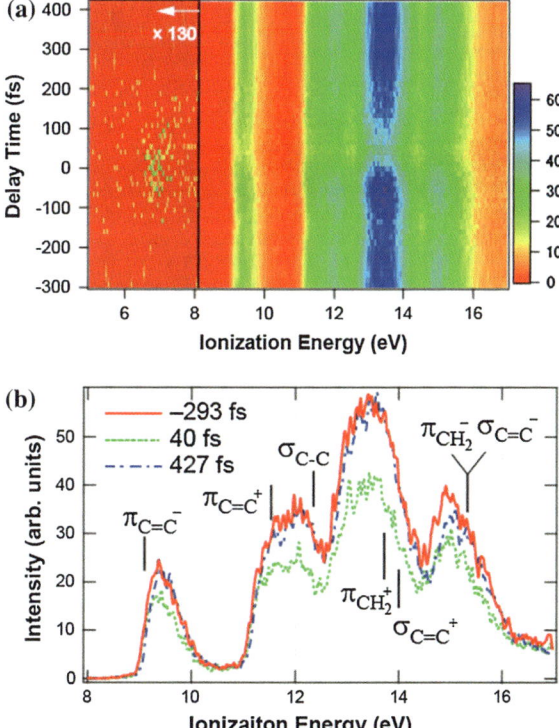

$\pi\pi^*$ state (1^1B_u). After the two-photon excitation to 2^1A_g, the molecule is relaxed rapidly to a region of the planar minimum energy structure, $(S_1)_{Cs-min}$, after passing through the conical intersection (CI) between S_2 and S_1, $(S_2/S_1)_{CI}$. Around $(S_1)_{Cs-min}$, the potential energy surface shows a very flat nature as shown in Fig. 2. In the $\pi^2\pi^{*2}$ excitation, the C=C bond becomes weak and lengthens to 1.5 Å, which is almost equivalent to the C–C single bond. Then, the planarity and the inversion symmetry of the molecule are broken by the twist of one terminal C=C bond. There is no barrier from $(S_1)_{Cs-min}$ to $(S_1)_{C1-min}$, and from $(S_1)_{C1-min}$ to the CI between S_1 and S_0, $(S_1/S_0)_{CI}$, indicating that the molecule easily goes back to the ground state through these key structures. At the MS-CASPT2 level, the binding energy for electron in the 2^1A_g state becomes smaller by 1.8 eV than in the 1^1A_g state, which decreases to 1.4 eV at the $(S_1)_{Cs-min}$ structure. The decrease in binding energy for electron due to photo-excitation is consistent with the experimentally observed value, 2.4 eV. Therefore, we attribute the observed excited state around 7 eV to the $(S_1)_{Cs-min}$ state. The 50-fs lifetime is due to the barrierless relaxation.

The MOs observed around 15 eV were found to be still shifted at 427 fs, while the HOMO band was not. The twist of the CH$_2$ bond in the excited state stimulates the antisymmetric C=C vibration. The theoretical calculation predicts that only the binding energy of $\sigma_{C=C}$ is largely shifted by this vibrational mode than those of the other MOs. The spectral shift of $\sigma_{C=C}^+$ directly indicates the structural deformation even in the ground state after the relaxation. Probing deep MOs by high harmonic pulses enables us to detect the local deformation of the molecular structure sensitively.

1,3-butadiene has various dissociation pathways upon photoexcitation by ultraviolet light. By detecting photoproduct ions, we can suspect the dissociation processes. However, the dissociation processes are still under cover. TRPES provides us unique opportunities to gain insight into the processes even in the longer

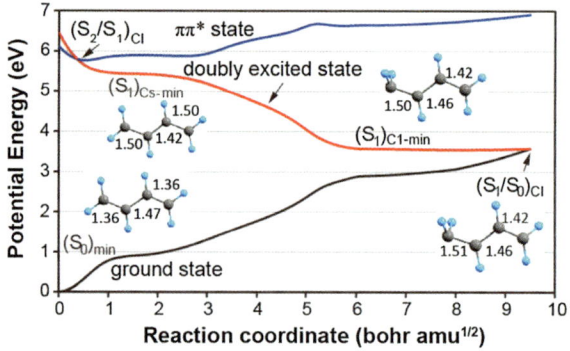

Fig. 2 Potential energy curve describing the relaxation processes in 1,3-butadiene along the reaction coordinate. The *red*, *blue*, and *black lines* are the curves for doubly excited state, $\pi\pi^*$ state, and the ground state, respectively. The *insets* are the molecular structures with bond length in Å at $(S_0)_{min}$, $(S_1/S_0)_{CI}$, $(S_1)_{Cs-min}$, and $(S_1)_{C1-min}$. The CI between S_2 and S_1 is indicated by $(S_2/S_1)_{CI}$

Fig. 3 Photoelectron
spectrogram in the picosecond
regime and the spectra of
1,3-butadiene at −67, 167,
and 1067 ps after pump and
1,2-butadiene

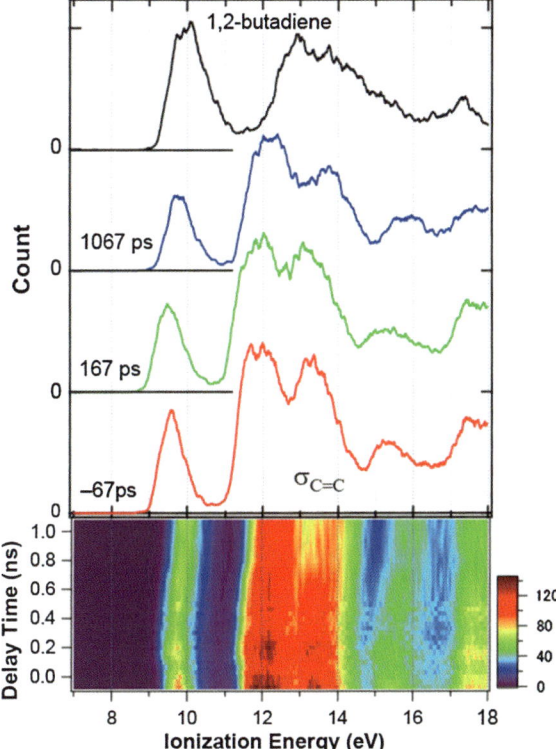

time ranges. Figure 3 shows the time-resolved photoelectron spectra in the picosecond regime. According to the previous work [3], the isomerization to 1,2-butadiene is the most probable relaxation pathway. Therefore, although the spectrum observed at 1.067 ns was still different from that of 1,2-butadiene, we suspect that the observed evolution of the spectrum is the isomerization process to 1,2-butadiene. In particular, the broadening of the MOs mainly consisting of $\sigma_{C=C}$ and σ_{CH} located between 14 and 18 eV suggests the recombination of the molecular bonds. TRPES is useful to observe the photochemical reactions.

References

1. A. Makida, et al. J. Phys. Chem. Lett. **5**, 1760 (2014).
2. H. Igarashi et al., Opt. Express **20**, 3725 (2012).
3. J.C. Robinson et al., J. Am. Chem. Soc. **124**, 10211 (2002).

Ultrafast and Photodissociation Dynamics of 1,2-Butadiene Studied by Photoelectron Spectroscopy

Ryo Iikubo, Takehisa Fujiwara, Taro Sekikawa, Yu Harabuchi and Tetsuya Taketsugu

Abstract Ultrafast and photodissociation dynamics of a cumulated dine molecule, 1,2-butadiene, were investigated by time-resolved photoelectron spectroscopy using a high harmonic. In contrast with 1,3-butadiene, coherent oscillation by stimulated Raman process was observed prior to photodissociation.

A conjugated polyene, consisting of alternating carbon-carbon double and single bonds, is one of the fundamental structures in the biological systems. In particular, the conjugation structures are often found in the light harvesting complex to transfer light energy to electrons. Therefore, it is interesting to investigate ultrafast relaxation of the excited states in the conjugated polyenes as a model system of the light harvesting complex. 1,3-butadiene has been investigated as the simplest conjugated polyene extensively. One of photochemical reactions in 1,3-butadiene, reported so far, is the isomerization to 1,2-butadiene before dissociation. Thus, in this work, we are interested in ultrafast and dissociation dynamics of 1,2-butadiene. Since 1,2-butadiene is a cumulative diene, the differences in the relaxation dynamics from that of 1,3-butadiene are also interesting issues.

To investigate the ultrafast and photodissociation dynamics, we employed time-resolved photoelectron spectroscopy (TRPES). PES probes the binding energies of molecular orbitals (MOs), depending on the chemical bonds inside a molecule. Hence, TRPES can detect the molecular rearrangement during the photo-physical and—chemical processes through the energy shifts and the appearance of new orbitals, which should be the advantage of TRPES over ionization spectroscopy observing only the photoproducts.

As a probe into MOs by TRPES, we used a single high harmonic selected by a time-delay compensated monochromator (TDCM) with the temporal duration

R. Iikubo · T. Fujiwara · T. Sekikawa (✉)
Department of Applied Physics, Hokkaido University, Kita 13 Nishi 8, Sapporo,
Kita-ku 060-8628, Japan
e-mail: sekikawa@eng.hokudai.ac.jp

Y. Harabuchi · T. Taketsugu
Department of Chemistry, Hokkaido University, Kita 10 Nishi 8,
Sapporo, Kita-ku 060-0810, Japan

© Springer International Publishing Switzerland 2015
K. Yamanouchi et al. (eds.), *Ultrafast Phenomena XIX*,
Springer Proceedings in Physics 162, DOI 10.1007/978-3-319-13242-6_11

44

preserved [1]. Since all harmonics are generated collinearly in high harmonic generation, it is necessary to select a single harmonic for PES. A TDCM separates the single harmonic order without the stretch of the pulse duration and the 19th harmonic (=42 nm) of a Ti:sapphire (TiS) laser was selected in this work. The sample in gas phase was excited by the second harmonic of a TiS laser (=400 nm) in the two-photon process. The time-resolved photoelectron spectra were recorded by a magnetic bottle photoelectron spectrometer by changing the optical delay between the pump and probe pulses.

Here, we would like to emphasize that the usage of a TDCM is advantageous over that of multilayer mirrors, selecting the harmonic order loosely, in that the observable spectral region is much wider than 10 eV. Because of this, we can gain insight into the structural changes by observing deeper MOs, which are inaccessible by visible and ultraviolet lights used so far.

In this work, we observed the photodissociation processes in the picosecond and ultrafast dynamics in the femtosecond region. Figure 1 shows (a) the three-dimensional plot of the photoelectron spectra between 11 and 18 eV in the picosecond region and (b) the spectra at −67, 700, and 1,067 ps after pump. The characters of MOs are also presented in Fig. 1b. The electron counts of $\sigma_{C\text{-}C}$ and C_{2s} bands, indicated by the blue and by the orange arrows in Fig. 1a, were found to decrease with delay time, although those from the highest occupied MO (HOMO) was almost constant. According to the ionization spectroscopy [2], 1,2-butadiene dissociates into methyl and methyl acetylene radicals as follows:

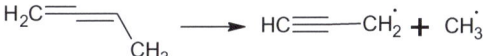

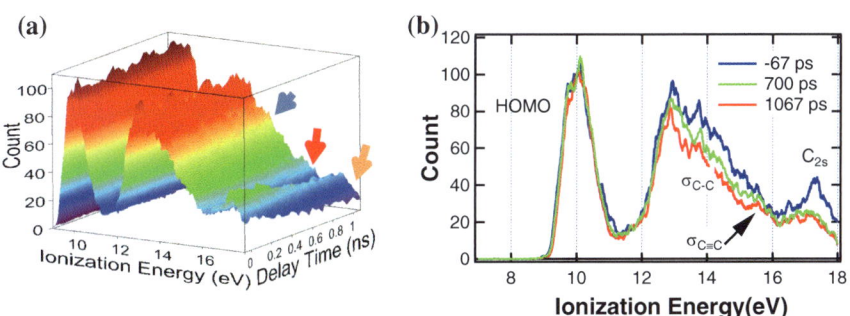

Fig. 1 **a** Three-dimensional plot of photoelectron spectra in the picosecond region. **b** Photoelectron spectra at –67, 700, and 1,067 ps after pump. The characters of molecular orbitals are shown

Therefore, the decrease in the σ_{C-C} band is attributable to the bond break and the formation of these two radicals. Our theoretical calculation predicts that the remaining methyl acetylene radical does not have any molecular orbitals around 14 eV and that the spectral shape of the C_{2s} band becomes broader, which is consistent with the experimental results. The time for photodissociation was 1 ns, estimated from the decay time of the C_{2s} band. When the dissociation takes place, the double bonds in the molecule transform into the triple bond in methyl acetylene radical, which is expected to appear in the photoelectron spectra. In fact, we observed a peak around 15.5 eV, appearing about after 0.8 ps indicated by the red arrow in Fig. 1a. Accordingly, our theoretical calculation predicts that the $\sigma_{C\equiv C}$ band appear around 15 eV. Therefore, we assign this band to the $\sigma_{C\equiv C}$ band in methyl acetylene radical formed upon photodissociation.

Figure 2a shows the three-dimensional plot of photoelectron spectra in the femtosecond region. The integrated electron numbers of the HOMO band and the transient structure observed around 7 eV were plotted as a function of delay time in Fig. 2b. The transient structure is the cross correlation between 400-nm photons and 19th harmonic and has a temporal duration of 90 fs. The intensity of the HOMO band oscillated with delay time in contrast with 1,3-butadiene, in which the depletion and the fast recovery with a time constant of 50 fs were observed [3]. Since the bands of all MOs oscillated in phase, the observed intensity oscillations were induced by stimulated Raman process in the ground state. In fact, Fourier transform of the oscillation of the HOMO band gives a frequency of 195 cm^{-1}, corresponding to an in-plane carbon-skeleton bending vibration.

The different dynamics between 1,3- and 1,2-butadiene in the femtosecond region is attributable to the difference in the absorption coefficient at 200 nm. Since 1,2-butadiene is a cumulated diene, the energy stabilization due to the conjugation is much smaller than 1-3-butadiene. This leads to the shift of the absorption to higher photon energies. Hence, at 200 nm, 1,2-butadiene was less excited and the stimulated Raman process in the ground state contributed more to the observed processes in the femtosecond region. On the other hand, in the picosecond region, the dissociation of a small number of the excited molecules was observed.

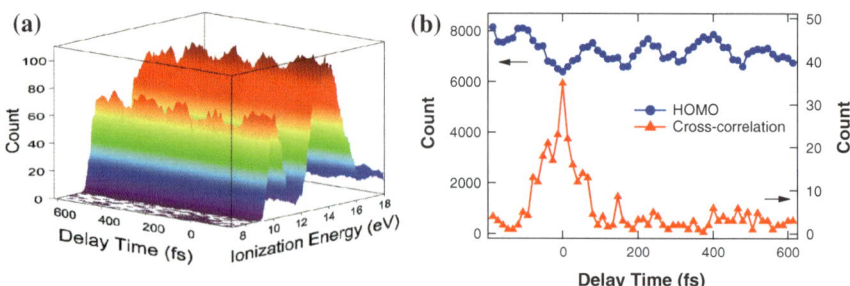

Fig. 2 **a** Three-dimensional plot of photoelectron spectra in the femtosecond region. **b** Time dependence of the integrated electron numbers of the HOMO band and the transient structure around 7 eV

References

1. H. Igarashi, A. Makida, M. Ito, and T. Sekikawa, Opt. Express **20**, 3725 (2012).
2. J. C. Robinson, W. Sun, S. A. Harris, F. Qi, and D. M. Neumark, J. Chem. Phys. **115**, 8359 (2001).
3. A. Makida, H. Igarashi, T. Fujiwara, T. Sekikawa, Y. Harabuchi, and T. Taketsugu, J. Phys. Chem. Lett. 5, 1760 (2014).

Direct Comparison of Multi-photon and EUV Single-Photon Probing of Molecular Relaxation Processes

T.J.A. Wolf, M. Koch, E. Sistrunk, J. Grilj and M. Gühr

Abstract We present a new setup for time-resolved photoelectron and photoion spectroscopy allowing for single-photon EUV or multi-photon NIR ionization. Comparison of the two different probe schemes reveals disagreements shedding light on the underlying advantages of different probes.

1 Introduction

Time-resolved photoelectron and photoion spectroscopy methods have proven to be a unique tool for investigating ultrafast processes in the excited states of organic molecules in the gas phase [1]. The method relies on probe photon energies overcoming the ionization potential of the molecular excited states. However, crystal-based frequency conversion techniques are limited to photon energies below 6 eV, which is often insufficient for a single-photon ionization process [2, 3]. The problem can in principle be circumvented by multi-photon or EUV single-photon ionization. In the case of multi-photon ionization, however, care has to be taken not to cover the molecular response by implications from the probe process, as will be shown in the following. With our recently completed setup [4], we can pursue both approaches. We employ either 800 nm (1.55 eV) or 14 eV pulses (9th harmonic) from high harmonic generation (HHG). The latter have sufficient photon energy to

T.J.A. Wolf (✉) · E. Sistrunk · M. Gühr
Stanford PULSE Institute, SLAC National Accelerator Laboratory,
Menlo Park, CA 94025, USA
e-mail: thomas.wolf@stanford.edu

M. Koch
Stanford PULSE Institute and Institute of Experimental Physics,
Graz University of Technology, Petersgasse 16, 8010 Graz, Austria, EU

J. Grilj
Stanford PULSE Institute and Laboratoire de Spectroscopie Ultrarapide,
Ecole Polytechnique Federale de Lausanne EPFL, 1015 Lausanne, Switzerland

© Springer International Publishing Switzerland 2015
K. Yamanouchi et al. (eds.), *Ultrafast Phenomena XIX*,
Springer Proceedings in Physics 162, DOI 10.1007/978-3-319-13242-6_12

ionize most organic molecules even from their ground states. Careful separation of signatures of the dynamical response from static background is, however, necessary.

As a first sample we investigated perylene. Since it is a fluorescence dye, only weak or no intramolecular vibrational redistribution (IVR) signatures can be expected within the first picoseconds after photoexcitation, which makes it an ideal sample to compare results from different ionization mechanisms.

2 Experimental

We use up to 4.5 mJ of a commercial Ti:Sapph 25 fs laser system for generation of EUV probe pulses in a HHG beamline which is described in [5]. Pump pulses at 400 nm are achieved by second harmonic generation. We use the combination of an indium filter and an MgF_2 coated aluminum mirror to select the 9th harmonic from the harmonic comb. By removing the indium filter from the beamline, samples can also be probed via multi-photon ionization at 800 nm. The focused pump and probe beams are temporarily and spatially overlapped in the interaction region of the time-of-flight spectrometer chamber, which is described in detail in [4] and can be operated in an ion and an electron mode. The time resolution is 70 ± 10 fs.

3 Results and Discussion

The time dependence of ion counts from *multi-photon ionization* at the mass peaks of the doubly charged perylene parent ion (126 u) and a singly charged fragment (250 u) are depicted in Fig. 1a. They show an exponential decay by more than 50 % with a time constant of (1.1 ± 0.1) ps to a level, which is stable for the rest of the investigated time-delay window (5 ps). This is in good agreement with the litera-ture-known internal vibrational redistribution (IVR) time scale in S_1 [6].

Figure 1b, top, shows the *single-photon* photoelectron spectrum from ionization at 14 eV (black). Additionally, a typical transient difference photoelectron spectrum obtained by subtracting spectra at a positive time delay from those at –660 fs delay is depicted (blue). It resembles a shift of the ground state spectrum by about the pump photon energy (3.1 eV, 400 nm). To investigate the time-dependence of the transient spectrum we analyzed the peak integrals and positions at 8 and 10 eV kinetic energy, as marked by bars in Fig. 1b, top. They do not show a time-dependent intensity change, broadening or peak shift as expected from the transient multi-photon photoionization data (see Fig. 1b, bottom).

The disagreement between the two datasets can be assigned to the difference in the single-photon compared to the multi-photon probe processes. Since the S_0 and S_1 minimum geometries differ, the ultrafast excitation induces nuclear dynamics. The absence of dynamics in the 14 eV photoelectron spectra points to a constant

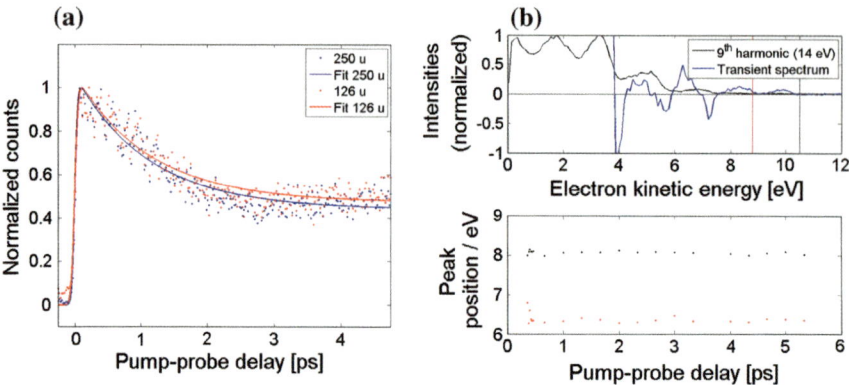

Fig. 1 **a** Time-dependence of the ion yield in the doubly charged perylene parent ion (126 u) and a singly charged fragment (250 u) together with an error function and exponential fit. **b** 14 eV photoelectron spectrum together with the transient spectrum after pre-excitation at 400 nm (*top*). In the latter spectra, slow photoelectrons were filtered out by applying a retardation voltage at the entrance of the flight tube. Additionally, the time dependence of the integrals of the transient peaks around 8 and 10 eV is visualized (*bottom*)

ionization potential from S_1 to the cationic states over the geometry range from Franck-Condon region to S_1 minimum. For the multi-photon probes we observed enhancement by intermediate resonant states in intensity-dependent measurements of the ion yield. A scan of the S_1 potential energy surface by TDDFT methods reveals that in contrast to the lowest cationic state these intermediate states experience a considerable shift in energy relative to S_1. This in turn strongly modulates the transition amplitude along the nuclear path in the S_1 state. Thus, in this case, time-resolved photoion spectroscopy with multi-photon ionization reveals a more modulated information about the excited state processes of the sample than time-resolved single-photon photoelectron spectroscopy. However, one has to keep in mind that the additional observations are due to resonant intermediate states, which are often not very well known. In the case of perylene, the excited state dynamics is sufficiently simple. Multi-photon probe results from more complex molecules showing intersystem crossing or internal conversion need to be thoroughly checked for intermediate resonances modulating the multi-photon ionization.

Acknowledgments This material is based on work supported by the U.S. Department of Energy, Office of Science, Office of Basic Energy Science, Chemical Sciences, Geosciences, and Biosciences Division. M. G. acknowledges funding via the Office of Science Early Career Research Program through the Office of Basic Energy Sciences, U.S. Department of Energy. T. W. acknowledges support from the German Academy of Sciences Leopoldina. M. K. acknowledges support form the Austrian Science Fund FWF (Schroedinger Grant) J3299-N20). J. G. acknowledges support from the European Commission under the Marie Curie Fellowship Programme.

References

1. Stolow, A, Underwood, J. G.: Time-Resolved Photoelectron Spectroscopy of Nonadiabatic Dynamics in Polyatomic Molecules., Advances in Chemical Physics **19**, 497–587 (2008).
2. Hudock, H. R., Levine, B. G., Thompson, A. L., Satzger, H., Townsend, D., Gador, N., Ullrich, S., Stolow, A., Martínez, T. J.: Ab Initio Molecular Dynamics and Time-Resolved Photoelectron Spectroscopy of Electronically Excited Uracil and Thymine. J. Phys. Chem. A **111**, 8500–8508 (2007).
3. Barbatti, M., Ullrich, S.: Ionization potentials of adenine along the internal conversion pathways. Phys. Chem. Chem. Phys. **13**, 15492–15500 (2011).
4. Koch, M., Wolf, T. J. A., Grilj, J., Sistrunk, E., Gühr, M.: Femtosecond photoelectron and photoion spectrometer with vacuum ultraviolet probe pulses. arXiv:1404.3039 (2014).
5. Grilj, J., Sistrunk, E., Koch, M., Gühr, M.: A Beamline for Time-Resolved Extreme Ultraviolet and Soft X-Ray Spectroscopy. J. Anal. Bioanal. Tech. S12:005. (2014).
6. Kiba, T., Sato, S., Akimoto, S., Kasajima, T., Yamazaki, I: Solvent-assisted intramolecular vibrational energy redistribution of S1 perylene in ketone solvents. J. Photochem. Photobiol., A **178**, 201–207 (2006).

Sub-4-fs Charge Migration in Phenylalanine

F. Calegari, D. Ayuso, L. Belshaw, A. Trabattoni, S. Anumula,
S. De Camillis, F. Frassetto, L. Poletto, A. Palacios, P. Decleva,
J. Greenwood, F. Martín and M. Nisoli

Abstract Charge migration initiated by isolated attosecond pulses was experimentally observed in the amino-acid phenylalanine. An oscillatory pattern in the yield of a doubly-charged ion was measured with periods below 4.5-fs.

The process of electron transfer in molecular complexes is of crucial importance in biochemistry since it triggers the first steps in a number of various biochemical processes [1]. Theoretical studies have pointed out that very efficient charge dynamics (called charge migration) can be driven by purely electronic effects, which precede any rearrangement of the nuclear skeleton and which can evolve on a temporal scale ranging from few femtoseconds down to tens of attoseconds [2–4].

F. Calegari · M. Nisoli (✉)
IFN-CNR, Piazza Leonardo da Vinci 32, 20133 Milan, Italy
e-mail: mauro.nisoli@polimi.it

F. Calegari
e-mail: daniele.brida@uni-konstanz.de

D. Ayuso · A. Palacios · F. Martín
Departamento de Química, Modulo 13, Universidad Autonoma de Madrid,
Cantoblanco, 28049 Madrid, Spain

L. Belshaw · S. De Camillis · J. Greenwood
Centre for Plasma Physics, School of Maths and Physics,
Queen's University Belfast, Belfast BT7 1NN, UK

A. Trabattoni · S. Anumula · M. Nisoli
Department of Physics, Politecnico di Milano, Piazza Leonardo da Vinci 32,
20133 Milan, Italy

F. Frassetto · L. Poletto
IFN-CNR, Via Trasea, 7, 35131 Padua, Italy

P. Decleva
Dipartimento di Scienze Chimiche e Farmaceutiche, CNR-IOM,
Università di Trieste, 34127 Trieste, Italy

F. Martín
Instituto Madrileno de Estudios Avanzados en Nanociencia, Cantoblanco,
28049 Madrid, Spain

© Springer International Publishing Switzerland 2015
K. Yamanouchi et al. (eds.), *Ultrafast Phenomena XIX*,
Springer Proceedings in Physics 162, DOI 10.1007/978-3-319-13242-6_13

In this work we report on a clear experimental measurement of charge migration in the amino acid phenylalanine, after attosecond excitation. Charge migration was evidenced in phenylalanine as an oscillatory evolution in the yield of a doubly-charged molecular fragment, with oscillation period below 4.5-fs.

In our experiments, charge migration was measured by using a two-color, pump-probe technique. Charge dynamics was initiated by isolated sub-300-as pulses [5], with photon energies in the spectral range between 17 and 35 eV, and subsequently probed by 4-fs, waveform-controlled near infrared (NIR) pulses, with central wavelength of 700 nm. A clean plume of isolated and neutral molecules of phenylalanine was generated by evaporation of the amino acid from a thin metallic foil heated by a CW diode laser. The parent and fragment ions produced by the interaction of the molecules with the pump and probe pulses were then collected by a linear time-of-flight device for mass analysis, where the metallic foil was integrated into one of the end electrodes [6]. Figure 1a, b shows the mass spectra measured by excitation with attosecond pulses only and NIR pulses only, respectively.

We have then measured the evolution of the yield of the doubly charged immonium ion (mass/charge = 60) as a function of the delay between the attosecond pump pulse and the NIR probe pulse (the immonium ion is formed by loss of the —COOH group). Figure 1c shows the measured dictation yield as a function of the pump-probe delay, on a 100-fs time scale. The experimental data display a rise time of 10 ± 2 fs and an exponential decay with time constants of 25 ± 2 fs (in agreement with the results reported in [7]). We then we increased the temporal resolution of the measurement by reducing the delay-step between pump and probe pulses from 3 to 0.5 fs. Figure 2 shows a 25-fs-wide zoom of the exponential decay, where an oscillation of the dictation yield is clearly visible.

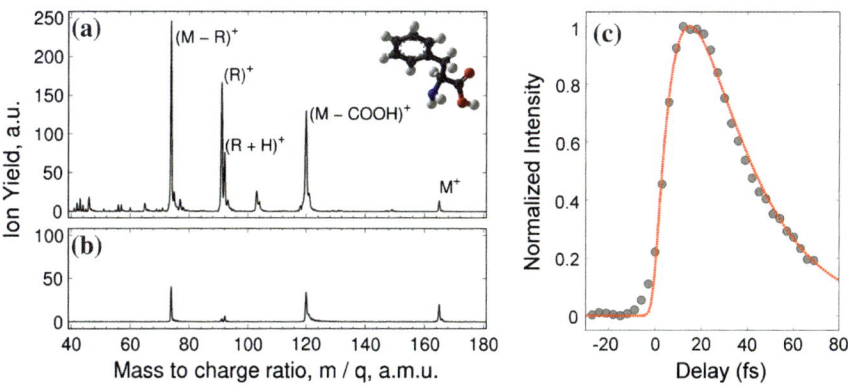

Fig. 1 Mass spectra measured from ionization of phenylalanine by **a** XUV pulses only and **b** NIR pulses only. M is the parent ion, R is the side-chain group. **c** Yield of doubly charged immonium ion (mass/charge = 60) versus pump-probe delay. The *dotted red line* is a fitting curve with an exponential rise time of 10 fs and an exponential relaxation time of 25 fs. The *inset* of panel (**a**) shows the structure of phenylalanine

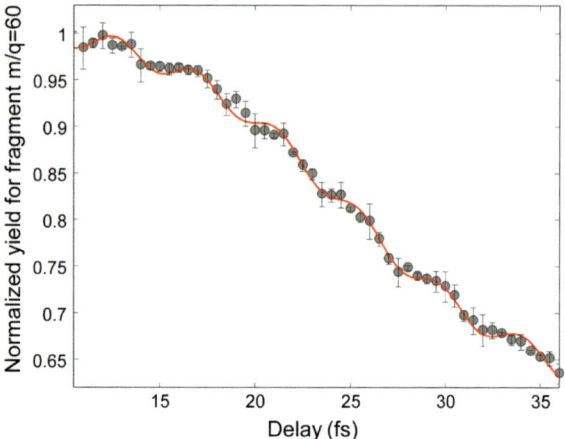

Fig. 2 Yield of doubly charged immonium ion (m/q = 60) as a function of the pump-probe delay measured with 0.5 fs temporal steps. The *red line* is the fitting curve discussed in the text

The corresponding fitting curve (red line in Fig. 2), which closely follows the measured points, is given by the sum of an exponential function (with a time constant of 25 fs) and a sinusoidal function, with frequency of 0.234 PHz (corresponding to a period of 4.3 fs), with lower and upper confidence bounds of 0.229 and 0.238 PHz, respectively, obtained from best fitting of the experimental data. We note that at short pump-probe delays (<10 fs) the experimental data can be fitted by the sum of two sinusoidal functions with frequencies of ∼0.14 and ∼0.3 PHz (corresponding to a period of 3.3 fs). This ultrafast dynamics can only be associated with purely electronic processes, thus constituting the first experimental measurement of charge migration in a biomolecule.

We have also performed theoretical calculations to describe the hole dynamics induced by an attosecond pulse similar to that used in the experiment. Due to the large bandwidth of the pulse, a manifold of ionization channels are open, thus leading to a superposition of many cationic states, i.e., to an electronic wave packet. For all open channels, the ionization amplitudes have been quantitatively determined. The evolution of the electronic wave packet has then been described by using a standard time-dependent density matrix formalism. The results of the numerical simulations clearly show the production of an ultrafast electron dynamics, characterized by oscillation frequencies in good agreement with the experimental result. We notice that only valence and inner-valence electrons are efficiently ionized by our XUV pulse, so that the observed dynamics is that of a delocalized hole.

Acknowledgments We acknowledge support from ERC (grants no. 227355 ELYCHE and 290853 XCHEM), from LASERLAB-EUROPE (no. 284464), from European COST Action CM1204 XLIC, the MICINN Project FIS2010-15127, the ERA-Chemistry Project PIM2010EEC-00751, the European Grants MC-ITN CORINF and MC-RG ATTOTREND 268284, the UK's STFC Laser Loan Scheme, the Eng. and Phys. Sciences Res. Council (grant EP/J007048/1), the Leverhulme Trust (RPG-2012-735), and the Northern Ireland Department of Employment and Learning.

References

1. J. Winkler *et al*, "Electron Transfer Through Proteins" in *Bioelectronics*, I. Willner, E. Katz, Eds. (Wiley-VCH, Weinheim, Germany, 2005), pp. 15-33.
2. L. S. Cederbaum, J. Zobeley, "Ultrafast charge migration by electron correlation," Chem. Phys. Lett. **307**, 205-210 (1999).
3. F. Remacle, R. D. Levine, "An electronic time scale in chemistry," PNAS **103**, 6793 (2006).
4. H. Hennig, J. Breidbach, L. S. Cederbaum, "Electron Correlation as the Driving Force for Charge Transfer: Charge Migration Following Ionization in *N*-Methyl Acetamide," J. Phys. Chem. A **109**, 409-414 (2005).
5. G. Sansone, E. Benedetti, F. Calegari, C. Vozzi, L. Avaldi, R. Flammini, L. Poletto, P. Villoresi, C. Altucci, R. Velotta, S. Stagira, S. De Silvestri, M. Nisoli, "Isolated Single-Cycle Attosecond Pulses," Science **314**, 443-446 (2006).
6. C.R. Calvert *et al.*, "LIAD-fs scheme for studies of ultrafast laser interactions with gas phase biomolecules," Phys. Chem Chem. Phys. **14**, 6289-6297 (2012).
7. L. Belshaw, F. Calegari, M. J. Duffy, A. Trabattoni, L. Poletto, M. Nisoli, and J. B. Greenwood, "Observation of ultrafast charge migration in amino acid," J. Phys. Chem. Lett. **3**, 3751-3754 (2012).

Recombination-Induced Autoionization Process in Rare-Gas Clusters

B. Schütte, M. Arbeiter, T. Fennel, F. Campi, M.J.J. Vrakking and A. Rouzée

Abstract We investigate electron-ion recombination to excited states in atomic clusters exposed to intense NIR and XUV pulses, which leads to a yet undiscovered autoionization mechanism as a consequence of multiple recombination processes.

1 Introduction

In the last two decades, a large interest was devoted to the investigation of the ionization mechanisms in clusters exposed to an ultrashort, intense light pulse in the near-infrared (NIR) and extreme-ultraviolet (XUV) region. While at the early stage of the interaction with the light pulse, the ionization mechanisms are very similar in atoms and clusters, i.e. multiphoton ionization processes in the NIR regime and single photon ionization for a XUV pulse, it was demonstrated that new electron emission mechanisms are observed in clusters at a later stage of the interaction. Indeed, as the electrons are stripped away by the laser field and the cluster is being charged, the increased Coulomb potential prevents other electrons from leaving the cluster. These quasifree electrons can move within the cluster and form a nanoplasma with the ions. In the case of an NIR pulse, field ionization was shown to play an important role where energy is transferred from the laser pulse to quasifree electrons in the nanoplasma, enabling them to drive electron impact ionization.

B. Schütte (✉) · M.J.J. Vrakking · A. Rouzée
Max-Born-Institut, Max-Born-Str. 2A, 12489, Berlin, Germany
e-mail: schuette@mbi-berlin.de

M. Arbeiter · T. Fennel
Institute of Physics, University of Rostock, Universitätsplatz 3,
18051 Rostock, Germany

F. Campi
Department of Physics, Lund University, P.O. Box 118
SE-221 00 Lund, Sweden

© Springer International Publishing Switzerland 2015
K. Yamanouchi et al. (eds.), *Ultrafast Phenomena XIX*,
Springer Proceedings in Physics 162, DOI 10.1007/978-3-319-13242-6_14

During the cluster expansion, quasifree electrons and ions in the nanoplasma may recombine. Here we demonstrate the recombination to high-lying atomic Rydberg states that are reionized by the DC detector electric field, a process known as frustrated recombination [1]. Recently, we could give experimental evidence for this effect by the detection of the corresponding electrons that have very low kinetic energies (meV to few tens of meV) [2]. In addition, we present time-resolved studies of recombination showing the formation of low-lying excited states. By using a weak 790 nm probe pulse following an intense XUV or NIR pump pulse, reionization of excited atoms from recombination (REAR) leads to additional peaks in the photoelectron spectra [3]. The overall ion yield is significantly enhanced by REAR even for large delays in the ns range, demonstrating the significance of electron-ion recombination processes. Also, as a consequence of multiple recombination processes, new autoionization processes in the cluster can take place, where one electron relaxes to the ground state, transferring the energy to a second electron that is emitted into the continuum. Here we show evidence for such an autoionization mechanism.

2 Experiment

A Ti:sapphire laser system delivering 35 mJ, 790 nm, 32 fs pulses at a repetition rate of 50 Hz is used for the experiment. XUV pulses are obtained by high-order-harmonic generation (HHG) employing a loose-focusing geometry. Part of the NIR beam is focused by a $f = 5$ m spherical mirror into a 15 cm long gas cell that is statically filled with Ar. After blocking the residual NIR light with a 200 nm thin Al filter, the HHG pulses are spectrally filtered and focused by a spherical multilayer mirror with a focal length of 75 mm into a velocity map imaging spectrometer, where it is crossed by a pulsed cluster beam. XUV peak intensities up to 5×10^{12} W/cm^2 are achieved. The generated ions and electrons are accelerated towards a microchannel plate/phosphor screen assembly, and the 2D momentum maps are recorded with a CCD camera. A small fraction of the 790 nm beam is split before the harmonic generation, and is later recombined with the harmonic beam to perform pump-probe experiments. Upon removal of the Al filter in the XUV path, it is also possible to perform NIR-pump NIR-probe experiments.

3 Results and Discussion

REAR is observed in the photoion and -electron momentum maps presented in Fig. 1, where the differences between the two-color XUV + NIR signals at a time delay of 5.2 ns and the yields obtained by the XUV pump pulse alone are displayed. In Fig. 1a, an additional ion contribution due to the NIR pulse is visible, with the

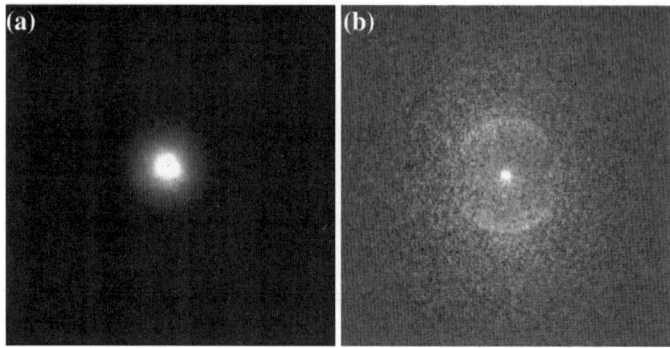

Fig. 1 **a** 2D Kr$^+$ ion momentum map from Kr$_{8000}$ clusters exposed to a 38 nm pump pulse with an intensity of 2×10^{12} W/cm^2 and a weak 790 nm probe pulse at a delay of 5.2 ns. The signal from the XUV pulse alone is subtracted. **b** Corresponding 2D electron momentum map showing ring structures attributed to REAR

ions having non-zero kinetic energies that are therefore identified as fragments formed during the cluster dissociation. Before recombination takes place, these fragments gain kinetic energy that is preserved until their reionization. In the electron momentum maps in Fig. 1b, clear ring structures are observed and attributed to single-photon ionization events of excited atoms formed by recombination in the nanoplasma with the weak NIR pulse. Nanoplasma formation is not limited to XUV ionization regimes, and therefore, a large number of recombination events is also observed when using an intense NIR pulse for cluster ionization. Although the NIR probe pulse is too weak to ionize a neutral cluster, it generates a large amount of ions and electrons from REAR that can even exceed the number of ions and electrons produced by the XUV pulse only. Due to the large number of excited atoms formed during the cluster expansion, a novel autoionization mechanism becomes possible, where one electron relaxes to the ground state of an atom, transferring the energy to a second electron, which can leave the atom with a substantial amount of kinetic energy. These two correlating electrons can either be located in one atom leading to autoionization, or in two atoms, similar to interatomic coulombic decay. The latter case was theoretically predicted in clusters, where excited atoms were formed by resonant laser excitation [4]. In an experiment after ionization of Ar clusters with strong NIR pulses, we observe a characteristic peak in the photoelectron spectrum close to the ionization potential of atomic Ar that is attributed to the autoionization following electron-ion recombination processes. In opposite to the studies carried out in [4], the autoionization mechanism discussed here does not involve resonant excitation, but is a consequence of excited atom formation in the nanoplasma. Therefore, it can be regarded as a general process that is independent from the ionization wavelength and is possible at photon energies far below the ionization potential. The current findings are important towards an improved understanding of the ionization dynamics in clusters and other extended systems.

References

1. Fennel, Th., Ramunno, L., and Brabec, Th.: Highly charged ions from laser-cluster interactions: local-field-enhanced impact ionization and frustrated electron-ion recombination. Phys. Rev. Lett. **99**, 233401 (2007).
2. Schütte, B., Arbeiter, M., Fennel, Th., Vrakking, M. J. J., and Rouzée, A.: Rare-gas clusters in intense extreme-ultraviolet pulses from a high-order harmonic source. Phys. Rev. Lett. **112**, 073003 (2014).
3. Schütte, B., Campi, F., Arbeiter, M., Fennel, Th., Vrakking, M. J. J., and Rouzée, A.: Tracing electron-ion recombination in nanoplasmas produced by extreme-ultraviolet irradiation of rare-gas clusters. Phys. Rev. Lett. **112**, 253401 (2014).
4. Kuleff, A. I., Gokhberg, K., Kopelke, S., and Cederbaum, L. S. Ultrafast interatomic electronic decay in multiply excited clusters. Phys. Rev. Lett. **105**, 043004 (2010).

X-Ray Magnetic Circular Dichroism Probed Using High Harmonics

Patrik Grychtol, Ofer Kfir, Ronny Knut, Emrah Turgut, Dmitriy Zusin, Dimitar Popmintchev, Tenio Popmintchev, Hans Nembach, Justin Shaw, Avner Fleischer, Henry Kapteyn, Margaret Murnane and Oren Cohen

Abstract We demonstrate the first generation and phase matching of circularly-polarized high harmonics, which are bright enough for X-ray magnetic circular dichroism measurements at the M absorption edges of the magnetic materials Fe, Co and Ni.

Circularly polarized light in the extreme ultraviolet (EUV) and soft X-ray regions of the electromagnetic spectrum is extremely useful for exploring chirality-sensitive light-matter interactions. X-ray magnetic circular dichroism (XMCD) makes it possible to extract detailed information about the magnetic state of matter and its interaction with phononic and electronic degrees of freedom on femtosecond time scales and nanometer length scales. Specifically, XMCD can be used to distinguish between spin and orbital contributions to the atomic magnetic moment in ferromagnetic materials, with element-specificity, which is not possible using ultrafast visible laser spectroscopy. To date, circularly polarized EUV and soft X-ray beams were restricted to large-scale electron storage facilities, such as synchrotrons and X-ray free electron lasers. Such facilities have great advantages of high peak and average powers in the X-ray region. However, drawbacks include experimental complexity, limited access and temporal resolution, as well as pump/probe jitter.

P. Grychtol (✉) · R. Knut · E. Turgut · D. Zusin · D. Popmintchev · T. Popmintchev · H. Kapteyn · M. Murnane
Department of Physics and JILA, University of Colorado, Boulder, CO 80309, USA
e-mail: p.grychtol@jila.colorado.edu

O. Kfir · A. Fleischer · O. Cohen
Solid State Institute and Physics Department, Technion, 32000 Haifa, Israel

R. Knut · H. Nembach · J. Shaw
Electromagnetics Division, National Institute of Standards and Technology, Boulder, CO 80305, USA

A. Fleischer
Department of Physics and Optical Engineering, Ort Braude College, 21982 Karmiel, Israel

© Springer International Publishing Switzerland 2015
K. Yamanouchi et al. (eds.), *Ultrafast Phenomena XIX*,
Springer Proceedings in Physics 162, DOI 10.1007/978-3-319-13242-6_15

Table-top short-wavelength sources based on high harmonic (HHG) up-conversion of femtosecond laser pulses represent a complementary and viable alternative to large scale sources, due to their unique ability to generate bright, broadband, ultrashort and coherent light from the UV to the keV region [1]. HHG sources have successfully exploited the resonantly enhanced magnetic contrast at the M absorption edges of the 3d ferromagnets Fe, Co and Ni in the EUV region to study element-specific dynamics in magnetic materials on femtosecond time scales on the tabletop [2, 3]. This new capability has opened up a wealth of opportunities for greater fundamental understanding of correlated phenomena. Specifically, novel insights on ultrafast (few-femtosecond) nanoscale phenomena were uncovered by capturing spin scattering and transport, as well as exchange interaction dynamics, in complex multi-species magnetic materials [4–6]. However, these investigations have so far been limited to linearly polarized HHG: to date, generating circularly-polarized harmonics has been highly inefficient, reducing both the photon flux and degree of circularity to a level that precludes scientific applications [7–9].

In this work, we present the first direct approach for generating bright circularly-polarized HHG under phase-matching conditions [10], based on a technique that was suggested almost two decades ago [11, 12] and that was recently demonstrated by Fleischer et al. [13]. Surprisingly, although for decades it was assumed that HHG from atoms was brightest when both the driving laser and HHG fields were linearly polarized, we show that circularly-polarized HHG can be as bright as linearly-polarized HHG. We also demonstrate the first XMCD measurements using a tabletop light source, in this case at the M absorption edges of Ni, thereby probing the magnetization component parallel to the light beam rather than in the sample plane. This experiment thus represents the first application of circularly-polarized HHG and the first measurement of the helicity of circularly-polarized harmonics.

In our experiment, we focus the fundamental and second harmonic of a Ti: sapphire laser amplifier with opposing helicities into a gas-filled hollow waveguide. As schematically depicted in Fig. 1, the polarization states of the two drivers can be

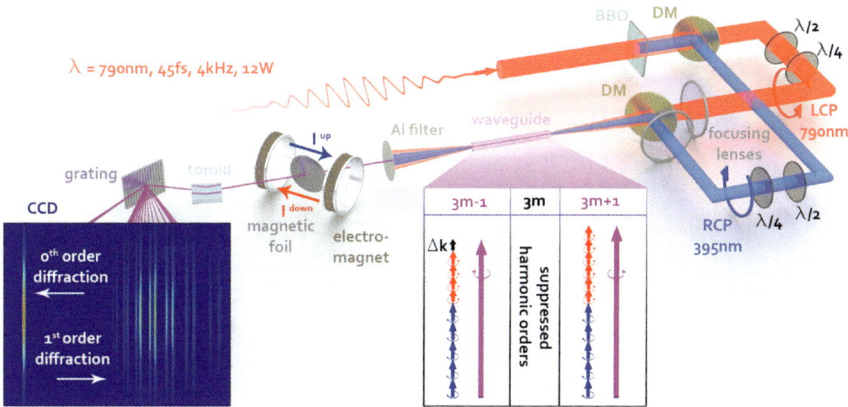

Fig. 1 Setup for generating phase-matched circularly-polarized harmonics [10]

independently adjusted in an interferometric Mach-Zehnder setup, thereby controlling the helicity of the generated harmonics. The threefold symmetry of the resultant bichromatic field and thus generation process gives rise to circularly polarized harmonics [10–13]: the electric fields of harmonic orders $q = 3n + 1$ have the same helicity as the fundamental, $q = 3n - 1$ rotate with the second harmonic, while harmonic orders $q = 3n$ are completely suppressed. Using Ar, Ne and He gas, we produce a circularly-polarized HHG spectrum that spans the entire M absorption edges of the 3d ferromagnets, with a photon flux comparable to that used in previous ultrafast element-selective magneto-optic HHG-based experiments [2–6]. Thus, the element specific magnetic state of materials can be probed simultaneously with spin and orbital selectively and with high signal-to-noise ratio, by exploiting XMCD.

Figure 2a plots HHG spectra from Ar at 90 Torr, in Ne at 650 Torr and in He at 900 Torr. Clearly, the $q = 3n$ harmonics are almost completely suppressed for all gas species. This feature is a clear indication that the HHG polarization is close to circular [13]. The detected photon flux is close to 10^9 photons/s. Next, we measured the HHG spectrum after passing through a magnetized 50 nm thick Ni foil, and extracted the XMCD asymmetry by flipping the in-plane magnetization of the sample. Figure 2b presents the HHG spectrum when the foil is magnetized either "up" or "down", labeled by I^{up} and I^{down}, respectively. The normalized XMCD asymmetry, $A = (I^{up} - I^{down})/(I^{up} + I^{down})$. is shown in Fig. 2c. The asymmetry of the $3n + 1$ and $3n - 1$ harmonics exhibit opposite signs, thereby proving that the helicity of the $3n + 1$ harmonics is indeed opposite to the $3n - 1$ harmonics. As the first bright source of circularly-polarized HHG, this work paves the way for investigating ultrafast circular dichroism of magnetic samples, chiral molecules and nanostructures.

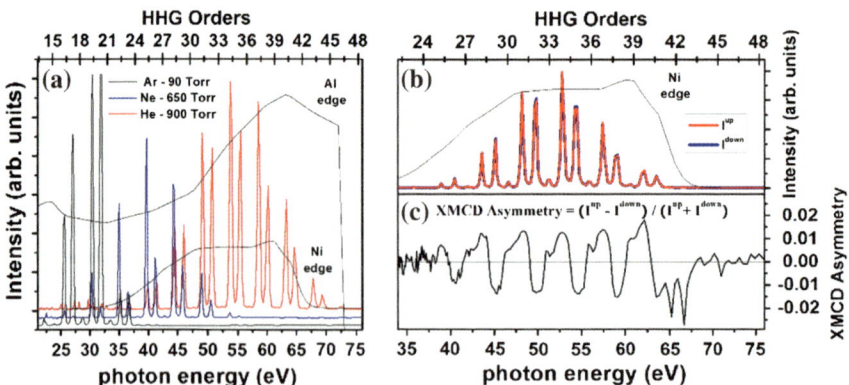

Fig. 2 **a** Spectra of bright circularly-polarized HHG in Ar, Ne and He and **b, c** XMCD of 50 nm Ni from circularly-polarized HHG in He

References

1. Popmintchev, *et al.,* Science **336**, 1287 (2012).
2. La-O-Vorakiat, *et al.*, Physical Review Letters **103**, 257402 (2009).
3. La-O-Vorakiat, *et al.*, Physical Review X **2**, 011005 (2012).
4. Mathias, *et al.*, Proceedings of the National Academy of Sciences **109**, 4792 (2012).
5. Rudolf, *et al.*, Nature Communications **3**, 1037 (2012).
6. Turgut, *et al.*, Physical Review Letters **110**, 197201 (2013).
7. Husakou, *et al.,* Optics Express **19**, 25346 (2011).
8. Vodungbo, *et al.,* Optics Express **19**, 4346 (2011).
9. Liu, *et al.*, Optics Letters **37,** 2415 (2012).
10. Kfir, *et al.*, Nature Photonics AOP (2014). doi:10.1038/nphoton.2014.293
11. Long, *et al.*, Physical Review A **52**, 2262 (1995).
12. Eichmann, *et al.*, Physical Review A **51**, R3414 (1995).
13. Fleischer, *et al.*, Nature Photonics **8**, 543 (2014).

Extreme Ultraviolet Transient Grating Measurement of Insulator-Metal Transition Dynamics in VO$_2$

Emily Sistrunk, Jakob Grilj, Jaewoo Jeong, Mahesh G. Samant, Alexander X. Gray, Hermann A. Dürr, Stuart S.P. Parkin and Markus Gühr

Abstract We demonstrate spectrally resolved transient grating (TG) spectroscopy in the extreme ultraviolet (EUV) near the M-edge of vanadium dioxide. Time-dependent and broadband EUV-TG measurements separate the index of refraction change due to the insulator to metal transition from purely acoustic effects.

1 Introduction

Multidimensional spectroscopy (MDS) has proven a powerful technique in chemistry and materials science, particularly in the infrared, visible and near-ultraviolet [1]. A form of MDS, transient grating (TG) spectroscopy has been highly successful

E. Sistrunk (✉)
Lawrence Livermore National Laboratory, Livermore, CA 94551, USA
e-mail: link7@llnl.gov

E. Sistrunk · J. Grilj
Stanford PULSE Institute, SLAC National Accelerator Laboratory, Menlo Park,
CA 94025, USA

J. Grilj
Laboratory de Spectroscopie Ultrarapide, Ecole Polytechnique Federale de Lausanne,
Lausanne, Switzerland

J. Jeong · M.G. Samant · S.S.P. Parkin
IBM Almaden Research Center, San Jose, CA 95120, USA

A.X. Gray · H.A. Dürr
Stanford Institute for Materials and Energy Sciences, SLAC National Accelerator Laboratory,
Menlo Park, CA 94025, USA

A.X. Gray
Department of Physics, Temple University, Philadelphia, PA 19122, USA

M. Gühr
SLAC National Accelerator Laboratory, Stanford PULSE Institute, Menlo Park,
CA 94025, USA
e-mail: mguehr@stanford.edu

© Springer International Publishing Switzerland 2015
K. Yamanouchi et al. (eds.), *Ultrafast Phenomena XIX*,
Springer Proceedings in Physics 162, DOI 10.1007/978-3-319-13242-6_16

64

in examining the electronic response of complex materials, electron-phonon coupling, and acoustic waves on surfaces [2]. In addition to being intrinsically background-free, TG experiments have proven highly sensitive to both surface and bulk acoustic waves and are widely used to characterize thin films. Few MDS experiments have been performed in the extreme ultraviolet (EUV) to study the acoustic response of a sample [3]. We further develop the technique to study the electronic sample response close to characteristic resonances in the EUV region. We demonstrate this for the example of an ultrafast photoinduced insulator-to-metal-transition (IMT) of vanadium dioxide (VO_2). The extensive literature documents a debate over the mechanism of the photoinduced IMT in VO_2 [4], which can also be induced thermally (heating above about 70 °C). The use of EUV TG spectroscopy for studying the photo-induced transition allows probing of the electronic structure with sensitivity to the vanadium M-edge during the IMT.

2 Experiment

The experimental geometry for the TG is shown in Fig. 1. High harmonics of an 800 nm, 30 fs, Ti:Sapphire laser are generated with a loose (~ 1.5 m) focusing geometry in an argon gas cell. The transient grating pump consists of two synchronized pulses of 800 nm light, crossed on the sample and impinging close to the surface normal. The grating period of ~ 15 μm is achieved with a total fluence of 15 mJ/cm^2, just below the damage threshold [5]. The high harmonic probe beam is refocused with a toroidal mirror, and probes the VO_2 sample surface at an incident angle of 22° before reaching the focus at the plane of the detector. Harmonics of order 15–25 (22–39 eV) are resolved in the first order diffracted signal and integrated using a back-thinned CCD camera. The average harmonic intensity is monitored through the current drawn by photoemission from the toroidal mirror and recorded to check source stability during long acquisitions [6].

3 Results

The time dependence of the transient response of VO_2 was studied with the sample below and above the thermally induced IMT around 70 °C. In the (heated) metallic phase, the pump is expected to drive phonons in the sample, but no IMT. The response of harmonic order 25 (HH25) is shown in Fig. 2 (green x) representative for all other probe wavelength exhibiting exactly the same transient behavior. The measured thermal grating has a rise-time of approx. 20 ps.

The colder room temperature sample exhibits two different temporal behaviors as a function of probe wavelength. Harmonic orders below HH25, (see e.g. HH21 (red O) in Fig. 2) display only the slow acoustic signal appearing in the hot measurements. Harmonics 25 (blue □) and wavelength shorter than this also exhibit

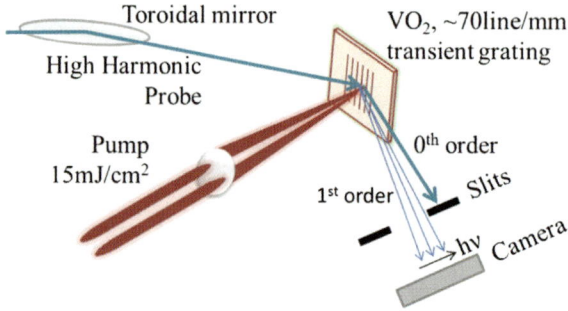

Fig. 1 Transient grating experiment geometry. A transient grating with ∼15 μm period is created on the VO$_2$ surface by two 800 nm beams. High harmonics generated in a high pressure gas cell are refocused using a toroidal mirror. First order diffraction from the sample surface is separated from the 0-order beam with slits and detected with a CCD camera

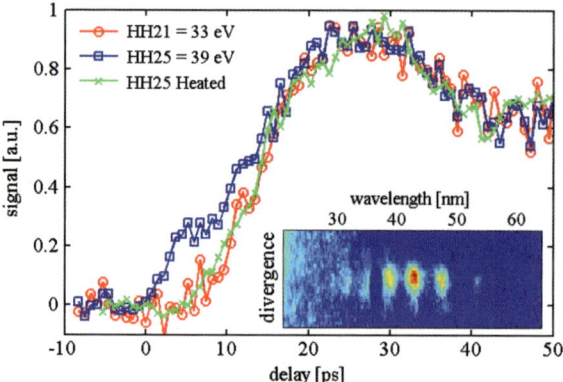

Fig. 2 Time dependence of integrated transient grating 1st order diffracted signal. Room temperature sample shows photon energy-dependent response: HH 21 (*red* ○) shows only slow rise from thermal expansion, while HH25 (*blue* □) exhibits fast electronic response. For the heated sample, HH25 (*green x*) no electronic response is observed. The *inset* shows the raw data, showing dispersed harmonics on the CCD camera

a much faster transient feature before the slow acoustic signal. The fast rise-time exhibited by the highest harmonic orders is resulting from a change in the refractive index as VO$_2$ undergoes the IMT.

The few picosecond rise-time is the resolution limit due to the grazing incidence of the probe. The highest-order harmonics have an increased electronic sensitivity since they are located at the onset of the vanadium M-edge [7].

We measure a maximal diffraction efficiency of order 10^{-2}, consistent with a heat induced sample expansion. Furthermore, we note that we are able to create a permanent grating due to sample damage by increasing the pump intensity. This produces a diffracted signal under the transient signal, opening the interesting opportunity for future heterodyne measurements. This in turn allows for signal amplification and decomposition in real and imaginary index of refraction.

By generating the pump's grating pattern with diffractive optics and utilizing a tilted wavefront scheme, we plan to compensate for the time-resolution limitations caused by the grazing incidence probing.

Acknowledgments This material is based upon work supported by the U.S. Department of Energy Science, Chemical Sciences, Geosciences and Biosciences Division. M.G. acknowledges funding via the Office of Science Early Career Research Program through the Office of Basic Energy Sciences, U.S. Department of Energy.

References

1. Ogilvie, J.P. and Kubarych, K.J.: Multidimensional electronic and vibrational spectroscopy: an ultrafast probe of molecular relaxation and reaction dynamics. In: Advances in Atomic, Molecular and Optical Physics, vol. 57, 249-321. Elsevier Inc., 2009.
2. Crimmins, T.F., Maznev, A.A., and Nelson, K.A.: Transient grating measurements of picosecond acoustic pulses in metal films. Appl. Phys. Lett. **74**, 1344-1346 (1999).
3. Tobey, R.I., *et al.*: Transient grating measurement of surface acoustic waves in thin metal films with extreme ultraviolet radiation. Appl. Phys. Lett. **89**, 091108 (2006).
4. Yang, Z., Ko, C., and Ramanathan, S.: Oxide Electronics Utilizing Ultrafast Metal-Insulator Transitions. Annu. Rev. Mater. Res. **41**, 337-367 (2011).
5. Sistrunk, E. *et al.*: Extreme Ultraviolet Transient Grating Spectroscopy of Vanadium Dioxide. In submission (2014). Pre-print available: arXiv.org/abs/1405.5964.
6. Grilj, J., Sistrunk, E., Koch, M., and Gühr, M.: A beamline for time-resolved extreme ultraviolet and soft x-ray spectroscopy. J. Anal. Bioanal. Tech. s12:005 (2014).
7. Martins, M., *et al.*: Open shells and multi-electron interactions: core level photoionization of the 3d metal atoms. J. Phys. B: At. Mol. Opt. Phys. **39**, R79-125 (2006).

Delayed Core-Level Photoemission from the van der Waals Crystal WSe$_2$

F. Merschjohann, S. Neb, P. Bartz, M. Hensen, C. Strüber,
S. Fiechter, N. Müller, W. Pfeiffer and U. Heinzmann

Abstract Attosecond time-resolved XUV streaking experiments are reported for cleaved WSe$_2$ surfaces. The photoemission from Se 3d and W 4f core levels occurs delayed by 50 as with respect to the valence band emission.

1 Introduction

The availability of single attosecond (as) XUV pulses allows investigating ultrafast electron dynamics on the attosecond time scale. Photoelectron wave packets generated by the XUV pulse are interacting with a simultaneously present IR streaking field. Measuring the photoelectron streaking for electrons emitted from different initial states as function of the delay between XUV and IR pulse allows determining the relative timing of the different emission processes. For a (110) tungsten surface Cavalieri et al. reported that the photoemission from tungsten 4f states is delayed by about 100 as with respect to the electron emission from the valence band [1]. The physical origin of this delay is not yet understood and controversial theoretical models coexist (for review see [2]). Recently, a study for Mg(0001) surfaces showed a vanishing delay between emission from Mg core levels and the valence band states [3] indicating that the degree of initial state localization has only minor influence on the photoemission dynamics. However, the identification of the physical mechanism responsible for the observed delays is obscured by the fact that (a) the streaking field distribution at the interface is unknown on the atomic length

F. Merschjohann · S. Neb · P. Bartz · M. Hensen · C. Strüber · N. Müller · W. Pfeiffer (✉) ·
U. Heinzmann
Fakultät für Physik, Universität Bielefeld, Universitätsstr. 25,
33615 Bielefeld, Germany
e-mail: pfeiffer@physik.uni-bielefeld.de

S. Fiechter
Institut für Solare Brennstoffe, Helmholtz Zentrum Berlin,
Hahn-Meitner-Platz 1, 14109 Berlin, Germany

© Springer International Publishing Switzerland 2015
K. Yamanouchi et al. (eds.), *Ultrafast Phenomena XIX*,
Springer Proceedings in Physics 162, DOI 10.1007/978-3-319-13242-6_17

scale, (b) the photoelectron emission occurs over a depth regime of several Å and varies substantially for different final states, and (c) any surface contaminations can alter the streaking effect. Hence, further as-time-resolved photoemission studies using other surfaces are needed to resolve the controversies related to the dynamics of the photoemission process. Here we report as-time-resolved photoemission from a van der Waals crystal that exhibits a completely different electronic structure compared to a bulk metal. The layered structure of the investigated solid yields element specific photoelectrons emitted from different but well defined depth and thus helps resolving the physical origin of temporal delays in photoemission.

2 Experimental Setup

The laser beam line is similar to the one used for the W(110) experiment [1] and the same UHV photoemission apparatus is used. Photoelectrons emitted under normal direction are detected using a field-free time-of-flight setup. WSe_2 crystals are cleaved under UHV conditions (10^{-10} mbar) and from the resulting surface XUV photoemission spectra are recorded over 1 day without showing degradation. After background subtraction the recorded photoemission spectra are fitted with a series of Gaussian peaks and the obtained peak positions are used to determine the streaking effect for the different spectral contributions.

3 Results and Discussion

The XUV photoemission spectrum is composed of a broad valence band emission centered at a kinetic energy E_{kin} of 78.8 eV, the emission from a Se s-band at $E_{kin} = 70$ eV, and the W 4f and Se 3d core level emission peaks at $E_{kin} = 51.7$ eV and $E_{kin} = 31.1$ eV, respectively. The feature appearing at $E_{kin} = 38$ eV is attributed to the filling of a Se 3d hole via a transition from Se s-band and related Auger electron emission from the valence band. From fitting the spectra recorded for different delay between IR and XUV pulse with a series of Gaussian peaks the energy positions of the various spectral components are determined. With exception of the Auger peak the photoemission components show pronounced streaking effects. The streaking effect for different initial states shows small phase shifts indicating that the electrons emitted by the attosecond XUV pulse experience the streaking field at slightly different times. The relative delay with respect to the emission from the valence band is determined by fitting the streaking curves. The obtained relative delays for 11 streaking measurements performed on 2 different samples for different times after cleaving the sample are summarized in Fig. 1b. Whereas there is no significant delay between the emission from the valence band and the Se s-band the photoelectrons from the Se 3d and W 4f are emitted with about 50 as delay. For about 30 h the behavior is unchanged indicating that surface

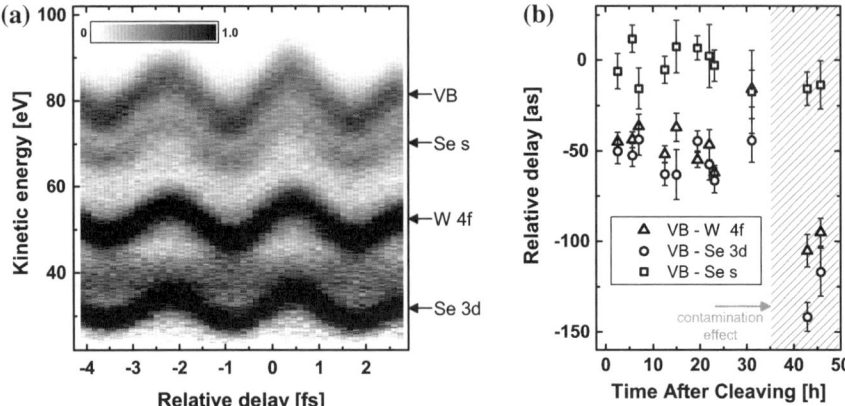

Fig. 1 Attosecond streaking from WSe$_2$ surface. **a** Contour plot of the photoemission spectra with background subtraction as function of delay between IR and XUV radiation. Positive delay indicates an earlier arrival of the XUV pulse. **b** Delay of different photoemission contribution relative to the valence band emission as function of the time after cleavage

contamination has no significant impact. For even longer exposure time the delay between core level and valence band emission increases. This increasing delay is explained by the growth of a surface contamination layer leading to longer transmission times of the photoelectrons emitted from core levels because of their lower kinetic energy compared to photoelectrons emitted from the valence band having the highest kinetic energy.

WSe$_2$ has a layered structure and each layer is composed of a single W layer sandwiched between two Se layers (± 1.67 Å apart). Neighboring W layers are separated by 6.48 Å and based on free electron propagation the emission from the second layer should be additionally delayed by about 150 and 120 as for the W 4f and the valence band electrons, respectively, and 195 and 130 as for the Se 3d and Se s-band electrons, respectively. No such delayed components can be identified in the streaking measurement and we therefore conclude that photoemission from the second WSe$_2$ layer is negligible. Thus, our measured 50 as delay between valence band and core level emission cannot be attributed to propagation effects. Because of the negligible role of surface contamination effects and the observation that the photoemission occurs from a single WSe$_2$ layer we conclude that properties of the photoemission initial and final states determine the observed delay. The created continuum electron wave packets originating from either localized core levels with different angular momentum or delocalized valence band states propagate in different effective potentials. As it was proposed in [4] this relates to relative phase shifts experimentally determined by means of spin-resolved photoemission and corresponding time delays that are accessible by means of attosecond time-resolved photoemission spectroscopy.

References

1. A. L. Cavalieri, N. Müller, T. Uphues, V. S. Yakovlev, A. Baltuska, B. Horvath, B. Schmidt, L. Blumel, R. Holzwarth, S. Hendel, M. Drescher, U. Kleineberg, P. M. Echenique, R. Kienberger, F. Krausz, and U. Heinzmann, Nature **449**, 1029-1032 (2007).
2. U. Heinzmann, *Attosecond Physics*, eds. L. Plaja, R. Torres, and A. Zaïr, eds. (Springer, Berlin Heidelberg, 2013) p. 231–253.
3. S. Neppl, R. Ernstorfer, E. M. Bothschafter, A. L. Cavalieri, D. Menzel, J. V. Barth, F. Krausz, R. Kienberger, and P. Feulner, Phys. Rev. Lett. **109**, 087401 (2012).
4. U. Heinzmann und J. H. Dil, J. Phys.: Condens. Matter. **24**, 173001 (2012).

Optimization of Quantum Trajectories Driven by Strong-Field Waveforms

S. Haessler, T. Balčiūnas, G. Fan, T. Witting, R. Squibb,
L. Chipperfield, A. Zaïr, G. Andriukaitis, A. Pugžlys,
J.W.G. Tisch, J.P. Marangos and A. Baltuška

Abstract We combine phase-locked femtosecond pulses with 1.5, 1.0 and 0.5 μm wavelength to shape optical cycles and experimentally realize the concept of the "perfect wave for high harmonic generation". This has far-reaching implications for attosecond spectroscopy.

OCIS codes (270.6620) Strong-field processes · (020.2649) Strong field laser physics · (140.7090) Ultrafast Lasers

Recently, with the generation and amplification of phase-locked more-than-octave spanning spectra, the capability for sub-cycle waveform shaping has been demonstrated [1] and applied for the sub-femtosecond confinement of field ionization to a single field crest [2]. Here we apply shaped optical cycles, synthesized from a 1.6-octave-spanning spectrum, to a more complex optimization task, the efficient acceleration and return of the ionized electron in HHG. As illustrated in Fig. 1, we address a fundamental bottleneck in HHG driven by sinusoidal driver waves: drive wavelength, λ, and intensity, I, combine to the highest achievable electron energy at recollision, $3.2U_p$, where $U_p \propto I \lambda^2$. The intensity is limited to a small range due to the combined effect of the exponentially growing ionization rate and saturation (unity ionization probability), while an increasing wavelength leads to a drop $\propto \lambda^{-6}$ of the conversion efficiency [3]. This is mainly due to the increased trajectory excursion duration, $\propto \lambda$, which causes additional quantum wave packet spreading and reduces the recollision amplitude. Here, applying shaped optical cycles we

S. Haessler (✉) · T. Balčiūnas · G. Fan · G. Andriukaitis · A. Pugžlys · A. Baltuška
Photonics Institute, Vienna University of Technology, Gusshausstrasse 27/387,
1040 Vienna, Austria
e-mail: stefan.haessler@ensta-paristech.fr

T. Witting · R. Squibb · A. Zaïr · J.W.G. Tisch · J.P. Marangos
Blackett Laboratory, Imperial College, London SW7 2AZ, UK

L. Chipperfield
Max Born Institute, Max-Born-Straße 2A, 12489 Berlin, Germany

© Springer International Publishing Switzerland 2015
K. Yamanouchi et al. (eds.), *Ultrafast Phenomena XIX*,
Springer Proceedings in Physics 162, DOI 10.1007/978-3-319-13242-6_18

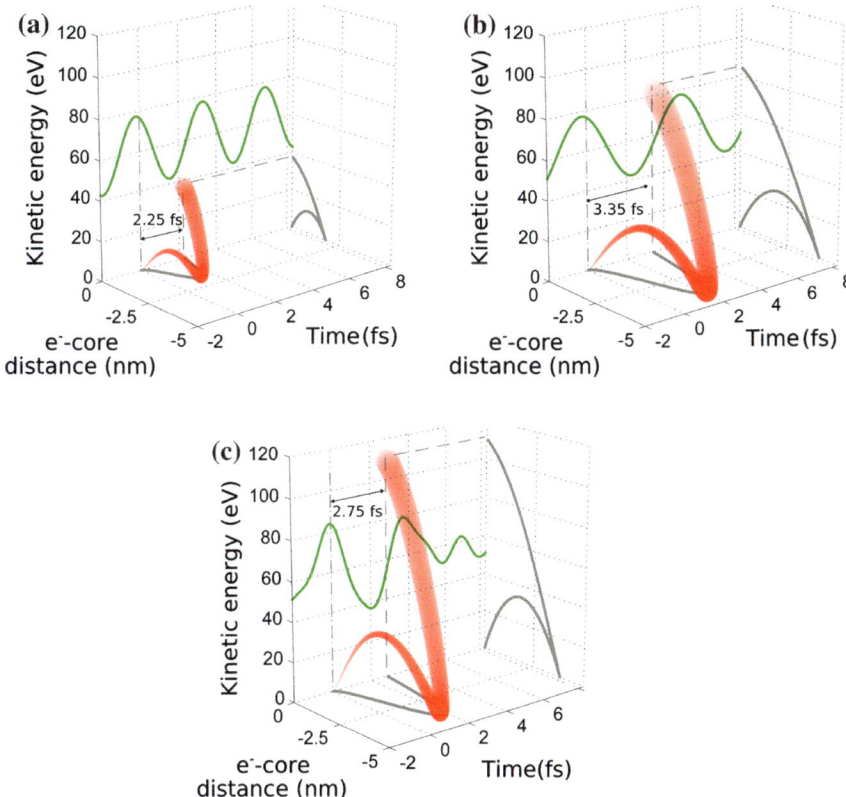

Fig. 1 Simulated optimization of HHG via the driving waveform. The driving electric field is shown by the *green line*. Both single-color drivers [1.03 μm (**a**) and 1.545 μm (**b**)] have an intensity of 1.2×10^{14} W cm^{-2}. The 3-color waveform [**c** combination of 1.03, 0.515 and 1.545 μm with relative intensities as in our experiments (Fig. 3) and optimal phase delays] has the same fluence within its 10.3-fs period as the single-color waves. The cutoff electron trajectory, leading to the highest recollision energy and thus highest emitted HHG photon energy is shown by points with radii that linearly increase with excursion time, symbolizing wavepacket spreading. The optimized 3-color-driver leads to enhanced acceleration of the continuum electron during a shorter excursion time, thus mitigating wavepacket spreading. Furthermore, the field strength launching the trajectories by tunnelling ionization is enhanced

tailor the single-atom quantum dipole to demonstrate the efficacy of the "perfect wave" concept [4] and so take an important step towards the removal of the above-mentioned bottleneck.

We coherently combine three color components: the $\lambda_1 = 1030$ nm from a CEP-locked Yb-based femtosecond laser amplifier, its second harmonic at $\lambda_2 = 515$ nm, and the $\lambda_3 = 1545$ nm signal wave from a white-light-seeded OPA, pumped by the 1030 nm laser. Superposing these three color components with controlled mutually locked phase delays, we realize a shot-by-shot stable Fourier synthesis of optical cycles, periodically repeated under a ≈180 fs envelope with >0.5 mJ total pulse energy. Previous control of HHG has been limited to two colors where these are generally a fundamental + second harmonic [5], which limits the scope for waveform shaping, or a pair of fields of incommensurate frequency with no possibility for systematic phase control [6]. The synthesized optical cycles were used to drive HHG in argon. We have studied how their shape governs the HHG emission by scanning the phase delays $\tau_2 = \phi_2 (\lambda_1 - \lambda_2) /2\pi c$ and $\tau_3 = \phi_{CEP} (\lambda_3 - \lambda_1)/2\pi c$ of the 515-nm and 1545-nm components, respectively, relative to the 1030-nm base component. Since we aim to observe the steering of the continuum electron wavepacket (EWP), we have normalized the measured HHG spectra by the squared recombination dipole matrix element for argon. Figure 2 shows these data as

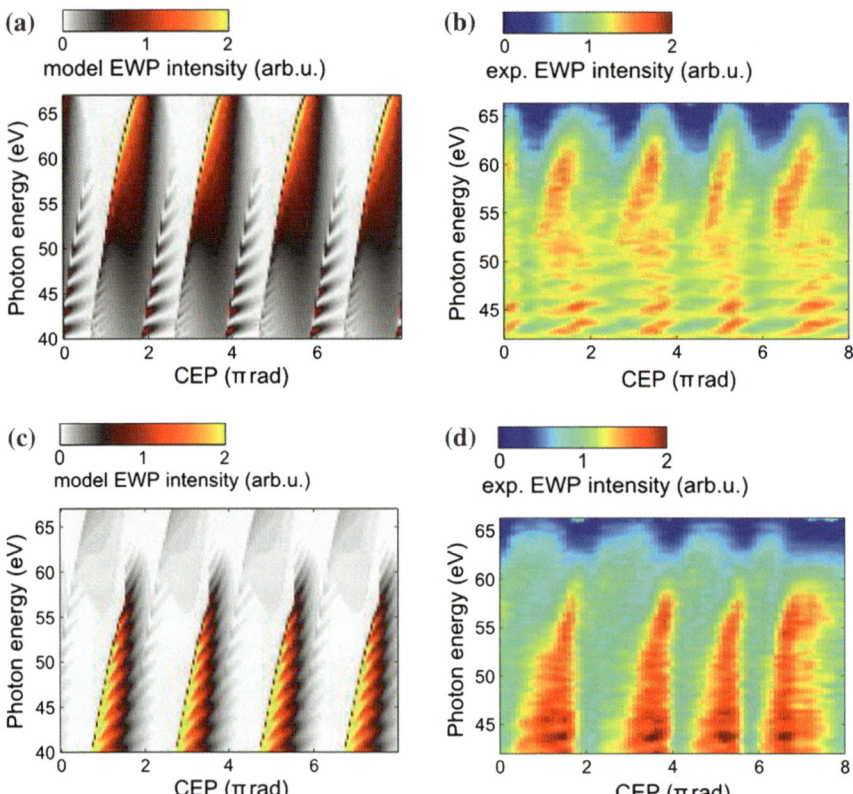

Fig. 2 Simulated **a** and measured **b** spectrogram for $\phi_2 = 1.2\pi$; simulated **c** and measured **d** spectrogram for $\phi_2 = 0.6\pi$. The simulations were performed for a single 10.3-fs long optical cycle using the intensities of the 3 color components as estimated for the measurements. Only the short trajectory family, which is best phase matched in a macroscopic medium, was included. In the experimental spectrograms, the harmonic peak modulation was removed to obtain the spectral envelope. The CEP-values in the experiment have an unknown offset, which could slip between scans for different ϕ_2. Note that ϕ_2 may be considered as jitter-free whereas ϕ_{CEP} jitters by ≥0.95 rad r.m.s

function of the laser CEP, ϕ_{CEP}, for two selected values of ϕ_2, together with EWP intensities simulated with the Lewenstein model using the quantum path analysis [7]. The agreement between the experiment and our single-atom simulations is good over all phase delays, which is proof of direct steering of electron quantum trajectories in HHG via our shaped optical cycles. The clear effect of both varied phase delays shows that the synthesized waveforms survive propagation in the HHG gas medium.

The calculated quantum trajectories show that the ionizing field strength launching the relevant trajectories are optimized, the recollision energies and thus the HHG photon energy cutoff are maximized, and the excursion duration of the relevant trajectories are kept short.

To demonstrate the usefulness of our shaped optical cycles for the cutoff and efficiency enhancement of HHG, we have repeated our experiments at higher driving intensity and thus in the practically most relevant interaction conditions close to saturation. Even under these conditions, we observe clear enhancement of HHG with the synthesized waveforms. Figure 3c shows a measured HHG spectrum generated by a selected optimal waveform. The spectral cutoff clearly lies above the 73 eV absorption edge of the Al-filter used. Figure 3a, b shows the HHG spectra generated by sinusoidal drivers. So whilst with the full available OPA output at 1545 nm a fairly high spectral cutoff is achieved, the HHG flux is very low. On the other hand, the laser output at 1030 nm with the same total pulse energy as the three-color waveform leads to saturated HHG with a cutoff below 60 eV, clearly showing the limitations of HHG driven by near-IR drivers. In comparison, the synthesized optimal waveform generates an HHG spectrum that unites high spectral intensities (>80 times increase compared to the mid-IR driver) with a cutoff well

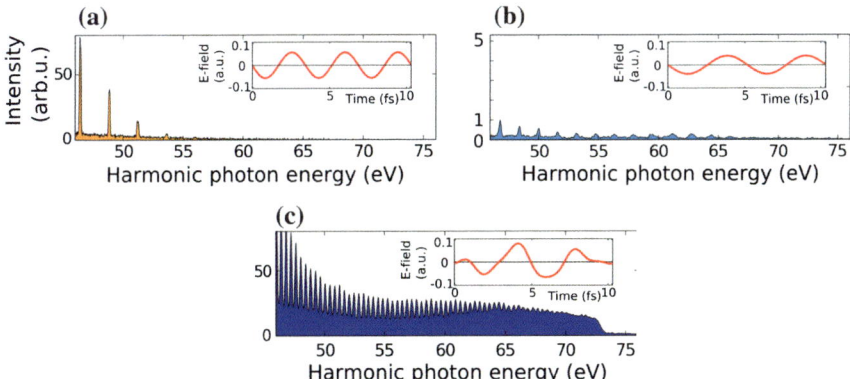

Fig. 3 Experimental HHG spectra generated in argon with different driver waves, shown in the insets. The spectral intensities are given on the same scale and are thus directly comparable. **a** 1030-nm driver with the same 0.54 mJ pulse energy as the three-color pulses, leading to the onset of saturation in HHG. **b** 1545-nm driver with the full OPA output pulse energy (0.35 mJ). **c** Chosen 3-colour driver ($\phi2 = 1.8\pi$ and $\phi_{CEP} = 0.3\pi$) giving the highest HHG signal between 60 eV and the Al-filter absorption edge

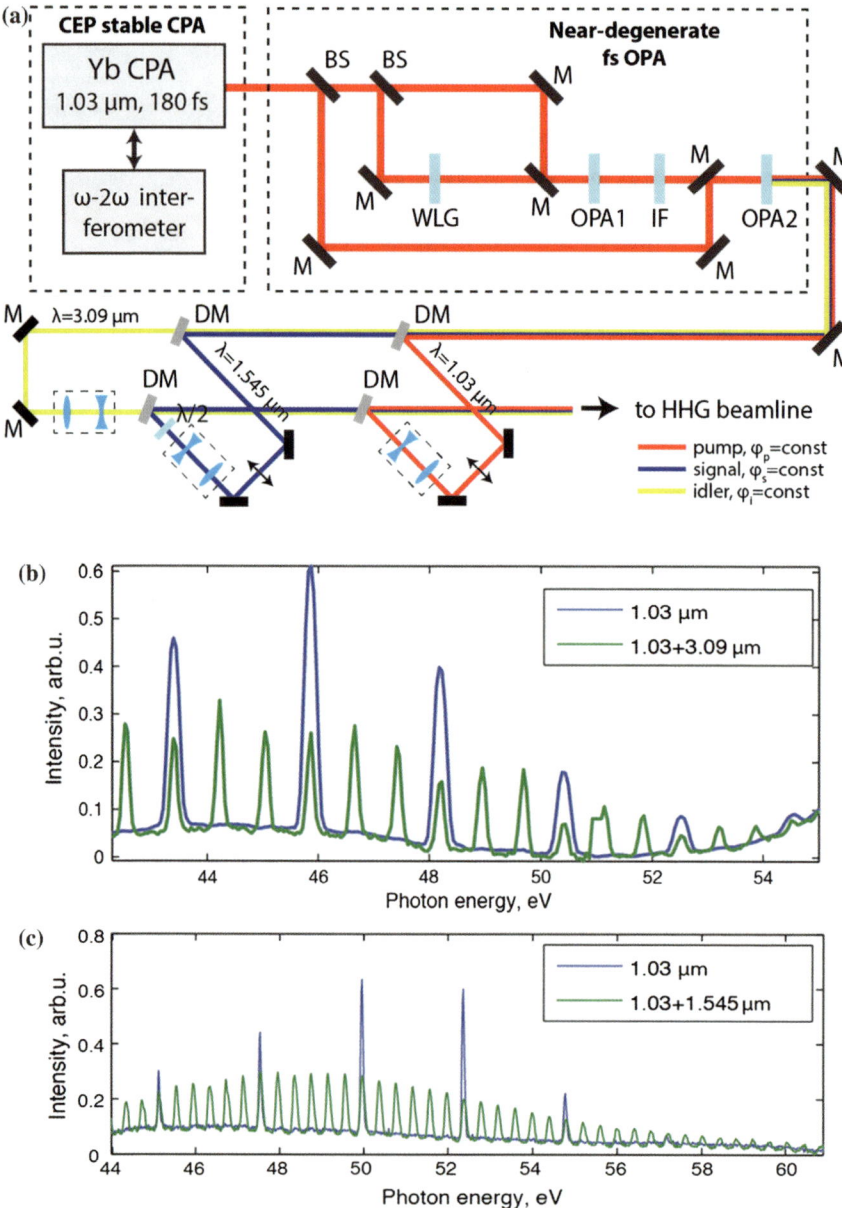

Fig. 4 Layout of our waveform synthesizer (**a**). Measured HHG spectra generated with the 1-µm component alone (*blue line*) and with the combination of 1 + 1.5 µm (**b**), and 1 + 3 µm (**c**)

beyond the saturation limit of the efficient near-IR driver. The denser harmonic comb spacing leads to even greater enhancement in the integral XUV flux (measured factor >140 in the 55–65-eV range). According to our simulations for the synthesized waveform, the dominant recollision events occur once per 10.3 fs (it would take a 6.2-μm single-color driver to realize the same periodicity), as opposed to once per 2.6 fs in the SWIR case. Consequently, we would expect a several hundred times enhancement in the flux per attosecond burst and thus great implications to future sources of high-energy (isolated) attosecond pulses.

Our demonstrated optimization holds significant potential for applications of HHG. Without sacrificing signal intensity for the self-probing of atoms and molecules [8], the desired higher bandwidths of the REWP can be attained even at moderate laser intensities, which is an important requirement for the study of larger molecules or clusters. The corresponding large temporal spacing of the attosecond pulses is of interest in gating techniques for selecting isolated attosecond pulses [9]. The reduced atto-chirp is highly advantageous for the generation of ever shorter and more intense attosecond pulses. It can be compensated with thinner, and thus less absorbing metal filters, thereby further increasing the attainable attosecond pulse energy on target.

Moreover, we report first indications of control over high harmonic generation via a synthesized driver waveform including a mid-IR color component. HHG spectra generated with two-color waveforms that consist of 1.545 μm + 3.09 μm and 1.03 μm + 3.09 μm combined colors along with the experimental scheme are shown in Fig. 4 b, c.

In summary, our results demonstrate a new example of many possible applications of cycle-shaped optical waveforms. Every process that is directly laser-field driven will benefit from the possibility of optimization via the shape of the optical cycle. We expect interesting possibilities to emerge in a broad range of laser-matter interaction regimes, involving, e.g., Brunel electrons whose field-driven trajectories can lead to THz-emission [10], plasma heating and HHG on plasma mirrors [11], or particle acceleration [12].

These studies were supported by the ERC (Projects No. CyFi 280202) and S.H. acknowledges support by the EU-FP7- IEF MUSCULAR

References

1. S.-W. Huang *et al.,* Nature Phot. **5**, 475 (2011).
2. A. Wirth *et al.,* Science **334**, 195 (2011).
3. A. Shiner *et al.,* PRL **103**, 073902 (2009).
4. L. Chipperfield *et al.,* PRL **102**, 063003 (2009).
5. I. Jong Kim *et al.,* PRL **94**, 243901 (2005).
6. F. Calegari *et al.,* Opt. Lett. **34**, 3125 (2009).
7. M. Lewenstein *et al.,* PRA **49**, 2117 (1994).
8. S. Haessler *et al., J. Phys. B* **44**, 203001 (2011).
9. E. Takahashi *et al.,* Nature Comms. **4**, 2691 (2013).
10. M. Kress *et al.,* Nature Phys. **2**, 327 (2006).
11. F. Quéré *et al.,* PRL **96**, 125004 (2006).
12. M. Veltcheva *et al.,* PRL **108**, 075004 (2012).

Phase-Matched Generation of High Order Harmonic for Study of Molecular Dynamics

L.V. Dao, K.B. Dinh and P. Hannaford

Abstract We present a pump-probe experiment based on the use of a second field to modulate the intensity and the spatial profile of the phase-matched high-order harmonics radiation for study of the dynamics of molecular gases.

1 Introduction

The wave-packet created by the laser field plays a fundamental role in understanding the quantum picture and provides a bridge between the quantum picture and the classical concept of the trajectory of a particle. The motion of the wave-packet reflects the time evolution of a coherent superposition of the system. Useful information regarding the discrimination and visualization of the wave-packet dynamics can be obtained from time-resolved photoelectron imaging because of the sensitivity of the photoelectron angular distribution to the electronic symmetry. The high order harmonic generation (HHG) process is a coherent recombination of wave-packets and can be used to study wave-packet properties [1]. In conventional pump-probe spectroscopy a strong laser pulse is used to excite a molecule and the variation of the optical properties following the recovery or excitation of the molecules is probed by a second delayed pulse. In a HHG experiment the propagation and phase of the fundamental and harmonic fields depend on the optical properties of the medium and are reflected in the observed harmonic intensity through the so-called phase mismatch factor. A second off-axis delayed pulse can be used to perturb the propagation process or to change the harmonic phase which leads to a variation of the total harmonic intensity and the harmonic beam profile [2].

L.V. Dao (✉) · K.B. Dinh · P. Hannaford
Centre for Quantum and Optical Science, Swinburne University of Technology,
Melbourne, Australia
e-mail: dvlap@swin.edu.au

© Springer International Publishing Switzerland 2015
K. Yamanouchi et al. (eds.), *Ultrafast Phenomena XIX*,
Springer Proceedings in Physics 162, DOI 10.1007/978-3-319-13242-6_19

We report the use of a second, off-axis, long-pulse, laser beam to control the HHG process and show that the dynamics of the excited wave-packet can be studied through the modulation of the HHG intensity and the variation of the beam profile. The experimental approach proposed here shows that the properties and dynamics of the quasi-bound electron wave-packet can be studied and characterized.

2 Experimental Results and Discussions

A 1 kHz multi-stage, multi-pass, chirped-pulse amplifier system producing up to 10 mJ pulses with a duration of 35 fs and centered at 810 nm is used in our experiment. The laser beam is split into two beams with an intensity ratio of first beam to second beam of about 4:1. A dispersion medium is added in the path of the second beam to give a pulse duration of ∼ 80 fs. The phase mismatch between harmonic and fundamental field $\Delta k = k_q - q k_0$ along the axis of the fundamental laser beam, which is reflected in the output of harmonic intensity, can be expressed as

$$\Delta k(z) = dispersion\ of\ the\ neutral\ medium + the\ plasma\ dispersion$$
$$+ (geometric\ term) + (dipole\ phase\ term) \tag{1}$$
$$+ (dispersion\ of\ highly\ excited\ medium)$$

The dispersion of the neutral medium is negative, and the plasma dispersion is positive. The geometric term is dependent on the configuration of the HHG setup. Electrons which have tunneled under the interaction of the laser field at the maximum electric field of a laser cycle can return to the atom after one cycle with zero kinetic energy and oscillate with the laser frequency in the next cycle. These electrons can be captured into high excited states after a few optical cycles of the pulses. We include the fifth term due to the dispersion of the highly excited medium to take account of excitation effects.

When the second beam is absent the HHG radiation is obtained along the optical-axis of the first beam. A good beam profile and high HHG intensity indicate that the phase mismatch is small for a short quantum pathway [3]. When the second laser field is applied the harmonic intensity varies with the delay time between the two beams because the phase-mismatch Δk is altered. The modulation of the harmonic intensity is very strong for parallel polarizations for delay times up to −300 fs. When an 80 fs hyperbolic secant second pulse is applied the ration between the first and second fields strength is ∼ 0.05 and 0.01 at delay time of −200 fs and −300 fs, respectively. The second field is not change the field strength of the first field but it is enough to modify the trajectory of free electron. A high modulation frequency, which is the same as the laser frequency, and a low modulation frequency (20–25 THz) can be seen in Fig. 1. The electrons are most likely produced

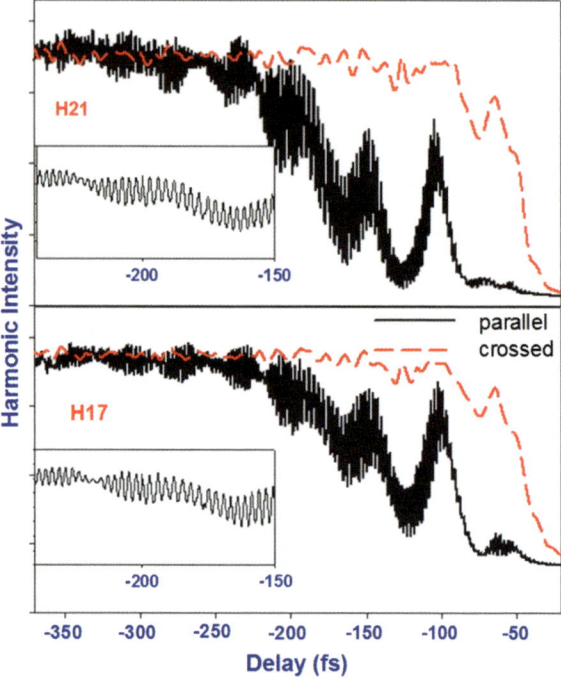

Fig. 1 Modulation of the intensity of harmonics H17 and H21. The inset shows a zoom of the region near the time delay of −200 fs

at the peak of the electric field but they return to the core with zero kinetic energy and trap into quasi-bound state. They are not contributed to HHG emission but their contributions to phase-matching cannot be neglect. The second (weak) pulse excites the quasi-bound and high excited wave-packet, which is likely a high excited Rydberg wave-packet, into the ionized continuum. The free electron wave-packets created by the first laser field, when they interact with the field before or after maximum, and the other free electron wave-packets created by the second laser field through the ionization of high–excited wave packet can interfere with each other. The ionization of the quasi-bound wave-packet and the interference of the two wave-packets lead to a change of the plasma dispersion. The interaction of the second laser field with the free electrons modifies the trajectory and re-scattering of the electron causes a change in the dipole phase. In the longitudinal direction or in the total harmonic intensity the contribution of the dipole phase to the phase-mismatch is smaller than that of the plasma dispersion. The Fig. 2 show the beam profile of H21 and H17 for oxygen at around −100 fs where the phase-mismatch is large and the harmonic intensity is low. Within one optical cycle we observe a shifting of the maximum with delay time away from the axis (for delay ∼ −108 fs) to an asymmetrical annular beam (around the optical axis of driving field) (for delay ∼ −109 fs) and then an abrupt change back to on-axis (for delay ∼ −110 fs). We can expect that the variation of the spatial profile depends on the atomic and molecular structures.

Fig. 2 the variation of beam profiles of H17 (**a**) and H21 (**b**) for oxygen at around −110 fs

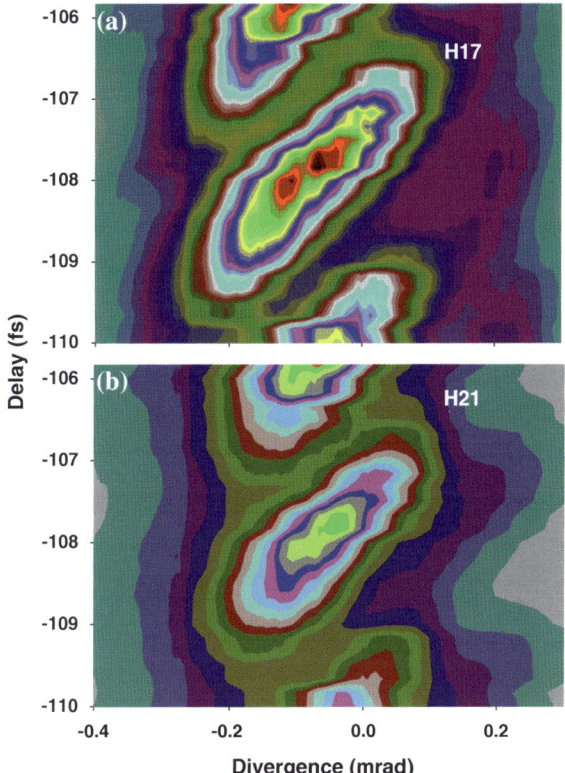

In conclusion, using a long off-axis pulse to modify the phase-matched harmonic intensity and the spatial distribution, the contribution of different electron trajectories can be revealed. We demonstrate that information on the quasi-bound state and the Rydberg state can be obtained from the high-order harmonic generation process. This experimental technique offers the possibility of studying the dynamics of the phase matching of an atom and the molecular structural dynamics with high time and spatial resolution.

References

1. O. Smirnova, Y. Mairesse, S. Patchkovskii, N. Dudovich, D. Villeneuve, P. Corkum, and M. Yu. Ivanov, "High harmonic interferometry of multi-electron dynamics in molecules", Nature 460, 972 (2009); H. J. Worner, J. B. Bertrand, D. V. Kartashov, P. B. Corkum, and D. M. Villeneuve, "Following a chemical reaction using high-harmonic interferometry", Nature 466, 604 (2010).
2. P. Balcou, P. Salieres, A. L'Huillier, and M. Lewenstein, "Generalized phase-matching conditions for high harmonics: The role of field-gradient forces", Phys. Rev. A 55, 3204–3210, (1997).

3. S. Teichmann, B. Chen, R. A. Dilanian, P. Hannaford, L. V. Dao, "Spectral characteristics across the spatial profile of a high-harmonic beam", J. Phys. D: Appl. Phys. 42, 135108 (2009); S. Teichmann, P. Hannaford, L.V. Dao, "Phase-matched emission of few high-harmonic orders from a helium gas cell" App. Phys. Lett. 94, 171111 (2009).

Multiphoton Transitions for Robust Delay-Zero Calibration in Attosecond Transient Absorption

J. Herrmann, M. Lucchini, S. Chen, M. Wu, A. Ludwig, L. Kasmi, K.J. Schafer, L. Gallmann, M.B. Gaarde and U. Keller

1 Introduction

Attosecond transient absorption spectroscopy plays a major role in the progress of attosecond science. A manifold of recent publications, either employing a single attosecond pulse (SAP) or a train of attosecond pulses (APT) in combination with an infrared (IR) pulse, provided detailed insight into the electron dynamics of atoms [1–4]. Due to the richness of the observed effects the correct calibration of the delay-zero in the experimental data is crucial for a useful interpretation.

Unfortunately, the low photon flux in the extreme ultraviolet (XUV) spectral region in combination with the absence of a suitable nonlinear material prevents the delay-zero calibration with a simple cross-correlation like in femtosecond pump-probe spectroscopy. Furthermore, the extraction of the delay-zero out of the experimental data by simply using e.g. the total absorption leads to incorrect results.

Here, we show the transient absorption of an APT overlapped with a moderately strong IR pulse around the first ionization threshold of helium (He). In the transmitted XUV radiation as a function of APT-IR delay we observe half- and quarter-laser-cycle oscillations. The appearance of the quarter-laser-cycle (4ω) oscillations

J. Herrmann (✉) · M. Lucchini · A. Ludwig · L. Kasmi · L. Gallmann · U. Keller
Department of Physics, Institute for Quantum Electronics, ETH Zurich,
8093 Zurich, Switzerland
e-mail: jens.herrmann@phys.ethz.ch

L. Gallmann
Institute of Applied Physics, University of Bern, 3012 Bern, Switzerland

S. Chen · M. Wu · K.J. Schafer · M.B. Gaarde
Department of Physics and Astronomy, Louisiana State University, Baton Rouge,
LA 70803, USA

© Springer International Publishing Switzerland 2015
K. Yamanouchi et al. (eds.), *Ultrafast Phenomena XIX*,
Springer Proceedings in Physics 162, DOI 10.1007/978-3-319-13242-6_20

allows us to experimentally define a precise and robust delay-zero. Our experimental results are supported by the solution of the time-dependent Schrö-dinger equation (TDSE).

2 Results

For the generation of the APT we use a Ti:sapphire-based laser amplifier system (repetition rate: 1 kHz, pulse duration: ≈25 fs, center wavelength: 789 nm) which is focused into a xenon gas jet. We separate a small fraction of the IR beam before the generation of the APT and send it on an independent beampath where we can introduce a delay with respect to the APT. After recombining APT and IR pulse, we focus both beams into a He gas target and detect the transmitted XUV radiation after filtering the residual IR radiation. The APT consists of higher-order harmonics (HHs) which are energetically centered around the first ionization threshold of He. The optical density of the He gas target is 0.79 for HH 17.

Figure 1a shows color-coded the transmitted XUV intensity as a function of photon energy and APT-IR delay τ. HHs 13 to 17 exhibit strong half-laser-cycle (2ω) oscillations. For a more detailed analysis we calculate the energy integrated absorption $\Pi(\tau)$. Figure 1b depicts $\Pi(\tau)$ for HH 13. In order to disentangle the different oscillation components we perform a delay-frequency analysis with a Gaussian-Wigner-transform into the frequency-delay domain (see Fig. 1c). This representation exhibits which frequency component appears at which APT-IR delay. Besides the strong 2ω-oscillation component at 0.76 PHz Fig. 1c reveals a 4ω-oscillation component at 1.52 PHz. These 4ω-oscillations were predicted theoretically by Chen et al. [5]. By integrating the frequency-delay representation along the frequency axis around 1.52 PHz we get the envelope of the 4ω-oscillations as a function of delay as shown in Fig. 1e for an IR intensity of $2.6 \cdot 10^{14}$ W/cm^2. This envelope has a symmetric shape with respect to the delay. The solution of the TDSE shows that the maximum of the envelope coincides with the delay-zero and is robust against changes of the gas density in the target and changes of the IR intensity. Figure 1f presents the calculated envelope for an intensity of $2.8 \cdot 10^{14}$ W/cm^2.

3 Conclusion and Outlook

We studied the interaction of an APT overlapped with a moderately strong IR pulse in He by means of attosecond transient absorption spectroscopy. A delay-frequency analysis allows us to identify 4ω-oscillations besides the already well-known 2ω-oscillations. The 4ω-oscillations originate from a multiphoton coupling of different HHs of the APT. Furthermore, we discussed how to use the 4ω-oscillations to

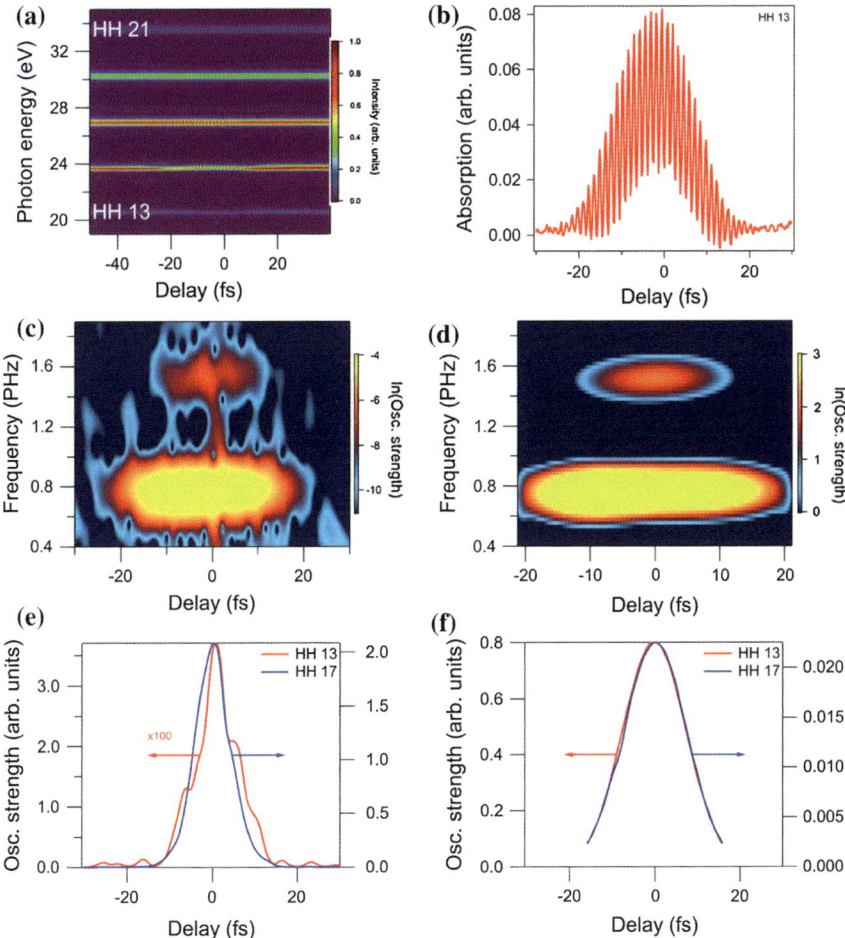

Fig. 1 **a** Transmitted XUV intensity as a function of APT/IR delay and photon energy at an IR intensity of $2.6 \cdot 10^{14}$ W/cm^2. **b** Energy-integrated absorption of HH 13. **c** Gaussian-Wigner transform of the energy-integrated absorption of HH 13 shown in (**b**). **d** Gaussian-Wigner transform of the calculated atomic response function. **e** Integration of the delay-frequency representation in the frequency domain around 1.52 PHz yielding the envelope of 4ω-oscillations of the experimental data in (**c**). **f** Envelope of the 4ω-oscillations calculated by solving the TDSE. The peak of the oscillations coincides with the delay-zero

robustly calibrate the delay-zero in an attosecond transient absorption experiment. By solving the TDSE we support our experimental findings. Finally, we discuss in detail the influence of the IR intensity on the sub-laser-cycle oscillations.

References

1. E. Goulielmakis, Z.-H. Loh, A. Wirth, R. Santra, N. Rohringer, V. S. Yakovlev, S. Zherebtsov, T. Pfeifer, A. M. Azzeer, M. F. Kling, S. R. Leone, and F. Krausz, Nature **466**, 739–743 (2010).
2. M. Holler, F. Schapper, L. Gallmann, and U. Keller, Phys. Rev. Lett. **106**, 123601 (2011).
3. M. Lucchini, J. Herrmann, A. Ludwig, R. Locher, M. Sabbar, L. Gallmann, and U. Keller, New J. Phys. **15**, 15 (2013).
4. M. Chini, X. Wang, Y. Cheng, Y. Wu, D. Zhao, D. A. Telnov, S.-I. Chu, and Z. Chang, Sci. Rep. **3**, 1105 (2013).
5. S. Chen, K. J. Schafer, and M. B. Gaarde, Opt. Lett. **37**, 2211–2213 (2012).

Infrared Double Optical Gating for Generating Submicrojoule Isolated Attosecond Pulses

Eiji J. Takahashi, Pengfei Lan and Katsumi Midorikawa

Abstract We experimentally demonstrate an infrared two-color polarization gating scheme for generating an intense isolated attosecond pulse using the multicycle laser. The output flux by the generalized infrared double optical gating was evaluated to be approximately 0.3 μJ/pulse, which is the highest value achieved by using the double optical gating method. The obtained submicrojoule continuum harmonic spectrum supports the generation of a pulse duration of sub-500 as.

1 Introduction

The generation of an isolated attosecond pulse (IAP) via high-order harmonic generation (HHG) essentially requires shortened pump pulses in a few-cycle pulse regime. To relax the required pulse duration for producing an IAP, several gating methods (see review [1]), such as polarization gating (PG), double optical gating (DOG), ionization gating, and two-color gating (TCG), have been proposed in the literature. To date, the maximum pulse energy of the IAP, as achieved by an inferred TCG (IR-TCG), reaches as high as 1.3 μJ [2], with a duration of 500 as. Recently, we have theoretically proposed a new gating method, called as the infrared double optical gating (IRDOG), to create IAPs using a multicycle CEP-unstabilized laser pulse. This scheme is based on the combination of the conventional PG and the recently reported optimized IR-TCG schemes [3]. In addition, the IRDOG scheme can be generalized to the case of elliptically polarized laser (e.g., $\varepsilon = 0.5$), called as the generalized-IRDOG (GIRDOG), which allows us to utilize a longer laser pulse of 42 fs for producing an IAP.

In this study, we have experimentally demonstrated GIRDOG for the first time. Furthermore, we have scaled up the generated HH output flux by combining a loose-focusing geometry. Science the GIRDOG scheme enables us to utilize a

E.J. Takahashi (✉) · P. Lan · K. Midorikawa
RIKEN Center for Advanced Photonics, 2-1 Hirosawa, Wako, Saitama 351-0198, Japan
e-mail: ejtak@riken.jp

© Springer International Publishing Switzerland 2015   87
K. Yamanouchi et al. (eds.), *Ultrafast Phenomena XIX*,
Springer Proceedings in Physics 162, DOI 10.1007/978-3-319-13242-6_21

phase matching technique under a neutral media, we have successfully demonstrated submicrojoules continuum HH. The obtained continuum HH spectrum by GIRDOG supports the generation of sub-500 as IAP at the XUV region.

2 High-Energy Continuum Harmonic Spectra

The GIRDOG field was generated by using a home built Ti:sapphire laser (30 fs, 50 mJ, and 10 Hz repetition rate). The schematic image of the experimental setup is shown in Fig. 1. The Output pulses from the Ti:sapphire laser were split into two beams using a partial reflecting mirror. The transmitted beam was used as the pump pulse for the OPA-1 with an energy of 2 mJ. The reflected Ti:sapphire laser pulse from the partial mirror was further split into two beams, while the corresponding transmitted beam was introduced to the BBO type-II at OPA-2. Subsequently, an output energy exceeding 7 mJ with 35 fs pulse width was achieved at a signal wavelength near 1.3 μm. This linearly polarized signal beam was combined to the PG field, and the corresponding GIRDOG field was generated.

The PG optics consists of a pair of birefringent quartz plates and a fused silica window. At the quartz plate, the linearly polarized input pulse is split into two orthogonal components with respect to the input polarization by the first quartz plate at 45°. The quartz plate also introduces a proper delay time of approximately 25 fs between the two pulses. Here the temporal jitter between the PG field and the 1.3 μm pulse becomes an important parameter, because the profile of the GIRDOG field depends on the relative phase of the TC interferometer. Therefore, we evaluate the relative phase variation with time in the optical path by injecting a continuous-wave (CW) laser beam of wavelength 547 nm into the TC interferometer. The spatial interference fringes on the superposed beam profile of the CW laser were recorded by using a CCD camera (exposure time: 10 μs with a repetition rate of

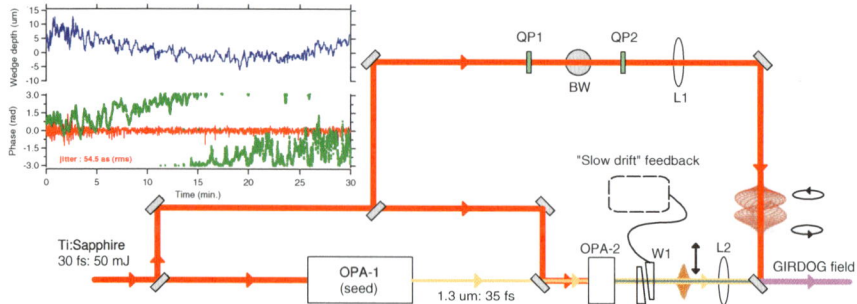

Fig. 1 GIRDOG setup. PG optics consists of a quartz plate (*QP1*), a Brewster window (*BW*), and a second quartz plate (*QP2*). The external IR seed pulses for the OPA-2 stage are produced by the generation of white-light continuum in OPA-1. L1: $f = 4500$ mm and L2: $f = 3500$ mm. Inset: Measured relative phase variation in the TC interferometer

10 Hz, which is synchronized with the pump laser) during 30 min. The phase of the spatial fringes at each recorded profile was converted to the relative delay and relative phase corresponding to the 1.3 μm field. The measured time evolution of the relative phase is shown in Fig. 1 (greed profile). In order to stabilize the temporal jitter in the TC interferometer, we used an active feedback circuit utilizing the information of the recorded relative phase. The red profile in Fig. 1 shows the time evolution of the relative phase with an active feedback obtained by controlling the insert depth of wedge plate (W1). We successfully suppressed the temporal jitter between the PG field and 1.3 μm pulse, which is induced by the fluctuation of optical path length. The temporal jitter of the TC interferometer was measured to be approximately 55 as r.m.s.

GIRDOG field was loosely focused onto a Xe gas target, which was statically filled in a gas cell of length 12 cm. The generated HH spectrum was measured by using a spectrometer with a microchannel plate. Figure 2 shows single-shot HH spectra obtained without gating (800 nm linearly polarized pulse) and GIRDOG. As expected, the absence of gating resulted in the generation of odd harmonics. These peaks merged to form a continuum for every laser shot under the GIRDOG field. The output flux of HH generated with GIRDOG was optimized at the Xe gas pressure of 1 Torr. The HH yield gradually increases with the gas pressure, suggesting a quadratic dependence of the harmonic yield as a function of gas pressure. This characteristic indicates that the optimized gas pressure satisfies the phase matching condition. Note that the profile of the XUV spectrum generated with GIRDOG changes shot-by-shot because the random changing of CEP. Extracting the cutoff HH (>17th HH), the corresponding pulse duration would be sub-500 as on the Fourier-transform limit. The output flux of HH by GIRDOG was evaluated to be approximately 0.3 μJ/pulse, which is the highest value achieved by using the DOG method [5].

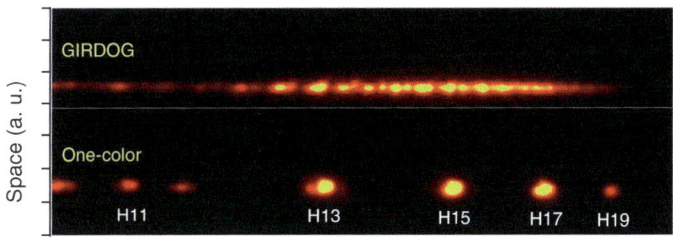

Fig. 2 Single-shot harmonic spectrum images obtained by one-color and GIRDOG

References

1. G. Sansone *et al.*, *Nat. Photonics* **5**, 655 (2011)
2. E. J. Takahashi *et al.*, Nat Commun. **4** 2691 (2013)
3. E. J. Takahashi *et al.*, Phys. Rev. Lett. **104** 233901 (2010)
4. P. Lan *et al.*, Phys. Rev. A **83** 063839 (2011)
5. Y. Wu *et al.*, Appl. Phys. Lett. **102** 201104 (2013)

Generation of Isolated Soft X-Ray Pulses Around the Carbon *K*-Edge Using CEP-Stabilized Few-Cycle IR Pulses

Nobuhisa Ishii, Keisuke Kaneshima, Kenta Kitano, Teruto Kanai, Shuntaro Watanabe and Jiro Itatani

Abstract We generate a 75-eV-wide continuum in the water window via the HHG process using CEP-stabilized, few-cycle IR pulses. A pressure dependence of harmonic spectra indicates sub-cycle deformation of the IR pulses in the process.

1 Introduction

Recent advances in the generation of intense few-cycle pulses at 800 nm from Ti: sapphire lasers have triggered many researches in strong-field physics such as high harmonic generation (HHG) and attosecond pulse generation. However, the maximum photon energy of attosecond pulses has been limited up to 200 eV. To extend the spectral range of attosecond pulses, new light sources with a longer wavelength are required. So far, extension of the cutoff has been demonstrated by relatively long femtosecond IR pulses without controlling their waveform [1]. Manipulation of the waveform of few-cycle laser pulses is a crucial point to access unprecedented time scales much shorter than the cycle period. Although many IR light sources have been developed so far [2], carrier-envelope phase (CEP)-dependent phenomena have not been observed at energy scales larger than 200 eV. We have developed a BiB_3O_6 (BIBO)-based optical parametric chirped pulse amplifier (OPCPA) [3], which is capable of delivering 9.0-fs, 0.55-mJ, CEP-stabilized optical pulses at 1,600 nm with a repetition rate of 1 kHz. In this work, we demonstrate CEP-dependent HHG in the water window using this IR light source [4].

N. Ishii (✉) · K. Kaneshima · K. Kitano · T. Kanai · J. Itatani
The Institute for Solid State Physics, The University of Tokyo,
5-1-5 Kashiwanoha, Kashiwa, Chiba 277-8581, Japan
e-mail: ishii@issp.u-tokyo.ac.jp

S. Watanabe
Research Institute for Science and Technology,
Tokyo University of Science, 2641 Yamasaki,
Noda, Chiba 278-8510, Japan

© Springer International Publishing Switzerland 2015
K. Yamanouchi et al. (eds.), *Ultrafast Phenomena XIX*,
Springer Proceedings in Physics 162, DOI 10.1007/978-3-319-13242-6_22

We also measure a dependence of half-cycle cutoffs (HCOs) [5] on the gas pressure, which indicates sub-cycle deformation of the waveform of the IR drive pulses in the HHG process.

2 CEP-Dependent HHG

HHs in the soft x-ray region were generated from neon gas irradiated by the IR pulses. The details of the light source and experimental setup are described elsewhere [3, 4]. We changed the CEP of the IR pulses by a Dazzler (FASTLITE) in the OPCPA. Figure 1a presents a collection of HH spectra recorded every 0.1π rad of the CEP shift at a backing pressure of 0.6 atm. Using a few-cycle Ti:sapphire laser, a similar CEP dependence has been observed in the extreme ultraviolet [5]. The CEP scan measurement shows that the cutoff energies depend significantly on the CEP where the highest cutoff energy reaches 325 eV. We found a good agreement between the HH spectra (Fig. 1a) and SFA-based simulation (Fig. 1b) assuming 10-fs, 1,600-nm IR

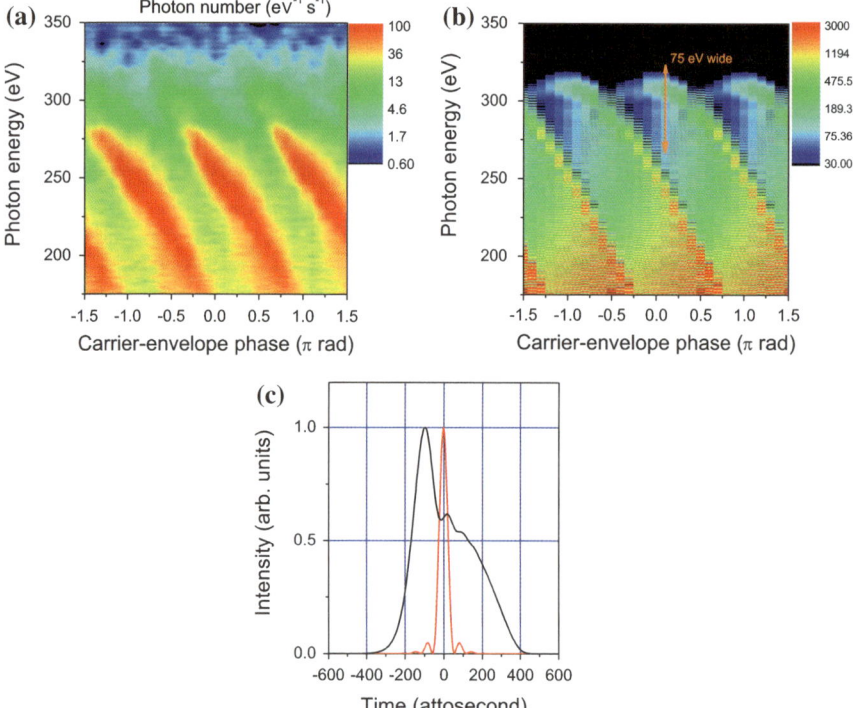

Fig. 1 **a** CEP-dependent HH spectra at a backing pressure of 0.6 atm. **b** Simulated CEP-dependent HH spectra. **c** Temporal profiles of the soft X-ray burst around the carbon K-edge with the chirp (*black line*) included in the calculation and without the chirp (*red line*)

pulses with a peak intensity of 3.8×10^{14} W/cm^2. Figure 1c shows the temporal profile of the attosecond burst, which was reconstructed by selecting out the spectral component with a 75-eV bandwidth as in Fig. 1b (orange arrow), assuming the spectral phase obtained from the simulation. The reconstructed pulse (Fig. 1c red line) would have a duration of about 300 attosecond, which has a bandwidth of supporting 60 attoseconds as shown by the black line in Fig. 1c.

3 Pressure Dependence of HH Spectra

We also investigated dependence of the HCOs on the target gas pressure. We recorded HH spectra by scanning the backing pressure from 0.2 to 2.0 atm at the fixed CEP (=0.2π rad) as shown in Fig. 2. By increasing the pressure from 0.2 to 0.4 atm, the on-axis intensity of HHs rapidly increased, which indicates the onset of self-guiding of the IR pulses [1]. The HH spectrum at 0.6 atm in Fig. 2 contains two peaks located at approximately 180 and 280 eV. These peaks can be attributed to be two HCOs from different electron trajectories [5]. The HCOs were shifted downward with different slopes when the gas pressure was increased. The difference in the slope for these HCOs indicates that the electric field within each half cycle is modulated differently from other half cycles by rapidly growing plasmas combined with self-phase modulation. The peak power of the IR pulses was 38 GW, which is about two times more than a critical power of 21 GW for self focusing when the pressure of neon is set at 1.0 atm. This two-dimensional plot allows us to retrieve sub-cycle waveform deformation of drive pulses in the HHG process.

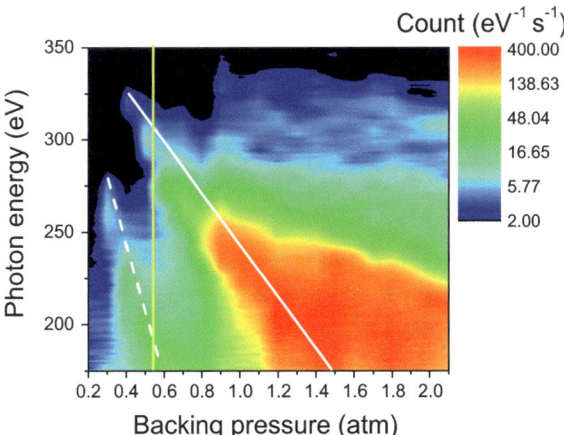

Fig. 2 Pressure dependence of HH spectra. The *white solid* and *dashed lines* are drawn along the spectral peaks. The *yellow line* is drawn at the gas pressure for self-focusing threshold (0.55 atm) when the critical power is identical to the peak power of the IR pulses (38 GW)

4 Conclusion

We applied the BIBO-based OPCPA system to generate the CEP-dependent HHs in the water window. Their spectra are well reproduced by the simulations, indicating the confinement of soft x-ray emission in a single recombination event around the carbon K edge. We observed systematic changes of HCOs when the gas pressure and CEP were varied, which indicate sub-cycle waveform deformation of drive pulses in the HHG process.

References

1. T. Popmintchev, *et al.*, Science **336**, 1287 (2012).
2. T. Fuji, *et al.*, Opt. Lett. **31**, 1103 (2006), D. Brida, *et al.*, Opt. Lett. **33**, 741 (2008), O. D. Mücke, *et al.*, Opt. Lett. **34**, 2498 (2009), B. E. Schmidt, *et al.*, Appl. Phys. Lett. **96**, 121109 (2010).
3. N. Ishii, *et al.*, Opt. Lett. **37**, 4182 (2012).
4. N. Ishii *et al.*, Nature Communications **5**, 3331 (2014).
5. C. A. Haworth, *et al.*, Nature Physics **3**, 52 (2007).

Bright Isolated Attosecond Soft X-Ray Pulses

M.-C. Chen, C. Mancuso, C. Hernández-García, F. Dollar,
B. Galloway, D. Popmintchev, B. Langdon, A. Auger, P.-C. Huang,
B.C. Walker, L. Plaja, A. Jaron-Becker, A. Becker, M.M. Murnane,
H.C. Kapteyn and T. Popmintchev

Abstract By driving the high harmonic generation process with multi-cycle mid-infrared laser pulses, we demonstrate bright isolated, attosecond soft X-ray pulses for the first time.

High harmonic generation (HHG) is the most extreme coherent nonlinear optical process in nature, making it possible to generate coherent beams that span from the EUV to $>$ keV photon energies. The ability to generate ultrashort femtosecond-to-attosecond pulses in the EUV region of the spectrum has already resulted in new understanding of charge and spin dynamics in molecules and materials on the fastest timescales relevant to function. To date, most HHG work has been done using multi-cycle 0.8 μm driving lasers, where HHG generally consists of a train of attosecond pulses in the EUV. Most schemes for creating isolated attosecond pulses rely on driving the process using either few-cycle 0.8 μm pulses, or complex polarization modulation, or a combination of multicolor fields. Furthermore, all these approaches to date have been limited to the EUV region of the spectrum because using near-IR Ti:Sapphire lasers limits the bright phased-matched upconversion to $<\approx 100$ eV photon energies.

M.-C. Chen (✉) · C. Mancuso · C. Hernández-García · F. Dollar · B. Galloway ·
D. Popmintchev · A. Jaron-Becker · A. Becker · M.M. Murnane · H.C. Kapteyn ·
T. Popmintchev
JILA and Department of Physics, University of Colorado at Boulder, Boulder,
CO 80309-0440, USA
e-mail: mingchang@mx.nthu.edu.tw

M.-C. Chen · P.-C. Huang
Institute of Photonics Technologies, National Tsing Hua University, Hsinchu 30013, Taiwan

C. Hernández-García · L. Plaja
Grupo de Investigación en Óptica Extrema, Universidad de Salamanca,
37008 Salamanca, Spain

B. Langdon · A. Auger
Kapteyn-Murnane Labs Inc, Boulder, CO 80301, USA

B.C. Walker
University of Delaware, Newark, DE 19716, USA

© Springer International Publishing Switzerland 2015   95
K. Yamanouchi et al. (eds.), *Ultrafast Phenomena XIX*,
Springer Proceedings in Physics 162, DOI 10.1007/978-3-319-13242-6_23

In this work, for the first time we generate isolated soft X-ray attosecond pulses at photon energies up to 180 eV, with a transform-limited pulse duration of 35 as and a predicted linear chirp of 300 as. Moreover, when mid-infrared lasers are used to drive the high harmonic generation process, the conditions for optimal bright soft X-ray generation naturally coincide with the generation of isolated attosecond pulses, that are also shorter in duration if compressed. Most surprisingly, advanced theory shows that long-duration, 10-cycle, driving laser pulses are required to generate isolated soft X-ray bursts efficiently, to mitigate group velocity walk-off between the laser and the X-ray fields that otherwise limit the conversion efficiency and the phase matching cutoff [1].

In our experiment, laser pulses at wavelengths of 0.8, 1.3, and 2.0 μm are generated using a Ti:Sapphire laser pumping an optical parametric amplifier (OPA) [2]. The pulse durations of all three wavelengths are adjusted to be ∼ 10 cycles in duration. Each beam is focused into a 2 mm long, Ar-filled cell, with the backing pressure and focus position both varied to obtain optimal phase matching. The temporal structure of the generated HHG beam is measured through a soft X-ray field autocorrelator. A fast Fourier transform of the field autocorrelation trace, captured by an X-ray CCD, is performed to obtain the HHG spectra. The attosecond field autocorrelation is sufficient to fully characterize how the temporal gating of phase matching scales with driving laser wavelength, by directly measuring the number of attosecond bursts contained in the bright HHG emission, i.e., a total of $2n - 1$ fringes will be measured if there are n bursts in the pulse train. Figure 1 shows that for all driving laser wavelengths, as the laser intensity is increased, the number of individual bursts in the attosecond pulse train decrease. The high laser intensity corresponds to optimal phase-matching at the highest photon energies possible in Ar, which generates the brightest HHG flux. Remarkably, for 2 μm lasers, when the optimal phase-matching laser intensity is used to achieve high HHG flux, the emission corresponds to an isolated attosecond pulse with a central photon energy of 140 eV, which is capable of supporting a ≈35 as transform-limited pulse.

To understand why the phase-matching temporal window shrinks as the driving laser wavelength is increased, we performed advanced 3D numerical macroscopic HHG simulations using the method presented in [1, 5]. To illustrate strong phase-matching temporal gating, as well as group velocity walk-off effects, Fig. 2 compares the HHG emission for single-cycle versus multicycle driving laser pulses (1.5- and 8-cycle pulses). The electric field of the laser is plotted at the entrance and the exit of the cell. The contribution of the neutral atom dispersion induces a phase shift on the front of the pulse, whereas the presence of the free-electron plasma induces a chirp on the trailing edge of the pulse. The combined effect confines the phase-matching window to a suboptical cycle duration (300 as) in the center of the pulse, where a bright isolated linearly chirped attosecond pulse is generated. As the 2-μm driving laser pulse duration is reduced from 8 to 1.5 cycles, there is a strong contribution due to group velocity walk-off that leads to a temporal delay of the envelope of the laser field as it propagates (Fig. 2a). As a result, the phase-matching window completely disappears for 1.5-cycle driving lasers, thus inhibiting bright

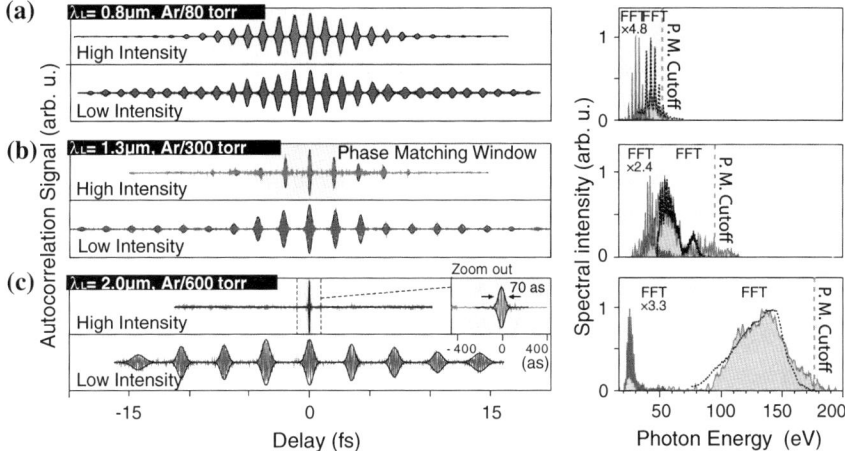

Fig. 1 Comparison of experimental HHG autocorrelation data (normalized) from Ar driven by ∼ 10 cycle laser pulses at **a** 0.8 μm, **b** 1.3 μm, and **c** 2 μm for high and low laser intensity conditions. *Left* field autocorrelation of the HHG field. The phase matching temporal window is highlighted in *gray*. Note that the bandwidth-limited pulse duration is half this coherence time. *Right* HHG spectra obtained using the fast Fourier transform of the field autocorrelation traces (filled-area plots), showing excellent agreement with the experimental spectra (*black dotted lines*) [1–4]. For longer laser wavelengths, the phase matching window is much narrower, leading to isolated attosecond pulses

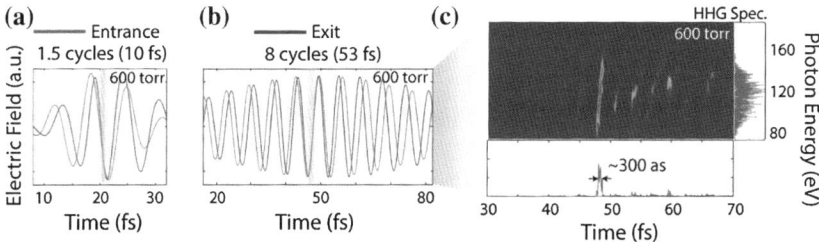

Fig. 2 Dependence of the phase-matching window (*gray*) on the number of driving laser cycles. The laser field is shown both at the entrance (*solid lines*) and exit (*dashed lines*) of a 2 mm gas cell at pressures of 600 torr for **a** 1.5 cycles and **b** 8 cycles FWHM. **c** A time-frequency analysis together with the HHG spectrum and temporal emission, as an isolated 300 as chirped pulse. Note that for a 1.5-cycle driving laser, at the high pressures required for bright HHG emission, phase matching is not possible due to group velocity walk-off

isolated as pulse generation. However, this group velocity walk-off can be mitigated by using multi-cycle driving lasers where the change in the electric field strength from cycle to cycle is negligible (Fig. 2b).

In summary, under optimal phase-matching conditions, long wavelength driving lasers naturally produce bright isolated attosecond soft X-ray pulses. Our work demonstrates a clear and straightforward approach for robustly generating bright isolated attosecond pulses of electromagnetic radiation throughout the soft X-ray region of the spectrum.

Acknowledgement The authors acknowledge support from the Department of Energy AMOs, by the Taiwan National Science Council and by the US National Science Foundation.

References

1. M-C. Chen, C. Mancuso, C. Hernandez-Garcia et al., "Generation of bright isolated attosecond soft X-ray pulses driven by multicycle midinfrared lasers," Proc. Natl. Acad. Sci. **111**, E2361 (2014).
2. T. Popmintchev, M-C. Chen, D. Popmintchev et al., "Bright Coherent Ultrahigh Harmonics in the keV X-ray Regime from Mid-Infrared Femtosecond Lasers," Science **336**, 1287–1291 (2012).
3. M-C. Chen, P. Arpin, T. Popmintchev et al., Bright, Coherent, Ultrafast Soft X-Ray Harmonics Spanning the Water Window from a Tabletop Light Source," Phys. Rev. Lett. **105**, 173901 (2010).
4. T. Popmintchev, M-C. Chen, A. Bahabad et al., Phase matching of high harmonic generation in the soft and hard X-ray regions of the spectrum," Proc. Natl. Acad. Sci. **106**, 10516–10521 (2009).
5. C. Hernandez-Garcia, JA. Pérez-Hernández, J Ramos et al., High-order harmonic propagation in gases within the discrete dipole approximation," Phys. Rev. A **82**, 033432 (2010).

Strong-Field-Enhanced Forward Scattering of High-Order Harmonics

Carles Serrat

Abstract We show that scattering of ultrashort XUV pulses from strong-field driven electron wave packets is enhanced as compared with normal weak scattering from bound or free electrons. We predict large XUV amplification in high-order harmonic generation. Recent measurements have corroborated the main effect that we report. Parametric amplification of attosecond pulse trains at 11 nm has been observed to occur only if the seed pulse-train is perfectly synchronized to the driving laser pulse in the amplifier as predicted here theoretically: J. Seres et al. Scientific Reports 4:4254 (2014).

We consider the interaction of an atom with an intense ultrashort infrared (IR) pulse combined with a weak attosecond pulse of central photon energy much higher than the principal atomic resonances [1, 2]. Specifically, we address the influence of the high-intensity IR field on the scattering of weak XUV radiation from atoms in high-order harmonic generation (HHG) processes. By solving the time-dependent Schrödinger equation in the single-atom strong-field approximation (SFA), we show that forward scattering is enhanced when the weak XUV attosecond pulse is synchronized with the amplitude maxima of the IR intense pulse. It is shown that this enhancement effect can be used for the amplification of coherent XUV radiation in the HHG cutoff spectral region to power levels that might be useful for most applications [3].

Figure 1 shows how the HHG yield is enhanced by the addition of the weak attosecond XUV pulse, and that this enhancement is governed by the delay at which the XUV pulse is added. The effect is more important for delays coinciding with higher values of the IR field, in such a way that the shape of the IR pulse is clearly reproduced. The enhancement calculated in Fig. 1 is given by $\int_{H_-}^{H_+} E_{IR+XUV}^2 dH \Big/ \int_{H_-}^{H_+} E_{IR}^2 dH$, where H_- and H_+ are the photon energies of the previous and next harmonics, respectively, with respect to the harmonic that coincides with the central wavelength of the XUV

C. Serrat (✉)
UPC—Universitat Politècnica de Catalunya, Colom 11, 08222,
Terrassa, Barcelona, Spain
e-mail: carles.serrat-jurado@upc.edu

© Springer International Publishing Switzerland 2015   99
K. Yamanouchi et al. (eds.), *Ultrafast Phenomena XIX*,
Springer Proceedings in Physics 162, DOI 10.1007/978-3-319-13242-6_24

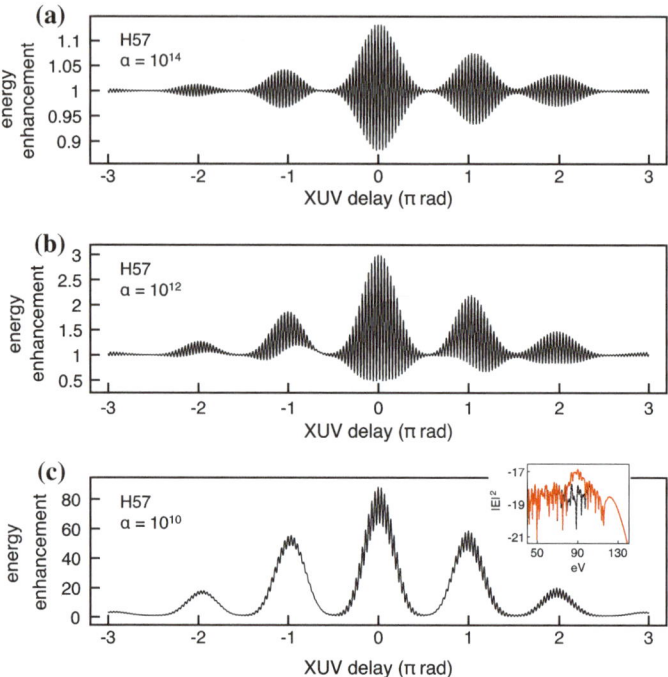

Fig. 1 Energy enhancement of the 57th harmonic of a 800 nm IR intense driving field in Helium, for different ratios between the IR and the XUV pulse peak intensities (α), as indicated, as a function of the delay between the IR and the XUV pulses. The delay is given in parts of half a cycle of the IR pulse, in radiants, so that 2π rad $\equiv \lambda_{IR}/c \approx 2.66$ fs. The *inset* in (**c**) shows the spectrum around the 57th harmonic, in the case of considering only the IR pulse (*black dotted line*) and for the combination of IR + XUV pulses (*red full line*)

pulse, so that the integral comprises the bandwidth of a single harmonic. E_{IR+XUV} is the HHG spectral field amplitude in the case that the interaction is performed by combining the IR and the XUV pulses, and E_{IR} corresponds to the case with the IR pulse alone. dH denotes the integration over the photon energy. The frequency-time analysis of the spectra reveals that the enhancement observed is due to XUV forward scattering from the non-stationary electronic wave packet promoted by the intense IR driving field, which is shown in Fig. 2.

Figure 3 shows the results obtained by implementing an iterative procedure. The harmonics yield produced by a strong IR pulse in a first HHG process is combined with the IR field and sent to second He target, and this process is repeated iteratively. The XUV delay is set to zero in each iteration, which means that the XUV pulse is added to the IR pulse as delayed as it comes out from the output of each interaction. We observe that the amplification of the yield becomes progressively centered at around 125 eV, and that the enhanced spectrum is roughly 50 eV wide at the 100th iteration. This shows that the spectral HHG region slightly above the cutoff is preferably enhanced, as in this region the harmonics yield is very low, and

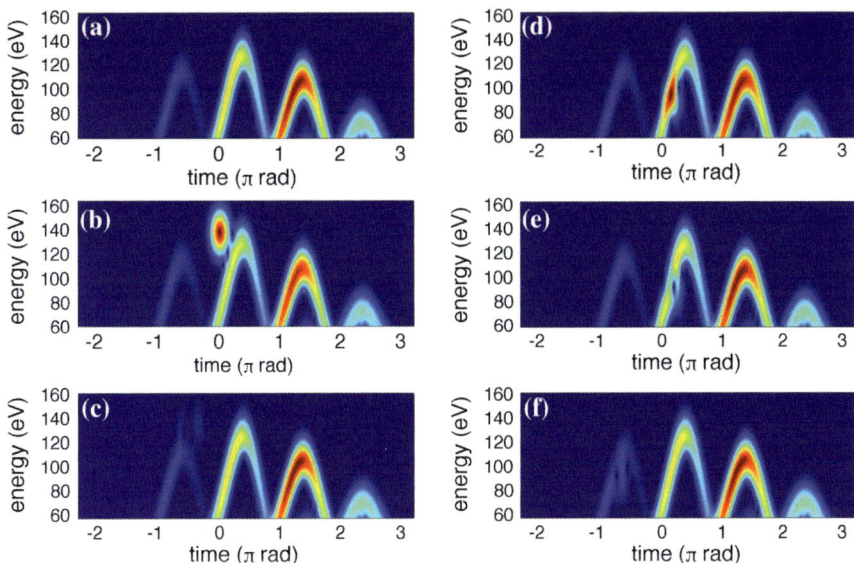

Fig. 2 Frequency-time analysis of some of the spectra shown in the figure at *left* with $\alpha = 10^{12}$. In **a** the He atom interacts only with the Gaussian IR pulse. In **b** and **c** the IR pulse is combined with its 87th harmonic at a delay of 0.014π and -0.48π rad, respectively. In **d**, **e** and **f** the IR pulse is combined with its 57th harmonic at the delays 0.206π, 0.223π and -0.6π rad respectively

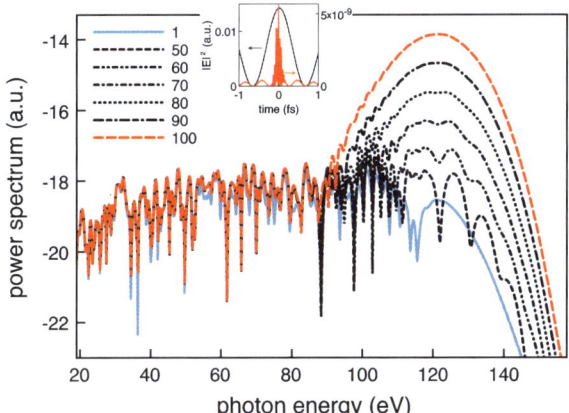

Fig. 3 XUV yield amplification obtained by filtering the harmonics spectrum at 80 eV with 120 eV spectral bandwidth. The filtered XUV is combined with the IR pulse as delayed as it comes out from the output of the interaction, in all iterations. The different lines correspond to the different iteration number, as indicated. The *inset* shows the filtered XUV pulse from the 100th iteration (*red full line*) compared with the input IR field (*black dotted line*)

as a consequence the scattered spectral components in the cutoff region initially dominate the amplification.

We conclude that the observed scattering enhancement effect can be used to strongly amplify the yield in the HHG cutoff region, which sets the basis for a new coherent X-rays source. We are currently extending this study in order to estimate quantitative results for particular experiments, which may include other interaction geometries and use different amplifying media.

References

1. C. Serrat, "Broadband Spectral Amplitude Control in High-Order Harmonic Generation", Appl. Sci. **2**, 816-830 (2012).
2. C. Serrat, "Broadband Spectral-Phase Control in High-Order Harmonic Generation", Phys. Rev. A **87**, 013825 (2013).
3. C. Serrat, "Coherent Extreme Ultraviolet Light Amplification by Strong-Field-Enhanced Forward Scattering", Phys. Rev. Lett. **111**, 133902 (2013).

Part II
Atomic, Molecular and Optical Sciences—Gas Phase

Non-adiabatic Effects in Electron Momenta

C. Hofmann, A.S. Landsman, A. Zielinski, C. Cirelli, A. Scrinzi
and U. Keller

Abstract In strong-field tunnel ionization of Helium, both adiabatic and fully non-adiabatic theoretical descriptions predict smaller final longitudinal electron momentum distributions than measured experimentally. Semiclassical simulations including an initial longitudinal momentum spread reproduce experimental values.

1 Strong Field Ionization

In many attosecond science methods, tunnel ionization in a strong laser field is the first step. Both adiabatic [1, 2] and non-adiabatic [3] models are applied to interpret experimental data. Common to all models is the debate on the longitudinal electron momentum directly at the tunnel exit. The assumption that it is equal to zero is usually adopted [4]. Depending on the phase of the field at the moment when an electron enters the continuum, it then gains a longitudinal momentum while propagating in the laser field, which leads to a final longitudinal momentum distribution.

For the adiabatic limit with Keldysh parameter [1] $\gamma \ll 1$, this acquired longitudinal momentum distribution has width

$$\sigma_{\parallel}^{A} = \sqrt{3\omega}/\sqrt{2\gamma^3(1-\varepsilon^2)}, \qquad \gamma = \omega\sqrt{2I_p(1+\varepsilon^2)}/F_0, \tag{1}$$

with ω denoting the laser frequency, ε the ellipticity, I_p the ionization potential and F_0^2 the peak field intensity. On the other hand, the non-adiabatic theory by PPT [3] predicts a spread with width

C. Hofmann (✉) · A.S. Landsman · C. Cirelli · U. Keller
Physics Department, ETH Zurich, 8093 Zurich, Switzerland
e-mail: chofmann@phys.ethz.ch

A. Zielinski · A. Scrinzi
Ludwig Maximilians Universität, Theresienstrasse 37, 80333 Munich, Germany

© Springer International Publishing Switzerland 2015
K. Yamanouchi et al. (eds.), *Ultrafast Phenomena XIX*,
Springer Proceedings in Physics 162, DOI 10.1007/978-3-319-13242-6_25

$$\sigma_{\parallel}^{NA} = \sqrt{\frac{\omega}{2c_x}}, \qquad \text{with } c_x = \frac{s_0(1-\varepsilon^2)}{(\varepsilon-s_0)(1-\varepsilon s_0)}\sqrt{\frac{s_0^2+\gamma^2}{1+\gamma^2}},$$

$$s_0 \text{ solution to: } \quad \operatorname{arctanh}\left(\sqrt{\frac{s^2+\gamma^2}{1+\gamma^2}}\right) = \frac{\varepsilon}{\varepsilon-1}\sqrt{\frac{s^2+\gamma^2}{1+\gamma^2}}. \tag{2}$$

It is worth noting that $\sigma_{\parallel}^{NA} > \sigma_{\parallel}^{A}$, therefore taking account of non-adiabatic effects leads to a wider momentum spread under the same experimental parameters.

2 Field Intensity Calibration

In strong field experiments, the exact intensity of the laser field must be calibrated based on the measured electron momenta and can not be determined independently to the desired accuracy. The most common and accurate in situ method is based on the final transverse momentum of electrons freed by an elliptically polarized field [5]. An important non-adiabatic effect is that the likeliest transverse momentum at the tunnel exit is non-zero [3]

$$p_{\perp,\text{initial}}^{NA} = \frac{\varepsilon F_0}{\omega\sqrt{1+\varepsilon^2}}\left((1-s_0/\varepsilon)\sqrt{\frac{1+\gamma^2}{1-s_0^2}}-1\right) \neq 0 \text{ au} \tag{3}$$

for $\varepsilon \neq 0$, while the adiabatic assumption yields a transverse momentum distribution centered around zero. This leads to larger transverse momentum predictions and consequently to reduced field calibration values for measurements when applying the non-adiabatic momentum-to-intensity mapping [6].

3 Quantitative Comparison of Theory and Experiment

The experiment is described in detail elsewhere [4]. The ion momenta during an ellipticity scan of strong field ionization of Helium were recorded in a COLTRIMS setup [7]. The calibrated laser field strengths are given by $F_0^{NA} = 0.14$ au (non-adiabatic) and $F_0^A = 0.151$ au (adiabatic) respectively. The measured longitudinal momentum spreads are considerably wider than the theoretical predictions (Fig. 1), even when the non-adiabatic formula (2) is calculated using the field strength from adiabatic calibration (red dotted). Using non-adiabatic theory [3] to calibrate the field strength compensates for the analytically wider non-adiabatic longitudinal spread, resulting in almost complete agreement in the predictions of both the non-adiabatic (red solid) and adiabatic (blue solid) cases.

Fig. 1 Experimental longitudinal momentum spread ○ [8] is wider than adiabatic (1) (*blue*) or non-adiabatic (2) (*red*) theoretical predictions. The best fitting initial momentum spreads ◆ are even larger than the initial transverse momentum (*black dashed*) [2]

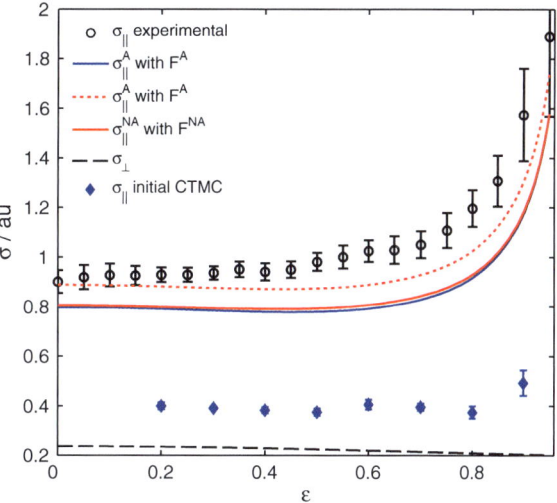

In classical trajectory Monte Carlo simulations [4] following adiabatic assumptions, an initial longitudinal momentum spread was varied to find the best fitting value to reproduce the final momentum distribution. Both experimental and simulated momentum distributions were analyzed by elliptical integration [4]. This new method to analyze angular momentum distribution is robust for any ellipticity and has been successfully applied in studying Coulomb effects for all ranges of polarization [9].

4 Conclusion and Outlook

Our findings show that independent of the uncertainty in the field calibration, current theoretical descriptions can not fully explain the observed longitudinal momentum spread. The initial longitudinal momentum spread that best reproduces the experimental distributions is of the order of twice the transverse momentum spread. Therefore, further theoretical work is necessary to more accurately model the spread of the electron wavepacket in strong field ionization.

Acknowledgements This research was supported by the NCCR MUST, funded by the Swiss National Science Foundation (SNSF), by ETH Research grant no. ETH-03 09-2, an SNSF equipment grant, an SNSF project grant, and by the ETH-FAST initiative.

References

1. L. V. Keldysh, Sov. Phys. JETP **20**, 1307 (1965)
2. N. B. Delone and V. P. Krainov, J. Opt. Soc. Am. B **8**, 1207–1211 (1991)
3. A. M. Perelomov, V. S. Popov, and M. V. Terent'ev, Sov. Phys. JETP **24**, 207–217 (1967)
4. C. Hofmann, A. S. Landsman, C. Cirelli, A. N. Pfeiffer, and U. Keller, J. Phys. B At. Mol. Opt. Phys. **46**, 125601 (2013)
5. A. S. Alnaser, X. M. Tong, T. Osipov, S. Voss, C. M. Maharjan, B. Shan, Z. Chang, and C. L. Cocke, Phys. Rev. A **70**, 23413 (2004)
6. R. Boge, C. Cirelli, A. S. Landsman, S. Heuser, A. Ludwig, J. Maurer, M. Weger, L. Gallmann, and U. Keller, Phys. Rev. Lett. **111**, 103003 (2013)
7. R. Dörner, V. Mergel, O. Jagutzki, L. Spielberger, J. Ullrich, R. Moshammer, and H. Schmidt-Böcking, Phys. Rep. **330**, 95–192 (2000)
8. A. N. Pfeiffer, C. Cirelli, A. S. Landsman, M. Smolarski, D. Dimitrovski, L. B. Madsen, and U. Keller, Phys. Rev. Lett. **109**, 83002 (2012)
9. A. S. Landsman, C. Hofmann, A. N. Pfeiffer, C. Cirelli, and U. Keller, Phys. Rev. Lett. **111**, 263001 (2013)

Determination of Absolute Cross-Sections of Nonresonant EUV-UV Two-Color Two-Photon Ionization of He

M. Fushitani, Y. Hikosaka, A. Matsuda, T. Endo, E. Shigemasa,
M. Nagasono, T. Sato, T. Togashi, M. Yabashi, T. Ishikawa
and A. Hishikawa

Abstract Single-shot photoelectron spectroscopy was performed for nonresonant EUV-UV two-color two-photon ionization of He. From data analysis on the shot-by-shot basis, the absolute cross-section was determined to be $\sigma^{(2)}$(59.7 nm, 268 nm) = 4.1(6) $\times 10^{-52}$ cm^4 s.

1 Introduction

Advances of free electron laser (FEL) technologies have made it possible to investigate various nonlinear phenomena in the EUV region [1]. Experiments performed by using optical lasers synchronized with the EUV-FEL provide fundamental information on how nonlinear optical processes take place by absorbing two different frequencies. In such processes, the simultaneous absorption depends not only on the field intensity of each laser pulse but also on the temporal overlapping between them. However, there is considerable instability in the synchronization due to technical limitations [2]. Such a timing jitter between the two pulses prevents from being kept the temporal overlapping optimal, and makes it difficult to evaluate the cross-section precisely. Here, we show that single-shot analysis enables us to extract

M. Fushitani (✉) · A. Matsuda · T. Endo · A. Hishikawa
Department of Chemistry, Nagoya University, Nagoya, Aichi 464-8602, Japan
e-mail: fusitani@chem.nagoya-u.ac.jp

M. Fushitani · Y. Hikosaka · E. Shigemasa · M. Nagasono · T. Sato · M. Yabashi ·
T. Ishikawa · A. Hishikawa
RIKEN, SPring-8 Center, Sayo, Hyogo 679-5148, Japan

Y. Hikosaka
Department of Environmental Science, Niigata University, Niigata, Niigata 950-2181, Japan

E. Shigemasa
Institute for Molecular Science, Okazaki, Aichi 444-8585, Japan

T. Togashi · M. Yabashi
Japan Synchrotron Radiation Research Institute, Sayo, Hyogo 679-5198, Japan

© Springer International Publishing Switzerland 2015
K. Yamanouchi et al. (eds.), *Ultrafast Phenomena XIX*,
Springer Proceedings in Physics 162, DOI 10.1007/978-3-319-13242-6_26

signals for the optimal overlap in time, thereby evaluating the cross-section in more accurate manner than those derived from average values. We determine the cross-section of non-resonant two-color two-photon ionization of He by comparing it to that of Ar by single photon ionization [3]. This study provides quantitative information on the new nonlinear optical processes induced by two different frequency components, which would be the basis of applications to multi-color experiments such as time-resolved studies.

2 Experimental

An EUV-FEL pulse (59.7 nm) obtained from the SCSS test accelerator facility at RIKEN Harima and an ultrashort UV laser pulse (268 nm) synchronized with the FEL were focused to gaseous He and Ar in the interaction region. The mean time-delay (Δt) between FEL and UV laser pulses was controlled by an optical stage with a resolution of 1 μm (Fig. 1a). Electrons formed by photoionization with FEL and UV laser pulses were detected shot-by-shot by a magnetic bottle type photoelectron spectrometer. The electron energy was calibrated with atomic lines of oxygen and Auger lines of xenon that were observed by irradiation with the 3rd order harmonics of FEL at 51 nm.

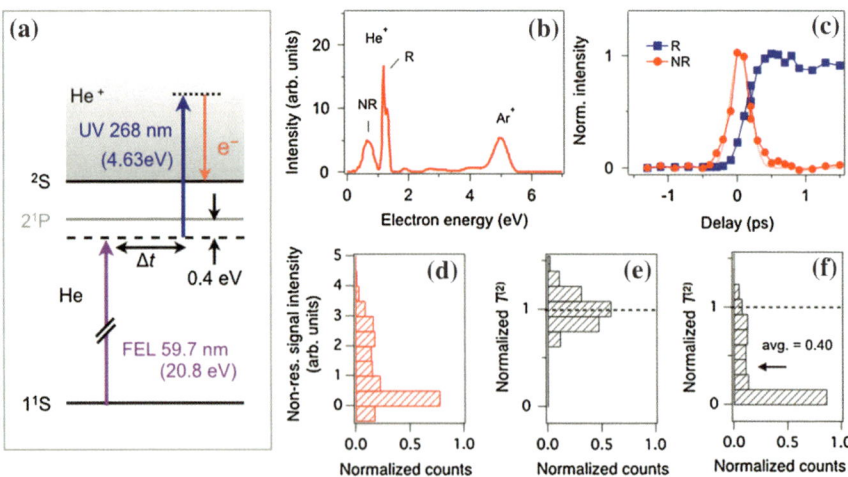

Fig. 1 **a** Energy diagram of two-color two-photon ionization of He using EUV-FEL and UV laser pulses. **b** Photoelectron spectrum at $\Delta t = 0$ ps. **c** Temporal behavior for resonant (R) and non-resonant (NR) components of the He peak. **d** Histogram of the NR signal at $\Delta t = 0$ ps. **e** Simulated histogram of $T^{(2)}(0)$ without the timing jitter. **f** Same as (**e**), but the timing jitter (0.33 ps) is included

3 Results and Discussion

Figure 1b shows a part of photoelectron spectrum (2,000 shots) at $\Delta t = 0$ ps. Photoelectron peak at 5 eV is due to electrons formed by one-photon ionization of Ar while He peaks appear about 1 eV. From the energy conservation the peak at 0.6 eV is assigned to photoelectrons produced by non-resonant (NR) two-color two-photon ionization of He. The peak at 1.3 eV is attributed to those by resonant (R) two-color two-photon ionization via the 2^1P state. The temporal behaviors of the NR and R peaks are plotted in Fig. 1c. The intensity of the R component increases until $\Delta t = 0.4$ ps and stays almost constant up to $\Delta t = 1.5$ ps, which is in line with the lifetime (560 ps) of the 2^1P state. In contrast, the NR peak exhibits the maximum at $\Delta t = 0$ ps. The full width at half maximum (FWHM) of the trace was obtained to be 0.36(3) ps by carrying out a least-squares fit to a Gaussian function. Since both FEL and UV laser pulse durations are about 100 fs, the temporal resolution primarily reflects the timing jitter (0.33 ps) between the FEL and UV laser pulses. This timing jitter prevents from keeping a stable temporal overlap in the time-resolved measurements, thereby resulting in accumulation of non-resonant signals at various time delays in the averaged spectra in Fig. 1b.

Figure 1d shows a distribution of the NR signals obtained from single-shot spectra at the nominal $\Delta t = 0$ ps. The broad distribution with a large contribution at the zero signal position reflects poor temporal overlapping between FEL and UV laser pulses mainly due to the timing jitter (δ). To understand the distribution in detail, we calculate a temporal overlapping $T^{(2)}(\delta)$ at $\Delta t = 0$ ps, which can be expressed as

$$T_{\text{EUV,UV}}^{(2)}(\delta) = \int_{-\infty}^{+\infty} f_{\text{EUV}}(t)\, f_{\text{UV}}(t - \delta)\,dt, \tag{1}$$

where $f(t)$ is a pulse envelope normalized by the peak maximum. Figure 1e shows a $T^{(2)}$ distribution without the timing jitter ($\delta = 0$). The distribution peaked at ~ 1, which reflects the intensity fluctuation of each FEL pulse that is simulated by using the partial coherence method [4]. On the other hand, the simulation including the timing jitter as shown in Fig. 1f is in good agreement with the experiment. From the comparison between Fig. 1d, f, the signal intensity $S_{\text{He}^+}^{(2)}$ for the non-resonant component at $\Delta t = 0$ ps is estimated to be 3.5 while $S_{\text{He}^+}^{(2)}$ is underestimated by a factor of 0.4 when the average value is adopted.

Since the cross-section of non-resonant two-color two-photon ionization of He in the present study can be expressed as [3],

$$\sigma_{\text{He}}^{(2)}(h\nu_i, h\nu_j) = \frac{S_{\text{He}^+}^{(2)}}{S_{\text{Ar}^+}^{(1)}} \frac{n_{\text{Ar}}^0}{n_{\text{He}}^0} \frac{T_i^{(1)}}{T_{i,j}^{(2)}(\Delta t)} \frac{h\nu_j}{I_{0,j}} \sigma_{\text{Ar}}^{(1)}(h\nu_i), \tag{2}$$

where $h\nu_i$, and $h\nu_j$ is photon energy of the pulse i (=EUV) and j (=UV), $S_{Ar^+}^{(1)}$ is the integrated intensity of the Ar peak at 5 eV, n_{Ar}^0/n_{He}^0 is the ratio of the number density of Ar and He, $T_i^{(1)}$ is the integral of the $f(t)$ for the EUV-FEL pulse, $I_{0,j}$ is the field intensity of the UV pulse, and $\sigma_{Ar}^{(1)}$ is the cross-section of Ar. By using (2) with the following parameters such as $n_{Ar}^0/n_{He}^0 = 1.2 \times 10^{-4}$, $T_{EUV}^{(1)}/T_{EUV,UV}^{(2)}(0) = 1.4$, $I_{0,UV} = 1.4 \times 10^{13}\,\mathrm{W/cm^2}$, $\sigma_{Ar}^{(1)} = 36.1$ Mb [5], the non-resonant cross-section is determined to be $\sigma^{(2)}(59.7\ \mathrm{nm},\ 268\ \mathrm{nm}) = 4.1(6) \times 10^{-52}\ \mathrm{cm^4\ s}$.

References

1. *e.g.* M. Yabashi *et al.*, J. Phys. B **46**, 164001 (2013).
2. A. Azima *et al.*, Appl. Phys. Lett. **94**, 144102 (2009).
3. M. Fushitani *et al.*, Phys. Rev. A **88**, 063422 (2013).
4. T. Pfeifer *et al.*, Opt. Lett. **35**, 3441-3443 (2010).
5. W. Chan *et al.*, Phys. Rev. A **46**, 149-171 (1992).

Attosecond Spatial Control of Electron Wave Packet Emission Dynamics

Li Zhang, Xinhua Xie, Stefan Roither, Yueming Zhou,
YanLan Wang, ChuanLiang Wang, Daniil Kartashov,
Markus Schöffler, Paul Corkum, Dror Shafir, Andrius Baltuška,
Igor Ivanov, Anatoli Kheifets, Peixiang Lu, XiaoJun Liu,
André Staudte and Markus Kitzler

Abstract Using orthogonally polarized two-color laser fields on neon and coincidence momentum imaging we gain access to the Coulomb influence in single ionization on sub-cycle times, and demonstrate a strong electron-electron anti-correlation in double ionization.

1 Introduction

Angström and attosecond control of free electron wave packets is one of the pinnacles of attosecond science. Orthogonally polarized two-color (OTC) laser fields allow to control the motion of field-ionizing electronic wave packets both in time

L. Zhang · X. Xie · S. Roither · D. Kartashov · M. Schöffler · A. Baltuška · M. Kitzler (✉)
Photonics Institute, Vienna University of Technology, 1040 Vienna, Austria
e-mail: markus.kitzler@tuwien.ac.at

Y. Zhou · P. Lu
School of Physics, Huazhong University of Science and Technology, and Wuhan National
Laboratory for Optoelectronics, Wuhan 430074, China

Y. Zhou · P. Lu
Key Laboratory of Fundamental Physical Quantities Measurement of Ministry of Education,
Wuhan 430074, China

Y. Wang · C. Wang · X. Liu
State Key Laboratory of Magnetic Resonance and Atomic and Molecular Physics, Wuhan
Institute of Physics and Mathematics, Chinese Academy of Sciences, Wuhan 430071, China

P. Corkum · A. Staudte
Joint Laboratory for Attosecond Science of the National Research Council and the University
of Ottawa, Ottawa K1A 0R6, Canada

D. Shafir
Department of Physics of Complex Systems, Weizmann Institute of Science,
76100 Rehovot, Israel

I. Ivanov · A. Kheifets
Research School of Physical Sciences, The Australian National University,
Canberra, Australia

© Springer International Publishing Switzerland 2015   113
K. Yamanouchi et al. (eds.), *Ultrafast Phenomena XIX*,
Springer Proceedings in Physics 162, DOI 10.1007/978-3-319-13242-6_27

and space [1]. In OTC pulses time and space are connected and thus an attosecond time scale is established in the polarization plane for both the emitted and the recolliding wave packets [2].

Here, we report on experiments that use OTC pulses for studying single and double ionization of neon using the COLTRIMS technique. By exploring the deviation of the measured electron momentum spectra from the time-to-angle mapping provided by the OTC field [2, 3] we gain access to the sub-cycle timing of the wave packet release from the momentum vector of emitted electrons. This allows us to study the influence of the parent ion's field on the trajectories of tunneling electrons during single ionization. In double ionization we demonstrate control over the correlated emission of two electrons in the intensity regime of nonsequential double ionization. We observe a strong electron-electron anti-correlation [5].

In the experiments we focused OTC pulses into a supersonic gas jet of neon atoms and measured the three-dimensional (3D) momentum spectra of the resulting singly charged ions and correlated electrons with a cold target recoil ion momentum spectroscopy (COLTRIMS) setup. The OTC pulses with a field as given by (1) were generated by combining a 46 fs, 800 nm (ω) laser pulse, and a 48 fs, 400 nm (2ω) pulse, in a collinear geometry at a rate of 5 kHz. The peak intensity in either color was $\hat{E}^2 = I_{800nm} = I_{400nm} = (1 \pm 0.1) \times 10^{14}$ W/cm^2. Temporal overlap of the two pulses was ensured by compensating for the different group velocities of the two colors with calcite plates and a pair of fused silica wedges. The latter were also used to vary the relative phase of the two colors with a precision of roughly 0.3 as. Further details of the optical and the COLTRIMS setup can be found in [5].

Figure 1a shows one cycle of an OTC pulse consisting of an 800 nm field, linearly polarized along x, and a superimposed, orthogonally polarized 400 nm field along z (blue) of equal peak intensity. The electric field of the OTC pulse is given by

$$\mathbf{E}(t) = \hat{E}[f_x(t)\cos(\omega t)\mathbf{e}_x + f_z(t)\cos(2\omega t + \Delta\varphi)\mathbf{e}_z] \tag{1}$$

with $\hat{E}$ the peak field strength, $f_{x,z}$ the pulse envelopes along the 800 and 400 nm direction, respectively, and $\Delta\varphi$ the relative phase between the two colors.

Figure 1b–f shows measured momentum distributions of electrons correlated with Ne$^+$ in the polarization plane of the OTC field. The spectra show that the electron emission direction is highly sensitive to the shape of the OTC field, featuring asymmetric emission patterns that vary with $\Delta\varphi$. Figure 1 furthermore illustrates the mapping of emission time to the angle in the laser polarization plane of electron momentum spectra that is established by the OTC pulses when the influence of the parent ion on the emitted electron wavepacket is neglected. The mapping is based on the classical relation $\mathbf{p} = -\mathbf{A}(t_0)$ obtained by neglecting the influence of the ion on the emitted electrons. Here $\mathbf{A}(t)$ is the laser vector potential and t_0 the electron birth time. This relation is widely used to interpret experimental electron spectra, often to extract sub-cycle timing information from them.

Wave packets detached by the OTC field within different laser quarter-cycles are observed in different momentum regions in the polarization plane [2, 3]. This is shown in Fig. 1b–f for different relative phases by the same color code as the field's quarter-cycles in Fig. 1a. A change in the phase of the 400 nm field relative to the 800 nm field changes the OTC field and therewith the driving force for the trajectories of tunnel ionized electron wave packets on sub-cycle time-scales, which leads to different time-to-momentum mapping for different $\Delta\varphi$ [2, 3].

For an in-depth analysis of the photoelectron momentum distributions beyond the classical relation $\mathbf{p} = -\mathbf{A}(t_0)$ we performed both a three-dimensional TDSE and a classical trajectory Monte Carlo (CTMC) simulation. Results of the CTMC simulations without the Coulomb field resemble classical spectra calculated by using the classical relation $\mathbf{p} = -\mathbf{A}(t_0)$ and taking the tunneling probability into account, but slightly differ from them by an accurate quantitative treatment of the initial electron momentum distribution. In order to understand the role of the ionic Coulomb potential in the photoelectron dynamics, we have performed another set of CTMC calculations in which the Coulomb potential $V(r) = -1/r$ has been included. In comparing the results of these two simulations it becomes evident that the Coulomb field of the parent ion strongly distorts the final electron momenta. The good qualitative agreement between the TDSE and the CTMC simulations verifies that the main features of the quantum calculation are captured very well in the classical picture. These results clearly show that the deviations of the measured photoelectron momentum distributions from the classical model (distortions and asymmetries) are caused by a purely classical effect, namely the additional driving force of the ionic Coulomb potential.

We now turn to nonsequential double ionization (NSDI) with OTC pulses. For these experiments the intensity has been increased to $I_{800nm} = I_{400nm} = (2 \pm 0.2) \times 10^{14}$ W/cm^2. Analysis of the spectra of the electron sum momentum vector in terms of its mean values and widths along p_x and p_z allows obtaining detailed insight into the correlated electron emission dynamics (Fig. 2a). Correlated two electron-emission (indicated in blue) results either in a large width and zero mean value, if it takes place symmetrically into both hemispheres (blue area), or in a narrow width

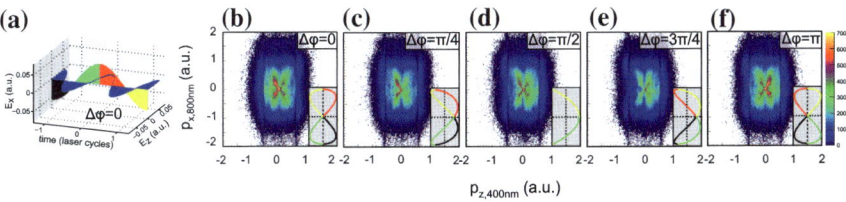

Fig. 1 **a** Separate electric fields of the 400 nm (*blue*) and 800 nm pulses. The *colors* encode quarter cycles of the 800 nm field. **b–f** Measured electron momentum distributions correlated to Ne$^+$ in the polarization plane of the OTC field (xz) with $|p_y| < 0.1$ a.u. for selected $\Delta\varphi$ from 0 to π. The *insets* illustrate the classical prediction $p = -A(t_0)$ of the electron momentum. The *color code* is illustrated in (**a**)

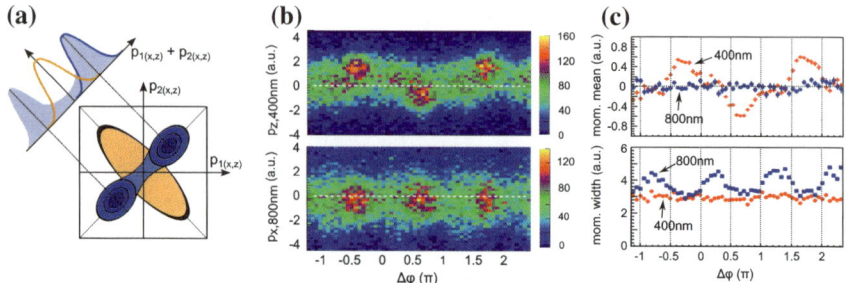

Fig. 2 a Signatures of correlated (*blue*) and anti-correlated (*orange*) two-electron emission dynamics in two-electron (*lower right*) and ion (*upper left*) momentum spectra. **b** Ne^{2+} p_z momentum (other directions integrated over) as a function of relative phase $\Delta\varphi$ (*upper panel*). Ne^{2+} p_x momentum (*lower panel*). **c** $\Delta\varphi$-dependent mean value of p_z (*red dots*) and p_x (*blue squares*). *lower panel*: width of p_z (*red dots*) and p_x (*blue squares*)

and large mean value if it happens dominantly into one hemisphere (blue thick line). Anti-correlated two electron-emission (indicated in orange) manifests itself as a narrow ion momentum spectrum with zero mean value (orange line). Figure 2b plots the momentum spectra of Ne^{2+} along the p_z and p_x momentum components as a function of $\Delta\varphi$, and Fig. 2c their widths and mean values. One scenario that can explain the small momentum width measured when the total double ionization probability maximizes, is that the two electrons are emitted into opposite $p_{x,800nm}$-hemispheres in a strongly anti-correlated emission scenario see Fig. 2a. For those phases where the spectral width is large, the emission happens in a correlated manner with emissions alternately into both hemispheres, which leads to spectra as sketched by the blue areas in Fig. 2a. Thus, our measurements demonstrate that by using OTC laser fields it is possible to control the electron-electron (anti-)correlation during NSDI by using $\Delta\varphi$ as the control parameter.

In conclusion, we used the COLTRIMS technique to investigate for the first time atomic single and double ionization by OTC laser fields. By exploiting the time to momentum space mapping provided by OTC fields we succeeded in gaining experimental access to the dynamics of emitted and recolliding electron wave packets on laser-sub-cycle times. We furthermore showed that by tuning the shape of the electric field of the OTC pulses on the sub-cycle scale it is possible to control the two electron-emission dynamics in nonsequential double ionization, and to dictate whether the two electrons are predominantly emitted in a correlated or anti-correlated manner.

Acknowledgements We acknowledge financial support by the Austrian Science Fund (FWF) under grants No. P21463-N22, No. P25615-N27 and No. SFB-F49 NEXTlite, and by a starting grant from the ERC (project CyFi).

References

1. M. Kitzler, M. Lezius. Spatial Control of Recollision Wave Packets with Attosecond Precision. *Phys. Rev. Lett.* **95**, 253001 (2005).
2. M. Kitzler *et al*. Optical attosecond mapping by polarization selective detection. *Phys. Rev. A* **76**, 011801 (2007).
3. M. Kitzler *et al*. Angular encoding in attosecond recollision. *New J. Phys.* **10**, 025029 (2008).
4. Y. Zhou *et al*. Correlated electron dynamics in nonsequential double ionization by orthogonal two-color laser pulses. *Opt. Expr.* **19**, 2301 (2011).
5. L. Zhang *et al*. Subcycle Control of Electron-Electron Correlation in Double Ionization. *Phys. Rev. Lett.* **112**, 193002 (2014).

Probing Elastic Rescattering Through Half-Cycle Cutoffs in Above-Threshold Ionization Spectra

Henning Geiseler, Nobuhisa Ishii, Keisuke Kaneshima, Teruto Kanai and Jiro Itatani

Abstract We observe photoelectron spectra from above-threshold ionization of xenon using carrier-envelope phase-stabilized few-cycle pulses at 1.6 µm. The signature of elastic scattering is imprinted on the spectra, and through careful analysis we successfully retrieve the electron-ion backscattering cross section in the energy range from 10 to 30 eV, where we find a resonance-like structure at 17.5 eV.

1 Introduction

Electron rescattering is an effect in strong-field physics, that opened the way to attosecond science. Atoms exposed to a strong optical field are tunnel-ionized, and the liberated electrons are subsequently accelerated in this field. Within the framework of the semi-classical three-step model [1], depending on the exact timing of ionization, the electron can return to the vicinity of the parent ion, and either recombine, or scatter elastically or inelastically. Controlling these sub-cycle processes via the waveform of optical fields enables experimental access to the attosecond time scale. When electrons are scattered elastically, they are accelerated by the optical field to kinetic energies up to $\sim 10U_p$, where U_p denotes the ponderomotive potential. This so called high-energy above-threshold ionization (HATI) allows experimental separation of high- from low-energy electrons, which result from ionization without rescattering. The signature of the rescattering event is imprinted on the angular- and energy-distributions of the high-energy electrons [2], which has been used, e.g., to extract angular-dependent scattering cross sections [3], and molecular structures [4]. Here, we use carrier-envelope phase (CEP)-stabilized few-cycle pulses to drive the rescattering process, which allows to precisely control the cutoff energy of

H. Geiseler (✉) · N. Ishii · K. Kaneshima · T. Kanai · J. Itatani
The Institute for Solid State Physics, The University of Tokyo, 5-1-5 Kashiwanoha,
Kashiwa, Chiba 277-8581, Japan
e-mail: geiseler@issp.u-tokyo.ac.jp

© Springer International Publishing Switzerland 2015
K. Yamanouchi et al. (eds.), *Ultrafast Phenomena XIX*,
Springer Proceedings in Physics 162, DOI 10.1007/978-3-319-13242-6_28

118

returning electrons. By varying the CEP, we spectrally resolve the elastic rescattering process in a single measurement, and extract the energy-dependent electron-ion backscattering cross section. By extension, the method can also be used to temporally resolve the rescattering process with attosecond resolution.

2 Experimental Setup

CEP-stabilized few-cycle pulses at 1.6 μm are readily provided by our optical parametric chirped-pulse amplification system [5]. The pulse energy is adjusted by an iris, and the pulses are focused into diffuse xenon gas positioned inside a vacuum chamber. The energy of generated photoelectrons is then analyzed within a cone of 1.2 msr solid angle around the polarization axis by a conventional time-of-flight spectrometer. Energy spectra are acquired while the CEP is slowly varied in steps of 0.05π rad, yielding two-dimensional spectral maps, as presented for two cases with different intensities in Fig. 1.

3 Analysis and Conclusions

When the CEP is varied, the amplitude in individual half-cycles of the optical pulses, and accordingly the kinetic energy of generated HATI electrons, are changed. This becomes obvious in Fig. 1, as the maximum electron energy displays a pronounced CEP-dependence. In the intermediate energy region, between 30 and 60 eV in Fig. 1a, and between 40 and 80 eV in Fig. 1b, another CEP-dependent structure becomes apparent, that is composed of contributions from two different effects. One is associated with the variation of tunnel-ionization rates, as the amplitude of the optical half-cycle, in which ionization occurs, is a function of the CEP. The other contribution is the imprint of the energy-dependent elastic

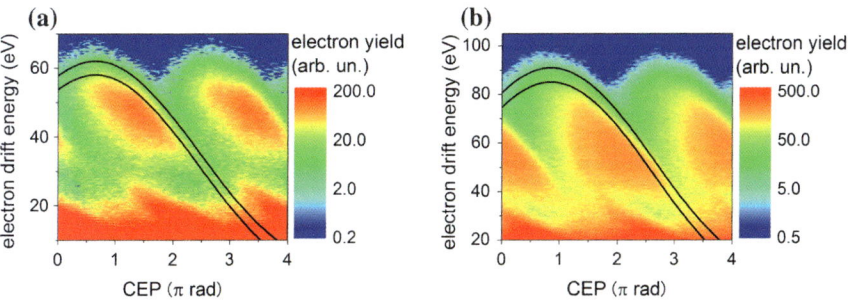

Fig. 1 Spectral maps obtained from ionization of xenon, taken at peak pulse intensities of **a** 2.6×10^{13} W cm^{-2} and **b** 4.0×10^{13} W cm^{-2}, respectively

scattering cross section of the ionic core, as the kinetic energy of the electrons at the moment of rescattering depends critically on the waveform in the optical cycle following ionization, which steers the electron motion. In order to isolate the effect of elastic rescattering, the two contributions have to be disentangled.

The analysis is complicated by the fact, that at least two separate electron trajectories have the same kinetic energy at the moment of rescattering (known as long and short trajectories). However, these trajectories converge near the maximum return energy, and for the cutoff energy only one trajectory remains. Therefore, we only focus on the electrons near the cutoff originating from ionization in one given optical half-cycle. The considered areas in Fig. 1a, b are marked by black lines. By modeling the effect of varying tunnel-ionization rates using the ADK formula [6], we subtract this contribution from the measured electron yield, and by following classical electron trajectories in an optical few-cycle field, a mapping of the electron rescattering energy onto the final energy is established. This procedure allows to extract the elastic backscattering cross section for electron-ion collisions in the energy range probed by the cutoff electrons. By combining the two measurements in Fig. 1, we obtain the cross section shown in Fig. 2. A resonance-like feature centered around 17.5 eV with a width of 6 eV emerges.

Using infrared fields at 1.6 μm, the electron rescattering energy can easily be extended to >100 eV, facilitating de Broglie wavelengths sufficiently short to resolve molecular structures. By experimentally suppressing one of the types of electron trajectories, also energy regions apart from the cutoff can be used to obtain a mapping of different electron return times onto the final energy distribution. Thus the method provides a way to conduct diffraction of single atoms or molecules with attosecond and Ångström resolution.

Fig. 2 Extracted elastic scattering cross section for electron-ion collisions of xenon in backscattering configuration, obtained from the data in Fig. 1

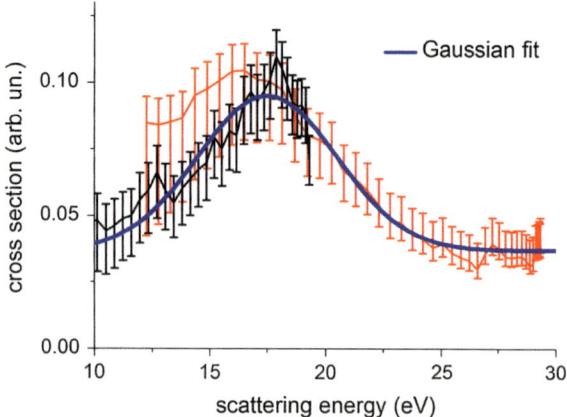

References

1. P.B. Corkum, Phys. Rev. Lett. **71**, 1994 (1993)
2. M. Meckel, D. Comtois, D. Zeidler, A. Staudte, D. Pavičić, H.C. Bandulet, H. Pépin, J.C. Kieffer, R. Dörner, D.M. Villeneuve, P.B. Corkum, Science **320**, 1478 (2008)
3. M. Okunishi, T. Morishita, G. Prümper, K. Shimada, C.D. Lin, S. Watanabe, K. Ueda, Phys. Rev. Lett. **100**, 143001 (2008)
4. C.I. Blaga, J. Xu, A.D. DiChiara, E. Sistrunk, K. Zhang, P. Agostini, T.A. Miller, L.F. DiMauro, C.D. Lin, Nature **483**, 194 (2012)
5. N. Ishii, K. Kaneshima, K. Kitano, T. Kanai, S. Watanabe, J. Itatani, Opt. Lett. **37**, 4182 (2012)
6. M.V. Amosov, N.B. Delone, V.P. Krainov, Sov. Phys. JETP **64**, 1191 (1986)

Experimental Signature of Light Induced Conical Intersections in Diatomics

Adi Natan, Matthew R. Ware and Philip H. Bucksbaum

Abstract We present evidence for the effect of light induced conical intersections in strong field photodissociation of H_2^+, manifested in modulations of angular distributions resulting from the topological singularity induced by the intense field.

Nature has an efficient and ultrafast mechanism to transfer energy without radiation when large molecules are excited. This mechanism is known as a Conical intersection (CI), and it takes place when electronic potential energy surfaces cross. CIs are commonly found in molecular systems and play an important role in molecular dynamics [1] because of the strong non-adiabatic coupling between electronic and rovibrational degrees of freedom. A hallmark of CI is the existence of a geometric or Berry's phase [2], which can generate observable interferences for parts of a wavepacket that encircles it. CIs cannot exist for free diatomic molecules as they require at least two degrees of freedom. Recently, it has been suggested [3, 4] that light induced conical intersections (LICI) can be realized for diatomic molecules interacting with strong laser fields. The required degrees of freedom are the internuclear separation distance r and the angle of the molecular axis with respect to the laser polarization θ. In theoretical studies that followed, it was shown that the realization of such LICI will have a strong impact on diatomic molecular dynamics [4, 5]. In this study, we show experimental evidence for the LICI phenomenon in a diatomic molecules undergoing strong field dissociation.

For diatomic molecules interacting with strong laser light, the Floquet formalism suggests that a two dressed potential curves would cross at some internuclear separation where the states are resonantly coupled by the laser field. The LICI is defined at this crossing point when the angle of the polarization of the laser with respect to the molecular axis is $\theta = \pi/2$. For other angles this crossing is avoided and the dynamics is adiabatic. For higher laser intensities, the LICI cone becomes narrower reflecting a more abrupt transition from adiabatic to non-adiabatic dynamics.

A. Natan (✉) · M.R. Ware · P.H. Bucksbaum
Stanford PULSE Institute, SLAC National Accelerator Laboratory,
2575 Sand Hill Rd, Menlo Park, California 94025, USA
e-mail: natan@stanford.edu

© Springer International Publishing Switzerland 2015
K. Yamanouchi et al. (eds.), *Ultrafast Phenomena XIX*,
Springer Proceedings in Physics 162, DOI 10.1007/978-3-319-13242-6_29

122

We'll focus on the influence a LICI has on the angle resolved H_2^+ photodissociation. We've analyzed the angular distributions of H_2^+ photodissociation fragments produced by a transform limited 30 fs pulses at 795 nm with a peak intensity of 2×10^{13} W/cm^2. For such peak intensity in H_2^+ we only need to consider the $1s\sigma_g$ and $2p\sigma_u$ potential curves, and this simplicity it attractive as it allows to disentangle the effect the LICI has from other strong field processes. The molecular ions were produced in a Nielsen source by ionizing H_2, and had a Franck-Condon distribution of vibrational states. Additional details on the experimental setup can be found here [6, 7]. We concentrate on angular distributions at specific kinetic energy releases (KER) from rovibrational states that were near the LICI. In Fig. 1a, we show the measured distributions that were obtained using narrow (20 meV) KER slices that correspond to vibration levels $v = 7, 8, 9$. We observe distinct modulations at $30°–60°$, that stand in contrast to the typical smooth $cos^{2n}(\theta)$ type distribution generated by strong field dissociation [6]. We believe the reason this has not been observed in previous experiments is related to the demanding resolutions needed in angle and energy. We found that bigger bin sizes in energy or angle will obscure the modulations observed. We also notice a flattening of the distribution toward $0°$, and explain it as saturation of the probability amplitude for near resonant levels, as also reported here [7].

Tracing how the LICI contributes to the interference features on the angular distributions was done by numerically solving the time-dependent Schrödinger equation both for nuclear and rotational degrees of freedom. Starting with a pure rovibrational state (v,j), we propagated the wavefunction as it interacted with the laser and calculated the dissociation probability, $P(\theta, t) = \int_0^\infty |\psi(r, \theta, t)|^2 W(r) dr$, by projecting the part of the dissociating wavefunction onto an absorbing boundary along the spatial axis, $W(r)$. Figure 1e-g show the distinct interference patterns in $P(\theta, t)$ for several initial rovibration states $(v = 7, 8, 9, j = 0)$. We shall now discuss the source of these interferences.

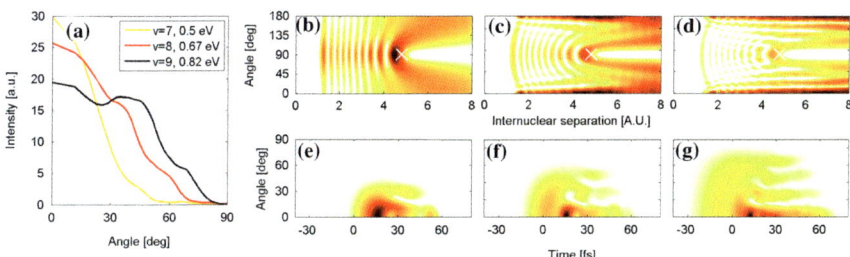

Fig. 1 **a** Measured dissociation yield versus angle of H_2^+. Different kinetic energy release values reveal modulations at different angles. Evolution of the calculated probability density initialized with $v = 9$ at **b** $t = -20$ fs, **c** $t = 0$ fs, and **d** $t = 15$ fs. Peak intensity of pulse happens at $t = 0$ fs and × marks the position of the LICI. e–g Calculated disassociation probability as function of time and angle for molecules that were initialized in the **e** $v = 7$, **f** $v = 8$ and **g** $v = 9$ vibrational eigenstates, in the ground rotational state $j = 0$

As the laser field increases in intensity, molecules dissociate via a typical adiabatic dynamics as well as promptly align via non-resonant Raman transitions that broaden the rotational distribution. Additionally, molecules are affected by the LICI, and non-adiabatic energy transfer to different rovibrational states takes place. As a result, there are at least two quantum paths for which the wavefunction can end up in the same final rovibrational state. Paths related to the LICI contribute a phase factor that creates interference. In a sense, the LICI operates as a quantum-point scatterer, mixing parts of the wavefunction across spatio-angular phase space and creating interferences in both the nuclear and rotation degrees of freedom. The simulation allows us to examine how the wavefunction propagates during the laser pulse, as shown in Fig. 1b–d. We start with an initial state $v = 9, j = 0$, where the spatial wavefunction is evenly distributed over all angles (not shown). However, 20 fs before the pulse peaks (Fig. 1b) we notice that the part of the population that was initially aligned with the laser (close to $0°$ or $180°$), has already started to dissociate, while parts closer to $\theta = \pi/2$ have accumulated. Later, when the pulse peaks (Fig. 1c) the population near the LICI scatters back to different angles, creating a spatio-angular interference pattern that bends around the LICI. This bending is just the superposition of the outgoing and the scattered parts of the wavefunction. Then, 15 fs after the pulse peaks, the wavefunction propagates almost freely, as the potential surfaces flatten, and the modulations in the angular distribution of the dissociating wavefunction are established (Fig. 1d).

Another aspect we capture in the simulation is how different initial states are affected by LICI. For example, in $v = 7$ only high rotational levels have apparent overlap in position with the LICI. Since these levels are most efficiently populated when the pulse peaks via Raman transitions, the non-adiabatic dynamics happens only for a very limited duration and the interference obtained is limited. However for $v = 9$, which is closest to resonance with the laser field, already at the leading edge of the pulse the LICI prevails, and a richer angular interference pattern appears.

In conclusion, LICI are intrinsic in any resonant strong field-matter interaction. It's controllability makes it attractive for novel spectroscopy techniques and research in quantum control, as recently shown in polyatomic systems [8].

Acknowledgments The authors would like acknowledge V. S. Prabhudesai, U. Lev, D. Zajfman and Y. Silberberg for collaborating on the original experiment from which the data was extracted. This research is supported through Stanford PULSE Institute, SLAC National Accelerator Lab by the U.S. Department of Energy, Office of Basic Energy Sciences.

References

1. W. Domcke and D.R.Yarkony, Annu. Rev. Phys. Chem. **63**, 325 (2012).
2. M. V. Berry, Proc. R. Soc. A, **392** 45 (1984).
3. N. Moiseyev et. al.,, J. Phys. B: At. Mol. Opt. Phys. **41** 221001 (2008).
4. M. Šindelka et. al.,, J. Phys. B: At. Mol. Opt. Phys. **44** 045603 (2008).

5. G. J. Halász, A. Vibók, N. Moiseyev, and L. S. Cederbaum, Phys. Rev. A **88**, 043413 (2013).
6. A. Natan *et. al.,* Phys. Rev. A, **86**, 043418 (2012).
7. V. S. Prabhudesai *et. al.,* Phys. Rev. A, **81**, 023401 (2010).
8. J. Kim, *et. al.,* J. Phys. Chem. A **116**, 2758-2763 (2012).

Sub-femtosecond Steering of Carbon Hydrogen Bonds

R. Siemering, M. Kübel, B. Bergues, A.S. Alnaser, M. Kling
and R. de Vivie-Riedle

Abstract During sub-fs double ionization of acetylene vibrational wavepackets are formed which contain the directional information on the targetted hydrocarbon bond. The mechanism for preferential deprotonation of individual bonds is demonstrated by quantum dynamical simulations.

1 Introduction

Carbon hydrogen bond cleavage is one of the most important dynamic reactions in chemistry, motivating considerable efforts to monitor and ultimately control the process. Various examples of efficient strong-field-induced proton ejection and hydrogen migration were reported [1–3]. They all reveal an extremely rapid hydrogen motion, making its real-time observation and control a challenge. Waveform-controlled few-cycle laser pulses have been successfully employed to manipulate the yields of ionization, fragmentation and hydrogen migration in hydrocarbons [1–3]. The full control of the reaction, however, would require also a directional control and has not been demonstrated yet. In our example of acetylene this means the control on which C-H bond is broken. We demonstrate the steering of proton ejection on sub-fs timescales prior to the breakup of the molecular

R. Siemering · R. de Vivie-Riedle (✉)
Department für Chemie, LMU München, 81377 Munich, Germany
e-mail: Regina.de_Vivie@cup.uni-muenchen.de

M. Kübel · M. Kling
Department für Physik, LMU München, 85748 Garching, Germany

B. Bergues · A.S. Alnaser · M. Kling
Max-Planck-Institut für Quantenoptik, Hans-Kopfermann-Str. 1,
85748 Garching, Germany

A.S. Alnaser
Department of Physics & Astronomy, King-Saud University,
Riyadh 11451, Saudi Arabia

© Springer International Publishing Switzerland 2015
K. Yamanouchi et al. (eds.), *Ultrafast Phenomena XIX*,
Springer Proceedings in Physics 162, DOI 10.1007/978-3-319-13242-6_30

126

dication. On the basis of quantum dynamical calculations, the experimental results are interpreted in terms of a novel sub-femtosecond control mechanism involving non-resonant excitation and phase-controlled superposition of vibrational states. The universal mechanism permits control over the directionality of chemical reactions in highly symmetric molecules.

2 CEP Dependent Asymmetry of the Deprotonation in Acetylene Dication

The electric field of few-cycle light pulses can be described as $E(t) = E_0(t) \cos(\omega t + \phi)$, where $E_0(t)$ is the amplitude envelope, ω the carrier frequency, and ϕ the carrier-envelope phase (CEP). In the experiments CEP-tagged reaction microscopy (REMI) was employed. The interaction of acetylene molecules with the intense light pulses results in the dissociative ionization and isomerization. We focus on the deprotonation after two consecutive ionization steps (Fig. 2, left). After population of the first excited state of the dication, the molecule breaks up into H^+ and C_2H^+, which are detected with the REMI in coincidence. The directional control of the deprotonation is then recorded by analyzing the H^+ fragments. The asymmetry parameter $A(\phi)$ with $N_{pos}(\phi)$ ($N_{neg}(\phi)$) the yield for positive (negative) momentum of H^+ (Fig. 1) is taken as a measure. The observed preferential CEP-dependent ejection of protons to the left and right of the laser polarization axis illustrates the sub-femtosecond steering of the hydrogen ejection.

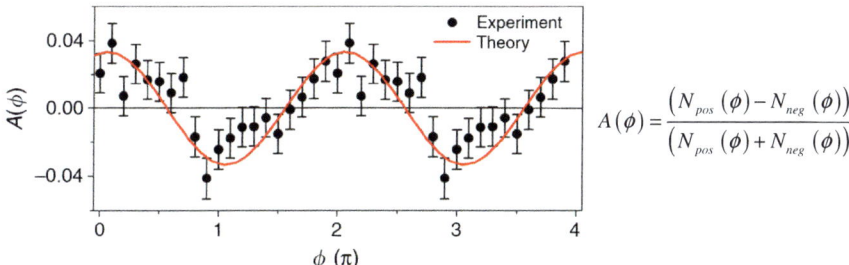

$$A(\phi) = \frac{\left(N_{pos}(\phi) - N_{neg}(\phi)\right)}{\left(N_{pos}(\phi) + N_{neg}(\phi)\right)}$$

Fig. 1 The asymmetry parameter, $A(\phi)$, integrated over all momenta for H^+ ions. The clear oscillation in the asymmetry for the directional emission of H^+ as a function of CEP demonstrates the CEP-control of the hydrogen emission direction

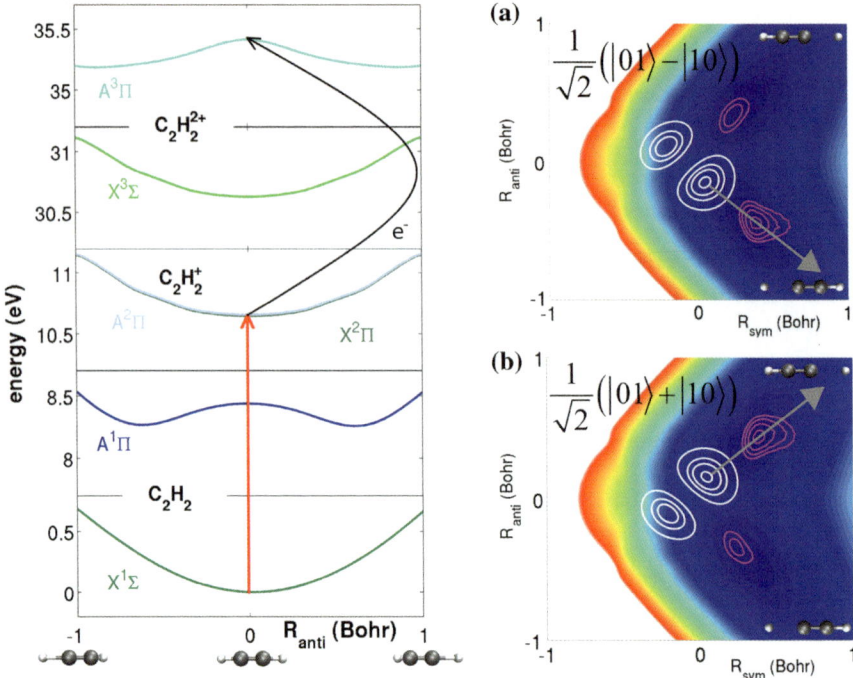

Fig. 2 *Left* 1-D cut of the states involved in the ionization and deprotonation of acetylene. Shown are ground (*green*) and first excited state (*blue*) of the neutral, the cation ground (*light blue*) and degenerate first excited state (*grey*), and the dication ground (*green*) and first excited state (*cyan*) along the anti-symmetric stretching coordinate R_{anti} at $R_{sym} = 0$. The *red vertical line* indicates the tunneling ionization, the *grey curved line* the second ionization step by electron recollision. *Right* 2D-surface of the first excited state of the dication along R_{sym} and R_{anti}. The *arrows* indicate the two directions for deprotonation. The control process is illustrated by wavepackets generated as superpositions of $|01\rangle$ and $|10\rangle$. The initial wavepacket is shown in *white*, a snapshot after 2.4 fs in *magenta*

3 Mechanism of the Directional Control via the CEP

The deprotonation of the dication is steered by a phase sensitive preparation of vibrational wavepackets. The precise preparation of vibrational superposition states has been established as the fundamental concept of molecular quantum computing [4]. In the present study it is the underlying mechanism for the observed asymmetry. The relevant coordinates for the deprotonation are the symmetric $|0m\rangle$ and the anti-symmetric $|n0\rangle$ C-H stretching mode with m, n the vibrational quantum numbers. The notation $|0m\rangle$ includes the time evolution factor $\exp(-iE_{0m}t/\hbar)$. The superposition state $1/\sqrt{2} (|0m\rangle \pm |n0\rangle)$ will induce a wavepacket motion along either one of the C-H bonds. When such a wavepacket is transferred onto a repulsive potential along the C-H bond, dissociation of either the left or the right H-atom will occur. To extract the control mechanism realized in the experiment, two-dimensional (2D) surfaces along these two coordinates (R_{sym}, R_{anti}) are calculated for all states involved. Exemplarily, the first excited state of the dication is shown in Fig. 2. The constraints for acetylene are that the symmetric mode $|0m>$ cannot be optically addressed by the applied IR-field, only the anti-symmetric mode $|n0>$ is IR-active. Our quantum dynamical analysis shows that the superposition is prepared in the neutral and cationic molecule and mainly from $m,n = 1$ states. The $|01>$ component is populated due to the field induced ionization, while the $|10>$ component is directly populated via the interaction with the non-resonant light field. Also a phase ($|10>$ exp ($i\phi_{CEP}t$)) is imprinted and can be controlled relative to the $|01>$ mode via the CEP (ϕ_{CEP}) of the light field. Variation of the CEP changes the phase and thus the sign in the superposition $1/\sqrt{2}$ ($|01> \pm |10>$). The superposition exhibits a modulation with a periodicity of 2π, which shows up in the asymmetry $A(\phi)$ for the directional emission of the H^+ ions and agrees well with the experimental results (Fig. 1).

In summary, the presented data support a new and very general coherent control scheme, where the direction of proton ejection in a symmetric hydrocarbon is steered through manipulating of the phases of the individual components of a vibrational wavepacket. The scheme can be transferred to control the direction of the proton migration and by tuning the IR pulse in resonance with the active mode the control can be enhanced significantly.

References

1. A.S. Alnaser, et al. "Momentum-imaging investigations of the dissociation of D2+ and the isomerization of acetylene to vinylidene by intense short laser pulses" J. Phys. B 39, 485-492 (2006).
2. X. Xie et al. "Attosecond-recollision-controlled selective fragmentation of polyatomic molecules" Phys. Rev. Lett. 109, 243001 (2012).
3. T. Okino, A. Watanabe, H. Xu, and K. Yamanouchi, "Ultrafast hydrogen scrambling in methylacetylene and methyl-d3-acetylene ions induced by intense laser fields" Phys. Chem. Chem. Phys. 14, 10640-10646 (2012).
4. R. de Vivie-Riedle and U. Troppmann, "Femtosecond lasers for quantum information technology" Chem. Rev. 107, 5082-5100 (2007).

Tabletop Imaging of Structural Evolutions in Chemical Reactions

Heide Ibrahim, Benji Wales, Samuel Beaulieu, Bruno E. Schmidt,
Nicolas Thiré, Emmanuel P. Fowe, Éric Bisson,
Christoph T. Hebeisen, Vincent Wanie, Mathieu Giguére,
Jean-Claude Kieffer, Michael Spanner, André D. Bandrauk,
Joseph Sanderson, Michael S. Schuurman and François Légaré

Abstract The first high-resolution molecular movie of proton migration in the acetylene cation is obtained using a tabletop multiphoton pump-probe approach— an alternative to demanding free-electron-lasers and other VUV light sources when ionizing from the HOMO-1.

OCIS codes (020.4180) Multiphoton processes · (100.0118) Imaging ultrafast phenomena

1 Introduction

Since the introduction of femto-chemistry, electron or X-ray diffraction have been the most commonly employed techniques to track nuclear rearrangement in molecules. Unfortunately, these techniques are largely insensitive to the more subtle and irregular structural changes that can occur within a single small molecule undergoing a chemical reaction. Pump-probe Coulomb Explosion Imaging (CEI)

H. Ibrahim (✉) · S. Beaulieu · B.E. Schmidt · N. Thiré · É. Bisson · V. Wanie · M. Giguére · J.-C. Kieffer · F. Légaré
Centre ÉMT, Institut National de la Recherche Scientifique,
1650 Boulevard Lionel-Boulet, Varennes, Quebec, QC J3X1S2, Canada
e-mail: Ibrahim@emt.inrs.ca

B. Wales · J. Sanderson
Department of Physics and Astronomy, University of Waterloo,
200 University Avenue West, Waterloo, ON N2L 3G1, Canada

E.P. Fowe · A.D. Bandrauk
Laboratoire de Chimie Théorique, Faculté des Sciences, Université de Sherbrooke,
2500, boul. de l'Université, Sherbrooke, QC J1K 2R1, Canada

C.T. Hebeisen · M. Spanner · M.S. Schuurman
National Research Council of Canada, 100 Sussex Dr, Ottawa, ON K1A 0R6, Canada

C.T. Hebeisen
Department of Physics, University of Ottawa, 150 Louis Pasteur, Ottawa,
ON K1N 6N5, Canada

© Springer International Publishing Switzerland 2015
K. Yamanouchi et al. (eds.), *Ultrafast Phenomena XIX*,
Springer Proceedings in Physics 162, DOI 10.1007/978-3-319-13242-6_31

allows observation of these changes on a femtosecond (fs) timescale with atomic resolution, so e.g. proton migration on the acetylene dication [1, 2].

Since laser driven tunnel ionization preferentially ionizes the highest occupied molecular orbital (HOMO) compared to lower lying orbitals, the dynamics within the cation typically involves mostly the electronic ground state—where a potential energy barrier of 2 eV prevents isomerization. To overcome this barrier it is thus necessary to populate the first electronically excited state which requires ionization from the HOMO-1 orbital. We present a table top approach to efficiently launch dynamics from ionization of the HOMO-1 of small organic molecules [3], which so far required VUV sources like free electron lasers (FEL) or high harmonics [4, 5]. Once on the first excited state of the acetylene cation, a proton can migrate from the linear acetylene cation ([HC=CH]+) to the vinylidene cation ([C=CH2]+).

While FELs are limited by repetition rate and fluctuations, our approach provides tremendous benefits in terms of statistics, accessibility and temporal resolution. Here we show that 266 nm ultrashort laser pulses are capable of initiating rich dynamics on excited states through multiphoton ionization. With our generally applicable tabletop approach that can be transferred to other small organic molecules, we have investigated two basic chemical reactions simultaneously: proton migration (shown in the first high resolution molecular movie) and C=C bond-breaking, triggered by multiphoton ionization. The experimental results are in excellent agreement with the timescales and relaxation pathways predicted by new and definitively quantitative ab initio trajectory simulations [3].

2 Results and Discussion

A pump pulse (266 nm, 32 fs, four photon absorption) ionizes the system to the acetylene cation and launches proton migration dynamics. A time delayed probe pulse (800 nm, 40 fs) further ionizes it to higher charged states. Its electric field stripes off electrons almost immediately, leaving positively charged fragments behind which undergo Coulomb explosion. They represent the molecule's geometric configuration at the arrival time of the probe pulse.

Newton plots of three correlated fragments (H^+, C^+ and CH^+) provide frames of the molecular movie of isomerization (Fig. 1)—starting from the linear geometrical shape of acetylene (·) on the left, via a transition state (+) to vinylidene (x) on the right hand side. The trans-configuration is reached after 20 fs, which matches a half-period of the trans-bent vibrational mode. At 40 fs the centre of mass reaches the classically calculated transition state, which agrees with timescales predicted by our semi-classical calculations for a population transfer to the ground state mediated by the conical intersection between ground and first excited state. A maximum in vinylidene is observed around 100 fs, after which the population maximum swings back to the acetylene side (with a vinylidene population minimum around 150 fs). This oscillation on the ground state continues, with the next vinylidene population maximum observed at 180 fs.

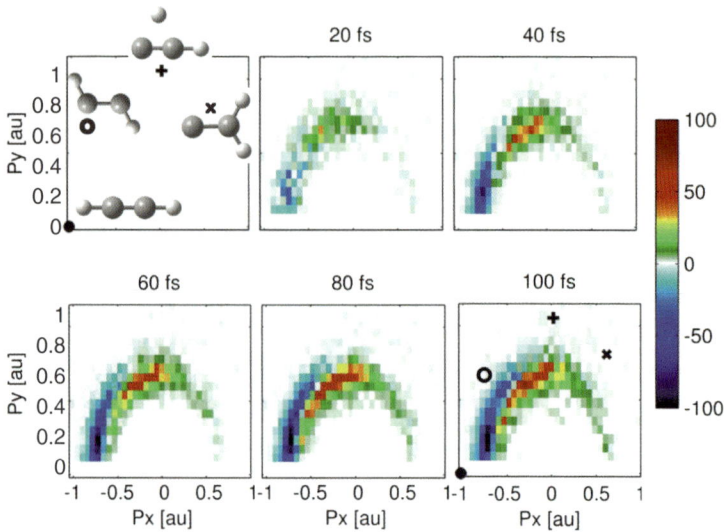

Fig. 1 The molecular movie of proton migration in the acetylene cation: Newton plots filtered for KER >13 eV. Symbols show classical calculations assuming Coulomb potentials for linear (·) and trans configuration (∘), transition state (+) and vinylidene (x), as given in the upper left corner. The plots show the evolution of H⁺ momenta (after subtracting the distribution at Δt = 0) from acetylene to vinylidene with increasing time delay from 0 to 100 fs, normalized to the integral. The distribution appears localized at 20 fs (*dark green data points*) and later spreads out. *White* corresponds to zero, *blue* to negative contributions (i.e. where the population originates) and *other colours* to positive contributions (i.e. where the population is going)

As shown in Fig. 2, the observed to and fro isomerization between acetylene (red) and vinylidene (blue) is in excellent agreement with the theoretical computations presented as solid lines. Due to careful investigation we are able to uniquely assign the involved electronic and charged states.

Additionally, new dissociation channels involving subsequent photon absorption on the cation have been observed from two fragment correlation ($CH^+ + CH^+$ and $C^+ + CH_2^+$). They show up in both, the acetylene and vinylidene molecule and are thus assigned to a C=C bond cleavage.

Merging the information from 1, 2, and 3 fragment correlations with theoretical simulations, our results present the most complete picture of population transfer on the excited state of the acetylene cation. If one aims to populate the excited states of charged molecules whose difference in ionization potentials is too large to be overcome by 800 nm photons, our experimentally rather simple approach provides an alternative to demanding VUV sources. It allows studying of excited states dynamics of e.g. the acetylene or ethylene cation and should be applicable to a great many of other small organic molecules.

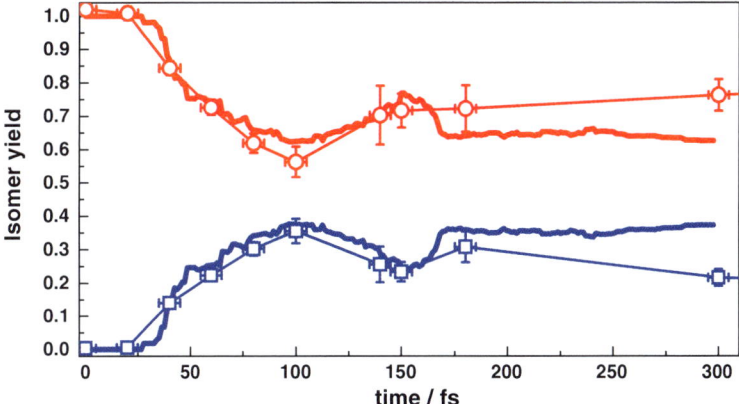

Fig. 2 Comparison of theory (*thick lines*) and experiment (*open symbols*) for acetylene (*red*) and vinylidene (*blue*) yield: Theoretical *curves* show the superposition of acetylene and vinylidene yield in both the ground state and excited states of the cation as a function of time. Experimental values are obtained by integrating the Newton plots of Fig. 1 for negative P_x (acetylene, *circles*) or positive P_x (vinylidene, *squares*), see [3]. Experimental data are corrected by offset and scaling factor to fit the theoretical points

References

1. A. Matsuda, M. Fushitani, E. J. Takahashi, and A. Hishikawa, Phys. Chem. Chem. Phys. **13**, 8697 (2011).
2. A. Hishikawa, A. Matsuda, M. Fushitani, and E. J. Takahashi, Phys. Rev. Lett. **99**, 258302 (2007).
3. H. Ibrahim, B. Wales, S. Beaulieu, B. E. Schmidt, N. Thiré, Emmanuel P. Fowe, É. Bisson, C. T. Hebeisen, V. Wanie, M. Giguére, J.-C. Kieffer, Michael Spanner, André D. Bandrauk, J. Sanderson, M. S. Schuurman, and F. Légaré, Nat. Commun. 5:4422 (2014).
4. Y. Jiang, A. Rudenko, O. Herrwerth, L. Foucar, M. Kurka, K. Kühnel, M. Lezius, M. Kling, J. van Tilborg, A. Belkacem, K. Ueda, S. Düsterer, R. Treusch, C. Schröter, R. Moshammer, and J. Ullrich, Phys. Rev. Lett. **105**, 1 (2010).
5. J. van Tilborg, T. K. Allison, T. W. Wright, M. P. Hertlein, R. W. Falcone, Y. Liu, H. Merdji, and A. Belkacem, J. Phys. B At. Mol. Opt. Phys. **42**, 81002 (2009).

Structure Dependence of Kinetic Energy Released in X-ray-Induced Fragmentation

Philip H. Bucksbaum, Chelsea E. Liekhus-Schmaltz, Ian Tenney and Vladimir S. Petrovic

Abstract We have analyzed transient structures in deuterated acetylene isomerized by 10 fs 400 eV X-rays, and probed by coulomb explosion with a second delayed X-ray pulse. Structural changes are revealed through fragment kinetic energy release.

X-ray induced isomerization experiments in several molecules were carried out by the AMO75113 collaboration[1] using X-ray pulse pairs from the newly commissioned soft X-ray split and delay at the LCLS X-ray free electron laser [1]. We have analyzed X-ray-induced fragmentation energies in deuterated acetylene data in the linear DCCD isomer and the vinylidene CCD_2 form to show how kinetic energy release correlates with isomerization.

The acetylene isomerization transition is a model for hydrogen migration in organic molecules. Relatively little is known about isomerization induced by photoionization of the 1 s shell. X rays above the carbon K-edge can induce core shell ionization followed by Auger relaxation, leading to the production of a molecular dication. Experiments at X-ray synchrotrons have shown strong evidence for X-ray dication formation accompanied by proton migration from the initial $HCCH^{2+}$ configuration to the $CCH2^{2+}$ vinylidene metastable isomer [2]. These studies cannot resolve the motion of the protons within the molecule prior to dissociation. Photoelectron-photoion coincidence measurements at synchrotrons can clock proton migration times, and have concluded that they are faster than 60 fs [3]. This is at odds with the ~ 2 eV isomerization barrier between acetylene and

---

[1] AMO75113 collaboration: V. Petrovic, C. Liekhus-Schmaltz, P. Bucksbaum, I. Tenney, T. Osipov, A. Belkacem, N. Berrah, R. Boll, C. Bomme, C. Bostedt, J. Bozek, S. Carron, R. Coffee, J. Devin, B. Erk, L. Fang, K. Ferguson, R. Field, L. Foucar, L. Frasinski, J.M. Glownia, M. Guehr, A. Kamalov, J. Krzywinski, H. Li, J. Marangos, T. Martinez, B. McFarland, S. Miyabe, B. Murphy, A. Natan, D. Rolles, A. Rudenko, A. Sanchez, M. Siano, E. Simpson, L. Spector, M. Swiggers, D. Walke, S. Wang, and T. Weber.

P.H. Bucksbaum (✉) · C.E. Liekhus-Schmaltz · I. Tenney · V.S. Petrovic
PULSE Institute, Stanford University, Stanford, CA 94305-4060, USA
e-mail: phbuck@Stanford.edu

© Springer International Publishing Switzerland 2015
K. Yamanouchi et al. (eds.), *Ultrafast Phenomena XIX*,
Springer Proceedings in Physics 162, DOI 10.1007/978-3-319-13242-6_32

134

vinylidene in the ground state of the dication, which suggests a longer isomerization time [4, 5].

Our studies used 10 fs, 400 eV X-ray pulse pairs. The first X-ray pulse ionizes a core electron and initiates the dynamics. The cation undergoes Auger relaxation in approximately 6 fs [5]. Deuteron migration dynamics is probed by a second X-ray pulse, which ionizes another core electron, leading to further Auger relaxation. The resulting tetracation is unstable, and dissociates promptly into several fragments. The three-layer delay-line anode detector can sort and resolve the position and arrival time of each ion fragment for up to 16 independent ions, therefore measuring the full momentum vectors. The pump and probe X-ray pulses were delayed by from 0 to 100 fs. The target gas density was maintained to a level where on the order of one molecule per shot was excited by two X-ray pulses (Fig. 1).

Data in this study include events where four charged dissociation products were produced from the same molecule. We found that an effective filter to eliminate backgrounds with false coincidences was to require that the four particles be singly charged, two deuterons and two carbon ions, and that the total vector sum of the momenta of all fragments was approximately 0, to within 10 % of the peak momentum. Figure 2 shows how this works in the case of only two fragments, where the display is simpler. We applied this to the four-dimensional product space of four particle coincidences.

The deuteron migration from acetylene to vinylidene can be detected by analyzing the individual ion momenta. For example, in this analysis we have used the following protocol: The four momentum vectors are $p_{C1}, p_{C2}, p_{H1},$ and p_{H2}. Our event selection filter ensures that the vector sum of all momenta is zero: $p_{C1} + p_{C2} + p_{H1} + p_{H2} = 0$. We then define the molecular axis in the lab frame as the unit vector $\hat{A} = (p_{C1} - p_{C2})/|p_{C1} - p_{C2}|$. Finally, we look at the projection of

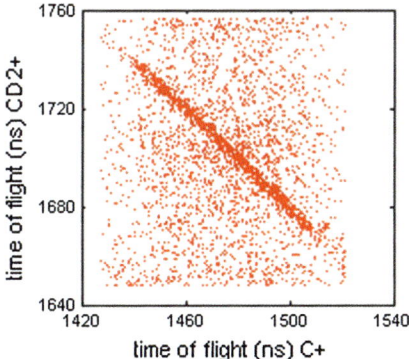

Fig. 1 Momentum correlation may be used to find coincidences corresponding to the isomerization pathway after the X-ray photon absorption. Here, coincidences of C^+ and CD_2^+ are found. For true coincidences the conservation of the momentum (along the time of flight axis in this case) results in the data lying on a *line* as shown above. In the study reported here, similar constraints were placed on four-particle coincidences that correspond to full dissociation of acetylene

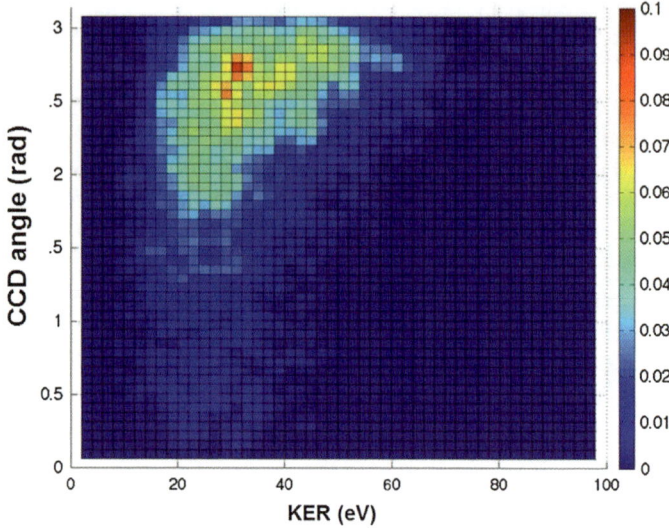

Fig. 2 Dependence of the kinetic energy released (KER) on the ejection angle of the deuteron. Angles below $\pi/2$ correspond to V-type structures. These have significantly lower average energy and a sharper energy distribution than the A-type dissociations

the deuteron momenta on $\hat{A}$. If the dot products $\hat{A} \cdot \boldsymbol{p}_{H1}$ and $\hat{A} \cdot \boldsymbol{p}_{H2}$ have different signs, then we identify that detected event as a molecule that has not undergone isomerization (the acetylene channel "A" or opposite-side deprotonation channel "P"). If, on the other hand, $\hat{A} \cdot \boldsymbol{p}_{H1}$ and $\hat{A} \cdot \boldsymbol{p}_{H2}$ have the same sign, then one deuteron has migrated to the other side of the molecule. This is the "V" channel that also includes the same-side deprotonation.

We find some striking physical features of the data sorted in this way. First the collection of A- and P-channel events have a total kinetic energy (KER) release distribution that is approximately 10 eV larger than V-channel events. This difference is so large that KER may be used as a method to separate non-isomerizing events from V-type events.

The ratio of V-type to non-isomerizing events is a strong function of the time delay between the pulses. The time-dependence is consistent with the LCLS X-ray autocorrelation time, and suggests that deuteron migration initiates on a time scale of approximately 20 fs or less following X-ray ionization.

Acknowledgments This research used the SLAC Linac Coherent Light Source, which is a DOE Office of Science User Facility. The members of the AMO75113 collaboration contributed advice and assistance. This research was supported by the National Science Foundation under Grant No. PHY-0649578.

References

1. AMO75113 collaboration, in preparation (2014).
2. J. Laksman, D. Céolin, M. Gisselbrecht, S. E. Canton, and S. L. Sorensen, The Journal of Chemical Physics 131 (2009) 244305.
3. T. Osipov, C. L. Cocke, M. H. Prior, A. Landers, T. Weber, O. Jagutzki, L. Schmidt, H. Schmidt-Böcking, and R. Dörner, Physical Review Letters 90 (2003) 233002.
4. D. Duflot, J. M. Robbe, and J. P. Flament, The Journal of Chemical Physics 102 (1995) 355.
5. T. S. Zyubina, Y. A. Dyakov, S. H. Lin, A. D. Bandrauk, and A. M. Mebel, The Journal of Chemical Physics 123 (2005) 134320.

Controlling Fragmentation Reactions of Polyatomic Molecules with Impulsive Laser Alignment

Xinhua Xie, Katharina Doblhoff-Dier, Huailiang Xu, Stefan Roither, Markus Schöffler, Daniil Kartashov, Sonia Erattupuzha, Tim Rathje, Gerhard G. Paulus, Kaoru Yamanouchi, Andrius Baltuška, Stefanie Gräfe and Markus Kitzler

Abstract We experimentally and theoretically demonstrate channel-selective control over strong-field induced fragmentation of a polyatomic molecule, acetylene, using impulsive laser alignment as the control mechanism.

Over the past decades control over the photodissociation dynamics of molecules has become a vivid area of research (see e.g. [1] for reviews). By applying temporally shaped femtosecond laser pulses, vibrational dynamics can be selectively excited and the molecular dynamics can be steered towards desired reaction products. However, only since very recently it has become possible to observe and control the ultrafast electronic dynamics in molecules. Control over the electronic dynamics of a molecule is typically achieved via control over the carrier-envelope phase (CEP) of few-cycle laser pulses [2].

X. Xie · S. Roither · M. Schöffler · D. Kartashov · S. Erattupuzha · A. Baltuška ·
M. Kitzler (✉)
Photonics Institute, Vienna University of Technology, Gusshausstrasse 27,
1040 Vienna, Austria
e-mail: markus.kitzler@tuwien.ac.at

K. Doblhoff-Dier · S. Gräfe
Institute for Physical Chemistry, Friedrich-Schiller University Jena,
07743 Jena, Germany

H. Xu
State Key Laboratory on Integrated Optoelectronics, Jilin University,
Changchun 130012, China

T. Rathje · G.G. Paulus
Institute of Optics and Quantum Electronics, Friedrich-Schiller University Jena,
07743 Jena, Germany

T. Rathje · G.G. Paulus
Helmholtz Institute Jena, 07743 Jena, Germany

K. Yamanouchi
Department of Chemistry, The University of Tokyo, 7-3-1 Hongo, Bunkyo-ku,
Tokyo 113-0033, Japan

© Springer International Publishing Switzerland 2015
K. Yamanouchi et al. (eds.), *Ultrafast Phenomena XIX*,
Springer Proceedings in Physics 162, DOI 10.1007/978-3-319-13242-6_33

Here, using the acetylene molecule, C_2H_2, as an example, we demonstrate that the alignment of the molecular axis relative to the laser polarization direction can serve as a control tool to determine not only the overall fragmentation yield but also the *relative* probability of individual reaction pathways in polyatomic molecules over a wide range of laser peak intensities and pulse durations. The method builds on the alignment-sensitive ionization yields from inner and outer valence orbitals and exploits the fact that selective ionization from specific inner valence orbitals allows for the controlled population of electronic surfaces in the ion associated with certain reaction pathways.

In our experiments we impulsively aligned [3] C_2H_2 molecules in an ultra-high vacuum chamber $(1.3 \times 10^{-10} \text{mbar})$ by applying a weak 795 nm 50 fs, linearly polarized laser pulse from a 5 kHz Titanium-Sapphire laser amplifier system [4]. As a probe (ionization) pulse, the remaining portion of the laser output, spectrally broadened and recompressed to a duration of 4.5 fs full width at half maximum (FWHM), was used. The time delay Δt between alignment pulse and probe pulse was precisely controlled using a piezo stage. The resulting momenta of the ions and ionic fragments are measured in coincidence by a reaction microscope [5] on a single shot basis, allowing channel-resolved examination of individual fragmentation.

The signature of the temporal evolution of the C_2H_2 rotational wavepacket induced by the alignment pulse is shown in Fig. 1 as a function of the delay between the alignment and probe laser pulses. The quarter and half revivals of C_2H_2 are clearly visible around 3.5 and 7 ps. The measured $\langle \cos^2(\theta) \rangle$ value was derived from a complete fragmentation channel of acetylene and shows that the maximum value of $\langle \cos^2(\theta) \rangle$ reaches 0.5 in our experiments.

The delay dependence of several fragmentation and ionization yields is shown in Fig. 2a, b for probe pulses with an intensity of $I = 4 \times 10^{14} \text{W}/\text{cm}^2$ and a pulse duration 4.5 fs (FWHM). The use of higher intensities (up to $7 \times 10^{14} \text{W}/\text{cm}^2$) and even longer pulses (up to 18 fs) does not change this picture qualitatively for any of

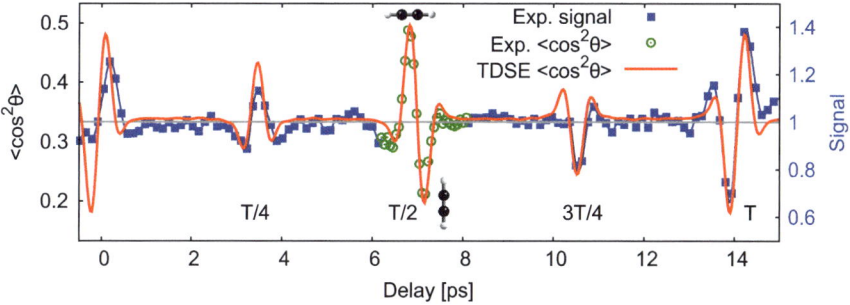

Fig. 1 The *blue filled squares* represent the yield of protons ejected with high energy from completely fragmenting acetylene molecules. The *red line* shows the calculated $\langle \cos^2(\theta) \rangle$ value as a function of time after application of the alignment pulse. Measured $\langle \cos^2(\theta) \rangle$ values for acetylene are shown by *green open circles* near the half revival structure around 7 ps

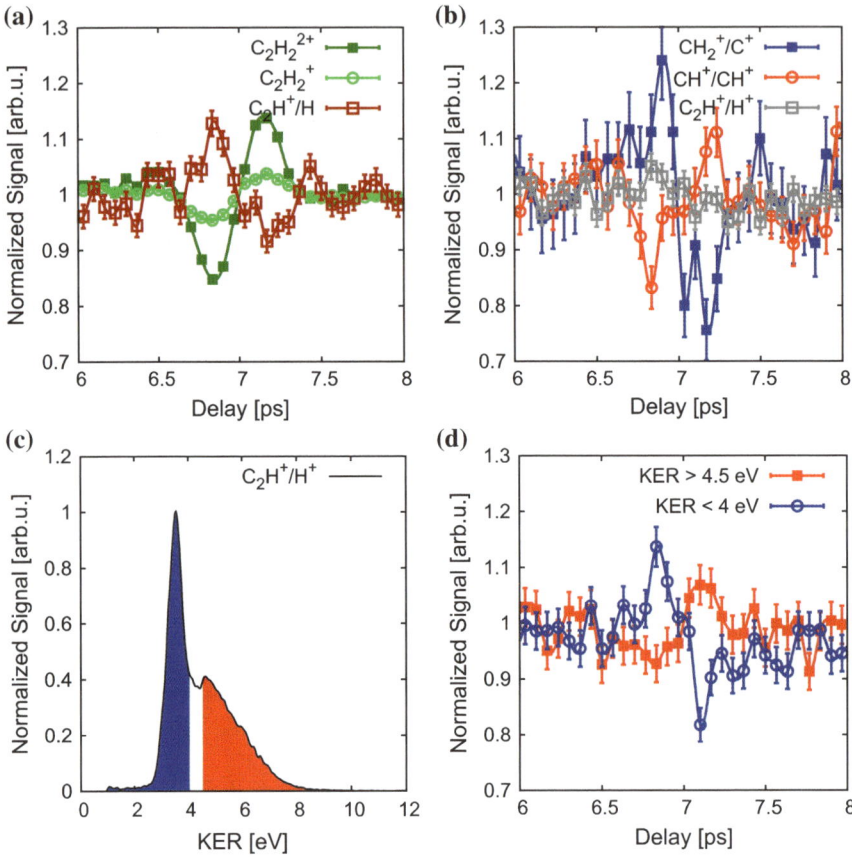

Fig. 2 **a** Normalized yield of the cation $C_2H_2^+$ (*light green*), the dissociation channel C_2H^+/H (*dark red*), and the dication $C_2H_2^{2+}$ (*dark green*) as a function of time delay. **b** Normalized yield of the fragmentation channels CH_2^+/C^+ (*blue*), CH^+/CH^+ (*red*), and C_2H^+/H^+ (*gray*). **c** KER spectrum of channel C_2H^+/H^+. **d** KER-resolved yield of channel C_2H^+/H^+ over Δt, with the selected KER spectral regions indicated in (**c**)

the observed channels. The yields of most channels feature a pronounced dependence on the time delay and hence on the relative orientation of the molecular axis to the laser polarization direction. An important feature of the measured fragmentation yields is that their delay-dependence differs for various examined fragmentation channels. While, for example, the yield of the fragmentation channel $C_2H_2^{2+} \rightarrow CH_2^+ + C^+$ peaks at parallel alignment, the yield of the channel $C_2H_2^{2+} \rightarrow CH^+ + CH^+$ peaks at perpendicular alignment. By adjustment of Δt it is thus possible to relatively enhance the yield of one channel with respect to the other (in our experiment by about 150 %). This finding clearly demonstrates the potential of molecular alignment to selectively enhance or suppress individual fragmentation channels of the same parent ion of polyatomic molecules on a very short time scale. The possibility of channel selective control is a novel feature as compared to previous attempts of controlling the fragmentation behavior of polyatomic molecules [2], where only the overall fragmentation yield, but not the relative yields of individual fragmentation channels could be controlled.

With the help of simulations that apply time-dependent density functional theory (TDDFT) we found that the demonstrated control is based on different angular-dependent ionization rates of inner and outer valence electrons. Ionization from different electronic states will result in the population of different electronic states in the dication: If two π-electrons are removed, the stable electronic ground state of the dication is reached and dissociation is inhibited. The removal of one π- and one σ-electron, however, puts the dication into an excited electronic state. Different electronic states feature, in general, different characteristic potential energy surfaces. The dissociation or isomerization processes that follow the ionization event may therefore, in turn, proceed along different specific nuclear degrees of freedom and as a result will end up in different fragmentation channels. Controlling the population of a specific excited electronic surface in the dication by determining the ratio of ionization from different orbitals (e.g. σ vs. π) using the molecular alignment as a knob, allows, thus, to control the yield of a certain fragmentation channel associated with this electronic surface. This is the essence of the control method demonstrated here.

Within this picture, using the angular dependent ionization rates obtained by the TDDFT simulations and by assuming a certain sequence of removing two electrons from the neutral molecule, it is possible to understand all experimentally observed modulations of the fragmentation yields with Δt [4]. A fragmentation channel that necessitates a more involved treatment, both experimentally and theoretically, is the deprotonation reaction with final products C_2H^+/H^+. The yield of this channel does not show any significant modulation with Δt (Fig. 2b). To understand this insensitivity to the alignment, we first note that the kinetic energy release (KER) distribution of the fragments features a clear double peak structure (Fig. 2c). This indicates the presence of (at least) two competing reaction pathways. By separating the events for low and high KER, it becomes apparent that the two reaction pathways have opposite alignment dependence (Fig. 2d): While the low KER part follows the predictions for ionization from one π- and one σ-orbital, the high KER part follows the predictions for electron removal from the π-system only. Only in the KER-integrated picture, where the two pathways contribute with nearly equal probability, the alignment dependence of the overall yield vanishes. From measurements with circular polarization and higher intensity we conclude that non-sequential processes, such as recollision-ionization and recollision-excitation, are important for fragmentation events in the high KER peak, while events in the low KER peak dominantly originate from sequential ionization. Molecular alignment in combination with fine-tuned laser parameters, thus, does not only allow to determine the relative yield of individual channels, but even the molecular pathway towards a certain final set of fragment ions.

In conclusion, we have introduced molecular alignment as a tool to control the yield of molecular fragmentations on the few-femtosecond time scale. We have demonstrated channel selectivity in different fragmentation reactions of a polyatomic molecule initiated by a strong laser field. The method exploits the presence of different ionization channels and the different angular dependence of the ionization probability from inner and outer valence orbitals for controlling the population of dissociative excited electronic surfaces associated with molecular reaction dynamics along specific pathways towards a desired set of fragment ions. Therefore, fragmentation control by molecular alignment is applicable to all (polyatomic) molecules that can be aligned and feature a distinctively different angular ionization probability from outer and inner valence orbitals.

Acknowledgments We acknowledge financial support by the Austrian Science Fund (FWF) under grants No. P21463-N22, No. P25615-N27 and No. SFB-F49 NEXTlite, and by a starting grant from the ERC (project CyFi).

References

1. M. Shapiro and P. Brumer, Quantum Control of Molecular Processes (Wiley, New York, 2012).
2. X. Xie *et al.* Attosecond-recollision-controlled selective fragmentation of polyatomic molecules, Physical Review Letters **109**, 243,001 (2012).
3. H. Stapelfeldt and T. Seideman, *Colloquium:* Aligning molecules with strong laser pulses, Rev. Mod. Phys. **75**, 543–557 (2003).
4. X. Xie *et al.* Selective Control over Fragmentation Reactions in Polyatomic Molecules Using Impulsive Laser Alignment Phys. Rev. Lett. **112**, 163003 (2014).
5. R. Dörner *et al.* Cold target recoil ion momentum spectroscopy: A 'momentum microscope' to view atomic collision dynamic, Physics Reports **330**, 95–192 (2000).

Intense Field Ionization of C_2H_2 and $^{12}C^{13}CH_2$ Aligned in Field-Free Space

Hirokazu Hasegawa, Yuki Ikeda, Kotaro Sonoda, Takahiro Sato, Atsushi Iwasaki and Kaoru Yamanouchi

Abstract Intense field ionization of nonadiabatically aligned C_2H_2 was investigated by a pump-probe technique. The yield of parent ions reveals that the ionization occurs preferentially when the molecular axis is perpendicular to the laser polarization direction.

1 Introduction

The interaction of molecules with an intense laser field induces a variety of dynamical processes, which are essentially triggered by ultrafast responses of electrons within a molecule to the intense laser field. However, in the gas phase, molecules rotate freely in space, and the information on the ionization processes in the molecular-fixed coordinate system could not be obtained straightforwardly because of the spatial averaging over the rotational angles of the molecular axes with respect to the laser polarization direction.

A molecular alignment is a promising technique free from this rotational averaging effect [1–3]. When linear molecules are irradiated with a linearly polarized ultrashort intense laser pulse, a rotational wave packet, which is a superposition of rotational eigenstates, is impulsively created by the interaction between the laser field and an induced dipole moment of molecules. After the interaction, molecules can be aligned parallel and perpendicular to the laser polarization direction periodically in field-free space. This phenomenon is called the nonadiabatic molecular alignment or the field-free molecular alignment [1–3]. In the present study, we

H. Hasegawa (✉) · K. Sonoda
Department of Integrated Sciences, Graduate School of Arts and Sciences,
the University of Tokyo, 3-8-1 Komaba, Meguro-ku 153-8902, Tokyo, Japan
e-mail: chs36@mail.ecc.u-tokyo.ac.jp

Y. Ikeda · T. Sato · A. Iwasaki · K. Yamanouchi
Department of Chemistry, School of Science, the University of Tokyo,
7-3-1 Hongo, Bunkyo-ku, Tokyo 113-0033, Japan

© Springer International Publishing Switzerland 2015
K. Yamanouchi et al. (eds.), *Ultrafast Phenomena XIX*,
Springer Proceedings in Physics 162, DOI 10.1007/978-3-319-13242-6_34

investigate the ionization process of C_2H_2 in an intense laser field by taking advantage of this nonadiabatic molecular alignment. In addition, we show that this method provides us with a powerful tool to extract spectroscopic information of the rotation of molecules in their electronic ground state.

2 Experiment

A femtosecond laser pulse (100 fs) generated by a Ti:Sapphire amplifier system is spilt into a pair of pulses by passing through a Michelson interferometer. One of the pulses called a pump pulse and the other called a probe pulse are used to align and ionize molecules, respectively. The temporal delay between the pump and probe pulses, τ, can be controlled by the delay stage in the pump beam arm in the interferometer. The pump and probe laser pulses are focused by a lens ($f = 200$ mm) on a supersonic molecular beam. The laser field intensity at the focal point is estimated to be 10 and 70 TW/cm^2 for the pump and probe pulses, respectively. A pulsed supersonic molecular beam of pure acetylene (C_2H_2) with the stagnation pressure of 1.1 atm was introduced into a vacuum chamber equipped with a Wiley-McLaren type time-of-flight mass spectrometer (TOF-MS). The nonadiabatic alignment of C_2H_2 is induced by the pump pulse. Parent ions generated by the probe pulse after the temporal delay are detected by the TOF-MS. The polarization of the pump pulse is set to be parallel to the TOF axis, and the polarization directions of the pump and probe pulses are set to be parallel to each other.

3 Results and Discussion

Figure 1 shows the dependence of the yield of $C_2H_2^+$ on the pump-probe delay, τ. The transient peaks are observed clearly at $\tau = 7.1$ and 14.2 ps. Since the rotational period T_{rot} of C_2H_2 is calculated to be $T_{rot} = 1/(2cB_0) = 14.2$ ps using the rotational

Fig. 1 The observed $C_2H_2^+$, $^{12}C^{13}CH_2^+$ yields, and the expectation value of $\sin^2\theta$ against the pump-probe delay

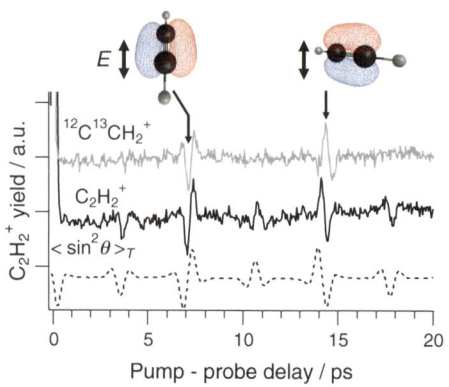

constant, $B_0 = 1.176608$ cm^{-1} [4], both the delay time for the first peak and the time difference between these two peaks correspond to a half of the rotational period, 7.1 ps, showing that these two peaks appear through the nonadiabatic molecular alignment.

In order to compare the observed ion yield with the spatial distribution of the molecular axis, $\langle \cos^2 \theta \rangle = \langle \psi(t) | \cos^2 \theta | \psi(t) \rangle$ is calculated by numerically solving a time-dependent Schrödinger equation, where θ represents the angle between the molecular axis and the polarization direction and $|\psi(t)\rangle$ is a rotational wave packet, $|\psi(t)\rangle = \sum_J A_J e^{i\delta_J} e^{-i\omega_J t} |J, M\rangle$, generated by the pump pulse defined using the probability amplitude A_J, the initial phase δ_J, the energy phase ω_J defined by $\omega_J = hcB_0 J(J+1)/\hbar$, and the rotational eigenstate $|J, M\rangle$. The $\langle \cos^2 \theta \rangle$ values calculated by assuming the rotational temperature of 50 K agree well with the peak positions in the observed signal, but the phases of these peaks are out of phase by 180°. The observed out-of-phase behavior implies that the ionization probability is higher when the laser polarization direction is perpendicular to the molecular axis. If we assume that the angular dependence of the ionization probability $W(\theta)$ is proportional to $\sin^2\theta$, the observed yield of $C_2H_2^+$ may be expressed as $\langle \sin^2 \theta \rangle = \langle \psi(t) | \sin^2 \theta | \psi(t) \rangle$.

The calculated $\langle \sin^2 \theta \rangle$ is plotted by the dashed curve as shown in Fig. 1, which reproduces well not only the positions of peaks but also the phase of the respective peaks. If the ionization proceeds by the tunneling ionization, $W(\theta)$ is expected to reflect the shape of the highest occupied molecular orbital (HOMO). Considering that the HOMO of C_2H_2 $1\pi_u$ as shown in the inset in Fig. 1, and its electron density distribution spreads along the perpendicular direction to the molecular z axis, the good agreement between the observed the $C_2H_2^+$ yield and the dashed curve in Fig. 1 suggests that the ionization proceeds via the tunneling ionization from the HOMO.

On the other hand, the Fourier transformation (FT) of the pump-probe signals of $C_2H_2^+$ and $^{12}C^{13}CH_2^+$ gives valuable information on molecular constants in the electronic ground state. Such FT spectra of molecules such as N_2, O_2, and N_2O were also reported previously [4, 5]. Figure 2 shows the FT spectra for $C_2H_2^+$ (lower trace) and $^{12}C^{13}CH_2^+$ (upper trace), in which the abscissa represents the frequencies in units of $cB_0 = 35.3$ GHz for $C_2H_2^+$ and $cB_0 = 34.4$ GHz for $^{12}C^{13}CH_2^+$. All peaks are assigned to the rotational energy differences between the rotational eigenstate $|J+2, M\rangle$ and $|J, M\rangle$, i.e., these peaks forms the S-branch of the rotational transitions. It should be noted that the spectra reflect both the rotational state distributions and the angular dependencies of the ionization probability. In the case of $C_2H_2^+$, a clear intensity alternation can be found between even Js and odd Js, which can be ascribed to the nuclear spin statistics of protons. Because the nuclear spin I of a proton is $I = 1/2$ and C_2H_2 belongs to the $D_{\infty h}$ point group, the statistical weight of the nuclear spin functions for the odd J states is three times larger than that for the even J states. Interestingly, the intensity alternation could not

Fig. 2 The Fourier transformed spectra of the observed ion signals

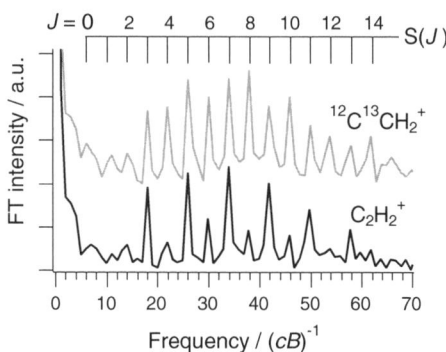

be recognized for $^{12}C^{13}CH_2^+$. This is because there is no center of symmetry in $^{12}C^{13}CH_2^+$. This FT analysis of the time-domain non-adiabatic rotational alignment signals is an efficient and universal technique to extract the information of rotational levels of molecules with high spectral resolution.

References

1. F. Rosca-Pruna and M. J. J. Vrakking, Phys. Rev. Lett., **87**, 153902 (4 pages) (2001).
2. H. Stapelfeldt and T. Seideman, Rev. Mod. Phys., **75**, 543-557 (2003).
3. T. Seideman and E. Hamilton, Adv. At. Mol. Opt. Phys., **52**, 289-329 (2005).
4. P. W. Dooley, I. V. Litvinyuk, K. F. Lee, D. M. Rayner, M. Spanner, D. M. Villeneuve, and P. B. Corkum, Phys. Rev. A, **68**, 023406 (12 pages) (2003).
5. Y. Gao, C. Wu, N. Xu, G. Zeng, H. Jiang, H. Yang, and Q. Gong, Phys. Rev. A, **77**, 043404(5 pages) (2008).

Ionization of Aligned O_2 by Intense Laser Pulse

Kotaro Sonoda, Hirokazu Hasegawa, Takahiro Sato,
Atsushi Iwasaki and Kaoru Yamanouchi

Abstract The dependencies of the yields of O_2^+ and O_2^{2+} on the angle between the polarization of the intense laser field and the molecular axis of O_2 were investigated by aligning O_2 molecules non-adiabatically by an ultrashort intense laser pulses. From the observed angular dependencies, the mechanisms of single and double ionization processes are discussed.

1 Introduction

Intense field ionization of molecules has been studied intensively in this decades, but it has been a difficult task to investigate the dependence of the ionization probability of a molecule on the angle between the molecular axis and the laser polarization direction because molecules rotate freely in space in the gas phase. This difficulty could be overcome by taking advantage of the process called the nonadiabatic molecular alignment [1].

The intense field ionization of nonadiabatically aligned diatomic molecules was investigated by the high harmonic generation (HHG) using diatomic molecules as a non-linear optical medium [2]. However, the generated high-order harmonics carries the information mostly on the angular dependence of the single ionization probabilities. On the other hand, through the angular dependence of the yield of the parent molecular ions generated by an intense laser pulse, the angular dependence of the multiple ionization can be investigated as well as that that of the single ionization. In the present study, by adopting the nonadiabatic alignment technique

K. Sonoda (✉) · H. Hasegawa
Department of Integrated Sciences, Graduate School of Arts and Sciences,
The University of Tokyo, 3-8-1 Komaba, Meguro-ku, Tokyo 153-8902, Japan
e-mail: 6017740755@mail.ecc.u-tokyo.ac.jp

T. Sato · A. Iwasaki · K. Yamanouchi
Department of Chemistry, School of Science, The University of Tokyo,
7-3-1 Hongo, Bunkyo-ku, Tokyo 113-0033, Japan

© Springer International Publishing Switzerland 2015
K. Yamanouchi et al. (eds.), *Ultrafast Phenomena XIX*,
Springer Proceedings in Physics 162, DOI 10.1007/978-3-319-13242-6_35

we investigate the angular dependence of the single and double ionization processes of O_2 by linearly polarized ultrashort intense laser pulses.

2 Experiment

A linearly polarized ultrashort laser pulse (110 fs, 800 nm, 10 Hz) generated from a Ti:sapphire laser amplifier system was split into a pair of pump and probe pulses by passing through a Michelson-type interferometer. A rotational wave packet of O_2 was created by the pump pulse, and after the temporal delay, O_2 was ionized by the probe pulse. The pump and probe pulses were focused into a Wiley-McLaren type time-of-flight mass spectrometer (TOF-MS) by a lens ($f = 200$ mm). The intensities of the pump and probe pulses were estimated to be 3×10^{13} and 1×10^{14} W/cm^2, respectively. The polarization directions of the pump and probe pulses were set to be parallel to the TOF axis. A pulsed supersonic molecular beam of a pure O_2 gas was introduced into the TOF-MS with the stagnation pressure of 3.2 atm. The generated ions at the laser focal spot were detected by a microchannel plate detector. The dependence of O_2^+ and O_2^{2+} yields on the pump-probe delay were obtained.

3 Results and Discussion

The delay dependence of the O_2^+ ion yield is shown in Fig. 1a as a function of the temporal delay τ_{delay} in units of the rotational period of O_2, $\tau_{rot} = 11.6$ ps. It is found that the sharp peaks appear at the interval of one-eighth of the rotational period, $\tau_{rot}/8 = 1.45$ ps, indicating that the ionization process is sensitively dependent on the molecular alignment with respect to the laser polarization direction.

In order to investigate the observed delay dependence of the O_2^+ ion yield, the expectation value of $\cos^2\theta$, $\langle\cos^2\theta\rangle = \langle\Psi(t)|\cos^2\theta|\Psi(t)\rangle$, was calculated by solving numerically the time-dependent Schrödinger equation, where θ is the angle between the molecular axis and the polarization direction and $|\Psi(t)\rangle$ is a rotational wave packet. The calculated $\langle\cos^2\theta\rangle$ shown in Fig. 1b has sharp peaks at the interval of one-fourth of the rotational period, $\tau_{rot}/4 = 2.90$ ps, which is not consistent with the observed interval of $\tau_{rot}/8$.

The observed ion yield should depend not only on the spatial alignment of the molecular axis of O_2 but also on the dependence of the ionization probability on the angle between the molecular axis and the laser polarization direction, $W(\theta)$. Therefore, $W(\theta)$ was calculated by the molecular Ammosov-Delone-Krainov (MO-ADK) theory [3] using the HOMO of O_2. As shown in Fig. 1c, the calculated $W(\theta)$ exhibits sharp peaks appearing at the interval of one-eighth of the rotational period, $\tau_{rot}/8 = 1.45$ ps, which is consistent with the observed ion yields. This result indicates that HOMO plays an important role in the single ionization process as has been reported previously [4].

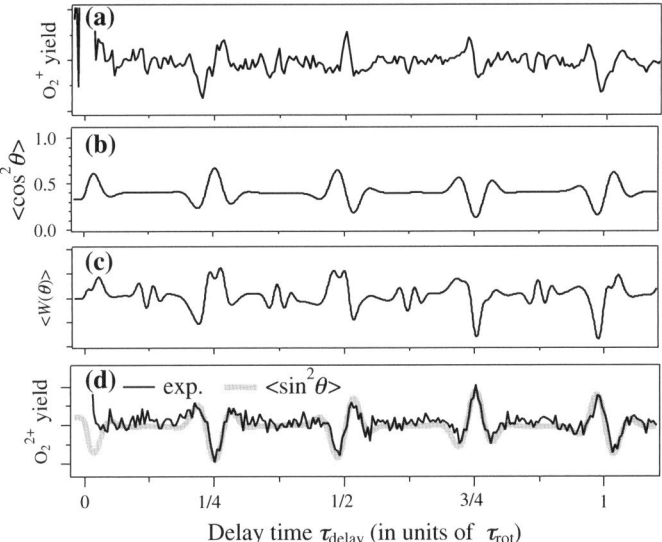

Fig. 1 The delay time dependence of **a** the observed O_2^+ yield, **b** $\langle\cos^2\theta\rangle$, **c** $\langle W(\theta)\rangle$ calculated by the MO-ADK theory with the HOMO of O_2, and **d** the O_2^{2+} yield (*thin solid curve*) and the calculated $\langle\sin^2\theta\rangle$ (*thick gray curve*) as a function of the delay time in units of $\tau_{rot} = 11.6$ ps

Contrary to the yield of O_2^+, the delay dependence of O_2^{2+} shown in Fig. 1d exhibits sharp peaks appearing at the interval of one-fourth of the rotational period, $\tau_{rot}/4$, of O_2. It is interesting to note that the peak profile is reproduced well by $\langle\sin^2\theta\rangle = 1 - \langle\cos^2\theta\rangle$, showing that O_2^{2+} is produced preferably when the laser polarization direction is perpendicular to the molecular axis. This delay dependence of the O_2^{2+} yield shows that MO-ADK calculations using only the HOMO of O_2 will not work and that the ejection of an electron or two electrons from the deeper molecular orbitals such as HOMO-1 and HOMO-2 needs to be considered in the double ionization process.

References

1. H. Stapelfeldt and T. Seideman, Rev. Mod. Phys. **75**, 543 (2003).
2. J. Itatani, D. Zeidler, J. Levesque, M. Spanner, D. M. Villeneuve, and P. B. Corkum, Phys. Rev. Lett. **94**, 123902 (2005).
3. X. M. Tong, Z. X. Zhao, and C. D. Lin, Phys. Rev. A **66**, 033402 (2002).
4. D. Pavičić, K. F. Lee, D. M. Rayner, P. B. Corkum, and D. M. Villeneuve, Phys. Rev. Lett. **98**, 243001 (2007).

Strong-Field Electronic Control of Multiple-Bond Breaking Dynamics in Ethylene

Xinhua Xie, Erik Lötstedt, Stefan Roither, Markus Schöffler, Daniil Kartashov, Kaoru Yamanouchi, Katsumi Midorikawa, Andrius Baltuška and Markus Kitzler

Abstract We experimentally and theoretically show that by controlling the dynamics of removing three electrons from ethylene with ultrashort intense laser pulses allows for the determination of the relative probabilities for fragmentation into two respectively three molecular moieties.

Manipulation of physical or chemical processes and control over their evolution, e.g., the breaking of a certain chemical bond in a polyatomic molecule, has been acknowledged as a crucial ability in both science and technology for decades. The nuclear motion involved in the breaking of a chemical bond in a molecule typically proceeds on time-scales from several femtoseconds to picoseconds. These dynamics are driven by the derivatives of the potential formed by the intra-molecular electron distribution, which can restructure on much faster, attosecond, time-scales. A suitable perturbation of the equilibrium bound electronic distribution, for example induced by ultrashort intense laser pulses, can therefore initiate nuclear motion towards a desired bond-breaking event. A relatively slow molecular fragmentation can thus be pre-determined on the much faster electronic time-scale. This has been demonstrated using the carrier-envelope offset phase of few-cycle laser pulses as the control parameter, e.g. in [1, 2], and by selective removal of electrons from either inner or outer valence orbitals based on their different shapes [3] or their different sensitivity to laser intensity and/or pulse duration [4].

So far, all successful demonstrations of electronically pre-determining molecular fragmentation with ultrashort intense laser fields considered only fragmentation

X. Xie · S. Roither · M. Schöffler · D. Kartashov · A. Baltuška · M. Kitzler (✉)
Photonics Institute, Vienna University of Technology,
Gusshausstrasse 27, 1040 Vienna, Austria
e-mail: markus.kitzler@tuwien.ac.at

E. Lötstedt · K. Midorikawa
Laser Technology Laboratory, RIKEN, 2-1 Hirosawa,
Wako, Saitama 351-0198, Japan

E. Lötstedt · K. Yamanouchi
Department of Chemistry, School of Science, the University of Tokyo,
Tokyo 113-0033, Japan

© Springer International Publishing Switzerland 2015
K. Yamanouchi et al. (eds.), *Ultrafast Phenomena XIX*,
Springer Proceedings in Physics 162, DOI 10.1007/978-3-319-13242-6_36

150

reactions, where the molecule is split into merely two moieties—even in these experiments that explored fragmentation control of polyatomic molecules. In this contribution we consider for the first time, to our knowledge, electronic pre-determination of molecular fragmentation into three moieties. Although this seems to be a minor enhancement by just one more moiety, the implications of this extension are vast and a number of different processes become possible that are not present for two-body fragmentations. First, a three-body fragmentation, which can only occur in a polyatomic molecules, takes place along (at least) two nuclear coordinates. Thus, the fragmentation dynamics must be necessarily described by molecular potential energy surfaces rather than only one-dimensional potential energy curves. Secondly, if a polyatomic molecule fragments into three moieties, the sequence and timing of the two involved fragmentation steps become important [5]: The two fragmentation steps can occur concertedly (simultaneous breaking of two bonds) or sequentially (one after another). For a sequential fragmentation dynamics it becomes additionally important which one of the two involved bonds breaks first.

We performed experiments in which we measured the three-dimensional momentum vectors of fragment ions from field ionized ethylene molecules in coincidence using a reaction microscope [3]. The laser beam, linearly polarized along the spectrometer axis and pulse durations ranging from 4.5 to 25 fs (FWHM) at a center wavelength of ~ 750 nm, was focused in an ultrahigh vacuum chamber ($\sim 1.3 \times 10^{-10}$ mbar) by a spherical Ag mirror with 60 mm focal length onto a super-sonic gas jet of ethylene.

Using momentum conservation in all three dimensions we identified from the measured data two two-body fragmentation and two three-body fragmentation channels from triply charged ethylene. Based on momentum correlation maps we also succeed in disentangling the fragmentation dynamics of the two three-body fragmentation channels, i.e., whether the three fragments are produced sequentially or concertedly. (1–6) summarize these results.

$$C_2H_4^{3+} \rightarrow C_2H_3^{2+} + H^+, \tag{1}$$

$$C_2H_4^{3+} \rightarrow CH_2^{2+} + CH_2^+, \tag{2}$$

$$C_2H_4^{3+} \rightarrow C_2H_2^+ + H^+ + H^+, \tag{3}$$

$$C_2H_4^{3+} \rightarrow C_2H_3^{2+} + H^+ \rightarrow C_2H_2^+ + H^+ + H^+, \tag{4}$$

$$C_2H_4^{3+} \rightarrow C_2H_3^{2+} + H^+ \rightarrow CH_2^+ + CH^+ + H^+, \tag{5}$$

$$C_2H_4^{3+} \rightarrow CH_2^{2+} + CH_2^+ \rightarrow CH_2^+ + CH^+ + H^+. \tag{6}$$

The information contained in (1–6) can be visualized in a three-dimensional coordinate system with the C–H and C–C stretching coordinates as the axes and the equilibrium configuration of the neutral ethylene molecule at the origin, see Fig. 1a. In this visualization the two-body fragmentation channels [channels (1) and (2)],

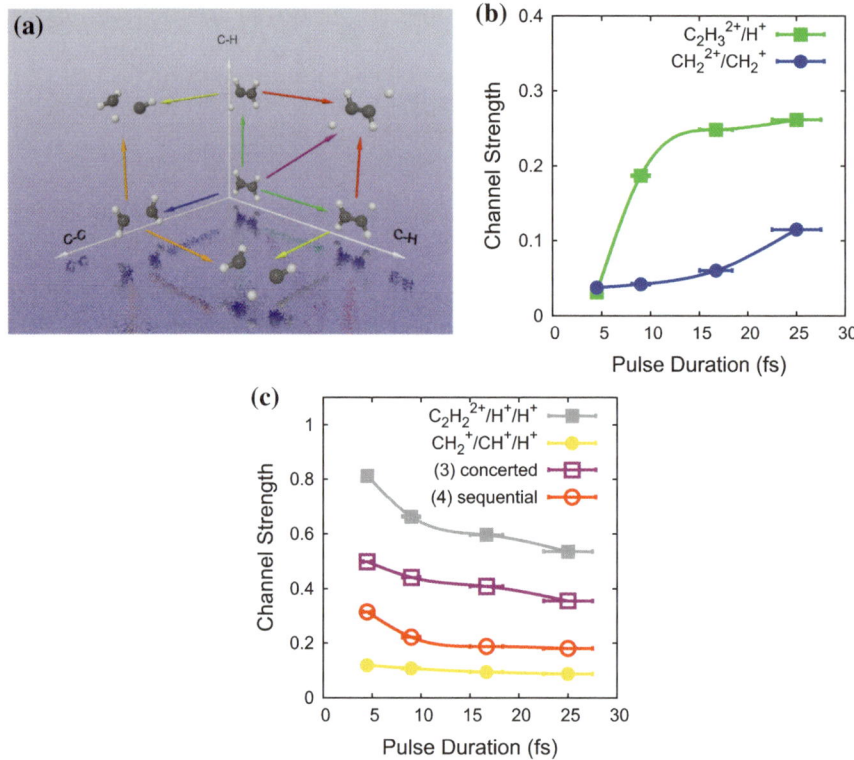

Fig. 1 a Schematic view of the fragmentation dynamics of all experimentally identified fragmentation reactions of triply ionized ethylene along the C–H and C–C stretching coordinates. **b** Relative yield of two-body fragmentation channels as a function of laser pulse duration. **c** Relative yield of three-body fragmentation channels as a function of laser pulse duration. Color coding in **b** and **c** is the same as that in **a**

where only one bond is broken, are depicted along the C–C and C–H stretching coordinates, respectively, and the three-body fragmentation channels, where two bonds are broken, appear in the planes of the coordinate system. A sequential three-body fragmentation dynamics proceeds along two different axes [channels (4), (5) and (6)], whereas the concerted three-body fragmentation dynamics of channel (3) proceeds along a diagonal, see Fig. 1a.

In Fig. 1b we show the measured normalized yield (channel strength) of the two-body fragmentation channels as a function of laser pulse duration. As can be seen, the channel strength dramatically and monotonically increases with the laser pulse duration. The yield of channel (1) increases by a factor of roughly 9 from 0.03 up to 0.26 when the laser pulse duration increases from 4.5 fs up to 25 fs. The fitting curve shows that the increasing slope is very steep between pulse durations 4.5 fs and about 12 fs, and the slope becomes small when the pulse duration is larger than 12 fs. This time scale of 12 fs is very close to the C–H vibrational period of 11 fs [7].

The yield of channel (2) increases by a factor of almost 3 from 0.04 up to 0.12 when the laser pulse duration increases from 4.5 fs up to 25 fs, and the trend shows that it will further increase for still longer pulse durations. On the other hand, as presented in Fig. 1c, the three-body fragmentation channels behave differently: The channel strength of all three-body fragmentation decreases with increasing laser pulse duration.

These results are somewhat counter-intuitive, as one would expect that for longer pulses more molecules can reach higher lying electronic states, which would lead to more three-body fragmentation products. However, our experiments show that with longer pulses more two-body fragmentations take place. With the help of quantum chemical simulations we can explain this somewhat counter-intuitive behaviour exemplarily for the C–H bond-breaking dynamics of channels (1), (3) and (4) as follows. Assuming that the three-body fragmentation reactions take place on the ground state of the trication, which we can rationalize by the measured distributions of the kinetic energy release and the angular distributions of the detected fragments, we can show that the dynamics of the removal of three electrons from the neutral determines, whether the molecule preferentially fragments into three or only two particles. The ground state potential energy surface (PES) of the ethylene trication is dissociative along the diagonal of the two C–H stretch coordinates, i.e., towards breakage of both C–H bonds, corresponding to channel (3). It is also dissociative along either C–H bond stretch coordinate, corresponding to the two-body fragmentation channel (1). Importantly, however, there exists a local minimum along the diagonal of the two C–H stretch coordinates, which needs to be overcome in order that fragmentation can take place along this direction. If now the formation of the trication by removing three electrons from the neutral takes place within only a very short duration, as it is the case for 4.5 fs pulses, the tricationic ground state PES is populated high up in energy close to C–H equilibrium geometry of the neutral, such that the local minimum can be overcome and concerted three-body fragmentation comprising the stretch of both C–H bonds can take place. For longer pulse durations, on the other hand, substantial C–H stretch can already take place during the process of electron removal on a dicationic PES towards the dicationic equilibrium geometry. As a consequence, for pulses longer than roughly 10 fs, the tricationic PES is populated in the region of the local minimum and three-body fragmentation becomes less likely, while the stretch motion of only one of the two C–H bonds, i.e. two-body fragmentation, becomes the dominant channel. Thus, a trapping of the nuclear dynamics on the ground state tricationic PES that depends on the dynamics of electron removal is behind the on first glance counter-intuitive observations in Fig. 1b, c.

In conclusion, we experimentally and theoretically investigated whether the three-body fragmentation dynamics of the ethylene trication can be determined by a suitable perturbation of the bound electron distribution using an intense ultrashort laser field. We found that by controlling the dynamics of electron removal using the laser pulse duration as a control parameter allows influencing the relative probability between fragmentation into two respectively three molecular moieties.

Acknowledgments We acknowledge financial support by the Austrian Science Fund (FWF) under grants No. P21463-N22, No. P25615-N27 and No. SFB-F49 NEXTlite, and by a starting grant from the ERC (project CyFi).

References

1. Kling *et al.* (Sub-)femtosecond control of molecular reactions via tailoring the electric field of light. *Physical chemistry chemical physics : PCCP* **15**, 9448 (2013).
2. X. Xie *et al.* Attosecond-Recollision-Controlled Selective Fragmentation of Polyatomic Molecules, Physical Review Letters **109**, 243001 (2012.
3. X. Xie, K. Doblhoff-Dier, H. Xu, S. Roither, M. S. Schöffler, D. Kartashov, S. Erattupuzha, T. Rathje, G. G. Paulus, K. Yamanouchi, et al., "Selective Control over Fragmentation Reactions in Polyatomic Molecules Using Impulsive Laser Alignment," Physical Review Letters **112**, 163003 (2014 a).
4. X. Xie, S. Roither, M. Schöffler, E. Lötstedt, D. Kartashov, L. Zhang, G. G. Paulus, A. Iwasaki, A. Baltuška, K. Yamanouchi, et al., "Electronic Predetermination of Ethylene Fragmentation Dynamics," Physical Review X **4**, 021005 (2014 b).
5. L. Zhang, S. Roither, X. Xie, D. Kartashov, M. Schöffler, H. Xu, A. Iwasaki, S. Gräfe, T. Okino, K. Yamanouchi, et al., "Path-selective investigation of intense laser-pulse-induced fragmentation dynamics in triply charged 1,3-butadiene," Journal of Physics B: Atomic, Molecular and Optical Physics **45**, 085603 (2012).
6. R. Dörner *et al.* Cold Target Recoil Ion Momentum Spectroscopy: a 'momentum microscope' to view atomic collision dynamics *Physics Reports* **330**, 95 (2000).
7. T. Shimanouchi, "Tables of molecular vibrational frequencies. Consolidated volume II," Journal of Physical and Chemical Reference Data **6**, 993 (1977).

Electronic Pre-determination of Ethylene Fragmentation Dynamics

Markus Kitzler, Xinhua Xie, Stefan Roither, Erik Lötstedt,
Markus Schöffler, Daniil Kartashov, Gerhard G. Paulus,
Atsushi Iwasaki, Andrius Baltuška and Kaoru Yamanouchi

Abstract We demonstrate, using ethylene, that controlling lower-valence ionization and field-driven excitation dynamics with ultrashort, intense laser pulses allows steering fragmentation reactions of polyatomic molecules along a certain pathway towards a specific set of fragment ions.

Strong laser fields can be used to both initiate and drive electronic dynamics in a polyatomic molecule on their natural, i.e., sub-femtosecond, time-scale. This potentially opens up the feasibility of controlling molecular fragmentation processes by influencing the intra-molecular dynamics of the electron cloud with strong laser-electric fields. Indeed, it was shown that the outcome of fragmentation processes in polyatomic molecules can be affected by the shape of the laser electric field [1]. In general, the key to controlling fragmentation and accompanying isomerization processes in molecular ions is to prepare them in specific dissociative states from which the reaction proceeds through a desired fragmentation *pathway* on multi-dimensional potential energy surfaces towards a certain set of final fragment products, called a *channel*. This can be achieved either during the ionization process, but also by subsequent excitation processes during the interaction with the laser field.

Here, we report experiments on the polyatomic molecule ethylene, C_2H_4, dedicated to disentangling the contributions of the ionization step and the subsequent field-driven excitation dynamics to different fragmentation pathways from the

M. Kitzler (✉) · X. Xie · S. Roither · M. Schöffler · D. Kartashov · A. Baltuška
Photonics Institute, Vienna University of Technology, Gusshausstrasse 27,
1040 Vienna, Austria
e-mail: markus.kitzler@tuwien.ac.at

E. Lötstedt · A. Iwasaki · K. Yamanouchi
Department of Chemistry, School of Science, The University of Tokyo,
Tokyo 113-0033, Japan

G.G. Paulus
Institute of Optics and Quantum Electronics, Friedrich-Schiller-University Jena,
07743 Jena, Germany

G.G. Paulus
Helmholtz Institute Jena, 07743 Jena, Germany

© Springer International Publishing Switzerland 2015   155
K. Yamanouchi et al. (eds.), *Ultrafast Phenomena XIX*,
Springer Proceedings in Physics 162, DOI 10.1007/978-3-319-13242-6_37

doubly charged ion. In particular, we show how these two excitation mechanisms that pre-determine the fragmentation reactions of the molecule, depend on laser pulse intensity and duration.

We observe that the relative importance of contributions to the fragmentation probability of a given *channel*, either from the ionization step or from field-driven excitations, strongly depends on both laser intensity and pulse duration. Moreover, we show that not only the probability of a given channel, but even the specific *pathways* that can be taken along the multitude of dissociative electronic states towards this channel, are dependent on the laser pulse parameters. Thus, by properly choosing pulse intensity and duration it becomes possible to steer the molecular dynamics along a desired pathway in the phase-space spanned by the nuclear coordinates and momenta towards a certain set of final fragment ions. This opens up new possibilities for controlling the outcome of fragmentation reactions of poly-atomic molecules in that it may allow to selectively enhance or suppress individual fragmentation channels, which was not possible in previous attempts of controlling the fragmentation behaviour of polyatomic molecules [1].

In our experiments, we measured in coincidence the three-dimensional momentum vectors of fragment ions resulting from the interaction of sub-5 and 25 fs (FWHM) laser pulses with an ethylene molecule using a COLTRIMS setup. By coincidence analysis of the measured data we identified for all pulse parameters the following two-body fragmentation channels of the ethylene dication:

$$C_2H_4^{2+} \rightarrow CH_2^+ + CH_2^+, \tag{1}$$

$$C_2H_4^{2+} \rightarrow C_2H_3^+ + H^+. \tag{2}$$

We observe that the angular distribution of fragment ions from channel (1) is anisotropic and peaks along the laser polarization direction, whereas that of channel (2) is isotropic, see Fig. 1. According to MO-ADK theory, the ionization rate from a certain orbital strongly depends on the relative orientation of the molecular orbital to the laser polarization direction and follows the shape of the electron density distribution of the respective orbital [2]. As a consequence, the anisotropic angular momentum distribution of channel (1) indicates, firstly, that the fragmentation process takes place from an excited ionic state of the ethylene dication formed by ionization of at least one HOMO-2 electron. Secondly, the strong anisotropy indicates that the fragmentation happens fast as compared to the rotational period of $C_2H_4^{2+}$ [3]. The isotropic angular momentum distribution of channel (2), in contrary, suggests molecular rearrangement after ionization in the course of the separation of two fragments, which takes much longer than the laser pulse duration. The fragmentation dynamics of channels (1) and (2), thus, take place apparently on different time-scales.

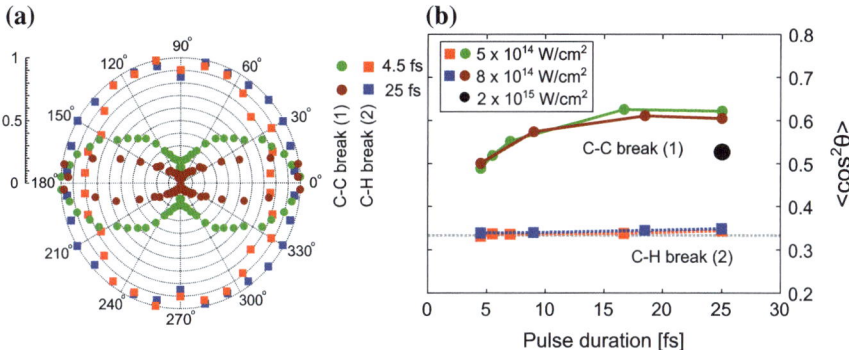

Fig. 1 a Normalized angular momentum distributions for channels (1) and (2). Polarization direction horizontal. **b** $\langle \cos^2 \theta \rangle$ over pulse duration for channels (1) and (2)

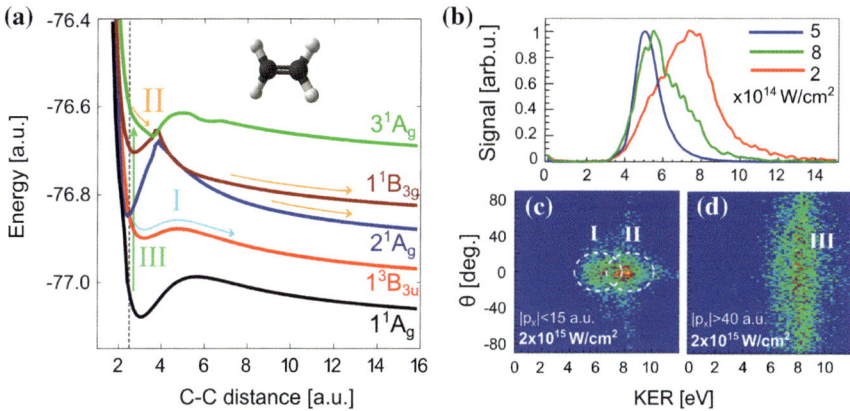

Fig. 2 a Calculated PESs of ground (1^1A_g, *black line*) and selected excited ionic states of the ethylene dication. The $^3B_{3u}$ state is formed by removing one HOMO electron and one HOMO-2 electron, the $^1B_{3g}$ by removing one HOMO-1 electron and one HOMO-2 electron, and the 2^1A_g and 3^1A_g states are formed by removing two HOMO-1 electrons and two HOMO-2 electrons, respectively. The vertical *black dashed line* indicates the C–C equilibrium distance of 2.53 a.u. The *coloured arrows* (labelled by I, II and III) refer to the three fragmentation pathways of channel (1), as discussed in the text. **b** Normalized KER spectra of fragmentation channel (1) for three different laser intensities as indicated. **c, d** Selections of the angular distributions over KER of CH_2^+ fragments from channel (1) measured at the indicated laser intensities for $|p_x| < 15$ a.u. (**c**) and $|p_x| > 40$ a.u. (**d**)

We can qualitatively explain our experimental results using the potential energy surfaces (PESs) of excited ionic states of the ethylene dication (Fig. 2a) and the kinetic energy release (KER) spectra for different laser peak intensities (Fig. 2b).

It can be clearly seen that with increasing laser intensity the mean value of the KER of channel (1) increases from around 5 eV up to 7 eV. The dependence of the KER of channel (1) on laser intensity indicates that the fragmentation of the C–C bond takes place via different pathways for different laser pulse parameters. As discussed above, this may involve the population of different initial dissociative (excited) states by ionization from different lower-valence orbitals, and/or field-induced excitations to other excited states by the action of the laser pulse. Three reaction pathways towards fragmentation into channel (1) (marked I, II and III in Fig. 2) are possible: Removal of one HOMO-2 electron and one HOMO electron prepares the dication in the $1B_{3u}$ state, dissociative along C–C direction (pathway I, marked by a cyan arrow in Fig. 2a). The KER of this pathway is estimated from the theoretical potential energy curves as about 5 eV, in very good agreement with the measured KER of region I. Removal of two electrons from the HOMO-2 can put the ethylene dication onto the 3^1A_g state. From there it will dissociate through crossings with the 2^1A_g and $^1B_{3g}$ states (Fig. 2a). As the initial state is populated by ionization from the HOMO-2, the angular distribution of the fragment ions will still be narrow along the laser polarization direction. However, fragmentations along this pathway II (orange arrows in Fig. 2a) will lead to a higher KER. Theory predicts values of 8 and 9 eV, respectively, for dissociation along these states, which fits the measured KER energy range for recorded events in region II very well (Fig. 2c). The fragments in region III (Fig. 2d), finally, show a KER distribution located around 8 eV, but their angular distribution is isotropic, indicating that the fragmentation starts from an excited ionic state formed by removal of electrons from the HOMO or HOMO-1. This would prepare the ion into the meta-stable ground state or a low excited state, which will not lead to breaking of the C–C center bond. Thus, the pathway leading to the fragment ions in region III must involve a field-induced excitation to the higher excited state 3^1A_g induced by the laser pulse. From there it dissociates along the same PESs as the pathway II leading to fragments with almost the same high KER as pathway II.

In conclusion, we demonstrated that the relative importance of the different molecular pathways along different dissociative electronically excited states, by which a particular set of final fragmentation products can be reached, strongly depends on the parameters of the laser pulse. Our work shows that selective population of excited ionic states by controlling intra-molecular electronic processes (in particular electron removal from lower-valence orbitals and non-adiabatic population transfer) with strong non-resonant laser fields is an efficient and general method for selectively enhancing or suppressing individual fragmentation channels.

Acknowledgments We acknowledge financial support by the Austrian Science Fund (FWF) under grants No. P21463-N22, No. P25615-N27 and No. SFB-F49 NEXTlite, and by a starting grant from the ERC (project CyFi).

References

1. Xinhua Xie *et al.* Attosecond-Recollision-Controlled Selective Fragmentation of Polyatomic Molecules. *Phys. Rev. Lett.* **109**, 243001 (2012).
2. X.M. Tong, Z. Zhao, and C. Lin. Theory of molecular tunneling ionization. *Phys. Rev. A* **66**, 033402 (2002).
3. A. Hishikawa, H. Hasegawa, and K. Yamanouchi. Hydrogen migration in acetonitrile in intense laser fields in competition with two-body Coulomb explosion. *J. El. Spec. Rel. Phen.* **141**, 195 (2004).

Long-Lived Neutral H_2 in Hydrogen Migration Within Hydrocarbon Dication

Katsunori Nakai and Kaoru Yamanouchi

Abstract First principles molecular dynamics calculations of energized $CH_3NH_2^{2+}$ and $CH_3CH_3^{2+}$ are performed to examine whether the formation of a long-lived neutral H_2 moiety identified in CH_3OH in our previous study [Nakai et al., *J. Chem. Phys.* **139**, 181103 (2013)] is a phenomenon commonly identified in energized hydrocarbon dications. A neutral H_2 moiety generated within a doubly charged parent ion is considered to play a crucial role in the hydrogen scrambling process in $CH_3CH_3^{2+}$, leading eventually to the ejection of H_3^+.

1 Introduction

When hydrocarbon molecules are exposed to an intense laser field whose field intensity exceeds $\sim 10^{13}$ W/cm^2, a variety of dynamical processes are induced such as multiple ionization, chemical bond rearrangement, and Coulomb explosion. In the series of our experimental studies on hydrocarbon molecules such as methanol [1] and ethane [2], we showed that the precursor hydrocarbon dications from which H_3^+ is ejected eventually have long lifetimes comparable with their rotational periods ($1 \sim 10$ ps) on the basis of the observed isotropic angular distributions of H_3^+. Furthermore, the yield ratio of the four kinds of isotopomers of triatomic hydrogen molecular ions generated from partially deuterated ethane dications $CH_3CD_3^{2+}$ was $H_3^+ : H_2D^+ : HD_2^+ : D_3^+ = 8 : 43 : 43 : 5.9$ [3], indicating that the lifetime of the precursor parent dication, $CH_3CD_3^{2+}$, is sufficiently long so that the three hydrogen atoms and the three deuterium atoms are scrambled almost statistically prior to a triatomic hydrogen molecular ion is ejected.

K. Nakai · K. Yamanouchi (✉)
Department of Chemistry, School of Science, The University of Tokyo,
Bunkyo-ku, Tokyo, Japan
e-mail: kaoru@chem.s.u-tokyo.ac.jp

© Springer International Publishing Switzerland 2015   160
K. Yamanouchi et al. (eds.), *Ultrafast Phenomena XIX*,
Springer Proceedings in Physics 162, DOI 10.1007/978-3-319-13242-6_38

In our recent theoretical study [4], we performed first-principles molecular dynamics trajectory calculations of energized CH_3OH^{2+} and found that the formation of a long-lived neutral H_2 moiety within CH_3OH^{2+} can be the origin of the long lifetime of CH_3OH^{2+}, from which H_3^+ is ejected eventually. In the present study, we performed the first-principles molecular dynamics trajectory calculations of the two energized dication species, $CH_3NH_2^{2+}$ and $CH_3CH_3^{2+}$, and examine whether a long-lived neutral H_2 moiety can also be formed within these dication species.

2 Results and Discussion

The relative yields of the decomposition pathways of $CH_3NH_2^{2+}$ and $CH_3CH_3^{2+}$ obtained by the present theoretical calculations are summarized in Table 1.

Snapshots of one of the trajectories of $CH_3CH_3^{2+}$ from which H_3^+ and $C_2H_3^+$ are eventually produced are shown in Fig. 1. First, two C–H chemical bonds in one of the two methyl groups are broken and a new H–H chemical bond is formed. As the next step, the resultant neutral H_2 moiety picks up an electron deficient H atom to form a triatomic H_3^+ moiety. It is interesting to note that the hydrogen migration proceeds within the rest of the molecule before the H_2 moiety picks up the electron deficient H atom.

The time evolutions of $|\vec{R}_1|$, the distance between the two H atoms forming the neutral H_2 within the dication, and $|\vec{R}_2|$, the distance between the center of the two H atoms and the center of the two C atoms, of one trajectory are shown in Fig. 2a. In the time range between 30 and 170 fs, $|\vec{R}_1|$ ($\sim 0.75\text{Å}$) oscillates at the frequency of around 1.25×10^{14} Hz, corresponding to the vibrational wavenumber of

Table 1 Relative yields (%) of the decomposition pathways of $CH_3NH_2^{2+}$, $CH_3CH_3^{2+}$, and CD_3OH^{2+} [4]

	$CH_3NH_2^{2+}$		$CH_3CH_3^{2+}$		CD_3OH^{2+} [4]	
No dissociation or hydrogen migration						
	$CH_3NH_2^{2+}$	26.7	$CH_3CH_3^{2+}$	4.7	CD_3OH^{2+}	0.5
Hydrogen migration						
	$CH_2NH_3^{2+}$	23.3	$CH_4CH_2^{2+}$	27.1	CD_2OHD^{2+}	0.2
Dissociation processes						
			$CH_3^+ + CH_3^+$	3.5		
	$H^+ + CH_4N^+$	10.8	$H^+ + C_2H_5^+$	3.5	$D^+ + CHD_2O^+$	67.6
					$H^+ + CD_3O^+$	0.4
	$H_2 + CH_3N^{2+}$	22.5	$H_2 + C_2H_4^{2+}$	52.9	$D_2 + CHDO^+$	11.9
	$H_3^+ + CH_2N^+$	16.7	$H_3^+ + C_2H_3^+$	8.2	$D_3^+ + COH^+$	13.7
					$HD_2^+ + CDO^+$	1.3

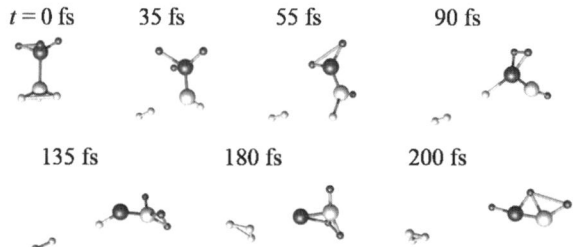

Fig. 1 Snapshots of $C_2H_6^{2+} \rightarrow H_3^+ + C_2H_3^+$. All the four atoms in the one of the two methyl groups are represented by *black balls*, while the four atoms in the other methyl group are represented by *white balls* for clarification

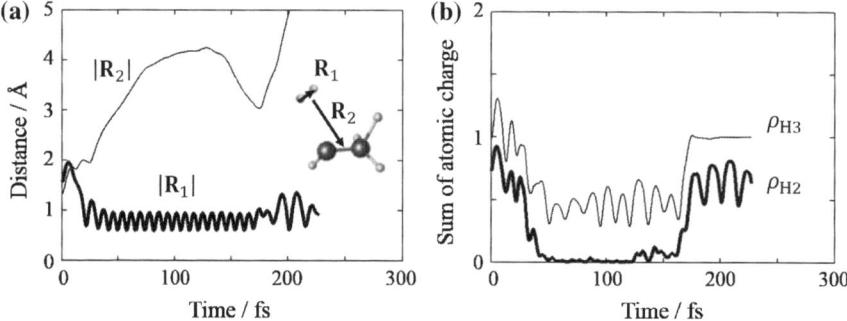

Fig. 2 a Time evolution of $|\vec{R}_1|$ and $|\vec{R}_2|$ for $CH_3CH_3^{2+} \rightarrow H_3^+ + CH_2CH^+$. The definition of the two vectors, $\vec{R}_1$ and $\vec{R}_2$ are shown in the inset. **b** Time evolution of (i) the sum of the total charge ρ_{H2} of the two H atoms forming a neutral H_2 moiety and (ii) the total charge of the three H atoms forming H_3^+ as a final product

4,175 cm^{-1}, and $|\vec{R}_2|$ increases to as large as 4.2 Å. This time evolutions of $|\vec{R}_1|$ and $|\vec{R}_2|$ show that the H_2 moiety exists for the relatively long period of time (~ 170 fs), during which it rotates and vibrates seemingly independently from the rest of the molecule.

As shown in Fig. 2b, the sum of the charge of the two H atoms ρ_{H2} is very close to zero during the period when the isolated H_2 moiety survives. Then, at $t = 180$ fs, the H_2 moiety picks up the H atom in the CH group to form an isolated H_3^+ having the net charge of $+1.0$.

The previous results we obtained for CH_3OH^{2+}[4] and those obtained in the present study for $CH_3NH_2^{2+}$ and $CH_3CH_3^{2+}$ suggest that the formation of a long-lived H_2 moiety within a doubly charged hydrocarbon molecule is a general phenomenon commonly seen in doubly charged energized hydrocarbon molecules.

References

1. T. Okino, Y. Furukawa, P. Liu, T. Ichikawa, R. Itakura, K. Hoshina, K. Yamanouchi, H. Nakano, Chem. Phys. Lett. **423**, 220-224 (2006).
2. K. Hoshina, Y. Furukawa, T. Okino, K. Yamanouchi, J. Chem. Phys. **129**, 104302 (2008).
3. R. Kanya, T. Kudou, N. Schirmel, S. Miura, K.-M. Weitzel, K. Hoshina, K. Yamanouchi, J. Chem. Phys. **136**, 204309 (2012).
4. K. Nakai, T. Kato, H. Kono, K. Yamanouchi, J. Chem. Phys. **139**, 181103 (2013).

Pump-Probe Photoelectron Imaging with 90-nm Excitation Pulses

Shunsuke Adachi, Motoki Sato, Yoshi-ichi Suzuki and Toshinori Suzuki

Abstract Pump-probe photoelectron imaging was performed with 90-nm excitation pulses. Quantum beat by coherent excitation of multiple Rydberg states in Kr, and photodissociation of CO_2 within a few ps from initially excited Rydberg state(s) were observed.

1 Introduction

The photoelectron ejection angle is an important observable in photoelectron spectroscopy. Time-resolved photoelectron imaging (TRPEI) provides valuable information on the electronic dynamics in photoexcited molecules. We have recently developed a laser system that generates 90-nm laser pulses through a third-harmonic generation process of deep UV driving pulses (270 nm, Ti:Sa third harmonic) [1]. This powerful (sub-mW) vacuum-UV laser source is ideal for TRPEI, which essentially requires numerous signal events to reconstruct reliable three-dimensional (3D) velocity distributions from observed 2D images. In the present report, pump-probe photoelectron imaging experiments for Kr and CO_2 samples are demonstrated, where the 90-nm pulses excite their Rydberg states that converge on their first ionic states.

S. Adachi (✉) · M. Sato · Y.-i. Suzuki · T. Suzuki
Department of Chemistry, Graduate School of Science, Kyoto University,
Kyoto 606-8502, Japan
e-mail: adachi@kuchem.kyoto-u.ac.jp

S. Adachi · T. Suzuki
RIKEN Center for Advanced Photonics, RIKEN, Wako 351-0198, Japan

© Springer International Publishing Switzerland 2015
K. Yamanouchi et al. (eds.), *Ultrafast Phenomena XIX*,
Springer Proceedings in Physics 162, DOI 10.1007/978-3-319-13242-6_39

2 Experiments and Results

The detail for the 90-nm laser source was discussed in the previous report [1]. The *on-target* pulse energy is 0.2 μJ at 1-kHz laser repetition rate. The wavelength of the 90-nm pulse was set to 91.2 nm (13.6 eV), which is slightly below the first ionic states of Kr (14.0 eV) and CO_2 (13.8 eV). The 90-nm pump pulse and another time-delayed 270-nm probe (i.e. ionization) pulse are focused onto a supersonic atomic (Kr) or molecular (CO_2) beam. The cross-correlation time of these pulses at the sample point was measured to be 80 fs by the nonresonant $(1 + 1')$ multiphoton ionization of Ar. The photoelectrons generated by the pump-probe photoionization are projected onto a 2D position-sensitive detector, and 3D photoelectron velocity distributions are reconstructed from the images using pBaseX method. Photoelectron kinetic energy and angular distribution in photoionization is expressed as follows:

$$I(E, \theta, t) = \frac{\sigma(E, t)}{4\pi} \{1 + \beta_2(E, t)P_2(\cos \theta) + \beta_4(E, t)P_4(\cos \theta)\}$$

where E, θ and t are photoelectron kinetic energy, electron ejection angle from laser polarization direction, and pump-probe time delay. $\sigma(E, t)$ and $\beta_n(E, t)$ represent photoelectron kinetic energy distribution and anisotropy parameters, respectively.

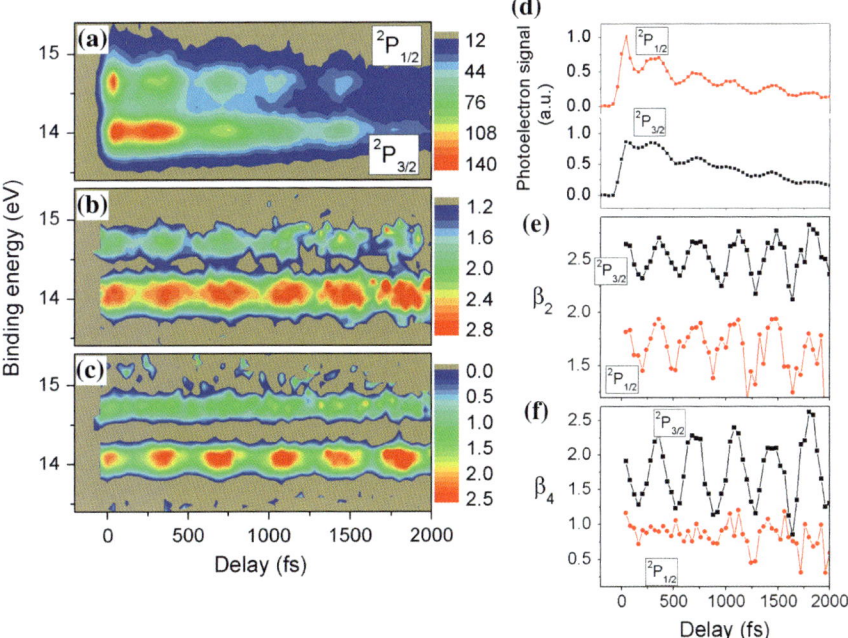

Fig. 1 Time-energy maps of **a** σ, **b** β_2, and **c** β_4 for Kr. **d–f** represent temporal profiles of σ, β_2, and β_4 for each final ionic state

Figure 1a–c depict time-energy maps of σ, β_2, and β_4 for Kr, and (d)–(f) represent temporal (i.e. spectrally-integrated) profiles of σ, β_2, and β_4 for each final ionic state ($^2P_{3/2}$ and $^2P_{1/2}$). The signal rise time (~ 80 fs) shown in Fig. 1d agrees well with the above-mentioned cross-correlation time. Since the 90-nm pulses with ~ 800 cm^{-1} spectral bandwidth excite two Kr Rydberg states ($4s^24p^5(^2P_{3/2})9s$ $[3/2]_1$ and $4s^24p^5(^2P_{3/2})7d[3/2]_1$) simultaneously, σ and β_n are temporally modulated with quantum beat frequency corresponding to the energy difference between the two Rydberg states (91 cm^{-1}; 370 fs). The initial phases of the β_2, and β_4 modulations are related to the total phase shift difference $\Delta\delta$ between f- and p-electrons. We have determined $\Delta\delta$ values for $^2P_{3/2}$ and $^2P_{1/2}$ channels to be -3.2 and -3.1 rad., respectively, both of which agree well with theoretically expected values.

Dissociation of CO_2 by solar vacuum-UV radiation is a fundamental photochemical process in planetary and environmental science. Furthermore, extreme-UV-induced molecular processes around/beyond their ionization energies are one of the research foci of attosecond electron dynamics. It is predicted that the manifold of coupled low-lying valence states acts as a "sink" for the optically bright Rydberg states and affects their dissociation lifetimes [2]. Figure 2a–c show time-energy maps of σ, β_2, and β_4 for CO_2, respectively. A very broad photoelectron spectrum (See Fig. 2d.) indicates there is large geometrical difference between the excited state(s) and ionic state. The ionic state (and Rydberg states converging on it) as well

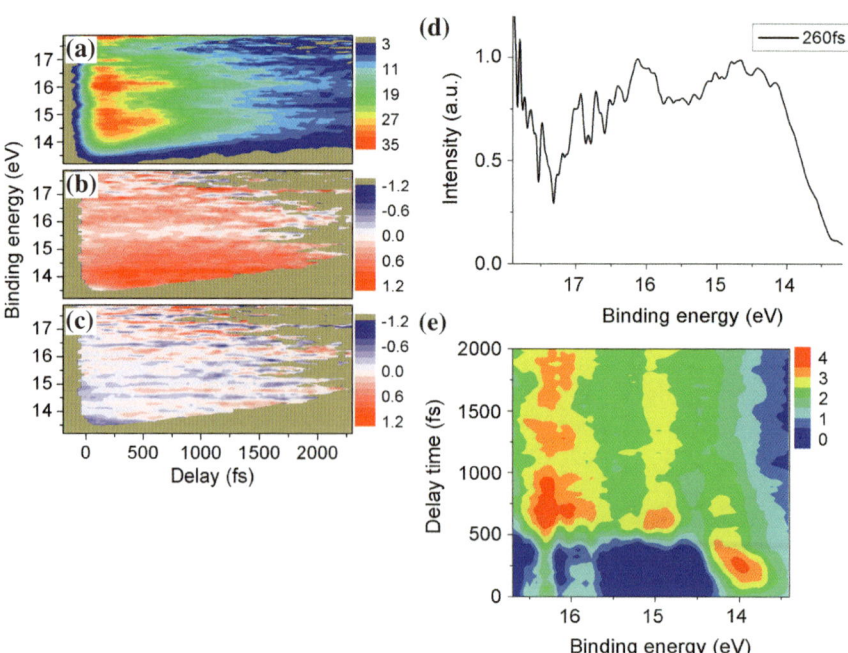

Fig. 2 Time-energy maps of **a** σ, **b** β_2, and **c** β_4 for CO_2. **d** A slice of σ map at 260 fs. **e** Minor contribution in σ map

as the ground state seems to have linear geometry [3], meanwhile the equilibrium geometries for the dissociative excited states are strongly bent according to an ion velocity-map imaging experiment [4]. Thus the observed very broad photoelectron spectrum represents photoionization from dissociative valence-like (pure or mixed valence) excited states. The dissociation time constant is determined to be $\tau = 700$ fs by global fit analysis, where no *spectral* change has been found. The global fit analysis also revealed that there is minor (probably Rydberg) contribution (Fig. 2e), where a short-lived vibrationally-cool peak appears at 13.8 eV, and it shifts and spreads toward higher binding energy in initial 500 fs. This is probably attributed to the crossing from an initially excited Rydgerg state to a valence state.

3 Conclusions

We have conducted time-resolved photoelectron imaging experiments in an unexplored spectral region. Quantum beat by coherent excitation of multiple Rydberg states in Kr, and photodissociation of CO_2 within a few ps from initially excited Rydberg state(s) were observed.

References

1. S. Adachi, T. Horio, and T. Suzuki, *Opt. Lett.* **37**, 2118 (2012)
2. S. Grebenschikov, *J.Chem. Phys.* **138**, 224106 (2013)
3. I. Reineck *et al.*, *Chem. Phys.* **78**, 311 (1983)
4. R. L. Miller *et al.*, *J.Chem. Phys.* **96**, 332 (1991)

Time-Resolved Coulomb Explosion Imaging of Ultrafast Fragmentation of CS_2 in Highly Charged States

Akitaka Matsuda, Eiji J. Takahashi and Akiyoshi Hishikawa

Abstract Time-resolved three-body Coulomb explosion imaging of dissociating CS_2 in few-cycle intense laser fields (9 fs, 4×10^{15} W/cm^2) is performed. It is revealed that the ultrafast fragmentation dynamics of CS_2 in highly charged states proceed in a different timescale depending on the charge state.

1 Introduction

Recent advances in laser technology enabled us to utilize few-cycle intense laser fields (<10 fs, $\sim 10^{15}$ W/cm^2) as a new probe for visualizing ultrafast structural deformation of molecules in course of chemical reaction. Molecules irradiated with few-cycle intense laser pulses promptly eject several electrons to form multiply charged ions, which subsequently undergo a rapid bond breaking process by strong Coulombic repulsion between the constituent atoms. The "Coulomb explosion" process provides a unique method to visualize the instantaneous structure of polyatomic molecules, because the momenta of the resultant fragment ions reflect sensitively the geometrical structure of the molecule at the time of the laser irradiation [1, 2].

However, for precise probing of the structural deformation of molecules, the effect of chemical bonding should be negligibly small compared to the Coulombic repulsion between the fragments so that the interactions between the fragment ions can be assumed to be purely Coulombic on the reconstruction of molecular structureHerewestudytheultrafastdynamicsoffragmentation(explosion)ofCS$_2$,

A. Matsuda · A. Hishikawa (✉)
Department of Chemistry, Graduate School of Science,
Nagoya University, Nagoya, Aichi 464-8602, Japan
e-mail: hishi@chem.nagoya-u.ac.jp

E.J. Takahashi
Attosecond Science Research Team, RIKEN Center for Advanced Photonics,
Wako, Saitama 352-0198, Japan

© Springer International Publishing Switzerland 2015
K. Yamanouchi et al. (eds.), *Ultrafast Phenomena XIX*,
Springer Proceedings in Physics 162, DOI 10.1007/978-3-319-13242-6_40

$CS_2^{z+} \rightarrow S^{p+} + C^{q+} + S^{r+}$ ($z = p + q + r$), in few-cycle intense laser fields by monitoring the three-body Coulomb explosion from a highly charged state so that the relative contribution of chemical bonding becomes less significant [3].

2 Experiment

A pair of 9 fs intense laser pulses (4×10^{15} W/cm^2) is employed as the pump and probe pulses (Fig. 1a). The pump laser pulse removes electrons to produce highly charged CS_2^{z+} ($z = 1$–6). The structural deformation associated with the fragmentation (explosion) of the molecular ion was then probed by the second laser pulse introduced with a time delay Δt by monitoring the three-body Coulomb explosion of CS_2^{6+}, $CS_2^{6+} \rightarrow S^{2+} + C^{2+} + S^{2+}$. The linearly polarized Ti:sapphire laser pulse (9 fs, 800 nm, 1 kHz) was introduced into a high-precision Michelson-type interferometer to obtain the pump and probe laser pulses, which were then focused onto an effusive molecular beam of CS_2 in an ultra-high vacuum chamber. The momenta of the fragment ions generated by the three-body Coulomb explosion were measured by coincidence momentum imaging technique [4]. The total kinetic energy release E_{kin} is calculated from the measured momenta, $E_{kin} = \Sigma |p_i|^2/(2m_i)$, where p_i and m_i is the momentum and the mass of the i-th fragment ion.

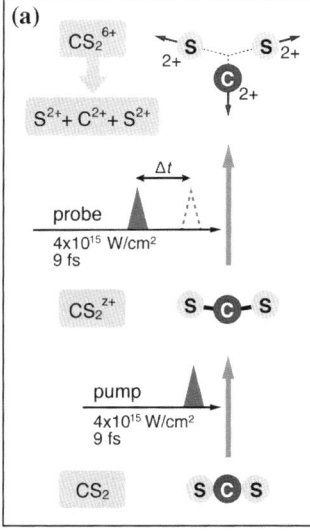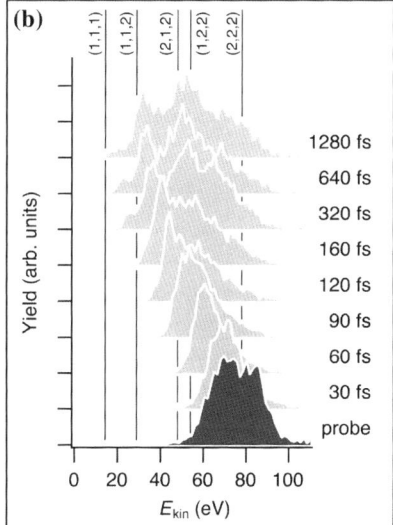

Fig. 1 **a** Scheme of pump-probe Coulomb explosion imaging employed in the present study. **b** Time evolution of the total kinetic energy release distribution. The *vertical lines* indicate peak kinetic energies for Coulomb explosion pathways observed with the pump pulse alone

3 Results and Discussion

The time evolution of the kinetic energy distribution is presented in Fig. 1b. At $\Delta t = 30$ fs, a clear shift of the peak position towards smaller energy is observed. Since the Coulomb repulsion energy is inversely proportional to the bond length, this indicates that the explosion proceeds within 30 fs after the irradiation of the pump pulse. This can be attributed to charge resonance enhanced ionization (CREI) [5] to CS_2^{6+} by the probe pulse, which explains the enhancement of the ionization rate by the charge localization near a critical distance. As the time delay increases, the peak of the kinetic energy release distribution further shifts to lower energy side due to stretching of the C–S bonds.

At a longer time delay ($\Delta t \geq 160$ fs), several sub-peaks appear in the kinetic energy distribution. These peaks are associated with different Coulomb explosion pathways induced by the pump pulse, which can be assigned by using the asymptotic energies observed at a sufficiently long time delay [6]. At $\Delta t = 1280$ fs, two peaks are observed at ~ 51 and ~ 32 eV, which well agree with that observed for pathways $(p, q, r) = (1, 2, 2)$, $(2, 1, 2)$ and $(1, 1, 2)$, respectively. The small shoulder observed at ~ 20 eV is identified as a contribution from pathway $(1, 1, 1)$. It is also shown that the timescale in which the asymptotic energy is reached depends on the explosion pathway: ~ 300 fs for pathway $(1, 2, 2)$ and $(2, 1, 2)$ and ~ 1 ps for pathway $(1, 1, 2)$. For pathway $(1, 1, 1)$, it is inferred that the dissociation along this pathway is too slow to reach the asymptotic region at this time delay since the shoulder peak is observed at a slightly higher energy compared to that expected (14.4 eV).

4 Conclusion

The ultrafast fragmentation of CS_2 in highly charged states was investigated by time-resolved Coulomb explosion imaging using few-cycle intense laser pulses. The explosion dynamics in different charge states (CS_2^{3+}, CS_2^{4+} and CS_2^{5+}) was simultaneously probed in real time showing that the explosion proceeds in different timescales depending on the charge state. By using a visible or ultraviolet ultrashort laser pulse as the pump along with the present technique, one can study ultrafast structural deformation during photochemical reactions of a variety of molecules to provide a deeper understanding of reaction processes.

References

1. A. Hishikawa, A. Matsuda, M. Fushitani, E. J. Takahashi, Phys. Rev. Lett. **99**, 258302 (2007).
2. A. Matsuda, M. Fushitani, E. J. Takahashi, A. Hishikawa, Phys. Chem. Chem. Phys. **13**, 8697 (2011).

3. A. Matsuda, E. J. Takahashi, A. Hishikawa, J. Electron Spectrosc. Relat. Phenom. **195**, 327 (2014).
4. H. Hasegawa, A. Hishikawa, K. Yamanouchi, Chem. Phys. Lett. **349**, 57 (2001).
5. I. Bocharova, R. Karimi, E. F. Penka, J. Brichta, P. Lassonde, X. Fu, J. Kieffer, A. D. Bandrauk, I. Litvinyuk, J. Sanderson, F. Légaré, Phys. Rev. Lett. **107**, 063201 (2011).
6. A. Hishikawa, M. Ueyama, K. Yamanouchi, J. Chem. Phys. **122**, 151104 (2005).

Initial Phase Shifts in the Quantum Beat Resulting from the Ultrafast Internal Conversion of Pyrazine

Yoshi-Ichi Suzuki and Toshinori Suzuki

Abstract We present a simple interpretation of the phase-shifted quantum beat resulting from the nonradiative transition from higher to lower electronic states of pyrazine, based on classical mechanics and harmonic potentials.

1 Introduction

Molecular photoabsorption spectra in the UV region are often broad and lacking in structure, since the highly excited states of polyatomic molecules generally undergo nonradiative decay to lower or ground states (Fig. 1a). Time-resolved photoelectron imaging [1] is a helpful means of investigating these ultrafast changes in electronic states, and we have previously employed this technique to observe a rapid change in the photoelectron angular distribution of the S_2 excited state in pyrazine on a 30 fs timescale [2–4]. This transition was interpreted as an internal conversion from the S_2 to S_1 states, an interpretation that was confirmed based on theoretical calculations [5]. It was also observed that the photoelectron signal acquired after 50 fs exhibited a quantum beat with a frequency of 560 cm^{-1} (Fig. 1b). This oscillation was assigned to the Q_{6a} mode (583 cm^{-1} in S_1), which has long been anticipated on a theoretical basis. The oscillation also exhibited a peculiar phase shift [4]. These results provide further evidence for the S_2-S_1 transition, since a phase shift other than π radians is not expected if the S_1 state is created by direct excitation from the S_0 state [6, 7]. It would be natural to conclude that the initial phase shift is related to the lifetime of the S_2 state. The purpose of this study was therefore to elucidate the relationship between the phase shift and the lifetime of the S_2 state, by assessing the potential energy curves of the excited states based on classical mechanics.

Y.-I. Suzuki (✉) · T. Suzuki
Department of Chemistry, Graduate School of Science, Kyoto University, Kyoto, Japan
e-mail: yoshi-ichi.suzuki@kuchem.kyoto-u.ac.jp

Y.-I. Suzuki · T. Suzuki
RIKEN Center for Advanced Photonics, RIKEN, Wako, Japan

© Springer International Publishing Switzerland 2015
K. Yamanouchi et al. (eds.), *Ultrafast Phenomena XIX*,
Springer Proceedings in Physics 162, DOI 10.1007/978-3-319-13242-6_41

172

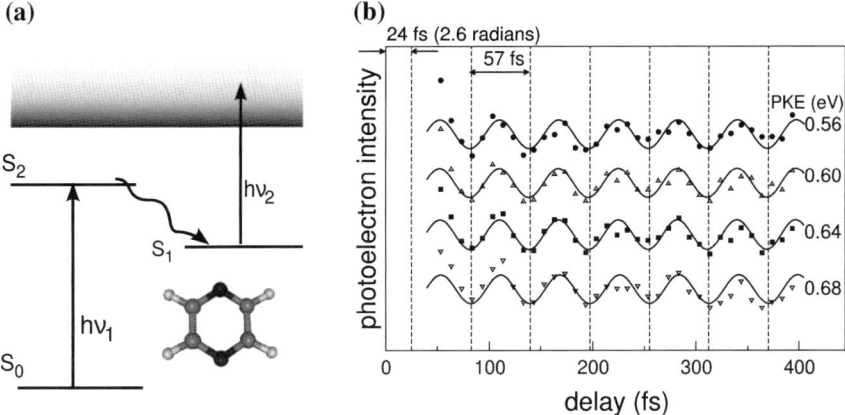

Fig. 1 **a** Schematic of the transitions observed in time-resolved photoelectron spectroscopy of pyrazine in a molecular excited state. **b** Energy-resolved photoelectron intensity values obtained from the two photon ionization of pyrazine as a function of the pump-probe delay [3]. *Vertical dotted lines* indicate the intensity minima

2 Results

Figure 2a presents schematic potential energy curves for the S_0, S_1, S_2 and D_0 states of pyrazine. The molecular structure is initially assumed to correspond to the equilibrium geometry of the S_0 state, following which the molecule is excited instantaneously to the S_2 state with no momentum. The geometry subsequently transitions to the equilibrium geometry of the S_2 state and then shifts to the S_1 surface suddenly and irreversibly at τ, the lifetime of the S_2 state. The phase shift of the quantum beat upon photoionization from S_1 to $D_0(v = 0)$ is affected by the relative positions of the equilibrium geometries of the S_0 and D_0 states. For the Q_{6a} mode of pyrazine (Fig. 2a), this phase shift is expected to be π. Classically, motion along the S_2 and S_1 potential curves is described, respectively, by $Q_2(t) = -Q_{eq2}\cos(\omega_2 t) + Q_{eq2}$ $(t < \tau)$ and $Q_1(t) = A\cos(\omega_1 t + \varphi + \pi) + Q_{eq1}$ $(t \geq \tau)$, where the equilibrium geometries and frequencies of the S_n state are denoted by Q_{eqn} and ω_n, respectively. Here φ is a function of τ, ω_1, ω_2, Q_{eq1} and Q_{eq2}. The φ can be obtained from $Q_2(t)$ and $dQ_2(t)/dt$ at $t = \tau$.

In the following discussion, we assume $\omega_1 = \omega_2$. Using theoretical values for Q_{eq1} and Q_{eq2} [8] and observed values for τ and ω_1, a phase shift of 2.6 radians is calculated, corresponding to a 25 fs delay time. This value agrees reasonably well with the observed delays of 18 fs in the total photoelectron signal and 24 fs in the energy-resolved photoelectron signal obtained at 0.6 eV (Fig. 1b).

In order to examine the sensitivity of the phase shift to the potential energy curves, we calculated the classical amplitudes of the oscillations of the Q_{6a} modes of the actual pyrazine molecule (Fig. 2a) and of a modified pyrazine model (Fig. 2b). In Fig. 2b, the equilibrium geometry of the S_2 state is intentionally moved

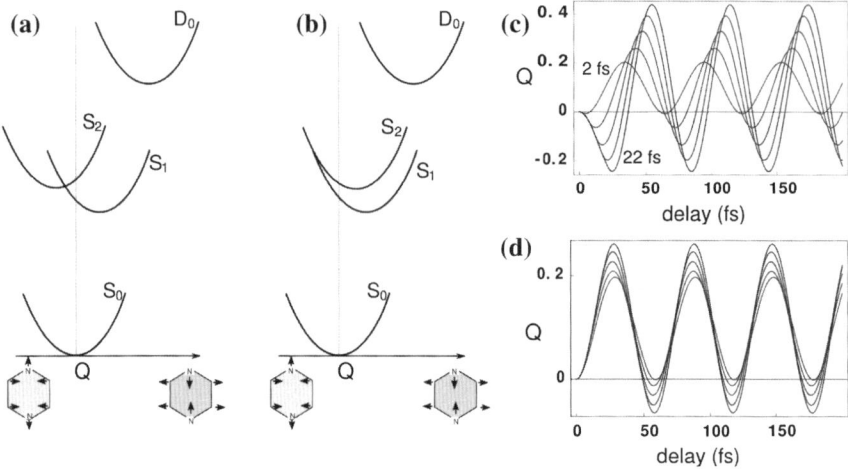

Fig. 2 Schematics of potential energy curves associated with the pyrazine Q_{6a} mode in **a** the actual pyrazine molecule and **b** a model of pyrazine in which the S_2 potential curve is intentionally moved. Vibrational motions for the potential curves are presented in **c** and **d** for the *curves* shown in (**a**) and (**b**), respectively, as a function of delay time and the lifetime of S_2, varied from 2 to 22 fs in 5 fs increments

to the same side as that of the S_1 state ($Q_{eq1} > 0$ and $Q_{eq2} > 0$) with respect to the S_0 minimum. Figure 2c, d show $Q(t)$ as a function of time, in which the S_2 state lifetime is varied from 2 to 22 fs in 5 fs increments. It is evident that, in the case of the actual pyrazine molecule, the beat pattern is sensitive to the lifetime and the phase shift is almost exactly proportional to the S_2 lifetime. In contrast, in the case of the model pyrazine (Fig. 2d), the S_2 lifetime has almost no effect on the phase shift. Therefore, we can conclude that the observed phase shift reflects the actual potential energy curves of the Q_{6a} mode of pyrazine. According to theory [8], the S_2 potential energy curve in the Q_1 mode of pyrazine is on the same side as the S_1 curve. Werner et al. performed theoretical simulations of pyrazine using "on the fly" ab initio nonadiabatic molecular dynamics [9] and the results suggested that there is no phase shift of the oscillation along the Q_1 mode. This is consistent with our model.

3 Conclusions

We determined the relationship between the initial phase of the quantum beat of excited pyrazine, the S_2 state lifetime and the potential energy curve parameters, assuming that the internal conversion takes place suddenly and irreversibly. This relationship represents a useful means of obtaining information on molecular dynamics near zero time delay, at which point the pump and probe pulses overlap temporally and the signal may contain intense and unwanted probe-pump signals and coherent artifacts. Since the excited state lifetime can also be estimated by other methods, our model can be used as a means of extracting information concerning the potential energy curves from the initial phase.

References

1. T. Suzuki, Annu. Rev. Phys. Chem., **57**, 555 (2006)
2. T. Horio, T. Fuji, Y. Suzuki, and T. Suzuki, J. Am. Chem. Soc., **131**, 10392 (2009)
3. Y. Suzuki, T. Fuji, T. Horio, and T. Suzuki, J. Chem. Phys., **132**,174302 (2010)
4. T. Suzuki and Y. Suzuki "Ultrafast Internal Conversion of Pyrazine via Conical Intersection" in Advances in multiphoton processes, ed. S. H. Lin, A. A. Villaeys, and Y. Fujimura, Vol. 21, Chap. 4 (World Scientific, Singapore, 2014)
5. Y. Suzuki and T. Suzuki, J. Chem. Phys., **137**, 194314 (2012).
6. P. M. Felker and A. H. Zewail, J. Chem. Phys., **82**, 2961 (1985).
7. T. Fuji, Y. Suzuki, T. Horio, and T. Suzuki, Chem. Asian J., **6**, 3028 (2011).
8. A. Raab, G. A. Worth, H.-D. Meyer, and L. S. Cederbaum, J. Chem. Phys., **110**, 936 (1999).
9. U. Werner, R. Mitrić, and V. Bonačić-Koutecký, J. Chem. Phys., **132**, 174301 (2010).

Ab Initio Quantum Dynamical Study on Ultrafast Nonradiative Transition Pathways of Pyrazine

Manabu Kanno, Yuta Ito, Noriyuki Shimakura, Shiro Koseki, Hirohiko Kono and Yuichi Fujimura

Abstract We theoretically verified the participation of optically dark $n\pi^*$ states other than S_1 in ultrafast internal conversion of pyrazine. Contrary to a recent semiclassical study, our quantum dynamical calculations demonstrated that their contributions are negligible.

1 Introduction

Pyrazine $C_4H_4N_2$ is a typical azabenzene that undergoes ultrafast nonradiative transitions. They come from the prominent vibronic couplings between two electronic excited states of different characters, i.e., the optically bright $\pi\pi^*$ S_2 ($^1B_{2u}$) and almost dark $n\pi^*$ S_1 ($^1B_{3u}$) states in the UV region [1]. Domcke and coworkers theoretically proposed that the ultrafast nonradiative transition from the S_2 to S_1 state takes place through their conical intersection (CI) [2]. Experimentally, Suzuki et al. measured the lifetime of the S_2 state (22 ± 3 fs) by pump-probe photoelectron

M. Kanno (✉) · H. Kono · Y. Fujimura
Department of Chemistry, Graduate School of Science,
Tohoku University, Sendai 980-8578, Japan
e-mail: kanno@m.tohoku.ac.jp

Y. Ito · N. Shimakura (✉)
Department of Chemistry, Niigata University,
Ikarashi Nino-Cho 8050, Niigata 950-2181, Japan
e-mail: shima@emeritus.niigata-u.ac.jp

S. Koseki
Department of Chemistry, Graduate School of Science and Research Institute
for Molecular Electronic Devices (RIMED), Osaka Prefecture University,
1-1 Gakuen-Cho, Naka-Ku, Sakai 599-8531, Japan

Y. Fujimura
Department of Applied Chemistry, Institute of Molecular Science and Center
for Interdisciplinary Molecular Science, National Chiao-Tung University,
Hsin-Chu 300, Taiwan

© Springer International Publishing Switzerland 2015
K. Yamanouchi et al. (eds.), *Ultrafast Phenomena XIX*,
Springer Proceedings in Physics 162, DOI 10.1007/978-3-319-13242-6_42

imaging spectroscopy and analyzed the observed spectrum under the condition that the nonradiative process from the S_2 to S_1 state occurs [3].

Recently, Werner et al. proposed new pathways via optically dark $n\pi^*$ ^1A_u and $^1B_{2g}$ states [4]. These dark states were theoretically predicted to be near the S_1 state but there have been no experimental evidence of their existence. Werner et al. calculated time-dependent populations of the lowest ^1A_u and $^1B_{2g}$ states in addition to those of the S_1 and S_2 states "on the fly" by combining the time-dependent density functional theory with Tully's stochastic fewest switches surface hopping procedure. Their interesting finding was that the S_2 lifetime was 21.1 fs, which agrees with the experimental value, while at the initial stage the population was preferentially transferred to both the ^1A_u and $^1B_{2g}$ states rather than the S_1 state.

In this study, we quantify the participation of the CIs between the optically bright S_2 state and the optically dark ^1A_u and $^1B_{2g}$ states in ultrafast internal conversion of pyrazine. "On-the-fly" methods are powerful to search for multi-mode nuclear dynamics but unable to properly describe coherent vibrational behaviors since they are based on classical mechanics. We employ the multireference configuration interaction (MRCI) method to obtain reliable excited-state potential energy surfaces (PESs) and the nuclear wave packet (WP) propagation method to solve the nuclear dynamics on the PESs coupled by CIs. The nuclear WP propagation method is applicable to ultrafast nonradiative transitions in pyrazine at the initial stage, in which restricted vibrational modes play an essential role.

2 Computational Outline

All electronic structure calculations were carried out with the 6-311++G** Gaussian basis set by using the quantum chemistry program MOLPRO. The geometry of pyrazine was optimized in the electronic ground state S_0 (^1A_g) at the complete-active-space self-consistent field (CASSCF) level of theory, followed by normal mode analysis. The active space comprised ten electrons distributed among eight orbitals (three π, three π^*, and two lone-pair orbitals). The lowest five singlet electronic states (S_0–S_4) were state-averaged with equal weights. To take dynamical electron correlation into account, the CASSCF energies were refined at the level of the internally contracted MRCI including single and double excitations (MRCISD).

Two-dimensional PESs were constructed within the subspaces defined by tuning and coupling modes, which span the branching plane of a CI. The vibrational ground-state wave function in the S_0 state was placed on the S_2 PES at the initial time $t = 0$. The subsequent time evolution of the Franck-Condon WP was performed in the diabatic basis to deal with nonadiabatic couplings by the split-operator method. The resultant diabatic WPs were converted to adiabatic WPs.

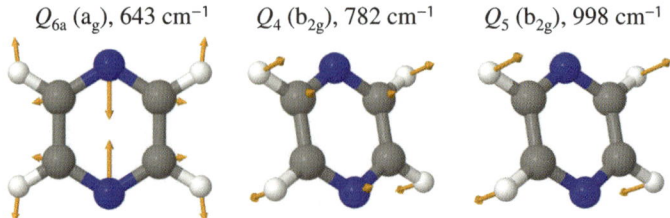

Q_{6a} (a_g), 643 cm^{-1} Q_4 (b_{2g}), 782 cm^{-1} Q_5 (b_{2g}), 998 cm^{-1}

Fig. 1 Calculated vibrational vectors and harmonic frequencies of selected normal modes in the S_0 state of pyrazine. The *white*, *black*, and *blue* balls represent hydrogen, carbon, and nitrogen atoms, respectively

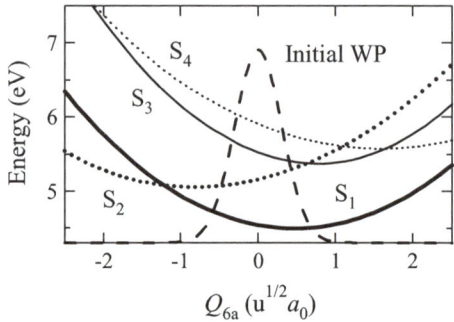

Fig. 2 One-dimensional MRCISD PESs of the S_1 ($^1B_{3u}$), S_2 ($^1B_{2u}$), S_3 (^1A_u), and S_4 ($^1B_{2g}$) states along the Q_{6a} mode. Also shown is the probability density of the Franck-Condon WP

3 Results and Discussion

We confirmed that the calculated results of geometrical parameters, excitation energies, and vibrational frequencies of pyrazine are in excellent agreement with the experimental data reported by Innes et al. [1]. For instance, the vibrational vectors of three selected normal modes are illustrated schematically in Fig. 1. For all of them, the calculated frequency is slightly higher than the experimental one but the difference is only less than 50 cm^{-1}.

The time-resolved measurements by Suzuki et al. suggested that the Q_{6a} (a_g) in-plane ring deformation mode (Fig. 1), where pyrazine maintains D_{2h} symmetry, is the dominant tuning mode for the $S_2 \leftarrow S_0$ photoabsorption [3]. The one-dimensional PESs of the S_1 ($^1B_{3u}$), S_2 ($^1B_{2u}$), S_3 (^1A_u), and S_4 ($^1B_{2g}$) states along the Q_{6a} mode are depicted in Fig. 2. The S_3 and S_2 states are degenerate at $Q_{6a} = 0.64$ u$^{1/2}a_0$, while the S_4–S_2 and S_2–S_1 crossings are located farther away at $Q_{6a} = 1.10$ and -1.21 u$^{1/2}a_0$, respectively. The tail of the Franck-Condon WP covers the S_3–S_2 crossing but the other crossings are outside the Franck-Condon region. The S_4 state is higher in energy than the S_3 state during the course of propagation to the S_2–S_1 crossing. Unless a

Fig. 3 Temporal behaviors in the population of the S_3 state in the $Q_{6a}-Q_4$ and $Q_{6a}-Q_5$ subspaces

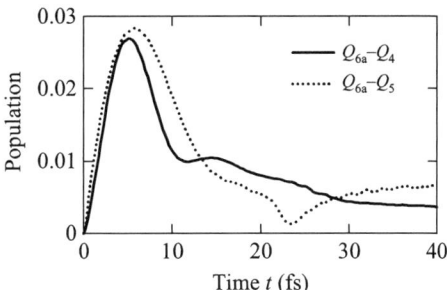

substantial population transfer to the S_3 state takes place by nonadiabatic transition, it is reasonable to conclude that the S_4 state plays a minor role in ultrafast internal conversion of pyrazine.

The Q_4 (b_{2g}) out-of-plane ring bending and Q_5 (b_{2g}) out-of-plane CH bending modes (Fig. 1) couple the S_3 (^1A_u) and S_2 ($^1B_{2u}$) states. These modes lower the symmetry of pyrazine to C_{2h} and the irreducible representation of the S_2 state to ^1A_u. Figure 3 shows the temporal behaviors in the population of the S_3 state obtained from the nuclear WP simulations on two-dimensional PESs within the $Q_{6a}-Q_4$ and $Q_{6a}-Q_5$ subspaces. In both cases, a small fraction of the population is transferred to the S_3 state but its maximum value is less than 0.03, which is much smaller than that in the semiclassical study by Werner et al. (~ 0.4) [4]. This indicates that nonadiabatic transitions to the S_3 and S_4 states are negligible. The dominant contribution to ultrafast internal conversion is electronic relaxation through the S_2-S_1 CI.

References

1. Innes KK, Ross IG, Moomaw WR (1988) Electronic States of Azabenzenes and Azanaphthalenes: A Revised and Extended Critical Review. J Mol Spectrosc 132:492-544.
2. Woywod C, Domcke W, Sobolewski AL, Werner H-J (1994) Characterization of the S_1-S_2 conical intersection in pyrazine using *ab initio* multiconfiguration self-consistent-field and multireference configuration-interaction methods. J Chem Phys 100:1400-1413.
3. Suzuki Y-I, Fuji T, Horio T, Suzuki T (2010) Time-resolved photoelectron imaging of ultrafast $S_2 \rightarrow S_1$ internal conversion through conical intersection in pyrazine. J Chem Phys 132:174302.
4. Werner U, Mitrić R, Suzuki T, Bonačić-Koutecký V (2008) Nonadiabatic dynamics within the time dependent density functional theory: Ultrafast photodynamics in pyrazine. Chem Phys 349:319-324.

The Ultrafast Wolff Rearrangement in the Gas Phase

Andreas Steinbacher, Sebastian Roeding, Tobias Brixner
and Patrick Nuernberger

Abstract The Wolff rearrangement of gas-phase 5-diazo Meldrum's acid is disclosed with femtosecond ion spectroscopy. Distinct differences are found for 267 nm and 200 nm excitation, the latter leading to even two ultrafast rearrangement reactions.

1 Introduction

Photoexcited 5-diazo Meldrum's acid (DMA) can undergo a Wolff rearrangement (WR) to form a ketene. The exact mechanism, stepwise or concerted, is still vividly discussed. While theoretical studies suggest a stepwise mechanism [1], pyrolysis [2] and liquid-phase studies [3] of DMA could not resolve the initial process. Hence, we investigate DMA with femtosecond (fs) pump–probe photofragment ion spectroscopy in the gas phase. Besides corroborating the theoretically predicted stepwise mechanism, we elucidate the crucial role of excess energy by exciting with 267 and also 200 nm pump pulses, leading to a second WR.

2 Experimental Results

The third (267 nm) or fourth (200 nm) harmonic of a Ti:Sa CPA system (35 fs, 800 nm, 1 kHz) are employed as pump pulses and recombined collinearly with the 800 nm probe (see Fig. 1, middle inset). Crystalline DMA molecules were expanded effusively into the vacuum chamber at 85 °C. Pump energies are adjusted such that no pump-only

A. Steinbacher · S. Roeding · T. Brixner
Universität Würzburg, Am Hubland, 97074 Würzburg, Germany

P. Nuernberger (✉)
Ruhr-Universität Bochum, 44780 Bochum, Germany
e-mail: patrick.nuernberger@rub.de

© Springer International Publishing Switzerland 2015
K. Yamanouchi et al. (eds.), *Ultrafast Phenomena XIX*,
Springer Proceedings in Physics 162, DOI 10.1007/978-3-319-13242-6_43

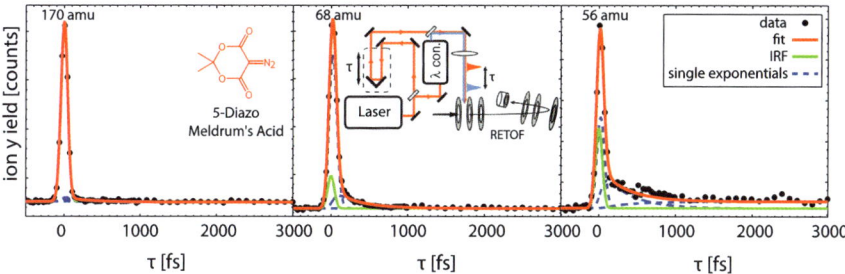

Fig. 1 Transients of selected mass peaks after 267 nm excitation (*black dots*). The result of the global fitting (*solid red lines*), the single exponential contributions (*dashed blue*), and the IRF (*solid green*) are presented. *Left inset* Structure of DMA. *Middle inset* Experimental setup

signal is generated. The probe generates ions by multiphoton ionization, which are detected in a home-built reflectron time-of-flight (RETOF) mass spectrometer.

In Fig. 1 the transient ion signals at 170, 68, and 56 amu after 267 nm excitation are presented. Prior to the WR in DMA, the diazo group is photolyzed off, which in theory could be traced by an ion signal at 142 amu corresponding to Carbene I and Ketene I (see inset in Fig. 2). Since this peak neither is present in the steady-state mass spectrum [4] nor in the pump–probe data due to fragmentation, the dynamics must be inferred from smaller fragments. In this context, the 68 amu signal is of special interest since it originates from Carbene I and Ketene I, but likely not from further reaction products. Likewise, the ion signals at 56 and 40 amu reflect the evolution of Carbene II and Ketene II (see inset in Fig. 2).

We fit the 16 dominant mass peaks of the 267 nm-excitation data simultaneously with a sequential model consisting of three time constants, convolved with a Gaussian instrument response function (IRF) with a FWHM of 104 fs, as obtained by a pump–probe cross-correlation. The time constants were determined to $\tau_1 = 27$ fs, $\tau_2 = 358$ fs, and $\tau_3 = \infty$, i.e., a permanent offset. The use of a time constant shorter than the IRF might seem arguable, but is needed to model our data appropriately.

As visible from the left panel in Fig. 1, the pump–probe signal of the parent ion (170 amu) follows the IRF, indicating that upon excitation, the DMA molecule is either instantly ionized or further fragmentation sets in. From the 68 amu transient (Fig. 1, middle panel) one can deduce that in addition to a signal following the IRF, also ion signals decaying with τ_1 and τ_2 contribute. The 27 fs time constant can be assigned to the WR from Carbene I to Ketene I. Corroborated by the evolution of the 56 amu peak (Fig. 1, right panel) which is associated to Carbene II, the lifetime of Ketene I can be inferred to be 358 fs. Moreover, the 56 amu signal exhibits a permanent offset (τ_3). Hence, we conclude that Ketene I loses a CO on a sub-picosecond time scale to form Carbene II, a reaction already proposed in [2].

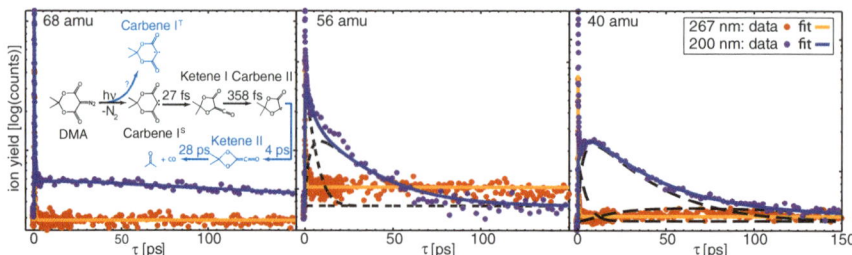

Fig. 2 Comparison of the 68, 56, and 40 amu transients for 200 (*violet dots*) and 267 nm (*red dots*) excitation. The result of the global fitting (*solid lines*) as well as the single exponential contributions (*dashed black lines*) are presented. *Inset* Scheme of the photochemical reactions monitored with 267 nm (*black*) and additional pathways upon 200 nm (*blue*) excitation, as inferred from our data

Theoretical studies reported [1] that higher-lying electronic states are involved in the WR of DMA. When exciting with 200 nm pulses, additional dynamics with time constants of $\tau_4 = 4$ ps and $\tau_5 = 28$ ps are observed. A comparison of the long-time behavior after 200 and 267 nm excitation is shown in Fig. 2. In the dynamics of the 68 amu ions (Fig. 2, left panel), a long-lasting contribution is present for 200 but not for 267 nm excitation, indicating that a subensemble of Carbene I is being transferred to a triplet state. This hinders the rearrangement and thus the species is long-lived [1]. Furthermore, pumping at 200 nm leads to picosecond dynamics in the 56 and 40 amu mass peaks (middle and right panel of Fig. 2). As put forward by the pyrolysis study [2], Carbene II can further undergo a WR to Ketene II. The additional excess energy upon 200 nm excitation enables this reaction. Hence, within 4 ps Ketene II is formed, which then fragmentizes to CO and acetone within 28 ps.

3 Conclusion

A comparison of the ion signal after excitation of DMA with two different excitation wavelengths reveals that the WR in the gas phase is an ultrafast stepwise process towards a ketene. The latter is only stable on a sub-picosecond time scale and reacts on to a further carbene species. If enough excess energy is provided, the carbene performs a second WR before further fragmentation sets in.

References

1. Bogdanova, A. and Popik, V. V.: Experimental and theoretical investigation of reversible interconversion, thermal reactions, and wavelength-dependent photochemistry of diazo Meldrum's acid and its diazirine isomer, 6,6-dimethyl-5,7-dioxa-1,2-diaza-spiro[2,5]oct-1-ene-4,8-dione. J. Am. Chem. Soc. **125**, 14153–14162 (2013)

2. Kammula, S. L., Tracer, H. L., Shevlin, P. B., and Jones, M.: Intramolecular decomposition of isopropylidene diazomalonate (diazo Meldrum's acid). J. Org. Chem. **42**, 2931–2932 (1977)
3. Rudolf, P., Buback, J., Aulbach, J., Nuernberger, P., and Brixner, T.: Ultrafast multisequential photochemistry of 5-diazo Meldrum's acid. J. Am. Chem. Soc. **132**, 15213–15222 (2010)
4. Steinbacher, A., Roeding, S., Brixner, T., and Nuernberger, P.: Ultrafast photofragment ion spectroscopy of the Wolff rearrangement in 5-diazo Meldrum's acid. Phys. Chem. Chem. Phys. **16**, 7290–7298 (2014)

Time-Resolved Photoelectron Spectroscopy and Ab Initio Multiple Spawning Studies of Hexamethylcyclopentadiene

T.J.A. Wolf, T.S. Kuhlman, O. Schalk, T.J. Martínez, K.B. Møller, A. Stolow and A.-N. Unterreiner

Abstract Time-resolved photoelectron spectroscopy and ab initio multiple spawning were applied to the ultrafast non-adiabatic dynamics of hexamethylcyclopentadiene. The high level of agreement between experiment and theory associates wavepacket motion with a distinct degree of freedom.

1 Introduction

The ultrafast excited state dynamics of small polyenes have been investigated in many studies either by experimental methods or via dynamical simulations. Only few, however, have used a combination of both approaches. In this contribution,

T.J.A. Wolf (✉) · A.-N. Unterreiner
Institute of Physical Chemistry, Karlsruhe Institute of Technology,
Karlsruhe, Germany
e-mail: thomas.wolf@stanford.edu

T.J.A. Wolf
PULSE Institute, Stanford University, Stanford, USA

T.S. Kuhlman · K.B. Møller
Department of Chemistry, Technical University of Denmark,
Kongens Lyngby, Denmark

O. Schalk
Stockholm University, Stockholm, Sweden

O. Schalk · A. Stolow
National Research Council, Ottawa, Canada

T.J. Martínez
PULSE Institute and Department of Chemistry,
Stanford University, Stanford, USA

A. Stolow
Department of Chemistry and Department of Physics, University of Ottawa,
Ottawa, Canada

© Springer International Publishing Switzerland 2015
K. Yamanouchi et al. (eds.), *Ultrafast Phenomena XIX*,
Springer Proceedings in Physics 162, DOI 10.1007/978-3-319-13242-6_44

we present a combination of time-resolved photoelectron spectroscopy (TRPES) and ab initio multiple spawning (AIMS) simulations for a profound understanding of excited state dynamics in polyenes. We aim at connecting phenomena observed in TRPES and AIMS simulations. In time-resolved photoelectron spectra we study the molecular dynamics of small polyenes by exciting the molecules with a pump pulse and following the dynamics through ionization with a probe pulse. The signature appearing immediately at time zero is often delayed throughout the spectrum. In dynamical simulations, depopulation of the initially excited electronic state is often observed to be preceded by an induction time [1]. In hexamethylcyclopentadiene ($CPDMe_6$), both effects are expected to be prominent because of an expected decrease of the relaxation time as compared to unsubstituted cyclopentadiene [2] due to the increased inertia. This makes $CPDMe_6$ an ideal benchmark molecule for the investigation of their connection.

2 Methods

TRPES data were recorded in a magnetic bottle-type spectrometer described elsewhere [4]. Pump pulses at 267 nm (2.6 µJ/pulse) and probe pulses at 320 nm (2 µJ/pulse) were set to magic angle and focussed into the interaction region. The cross-correlation was 130 ± 10 fs. $CPDMe_6$ was supersonically expanded into vacuum by a pulsed Even-Lavie valve using 3 bar helium as a backing gas.

The details of the employed theoretical methods are described in [3]. Briefly, dynamical simulations were carried out employing an in-house code, which combines AIMS dynamics with MS-MR-CASPT2/6-31G** electronic structure calculations. The latter were done with the MOLPRO2006.1 program package [5] employing a complete active space consisting of four π orbitals and four electrons. The methyl substituents of $CPDMe_6$ were approximated as hydrogens with a mass of 15 amu. 38 trajectory basis functions were propagated for ≈ 300 fs, whereas initial conditions were sampled from a Wigner distribution in the electronic ground state. From the trajectories, ionization energies and cross-sections were calculated employing Dyson orbitals. To simulate time-resolved photoelectron spectra, they were convoluted with Gaussians to account for the limited energy and time resolution of the experiment.

3 Results and Discussion

Figure 1a shows the experimental TRPES data of $CPDMe_6$. They can be partitioned in regions with contributions from ionization with one (left of yellow line) or two photons (right of yellow line). Observable features of both regions refer to the same

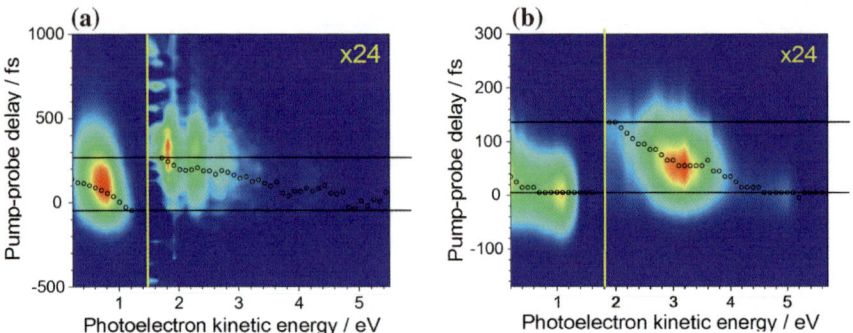

Fig. 1 a Experimental and **b** simulated time-resolved photoelectron spectra. *Red* refers to high, *blue* to low photoelectron intensities. Note the high qualitative agreement of spectra (**a**) and (**b**) and the disagreement in time scales

underlying dynamics. The photoelectron bands appear earliest in the high energy part of both regions and are delayed and broadened in time throughout the photoelectron spectrum. The coefficients of a fit quantifying the delay are inserted as black circles.

The AIMS simulations of the excited state dynamics reveal that population transfer to the ground state is preceded by an induction period of 108 fs. As projections of the simulated wavepacket onto different nuclear degrees of freedom of the molecule show, the induction period can be connected to a torsion around one of the double bonds in $CPDMe_6$, which also induces the mandatory geometry change to access the conical intersection with the ground state. TRPES data simulated from the AIMS results are depicted in Fig. 1b. Like in the experimental data, they show a delay and broadening of the photoelectron bands, which is again indicated by fit coefficients. The high qualitative agreement of the spectra indicates that the simulation reflects very well the relevant properties of the nuclear wavepacket, which was probed by the experiment. Their quantitative disagreement on the time scales is attributed to the approximation of the methyl groups as heavy hydrogens.

To connect the spectral delays in experimental and simulated TRPES data to the torsion degree of freedom, which is responsible for the induction period, a simple stepladder-type model is employed [3, 6]. It treats the wavepacket evolution as a time-dependent distribution of population density onto a one-dimensional series of discrete steps between the Franck-Condon region and the conical intersection and thereby reduces the description of the dynamics to the torsion degree of freedom. By providing each step with a spectral signature, it is also able to describe the delays in the time-resolved photoelectron spectra. Furthermore, induction period and depopulation can together be quantified by a single characteristic time parameter $t_{ch} = 140$ fs in the simulated data. In the case of the experimental spectra, the fit yields $t_{ch} = 540$ fs. Thus, the time scale of the dynamics as well as the spectrally dependent delay can be directly associated with the double bond torsion.

This finding seems of more general validity for excited state dynamics of small polyenes. The results emphasize the potential of the combined approach of TRPES and AIMS for a profound understanding of excited state dynamics.

References

1. Kuhlman, T. S., Glover, W. J., Mori, T., Møller, K. B., Martínez, T. J.: Between Ethylenes and Polyenes - The Nonadiabatic Dynamics of cis-dienes. Faraday Discuss. **157**, 193 (2012).
2. Schalk, O., Boguslavskiy, A. E., Stolow, A.: Substituent Effects on Dynamics at Conical Intersections: Cyclopentadienes. J. Phys. Chem. A **114**, 4058 – 4064 (2010).
3. Wolf, T. J. A., Kuhlman, T. S., Schalk, O., Martínez, T. J., Møller, K. B., Stolow, A., Unterreiner, A.-N.: Hexamethylcyclopentadiene: time-resolved photoelectron spectroscopy and ab initio multiple spawning simulations. Phy. Chem. Chem. Phys. **16**, 11770 (2014).
4. Lochbrunner, S., Larsen, J. J., Shaffer, J. P., Schmitt, M., Schultz, T., Underwood, J. G., Stolow, A.: Methods and applications of femtosecond time-resolved photoelectron spectroscopy. J. Electron Spectrosc. Relat. Phenom. **112**, 183 – 198 (2000).
5. Werner, H.-J. et al., Molpro, version 2006.1, a package of ab initio programs, see http://www.molpro.net.
6. Møller, K. B., Zewail, A. H., Kinetics modeling of dynamics: the case of femtosecond-activated direct reactions. Chem. Phys. Lett. **351**, 281 – 288 (2002).

Laser-Assisted Electron Diffraction for Probing Femtosecond Nuclear Dynamics of Gas-Phase Molecules

Yuya Morimoto, Reika Kanya and Kaoru Yamanouchi

Abstract By detecting 1 keV electrons scattered by CCl_4 in a femtosecond laser field, we observed laser-assisted electron diffraction images with which we can probe ultrafast nuclear dynamics of gas-phase molecules with <10 fs and ~ 0.01 Å resolutions.

1 Introduction

When an electron is scattered by an atom or a molecule in a laser field, the electron can change its kinetic energy by multiples of the photon energy ($h\nu$). This scattering process is called laser-assisted electron scattering (LAES) or free-free transition of electrons. Recently we proposed an ultrafast electron diffraction method called laser-assisted electron diffraction (LAED) [1] to determine the instantaneous structure of polyatomic molecules by taking advantage of the LAES process. Because the LAES process occurs only in the presence of the laser field, the temporal resolution of the LAED method can be as high as the laser pulse duration (<10 fs), which is more than two orders of magnitude shorter than a few picoseconds achieved by the pulsed gas-phase electron diffraction method [2]. In the present study, we recorded an LAED pattern of gas-phase CCl_4, i.e., an electron diffraction pattern appearing through the interference among LAES electrons

Y. Morimoto (✉) · R. Kanya · K. Yamanouchi
Department of Chemistry, School of Science, The University of Tokyo,
7-3-1 Hongo, Bunkyo-ku, Tokyo 113-0033, Japan
e-mail: morimoto@chem.s.u-tokyo.ac.jp

R. Kanya
e-mail: kanya@chem.s.u-tokyo.ac.jp

K. Yamanouchi
e-mail: kaoru@chem.s.u-tokyo.ac.jp

© Springer International Publishing Switzerland 2015
K. Yamanouchi et al. (eds.), *Ultrafast Phenomena XIX*,
Springer Proceedings in Physics 162, DOI 10.1007/978-3-319-13242-6_45

scattered by the respective atoms within a molecule at the energy shifts (ΔE) of $+h\nu$. The observed LAED pattern of CCl_4 is well reproduced by the numerical simulation in which the field-free geometrical structure of CCl_4 is adopted.

2 Experiment

A linearly polarized output of a Ti:sapphire chirped pulse amplification laser system ($\lambda = 800$ nm, $\tau = 520$ fs) was focused into a vacuum chamber. The peak field intensity was 6×10^{11} W/cm^2. A part of the laser output was split and converted into UV pulses ($\lambda = 266$ nm, $\tau = 15$ ps) by a pair of BBO crystals. The UV pulses were focused onto a gold photocathode of an electron gun to generate electron pulses. The electron pulses were accelerated to 1 keV and crossed both an effusive CCl_4 beam and the laser beam at right angles (Fig. 1a). The kinetic energy and the scattering angle of the scattered electrons were resolved by a toroidal energy analyzer and projected onto a position sensitive detector (Fig. 1b). The further details of the apparatus have been described elsewhere [3, 4].

3 Results and Discussion

The angular distribution of one-photon LAES signals ($\Delta E = +h\nu$) is shown in Fig. 2a as filled circles. In the angular distribution, we observed a clear interference structure with a minimum around 5.5° and a maximum around 9.0°. In order to confirm the origin of the interference pattern, we conducted a numerical simulation based on the Kroll-Watson approximation [5] and on the independent atomic model (IAM) [6] with corrections of the polarization effect induced by the incident

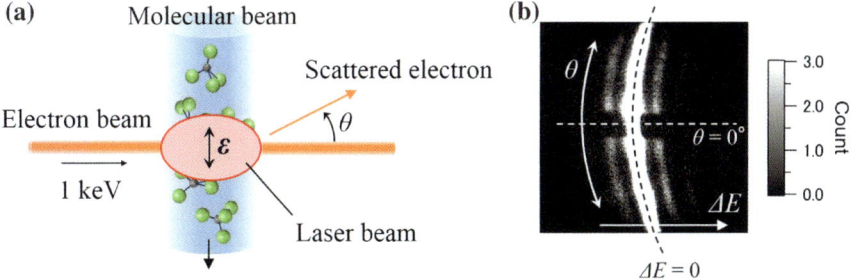

Fig. 1 **a** The schematic of the LAED experiment. The laser beam axis is set to be normal to the plane of the figure. The direction of the laser field polarization (ε) is vertical to the electron beam. **b** The image of the scattered electron signals. The one-photon LAES signals at $\Delta E = \pm h\nu$ are observed on both sides of the intense elastic signals at $\Delta E = 0$

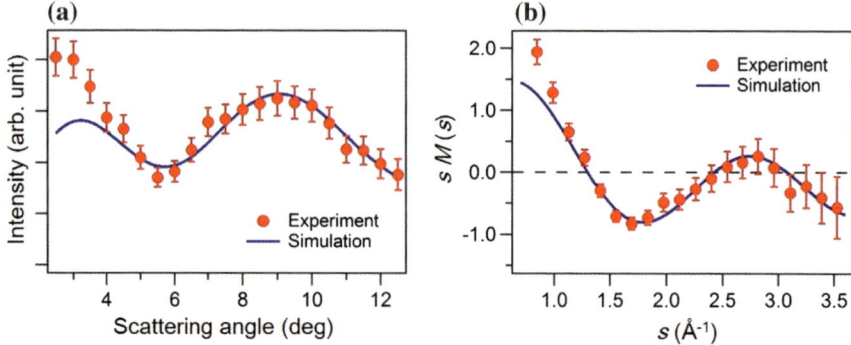

Fig. 2 **a** The LAED pattern of CCl_4 at $\Delta E = +h\nu$. **b** The modified molecular scattering intensity calculated from the LAED pattern

electrons and the chemical bonding effect. We adopted the geometrical parameters of CCl_4 at room temperature [7] determined by the conventional gas-phase electron diffraction (GED) method. The details of the simulation have been given in [4]. The simulated LAES angular distribution represented by a solid curve in Fig. 2a is in good agreement with the experimental distribution.

The comparison of the experimental results with simulation can also be made using a modified molecular intensity, $sM(s)$, which has been commonly used in determining geometrical structures of molecules by the conventional GED method [6]. In Fig. 2b, $sM(s)$ calculated from the observed LAED pattern are shown with filled circles and the simulated $sM(s)$ based on IAM as a solid curve. The agreement between the experimentally obtained $sM(s)$ and the simulated $sM(s)$ shows that the geometrical structure of CCl_4 at the moment of the laser irradiation can be determined from the analysis of LAED patterns with high precision.

4 Conclusion

We reported the observation of the LAED pattern of CCl_4 at $\Delta E = +h\nu$. The observed LAED pattern and the experimentally obtained $sM(s)$ are reproduced well by the numerical simulations in which the field-free geometrical structure of CCl_4 is adopted. The present study confirms that ultrafast nuclear dynamics can in principle be probed in real time with high temporal (<10 fs) and spatial (~ 0.01 Å) resolutions if molecules are pumped by an ultrashort laser pulse and are probed by the LAED method.

References

1. R. Kanya, Y. Morimoto, and K. Yamanouchi, Phys. Rev. Lett. **105**, 123202 (2010).
2. R. Srinivasan, V. A. Lobastov, C. Y. Ruan, and A. H. Zewail, Helv. Chim. Acta **86**, 1761 (2003).
3. R. Kanya, Y. Morimoto, and K. Yamanouchi, Rev. Sci. Instrum. **82**, 123105 (2011).
4. Y. Morimoto, R. Kanya, and K. Yamanouchi, J. Chem. Phys. **140**, 064201 (2014).
5. N. M. Kroll and K. M. Watson, Phys. Rev. A **8**, 804 (1973).
6. K. Yamanouchi, *Quantum Mechanics of Molecular Structures* (Springer, Heidelberg, 2012).
7. Y. Morino, Y. Nakamura, and T. Iijima, J. Chem. Phys. **32**, 643 (1960).

Filament-Driven Lasing Action for Combustion Diagnosis

Huailiang Xu, Wei Chu, Helong Li, Jielei Ni, Bin Zeng, Jinping Yao,
Haisu Zhang, Guihua Li, Chengrui Jing, Hongqiang Xie,
Kaoru Yamanouchi and Ya Cheng

Abstract We report on the lasing action for the combustion intermediate of CN in an ethanol/air flame by femtosecond laser excitation. It is confirmed that the lasing action results from amplified spontaneous emission with the population inversion achieved in the femtosecond-laser-induced plasma filament.

1 Introduction

Ultrafast femtosecond laser-induced lasing actions via amplified spontaneous emission (ASE) or seed amplification in the plasma filaments have been a subject of increasing interest in recent years because of its potentials for remote applications in standoff spectroscopy and remote sensing [1–3]. In air, it was shown that when an plasma filament is formed by a powerful femtosecond laser pulse, population inversion of neutral nitrogen molecules (i.e., N_2) for the $C^3\Pi_u - B^3\Pi_g$ transition and that of nitrogen molecular ions (i.e., N_2^+) for the $B^2\Sigma_u^+ - X^2\Sigma_g^+$ transition could be established although their underlying mechanisms are totally different [4, 5]. It was further demonstrated that the population inverted N_2 and N_2^+ systems in

H. Xu (✉) · W. Chu · H. Li
State Key Laboratory on Integrated Optoelectronics, College of Electronic Science,
and Engineering, Jilin University, Changchun 130012, China
e-mail: huailiang@jlu.edu.cn

W. Chu · J. Ni · B. Zeng · J. Yao · H. Zhang · G. Li · C. Jing · H. Xie · Y. Cheng
State Key Laboratory of High Field Laser Physics, Shanghai Institute of Optics
and Fine Mechanics, Chinese Academy of Sciences, Shanghai 201800, China
e-mail: ya.cheng@siom.ac.cn

K. Yamanouchi
Department of Chemistry, School of Science, The University of Tokyo,
7-3-1 Hongo, Bunkyo-ku, Tokyo 113-0033, Japan

© Springer International Publishing Switzerland 2015
K. Yamanouchi et al. (eds.), *Ultrafast Phenomena XIX*,
Springer Proceedings in Physics 162, DOI 10.1007/978-3-319-13242-6_46

ambient air can be seeded by self-generated white light continuum or harmonics during the filamentation of the pump laser beams [2, 5, 6]. Based on such approach, the µJ output energy of the self-seeded laser has been reported [5].

In the current contribution, we show that when a femtosecond filament is formed in a laminar ethanol/air flame on an alcohol burner, clean fluorescence emissions from free radicals CH, CN, NH, OH, and C_2, as well as atomic C and H, can be obtained, and their intensities vary as functions of the position of interaction of the filament with the flame along the vertical axis of the central combusting flow, showing the fluorescence signal intensity can be used to characterize the concentration of combustion intermediates [7]. Furthermore, it is found that the fluorescence emission from some specific species such as CN can be amplified by observing the backward fluorescence intensity as a function of the plasma length in the flame, opening up a new way to enhance the signal-to-noise ratio in combustion diagnosis by using fluorescence techniques [8].

2 Experimental Results and Discussion

Experiments were performed with 1 kHz and 800 nm Ti:Sapphire femtosecond laser pulses. The laser beam with a 0.5 mJ pulse energy was focused by a lens of 20 cm into the ethanol/air flame on an alcohol burner to generate a single filament with the length of ~ 1 cm. A moving stage was used to raise or lower the alcohol burner to control the position of interaction of the flame with the short filament. The fluorescence emitted from the filament was collected at a right angle to the laser propagation direction, and focused onto the entrance slit of a spectrometer (Andor Shamrock SR-303i) and detected by a gated intensified charge coupled device (ICCD, Andor iStar). Figure 1a shows the filament-induced spectrum of the ethanol/air flame in ambient atmosphere measured from the side of the filament. The spectral bands are assigned to free radicals of C_2, CH, CN, N_2, NH and OH, and atomic species of C and H, showing that the species in the alcohol burner flame are very rich consisting of not only atoms but also molecules. To demonstrate the ability of FINS in mapping the concentration distribution of combustion intermediates in flames, we carried out the measurements of filament-induced spectra at different positions of the flame along the dash-arrow line of the inset (a) in Fig. 1. It is found that the intensities of the four species of C, CH, CN, and C_2 vary as a function of the filament position in the flame, and found the signal intensities of all the four species vary, which reflects the concentration distribution of the four intermediate species in the flame.

Furthermore, we generated the filament in the flame on an alcohol burner array by focusing the laser beam with an $f = 40$ cm lens. The laser energy was 1.6 mJ/ pulse. The total length of the alcohol burner array with five burner wicks was approximately 40 mm. In this measurement, the fluorescence emitted from the filament was collected and collimated in the backward direction. It was noticed that the main features of the measured backward spectrum were consistent with that

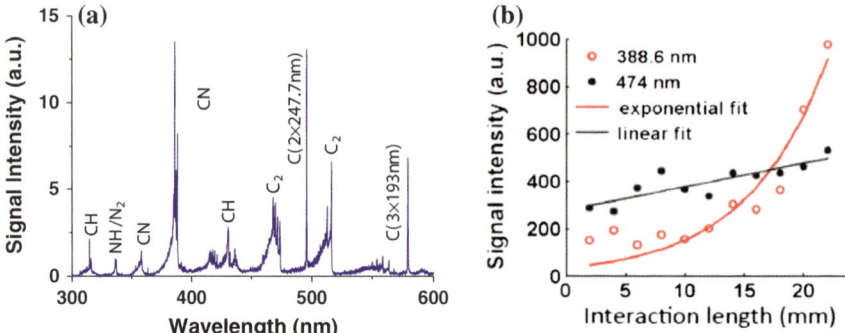

Fig. 1 **a** Spectrum of the ethanol/air flame on an alcohol burner by the excitation of a femtosecond filament. *Inset* Picture of the flame and the interaction positions marked by the *dash lines*; **b** Measured (*circle*) and fitted (*solid line*) dependence of the emission intensity of CN at 388.6 nm and that of C_2 at 474 nm on the length of the filament in the flame

measured previously from the side of the laser-induced flame filament. However, the emission at $\sim$388 nm for CN was strikingly enhanced when compared to the spectral bands of other species. By changing the length of the filament interacting with the flame by moving the burner array along the filament direction, we plotted the dependence of the peak intensities of CN at 388 nm and C_2 at 474 nm on the interaction length of the filament in the flame in Fig. 1b. It can be seen that the signal intensity of CN increases exponentially, but that of C_2 increases linearly when the length of the filament interacting with the flame becomes longer, confirming that there exists gain only for CN in the flame filament.

3 Summary

We have demonstrated the generation of ASE lasing action for the spectral band of CN at $\sim$388 nm in the ethanol/air combustion flame. This finding shows that lasing actions triggered by femtosecond laser filaments can be employed for sensing specific species in combustion and explosion processes.

References

1. Q. Luo, W. Liu, and S. L. Chin, Appl. Phys. B **76**, 337-340 (2003).
2. J. Yao, B. Zeng, H. Xu, G. Li, W. Chu, J. Ni, H. Zhang, S. L. Chin, Y. Cheng, and Z. Xu, Phys. Rev. A **84**, 051802(R) (2011).
3. A. Dogariu, J. B. Michael, M. O. Scully, and R. B. Miles, Science **331**, 442 (2011).
4. J. Ni, W. Chu, C. Jing, H. Zhang, B. Zeng, J. Yao, G. Li, H. Xie, C. Zhang, H. Xu, S. L. Chin, Y. Cheng, and Z. Xu, Opt. Express **21**,8746-8752(2013).

5. D. Kartashov, S. Ališauskas, A. Baltuška, A. Schmitt-Sody, W. Roach, P. Polynkin, Phys. Rev. A **88**, 041805 (2013).
6. W.Chu, G. Li, H. Xie, J. Ni, J. Yao, B. Zeng, H. Zhang, C. Jing, H. Xu, Y. Cheng, and Z. Xu, Laser Phys. Lett. **11**, 015301(2014).
7. H. Li, H.L. Xu, B. Yang, Q. Chen, T. Zhang, H. Sun, Opt. Lett. 38, 1250 (2013).
8. W. Chu, H. Li, J. Ni, B. Zeng, J. Yao, H. Zhang, G. Li, C. Jing, H. Xie, H. Xu, K.Yamanouchi, and Y. Cheng, Sensors and Actuators B at press (2014).

Part III
Solid State Physics and Chemistry

Anomalous Phase Change in $[(GeTe)_2/(Sb_2Te_3)]_{20}$ Superlattice Observed by Coherent Phonon Spectroscopy

K. Makino, Y. Saito, K. Mitrofanov, J. Tominaga, A.V. Kolobov, T. Nakano, P. Fons and M. Hase

Abstract The temperature-dependent ultrafast coherent phonon dynamics of topological $(GeTe)_2/(Sb_2Te_3)$ super lattice phase change memory material was investigated. By comparing with Ge-Sb-Te alloy, a clear contrast suggesting the unique phase change behavior was found.

1 Introduction

Phase change materials including $Ge_2Sb_2Te_5$ (GST) are commonly used as a recording film for optical disks and nonvolatile electrical memories because they have significant contrast in the optical and electrical properties depending on the two stable phases (SET or RESET). Recently, GST super lattice (SL) [1] consisting of GeTe sub-layers and Sb_2Te_3 sub-layers has received considerable attention due to its topological property [2] as well as low SET-RESET switching energy. In GST SL, Ge atoms are restricted to displace in the SL interfaces between tetrahedral sites and octahedral sites during the SET-RESET operation. It is therefore considered that phase change process is different from that in GST alloys. In this study, coherent phonon spectroscopy was performed for GST SL and GST alloy in order to reveal the difference in phase change mechanism from the point of view of ultrafast lattice dynamics.

K. Makino (✉) · Y. Saito · K. Mitrofanov · J. Tominaga · A.V. Kolobov · T. Nakano · P. Fons · M. Hase
Nanoelectronics Research Institute, National Institute of Advanced Industrial Science and Technology, Tsukuba Central 4, Higashi 1-1-1, Tsukuba 305-8562, Japan
e-mail: k-makino@aist.go.jp

© Springer International Publishing Switzerland 2015   199
K. Yamanouchi et al. (eds.), *Ultrafast Phenomena XIX*,
Springer Proceedings in Physics 162, DOI 10.1007/978-3-319-13242-6_47

2 Experiment

We employed an optical pump-probe measurement to excite and probe local atomic motions using optical pulses with 20 fs duration and 850 nm wavelength from a Ti: sapphire laser oscillator. The sample investigated was $[(GeTe)_2/(Sb_2Te_3)]_{20}$ (GST SL) which is theoretically predicted to be a 3D topological insulator. GST alloy film was also investigated for comparison. Both samples in REST state were fabricated on Si (100) substrates by helicon RF magnetron sputtering. Time-resolved reflectivity change ($\Delta R/R$) was recorded at room temperature (RT, 25 °C) and 180 °C. Here, the phase change temperatures to SET state are 170 °C for GST SL and 150 °C for GST alloy.

3 Result

Figure 1 shows the $\Delta R/R$ signals as a function of pump-probe time delay for GST SL and GST alloy obtained at RT and 180 °C. In GST SL, the intensity of the oscillation originated from coherent phonons is preserved even at high temperature. In GST alloy, on the other hand, the oscillation is strongly damped at high temperature. This is the most significant difference reflecting the difference of phase change process as discussed later.

Them, Fourier transformed (FT) spectra were obtained from the $\Delta R/R$ signals for both samples measured at RT and 180 °C. In GST alloy, mainly two peaks were observed at 3.72–4.75 THz and the dramatic decrease in the intensity of these mode was clearly seen at 180 °C. In GST SL, in contrast, mainly three modes are observed at 2.11, 3.40 and 5.07 THz. Interestingly, the temperature dependence on the FT intensity are small compared to GST alloy.

Fig. 1 Time-resolved $\Delta R/R$ signals for GST SL and GST alloy measured at RT (*solid lines*) and 180 °C (*dashed lines*)

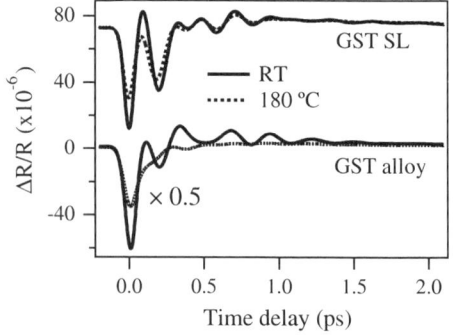

4 Discussion

The observed phonon modes can be assigned based on the phase change model and previous reports [3, 4] as follows. The lowest frequency mode at 2.11 THz and the highest frequency mode at 5.07 THz are reasonably attributed to the two different A$_1$ mode of Sb$_2$Te$_3$ sub-layers. However, the symmetry of 3.40 THz mode cannot be assigned by current result. There are two candidate that are the E$_g$ mode of Sb$_2$Te$_3$ sub-layers and the A$_1$ mode of GeTe sub-layers. The small attenuation of coherent phonons observed in GST SL at high temperature suggests that rearrangement of structure by the phase change is small compared with GST alloy. The phase change process in GST SL is considered to be different from a conventional amorphous-to-crystalline phase change which is applicable to GST alloy [1]. Therefore, the clear contrast between the two GST samples observed at high temperature presumably speculates the difference in phase change mechanisms. This result is very important for better understanding of the low energy phase switching property in GST SL [1].

5 Conclusion

In conclusion, temperature dependent coherent phonon spectroscopy was carried out in GST SL and GST alloy films. The clear difference in coherent phonon intensity between these two samples was observed at high temperature above the phase change temperature. This result presumably reflects the difference in phase change mechanism between GST SL and GST alloy.

References

1. Simpson, R.E., Fons P., Kolobov A.V., Fukaya T., Krbal M., Yagi T., Tominaga J.: Interfacial phase-change memory. Nature Nanotech. **6**, 501–501 (2011)
2. Tominaga J., Simpson R.E., Fons P., Kolobov A.V.: Electrical-field induced giant magneto-resistivity in (non-magnetic) phase change films. App. Phys. Lett. **99**, 152105 (2011)
3. Hase M., Miyamoto Y., Tominaga J.: Ultrafast dephasing of coherent optical phonons in atomically controlled GeTe/Sb$_2$Te$_3$ superlattices. Phys. Rev. B **79**, 174112 (2009)
4. Först M., Dekorsy T., Trappe C., Laurenzis M., Kurz H., Bechevet B.: Phase change in Ge2Sb2Te5 films investigated by coherent phonon spectroscopy. Appl. Phys. Lett. **77**, 1964–1966 (2000)

Dynamical Coupling of Rabi Oscillation to Coherent Phonon in Semiconductor Microcavities

K. Mizoguchi, S. Yoshino and G. Oohata

Abstract We report on the dynamical coupling between Rabi oscillation and coherent phonon in CuCl semiconductor microcavities, which induces the time-dependent frequency-shift of the coherent phonon mode driven by Rabi oscillation.

1 Introduction

The study on the dynamical coupling between two oscillations is one of the interesting subjects in ultrafast phenomena [1–4]. When two oscillations at two discrete states coupled to each other are resonant in frequency, it is well known that coupled modes appear, which show an anti-crossing behavior. For example, when the quantum beats which originate from quantum interference between two excited states are coupled to coherent phonons, the coherent phonons are remarkably enhanced and the coupled modes are observed [3, 4]. Since Rabi oscillations observed in semiconductor microcavities also originate from quantum interference between two polariton states, it is expected that the Rabi oscillations will be coupled to coherent phonons through the polarization interaction between the coherent phonon and the excitonic components of polaritons. In this study, we have investigated the coupling dynamics between the Rabi oscillations and coherent phonons in the CuCl semiconductor microcavity.

K. Mizoguchi (✉) · S. Yoshino · G. Oohata
Department of Physical Science, Osaka Prefecture University,
1-1 Gakuen-Cho, Naka-Ku, Osaka, Sakai 599-8531, Japan
e-mail: k.mizoguchi@p.s.osakafu-u.ac.jp

© Springer International Publishing Switzerland 2015
K. Yamanouchi et al. (eds.), *Ultrafast Phenomena XIX*,
Springer Proceedings in Physics 162, DOI 10.1007/978-3-319-13242-6_48

2 Experiment

The sample used was CuCl semiconductor microcavities with distributed Bragg reflectors consisting of $PbCl_2/AlF_3$ multilayers on Al_2O_3 substrates [5]. In a fabricated CuCl microcavity, a cavity layer with the thickness of a cavity length ($L_{cav} = \lambda$) consists of a CuCl active layer with the thickness of $L_{act} = \lambda/8$ sandwiched by AlF_3 spacer layers. Here, λ is given by λ_{ex}/n_b, λ_{ex} is the resonant wavelength of the Z_3 exciton of CuCl in vacuum ($\lambda_{ex} = 387$ nm), and n_b is the background refractive index. In the transmittance spectra of the CuCl microcavity obtained at various incident angles of white light, three peaks of the cavity polariton modes varying with the incident angle have been observed. Here, these peaks are called the lower polariton branch (LPB), the middle polariton branch (MPB) and the upper polariton branch (UPB) in energy order. The peak energies of the cavity polariton modes are plotted as a function of incident angle in Fig. 1a. Three solid curves are the dispersion relationships of the cavity polariton fitted to the experimental results. We focused on the Rabi oscillation between the MPB and LPB, because the energy deference between the MPB and LPB can be controlled over the range including the longitudinal optical (LO) phonon energy of CuCl ($E_{LO} = 26$ meV, $\nu_{LO} = 6.3$ THz) by changing the incident angle of pump pulses in a pump-probe technique. Time-domain signals were measured at 10 K by a transmission-type electro-optic (EO) sampling method with second harmonic pulses of a mode-locked Ti:sapphire pulse laser delivering about 90-fs pulse. The center energy of the laser pulses was tuned to 3.207 eV, which was the central energy between the LPB and MPB modes, shown in Fig. 1a.

3 Experimental Results and Discussion

The time-resolved transmission changes of the CuCl microcavity observed at the various incident angles of pump pulses are shown in Fig. 1b. The two oscillatory components, which show the strong oscillation with the fast decay time and the weak oscillation with the long decay time, are observed. The short-lived strong oscillation results from the Rabi oscillation between the MPB and LPB, because the period of the strong oscillation changes with the incident angle. On the other hand, the long-lived weak oscillation is related to the coherent LO phonon driven by Rabi oscillation, because the frequency of the weak oscillation mode observed in the time-partitioning Fourier transformed spectra after 1 ps is located at about 6.3 THz and the amplitude of the weak oscillation is strengthened with increasing the amplitude of Rabi oscillation.

Now, we consider the coupling between Rabi oscillations and coherent phonons. When we assume that the pump pulse excites only the Rabi oscillation and the coherent phonon is generated by the Rabi oscillation under the weak coupling condition, the oscillation $f(t)$ generated by the coupling between two oscillations is

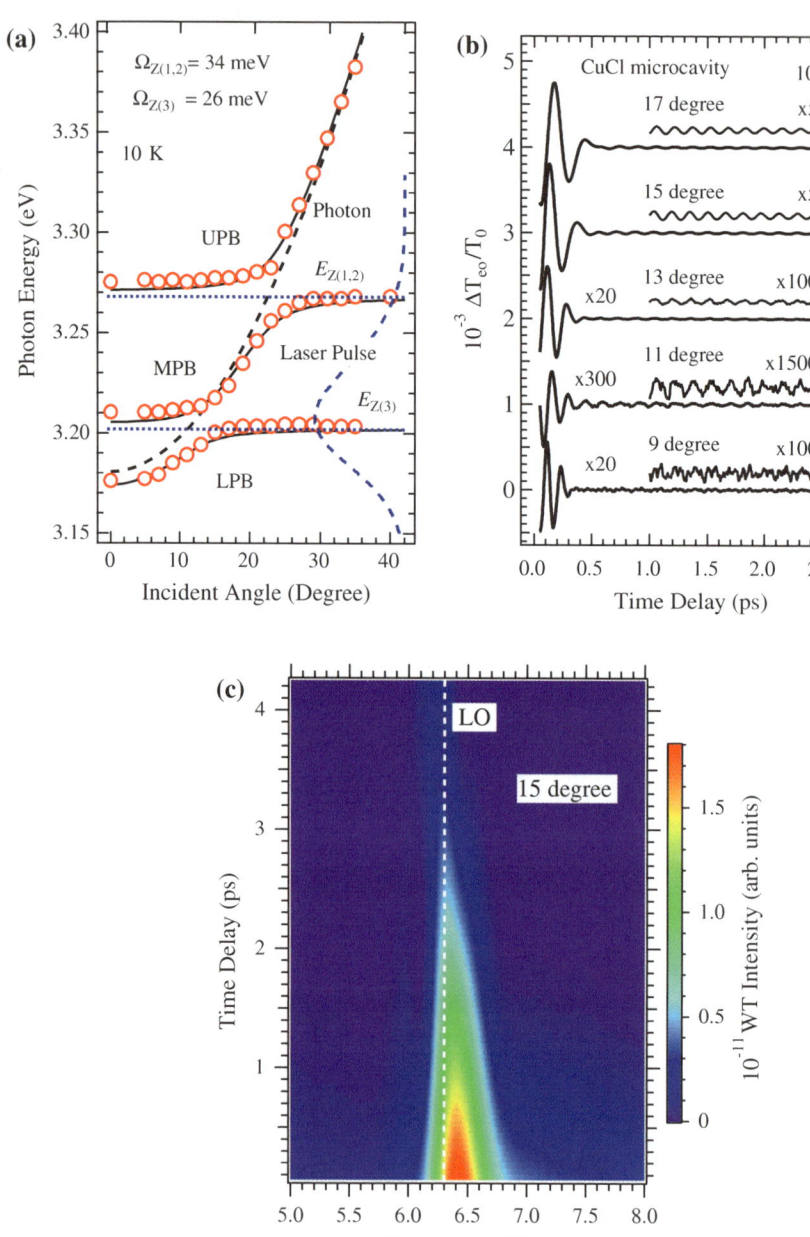

◀**Fig. 1** **a** Incident-angle dependence of the peak energies of the cavity polariton modes in the CuCl microcavity (*open circles*) and dispersion relationship of the cavity polaritons (*solid curves*). The *dotted horizontal lines* denote the energies of the Z_3 and $Z_{1,2}$ excitons: $E_{Z(3)} = 3.202$ and $E_{Z(1,2)} = 3.268$ eV, respectively, and the *black dashed curve* indicates the dispersion relationship of the cavity photon. The *blue dashed curve* shows the laser pulse spectrum. The estimated Rabi splitting energies for Z_3 and $Z_{1,2}$ excitons are $\Omega_{Z(3)} = 26$ meV and $\Omega_{Z(1,2)} = 34$ meV, respectively. **b** Time-resolved transmission changes of the CuCl microcavity obtained at various incident angles of pump pulses. **c** Image plot of the wavelet transformation (WT) of the coupled oscillation extracted from time domain signals observed at the incident angle of 1°. The *vertical dotted line* indicates the LO phonon frequency of CuCl

given by $f(t) = f_{RO}(t) + f_C(t)$, where $f_{RO}(t)$ and $f_C(t)$ indicate the Rabi oscillation driven by pump pulses and the coupled oscillation between the Rabi oscillation and coherent phonon, respectively. Then, to reveal the time evolution of the coupled oscillation, we numerically subtracted the Rabi oscillation signals from the observed time-domain signals, and performed the wavelet transformation (WT) of the subtracted time-domain signals. The image plot of the WT of the coupled oscillation obtained at the incident angle of 15° is shown in Fig. 1c. It is clear that the frequency of the coupled oscillation mode is shifted to the phonon frequency with increasing the time delay; namely, the coupled oscillation shows the time-dependent frequency shift. This result demonstrates that the coupled oscillation mode exists as a coupling state between coherent phonon and Rabi oscillation before Rabi oscillation is relaxed, and the coherent phonon component of the coupled oscillation mode remains after the relaxation of Rabi oscillation.

References

1. J. Shah, *Ultrafast Spectroscopy of Semiconductors and Semiconductor Nanostructures*, (Springer, Berlin, 1996)
2. T. Dekorsy, G. C. Cho, and H. Kurz, *Coherent Phonons in Condensed Media*, In M. Cardona and G. Güntherodt, (Ed.) *Light Scattering in Solids* VIII, (Springer, Berlin, 2000), Chap. 4.
3. O. Kojima, K. Mizoguchi and M. Nakayama: Coupling of coherent longitudinal optical phonons to excitonic quantum beats in GaAs/AlAs multiple quantum wells. Phys. Rev. **B 68**, 155325 (2003).
4. K. Mizoguchi, *et al.*: Coupled mode of the coherent optical phonon and excitonic quantum beat in GaAs/AlAs multiple quantum wells. Phys. Rev. **B 69**, 233302 (2004).
5. S. Yoshino, *et al.*: Optical properties of CuCl microcavities with fluctuations in their refractive index profiles along the cavity structures. Phys. Rev. **B 88**, 205311 (2013).

Coherent Control Over Two-Dimensional Lattice Vibrational Trajectories in α-Quartz Using Polarization Pulse Shaping

Masaaki Sato, Takuya Higuchi, Makoto Kuwata-Gonokami and Kazuhiko Misawa

Abstract Polarization pulse shaping was applied to control the trajectory of two-dimensional vibrational motion in α-quartz. Polarization-twisting pulses were used to impart pseudorotational motion in a degenerate E-symmetry optical phonon mode selectively through impulsive stimulated Raman scattering.

1 Introduction

Polarization-shaped pulses with arbitrary vectorial electric fields as a function of time have attracted considerable interest for manipulating spatially anisotropic or chiral materials [1]. One of the targets is to optically control two-dimensional vibrational trajectories. An earlier study [2] used pairs of pulses with different polarizations to impart momentum impulsively and selectively in orthogonal directions in α-quartz crystals. However, the only adjustable parameter was the time delay between the paired pulses, which simultaneously determines the relative phase between the vibrational motions along the two orthogonal axes of the coordinate and the beat frequency for the repetitive excitation resonant in a particular mode.

We recently implemented polarization pulse shaping for arbitrary vectorial fields by employing a Fourier-synthesis-based optical pulse shaper. We controlled the instantaneous intensity and polarization state with the pulse shaper and converted the tailored pulses into terahertz waves using a nonlinear crystal [3]. In the present

M. Sato · K. Misawa (✉)
Tokyo University of Agriculture and Technology, 2-24-16 Naka-Cho, Koganei 184-8588, Japan
e-mail: kmisawa@cc.tuat.ac.jp

T. Higuchi · M. Kuwata-Gonokami
The University of Tokyo, 7-3-1 Hongo, Bunkyo-Ku, Tokyo 113-8656, Japan

© Springer International Publishing Switzerland 2015
K. Yamanouchi et al. (eds.), *Ultrafast Phenomena XIX*,
Springer Proceedings in Physics 162, DOI 10.1007/978-3-319-13242-6_49

study, we were successful in selectively driving a desired pseudorotational motion of a single vibrational mode in an α-quartz crystal using polarization-twisting pulses traveling along the three-fold axis of the crystal.

2 Methods

We used polarization-twisting pulses in which the direction of linear polarization was rotated at a constant angular frequency of ω over the intensity envelope, as depicted in Fig. 1. Polarization-twisting pulses were generated as a superposition of right- and left-circularly polarized pulses which are linearly chirped and temporally displaced [3]. The dynamic polarization states of the shaped pulses can be represented by the intensity envelope $I(t)$, polarization azimuth angle $\phi(t)$, and ellipticity $\eta(t)$. $\phi(t)$ evolves in time as $\phi(t) = \omega t$ and η is kept 0 in case of the polarization-twisting pulses. The temporal duration of the intensity envelope was set to be longer than the vibrational periods by adding a sufficiently large chirp.

The α-quartz crystal has a C_3 axis of symmetry that yields three normal modes; one is the breathing mode with A symmetry and the other two are degenerate E-modes. The displacements of the A-mode $q_A(t)$ and E-modes $q_{E_a}(t)$ and $q_{E_b}(t)$ can be written in the following simple form owing to the three-fold crystal symmetry [4].

$$q_A(t) = \int_{-\infty}^{t} R_A(t - \tau)I(\tau)\mathrm{d}\tau$$

$$q_{E_a}(t) = \int_{-\infty}^{t} R_E(t - \tau)I(\tau)\cos 2\omega\tau\mathrm{d}\tau, \qquad q_{E_b}(t) = -\int_{-\infty}^{t} R_E(t - \tau)I(\tau)\sin 2\omega\tau\mathrm{d}\tau$$

The breathing A-mode is not efficiently excited because the convolution of the vibrational response function $R_A(t - \tau)$ with the intensity envelope $I(t)$ is smoothed out because of the longer pulse duration. On the other hand, the E-mode vibration

Fig. 1 Pump-probe measurements with polarization-twisting pulse

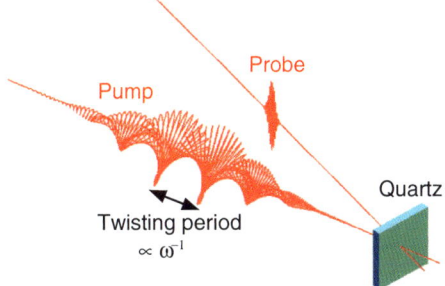

Fig. 2 Fourier-transformed Raman spectra for excitations with polarization-twisting pulses at $\Omega = 128$ cm^{-1} (*red*) and 402 cm^{-1} (*blue*), as well as with Fourier-transform limited pulses (*green*)

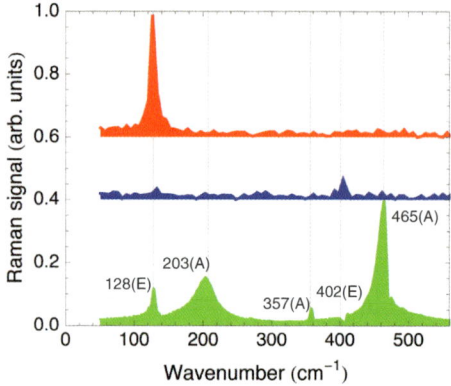

was resonantly enhanced with an angular frequency of the electric polarization vector of $\Omega = 2\omega$. The direction of the E-mode vibration was determined by the rotational direction of the polarization vector.

3 Results

Impulsive stimulated Raman scattering (ISRS) was used to study mode selection by polarization-shaped pulses in the time domain. By Fourier-transform limited (FTL) pulses, A- and E-modes are simultaneously excited. The vibrational frequencies of 128, 203, 357, and 465 cm^{-1} observed in the Raman spectra converted from the ISRS transients are in good agreement with previously reported values [5].

The red and blue curves in Fig. 2 represent the Raman spectra excited by polarization-twisting pulses at tuning frequencies of 128 and 402 cm^{-1}, respectively. The Raman spectrum obtained with FTL pulses is shown in green. A well-defined spectral peak was observed in each Raman spectrum just around the tuning frequency shown on the red and blue curves. Both peaks show resonance with the pseudorotational motion in α-quartz. All other A-modes were eliminated as designed. With 402 cm^{-1} polarization-twisting pulses, a small peak appeared at 401 cm^{-1} on the blue curve. This 401 cm^{-1} mode has a very weak Raman cross section, and it is usually hidden behind the tail of the much stronger A-mode at 464 cm^{-1}.

4 Discussion and Conclusion

Using a polarization-twisting pulse, two-dimensional pseudorotation of the E-mode was selectively enhanced, and the breathing motion of the A-mode was strongly suppressed. If arbitrarily-shaped excitation waveforms are used, more complex vibrational trajectories are also controlled. This technique is applicable to a broad class of degenerate modes, such as electron spin coherence, which can also be induced in different polarization directions by pulses of different polarizations.

References

1. M. Sato, T. Suzuki, and K. Misawa, Rev. Sci. Instrum. **80**, 123107 (2009).
2. M. M. Wefers, H. Kawashima, and K. A. Nelson, J. Chem. Phys. **108**, 10248 (1998).
3. M. Sato, T. Higuchi, N. Kanda, K. Konishi, K. Yoshioka, T. Suzuki, K. Misawa and M. Kuwata-Gonokami, Nature Photonics **7**, 724-731 (2013).
4. T. Higuchi, H. Tamaru, and M. Kuwata-Gonokami, Phys. Rev. A **87**, 013808 (2013).
5. J. F. Scott, and S. P. S. Porto, Phys. Rev., **161**, 901 (1967).

Ultrafast Lattice Dynamics of Phase-Change Materials Monitored by a Pump-Pump-Probe Technique

Muneaki Hase, Paul Fons, Kirill Mitrofanov, Alexander V. Kolobov and Junji Tominaga

Abstract We explore an ultrafast structural transformation in a Ge_2Te_2/Sb_2Te_3 superlattice, using pump-pump-probe spectroscopy. The coherent phonon spectra exhibit complex structural dynamics upon photo-excitation, being described as a mixing of two different Ge bonding configurations.

1 Introduction

$Ge_2Sb_2Te_5$ (GST) is one of the highest-performance optical recording media among commercially available phase-change materials [1]. Recently, in order to reduce both the speed and energy required for switching, interfacial phase change memory (iPCM) has been proposed. iPCM is a superlattice (SL) structure consisting of GeTe and Sb_2Te_3 layers [2]. A question arising from the dynamics of the phase change in GST and iPCM is how fast the phase transformation between the amorphous (referred to as RESET) and the crystalline (referred to as SET) phases occurs. Motivated by understanding the mechanism of the rapid phase change process, thermal and non-thermal crystallization of GST and iPCM films was examined using a femtosecond pump-probe technique and it was found that the appearance of the coherent phonons was significantly modified upon the phase change [3, 4]. Despite the recent activity investigating nonthermal phase change in GST alloys and iPCM [4, 5], the mechanism of the nonthermal phase change remains largely unknown, especially on sub-picosecond time scales under strong photo-excitation. Here we focus on the femtosecond order phase transformation between the SET and RESET phases under

M. Hase (✉)
Institute of Applied Physics, University of Tsukuba,
1-1-1 Tennodai, Tsukuba 305-8573, Japan
e-mail: mhase@bk.tsukuba.ac.jp

P. Fons · K. Mitrofanov · A.V. Kolobov · J. Tominaga
Nanoelectronics Research Institute, National Institute of Advanced Industrial Science
and Technology, Tsukuba Central 4, 1-1-1 Higashi, Tsukuba 305-8562, Japan

© Springer International Publishing Switzerland 2015   210
K. Yamanouchi et al. (eds.), *Ultrafast Phenomena XIX*,
Springer Proceedings in Physics 162, DOI 10.1007/978-3-319-13242-6_50

strong photo-excitation with fluences corresponding to the excitation of several percent of the valence electrons, in which 'hardening' of the phonon frequency, instead of usual softening of the phonon frequency, is observed in a prototypical iPCM, Ge_2Te_2/Sb_2Te_3 SL.

2 Experimental

The samples used were Ge_2Te_2/Sb_2Te_3 SL thin films grown on a silicon (100) substrate by using a helicon-wave RF magnetron sputtering machine. To measure time-resolved reflectivity change of the sample, 40 fs-width optical pulses (wavelength = 800 nm) from a Ti:sapphire regenerative amplifier system were utilized. A pair of the pump-pulses was generated through a Michelson interferometer, in which the separation time (Δt) of the double pump-pulses was precisely controlled. Using a sequence of a double pump-pulse and a probe pulse, pump-pump-probe spectroscopy was carried out, in which the prepump-pulse ($P_1 = 10.6$ mJ/cm^2) excited the electrons from bonding into the excited anti-bonding state, followed by generation of a coherent phonon in the excited state by another weak pump ($P_2 = 6.9$ mJ/cm^2), which was monitored by the probe pulse ($P_3 = 0.2$ mJ/cm^2).

3 Results and Discussion

Figure 1(a) shows the time-resolved reflectivity signal observed at room temperature for a Ge_2Te_2/Sb_2Te_3 SL film in the SET phase. After the transient electronic response due to the excitation of carriers, coherent phonons with a few picoseconds relaxation time appear without prepump-pulse excitation (see the lower part of Fig. 1a). When the prepump-pulse is applied, the relaxation time of the coherent phonon changes significantly depending on the separation time (Δt); the coherent phonon exhibits a strongly damped oscillation at $\Delta t = 290$ fs. As shown in Fig. 1b, Fourier transformed (FT) spectra taken in the SET state without prepump-pulse excitation exhibit a single peak at ≈ 3.48 THz, which is consistent with that of the optical mode of the SET phase [3, 4]. The FT spectra taken only from the time-domain data ($\Delta t = 290$ fs) after the P_2 pump arrival explore the double peak structure peaking at ≈ 2.55 and ≈ 3.70 THz. This new peak position at 2.55 THz changes to higher frequency with increasing Δt and disappears at $\Delta t = 2,320$ fs. The new peak at 2.55 THz is interpreted as a result of phonon softening by electronic excitation. Most importantly, the main peak at 3.48 THz 'blue-shifts' by 0.22 THz. This blue-shift implies the RESET phase appears under the strong photo-excitation since the frequency of 3.70 THz is very close to the frequency of the RESET phase (3.74 THz). Based on ab initio molecular dynamics simulations, Li et al. reported that in the early stage of the phase change from the SET to RESET phase in GST alloy the coordination number of Ge atoms changed from the original

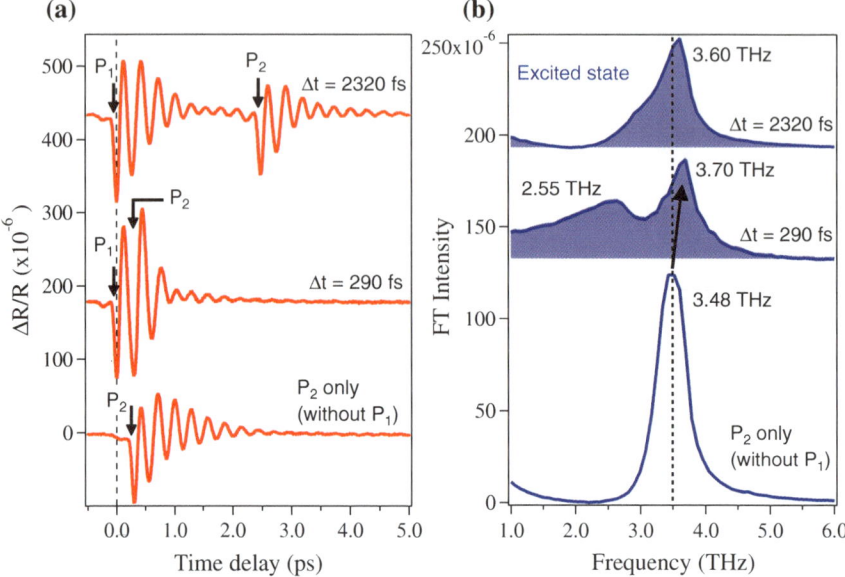

Fig. 1 a Transient reflectivity trace observed in the SET phase of a SL film at $\Delta t = 290$ and 2,320 fs. The result for the case without a prepump (P_1) pulse is shown at the bottom for reference. **b** FT spectra obtained from the time-domain data in **a**

six-fold into a mixture of five-fold and four-fold coordinations within 450 fs [6]. Therefore, the characteristic double-peak FT spectrum observed at ultrafast time spans after photoexcitation implies a mixture of two different Ge bonding configurations.

4 Summary

In summary, we have experimentally explored the ultrafast 'blue-shift' of the coherent optical phonon in iPCM films using a pump-pump-probe sequence. The transient phase characterized by a double-peak FT spectrum is interpreted as being due to a mixture of two different Ge bonding configurations, which relax within a few picoseconds.

Acknowledgement This work was supported by X-ray Free Electron Laser Priority Strategy Program, entitles "Lattice dynamics of phase change materials by time-resolved X-ray diffraction" (NO. 12013011 and 12013023), from the MEXT of Japan.

References

1. M. Wuttig and N. Yamada, "Phase-change materials for rewritable data storage," Nature Mater. **6**, 824–832 (2007).
2. R. E. Simpson *et al.*, "Interfacial phase-change memory," Nature Nanotech. **6**, 501–505 (2011).
3. M. Först *et al.*, "Phase change in $Ge_2Sb_2Te_5$ films investigated by coherent phonon spectroscopy," Appl. Phys. Lett. **77**, 1964–1966 (2000).
4. K. Makino, J. Tominaga, and M. Hase, "Ultrafast optical manipulation of atomic arrangements in chalcogenide glass memory materials," Opt. Exp. **19**, 1260–1270 (2011).
5. P. Fons *et al.*, "Photoassisted amorphization of the phase-change memory alloy $Ge_2Sb_2Te_5$," Phys. Rev. B. **82**, 041203(R) (2010).
6. X. B. Li *et al.*, "Role of electronic excitation in the amorphization of Ge-Sb-Te alloys," Phys. Rev. Lett. **107**, 015501 (2011).

Enhancement of Superconducting Coherence in $YBa_2Cu_3O_x$ by Resonant Lattice Excitation

Daniele Nicoletti, W. Hu, S. Kaiser, C.R. Hunt, I. Gierz, M. Le Tacon, T. Loew, B. Keimer and A. Cavalleri

Abstract By using femtosecond pulses in the mid-infrared, we resonantly excite an infrared-active phonon mode in the high-temperature superconductor $YBa_2Cu_3O_x$. The electronic properties of the driven state, probed with ultra-broadband time-domain terahertz spectroscopy, are highly unconventional.

Non-linear resonant excitation of infrared-active vibrational modes of the crystal lattice has been proven to melt magnetic or orbital orders and to induce insulator-to-metal transitions in complex solids [1, 2]. In high-T_c superconducting cuprates, this technique was used to remove the competing charge- and spin-order in the so-called stripe phase in $La_{1.675}Eu_{0.2}Sr_{0.125}CuO_4$, thus transiently inducing superconductivity (at temperatures as high as 20 K) in a material which is not superconducting at equilibrium [3]. In the present experiment, we show that a large-amplitude modulation of the apical oxygen positions (see Fig. 1a) in $YBa_2Cu_3O_x$ can promote highly unconventional electrodynamics, which can be captured with ultra-broadband time-domain terahertz spectroscopy.

Superconductors at equilibrium display two characteristic physical properties: zero DC resistance and the expulsion of static magnetic fields. The first of these properties manifests itself as a positive imaginary part of the optical conductivity $\sigma_2(\omega)$ that diverges at low frequency as $1/\omega$. In high-T_c cuprates, the layered structure gives rise to additional c-axis excitations of the superfluid, with the appearance of one or more longitudinal Josephson plasma modes due to tunneling of Cooper pairs between capacitively coupled superconducting planes. In $YBa_2Cu_3O_x$, a bi-layer cuprate, two longitudinal plasma modes are found [4], reflected

D. Nicoletti (✉) · W. Hu · S. Kaiser · C.R. Hunt · I. Gierz · A. Cavalleri
Max Planck Institute for the Structure and Dynamics of Matter, c/o DESY - Building 99 (CFEL) Luruper Chaussee 149, 22761 Hamburg, Germany
e-mail: daniele.nicoletti@mpsd.mpg.de

M. Le Tacon · T. Loew · B. Keimer
Max Planck Institute for Solid State Research, Stuttgart, Germany

A. Cavalleri
Department of Physics, Oxford University, Clarendon Laboratory, Oxford, UK

© Springer International Publishing Switzerland 2015
K. Yamanouchi et al. (eds.), *Ultrafast Phenomena XIX*,
Springer Proceedings in Physics 162, DOI 10.1007/978-3-319-13242-6_51

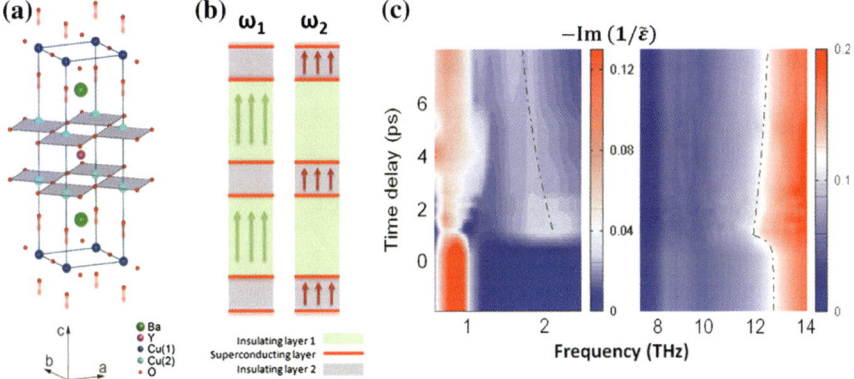

Fig. 1 **a** Crystal structure of $YBa_2Cu_3O_x$ and sketch of the optically-driven distortion of the apical oxygen. **b** Josephson plasma modes of the bi-layer structure. **c** Transient energy loss function of $YBa_2Cu_3O_{6.5}$ after mid-infrared excitation at 10 K. The dynamics of the Josephson plasma modes is displayed by exponential fits as a function of pump-probe time delay

by two peaks in the energy loss function at $\omega_1 \sim 1$ and $\omega_2 \sim 14$ THz (Fig. 1b, c, negative time delays). Within each family of cuprates, the mode frequency quantifies the strength of the Josephson coupling between pairs of CuO_2 layers.

Here, we measure the transient c-axis optical properties of $YBa_2Cu_3O_x$ ($x = 6.45$, 6.5, 6.6) after photo-excitation, both below and above the superconducting transition temperature T_c [5, 6]. Mid-infrared pump pulses of ~ 300 fs duration, polarized along the c direction and tuned to ~ 20 THz frequency, were made resonant with the infrared-active distortion shown in Fig. 1a. Such pump pulses were generated by difference-frequency mixing in an optical parametric amplifier and focused onto the samples with a maximum fluence of 4 mJ/cm^2, corresponding to peak electric fields up to ~ 3 MV/cm. At these strong fields, the apical oxygen positions are driven in an oscillatory way by several percent of the equilibrium unit cell [5]. The solid was then investigated with broadband THz probe pulses generated both by optical rectification in ZnTe (0.5–2.5 THz) and by gas ionization (1–14 THz).

The transient optical properties of $YBa_2Cu_3O_x$ were extracted from measurements of the amplitude and phase of the reflected electric field after photo-excitation, taking into account the pump-probe penetration depth mismatch [5]. In Fig. 1c, we report the changes in the inter- and intra-bilayer Josephson coupling strength by plotting the time- and frequency-dependent energy loss function of $YBa_2Cu_3O_{6.5}$ in the superconducting state [6]. The peak at $\omega_1 \sim 1$ THz, which reflects the inter-bilayer plasma mode at equilibrium, reduces in amplitude after photo-excitation, as a second higher-frequency peak appears at ~ 2 THz. Simultaneously, the intra-bilayer mode, which at equilibrium is observed at $\omega_2 \sim 14$ THz, shifts to the red. All spectral changes relax back to equilibrium within ~ 7 ps.

A qualitatively similar response to that described above is found in the same material above T_c (not shown). After excitation, a peak in the loss function appears

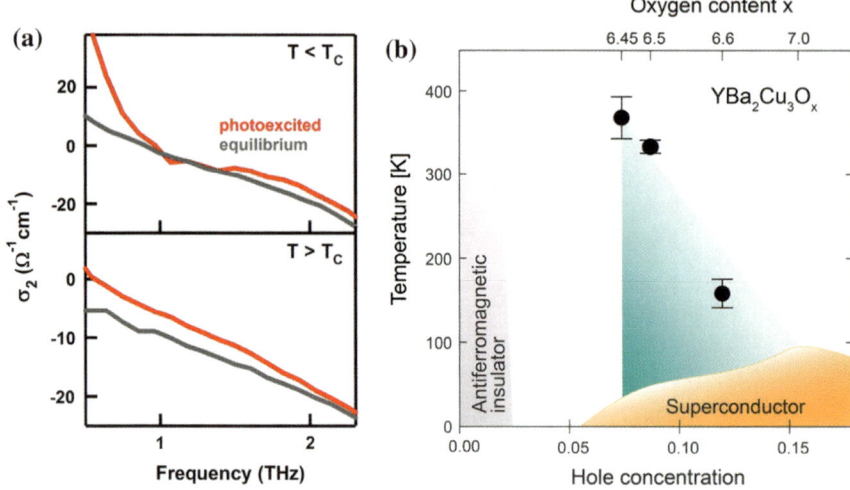

Fig. 2 a Transient increase in the imaginary part of the optical conductivity at low frequency, measured 0.8 ps after photo-excitation in YBa$_2$Cu$_3$O$_{6.5}$, both *below* and *above* T_c = 50 K (at 10 and 100 K, respectively). **b** Temperature-doping phase diagram of YBa$_2$Cu$_3$O$_x$. The region affected by optically-enhanced coherence (*shaded*) has been determined by estimating the maximum temperature at which a divergence in $\omega\sigma_2(\omega)|_{\omega\rightarrow0}$ could be detected for each doping level (*black circles*)

at ~ 2 THz, while a red-shift of spectral weight is detected around ~ 14 THz. These observations are consistent with an optically-driven transfer of coupling strength from the bi-layers towards the inter-bilayer region, occurring both below and above T_c.

In Fig. 2a, we report the imaginary part of the optical conductivity of YBa$_2$Cu$_3$O$_{6.5}$ after excitation. Since the superfluid density at equilibrium is quantified by $\omega\sigma_2(\omega)|_{\omega\rightarrow0}$, the increase in the slope of $\sigma_2(\omega)$ at low frequency indicates a photo-induced enhancement of superfluid density in the superconducting state (upper panel). Above T_c (lower panel) $\omega\sigma_2(\omega)|_{\omega\rightarrow0}$ also exhibits an increase, turning positive and diverging down to the lowest measured frequency. This behavior, combined with the other optical constants, is consistent with the response of a conductor with anomalously high mobility, which would be highly unusual for incoherent transport in oxides. An interpretation based on a transient supercon-ducting state above T_c is in our view more plausible [6].

The phase diagram of YBa$_2$Cu$_3$O$_x$ is displayed in Fig. 2b. The shaded region indicates the non-equilibrium high-mobility phase induced by resonant lattice excitation. Remarkably, this extends even above room temperature at the lowest doping levels and tracks surprisingly well the onset temperature of the so-called pseudogap state in the equilibrium phase diagram of high-T_c cuprates [7].

All the above observations are compatible with previous reports of an inho-mogeneous state above T_c, which would retain important properties of a super-conductor [8]. In such scenario, mid-infrared light might melt a competing order, or dynamically synchronize the inter-layer phase [3, 9]. The transient redistribution of coherence demonstrated here could lead to new strategies to enhance superconductivity in the steady state.

References

1. M. Först, C. Manzoni, S. Kaiser, *et al.*, Nature Physics **7**, 854 (2011).
2. M. Rini, R. I. Tobey, N. Dean, *et al.*, Nature **449**, 72 (2007).
3. D. Fausti, R. I. Tobey, N. Dean, *et al.*, Science **331**, 189 (2011).
4. P. W. Anderson, Science **268**, 1154 (1995).
5. S. Kaiser, C. R. Hunt, D. Nicoletti, *et al.*, Phys. Rev. B **89**, 184516 (2014).
6. W. Hu, S. Kaiser, D. Nicoletti, *et al.*, Nature Materials **13**, 705 (2014).
7. T. Timusk and B. Statt, Rep. Prog. Phys. **62**, 61 (1999).
8. K. K. Gomes, A. N. Pasupathy, A. Pushp, *et al.*, Nature **447**, 569 (2007).
9. V. J. Emery and S. A. Kivelson, Nature **374**, 434 (1995).

Coherent Phonon Dynamics in Singlet Fission of Rubrene Single Crystal

**Kiyoshi Miyata, Shunsuke Tanaka, Toshiki Sugimoto,
Kazuya Watanabe, Takafumi Uemura, Jun Takeya
and Yoshiyasu Matsumoto**

Abstract We have observed coherent phonons of singlet and triplet exciton states in a rubrene single crystal at 35 K. Our results suggest that the instantaneous triplet exciton formation through the singlet fission of the rubrene single crystal generates coherent phonons on the potential energy surface of a triplet exciton.

1 Introduction

Singlet fission (SF) in organic materials, the fission of a singlet exciton (S_1) into two triplet excitons ($2T_1$) within picoseconds time scale, has recently attracted much attention because of their potential to boost the light conversion efficiency of organic photovoltaic devices [1]. A better understanding of the microscopic mechanisms of SF is necessary to improve SF yields.

Recently, the nearly instantaneous formation of triplet pair states, $^1(T_1T_1)$, has been observed in the SF of tetracene and pentacene thin films by use of two-photon photoemission spectroscopy [2, 3]. And the rate of SF in tetracene and pentacene thin films shows no temperature dependence. In contrast, the SF of rubrene is supposed to be thermally-activated; the SF rate shows substantial temperature dependence [4]. This suggests that the mechanism of SF in a rubrene single crystal is different from those of tetracene and pentacene. Thus, to understand the SF mechanism in a rubrene single-crystal, it is vital to investigate whether or not the formation of triplet pair state occurs nearly instantaneously. In this report, we describe the SF process in a rubrene single-crystal at 35 K observed with ultrafast

K. Miyata · S. Tanaka · T. Sugimoto · K. Watanabe · Y. Matsumoto (✉)
Department of Chemistry, Graduate School of Science, Kyoto University,
Kyoto 606-8502, Japan
e-mail: matsumoto@kuchem.kyoto-u.ac.jp

T. Uemura · J. Takeya
Department of Frontier Sciences, The University of Tokyo, Kashiwa,
Chiba 277-8561, Japan

© Springer International Publishing Switzerland 2015
K. Yamanouchi et al. (eds.), *Ultrafast Phenomena XIX*,
Springer Proceedings in Physics 162, DOI 10.1007/978-3-319-13242-6_52

pump-probe spectroscopy, focusing on rapid $^1(T_1T_1)$ formation. We discuss the mechanisms of SF in a rubrene single crystal on the basis of coherent phonon oscillations observed.

2 Experimental Methods

The pump-probe measurements were done by using home-built noncollinear optical parametric amplifiers (NOPA) pumped by a second harmonics of outputs of a Ti: Sapphire regenerative amplifier (1 kHz) [5]. The NOPA output at 2.1–2.4 eV (40 fs) was used for pumping a rubrene crystal at the absorption band edge. Another NOPA was used for the probe light in the range from 1.65 to 2.5 eV (40 fs), and outputs of Ti:Sapphire laser were used for probe (120 fs) at 1.56 eV. The probe pulses were detected by a photodiode equipped with a monochromator. The rubrene single crystal grown by physical vapor transport was cooled down to 35 K using a He cryostat mounted in a high vacuum cell. The polarization directions of pump and probe pulses were set to be parallel to b-axis of the crystal. Pump power was kept low enough to avoid multiple photon absorption and heating induced by absorption of pump pulses.

3 Results and Discussion

Figure 1a shows transient absorption (TA) spectra of rubrene single-crystal at room temperature as a function of pump-probe delay. While a broad feature around 2.0 eV due to S_1–S_n transition decays within a few tens of ps (Fig. 1b), the absorptions at 1.55 and 2.45 eV due to T_1–T_n transition grows (Fig. 1c). The rise of T_1 and the decay of S_1 are clearly suppressed at 35 K (Fig. 1b, c).

Oscillatory signals due to coherent phonons (CPs) were observed in temporal profiles of transient absorption signals at 35 K. At 2.4 eV, we observed the oscillatory signals due to a CP with a frequency of 83 cm^{-1} (not shown); this is ascribed to the CP on a ground state potential energy surface (PES). Oscillatory features observed at 2.06 and 1.56 eV are shown in Fig. 2. At 2.06 eV in the S_1–S_n transition band, the oscillatory feature with a center frequency of 78 cm^{-1} dephases with a time constant of ~ 1 ps. We ascribed this oscillatory feature to CPs on the S_1 PES; this is consistent with the oscillation observed at 1.2 eV in a band that has been assigned to S_1–S_n transition [6]. At 1.56 eV in the T_1–T_n transition band, the oscillatory feature with its dephasing time longer than 5 ps was observed and its frequency was 124 cm^{-1}. The feature is clearly distinct from the oscillations observed on the ground and S_1 exciton PESs. Therefore, we assigned this oscillation to the CPs on the T_1 PES.

This implies that the coupling between S_1 and $^1(T_1T_1)$ is so strong that nearly instantaneous triplet formation occurs in the rubrene single crystal even at low

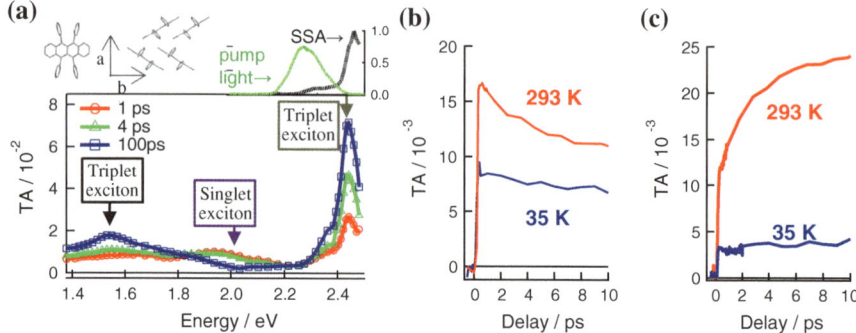

Fig. 1 a Transient absorption (TA) spectra observed at room temperature, a steady state absorption (SSA) spectrum, and the spectrum of pump light. The molecular structure and crystalline structure of a rubrene single-crystal are depicted in the *inset*. **b** Temporal profiles of TA at 2.06 eV and **c** 1.56 eV

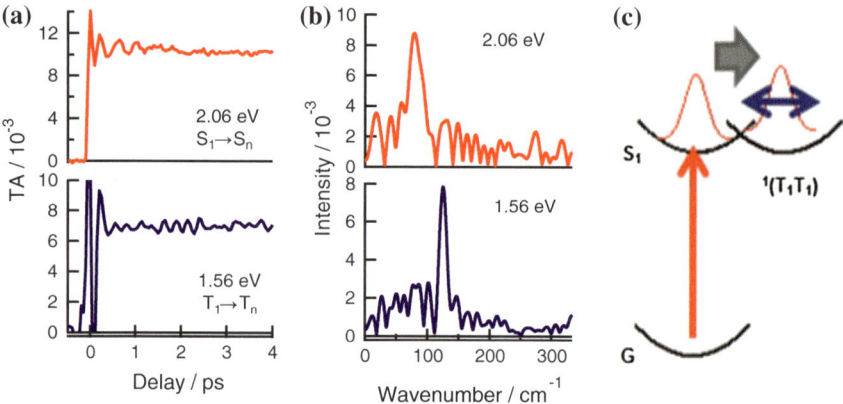

Fig. 2 a Temporal profiles of transient absorption observed at 1.56 and 2.06 eV at 35 K and **b** Fourier spectra of the oscillatory features in (**a**). **c** A schematic diagram of coherent phonon generation on T_1 PES

temperature; this rapid formation generates coherent nuclear motions on the triplet exciton PES (Fig. 2c). Therefore, in addition to the slow thermally activated fission, nearly instantaneous triplet generation occurs in the rubrene single crystal as in tetracene and pentacene thin films.

Acknowledgments This work was supported by the Grant-in-Aid for Scientific Research (A) from the Japanese Society for the Promotion of Sciences (Grant No. 22245001), Grant-in-Aid for Challenging Exploratory Research from Japan Society for the Promotion of Science (Grant No. 24655011), Kyoto University Global COE program, Grants for Excellent Graduate Schools, MEXT, Japan, and the program of Network of Joint Research Center for Advanced Materials and Devices.

References

1. M. B. Smith and J. Michl, *Chem. Rev* 110, 6891 (2010).
2. W.–L. Chan, M. Ligges, A. Jailaubekov, L. Kaake, L. Miaja-Avila, and X.–Y. Zhu, *Science,* 334, 1541 (2011).
3. W.–L. Chan, M. Ligges, and X.–Y. Zhu, *Nat. Chem.,* 4, 840 (2012).
4. L. Ma, K. Zhang, C. Kloc, H. Sun, C. Soci, M. E. Michel-Beyerle, and Gagik G. Gurzadyan, *Phys. Rev. B,* 87, 201203(R) (2013).
5. K. Watanabe, K. Inoue, I. F. Nakai, M. Fuyuki, and Y. Matsumoto, Phys. Rev. B, 80, 075404 (2009).
6. S. Tao, N. Ohtani, R. Uchida, T. Miyamoto, Y. Matsui, H. Yada, H. Uemura, H. Matsuzaki, T. Uemura, J. Takeya, and H. Okamoto, *Phys. Rev. Lett.,* 109, 097403 (2012).

Acceleration of Ultrafast Singlet Fission in Aza-Derivative of TIPS-Pentacene

T. Buckup, J. Herz and M. Motzkus

Abstract We unveil a new general channel for formation of triplet states via singlet fission in TIPS-pentacene derivatives by probing sub 100 fs dynamics in the near infrared spectral region with transient absorption.

1 Introduction

The efficiency of organic materials acting as single junction solar cells depends on the formation of two triplet states via the fission of a singlet state [1]. In this regard, a singlet fission (SF) occurring on an ultrashort time scale is an advantage since competing channels will be not populated, potentially leading to very high quantum yields. Pentacene (Pn) and its derivatives are very attractive compounds in organic electronics [2], especially because of their high efficient formation of triplets via SF. TIPS-pentacene (TIPS-Pn) is one important derivative since it enhances orbital overlap due to its side groups, leading to quantum yields of 144 % through the formation of triplet states [3]. The SF dynamics has been suggested to take place on a time scale of 1 ps. This contrasts to SF in unsubstituted pentacene, which reaches a quantum yield of 160–180 %, but takes place on a much shorter time scale of just about 80 fs [2]. Here we apply transient absorption with high time resolution to TIPS-Pn and a new derivative in a spin-coated thin-film. It is shown that the formation of triplet states via SF in these compounds is much faster than observed previously. In order to avoid overlapping contributions in the visible spectral

T. Buckup · J. Herz · M. Motzkus (✉)
Physikalisch-Chemisches Institut, Ruprecht-Karls-Universität Heidelberg,
69120 Heidelberg, Germany
e-mail: marcus.motzkus@pci.uni-heidelberg.de

T. Buckup
e-mail: tiago.buckup@pci.uni-heidelberg.de

© Springer International Publishing Switzerland 2015
K. Yamanouchi et al. (eds.), *Ultrafast Phenomena XIX*,
Springer Proceedings in Physics 162, DOI 10.1007/978-3-319-13242-6_53

region, we additionally probe the formation of triplet states in the near infrared (NIR) region. Moreover, we show that the carbon to nitrogen substitution can accelerate SF in diaza-TIPS-Pn by almost a factor of two [4].

2 Experimental Details

Transient absorption measurements were performed at 1 kHz repetition rate using a short pulse generated in non-collinear optical parametric amplifier (at about 600 nm) as pump pulse and a white light supercontinuum as probe pulse. Pump duration was about 20 fs. Probe light in the visible was generated via tight focusing of few microjoules at 800 nm in a 2 mm sapphire crystal. TIPS-Pn was obtained from Sigma-Aldrich and used without further purification. Diaza-TIPS-Pn has been synthesized according to [5]. Thin films were prepared via spin coating, leading to a thickness of ca. 100 nm.

3 Results and Discussion

The transient absorption signal of TIPS-Pn and diaza derivative shows similar features (Fig. 1). The visible region is dominated by a strong and broad excited state absorption (ESA). In TIPS-Pn, the ESA band has a shoulder at about 460–490 nm, which decays fast while the maximum at about 530 nm rises. Such a transient evolution of the shoulder and the maximum of the ESA cannot be clearly observed

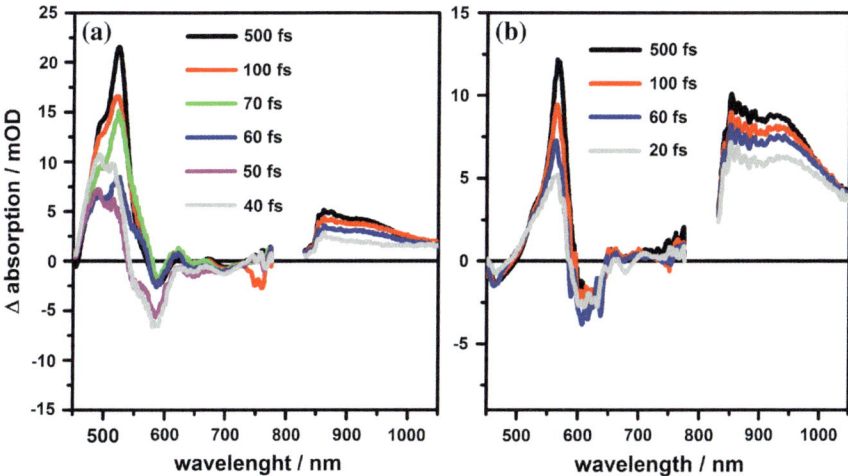

Fig. 1 Transient absorption spectra at several probe delay times for **a** TIPS-Pn and **b** Diaza-TIPS-Pn

for diaza, due to the faster and overlapping dynamics. For TIPS-Pn, the band with a maximum at 530 nm is usually identified as the triplet band, while the shoulder at 490 nm is originated from a singlet absorption [3].

A careful analysis and fitting of selected transients at the maximum of the ESA and its shoulder unveils an ultrafast rise and decay time, respectively (Fig. 2a, b). For TIPS-Pn, the transient at 490 nm decays with just 120 fs while at 530 nm it rises with a time constant of just 200 fs, instead of 1 ps as observed previously. For the diaza derivative, the rise time of the triplet band is much faster with a time constant of about 100 fs. In order to avoid the overlap between the singlet and triplet absorption bands in the visible, the dynamics in the NIR was probed. For pentacene, bands in the NIR at about 900 nm were assigned exclusively to triplet ESA [6]. The NIR bands (Fig. 2c) show a rise time of 200 fs for TIPS-Pn and 100 fs for diaza, agreeing with the results observed in the visible spectral region. The relative amplitude of the NIR to visible absorption bands further indicates a high efficient formation of triplets for diaza when compared with TIPS-Pn.

Moreover, a global target analysis (not shown) was performed, taking into account several models. A sequential model with an ultrafast triplet formation fitted well the data for both TIPS-Pn and diaza. In this regard, no additional loss channels from the initially populated singlet state were required to fit the data.

Summing up, transient absorption of TIPS-pentacene derivatives shows that the formation of triplets is much faster than observed before. In this process, loss channels are not required to explain the dynamics. Such an ultrafast singlet fission can now explain the high quantum yield of TIPS-pentacene. Moreover, substitution of carbon by nitrogen atoms in diaza-TIPS-pentacene can accelerate the SF process by a factor of two, making it a promising new compound for organic electronics applications.

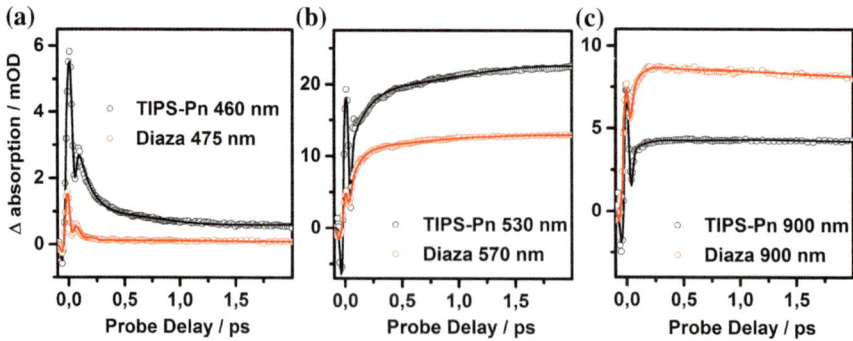

Fig. 2 Transients at selected detection wavelengths with corresponding fits. **a** TIPS-Pn at 460 nm and diaza-TIPS-Pn at 475 nm. **b** TIPS-Pn at 530 nm and diaza-TIPS-Pn at 570 nm. (**c**) TIPS-Pn and diaza-TIPS-Pn at 900 nm

References

1. M.B. Smith and J. Michl, Chemical Reviews **110,** 6891 (2010).
2. C.D. Dimitrakopoulos and P.R.L. Malenfant, Advanced Material. **14,** 99 (2002).
3. C. Ramanan et al., Journal of the American Chemical Society **134,** 386 (2011).
4. J. Herz et al., Journal of Physical Chemistry Letters **5,** 2425 (2014).
5. A.L. Appleton et al., Nature Communications **1,** 91 (2010).
6. M.W.B Wilson et al., Journal of the American Chemical Society **133,**11830 (2011).

Vibrational Coherence Reveals the Role of Dark Multiexciton States in Ultrafast Singlet Exciton Fission

Artem A. Bakulin, Sarah E. Morgan, Jan Alster, Dassia Egorova, Alex Chin, Donatas Zigmantas and Akshay Rao

Abstract We use 2D electronic photon-echo spectroscopy to study ultrafast singlet exciton fission in pentacene. Our observations and analysis of vibronic coherences provide insight to the role played by dark multiexcitonic states in mediating fission.

OCIS codes (320.7130) Ultrafast processes in condensed matter · Semiconductors · (350.6050) Solar energy

Singlet exciton fission has been attracting increasing attention recently due to its potential for use in excitonic solar cells, enabling the generation of two electron–hole pairs per photon absorbed. This allows better use of high-energy photons in the solar spectrum and could enable solar cells to overcome the Shockley-Queisser limit on the power conversion efficiency of a single junction cell [1]. The first steps in the process have been taken with all organic cells based on pentacene that show external quantum efficiencies above 126 %, the highest for any solar technology till date [2]. The basic equation describing fission can be written as:

$$S_0 + S_1 \overset{\rightarrow}{\rightarrow} (TT) \overset{\rightarrow}{\rightarrow} T_1 + T_1$$

A.A. Bakulin (✉)
FOM Institute AMOLF, 1098 XG Amsterdam, The Netherlands
e-mail: a.bakulin@amolf.nl

S.E. Morgan · A. Chin · A. Rao
Cavendish Laboratory, University of Cambridge, JJ Thomson Avenue,
Cambridge CB30HE, UK
e-mail: ar525@cam.ac.uk

J. Alster · D. Zigmantas
Department of Chemical Physics, Lund University, P.O. Box 124
22100 Lund, Sweden

D. Egorova
Institut für Physikalische Chemie, Christian-Albrechts-Universität zu Kiel,
Olshausenstr. 40, 24098 Kiel, Germany

© Springer International Publishing Switzerland 2015
K. Yamanouchi et al. (eds.), *Ultrafast Phenomena XIX*,
Springer Proceedings in Physics 162, DOI 10.1007/978-3-319-13242-6_54

Here S_0 is a chromophore in the ground state, S_1 a chromophore in the excited singlet state, T_1 the molecular triplet state. $^1(TT)$ is a doubly excited state that can be described as a pair of triplets coherently coupled to form an overall spin singlet. The $^1(TT)$ is the theoretically required multiexcitonic intermediate state in the fission process, however it is optically dark and has a very short lifetime. For these reasons no experimental data about the electronic structure of this state have yet been presented. This combined with a lack of information about the role of vibrational coupling mean that a comprehensive mechanistic picture of the efficient singlet fission is currently lacking. Here we study pentacene, a model system for ultrafast fission [3], using ultrafast 2D electronic photon echo (2DPE) spectroscopy. We demonstrate that this technique allows us to study not only previously 'dark' multiexcitonic $^1(TT)$ state but also the underlying vibrational and electronic dynamics that mediates transfer between $S_0 S_1$ and $^1(TT)$.

Figure 1a presents three time slices showing the evolution of the real-part of the 2D spectra of a polycrystalline pentacene film. On the right the pump-probe spectra at the same time delay are shown for comparison. The 2D spectra are dominated by a large positive peak on the diagonal at 1.83 eV, which corresponds to ground-state bleach (SGB) of the main excited singlet transition and at short times (<90 fs) also includes a contribution from stimulated emission (SE) from the singlet exciton. At 30 fs a PIA feature, seen as negative peak at 1.87 eV probe energy, is present. We assign this feature to the PIA of singlet excitons in agreement with previous pump-probe measurements. At longer waiting times, this PIA is lost as singlets undergo fission and a new PIA feature at 1.65 eV probe energy appears, which corresponds to the transition from the lowest triplet state T_1 to the T_2 state. Importantly, no feature is observed at pump energy of 1.72 eV where the $^1(TT)$ state is predicted to be. This indicates that the state can not excited directly from the ground state.

The quality of the 2D spectra obtained allows us to reveal not just the kinetics of the triplet state formation but also to map and investigate the role of intermediate states. For this we analysed the coherent beatings observed in the 2DPE data along the evolution-time axis. The spectrum of these beatings is shown in the Fig. 1b. Figure 2 shows 2D FT 'beating-maps' for frequencies ranging from 170   to

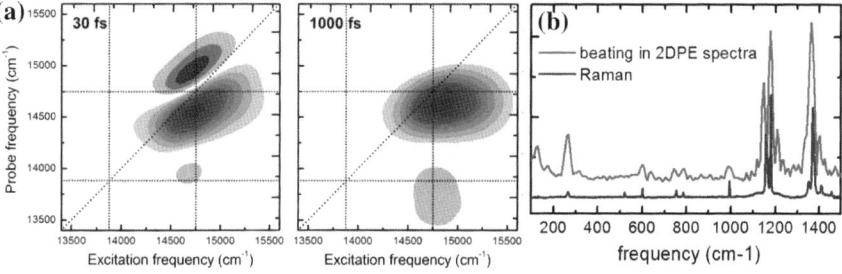

Fig. 1 **a** 2DPE real-part spectra of pentacene molecular crystal at different evolution times. **b** Spectrum of evolution-time oscillations observed in 2DPE data compared to the Raman spectrum of pentacene

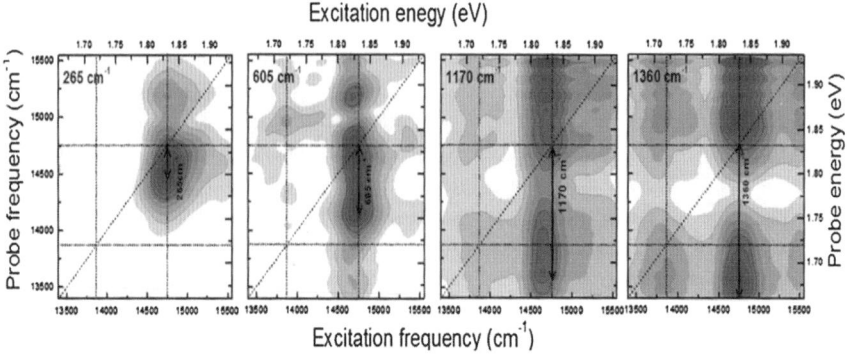

Fig. 2 The 2D maps of beating in 2DPE spectra corresponding to the strongest oscillatory features observed in Fig. 1b

1,500 cm^{-1}. For all the 2D beating-maps presented the most intense oscillatory component is found at $\omega_{excitation} = \omega_{probe} = 1.83$ eV, i.e. the position of the singlet exciton, $\omega_{singlet}$. It can also be seen that peaks occur that are separated from this main peak by the energy of the vibrational mode, as indicated by the arrows in the figure. For example, peaks can be seen at probe frequency $\omega_{singlet} - 605$ cm^{-1} in Fig. 1b and at probe frequency $\omega_{singlet} -1,170$ cm^{-1}. These and additional peaks that occur below and to the right of the diagonal are caused by the coupling of vibrational modes to an electronic transition, in this case the singlet, $\omega_{singlet}$.

Quite surprisingly, the higher energy beating-maps at 1,170 and 1,360 cm^{-1} an additional peaks at the excitation frequency of 13,900 cm^{-1} are observed,. This peak is absent from the lower energy beating-maps and is not predicted within the displaced-oscillator theoretical framework, within which peaks should not appear at lower energies on the diagonal [4, 5]. The position of the new peak is at $\omega_{excitation} = 1.72$ eV, i.e. at the expected position for the ^{1}(TT) state. We note that no state was observed here in the absorptive 2D maps shown in Fig. 1a. Thus suggests that the previously 'dark' ^{1}(TT) state is being revealed via vibrational coherence because there are transitions with strong dipoles contributing to the 4-wave mixing signals observed. The presented work thus demonstrates a novel method of revealing and studying such 'dark' states, not accessible via conventional optical methods.

Interestingly, the dark ^{1}(TT) state only contributes to the vibrational coherence when the vibrational mode can bridge the energy gap between S$_1$ and ^{1}(TT). This suggests that it is vibronic rather than direct electronic coupling that strongly mixes ^{1}(TT) and S$_1$, allowing for ultrafast fission.

References

1. A. Rao et al., Journal of the American Chemical Society, **132**, 12698 (2010).
2. D. N. Congreve et al., Science, **340**, 6130, 334-337, (2013).
3. M. W. B. Wilson et al., Accounts of Chemical Research, **46**, 1330–1338 (2013).
4. D. Egorova, Chem. Phys. **347** 166 (2008).
5. V. Butkus et al., Chemical Physics Letters, **545**, 40–43 (2012).

Ultrafast Carriers Dynamics in Silicon: A Joint Experimental and Theoretical Study

S. Dal Conte, D. Sangalli, A. Marini, G. Cerullo and C. Manzoni

Abstract We investigate the carriers dynamics in bulk silicon using pump-probe spectroscopy. The experimental results are compared with theoretical calculations which combine for the first time the non-equilibrium Green's functions theory with ab initio methods.

Semiconductor nanostructures are the most promising building blocks for future electronic devices. In the light of a continuous progress in the realization and miniaturization of semiconductor devices, ultrafast carriers dynamics is a research field of paramount importance for these materials. The transient dynamics in silicon was extensively studied by different time-resolved experimental techniques, such as single color pump-probe reflectivity [1], transient grating diffraction and time resolved photoemission [2]. However the limited spectral range explored by pump-probe measurements and the rather complicated silicon band structure prevents to deeply understand the physical origin of the transient variation of silicon optical properties. Therefore a quantitative understanding of all the relaxation processes taking place after the excitation of a femtosecond laser pulse, is still lacking.

Here we use an optical pump-supercontinuum-probe technique to investigate the dynamics of the silicon dielectric function over an extended energy range. These experimental results are compared with theoretical calculation which combines for the first time the non-equilibrium Green's functions theory with ab initio methods, like Density Functional Theory [3, 4].

The silicon electronic band structure is well understood both theoretically and experimentally. Silicon is an indirect gap semiconductor: direct optical transitions from the top of the valence band to the bottom of the conduction band occur for electronic states at the center of the Brillouin zone (Γ point) and close to the

S. Dal Conte (✉) · C. Manzoni
IFN-CNR, Piazza L. da Vinci 32, 20133 Milan, Italy
e-mail: stefano.dalconte@polimi.it

S. Dal Conte · G. Cerullo · C. Manzoni
Dipartimento di Fisica, Politecnico di Milano, Piazza L. da Vinci 32, 20133 Milan, Italy

D. Sangalli · A. Marini
ISM-CNR, Monterotondo Stazione, Italy

© Springer International Publishing Switzerland 2015
K. Yamanouchi et al. (eds.), *Ultrafast Phenomena XIX*,
Springer Proceedings in Physics 162, DOI 10.1007/978-3-319-13242-6_55

L critical point. Below the threshold of the direct optical transition, which is ≈3.3 eV, it is still possible for the valence band electrons to be promoted to the conduction band through excitation processes mediated by phonon scattering into the L and X valleys.

In this work we report on ultrafast transient reflectivity experiment in bulk silicon. Our pump-probe setup is powered by a regeneratively amplified Ti:sapphire laser system, providing 150 fs, 800 nm pulses at 1 kHz repetition rate. The electronic carriers are excited to the conduction band by the second harmonic (3.1 eV) and are probed by a broadband delayed pulse, in a spectral range between 1.7 and 3.6 eV. The probe pulse is generated by spectral broadening of a fraction of the fundamental light in a CaF_2 plate. Upon pumping with the fluence of 1 mJ/cm^2, we measure the transient reflectivity response reported in Fig. 1. The trace displays very rich temporal and spectral dynamics due to the interplay between different physical processes. An intense and narrow negative band peaked around 3.5 eV is ascribed to the bleaching of the interband direct transition by Pauli blocking due to the pump induced depletion of states at the top of the valence band around Γ. The negative signal below 3 eV, increasing at lower probe energies, arises from the bleaching of the phonon mediated optical transitions. This interpretation is confirmed by the different temporal behavior below (2 eV) and above (3.5 eV) the direct transition threshold (Fig. 1c). This experimental result is consistent with the fact that direct optical transitions are Pauli blocked only by the photoinduced holes around Γ while, in the case of indirect transitions, the hot electrons relaxing towards the minima of the conduction band give an additional contribution to this process.

The DFT band structure is numerically computed within the standard LDA approximation applying a quasiparticle correction of 0.8 eV. The out of equilibrium Green function, constructed using quasiparticle energies and DFT wave-functions, is used to study the non-equilibrium dynamics of the system following a laser pulse excitation close to the experimental condition (pulse width = 110 fs and pulse energy = 3.3 eV). In this case electrons are directly injected from the valence to the conduction band, around Γ, and then relax towards the conduction band minimum X due to electron-electron scattering, described within the out-of equilibrium GW approximation, and the electron-phonon scattering, described starting from the Fan Self-energy.

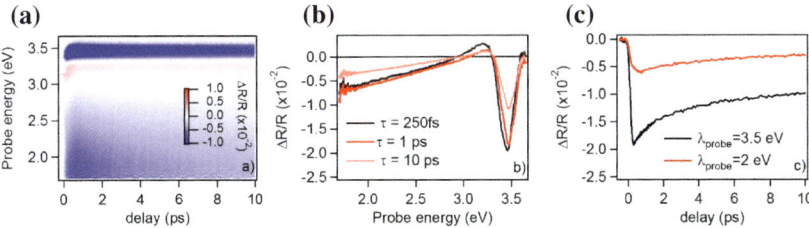

Fig. 1 a Time and frequency resolved reflectivity measurements carried on a sample of monocrystalline silicon at T = 300 K (pump fluence equal to 1 mJ/cm^2) **b** Energy resolved spectra at different delay times **c** Temporal traces taken at photon energy at respectively 3.5 and 2 eV

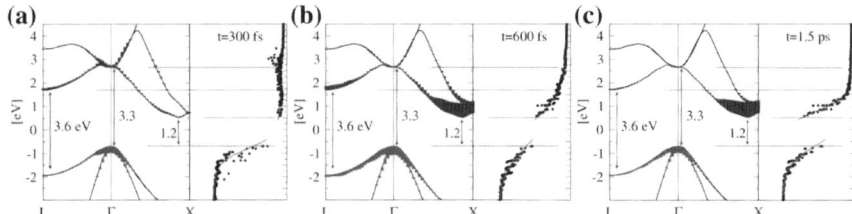

Fig. 2 a First principles carriers dynamics in bulk silicon under a 110 fs laser pulse. The electrons and holes density is represented on the silicon band structure and projected on the energy axes to monitor the creation of two Fermi distributions with decreasing temperatures. Three temporal snapshots are here represented at t = 300 fs, before the maximum of the laser pulse, at t = 600 fs (**b**) and finally at t = 1.5 ps (**c**)

Figure 2 sketches the non-equilibrium carriers dynamics in bulk silicon. The electrons, initially injected close to the Γ point (Fig. 2a), relax towards the two minima L and X on a very fast timescale. Thereafter the population in L reaches its maximum at a delay time of ≈600 fs (Fig. 2b), while the complete relaxation towards the conduction band minimum takes more than 1 ps (Fig. 2c).

These preliminary numerical results can be used to compute the silicon transient dielectric function and then be compared with the experiments. This joint experimental and theoretical study of the transient silicon optical properties could pave the way to the quantitative understanding of several non-equilibrium physical phenomena in semiconductors.

Acknowledgments Futuro in Ricerca grant No. RBFR12SW0 J of the Italian Ministry of Education, University and Research.

References

1. Sabbah, A. J. *et al.* Femtosecond pump-probe reflectivity study of silicon carrier dynamics. Phys. Rev. B **66**, 165217 (2002)
2. Ichibayashi, T. *et al.* Ultrafast relaxation of highly excited hot electrons in Si: Roles of the LX intervalley scattering. Phys. Rev. B **84**, 235210 (2011)
3. Attaccalite, C. *et al.* Real-time approach to the optical properties of solids and nanostructures: Time-dependent Bethe-Salpeter equation. Phys. Rev. B **84**, 245110 (2011)
4. Marini, A. Competition between the electronic and phonon mediated scattering channels in the out-of-equilibrium carrier dynamics of semiconductors: an **ab initio** approach. J. Phys. Conf. Ser. **427**, 012003 (2013)

Ultrabroadband Infrared Pump-Probe Spectroscopy Using Chirped-Pulse Upconversion

Hideto Shirai, Tien-Tien Yeh, Yutaka Nomura, Chih-Wei Luo and Takao Fuji

Abstract We have demonstrated infrared pump-probe spectroscopy using chirped-pulse upconversion with a nonlinear mixing in a gas. The ultrabroadband mid-infrared spectrum spanning from 200 to 5,000 cm^{-1} was upconverted by using a 0.4 ps chirped pulse to visible wavelength generation in gas. This technique allow for the high speed and high sensitivity detection in ultrabroadband mid-infrared region. Ultrafast dynamics of free-carrier in Ge was clearly observed in the range 200−5,000 cm^{-1}.

Ultrabroadband mid-infrared (MIR) coherent light is highly attractive for studies in molecular science and semiconductor physics since a number of molecular vibrations and intraband transition of free-carriers have resonance in this wavelength region. Such a light source has been provided with nonlinear frequency conversion through two-color filamentation in gases [1]. Recently, the light source have been used for a femtosecond pump-probe spectroscopy [2].

Single-shot detection of MIR spectra with a dispersive spectrometer is essential for fast acquisition at the pump-probe spectroscopy. However, one has to suffer from low pixel count, reduced sensitivity, and high cost of multichannel infrared detectors. Chirped-pulse upconversion (CPU) is a technique to solve the problem. A MIR pulse is converted to visible with a nonlinear frequency mixing between the MIR pulse and a chirped pulse, and the visible spectra are detected with a high

H. Shirai (✉) · Y. Nomura · T. Fuji
Institute for Molecular Science, 38 Nishigonaka, Myodaiji, Okazaki 444-8585, Japan
e-mail: shirai@ims.ac.jp

Y. Nomura
e-mail: nomura@ims.ac.jp

T. Fuji
e-mail: fuji@ims.ac.jp

T.-T. Yeh · C.-W. Luo
Department of Electrophysics, National Chiao Tung University, Hsinchu 300, Taiwan
e-mail: yahahaha0723.cv96@nctu.edu.tw

C.-W. Luo
e-mail: cwluo@mail.nctu.edu.tw

© Springer International Publishing Switzerland 2015
K. Yamanouchi et al. (eds.), *Ultrafast Phenomena XIX*,
Springer Proceedings in Physics 162, DOI 10.1007/978-3-319-13242-6_56

233

quality optical multichannel analyzer [3]. Very recently, Ultrabroadband detection of infrared spectra with CPU has been demonstrated by using a gas medium for the frequency conversion [4]. In this contribution, we report an application of the ultrabroadband CPU to infrared pump-probe spectroscopy for the investigation of ultrafast carrier dynamics in Ge. We have succeeded in measuring free-carrier dynamics of Ge in an entire MIR range (200–5,000 cm^{-1}).

The ultrabroadband infrared pump-probe spectroscopy with CPU was realized with the system shown in Fig. 1. The light source was based on a Ti:sapphire multi-pass amplifier system (800 nm, 30 fs, 0.85 mJ at 1 kHz, Femtopower CompactPro, FEMTOLASERS). The output pulse was split into three, the first pulse was for infrared generation, the second pulse was the pump pulse for the pump-probe spectroscopy, and the third pulse was used as a chirped pulse.

The ultrabroadband infrared pulse for the probe pulse at the pump-probe spectroscopy was generated by using four-wave mixing of fundamental and second harmonic of Ti:sapphire amplifier output through filamentation in nitrogen. The pulse duration and spectral range of the infrared pulse were 7 fs and 200–5,000 cm^{-1}, respectively. The infrared pulse was reflected by several substrates coated with indium tin oxide ($t = 300$ nm) to reduce the residual visible beam along the infrared pulse. The infrared pulse was focused onto a sample, a Ge substrate, by using a concave mirror ($f = 0.75$ m). The diameter of the infrared beam on the sample was 0.4 mm.

The pump pulse (800 nm, 30 fs) was collimated down to 4 mm diameter and overlapped to the infrared pulse on the sample. For shot-to-shot data acquisition of reflectivity-change signals, every second pump pulse was blocked by using a mechanical chopper, which is synchronized with a half-frequency of the repetition rate of the laser pulse train.

The third beam was chirped by passing through several dispersive media, BK7 and ZnSe substrates. The pulse duration of the chirped pulse became ~ 0.4 ps, which corresponds to the frequency resolution of 41 cm^{-1} at the upconversion spectroscopy. The chirped pulse was combined with the mid–infrared pulse

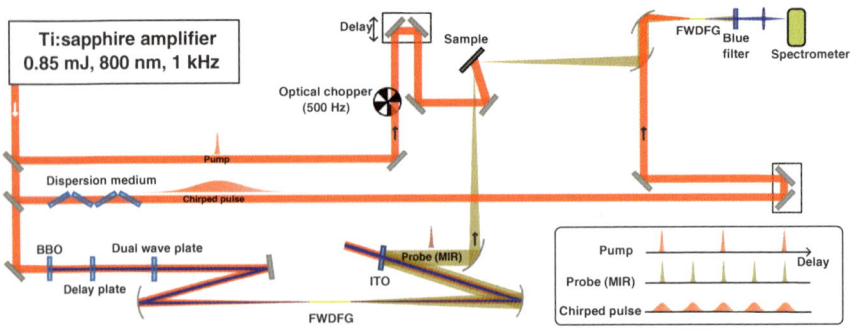

Fig. 1 Schematic illustration of ultrabroadband mid-infrared pump-probe spectroscopy with CPU

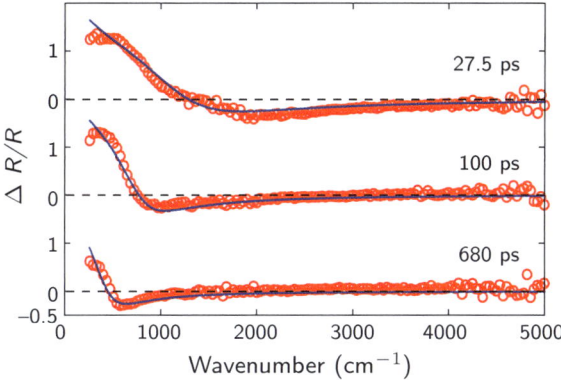

Fig. 2 Measured ultrafast reflectivity-change signal $\Delta R/R$ of Ge for the excitation density 135 µJ/ cm^2 at several delay time. The *circles* are the experimental data and the *solid lines* are analytical fits to data using the Drude model

reflected by the sample through a mirror with a hole. The combined beam was focused into nitrogen with a parabolic mirror ($f = 50$ mm) and upconverted into a visible beam (ω_2, 400–500 nm) through four-wave difference frequency generation (FWDFG, $\omega_1 + \omega_1 - \omega_0 \rightarrow \omega_2$) of the chirped pulse ($\omega_1$) and the infrared pulse (ω_0). The FWDFG signal generated at a fixed delay between the infrared and chirped pulses was sent to a spectrometer with a 1,600 × 400 pixel EMCCD camera (SP-2358 with ProEM + 1,600, Princeton Instruments). The upconverted spectrum was obtained by accumulated 100 pairs of with/without-pump infrared spectra at each delay time between the pump and probe pulses to achieve a reasonable signal-to-noise ratio.

Figure 2 shows the ultrafast reflectivity-change signal, $\Delta R/R$, of Ge for the excitation density of 135 µJ/cm^2 at several delay time. The cross–phase modulation due to CPU was removed using simple calculation [5]. The solid lines were fitted by using the Drude model. The positive and negative reflectivity-changes due to photoinduced free-carriers was clearly observed in the range from far-infrared (200 cm^{-1}) to near infrared (5,000 cm^{-1}).

References

1. T. Fuji and T. Suzuki, "Generation of sub–two–cycle mid–infrared pulses four–wave mixing through filamentation in air," Opt. Lett. **32**, 3330–3332(2007)
2. C. Calabrese, A. M. Stingel, L. Shen, and P. B. Petersen, "Ultrafast continuum mid-infrared spectroscopy: probing the entire vibrational spectrum in a single laser shot with femtosecond time resolution," Opt. Lett. **37**, 2265–2267(2012)
3. J. Knorr, P. Rudolf, and P. Nuernberger, "A comparative study on chirped–pulse upconversion and direct multi–channel mct detection," Opt. Express **21**, 30693–30706(2013)

4. Y. Nomura, Y. T. Wang, T. Kozai, H. Shirai, A. Yabushita, C. W. Luo, S. Nakanishi, and T. Fuji, "Single-shot detection of mid-infrared spectra by chirped-pulse upconversion with four-wave difference frequency generation in gases," Opt. Express **21,** 18249–18744(2013)
5. K. F. Lee, P. Nuernberger, A. Bonvalet, and M. Joffre, "Removing cross-phase modulation from midinfrared chirped-pulse upconversion spectra," Opt. Express **17,** 18738–18744(2009)

Investigation of Laser-Induced Currents in Large-Band-Gap Dielectrics

Sabine Keiber, Tim Paasch-Colberg, Alexander Schwarz,
Olga Razskazovskaya, Elena Fedulova, Özge Sağlam,
Clemens Jakubeit, Shawn Sederberg, Péter Dombi,
Nicholas Karpowicz and Ferenc Krausz

Abstract Applying few-cycle laser pulses to dielectrics reversibly and rapidly increases their polarizability, allowing for the switching of currents at the frequency of light (Schiffrin et al., Nature 493:70–74, 2013). We report on the dependence of these ultrafast currents on material band gap and sample geometry.

1 Introduction

Recent experiments investigating electron dynamics on the ultrashort timescale in solids have shown exciting results from the generation of high harmonics [2] to the observation of Bloch oscillations [3]. These experiments have only recently become possible due to the advancement of ultrafast laser sources, which allow material properties to be probed at field strengths of a few V/Å, far above the dc damage threshold.

The detection of ultrafast currents which are sensitive to the carrier envelope phase (CEP) of the applied laser pulses has opened the door to new metrology techniques [4] and is a first step towards PHz optoelectronics [1]. This work aims to

S. Keiber (✉) · T. Paasch-Colberg · A. Schwarz · O. Razskazovskaya · E. Fedulova ·
C. Jakubeit · S. Sederberg · P. Dombi · N. Karpowicz · F. Krausz
Max-Planck-Institut für Quantenoptik, Hans-Kopfermann-Strasse 1,
85748 Garching, Germany
e-mail: sabine.keiber@mpq.mpg.de

S. Keiber · A. Schwarz · F. Krausz
Fakultät für Physik, Ludwig-Maximilians-Universität, Am Coulombwall 1,
85748 Garching, Germany

Ö. Sağlam
Physik-Department, Technische Universitat München, James-Franck-Strasse,
85748 Garching, Germany

P. Dombi
MTA "Lendület" Ultrafast Nanooptics Group, Wigner Research Centre for Physics,
Konkoly Thege Miklós út 29-33, Budapest, XII H-1121, Hungary

© Springer International Publishing Switzerland 2015
K. Yamanouchi et al. (eds.), *Ultrafast Phenomena XIX*,
Springer Proceedings in Physics 162, DOI 10.1007/978-3-319-13242-6_57

provide a better understanding of the nature of the detected current and its
dependence on material properties as well as sample geometry.

2 Experimental Setup and Results

Visible/near infrared laser pulses with 400 µJ energy are generated at a repetition
rate of 3 kHz by a customized Ti:Sapphire chirped-pulse amplifier system. The
pulse duration of the CEP-stable pulses is 4 fs (FHWM), corresponding to about 1.5
optical cycles at the carrier wavelength of 760 nm. The linearly polarized beam is
focused with an off-axis-parabolic mirror to a diameter of about 60 µm ($1/e^2$) onto
the sample which consists of a 250 µm thick dielectric substrate coated with a
200 nm silver layer on both sides (see Fig. 1). The current between the electrodes is
amplified and filtered from the CEP-independent background using a lock-in
amplifier referenced to a modulation of the CEP. No bias voltage is applied to the
electrodes.

When changing the CEP of the applied laser pulse by scanning a pair of fused
silica wedges, we record oscillations in the transferred charge per pulse (see
Fig. 2a). The amplitude of the charge oscillation varies depending on the dielectric
material used in the measurement. For investigating how the transferred charge
depends on band gap (E_g), three crystalline dielectrics with the space group $Fm\bar{3}m$
were chosen: CaF_2 (E_g = 12.1 eV), BaF_2 (E_g = 9.1 eV) and MgO (E_g = 7.8 eV). As
can be seen in Fig. 2b), the critical field strength, at which the signal sets in,

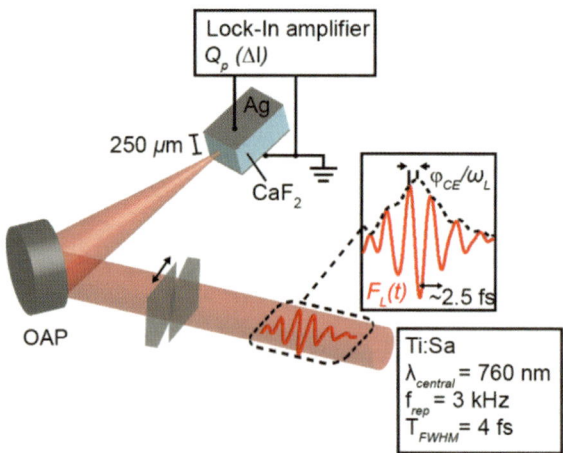

Fig. 1 Experimental setup, adapted from [4]. The edge of a 250 µm thick dielectric substrate is
illuminated by few-cycle Ti:Sapphire pulses at a repetition rate of 3 kHz. The beam is polarized
perpendicularly to the gap formed by two silver electrodes and focused down to about 60 µm ($1/e^2$),
providing electric field strengths of a few V/Å. The CEP-dependent transferred charge is measured
with a lock-in amplifier

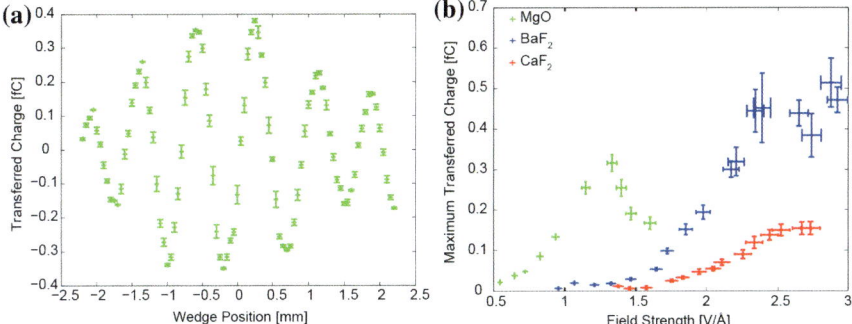

Fig. 2 Laser-induced transferred charge. **a** The transferred charge per pulse averaged over 1,500 shots shows a clear dependence on the position of the pair of fused silica wedges used to adjust the carrier envelope phase of the pulse. **b** The increase of the maximum transferred charge with field strength is shown for MgO, BaF$_2$, and CaF$_2$. The signal amplitude as well as the critical field strength at which the signal sets in depends on the band gap of the material

increases with band gap. The signal decrease in MgO around 1.5 V/Å coincides with the onset of ablation of the material in this sample.

In contrast to previous experiments [1, 4], the electrode gap in the reported measurements is much larger than the focus size. This raises the question how the laser-induced charge separation can be observed in the electronic circuit which has a response time of a few μs. Although our results do not allow a definite answer at this point, they suggest that trapping at the metal-dielectric interface does not play the predominant role in the current detection.

3 Conclusion and Outlook

We have shown that the amplitude of laser-induced ultrafast currents strongly depends on the band gap of the active material. In addition, the recent sample geometry allows for the detection of transferred charges for electrode gap sizes much bigger than the laser focus.

In a next step, we plan to use an infrared optical parametric amplifier (OPA) to control currents inside of solids. It operates at 2.1 μm carrier wavelength with a repetition rate of 3 kHz and can provide CEP-stable pulses with a sub-two-cycle duration and up to 1.2 mJ pulse energy [5]. By reducing the photon energy, the influence of low-order nonlinear effects is reduced, while at the same time the damage threshold of most materials is increased. This will enable us to investigate optically induced currents in semiconductor materials, which could be appealing for future optoelectronics applications.

References

1. A. Schiffrin *et al.*, Nature **493**, 70–74 (2013)
2. S. Ghimire *et al.*, Nature Physics **7**, 138–141 (2011)
3. O. Schubert *et al.*, Nature Photonics **8**, 119–123 (2014)
4. T. Paasch-Colberg *et al.*, Nature Photonics **8**, 214–218 (2014)
5. Y. Deng *et al.*, Optics Letters **37**, 23, 4973–4975 (2012)

Field-Induced Dynamics of Correlated Electrons in LiH and NaBH$_4$

Vincent Juvé, Marcel Holtz, Flavio Zamponi, Michael Woerner,
Thomas Elsaesser and Andreas Borgschulte

Abstract Femtosecond x-ray powder diffraction maps electron density in response to a strong electric field. In LiH, electron correlations lead to an electron transfer from Li to H while NaBH$_4$ shows a transfer from BH$_4^-$ to Na$^+$.

1 Introduction

Field-driven physical processes play a key role for the electronic and optical properties of condensed matter and should allow for steering charge transport on the time scale of the optical cycle. To distinguish local electron displacements from a real-space electron flow, insight into the spatial electron distribution on ultrashort time scales is required. Recently, we have shown with the help of femtosecond x-ray diffraction that a strong non-resonant laser field induces transient optical polarizations which are connected with a spatial redistribution of electronic charge in the concomitant "virtual" or mixed quantum state [1]. In ionic crystals, the electronic states at the valence band maximum are typically localized on the negative ions and those at the conduction band minimum on the positive ions. When applying the external field, the simplest picture predicts a quasi-instantaneous, i.e. fully reversible, electron transfer from a negative to a neighboring positive ion. This picture is expected to break down whenever strong Coulomb correlations exist between valence electrons. Here, we demonstrate the strong influence of Coulomb correlations on the field-induced response of the model system lithium hydride (LiH), the most elementary heteronuclear solid. Electron density maps derived from

V. Juvé (✉) · M. Holtz · F. Zamponi · M. Woerner · T. Elsaesser
Max-Born-Institut für Nichtlineare Optik und Kurzzeitspektroskopie, 12489 Berlin, Germany
e-mail: juve@mbi-berlin.de

A. Borgschulte
Swiss Federal Laboratories for Materials Testing and Research,
Laboratory for Hydrogen and Energy, EMPA, 8600 Dübendorf, Switzerland

© Springer International Publishing Switzerland 2015
K. Yamanouchi et al. (eds.), *Ultrafast Phenomena XIX*,
Springer Proceedings in Physics 162, DOI 10.1007/978-3-319-13242-6_58

femtosecond x-ray powder diffraction patterns reveal a net transfer from Li to H, in sharp contrast to the anticipated ionic behavior. Such results are in line with theoretical calculations of the bandstructure of correlated electrons.

2 Results

In our experiments, we study crystalline powders of LiH and sodium boron-hydride (NaBH$_4$), both crystallizing in the rock-salt (NaCl) structure. In a pump-probe approach, the 800 nm excitation pulse provides a strong off-resonant optical field and the material's response is mapped by diffracting synchronized hard x-ray pulses (Cu Kα, wavelength 0.154 nm) from the powder sample [2]. The pump wavelength of 800 nm is well below the bandgap of the materials and, on the other hand, above all vibrational transitions. The peak intensity corresponds to a peak electric field of 1 GV/m. Several Debye-Scherrer diffraction rings were recorded simultaneously as a function of the time delay between the pump and the x-ray probe in order to monitor the diffracted intensities changes of the different rings. The introduction of a chopper in the pump arm allows us to determine the relative change of the diffracted intensity of each individual ring and, moreover, reduces the source noise fluctuations down to the shot noise of x-ray photon detection.

We observed intensity changes on several Debye-Scherrer rings which occur in time around delay zero, i.e., when both the pump and probe beams are temporally overlapped. The angular positions of the rings remain unchanged. The diffraction ring which corresponds to the (111) reflection, shows a decrease of intensity on the order of 1 % for LiH (Fig. 1a), whereas the same reflection shows an increase of a few percent for NaBH$_4$ (not shown). The extracted structure factors $\Delta F_{hkl}(t)$ from the intensities changes $\Delta I_{hkl}(t)$ are used to reconstruct the temporal evolution of the electronic density $\Delta \rho$ (x,y,z,t) by means of the Maximum Entropy Method [3]. Surprisingly, the LiH results show a shift of electronic charge from the cation Li$^{0.5+}$ to the anion H$^{0.5-}$ (Fig. 1b), whereas in NaBH$_4$ we measured a charge transfer from the anion to the cation (not shown). This unexpected result means that LiH becomes more ionic upon application of the external field, a behavior in contrast to NaBH$_4$ and the previously studied LiBH$_4$ [1].

The experimental result is reproduced by a theoretical analysis where the response to the external field is treated in the Coulomb-Hole-plus-Screened-Exchange (COHSEX) formalism [4, 5]. In contrast to the simpler Hartree-Fock picture, the COHSEX formalism includes electron correlations beyond the single-particle level. By properly incorporating the inhomogeneous screening of the electron-electron interaction, the COHSEX calculations predict the experimentally observed electron transfer from the cation Li$^{0.5+}$ to the anion H$^{0.5-}$ in LiH (Fig. 1c), a behavior not reproduced by a Hartree-Fock calculation. Such findings underline the role of electronic correlations in LiH on the electron dynamics induced by strong optical fields and, vice versa, show that ultrafast x-ray diffraction represents a key method for investigating charge correlations.

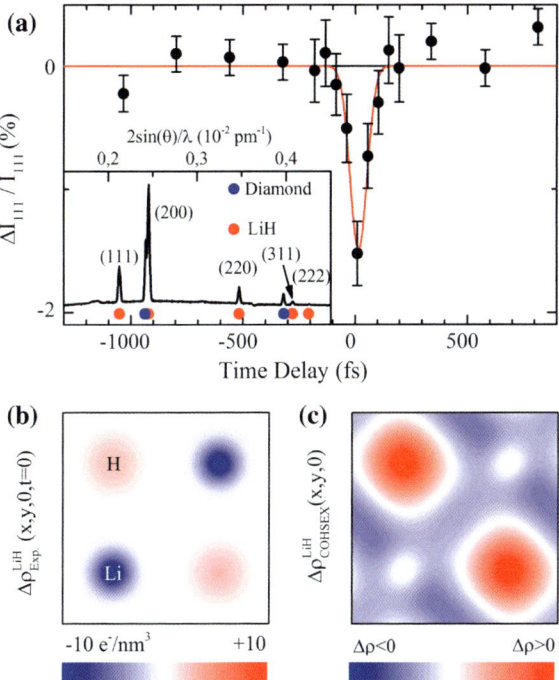

Fig. 1 **a** Relative change of the number of photons diffracted off the (111) plane as a function of the time delay between the 800 nm pump and the x-ray probe. *Inset* Steady-state powder diffraction of lithium hydride. The *diamond* diffraction peaks arise from the sample holder. **b** Experimental electron density change at time delay zero in the lithium-hydrogen plane. **c** Calculated electron density change at time delay zero in the lithium-hydrogen plane as calculated within the COHSEX framework

References

1. J. Stingl, F. Zamponi, B. Freyer, M. Woerner, T. Elsaesser, and A. Borgschulte, "Electron Transfer in a Virtual Quantum State of LiBH$_4$ Induced by Strong Optical Fields and Mapped by Femtosecond X-Ray Diffraction", *Phys. Rev. Lett.* **109**, 147402 (2012).
2. V. Juvé, M. Holtz, F. Zamponi, M. Woerner, T. Elsaesser, and A. Borgschulte, "Field-driven dynamics of correlated electrons in LiH and NaBH4 revealed by femtosecond x-ray diffraction", *Phys. Rev. Lett.* **111**, 217401 (2013).
3. C. J. Gilmore, "Maximum Entropy and Bayesian Statistics in Crystallography: a Review of Practical Applications", *Acta Crystallogr. Sect. A* **52**, 561 (1996).
4. L. Hedin, "New Method for Calculating the One-Particle Green's Function with Application to the Electron-Gas Problem", *Phys. Rev.* **139**, A796 (1965).
5. M. S. Hybertsen and S. G. Louie, "Electron correlation in semiconductors and insulators: Band gaps and quasiparticle energies", *Phys. Rev. B* **34**, 5390 (1986).

Spontaneous Formation of Correlated Charge Coherence Induced by 1.5-Cycle Pulse in 1-D Organic Metal (TMTTF)$_2$AsF$_6$

T. Ishikawa, Y. Sagae, Y. Naito, J. Ichimura, Y. Kawakami,
H. Itoh, K. Yamamoto, K. Yakushi, S. Ishihara, T. Sasaki,
K. Yonemitsu and S. Iwai

Abstract Ultrafast optical response of organic metal (TMTTF)$_2$AsF$_6$ induced by 1.5 cycle (7 fs) infrared pulse was investigated. Intense coherent oscillation of correlated charge (18 fs) grows in the time scale of 50 fs, reflecting the spontaneous-formation of the electronic coherence before the electronic thermalization.

1 Introduction

Since the first observation of the increase of electron temperature in inorganic metals, ultrafast optical responses in metal have been discussed in terms of the increase of an electron temperature in the framework of two-temperature model [1]. On the other hand, in strongly correlated materials, ultrafast charge dynamics have been investigated in insulators such as Mott insulator and charge order (CO) [2–4], because various ultrafast changes in the conducting and/or magnetic natures can be

T. Ishikawa · Y. Sagae · Y. Naito · J. Ichimura · Y. Kawakami · H. Itoh · S. Ishihara · S. Iwai (✉)
Department of Physics, Tohoku University, Sendai 980-8578, Japan
e-mail: s-iwai@m.tohoku.ac.jp

H. Itoh · T. Sasaki · S. Iwai
JST-CREST, Sendai 980-8578, Japan

K. Yamamoto
Department of Applied Physics, Okayama Science University, Okayama 700-0005, Japan

K. Yakushi
Toyota Physical and Chemical Research Institute, Nagakute 480-1192, Japan

T. Sasaki
Institute for Material Research, Tohoku University, Sendai 980-8577, Japan

K. Yonemitsu
Department of Physics, Chuo University, Tokyo 112-8551, Japan

© Springer International Publishing Switzerland 2015
K. Yamanouchi et al. (eds.), *Ultrafast Phenomena XIX*,
Springer Proceedings in Physics 162, DOI 10.1007/978-3-319-13242-6_59

244

Fig. 1 Schematic illustration
of **a** charge ordering phase
and **b** metallic phase in
$(TMTTF)_2AsF_6$

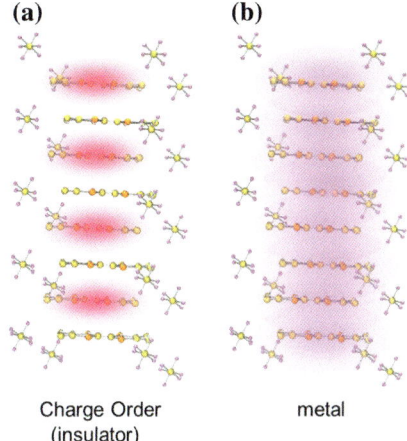

(a)   **(b)**

Charge Order   metal
(insulator)

realized. However, there are few studies on correlated metals. Charge dynamics
reflecting the strong electron correlation are expected to be captured before the
electronic thermalization is completed.

The organic salt $(TMTTF)_2AsF_6$ (TMTTF; tetramethyl tetrathiafulvalen)
are well known strongly correlated 1-D metal or Tomonaga-Luttinger liquid at
>250 K (Fig. 1b), although this compound shows the ferroelectric CO below 102 K
(Fig. 1a). In this study, we have observed the ultrafast coherent dynamics reflecting
the correlated charge motion before the thermalization of the electron system by
using the 7 fs, 1.5-cycle near infrared pulse.

2 Experiment

Super broadband infrared spectrum covering 1.2–2.3 µm was obtained by focusing
the idler pulse (1.7 µm, carrier- envelope phase is self-stabilized) from the optical
parametric amplifier into the hollow-fiber in Kr filled chamber. Pulse compression
was performed using active mirror and chirped mirror. Pulse width evaluated from
the second harmonic generation autocorrelation is 7 fs which corresponds to
1.5-optical cycle. Spectrum of the 7 fs pulse for pump and probe are shown as a
dotted curve in Fig. 2a.

3 Results and Discussions

Figure 2 shows the steady state reflectivity (R) (Fig. 2a), and transient reflectivity
$\Delta R/R$ (open circles in Fig. 2b) at 300 K, respectively. Decrease of R at 0.2–0.75 eV
was observed immediately after the excitation by 150 fs, 0.89 eV pulse. As shown

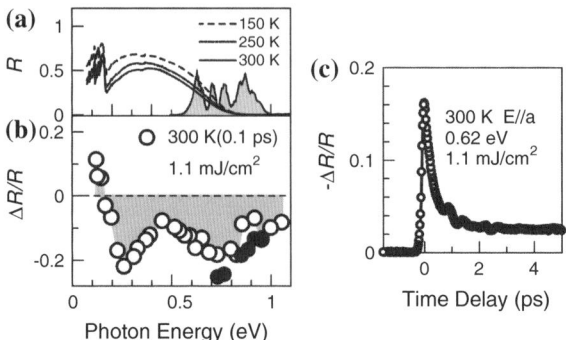

Fig. 2 **a** Steady state reflectivity (*R*) at various temperatures. **b** Transient reflectivity (Δ*R/R*) at 0.1 ps (300 K) measured by 100 fs pulse (*open circles*). Δ*R/R* spectrum measured by 7 fs pulse at 30 fs (*closed circles*). **c** Time evolution of Δ*R/R* (measured by 100 fs pulse)

in the time profile in Fig. 2c, a build-up time and a recovery time of Δ*R/R* are in the time scale of <100 fs, and <1 ps, respectively. Such ultrafast responses are attributable to the increase of electron temperature up to 700 K and subsequent decrease induced by electron–phonon interactions.

Closed circles in Fig. 2b shows the Δ*R/R* spectrum measured using 7 fs pulse. Figure 3 shows the time evolution of Δ*R/R* (Fig. 3a) measured at 0.83 eV, oscillating component (Fig. 3b) and the time resolved spectra obtained from the wavelet (WL)

Fig. 3 **a**, **b** Time evolutions of Δ*R/R* (**a**) at 0.83 eV and the oscillating component (**b**) after excitation by 7 fs pulse, respectively. **c**, **d** Steady state optical conductivity spectrum (**c**) and transient spectra obtained by the wavelet analysis of the oscillating component (**d**)

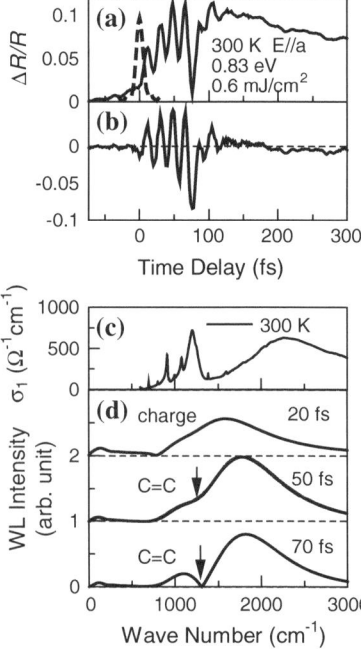

analysis (Fig. 3d). (i) Intense oscillation with a period of 18 fs $(1,850\ cm^{-1})$ was observed in the timescale of <100 fs in Fig. 3a, b. The broad spectra in Fig. 3d reflects the fast damping of the oscillation. This short-lived 18 fs oscillation is attributable to the coherent intermolecular charge motion, because the oscillating energy in Fig. 3d corresponds to the intermolecular charge transfer excitation in the optical conductivity spectrum (Fig. 3c). (ii) complicated waveform in time profile (Fig. 3b) and spectral dip (Fig. 3d) reflecting the destructive interference between the charge and intra-molecular C=C (v_4) vibration shows the beginning of the electron-molecular vibration (EMV) coupling. Such coherent electron and electron-vibration (EMV) dynamics are beyond the two-temperature model which has been used for describing the ultrafast optical responses in the metallic state [1]. In particular, the electron oscillation grows in the time scale of 50 fs as shown in Fig. 3a, b after the excitation by 0.89 eV pulse. Considering that the 0.89 eV excitation can generate the charges with large excess energy (~ 0.8 eV) which are analogous to the free-carriers [5], the growth of the oscillation indicates that the coherence of the correlated charge is spontaneously organized after the excitation. The formation of the electronic coherence observed here is considered to be the elementary dynamics in correlated electron system.

4 Summary

We have observed the ultrafast coherent dynamics reflecting the correlated charge motion before the thermalization of the electron system in $(TMTTF)_2AsF_6$ by using the 7 fs, 1.5-cycle near infrared pulse. Spontaneous formation of the electronic coherence is induced in the time scale of 50 fs.

References

1. R.W. Schoenlein, W.Z. Lin, J.G. Fujimoto, G.L. Eesley, Phys. Rev. Lett. **58**, 1680 (1987)
2. S. Iwai, K. Yamamoto, A. Kashiwazaki, F. Hiramatsu, H. Nakaya, Y. Kawakami, Y. Yakushi, H. Okamoto, H. Mori, Y. Nishio, Phys. Rev. Lett. **98**, 097402 (2007)
3. Y. Kawakami, T. Fukatsu, Y. Sakurai, H. Unno, H. Itoh, S. Iwai, T. Sasaki, K. Yamamoto, K. Yakushi, K. Yonemitsu, Phys. Rev. Lett. **105**, 246402 (2010)
4. Y. Kawakami, S. Iwai, T. Fukatsu, M. Miura, N. Yoneyama, T. Sasaki, and N. Kobayashi, Phys. Rev. Lett. **103**, 066403 (2009)
5. H. Gomi, A. Takahashi, T. Tastumi, S. Kobayashi, K. Miyamoto, J. Lee, M. Aihara, J. Phys. Soc. Jpn. **80**, 034709 (2011)

Coherent Dynamics of Structural Symmetry During the Ultrafast Melting of a Charge Density Wave

T. Huber, S.O. Mariager, A. Ferrer, H. Schaefer, J.A. Johnson,
S. Gruebel, A. Luebcke, A. Caviezel, L. Huber, T. Kubacka,
C. Dornes, C. Laulhe, S. Ravy, G. Ingold, P. Beaud, J. Demsar
and S.L. Johnson

Abstract We use time-resolved hard X-ray diffraction to directly follow the dynamics of structural symmetry change during the ultrafast melting of a charge density wave. We observe a transient recovery of the periodic lattice distortion on a sub-picosecond timescale.

1 Introduction

In strongly correlated electron systems, the coupling of the lattice to long-range electronic order can lead to a break of structural symmetry upon cooling below the critical temperature. In prototypical charge density wave (CDW) compounds, this leads to a low temperature periodic lattice distortion with a modulation vector reproducing the Fermi-nesting vector of the system [1]. The ultrafast melting of

T. Huber (✉) · A. Ferrer · L. Huber · T. Kubacka · C. Dornes · S.L. Johnson
Institute for Quantum Electronics, Physics Department, ETH Zurich,
8093 Zurich, Switzerland
e-mail: tihuber@phys.ethz.ch

S.O. Mariager · A. Ferrer · J.A. Johnson · S. Gruebel · A. Luebcke · A. Caviezel · G. Ingold ·
P. Beaud
Swiss Light Source, Paul Scherrer Institut, 5232 Villigen PSI, Switzerland

H. Schaefer · J. Demsar
Physics Department, Universitaet Konstanz, 78457 Konstanz, Germany

J. Demsar
Institute of Physics, Ilmenau University of Technology, 98693 Ilmenau, Germany

A. Luebcke
Laboratoire de Spectroscopie Ultrarapide, EPF Lausanne, 1015 Lausanne, Switzerland

C. Laulhe · S. Ravy
Synchrotron SOLEIL, L'Orme Des Merisiers, Saint-Aubin, BP 48, 91192 Gif-Sur-Yvette
Cedex, France

C. Laulhe
Université Paris-Sud, 91405 Orsay Cedex, France

© Springer International Publishing Switzerland 2015
K. Yamanouchi et al. (eds.), *Ultrafast Phenomena XIX*,
Springer Proceedings in Physics 162, DOI 10.1007/978-3-319-13242-6_60

electronic order driven by femtosecond laser pulses has been studied in various systems [2–5], but we still lack crucial direct information about the dynamics of structural symmetry associated with CDW melting. Here we show that with sufficient space and time resolution in a time-resolved hard X-ray experiment, we can observe the coherent structural dynamics associated with CDW formation. As a model system, we choose the prototypical one-dimensional CDW compound $K_{0.3}MoO_3$ (usually termed Blue Bronze), widely regarded as a textbook system for Fermi-nesting driven CDW formation. In $K_{0.3}MoO_3$, an incommensurate CDW and an accompanying periodic lattice distortion occurs below $T_c = 183$ K [6, 7].

2 Experimental

The measurements were carried out at the hard X-ray slicing source FEMTO at the Swiss Light Source (SLS) of the Paul-Scherrer-Institut in Villigen, Switzerland [7]. We employed a pump-probe scheme, using 1.5 eV pump pulses to excite a single crystal $K_{0.3}MoO_3$ cooled well below the metal-to-CDW transition temperature to T = 95 K. In the low temperature phase, the periodic lattice modulation leads to superlattice (SL) reflections that can be directly measured with the 7 keV X-ray pulses provided by the synchrotron slicing source. The X-ray pulse duration (FWHM) is around 100 fs, assuring sufficient time-resolution to resolve all relevant collective excitations of the CDW ground state. The measurements were carried out in a grazing incidence geometry, allowing for a close match of pump and probe penetration depth. The time evolution of the measured SL reflection gives a direct view of the nonequilibrium structural dynamics and is not distorted by electronic relaxation processes right after intense photoexcitation.

3 Results and Discussion

In Fig. 1, we present the measured coherent structural dynamics after photoexcitation for all relevant excitation fluences. For fluences below the melting threshold of the CDW condensate (F = 0.3 mJ/cm^2), we can directly follow the structural dynamics associated with the collective amplitude mode (AM) of the system. The relaxation timescale of the non-oscillatory signal is $t_{disp} = 3$ ps. Upon increasing the fluence to F = 1 mJ/cm^2, t_{disp} increases to around 10 ps, and we do not observe a clear signature of the AM. The timescale of the initial drop of SL diffraction intensity is around 100 fs, limited by the experimental time resolution. For higher fluences (F = 3.7 mJ/cm^2) well above the melting threshold [3], we observe a transient recovery of the SL diffraction intensity, peaking at a delay of around t = 350 fs. After the transient recovery, we do not observe a clear relaxation of the SL diffraction signal on the measured timescale. The background level for time delays t >0.5 ps with increasing fluence saturates at a background level of $I(t)/I_0 \sim 0.4$ that we

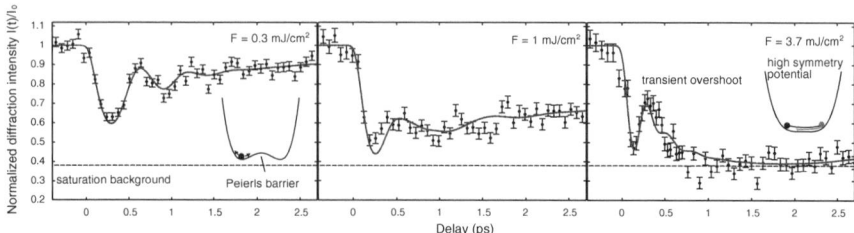

Fig. 1 Time evolution of (1 3.25–0.5) SL peak diffraction intensity, normalized to the equilibrium diffraction intensity I_0. The dashed line depicts the saturation background. Error bars correspond to photon counting statistics. Lines are fits to the model as described in the text. The low fluence data shows a clear signature of the soft mode of the system at 1.62 THz. For the highest fluence, we observe a transient overshoot

attribute to diffraction of non-excited surface regions of the cleaved single crystal. A striking property of the transient recovery is its timescale, which is faster than dynamics that could be associated with the AM.

We can model the results as a coherent motion along the structural coordinate of the Peierls distortion in a fluence dependent double-well potential: For low fluences, the system oscillates around low-temperature equilibrium positions. For fluences above the melting threshold, the Peierls barrier collapses immediately, leading to a coherent motion in a high symmetry potential landscape. The measured diffraction signal in our experimental geometry then results from an incoherent superposition of the signal with different excitation levels of the inhomogeneously excited crystal. In Fig. 1, we show fits to the data according to the model, where the only fluence dependent parameters are the height of the Peierls barrier and the relaxation timescale t_{disp}. To account for the evolution of the transient overshoot after the first peak, we introduce a phenomenological time-dependent damping parameter describing the damping of the motion along the distortion coordinate in the transient state after intense photoexcitation.

The agreement with the model indicates that during the ultrafast melting of a CDW, the structural dynamics are determined by the properties of the high-symmetry phase and not by the lattice modes of the initial state. This could have wide implications for the understanding of materials with similar periodic lattice modulations that are the result of a coupling of the lattice to electronic order.

Ultimately, the results help explain how the structural symmetry dynamics associated with so-called nonthermal phase transitions can be fast, as opposed to the much slower behavior observed in adiabatic soft-mode phase transitions.

References

1. G. Gruner, "Density waves in solids" (Addison-Wesley, 1994).
2. M. Eichberger et al., "Snapshots of cooperative atomic motions in the optical suppression of charge density waves", Nature **468**, 799 (2010).

3. A. Tomeljak et al., "Dynamics of Photoinduced Charge-Density-Wave to Metal Phase Transition in $K_{0.3}MoO_3$", Phys. Rev. Lett. **102**, 066404 (2009).
4. F. Schmitt et al., "Transient electronic structure and melting of a charge density wave in $TbTe_3$", Science **321**, 1649 (2008).
5. E. Moehr-Vorobeva et al., "Nonthermal Melting of a Charge Density Wave in $TiSe_2$", Phys. Rev. Lett. **107**, 36403 (2011).
6. G. Travaglini, P. Wachter, "The blue bronze $K_{0.3}MoO_3$: A new one-dimensional conductor", Solid State Comm. **37**, 599 (1981).
7. J.P. Pouget et al., "Structural study of the charge-density-wave phase transition of the blue bronze: $K_{0.3}MoO_3$", Journal de Physique **46**, 1731 (1985)
8. P. Beaud et al., "Spatiotemporal Stability of a Femtosecond Hard–X-Ray Undulator Source Studied by Control of Coherent Optical Phonons" Phys. Rev. Lett. **99**, 174801 (2007).

10 fs Dynamics of Photoinduced Magnetic Transition in Double-Layered Charge Ordering in LuFe₂O₄ Under Interlayer Excitation

Y. Sagae, K. Yamada, T. Ishikawa, K. Itoh, H. Itoh, T. Sasaki,
T. Nagata, J. Kano, T. Kambe, S. Ishihara, N. Ikeda and S. Iwai

Abstract Photoinduced growth of the antiferromagnetic state in the ferrimagnetic phase were demonstrated in double layered Fe oxide $LuFe_2O_4$ by 12 fs infrared pulse under inter-layer excitation. Inter-layer charge imbalance successively induces the changes of charge/magnetic structures interacting with local Fe-O stretching and inter-layer sliding phonons through the exchange interaction.

1 Introduction

Optical control of magnetic properties in strongly correlated system through the charge-spin interactions such as double-exchange interaction and Dzyaloshinsky-Moriya interactions attract much attention in perovskite manganese. On the other hand, the charge ordering (CO) in double-layered $LuFe_2O_4$ enables us to expect another mechanism of photoinduced magnetic response. In $LuFe_2O_4$, as shown in Fig. 1a, the LuO_2 and doubled FeO layers having a triangular lattice stack along the c-axis. While average valence of Fe ions is 2.5^+, Fe^{2+}/Fe^{3+} superstructure appears in the FeO layers with an intra-layer three-fold periodicity. Below $T_{CO3D} \sim 330$ K, the CO develops along the c-axis and an inter-layer arrangement of the charge pattern is locked, in other words, three-dimensional (3-D) CO appears. Consequently, macroscopic polarization or ferroelectricity are expected to be appeared from periodical

Y. Sagae · K. Yamada · T. Ishikawa · K. Itoh · H. Itoh · S. Ishihara · S. Iwai (✉)
Deaprtment of Physics, Tohoku University, Sendai 980-8578, Japan
e-mail: s-iwai@m.tohoku.ac.jp

H. Itoh · T. Sasaki · S. Iwai
JST-CREST, Sendai 980-8578, Japan

T. Sasaki
Institute for Material Research, Tohoku University, Sendai 980-8577, Japan

T. Nagata · J. Kano · T. Kambe · N. Ikeda
Department of Physics, Okayama University, Okayama 700-8530, Japan

© Springer International Publishing Switzerland 2015
K. Yamanouchi et al. (eds.), *Ultrafast Phenomena XIX*,
Springer Proceedings in Physics 162, DOI 10.1007/978-3-319-13242-6_61

Fig. 1 a Crystal structure of LuFe$_2$O$_4$ consisting of FeO double-layers and LuO$_2$ blocks. **b** Schematic illustration of magnetic transition. **c** Steady state reflectivity (R) at 150 K. **d** Temperature dependence of R at 0.77 eV and at 0.31 eV

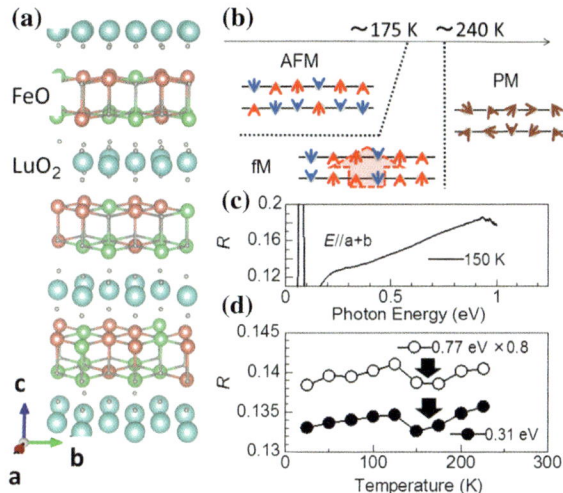

charge imbalance between neighboring FeO layers [1, 2]. It is noteworthy that such 3-D CO also arise the magnetic transitions; i.e., Ferrimagnetic (fM) ordering sets in below T_N = 240 K. Antiferromagnetic (AFM) ordering grows in fM phase below T_{LT} = 175 K through the charge-spin coupling in double layered structure (Fig. 1b) [3]. In this compound, optical responses in visible [4] and THz [5] region have been investigated. However, ultrafast optical response in mid-IR has not been measured, although fM-AFM change markedly affects the mid-IR reflectivity as shown below.

2 Experiment

3-cycle 12 fs pulse in the 1.2–1.8 μm wavelength region was generated in optical parametric amplifier using type I BBO with degenerate configuration and chirped mirror compressor. Pulse width evaluated from the FROG pattern is 12 fs which corresponds to 3-optical cycle. The time resolution of the reflection detected pump-probe measurement is 15 fs [6].

3 Results and Discussions

Figure 1c shows the steady state reflectivity (R) spectrum at 150 K for E//c polarization. As shown in Fig. 1d, R shows the anomalous change between 150–200 K for E//c (arrows). Therefore, R in mid-IR region reflects the change of the magnetic property, i.e., the growth of the AFM state in the fM phase.

Fig. 2 a $\Delta R/R$ spectrum
for $E_{pump}//c$ (0.3 mJ/cm^2),
$E_{probe}//c$ at 150 K **b** $\Delta R/R$
spectrum for $E_{pump}//a + b$
(0.3 m/cm^2), $E_{probe}//c$ at
150 K. Spectral differences
$[R(100\ K)-R(150\ K)]/R$
(150 K) **(a)** and $[R(200\ K)-R$
(150 K)]/R(150 K) **(b)** are
also shown by solid curves.
c, d Temperature
dependences of $\Delta R/R$ for
$E_{pump}//c$, $E_{probe}//c$ **(c)** and
for $E_{pump}//a + b$, $E_{probe}//c$ **(d)**

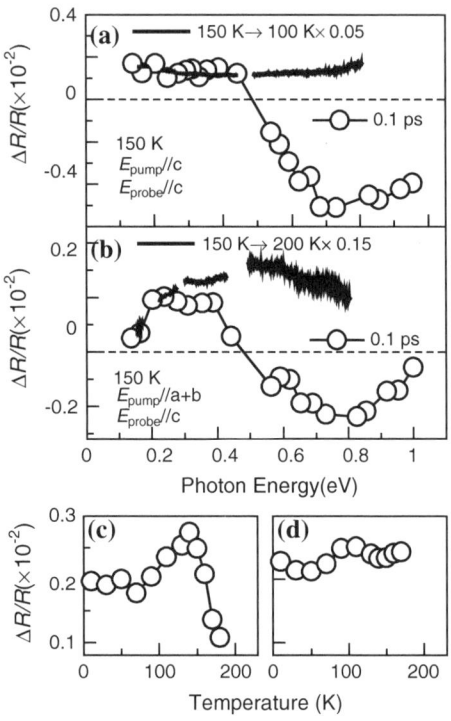

Figure 2a, b show transient reflectivity $\Delta R/R$ at 150 K for $E_{pump}//c$ (a) and for $E_{pump}//a + b$ (b) measured at 0.1 ps after excitation of 100 fs pulse. Excitation energy 0.89 eV corresponds to the Fe^{2+}-Fe^{3+} CT excitation for both polarizations. For both excitation polarization, $\Delta R/R$ is positive at low energy (<0.5 eV), whereas $\Delta R/R$ is negative at high energy region (>0.5 eV). As shown in Fig. 2c, anomalous change in $\Delta R/R$ signal at 150–200 K indicates that the increase of R is closely related to the change of the magnetic property. Such anomaly in $\Delta R/R$ cannot be observed for $E_{pump}//a + b$.

The spectral shape of the reflectivity increases is analogous to that $R(100\ K)–R$ (150 K)/R(150 K) (solid curve in Fig. 2a), reflecting the growth of the AFM in fM phase. On the other hand, $\Delta R/R$ at 300 fs is analogous to the $[R(200\ K)–R(150\ K)]/$R(150 K)(solid curve in Fig. 2b), indicating the temperature rise.

Figure 3 shows the time evolution of $\Delta R/R$ at 0.64 eV (Fig. 3a) measured by using 12 fs pulse and the oscillating component (Fig. 3b). The time profile exhibits instantaneous rise and 150 fs decay. Oscillating structures with the frequencies of 900, 620 and 190 cm^{-1} were also observed. According to the assignment for the IR and Raman spectra, these oscillations are attributable to the overtone of local Fe-O stretching mode (//c), Fe A_g (//c) mode, and E_g (//a + b) mode, respectively. In particular, A_g (//c) and E_g (//a + b) modes have been known to be coupled to the fM-AFM crossover. It is noteworthy that the decay time constant of the 900 cm^{-1}

Fig. 3 a Time evolution of
$\Delta R/R$ at 150 K(0.26 mJ/cm^2)
measured by using 12 fs pulse
for $E_{pump}//c$, $E_{probe}//c$.
b Oscillating components

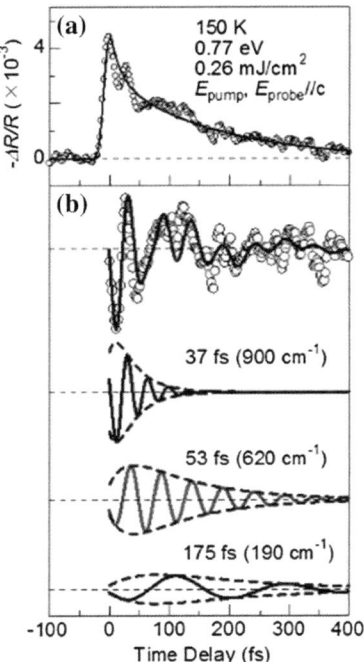

(35 fs) corresponds to the rise time of 620 cm^{-1}, and the decay of 620 cm^{-1} (100 fs) is approximately equal to the rise of 190 cm^{-1} mode. Such successive appearance of the coherent phonons reflects the interactions between the change of the charge/magnetic properties triggered by the inter-layer CT excitation and the phonons; i.e., (i) charge imbalance is locally induced, (ii) such imbalance is extended and stabilized in the double layer, (iii) magnetic ordering changes from fM to AFM. Here, it should be emphasized that the decay time of the 620 cm^{-1} mode 100 fs (= rise of 190 cm^{-1} mode) roughly corresponds to the energy scale of the exchange interaction and/or the spin-orbit interaction.

4 Summary

Photoinduced growth of the anti-ferromagnetic state in the ferrimagnetic phase was demonstrated by 100 fs mid-infrared transient reflectivity measurement. Furthermore, the fast dynamics were also investigated using 12 fs infrared pulse. Successive phonon dynamics interacting with the changes of the charge/magnetic properties have been clarified.

References

1. N. Ikeda, H. Ohsumi, K. Ohwada, K. Ishii, T. Inami, K. Kakurai, Y. Murakami, K. Yoshii, S. Mori, Y. Horibe, and H. Kito, "Ferroelectricity from iron valence ordering in the charge-frustrated system $LuFe_2O_4$", Nature 436, 1136(2005).
2. S. Ishihara, "Electronic ferroelectricity and frustration", J. Phys. Soc. Jpn., 79, 011010(2010).
3. J. Lee, S. A. Trugman, C. D. Batista, C. L. Zhang, D. Talbayev, X. S. Xu, S. -W. Cheong, D. A. Yarotski, A. J. Taylor, and R. P. Prasankumar, "Probing the interplay between quantum charge fluctuations and magnetic ordering in $LuFe_2O_4$", Sceintific Reports, 3, 2654(2013).
4. J. Bourgeois, G. Andre, S. Petit, J. Robert, M. Poienar, J. Rouquette, E. Elkaim, M. Hervieu, A. Maignan, C. Martin, and F. Damay, "Evidence of magnetic phase separation in $LuFe_2O_4$", Phys. Rev. B86, 024413(2012).
5. H. Itoh, K. Itoh, K. Anjo, H. Nakaya, H. Akahama, D. Ohishi, S. Saito, T. Kambe, S. Ishihara, N. Ikeda, and S. Iwai, "Ultrafast melting of charge ordering in $LuFe_2O_4$ probed by terahertz spectroscopy", J. Lumin. 149-151, 133(2013).
6. Y. Kawakami, T. Fukatsu, Y. Sakurai, H. Unno, H. Itoh, S. Iwai, T. Sasaki, K. Yamamoto, K. Yakushi, and K. Yonemitsu, "Early-Stage Dynamics of Light-Matter Interaction Leading to the Insulator-to-Metal Transition in a Charge Ordered Organic Crystal", Phys. Rev. Lett. 105, 246402(2010).

Magnetically Induced Lattice Dynamics in a Magnetoelectric Antiferromagnet Cr$_2$O$_3$

T. Nishimoto, T. Moriyasu and T. Kohmoto

Abstract We studied the optically induced lattice dynamics in a magnetoelectric antiferromagnet Cr$_2$O$_3$ by polarization spectroscopy. The observed divergence behavior of the relaxation rate at the Néel temperature suggests the correlation between lattice and spin fluctuations.

1 Introduction

In recent years, various types of multiferroic materials have been discovered, in which magnetic order and ferroelectricity coexist. Many of them are antiferromagnets with spiral spin structures, and their giant magnetoelectric effects have been attracting attention.

Chromium oxide Cr$_2$O$_3$ is known to show the linear magnetoelectric effect [1], in which electric polarization is induced in proportion to an applied magnetic field and magnetization is induced in proportion to an applied electric field. Although the linear magnetoelectric effect was first observed in Cr$_2$O$_3$ a half century ago [2, 3], recently the room temperature magnetoelectric effect has been attracting attention again [4, 5].

In this report, we studied the optically induced lattice dynamics in a Cr$_2$O$_3$ single crystal by using a transient birefringence measurement with the pump-probe technique.

2 Sample and Experiment

Cr$_2$O$_3$ has a corundum structure with a threefold symmetry axis. Cr^{3+} ions align along the symmetry axis and are surrounded by distorted octahedra of O^{2-} ions. The spins on Cr^{3+} ions shows an antiferromagnetic order below the Néel temperature $T_N = 307$ K.

T. Nishimoto · T. Moriyasu · T. Kohmoto (✉)
Graduate School of Science, Kobe University, Kobe 657-8501, Japan
e-mail: kohmoto@kobe-u.ac.jp

© Springer International Publishing Switzerland 2015
K. Yamanouchi et al. (eds.), *Ultrafast Phenomena XIX*,
Springer Proceedings in Physics 162, DOI 10.1007/978-3-319-13242-6_62

A lattice distortion is generated by a linearly polarized pump pulse (805 nm, 0.2 ps). The induced anisotropy of the refractive index, or linear birefringence, is detected by a polarimeter with a quarter-wave plate as the change in the polarization of a linearly polarized probe beam (900 nm, 0.2 ps). The polarization plane of the probe beam is tilted by 45° from that of the pump beam, and the ellipticity of the transmitted probe beam caused by the birefringence is monitored. The thickness of the sample is 0.25 mm.

3 Result and Discussion

The optically induced lattice distortion signal observed in the picosecond region is shown in Fig. 1. Oscillation components of ~ 0.1 and ~ 0.3 THz are found in the lattice distortion signals at low temperatures. These components disappear above 50 K. The magnon frequency of 0.17 THz has been reported in an experiment of antiferromagnetic resonance [6]. However, in our experiment, no magnon signal was observed with circular pumping and Faraday-rotation detection. The origin of the observed oscillation components is unknown at present.

The optically induced lattice distortion signals observed in the nanosecond and subnanosecond regions are shown in Fig. 2. The observed lattice distortion signal is the sum of two components. One has a long relaxation time, and the other has a shorter relaxation time and a small opposite-sign amplitude.

The temperature dependence of the relaxation rate $1/\tau$ obtained from the long-relaxation-time component is shown in Fig. 3. As the temperature increases, its relaxation rate shows a divergence behavior towards the Néel temperature. The solid line in Fig. 3 represents the relation $1/\tau = (T_N - T)^{-n}$ with $n \sim 0.6$.

In the ferroelectrics of the displacement type, oscillations of soft phonon modes appear. In the ferroelectrics of the order-disorder type, the relaxation rate is expected to decrease at the transition temperature (critical slowing down). However, the relaxation rate observed in Cr_2O_3 increases at T_N. The experimental result in Fig. 3 shows the lattice fluctuation also increases at T_N where the spin fluctuation

Fig. 1 Optically induced lattice distortion signal in the picosecond region observed at 7 K

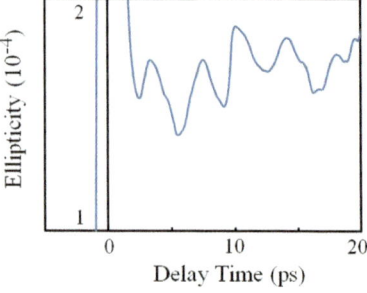

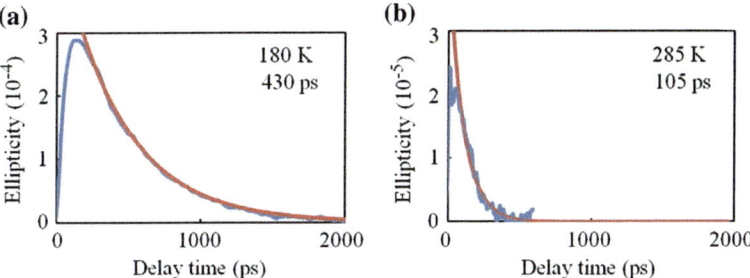

Fig. 2 Optically induced lattice distortion signal in the nanosecond and subnanosecond regions observed at **a** 180 and **b** 285 K

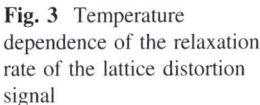

Fig. 3 Temperature dependence of the relaxation rate of the lattice distortion signal

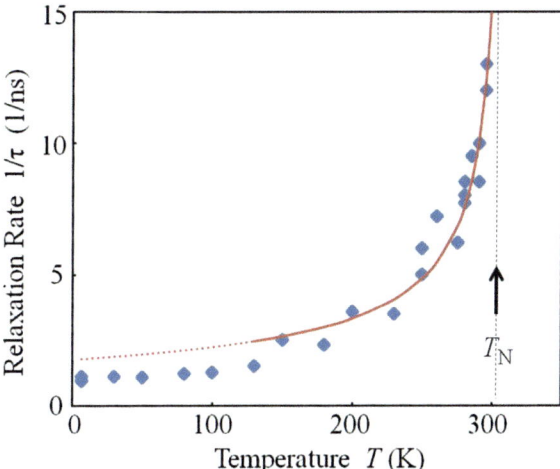

increases and suggests the correlation between lattice and spin fluctuations. The observed dynamics is considered to be a phenomenon peculiar to magnetoelectric materials and may be called a dynamic magnetoelectric effect.

References

1. I. E. Dzyaloshinskii, Zh. Exp. Teor. Fiz. 37, 881-882 (1959) [Sov. Phys. JETP **10**, 628-629 (1959)].
2. D. N. Astrov, Zh. Exp. Teor. Fiz. 38, 984-985 (1960) [Sov. Phys. JETP **11**, 708-709 (1960).
3. G. T. Rado and V. J. Folen, Phys. Rev. Lett. **7**, 310 (1961).
4. A. Iyama and T. Kimura, Phys. Rev. B **87**, 180408(R) (2013).
5. T. Satoh, B. B. V. Aken, N. P. Duong, T. Lottermoser, and M. Fiebig, Phys. Rev. B **75**, 155406, (2007).
6. G. S. Heller, J. J. Stickler, and J. B. Thaxter, J. Appl. Phys. **32**, S307 (1961).

Coherent Magnetism: Pushing the Limits of Spin-Photon Interaction

M. Barthelemy, M. Sanches Piaia, M. Vomir, H. Vonesh and J.-Y. Bigot

Abstract We report on magneto-optical coherent dynamics measurements on a 9 fs time scale. We show that the laser field induces an anisotropy, driving the magnetization out of equilibrium. The corresponding mechanism is the spin-orbit interaction.

OCIS codes 320.0320 · 320.7130 · 300.6290

1 Introduction

The question of how fast can one modify the magnetic state of magnetic materials is important in the context of controlling the recording process in magnetic media [1]. It has been previously shown that coherent magnetism can result from the strong spin-photon interaction, mediated by the spin-orbit coupling [2]. This finding has been deduced from magneto-optical Faraday measurements in Ni and CoPt ferromagnetic thin films using 50 fs pulse duration. Alternatively, it has been shown that the in-plane magnetization of Garnet films can be controlled optically, resulting in a motion of precession [3]. Its origin has been attributed to a magneto-electric effect that is a coherent mechanism corresponding to an induced anisotropy δH_a which modifies the effective field sensed by the material. Here, we explore the limits of this coherent spin-photon interaction by performing time resolved Magneto-Optical Four Wave Mixing (MO-FWM) in a ferrimagnetic Garnet film with 9 fs pulses. We propose a simple model based on the spin-orbit interaction coupled to the laser field which was previously used to explain the coherent Faraday measurements in metals [4].

M. Barthelemy · M. Sanches Piaia · M. Vomir · H. Vonesh · J.-Y. Bigot (✉)
Institut de Physique et Chimie Des Matériaux de Strasbourg, Unité Mixte 7504 CNRS,
Université de Strasbourg, BP. 43, 23 Rue Du Loess, 67034 Strasbourg Cedex, France
e-mail: bigot@unistra.fr

© Springer International Publishing Switzerland 2015
K. Yamanouchi et al. (eds.), *Ultrafast Phenomena XIX*,
Springer Proceedings in Physics 162, DOI 10.1007/978-3-319-13242-6_63

260

2 9 fs Magneto Optical Four Wave Mixing Experiment

The experiment is sketched in the inset of Fig. 1. It consists in a degenerate four wave mixing configuration performed at 800 nm using 9 fs pulses with 80 MHz repetition rate. $\mathbf{k}_p$ (respectively $\mathbf{k}_s$) corresponds to the pump (respectively probe) wave vectors. Note that in such configuration both laser fields can have similar amplitudes. The sample is a 7 μm thick bismuth doped iron garnet (BIG) film and it has been chosen for its strong spin orbit coupling, giving rise to a large magneto optical response in the visible and near IR. A static magnetic field $\mathbf{H}_0$ is applied perpendicular to the sample plane ($\phi = 0°$ or $180°$ being the angle between $\mathbf{H}_0$ and the normal to the sample). The magnetic signal is obtained from the difference between the measurements for opposite directions of $\mathbf{H}_0$.

The MO-FWM signal S^ϕ_{2kp-ks} shown on Fig. 1a is obtained by analyzing the polarization state of the coherent four wave mixing emission in the $2\mathbf{k}_p-\mathbf{k}_s$ direction, using a polarization bridge. Note the opposite rotation for the complementary angles of $\mathbf{H}_0$. The differential signal $\Delta S^\phi_{2kp-ks} = S^{180°}_{2kp-ks} - \Delta S^{0°}_{2kp-ks}$ is shown on Fig. 2b. The coherent dynamics of the charges is expected to decay with a characteristic time T_2, corresponding to the dephasing time of charges with respect to the laser field. By analogy, the magnetic field dependent coherent emission (Fig. 1b) corresponds to a magneto optical dephasing time T_2^{MO}. The asymmetry of the ΔS^ϕ_{2kp-ks} allows to retrieve $T_2^{MO} = 2.8$ fs assuming an inhomogeneously broadened medium, due to the multi-domain structure of Garnet on the spatial scale of the focused beams (~ 30 μm diameter). This is confirmed by a detailed analysis of the signals emitted in the two directions $2\mathbf{k}_p$-$\mathbf{k}_s$ and $2\mathbf{k}_s$-$\mathbf{k}_p$.

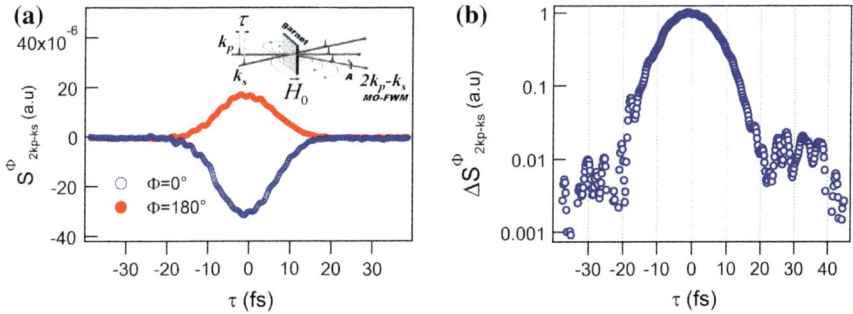

Fig. 1 a Two beams MO-FWM signals for two opposite directions of $\mathbf{H}_0$ ($\phi = 0°$ *blue opened circles* and $\phi = 180°$ *red closed circles*). *Inset* sketch of the two beams MO-FWM configuration. **b** MO-FWM given by $\Delta S^\phi_{2kp-ks} = S^{180°}_{2kp-ks} - S^{0°}_{2kp-ks}$ in log scale and relaxation time T_2^{MO}

Fig. 2 Magneto-optical
rotation. Population dynamics
(*black*). Coherent dynamics:
Pump polarization coupling
(*red*) and pump perturbed free
induction decay (*blue*)

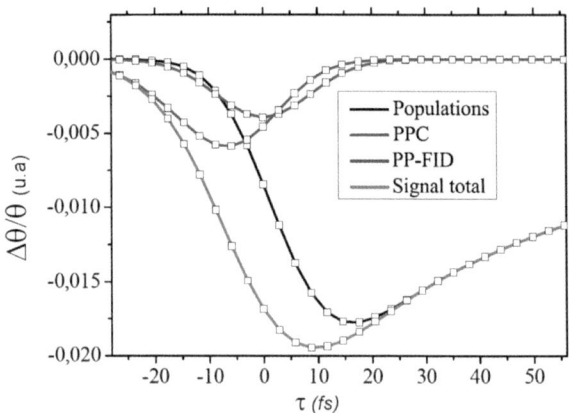

3 Microscopic Origin of the Magneto-optical Coherent Emission

To understand the origin of the coherent magneto optical dynamics, and distinguish
it from effects associated to the population dynamics, like thermal equilibrium and
magnetization precession, it is necessary to separate the transverse from longitudinal magneto-optical processes. It is best made using the density matrix formalism
[5] and considering a simplified Hamiltonian. For the case of a single electron with
charge e mass m momentum $\mathbf{p}$ and spin s excited in hydrogenic quantum states and
interacting with the laser and magnetic fields, also taking into account the spin-orbit
interaction, one has:

$$H_{\text{int}} = \frac{e}{m}\vec{\pi}_{\pm}\vec{A}_L \text{ with } \vec{\pi}_{\pm} = \left(\vec{p} + \frac{e}{c}\vec{A}_M + \vec{s} \times \vec{\nabla}V_{ion}\right)_{\pm}$$

where A_L and A_M are the vector potentials of the laser electromagnetic field and of
the magnetic field respectively; V_{ion} is the scalar potential of the ion. The spin-orbit
coupling term $\vec{s} \times \vec{\nabla}V_{ion}$ plays a fundamental role in the optical control of the
magnetic states. It is a relativistic term present in the quantum electrodynamics of
the spin-photon interaction. It can be simply incorporated in the time dependent
third order nonlinear response of a 8-level hydrogenoid model system. The corresponding third-order non linear polarization can be expressed in terms of the
rotation θ and the ellipticity ε responses either in the Faraday or in the Four Wave
Mixing configurations. It allows distinguishing clearly between the population and
coherent contributions to the magneto optical signals as usually obtained for the
charge dynamics. This is exemplified in Fig. 2 which shows the Faraday rotation
obtained with laser pulses of 10 fs, resonant with the main transition. It can be
decomposed into two coherent contributions, the pump polarization coupling and
the pump-perturbed free induction decay having an electronic dephasing time
$T_2 = 10$ fs and a population contribution with a lifetime $T_1 = 100$ fs. While the

coherent contribution has its counter part in the MO-FWM signal, the population is absent for the two-beam configuration.

In conclusion, the coherent MO-signals observed in transparent ferrimagnetic media reveal the importance of the spin-orbit interaction both for the dephasing dynamics of the charges and spins states at the femtosecond time scale and for the control of the precession at the time scale of tens of picoseconds.

Acknowledgement JYB thanks the European Research Council for financial support: project Atomag, ERC-2009-AdG-20090325#247452.

References

1. J.-Y. Bigot, M. Vomir, "Ultrafast magnetization dynamics of nanostructures", Ann. Phys. 1–29 (2012).
2. J.-Y. Bigot, M. Vomir & E. Beaurepaire, *"Coherent ultrafast magnetism induced by femtosecond laser pulses"*, Nat. Phys. **5**, 515–520 (2009).
3. F. Hansteen, A. Kimel, A. Kirilyuk, Th. Rasing, *"Nonthermal ultrafast optical control of the magnetization in garnet films"* Phys. Rev. B, **73**, 014421 (2006).
4. H. Vonesh and J.-Y. Bigot, *"Ultrafast spin-photon interaction investigated with coherent magneto-optics"*, Phys. Rev. B, **85**, 180407(R), (2013).
5. S. Mukamel, *"Principles of Nonlinear Optical Spectroscopy"*, Oxford Series on Optical and Imaging Sciences, Oxford University Press (1999).
6. J.-Y. Bigot *"Femtosecond magneto-optical processes in metals"*, Compte rendu de l'Académie des sciences, **2**, Issue 10, 1483–1504, (2001).

Quantum Droplets of Electrons and Holes in GaAs Quantum Wells

S.T. Cundiff, A.E. Almand-Hunter, H. Li, M. Mootz, M. Kira
and S.W. Koch

Abstract We present evidence for electron-hole quantum droplets in GaAs quantum wells using spectrally-resolved transient-absorption spectroscopy. Quantum droplets have a correlation function characteristic of a liquid, but have quantized binding energy, unlike macroscopic droplets.

Many-body dynamics can be greatly simplified by the identification of appropriate quasiparticles. In semiconductors near the ground state, interactions between free electrons and the crystal lattice can be neglected by considering quasiparticle electrons and holes with modified mass and, for holes, charge. Electrons and holes can bind to form excitons [1], and two opposite-spin excitons can bind to form biexcitons. In indirect semiconductors, such as silicon and germanium, macroscopic droplets of electrons and holes in thermal equilibrium with an electron-hole plasma have been observed [2]. Evidence for polyexcitons has also been reported [3]. We present experimental and theoretical evidence for a new quasiparticle, which we call a quantum droplet [4]. A quantum droplet is a bound state of a small number of electrons and holes with no pairwise correlations, as in a polyexciton, but instead a two-particle correlation function characteristic of a liquid. Quantum droplets can form in GaAs quantum wells (QWs) under sufficiently high excitation density, and exist on a picosecond time scale, long before the system has reached thermal equilibrium.

In the spectrally-resolved transient-absorption experiment, a high-intensity pump pulse resonantly excites the heavy-hole (HH) QW exciton in GaAs multiple quantum well sample. The number of photons in the pump pulse, N_{pump}, is systematically varied while the average number is locked with a feedback circuit. A low-intensity probe pulse, with opposite circular polarization, arrives after a time

S.T. Cundiff (✉) · A.E. Almand-Hunter · H. Li
JILA, National Institute of Standards and Technology & University of Colorado, Boulder,
CO 80309-0440, USA
e-mail: cundiff@jila.colorado.edu

M. Mootz · M. Kira · S.W. Koch
Department of Physics, Philipps-University Marburg, Renthof 5, 35032 Marburg, Germany
e-mail: kira@Staff.Uni-Marburg.DE

© Springer International Publishing Switzerland 2015
K. Yamanouchi et al. (eds.), *Ultrafast Phenomena XIX*,
Springer Proceedings in Physics 162, DOI 10.1007/978-3-319-13242-6_64

264

delay of zero to several picoseconds. We measure the spectrum of the transmitted probe light and calibrate the probe transmission $T(\omega)$ in absolute units.

Figure 1a shows the probe $1 - T(\omega)$ for a probe delay of 2 ps over a range of N_{pump}. The HH resonance is a peak at $\hbar\omega = E_{HH}$. As N_{pump} is increased, the HH resonance blue shifts because of bandgap renormalization, screening, and Pauli blocking of the low-energy states [5–7]. At the same time, the lower-energy resonance, which we naively identify as a biexciton, red shifts, indicating an increase in the binding energy E_{bind}. However, this binding energy should respond similarly to the HH binding energy. The unexpected red shift suggests that the lower-energy resonance cannot be due to biexcitons alone.

More information about the many-body aspects of the sample response is obtained by projecting out the part of the measured spectra caused by correlations of three or more photons (i.e., higher-order correlations than biexcitons) [7]. Figure 1b shows the resulting differential spectra, for a probe delay of 8 ps. As in the unprojected data, the resonance red shifts with increasing N_{pump}, but the projection reveals a stepped behavior between quantized energy levels.

Quantum droplets explain both the red shift of the resonance, and the quantized steps evident in Fig. 1b. The electron-hole correlation function of a quantum droplet is shown in Fig. 2a. The correlation function has a sharp peak in the middle because of the attractive Coulomb interaction between electrons and holes. The rings in the correlation function are separated approximately by the mean electron-hole separation in the droplet, and vanish beyond the droplet-plasma boundary at radius R. Figure 2b shows the calculated binding energy of droplets as a function of electron-hole density, which closely matches the measured binding energy from Fig. 1b.

To test the identification of this resonance as quantum droplets, we performed two control experiments. First, we repeated the transient absorption measurement

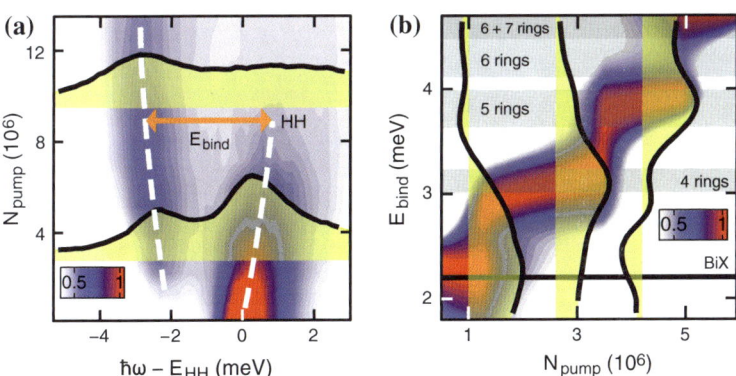

Fig. 1 **a** Probe $1 - T(\omega)$ for a delay of 2 ps as a function of pump photon number. *White dashed lines* track the peaks of the two main resonances. The HH exciton *blue* shifts with increasing number of pump photons. The lower-energy resonance unexpectedly *red* shifts. **b** Projection of probe $1 - T(\omega)$ to a "slanted-Schrödinger's-cat" state, which enhances the contribution of ≥ 3-photon correlations. The binding energy increases in a series of quantized steps

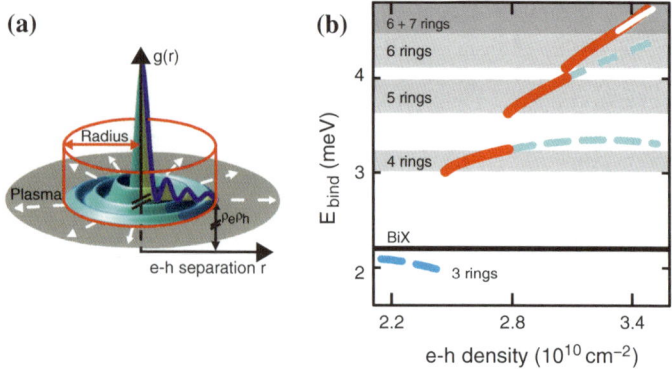

Fig. 2 a Electron-hole correlation $g(r) = \Delta g(r) + \rho_e \rho_h$ for a quantum droplet, as a function of e-h separation r. **b** Calculated E_{bind} for quantum droplets, as a function of e-h density

with co-circular polarizations for pump and probe. Thus, selection rules prohibit excitation of biexcitons or polyexcitons, but we still see the quantum droplet resonance. Second, we raised the sample temperature to 70 K, where phonon scattering should prevent the formation of biexcitons or quantum droplets. In this case, the quantum droplet signature is absent from both the recorded data and the projection. Thus, we conclude that all observations are consistent with identifying the low-energy resonance of Fig. 1a as a quantum droplet.

References

1. Elliott, R. J.: Intensity of optical absorption by excitons, Phys. Rev. **108**, 13841389 (1957)
2. Suzuki T., Shimano R.: Time-resolved formation of excitons and electron-hole droplets in si studied using terahertz spectroscopy, Phys. Rev. Lett. **103**, 057401 (2009)
3. Turner D.B., Nelson K.A.: Coherent measurements of high-order electronic correlations in quantum wells, Nature **466**, 10891092 (2010)
4. Almand-Hunter, A. E., Li, H., Cundiff, S.T., Mootz, M., Kira, M., Koch, S.W.: Quantum droplets of electrons and holes, Nature **506**, 471-475 (2014)
5. Khitrova, G., Gibbs, H.M., Jahnke, F., Kira, M., Koch, S.W.: Nonlinear optics of normal-mode-coupling semiconductor microcavities, Rev. Mod. Phys. **71**, 15911639 (1999)
6. Smith, R. P., Wahlstrand, J. K., Funk, A. C., Mirin, R. P., Cundiff, S. T., Steiner, J. T., Schafer, M., Kira, M., Koch, S. W.: Extraction of many-body configurations from nonlinear absorption in semiconductor quantum wells, Phys. Rev. Lett. **104**, 247401 (2010)
7. Kira M., Koch, S.W., Smith, R.P., Hunter, A.E., Cundiff, S.T.: Quantum spectroscopy with Schrödinger-cat states, Nat. Phys. **7**, 799804 (2011)

Exciton Dynamics in Cu-Doped InAs Colloidal Quantum Dots

Chunfan Yang, Itay Gdor, Yorai Amit, Adam Faust, Uri Banin
and Sanford Ruhman

Abstract Femtosecond transient absorption spectroscopy has been used to investigate the exciton dynamics in native and Cu-doped InAs quantum dots from three respects: (1) hot exciton cooling; (2) Auger recombination; (3) absorption cross section.

1 Introduction

Doping is an attractive way to control the physical and chemical properties of colloidal quantum dots (QDs). It can change the Fermi level by increasing electron or hole concentrations, and alter other physical characteristics which can broaden their technological applications in micro- and optoelectronics [1]. It is important to understand the role played by dopants in carrier dynamics and photophysics of QDs. Our recent work focus on the ultrafast exciton dynamics in Cu-doped InAs quantum dots. Cu acts as an electron donor leading to n-type doping [2]. Carrier dynamics are studied as function of the concentration of metal dopant using broadband near IR femtosecond transient absorption spectroscopy.

C. Yang (✉) · I. Gdor · S. Ruhman
The Institute of Chemistry, Hebrew University, Jerusalem 91904, Israel
e-mail: chunfan.yang@mail.huji.ac.il

Y. Amit · A. Faust · U. Banin
The Institute of Chemistry and the Center for Nanoscience and Nanotechnology,
Hebrew University, Jerusalem 91904, Israel

© Springer International Publishing Switzerland 2015   267
K. Yamanouchi et al. (eds.), *Ultrafast Phenomena XIX*,
Springer Proceedings in Physics 162, DOI 10.1007/978-3-319-13242-6_65

2 Experiments

Collodial InAs QDs and doped InAs QDs are synthesized following a well-established method. We note that the symbol "Cu Number" (for instance Cu 250) reported in the manuscript correspond to the ratio Cu impurities per QD for the doping reaction in solution. The actual Cu concentration per QD is determined by ICP-AES (Inductively Coupled Plasma Atomic Emission Spectroscopy) Measurements, which gives the loading ratio is about 10–25 % [1, 2].

Figure 1 shows the transient absorption (TA) spectra of different doped QDs at different time delays. When irradiated at 640 nm, the bleaching at the band edge, which is attributed to the 1S-1S transition, is observed. 1S transient-absorption signal exhibits a fast build up process in the initial time delays, which is assigned to hot exciton cooling process. The descent of these hot excitons from the initial excited state in conductive band decay to the lowest 1S state by intraband relaxation [3]. The buildup is fitted to an exponential rise, and the associated time constants decrease monotonically with the rise in dopant concentrations.

After initial rapid evolution, a gradual decay is observed at longer delays (t > 2 ps), which is expected to be the exciton annihilation, shown in Fig. 2a. At low fluence, the bleach exhibits a slow decay which is assigned to the single exciton decay. As the fluence is increased, rapid bleach decay due to Auger recombination of the multi-excitons is observed along with an incessant slow decay of the single exciton [3]. In different doping level QDs, the Auger recombination of multiple excitons does not show dramatically different, and the associated time constants are fitted about 12 ps for samples irrespective of level of doping. However at high

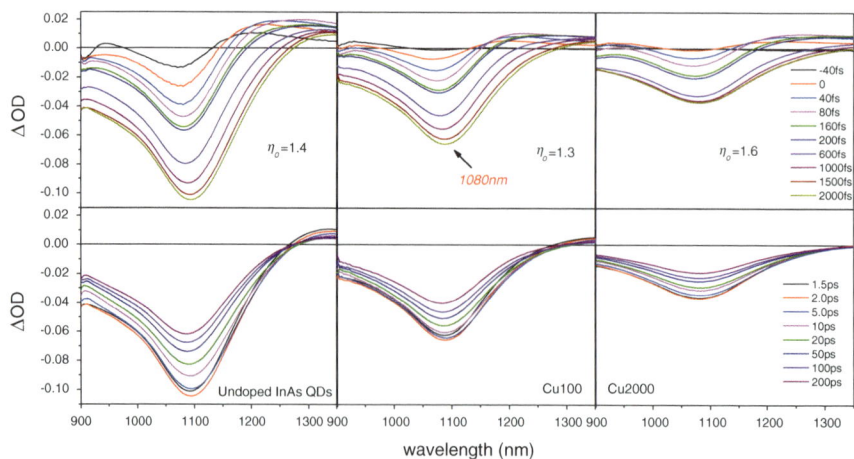

Fig. 1 Transient spectra at different time delays after excitation at 640 nm in different doped InAs QDs. Different time delays ranges, 0–2 ps, 1.5–200 ps are presented above and below respectively, with a column for each QDs; Pump photon fluxes are controlled to deposit similar exciton per QDs (η_0) on average at the front surface for all sample

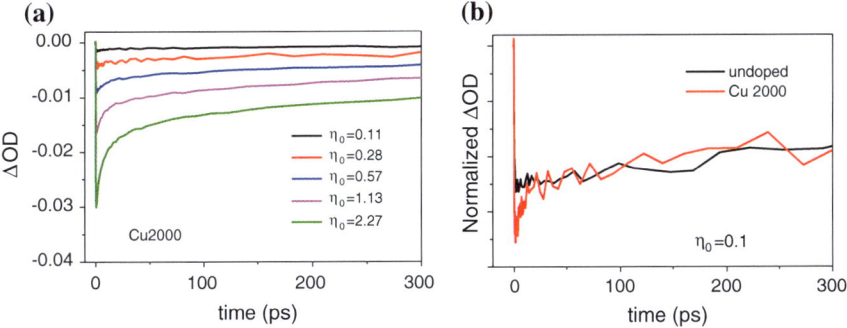

Fig. 2 **a** Pump intensity dependence TA dynamics pumped at 800 nm exhibit both single and multiple exciton dynamics in Cu2000 QDs. **b** Even pumping at lowest intensity we have measured for which less than one exciton is generated per QDs, Cu2000 show fast compare to the undoped ones

doping levels (Cu1000, Cu2000) a fast decay component appears, even when pumping at lowest intensity we have measured for which less than one exciton is generated per QDs, and this fast decay are assigned to the mono-exciton decay in this situation displayed in Fig. 2b.

To further probe doping effects on the photo-physics, a comparison of QD absorption cross sections and the bleach induced by a single relaxed exciton was conducted for doped and undoped samples. To do that, two measurements were performed. In the first, the distribution of exciton number states was measured as a function of pump fluence well above the band edge [4]. This is obtained from comparing band bleach residual after Auger recombination with the bleach induce by a single exciton after photo excitation at 640 nm. Assuming Poisson statistics for multiple absorptions and the known fluence, this ratio can be simulated to produce $\sigma_{640\ nm}$, and this cross section is essentially unchanged by doping. The bleach induced by a single relaxed exciton at the band edge can be determined by measuring the absolute value of band edge bleach ~ 2 ps after excitation, as a function of the density of absorbed photons. Unlike the cross section for absorption, the band edge cross section (σ_{bg}) for a single exciton is reduced dramatically by the doping with Cu.

3 Conclusions

We have investigated ultrafast exciton dynamics in native and Cu-doped InAs quantum dots by femtosecond transient absorption spectroscopy. And our experimental results show that:

(1) Hot exciton cooling is accelerated as doping is increased.
(2) Similar relaxation Auger recombination rates demonstrate that the dopant band electrons are not involved in carrier dynamics in this regime. The fast mono-exciton decay observed in heavily doped QDs requires further study.

(3) The cross sections derived from the band edge bleach gradually decrease to as increasing the doping levels, while the cross section obtained from the distribution of exciton number at 640 nm doesn't change. This shows drastically reduction of the bleach contributed by a single relaxed exciton at the band edge, indicating an effective increase in the band edge degeneracy due to the doping.

References

1. D. Mocatta et al (2011) Heavily Doped Semiconductor Nanocrystal Quantum Dots. Science 332:77-81
2. Yorai Amit et al (2013) Unraveling the Impurity Location and Binding in Heavily Doped Semiconductor Nanocrystals: The case of Cu in InAs Nnaocrystals. J. Phys. Chem. C 117:13688-13696
3. Victor I. Klimov (2007) Spectral and Dynamical Properties of Multiexcitons in Semiconductor Nanocrystals. Annu Rev Phys Chem 58:653-73
4. M. Ben-Lulu et al (2008) On the Absence of Detectable Carrier Multiplication in a Transient Absorption Study of InAs/CdSe/ZnSe Core/Shell1/Shell2 Quantum Dots. Nano Lett 8(4): 1207-1211

Rabi Oscillations in an InAs Quantum Dot Ensemble Observed in Pre-pulse 2D Coherent Spectroscopy

T. Suzuki, R. Singh, I.A. Akimov, M. Bayer, D. Reuter, A.D. Wieck and S.T. Cundiff

Abstract We have observed Rabi oscillations in an InAs quantum dot ensemble by using optical pre-pulse 2D coherent spectroscopy. The polarization for 2D coherent spectroscopy is set to be cross-linear in order to obtain biexciton signal, which enables us to distinguish the signals from the ground and excited states. Furthermore, the spectral domain in 2D can reveal the coherent evolution in an inhomogeneously broadened ensemble. With increasing pre-pulse intensity, the signals attributed to the ground and excited states exhibit the sinusoidal decrease and increase, respectively. The observed excitation behavior is well reproduced by a damped oscillation model. From the fitting the pulse area achieved in this work is deduced to be 0.41π and the dipole moment is estimated as 29 Debye.

1 Introduction

A semiconductor quantum dot (QD) is one of the most promising candidates for use as a qubit in quantum information devices because the large dipole moments μ and the discrete level system in QD enable us to control the light-matter interaction coherently. Optical Rabi oscillations are a fundamental demonstration of coherent

T. Suzuki (✉) · R. Singh · S.T. Cundiff
JILA, University of Colorado & National Institute of Standards and Technology, Boulder, CO 80309-0440, USA
e-mail: takeshi.suzuki@jila.colorado.edu

S.T. Cundiff
e-mail: cundiff@jila.colorado.edu

I.A. Akimov · M. Bayer
Experimentelle Physik 2, Technische Universität Dortmund, 44221 Dortmund, Germany
e-mail: manfred.bayer@tu-dortmund.de

D. Reuter · A.D. Wieck
Lehrstuhl fuer Angewandte Festkoerperphysik, Ruhr-Universitaet Bochum, Universitaetsstrasse 150, 44780 Bochum, Germany
e-mail: andreas.wieck@ruhr-uni-bochum.de

© Springer International Publishing Switzerland 2015
K. Yamanouchi et al. (eds.), *Ultrafast Phenomena XIX*,
Springer Proceedings in Physics 162, DOI 10.1007/978-3-319-13242-6_66

manipulation, which manifest themselves as the sinusoidal dependence of the population inversion in a two level system on the pulse area $\int_{-\infty}^{\infty} dt \mu E(t)/\hbar$, where $E(t)$ is an envelope function of the driving electric field. Compared to the demonstrations of coherent control for single QDs [1], there have been few studies for ensembles of QDs [2, 3]. 2D coherent spectroscopy (2DCS) has the advantage of being able to clearly distinguish between contributions from the ground and excited states in an optically excited QD ensemble, which is discussed later.

In this work, we have investigated Rabi oscillations in an InAs QD ensemble by using pre-pulse 2DCS, which is a pump-probe experiment, where the pre-pulse is the pump and the 2DCS works as the probe. 2DCS is a three-pulse transient four wave mixing (TFWM) technique with the addition of interferometric stabilization of the pulse delays. Three pulses are incident on the sample in a rephasing (photon echo) time-ordering and the TFWM signal is interfered with a reference and spectrally-resolved while the delay between the first two pulses is varied. A Fourier transform of the signal with respect to this delay generates a 2D spectrum, in which the excitation and emission energies are correlated along the vertical and horizontal axes, respectively. The sample investigated is a self-assembled InAs/GaAs QD ensemble, consisting of 10 quantum-mechanically-isolated epitaxially-grown layers that are thermally annealed post-growth at 900 °C for 30 s, which results in a 100 meV in-plane confinement. The sample was unintentionally doped during growth, resulting in approximately half of the QDs being charged with a hole. The lowest exiton transition is four-fold degenerate with two optically-active (bright) states that are coupled through the exchange interaction, forming an eigenbasis of horizontally (H) and vertically (V) polarized states for asymmetrically shaped QDs (Fig. 1a). All of the measurements are performed with the sample cooled at 10 K in a sample in vapor flow cryostat.

2 Results and Discussion

A cross-linearly polarized excitation and detection scheme is used, i.e. VHHV for A, B, C and the TFWM signal respectively, which suppresses neutral exciton signals. Figure 1e shows a typical 2D rephasing amplitude spectrum without pre-pulse. The spectrum features two peaks that are inhomogeneously broadened along the diagonal direction due to QD size dispersion. The peak labeled D arises from the trion nonlinear response, whereas the peak labeled B1 arises from the biexciton nonlinear response and is red-shifted along the emission energy by the biexciton binding energy corresponding to the quantum pathway shown in Fig. 1b. The polarization for the pre-pulse is set along the V direction, which creates the superposition between the ground state and the V exciton state, and other quantum paths starting with population in the V exciton state (Fig. 1c, d) contribute to the 2D signal. Figure 1f shows a 2D rephasing amplitude spectrum at a prepulse intensity of 8.8 μ J/cm². Compared to Fig. 1e, the B1 signal becomes weaker, due to the quantum path shown in Fig. 1c, which has the opposite sign to the signal shown

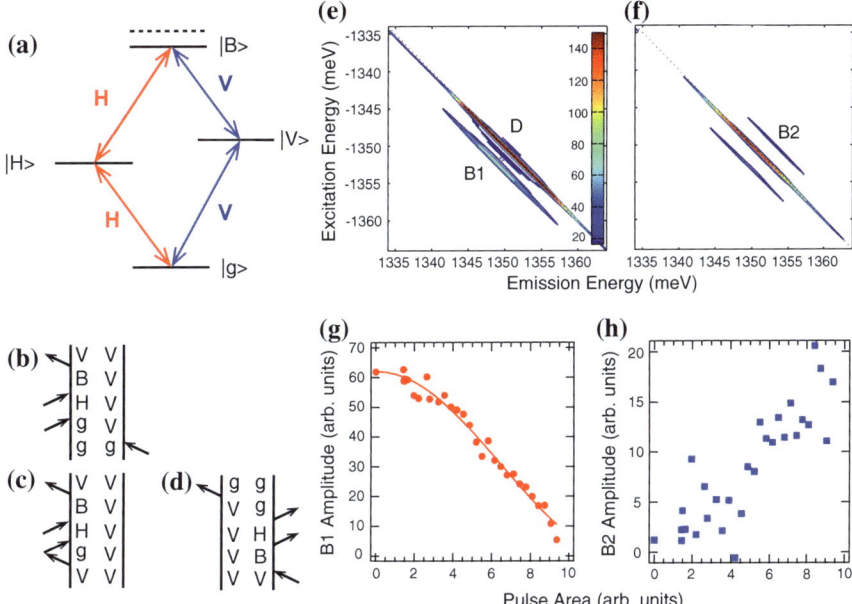

Fig. 1 **a** The energy level diagram showing the horizontal (H) and vertical (V) exciton and biexciton (B). The quantum pathway for VHHV polarization starting with population in the ground state (**b**) and the V exciton state (**c** and **d**). 2D rephasing amplitude spectrum for an InAs self-assembled QD ensemble at prepulse intensity of **e** 0 and **f** 8.8 μ J/cm². The *colorbar* in **e** shows the scales for both (**e**) and (**f**). Amplitude for peak B1 (**g**) and B2 (**h**) as a function of pulse area which is proportional to the square root of pre-pulse intensities

in Fig. 1b. Additionally, a peak labeled B2 arises from the biexciton nonlinear response from the V exciton state corresponding to Fig. 1d and is red-shifted along excitation energy by the biexciton binding energy. Using 2DCS, the biexciton nonlinear responses starting with population in the ground state and the V exciton state can be clearly separated, enabling us to estimate the ratio of the population in the V exciton state to the ground state quantitatively in a QD ensemble. Figure 1g, h show the amplitude of peaks B1 and B2 in red circles and blue squares, respectively, as a function of pulse area which is proportional to the square root of pre-pulse intensity. B1 gets weaker, while B2 gets stronger with increasing pulse area. The solid red line in Fig. 1g is a fit to a damped oscillator model given by $A \exp(-\gamma\sqrt{I})\left(\cos(a\sqrt{I}) + 1\right)$, where γ is a decay of coherence, I is pre-pulse intensity, and A and a are the scaling parameters. From the fitting, we estimate that a pulse area of up to 0.41 π is achieved in our experiment, from which the dipole moment is deduced as 29 Debye. Work on increasing the pulse area to study highly excited coherent state is ongoing.

In summary, we have observed Rabi oscillations in an InAs QD ensemble using pre-pulse 2DCS. The 2D spectra can clearly separate the upper state's signal from the lower state, enabling us to monitor the coherent evolution despite the inhomogeneous broadening.

References

1. A. J. Ramsay, "A review of the coherent optical control of the exciton and spin states of semiconductor quantum dots", Semicond. Sci. Technol. **25**, 103001 (2010)
2. P. Borri, *et al.*, "Rabi oscillations in the excitonic ground-state transition of InGaAs quantum dots", Phys. Rev. B **66**, 081306(R) (2002)
3. M. Kujiraoka, *et al.*, "Optical Rabi oscillations in a quantum dot ensemble", Appl. Phys. Express **3**, 092801 (2010)

Slow Electron Cooling Dynamics of Highly Luminescent CdS$_x$Se$_{1-x}$ Alloy Quantum Dot

Partha Maity, Tushar Debnath and Hirendra Nath Ghosh

Abstract Ultrafast Electron cooling dynamics of highly luminescent oleic acid caped CdS$_x$Se$_{1-x}$ alloy quantum dot (QD) is investigated by femtosecond transient absorption studies and found to be much slower as compared to pure CdSe and CdS QDs.

1 Introduction

Alloy nanocrystals [1, 2] quantum dots are important class of composite materials due to their enormous applications like photovoltaic performances [3], coherent emitter, biological imaging, plasmon wave guide and magneto optical devices etc. The physical and optical properties of the alloy semiconductor quantum dots depend on both size of the nanocrystals as well as the composition of the constituents. As a result composition of nanocrystals plays an extra degree of freedom towards selecting enviable properties for nanostructure designing purpose. The size dependent band structure of the semiconductor quantum dots appear due to their strong confinement which is governed by the size quantization effect (SQE). Thus, by changing the composition of the constituents, one can achieve next contrivance for altering physical and optical properties of the nanocrystals. To obtain higher efficiency in quantum dot based solar cell, it is important to separate the electron and hole pair before their exciton-exciton annihilation which occur in sub picoseconds time scale. Higher emission lifetime and emission quantum yield for alloy QD is reported in literature [4, 5], however no reports are available on charge carrier dynamics in ultrafast time scale. In the present investigation we are reporting optical and photo-physical

P. Maity · T. Debnath · H.N. Ghosh (✉)
Radiation and Photochemistry Division, Bhabha Atomic Research Centre,
Mumbai 400 085, India
e-mail: hnghosh@barc.gov.in

P. Maity
e-mail: pmaity@barc.gov.in

© Springer International Publishing Switzerland 2015
K. Yamanouchi et al. (eds.), *Ultrafast Phenomena XIX*,
Springer Proceedings in Physics 162, DOI 10.1007/978-3-319-13242-6_67

properties of newly synthesized $CdS_{0.7}Se_{0.3}$ alloy quantum dot by steady state and time-resolved absorption and emission spectroscopy. Femtosecond transient absorption measurement has been carried out to comprehend charge carrier dynamics like carrier cooling and charge recombination dynamics in ultrafast time scale.

2 Results and Discussion

Figure 1 shows excitonic absorption at 595 nm and excitonic emission at 617 nm for $CdS_{0.7}Se_{0.3}$ alloy QD which are shifted to red region of the spectrum as compared to that of pure CdS QD of similar size. Photoluminescence quantum yield determined to be 0.7, which indicates that the electrons and holes are more confined in the electronic states of the nanocrystal. Excited emission life time measured to be more than 20 ns (Fig. 1, left panel, inset) for alloy QD as compared pure CdS (τ_{em} = 12 ns) and CdSe (τ_{em} = 13 ns). Figure 1 (right panel) depicts the transient absorption spectra of $CdS_{0.7}Se_{0.3}$ alloy QD with two negative absorption band at 600 and 500 nm which can be attributed to ($1S_e - 1S_{3/2}$ (1S) and $1P_e - 1P_{3/2}$ (1P)) excitonic bleach respectively. It is interesting to see that appearance and disappearance time for both the excitonic bleach peaks are different. Growth and recovery dynamics for both the excitonic bleach are shown in the left panel of Fig. 2. It is interesting to see that both growth and recovery time for 1P exciton bleach is much faster as compared to 1S exciton bleach. Slower bleach growth determined to be 8 ps at 1S excitonic position can be attributed electron cooling dynamics of photo-excited electron in the conduction band of alloy QD as shown in the scheme below (Fig. 2, right panel).

At 400 nm laser pump excitation the highest electronic state of the alloy QD will be populated as shown in the scheme. Now to understand the carrier dynamics (both cooling and recombination) it is very important to compare the bleach recovery kinetics of the alloy QD at two excitonic position as well as with the pure CdSe QD.

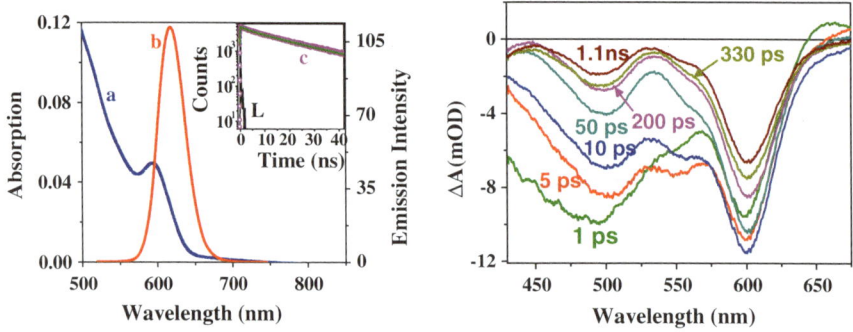

Fig. 1 *Left-Panel* steady state *a* optical absorption and *b* emission spectra of $CdS_{0.7}Se_{0.3}$ alloy QD. *Inset c* time resolved emission decay trace of alloy QD, λ_{ex} = 445 nm and λ_{em} = 617 nm, '**L**' stands for excitation lamp profile. *Right panel* transient absorption spectra of that $CdS_{0.7}Se_{0.3}$ alloy QD recorded at different time delay after excitation at 400 nm laser light

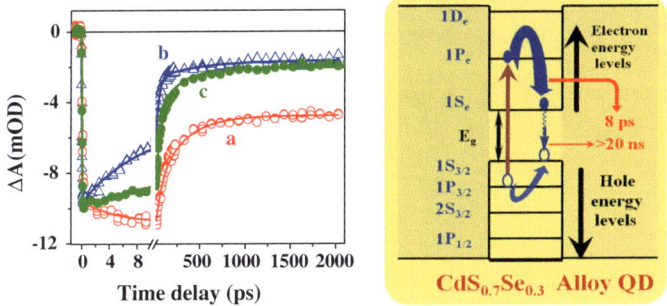

Fig. 2 *Left-Panel* bleach recovery kinetics at *a* 600 nm (1S exciton), *b* 500 nm (1P exciton) for CdS$_{0.7}$Se$_{0.3}$ QD and at *c* 626 nm for 1S exciton for CdSe QD, *Right-Panel* schematic presentation of carrier cooling/relaxation dynamics in CdS$_{0.7}$Se$_{0.3}$ alloy quantum dot

Here we would like to mention that we have compared the dynamics of the alloy with pure CdSe QD not with pure CdS QD because the exciton position of the alloy QD is very close to CdSe QD and at the same time at 400 nm pump excitation the higher electronic state excitation of the CdS QD is not possible (first exciton of the CdS QD of similar size appear at 410 nm). The first excitonic (1S) bleach recovery dynamics of the alloy at 600 nm (Fig. 2 trace a) which has bleach growth and can be fitted bi-exponentially with time constants $\tau_1 = 100$ fs (80 %), $\tau_2 = 8$ ps (20 %) and it recovers multi-exponentially with time constants $\tau_1 = 60$ ps (40 %), $\tau_2 = 300$ ps (15 %) and $\tau_3 > 2.2$ ns (45 %). On the other hand the upper excitonic bleach (1P, scheme) kinetics for the alloy QD at 500 nm (Fig. 2 trace b) can be fitted with pulse-width limited growth (~ 100 fs) and multi-exponential recovery with time constants $\tau_1 = 6$ ps (52 %), $\tau_2 = 35$ ps (24 %), $\tau_3 = 200$ ps (8 %) and $\tau_4 > 2.2$ ns (16 %). We have monitored the bleach recovery kinetics at 1S excitonic position for both for CdSe QD (Fig. 2c) and alloy QD to compare charge carrier and recombination dynamics. Bleach recovery kinetics for 1S excitonic position of CdSe QD can be fitted with 400 fs growth time which can be attributed to electron cooling dynamics from upper excitonic state to 1S excitonic state. Recovery kinetics of 1S exciton for CdSe QD can fitted with multi-exponential time constants of $\tau_1 = 12$ ps (40 %), $\tau_2 = 70$ ps (30 %), $\tau_3 = 400$ ps (10 %) and $\tau_4 > 2.2$ ns (20 %). It is interesting to see that cooling time in alloy QD is much slower (8 ps) as compared to CdSe QD (400 fs). In addition to that charge recombination time was found to much slower in alloy QD as compared to pure CdSe QD.

3 Conclusion

In conclusion, we have synthesized CdS$_{0.7}$Se$_{0.3}$ alloy QD which has very high emission quantum yield (70 %) and long radiative life time. Ultrafast transient absorption studies reveal that electron cooling time for CdS$_{0.7}$Se$_{0.3}$ alloy QD is extremely slow (8 ps) as compared to both CdS and CdSe QDs (400 fs). Charge recombination time found to be exceedingly slower in alloy QD as compared to pure QDs. To the best of our knowledge we are reporting for first time the slow electron cooling of $CdS_{0.7}Se_{0.3}$ alloy QD. Longer excited state life time, slower electron cooling dynamics for alloy QD confirms the reality of design and development of higher efficient QD solar cell.

References

1. Swafford, L. A.; Weigand, L. A.; Bowers, M. J.; . McBride, J. R.; Rapaport, J. L.; Watt, T. L.; Dixit, S. K.; Feldman, L. C.; Rosenthal, S. J. "Homogeneously Alloyed CdSxSe1-x Nanocrystals: Synthesis, Characterization, and Composition/Size-Dependent Band Gap" J. Am. Chem. Soc. **2006**, *128*, 12299.
2. Ki, B. W.; Padilha, L. A.; Park, Y.; McDaniel, H.; Robe, I.; Pietryga, J. M.; Klimov, V. I.; "Controlled Alloying of the Core/Shell Interface in CdSe/CdS Quantum Dots for Suppression of Auger Recombination". ACS Nano, **2013**, *7*, 3411.
3. Kim, J.P, Christians J A.; Choi, H.; Krishnamurthy, S.; Kamat, P. V., "CdSeS Nanowires: Compositionally Controlled Band Gap and Exciton Dynamics". J. Phys. Chem. Lett. **2014**, *5*, 1103.
4. Sarma, D. D.; Nag, A.; Santra, P. K.; Kumar, A.; Sapra, S.; Mahadevan, P. "Origin of the Enhanced Photoluminescence from Semiconductor CdSeS Nanocrystals". J. Phys. Chem. Lett. **2010**, *1*, 2149.
5. Boldt, K.; Kirkwood, N.; Beane, G. A.; Mulvaney, P. "Synthesis of Highly Luminescent and Photo-Stable, Graded Shell CdSe/Cd$_x$Zn$_{1-x}$S Nanoparticles by In Situ Alloying". Chem. Mater. **2013**, *25*, 4731.

Ultrafast Dynamics Related to Spin Crossover Processes in Single Crystal [Fe^{II}(bpy)$_3$](PF$_6$)$_2$

R.L. Field, L. Liu, C. Lu, Y. Jiang, W. Gawelda and R.J.D. Miller

Abstract Transient absorption spectroscopy is used to characterize the ultrafast spin crossover process in iron(II)-tris(bipyridine)-bis(hexafluorophosphate) single crystals. Preliminary data shows evidence of instrument response limited excited state absorption, long-lived (>1 ns) ground state bleach, and oscillatory signals on multiple time scales.

1 Introduction

Transition metal complexes (TMC) are molecules in which central metal atoms or ions are bound to a surrounding array of organic molecules (ligands). Some TMCs have been observed to undergo a phenomenon known as spin crossover (SCO) where a transition from a low-spin (LS) state to a high-spin (HS) state is induced upon photoexcitation or temperature/pressure change [1].

Iron(II)-tris(bipyridine) is a complex that has been extensively studied with regard to the SCO phenomenon. Studies have used transient absorption (TA) and fluorescence spectroscopy in the visible [2], UV [3], and X-ray regimes to elucidate to relaxation cascade leading to SCO in aqueous iron(II)-tris(bipyridine) ([Fe^{II}(bpy)$_3$]$^{2+}$) [4, 5]. Using pump wavelengths in the visible range, these studies

R.L. Field · L. Liu · C. Lu · R.J.D. Miller (✉)
Departments of Chemistry and Physics, University of Toronto,
80 St. George Street, Toronto, ON M5S 3H6, Canada
e-mail: dwayne.miller@mpsd.mpg.de

R.L. Field · L. Liu · Y. Jiang · R.J.D. Miller
Centre for Free Electron Laser Science, Max Planck Institute for the Structure
and Dynamics of Matter, Bld. 99, Luruper Chaussee 149, 22761 Hamburg, Germany

W. Gawelda
European XFEL, Albert-Einstein-Ring 19, 22761 Hamburg, Germany

© Springer International Publishing Switzerland 2015   279
K. Yamanouchi et al. (eds.), *Ultrafast Phenomena XIX*,
Springer Proceedings in Physics 162, DOI 10.1007/978-3-319-13242-6_68

have shown that the complex is excited from the LS ground state to a metal-to-ligand-charge-transfer singlet state (^{1}MLCT) from which it rapidly (20 fs) populates a triplet state (^{3}MLCT) via intersystem crossing (ISC). The system then relaxes within 120 fs, again via ISC, to the HS quintet state from which the LS ground state is repopulated with a time constant of 665 ps [2, 4]. Additionally, using a 520 nm pump, oscillations with a 254 fs period were observed in the excited state absorption (ESA) region of the UV TA spectrum [3]. These oscillations were ascribed to a vibrational wave packet on the HS surface. Iron(II)-tris(bipyridine)-bis (hexafluorophosphate) ($[Fe^{II}(bpy)_3](PF_6)_2$) crystals in powdered form have also been studied using X-ray diffraction [6].

The extremely fast spin transitions observed defy conventional descriptions of spin-orbit coupling. The nuclear motions coupled to the MLCT band are thought to play a role. Femtosecond electron diffraction studies are in progress to separate the nuclear dynamics from electronic effects. However, the excited state dynamics must be characterized within the single crystal environment needed for adequate spatial resolution to resolve this issue. Crystal contacts and the difference in medium polarizability could affect these dynamics. Here, we present preliminary data of the first TA studies of single crystal $[Fe^{II}(bpy)_3](PF_6)_2$.

2 Experimental Methods

The data presented were collected using a home-built pump-probe spectrometer. The setup used a commercial Ti:Sapphire regenerative amplifier laser system (Coherent Legend) outputting 40 fs pulses at 1 kHz, centered at 800 nm. A fraction of the output beam was used to generate a 400 nm pump beam via second harmonic generation. This beam was passed through an acousto-optic pulse shaper used to compress the pulses to near the transform limit. A 52 fs pulse duration was retrieved using a home-built TG FROG spectrometer. The pump fluence used in the data shown was approximately 6 mJ/cm^2. A white-light continuum ranging from 375 to 625 nm was used as a probe. The continuum was generated using a small fraction of the 800 nm beam focused into a 2 mm thick CaF_2 window. The instrument response function was estimated to be 130 fs by fitting the measured Kerr effect signal in a glass microscope slide.

Optical choppers were used in both the pump and probe beams, with the probe beam being chopped at half the frequency of the pump beam. This allowed for the signal from pump scattering to be isolated and largely removed. All pump-probe data shown were corrected to compensate for group velocity dispersion in the probe beam. The $[Fe^{II}(bpy)_3](PF_6)_2$ samples were microtomed to 200 nm in thickness (except for the data shown in Fig. 1d, as indicated) and placed on 1 mm glass microscope slides for measurement.

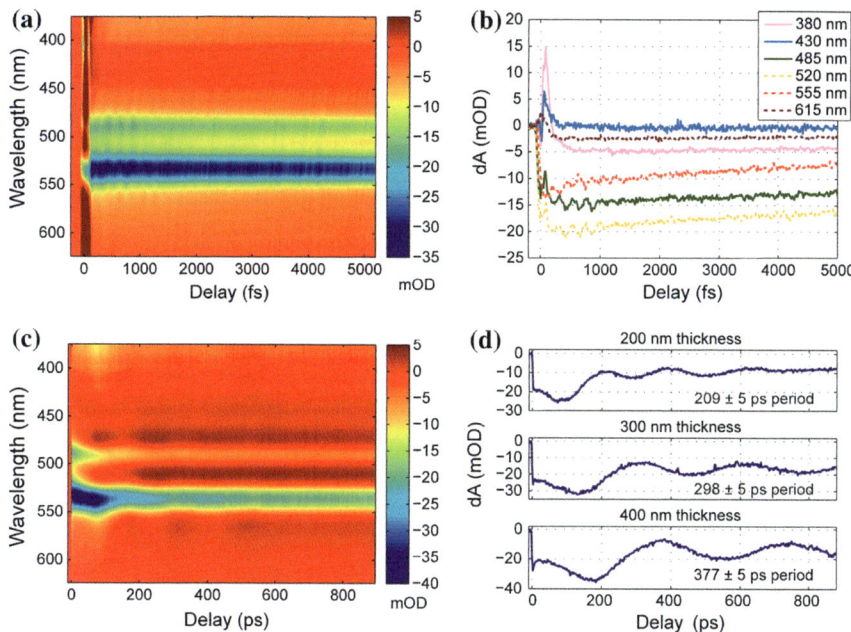

Fig. 1 **a** TA spectrum of [Fe^{II}(bpy)$_3$](PF$_6$)$_2$ at short times after excitation. **b** Short-time kinetic traces at selected wavelengths with substrate response subtracted. **c** TA spectrum at long times after excitation. **d** Long-time kinetic traces at 550 nm using samples with different thicknesses

3 Results and Discussion

Preliminary TA data are shown for short and long times after excitation in Fig. 1a, c respectively. Spectral signatures of excited state absorption (ESA) are present between 375–460 and 600–625 nm. The decay rate of the ESA signal is instrument response limited, which is consistent with the decay of the ^{1}MLCT or ^{3}MLCT states (20 and 120 fs, respectively) found in [2, 4]. Strong ground state bleach (GSB) is observed between 375–400 and 470–560 nm. The long-time scan shows that the GSB signal persists well beyond 850 ps. However, the sample almost fully recovers to the ground state within 1 ms, as evidenced by the negligible background signal before the zero delay. We note that the kinetic trace at 430 nm (Fig. 1b) decays to zero after the initial fast ESA. This suggests the presence of a long-lived absorptive component overlapping with the GSB. In contrast to [2], a long-lived absorptive feature in the red (>600 nm) previously attributed to the HS state is absent. This difference may be related to the effects of crystal contacts on the bond displacement associated with the HS transition.

In addition to the above, the kinetic traces of the short-time scan show clear oscillations in the GSB signal. Fourier analysis of these oscillations gives a period of 223 ± 30 fs. Although the cause of these oscillations is yet to be determined, we note that the period is similar to that assigned to vibrational wave packets under solution phase conditions (254 fs) in [3]. In the long-time scan, oscillations with time scales on the order of hundreds of picoseconds are observed. Using samples of different thicknesses, we note that the period of the oscillation varies linearly with the crystal thickness along the optical axis (Fig. 1d). These oscillations can therefore be attributed to acoustic phonons oscillating between the crystal surfaces. Further analysis and TA experiments in the UV are required to fully elucidate the states involved and the nature of the fast oscillations observed.

References

1. P. Gutlich, A. Hauser, H. Spiering, Angew. Chem. Int. Edit., **33**, 2024 (1994).
2. W. Gawelda, A. Cannizzo, V. T. Pham, F. Mourik, C. Bressler, M. Chergui, J. Am. Chem. Soc. **129**, 8199 (2007).
3. C. Consani, M. Prémont-Schwarz, A. ElNahhas, C. Bressler, F. Mourik, A. Cannizzo, M. Chergui, Angew. Chem. **121**, 7320, (2009).
4. C. Bressler et al., Science, **323**, 489, (2009).
5. W. Zhang et al., Nature, **509**, 345, (2014).
6. B. Freyer, F. Zamponi, V. Juvé, J. Stingl, M. Woerner, T. Elsaesser, M. Chergui, J. Chem. Phys., **138**, 144504 (2013).

Femtosecond Electron Diffraction Study of the Spin Crossover Dynamics of Single Crystal [Fe(PM-AzA)₂](NCS)₂

Yifeng Jiang, Lai Chung Liu, Henrike M. Müller-Werkmeister,
Meng Gao, Cheng Lu, Dongfang Zhang, Eric Collet
and R.J. Dwayne Miller

Abstract The atomic motions involved in spin crossover photo-switching dynamics of single crystal [Fe(PM-AzA)₂](NCS)₂ are investigated by femtosecond electron diffraction (FED). The experiment was performed with an ultrabright femtosecond electron source using 8.5×10^4 electrons per pulse with 400 fs temporal instrument response function.

1 Introduction

Spin crossover (SCO), a conversion from low spin (LS) ground state to high spin (HS) excited state (or visa versa) due to temperature change or light absorption [1, 2], has been extensively studied for its potential applications in optical memory and photo switchable devices [3]. Among the compounds with SCO dynamics, the group of ferrous Fe(II) metal compounds with an FeN_6 coordination environment is the largest [4]. After the Fe(II) compounds absorb light, the LS $^1A_1(t_{2g}^6)$ ground state changes to the HS $^5T_2(t_{2g}^4e_g^2)$ excited state by 2 paired electrons in the t_{2g} sublevel entering the e_g sublevel as unpaired electrons. As the Fe is bonded to the ligand by N atoms, the Fe-N bond distance in the HS state is elongated by ~ 0.2 Å

Y. Jiang · D. Zhang · R.J.D. Miller (✉)
Centre for Free Electron Laser Science, Department of Chemistry and Physics, Centre for Free Electron Laser Science, Max Planck Institute for the Structure and Dynamics of Matter, University of Hamburg, Notkestrasse 85, 22607 Hamburg, Germany
e-mail: dwayne.miller@mpsd.mpg.de

L.C. Liu · H.M. Müller-Werkmeister · M. Gao · C. Lu · R.J.D. Miller
Departments of Chemistry and Physics, University of Toronto, 80 St. George Street, Toronto, ON M5S 3H6, Canada

E. Collet
Institut de Physique de Rennes, Campus de Beaulieu, University Rennes 1,
UMR 6251 UR1-CNRS,Bat 11A Campus de Beaulieu, Rennes, France

© Springer International Publishing Switzerland 2015
K. Yamanouchi et al. (eds.), *Ultrafast Phenomena XIX*,
Springer Proceedings in Physics 162, DOI 10.1007/978-3-319-13242-6_69

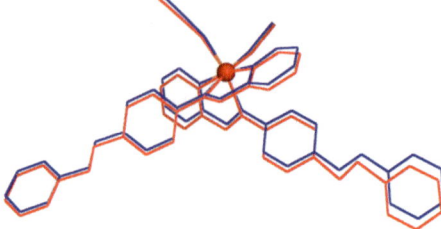

Fig. 1 For single crystal [Fe(PM-AzA)$_2$](NCS)$_2$, the Fe-N bond length of the low spin state is 2 Å and it increases 0.2 Å upon conversion to the high spin state [6]. The Fe atom is represented by the *large ball*

concomitant with the transfer of two electrons to the antibonding e_g orbital and loss of π-backbonding from the t_{2g} [5] orbital. Therefore, the molecular structure rearranges between LS and HS states as shown in Fig. 1.

Cis-bis(thiocyanato)-bis(N-2′-pyridyl methylene)-4-(phenylazo) aniline iron(II) ([Fe(PM-AzA)$_2$](NCS)$_2$) is a model system for Fe(II) metal compounds. Single crystal [Fe(PM-AzA)$_2$](NCS)$_2$ has been studied by femtosecond optical pump-probe reflectivity [6]. Figure 2 illustrates the spin crossover process of single crystal [Fe(PM-AzA)$_2$](NCS)$_2$, which has been identified as a two-step photo-switching process with a fast relaxation of short-lived intermediate states (INT) and a slow vibrational cooling of the photoinduced high spin state. Additionally, X-ray diffraction with 100 ps time resolution studies on the SCO compound [(TPA) Fe$^{(III)}$(TCC)]PF$_6$ indicate that molecules are locally photoswitched with structural reorganization at constant volume [7, 8]. However, the photo-induced SCO structural behavior of single crystal [Fe(PM-AzA)$_2$](NCS)$_2$ on the sub 10 ps time scale is still unknown. Here we use FED to study the structural dynamics of [Fe(PM-AzA)$_2$](NCS)$_2$ spin crossover in single crystal form. Here we use a ultrabright FED source to closely track structural signatures upon photoconversion from the LS to the HS state.

Fig. 2 A simplified energy level diagram of single crystal [Fe(PM-AzA)$_2$](NCS)$_2$ has been deduced from femtosecond optical pump-probe reflectivity studies [6]

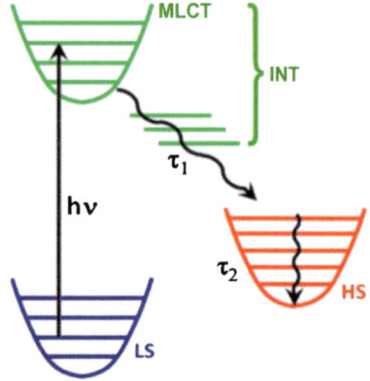

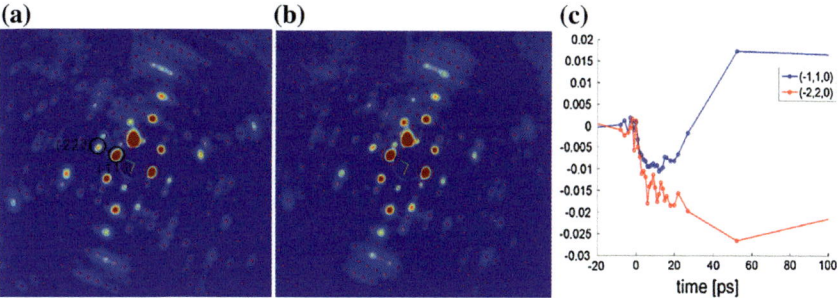

Fig. 3 **a** Diffraction pattern of LT state of $[Fe(PM-AzA)_2](NCS)_2$ was obtained at 160 K. The signal is an average over 10 electron pulses. **b** Diffraction pattern of HT state of $[Fe(PM-AzA)_2]$ $(NCS)_2$ was obtained at 300 K. **c** Relative intensity change of selected diffraction spot $(-1, 1, 0)$ and $(-2, 2, 0)$ from **a** where clear ultrafast structural processes are resolved

2 Experimental Methods

In this study, a 95 keV hybrid DC-RF electron source is employed to generate ultrabright femtosecond electron pulses as an ultrafast structural probe [9]. We generate 8.5×10^4 electrons per pulse with a spot size of 300 μm diameter and a repetition rate of 100 Hz. To investigate the SCO dynamics of $[Fe(PM-AzA)_2]$ $(NCS)_2$, we employed 60 fs pump pulses centered at 800 nm. At the sample position, the pulse energy was 8 μJ per pulse and the laser beam diameter was 520 μm. The incident excitation fluence was 4.2 mJ/cm^2.

3 Results and Discussion

Figure 3 shows preliminary results of the FED study. The diffraction pattern of the LS state of $[Fe(PM-AzA)_2](NCS)_2$ is shown in Fig. 3a, b shows the diffraction pattern of the HS state. The important point is the high quality of the diffraction data for an organic system. We have studied several crystalline orientations and have conducted in parallel fs transient absorption studies of these same single crystals to provide complementary information on the excited state dynamics to enable separation of the electronic and nuclear contributions to the spin cross over dynamics. Figure 3c shows the picosecond structural dynamics out to 100 ps for two diffraction orders. The full structural dynamics reconstructed from these studies are outside the scope of this brief report and will be reported elsewhere.

Acknowledgements We thank J.F. Létard for providing the samples. This work was supported by Max Planck Institute for the Structure and Dynamics of Matter, University of Hamburg, DESY, Marie Curie Actions, University of Toronto and the NSERC.

References

1. P. Gutlich *et al.*, Top. Curr. Chem. 233, 1 (2004).
2. A. Hauser, Top. Curr. Chem. 234, 155 (2004).
3. J. F. Letard *et al.*, Top. Curr. Chem. 235, 221 (2004).
4. A. Halcrow, Polyhedron 26, 3523 (2007).
5. W. Gawelda *et al.*, J. Am. Chem. Soc. 129, 8199 (2007).
6. Andrea. Marino *et al.*, Polyhedron 66 (2013) 123-128.
7. Eric Collet, *et al.*, Phys. Chem. Chem. Phys., 2012, 14, 6192–6199.
8. Eric Collet, *et al.*, Chem.Eur. J., DOI: 10.1002/chem.201103048.
9. M. Gao *et al.*, Opt. Express 20, 12048 (2012).

Ab Initio Solution of Structural Dynamics with Ultrafast Electron Diffraction and Charge Flipping

Lai Chung Liu, Yifeng Jiang, Cheng Lu, Meng Gao, Manabu Ishikawa, Hideki Yamochi and R.J. Dwayne Miller

Abstract Ultrafast electron diffraction is used to probe the photoinduced structural dynamics of single crystal $(EDO\text{-}TTF)_2PF_6$ with femtosecond time resolution. Structure factor phases at key time points are solved ab initio using the charge-flipping method.

OCIS codes 320.2250 Femtosecond phenomena · (320.7130) Ultrafast processes in condensed matter, including semiconductors

1 Introduction

Diffraction methods have played an important role in determining the structure of molecules and crystals over the past 100 years. The advent of coherent femtosecond X-ray and electron pulse sources has enabled structural exploration of materials with a low radiation-damage threshold, as well as short-lived intermediate states in chemical reactions [1]. However, the general "phase problem" inherent to crystallography remains a challenge. Conventional diffraction studies can only measure the modulus of the structure factors; without the complex phase, the data cannot be easily inverted from momentum space to position space. A solution is to collect data at multiple sample orientations and apply constrained refinement of a rough structure

L.C. Liu · C. Lu · M. Gao · R.J. Dwayne Miller (✉)
Departments of Chemistry and Physics, University of Toronto,
80 St. George Street, Toronto, ON M5S 3H6, Canada
e-mail: dwayne.miller@mpsd.cfel.de

L.C. Liu · Y. Jiang · M. Gao · R.J. Dwayne Miller
Max Planck Institute for the Structure and Dynamics of Matter,
CFEL (Bld. 99), Luruper Chaussee 149, 22761 Hamburg, Germany

M. Ishikawa · H. Yamochi
Research Center for Low Temperature and Materials Sciences,
Kyoto University, Kyoto 606-8501, Japan

© Springer International Publishing Switzerland 2015
K. Yamanouchi et al. (eds.), *Ultrafast Phenomena XIX*,
Springer Proceedings in Physics 162, DOI 10.1007/978-3-319-13242-6_70

against the three-dimensional set of intensities [2, 3]. Recently, an iterative algorithm called Charge Flipping (CF) has been developed to offer a simpler, more efficient, and ab initio method for solving the phase problem [4, 5]. It does so by modifying a random scattering function in momentum and position space alternatively to generate a highly localised and positive-definite solution with structure factor magnitudes matching given data.

As a proof of concept, we present preliminary work that combines the high-source intensity and sub-picosecond time resolution of our ultrafast electron diffraction (UED) setup with the CF method to directly reconstruct the time-dependent structure of $(EDO-TTF)_2PF_6$ (EDOP). This molecular system undergoes a photoinduced phase transition and exhibits femtosecond structural dynamics [6]. We have chosen it as our model system because it is an effective stand-in for more complex ones and has a multi-picosecond transient intermediate state whose structure has been determined from a refinement model alone [7].

2 Experimental Methods

Electron diffraction patterns were collected using our 95 kV DC electron gun with RF compression. A 270 nm laser pulse back-illuminates a 20 nm thick gold photocathode to release a high density electron bunch which is then accelerated to a kinetic energy of 95 keV and recompressed by 3-GHz RF cavity before arriving at the sample position with a duration of 430 fs.

Monocrystalline EDOP were grown and ultramicrotomed to 100-nm thick slices. The 500×500-μm samples are then mounted on standard TEM copper meshes. They are in turn placed on a sample holder assembly cooled to 230 K with a liquid nitrogen cold finger. Multiple crystal orientations are sampled using the rotation stage that supports the sample holder. Photoexcitation of the samples is provided by a pump pulse (wavelength: 800 nm, duration: 60 fs, repetition rate: 10 Hz, excitation fluence: 0.55 mJ/cm^2, spot size: 400 μm FWHM) from a Ti:Sapphire regeneratively amplified laser system. The polarisation was linear and chosen to be along the stacking axis of EDOP. The incident angle is varied to ensure that the excited fraction (~ 10 %) is constant over the sampled orientations. The entire setup is held under a background vacuum pressure of 10^{-8} mbar.

The CF method is based on the assumption that distribution of scattered electron in momentum space is proportional to the Fourier transform of the electrostatic potential function. When the electron path length in the sample is long, there could be multiple scattering events and this proportionality would no longer be exact. In addition, the sample surface area is limited and does not allow arbitrarily high longitudinal rotations. As such, sample rotation would need to be minimised. We have performed simulations to determine the minimal momentum space coverage (i.e. orientation sampling) to ensure reasonable convergence of the CF algorithm. Calculations are also being done to study the effect of increased path length under sample rotation on CF structure solutions.

3 Preliminary Results

Using a low-temperature EDOP structure model from previous measurements, model calculations of kinematic electron diffraction show that full coverage of momentum space is not necessary for the CF algorithm. For a spatial resolution of 1 Å and a coverage as low as 50 % (i.e. ±30° longitudinal rotation), convergence is achieved without any additional constraints and yields a structure solution which is comparable to the starting structure. The lower bound is reached and convergence fails when there is insufficient structure factor information perpendicular to the 0° plane to constrain the solution along the transverse direction.

We are focusing on three key points in the phase transition of EDOP: initial low-temperature state at t < 0 ps, the transient intermediate state at t = 3–10 ps, and the final high-temperature state t > 1 ns. For the first and last cases, diffraction measurements have been done with longitudinal rotations up to 40° (see Fig. 1a); the scattering electrostatic potentials were reconstructed and found to be consistent with the known ones (see Fig. 1b). By taking the peak centroids in the potential as atomic positions, spatial resolutions of ∼1.3 and ∼2.1 Å were found along the directions longitudinal and transverse to the 0° plane.

Data collection is in progress to capture a similar data set for the intermediate state. Here, the challenge is to maintain subpicosecond time resolution while maintaining a fairly constant excitation density and obtaining sufficiently a high signal-to-noise level to extract useful structure factor intensities. When this is achieved, a full reconstruction of the transient molecular structure, comparable to the unexcited ones, would be obtained.

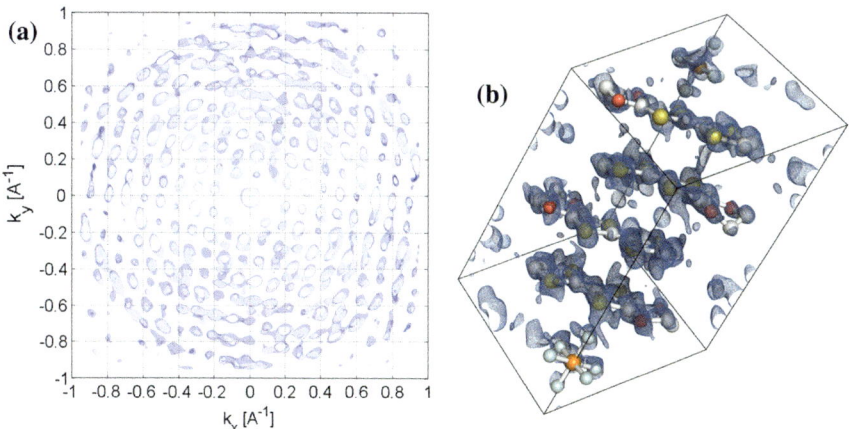

Fig. 1 **a** Top view of 3D isosurface plot of measured electron diffraction intensities in momentum space (k_x, k_y-plane) and sample rotation angle (z-axis). **b** 3D isosurface plot of electrostatic potential of low-temperature of EDOP solved *ab initio* using our UED data and the CF algorithm, overlaid over the molecular structure from published data [6]

References

1. R. J. Dwayne Miller, "Femtosecond Crystallography with Ultrabright Electrons and X-rays: Capturing Chemistry in Action", Science **343**, 1108–1116 (2014).
2. George M. Sheldrick, "A short history of SHELX", Acta Cryst. A**64**, 112-122 (2008).
3. Friedrich Schotte, Manho Lim, Timothy A. Jackson, Aleksandr V. Smirnov, Jayashree Soman, John S. Olson, George N. Phillips Jr., Michael Wulff, and Philip A. Anfinrud, "Watching a Protein as it Functions with 150-ps Time-Resolved X-ray Crystallography", Science **300**, 1944-1947 (2003).
4. Gábor Oszlányi and András Sütö, "Ab Initio structure solution by charge flipping", Acta Cryst. A**60**, 131–141 (2004).
5. Jinsong Wu, Kurt Leinenweber, John C. H. Spence, and Michael O'Keeffe, "Ab Initio phasing of X-ray powder diffraction by charge flipping", Nature Mater. **5**, 647-652 (2006).
6. Akira Ota, Hideki Yamochi, and Gunzi Saito, "A novel metal-insulator phase transition observed in (EDO-TTF)$_2$PF$_6$", J. Mater. Chem. **12**, 2600–2602 (2002).
7. Meng Gao, Cheng Lu, Hubert Jean-Ruel, Lai Chung Liu, Alexander Marx, Ken Onda, Shin-ya Koshihara, Yoshiaki Nakano, Xiangfeng Shao, Takaaki Hiramatsu, Gunzi Saito, Hideki Yamochi, Ryan R. Cooney, Gustavo Moriena, Germán Sciaini, and R. J. Dwayne Miller, "Mapping molecular motions leading to charge delocalization with ultrabright electrons", Nature **496**, 343–346 (2013).

Laser Streaking of Free-Electron Pulses at 25 keV

A. Gliserin, F.O. Kirchner, M. Walbran, F. Krausz and P. Baum

Abstract We demonstrate an optical-field-driven streak camera for temporal characterization of ultrashort free-electron pulses with sub-ångström de Broglie wavelength. This metrology reveals duration, chirp, and coherence of diffraction-capable electron pulses and offers attosecond resolution.

1 Introduction

Ultrafast electron microscopy and diffraction [1, 2] can visualize atomic motion in space and time. However, the temporal resolution is limited to about 150 fs due to space charge broadening and electron dispersion. These problems can be overcome by combining single-electron pulses with a microwave cavity for temporal compression, offering the possibility to achieve a resolution in the few-femtosecond range or even below [3]. Eventually, this may allow visualizing primary reactions dynamics in condensed matter and, on the long run, the dynamics of atomic-scale electron densities [4].

However, approaching this novel regime necessitates an electron pulse metrology with sufficient temporal resolution. Characterization by ponderomotive scattering [5] requires high-intensity laser pulses and the resolution is limited to about 100 fs. Here, we present an optical-field-driven streak camera for temporal characterization of diffraction-capable free-electron pulses [6]. Laser pulses with nJ energy provide a sufficient field strength and the temporal resolution is determined by the optical-field transients, i.e. in the sub-cycle regime.

A. Gliserin · F.O. Kirchner · M. Walbran · F. Krausz · P. Baum (✉)
Ludwig-Maximilians-Universität München, Am Coulombwall 1,
85748 Garching, Germany
e-mail: peter.baum@lmu.de

A. Gliserin · F.O. Kirchner · M. Walbran · F. Krausz · P. Baum
Max-Planck-Institute of Quantum Optics, Hans-Kopfermann-Str. 1,
85748 Garching, Germany

© Springer International Publishing Switzerland 2015
K. Yamanouchi et al. (eds.), *Ultrafast Phenomena XIX*,
Springer Proceedings in Physics 162, DOI 10.1007/978-3-319-13242-6_71

2 Concept, Experiment, Results

Optical streaking is based on a rapid transition of the electrons into or out of a controlled optical field within a transition time below the duration of an optical cycle. This is the principle of the attosecond streak camera [7], where photoelectrons are released by attosecond extreme-ultraviolet pulses in the presence of an optical streaking field. The rapid transition (or "birth") into the optical field imprints a momentum modulation onto the electrons, depending on the phase of the optical field.

We adapt this concept to 25 keV free-electron pulses by reflecting the streaking laser pulse off a 50-nm free-standing aluminum foil, creating a maximum longitudinal streaking field of about 1.8×10^9 V/m, while the electron pulses are partially transmitted (see Fig. 1). The optical field is shielded by the foil within a fraction of the wavelength, ensuring a sub-cycle transition of the electron pulse out of the field and therefore a momentum modulation. The noncollinear incidence angles of the two beams are chosen to match the sweeping velocity of the optical phase and electron pulse's group velocity along the foil's surface. A home-built time-of-flight spectrometer records the change of the electron's kinetic energy after the interaction as a function of the laser-electron delay time.

In optical field streaking, the maximum energy gain (or loss) of the electrons is proportional to the peak field strength [7]. The 800-nm streaking laser pulses have a full-width-at-half-maximum (FWHM) duration of 50 fs (intensity envelope) and their carrier-envelope phase is not stabilized. Also, the electron pulses used here are

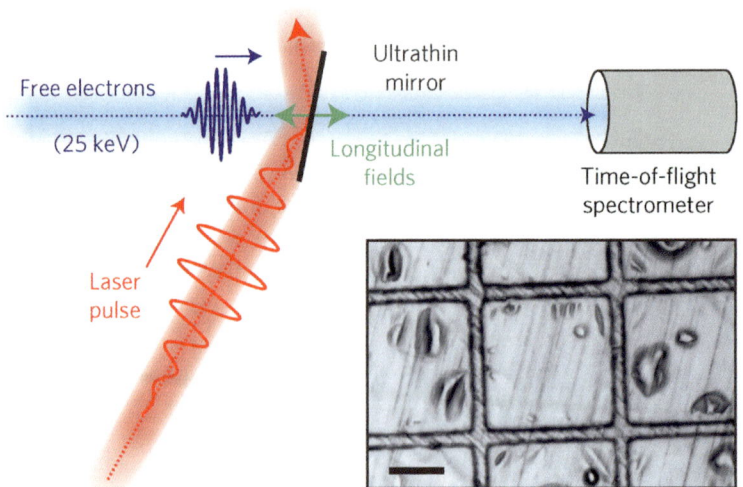

Fig. 1 Concept of the optical-field-driven streak camera for freely propagating electron pulses at diffraction-capable central energy (tens of keV). The key element is an ultrathin metal mirror (*black*) that is transparent to the electrons. The *inset* shows our foil, made of 50-nm thick aluminum (*scale bar*, 100 μm)

much longer than an optical cycle. Therefore, a streaking spectrogram (energy spectrum as a function of delay time) yields a cycle-averaged cross-correlation between the electron pulse and the envelope of the laser field. The width of the cross-correlation at a particular energy gain is a convolution between the electron pulse duration and that portion of the field envelope that is sufficiently strong for inducing this energy gain. Thus, for higher energy gain values, the electron pulse is sampled with an effective probe of decreasing duration [6].

Figure 2, left panel, shows a streaking spectrogram of dispersed uncompressed electron pulses with about one electron per pulse. A spectral interference with a period of about 1.6 eV, corresponding to the photon energy, is visible. This is a consequence of the electron's longitudinal coherence extending over more than one optical cycle, causing different parts of the electron's wave function leaving the field with the same final energy in subsequent optical cycles. The measured maximum energy gain is about 65 eV, in agreement with calculations based on the peak streaking field [6]. The width of the cross-correlation converges to (360 ± 20) fs FWHM at above ~ 10 eV of energy gain. This denotes the electron pulse duration. The tilt of the interference features indicates a linear chirp of ~ 0.9 ps/eV.

Figure 2, middle and right panels, show the results for electron pulse consisting of more than one electron. The increasing tilt, best visible in the loss feature at zero energy gain, reveals an increasing chirp of the electron pulses. Also, the pulses

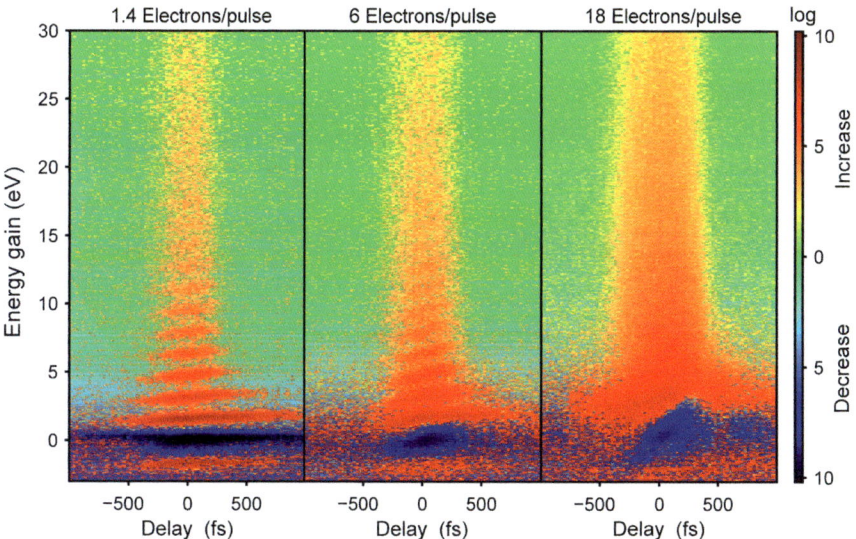

Fig. 2 Laser streaking results. Negative delay times denote an earlier arrival of the electron pulse. The unstreaked electron spectra have been subtracted in order to enhance the laser-induced changes. *Left panel* Dispersed electron pulse without microwave compression. The duration is (360 ± 20) fs FWHM and a linear chirp is evident. *Middle/right panel* Increasing the electron density causes additional dispersion, decoherence by multi-electron interactions, and pulse broadening

become longer. The decrease of interference contrast is probably due to multi-electron dynamics causing single-electron wavepacket localization.

Extension of this metrology technique to few-femtosecond or attosecond free-electron pulses is straightforward if applying few-cycle laser pulses with stabilized carrier-envelope phase. Simulations [6] reveal a transition between the coherent interference regime and field-resolved sampling, like in conventional attosecond streaking.

3 Conclusions

The reported optical-field-driven streak camera constitutes a powerful tool for electron pulse characterization in the time and energy domain, with sub-femtosecond resolution. Pump-probe single-electron diffraction seems feasible [8]; together with the here reported pulse metrology this provides a perspective for advancing ultrafast electron diffraction into novel resolution regimes in space [9] and time [4].

References

1. D. J. Flannigan and A. H. Zewail, Acc. Chem. Res. 45, 1828 (2012).
2. G. Sciaini and R. J. Dwayne Miller, Rep. Prog. Phys. 74 (2011).
3. P. Baum, Chem. Phys. 423, 55–61 (2013).
4. P. Baum, J. Phys. B 47, 124005 (2014).
5. M. Gao, H. Jean-Ruel et al., Opt. Express 20, 12048 (2012).
6. F. O. Kirchner, A. Gliserin, F. Krausz, P. Baum, Nature Photonics 8, 53 (2014).
7. M. Drescher, M. Hentschel et al., Nature 419, 803–807 (2002).
8. S. Lahme, C. Kealhofer, F. Krausz, P. Baum, Struct. Dyn. 1, 034303 (2014).
9. J. Hoffrogge, J. P. Stein, M. Krüger, M. Förster, J. Hammer, D. Ehberger, P. Baum, P. Hommelhoff, J. Appl. Phys. 115, 094506 (2014).

Ultrafast Single-Electron Diffraction

A. Gliserin, S. Lahme, M. Walbran, F. Krausz
and P. Baum

Abstract We report compression of single-electron pulses using microwave fields to 28-fs duration (FWHM), characterized by laser streaking. Atomic resolution is achieved in diffraction. A time-resolved study on fibrous graphite polycrystals without the compressor reveals the practical feasibility of pump-probe diffraction with single electrons.

1 Introduction

Ultrafast electron microscopy and diffraction [1, 2] can provide a four-dimensional visualization of atomic motion in space and time. Currently, the temporal resolution is limited by space charge effects and dispersion to about 160 fs [3, 4]. This is often not good enough to observe primary processes in molecules or condensed matter, i.e. the half periods of vibrational modes or optical phonons. Single-electron pulses are intrinsically immune to Coulomb repulsion and can in principle have few-fs duration and below [5, 6]. Realization of that potential, however, requires a careful management of electron dispersion, similar to the handling of chirp in femtosecond optics. Furthermore, single-electron diffraction comes at a price: The high pump-probe repetition rate required for obtaining data in reasonable time restricts the investigations to reversible processes. Also, thermal load on the sample must be

A. Gliserin · S. Lahme · M. Walbran · F. Krausz · P. Baum (✉)
Ludwig-Maximilians-Universität München, Am Coulombwall 1,
85748 Garching, Germany
e-mail: peter.baum@lmu.de

A. Gliserin · S. Lahme · M. Walbran · F. Krausz · P. Baum
Max-Planck-Institute of Quantum Optics, Hans-Kopfermann-Str. 1,
85748 Garching, Germany

© Springer International Publishing Switzerland 2015
K. Yamanouchi et al. (eds.), *Ultrafast Phenomena XIX*,
Springer Proceedings in Physics 162, DOI 10.1007/978-3-319-13242-6_72

handled. Here we report on our progress in overcoming these hurdles, reporting (i) 28-fs electron pulses (FWHM) characterized by field-induced streaking [7], and (ii) a first time-resolved diffraction study with entirely single-electron methodology [8].

2 Single-Electron Pulse Compression

Our experiment is sketched in Fig. 1. A femtosecond laser produces single-electron pulses, which are accelerated to 25–100 keV by a static electric field. The beam is collimated by a well-aligned magnetic lens [9] with 'isochronic' properties [10].

The initial energy bandwidth after photoemission translates to dispersion in time during propagation. A microwave cavity with optically enhanced jitter suppression [11] is applied to reshape the single-electron phase space from the temporal into the energetic domain, hence shortening the pulses at the cost of an increase in energy bandwidth [12]. The electron pulses can become shorter than the laser pulses used initially for photoemission.

Pulse characterization is achieved by using an optical field for streaking the electron pulses in the energy domain [7], similar to attosecond XUV streaking. The shortest electron pulses we achieved so far in our laboratory have a cross-correlation width of 15 fs (rms) corresponding to an electron pulse duration of 12 ± 2 fs (rms) or 28 ± 5 fs (FWHM). These are the shortest diffraction-capable electron pulses that we know of. Diffraction from the ground state of a molecular proton transfer switch compound, N-(triphenylmethyl)-salicylidenimine, demonstrates the high transverse coherence [13] and ability to obtain atomic-scale resolution with compressed single-electron pulses. Also, electron-energy-loss spectra (EELS) were obtained successfully.

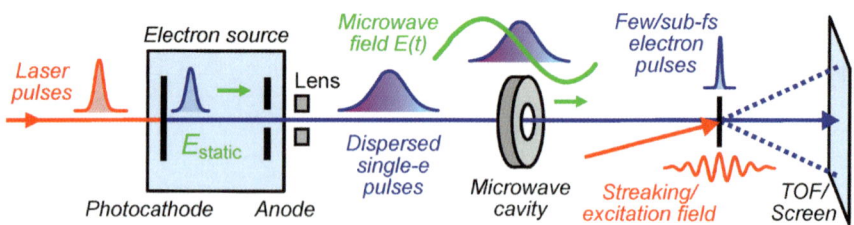

Fig. 1 Experimental concept for single-electron diffraction at ultimate temporal resolution, 28 fs. Key elements are a single-electron source, a microwave compressor and a laser streaking pulse metrology

3 Time-Resolved Single-Electron Diffraction

Roughly 10^7 incoming electrons are required for a suitable diffraction image and about 10–100 different pump-probe delays must be scanned for a complete, dynamical picture. In the single-electron regime, an acceptable overall measurement time can only be achieved at rather high pump-probe repetition rates, potentially causing problems with average power and reversibility.

Here we report on our first success with using single-electron pulses, intrinsically space-charge-free over the entire trajectory right from the source, for recording time-resolved diffraction of a realistic sample, here a fibrous polycrystalline thin-film made from highly-ordered pyrolytic graphite (HOPG). An extremely fine TEM mesh is used as a support structure for heat removal. The graphite layers are tilted by about 20° in order to split the diffraction rings into arcs with in-plane and out-of-plane contributions. Thanks to the good spatial coherence of single-electron pulses [13], distinct Bragg spots are discernible, corresponding to selected-area diffraction from individual graphite grains within the 25-nm thick film.

Figure 2 shows the time-resolved intensity of 27 selected spots, partially obtained with ten-electron pulses at 128 kHz or alternatively with genuine single-electron pulses at 256 kHz [8], at the natural few-hundred femtosecond resolution of single-electrons without the microwave compressor [7].

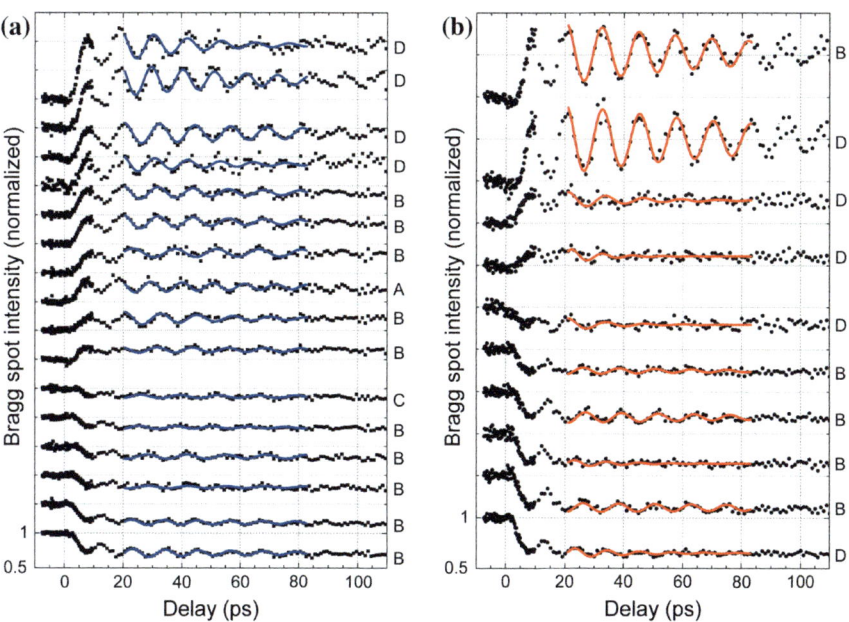

Fig. 2 Time-resolved single-electron diffraction on fibrous graphite polycrystals produces individual Bragg spots that reveal coherent acoustic phonons of individual graphite grains

Laser-induced excited carriers first relax into strongly coupled optical phonons (SCOP) and subsequently thermalize with all phonons, causing significant thermal expansion along the c-axis. Shifts of the grain-specific rocking curves cause Bragg spot intensities to increase or decrease. The data hence reveals the time-resolved c-axis dynamics of each grain individually. The blue and red lines are fits to an oscillator model with a driving term related to a two-temperature model. The results reveal a rather homogeneous oscillation period and film thickness (25 ± 1 nm), but also a surprisingly wide distribution of damping times (25–300 ps) and initial intensity changes (2–12 ps), indicating nonlinear influences of the rocking curve and strain for assessing the early structural dynamics of polycrystalline graphite [8].

4 Conclusions

These results show that time-resolved single-electron diffraction meets the requirements of realistic samples. Combined with microwave compression, it is a promising approach for visualizing structural dynamics in the few-fs regime and eventually, on the long run, even below [14]. A next step will be the application of a tip-based electron source [15] for selected-area diffraction from more complex materials.

References

1. D. J. Flannigan and A. H. Zewail, Acc. Chem. Res. 45, 1828 (2012).
2. G. Sciaini and R. J. D. Miller, Rep. Prog. Phys. 74 (2011).
3. T. van Oudheusden, P. L. E. M. Pasmans, S. B. van der Geer, M. J. de Loos, M. J. van der Wiel, O. J. Luiten, Phys. Rev. Lett. 105, 264801 (2010).
4. M. Gao, Y. Jiang, G. H. Kassier, R. J. D. Miller, Appl. Phys. Lett. 103, 033503 (2013).
5. E. Fill, L. Veisz, A. Apolonski, F. Krausz, New J. Phys. 8 (2006).
6. P. Baum, Chem. Phys. 423, 55–61 (2013) .
7. O. Kirchner, A. Gliserin, F. Krausz, P. Baum, Nature Photonics 8, 53 (2014).
8. S. Lahme, C. Kealhofer, F. Krausz, P. Baum, Struct. Dyn. 1, 034303 (2014).
9. D. Kreier, D. Sabonis, P. Baum, J. Optics 16, 075201 (2014).
10. C. Weninger and P. Baum, Ultramicroscopy 113, 145–151 (2012).
11. A. Gliserin, M. Walbran, P. Baum, Appl. Phys. Lett. 103, 031113 (2013).
12. A. Gliserin, A. Apolonski, F. Krausz, P. Baum, New J. Phys. 14, 073055 (2012).
13. F. O. Kirchner, S. Lahme, F. Krausz, P. Baum, New J. Phys. 15, 063021 (2013).
14. P. Baum, J. Phys. B 47, 124005 (2014) .
15. J. Hoffrogge, J. P. Stein, M. Krüger, M. Förster, J. Hammer, D. Ehberger, P. Baum, P. Hommelhoff, J. Appl. Phys. 115, 094506 (2014).

Part IV
Interfacial, Surface, Thin Films and Carbon Nanotubes

Hydrated Phospholipid Surfaces Probed by Ultrafast 2D Spectroscopy of Phosphate Vibrations

Rene Costard, Ismael A. Heisler and Thomas Elsaesser

Abstract Phosphate stretching vibrations probe interfacial dynamics in hydrated phospholipids. Two-dimensional spectra in the $1,000–1,300$ cm^{-1} range reveal structural fluctuations on a 300 fs time scale while water-phosphate hydrogen bonds persist for longer than 10 ps.

The interaction of phospholipids with water plays a key role for the structure and function of biological membranes. Interfacial water molecules hydrate the charged phospholipid headgroups at the membrane's surface where the interplay of long-range Coulomb polarization forces and local hydrogen bonds determines the molecular arrangement [1]. While the time-averaged equilibrium structures of hydrated phospholipids have been studied in substantial detail, insight into their dynamics on the time scale of vibrational motions and structural fluctuations, processes relevant for biological function, is limited. Femtosecond infrared spectroscopy has mainly addressed water stretching vibrations of hydrated membranes [2] and reverse micelles, an approach inherently connected with a spatial averaging over all water environments, i.e., not allowing for a selective observation of interfacial water. Here, we introduce the phosphate stretching vibrations of phospholipid head groups as novel specific interfacial probes to study ultrafast dynamics of the phospholipid-water interface in a wide range of hydration levels. Applying two-dimensional (2D) photon-echo spectroscopy in the frequency range from $1,000$ to $1,300$ cm^{-1}, we map subpicosecond structural fluctuations of the hydrated membrane surface and show that water-phosphate hydrogen bonds persist for longer than 10 ps.

As a model system, reverse dioleoylphosphatidylcholine (DOPC) micelles containing a nanopool of water molecules (Fig. 1a) were prepared in benzene solution for different hydration levels w_0, where w_0 represents the ratio of the water to DOPC concentration [3, 4]. The infrared absorption spectra of reverse micelles at different w_0 display the asymmetric and symmetric $(PO_2)^-$ stretch absorption around $\nu_{AS} = 1,250$ cm^{-1} and $\nu_S = 1,095$ cm^{-1} (Fig. 1b), both showing a red-shift

R. Costard · I.A. Heisler · T. Elsaesser (✉)
Max-Born-Institut f. Nichtlineare Optik Und Kurzzeitspektroskopie, 12489 Berlin, Germany
e-mail: elsasser@mbi-berlin.de

© Springer International Publishing Switzerland 2015   301
K. Yamanouchi et al. (eds.), *Ultrafast Phenomena XIX*,
Springer Proceedings in Physics 162, DOI 10.1007/978-3-319-13242-6_73

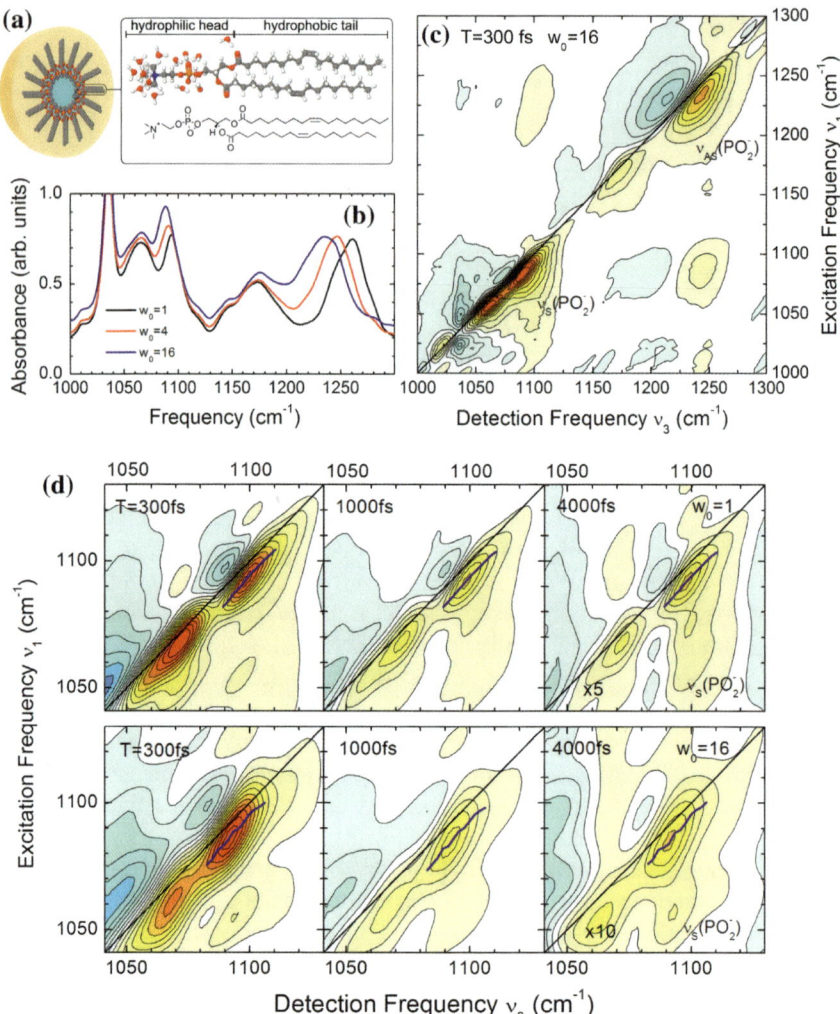

Fig. 1 a Chemical structure of the phospholipid DOPC and schematic of a DOPC reverse micelle containing a water nanopool. **b** Infrared absorption spectra of DOPC reverse micelles for hydration levels of $w_0 = 1$, 4, and 16. Both the asymmetric $(PO_2)^-$ stretch vibration ν_{AS} around 1,250 cm^{-1} and the symmetric $(PO_2)^-$ stretch vibration ν_S around 1,095 cm^{-1} undergo a redshift with increasing hydration level. **c** Absorptive 2D infrared spectrum of DOPC reverse micelles at $w_0 = 16$ and a population time T = 300 fs. *Yellow-red* contours are positive signal, *blue* contrours are negative signals. The phosphate stretch contributions are marked. **d** Series of 2D spectra of the symmetric $(PO_2)^-$ stretch vibration ν_S at $w_0 = 1$ (*top row*) and $w_0 = 16$ (*bottom row*). There is a minor reshaping with population time T, as also evident from the center lines shown in *blue*

with increasing water content. Femtosecond 2D photon-echo spectroscopy was implemented for the first time in the $1,000$–$1,300$ cm^{-1} range by generating two phase-locked pulse pairs with the help of diffractive optics and using heterodyne detection based on frequency-domain interferometry [5]. A 2D spectrum recorded with a $w_0 = 16$ (fully hydrated) sample at a population time $T = 300$ fs is presented in Fig. 1c. The 2D lineshapes of ν_{AS} at an excitation frequency $\nu_1 = 1,235$ cm^{-1} and of ν_S at $\nu_1 = 1,090$ cm^{-1} consist of a positive component (yellow red) due the $v = 0$ to 1 transition of the oscillator and a negative component (blue contours) due the $v = 1$ to 2 transitions. All contours are elongated along the diagonal $\nu_1 = \nu_3$, indicating a substantial inhomogeneous broadening that reflects the structural heterogeneity of the phospholipid-water interface. The other 2D peaks originate from other fingerprint vibrations in this range as has been discussed in [5]. In Fig. 1d, we show 2D spectra of the symmetric $(PO_2)^-$ stretch mode ν_S which has a 1–1.5 ps population lifetime, for hydration levels $w_0 = 1$ and 16. Even for full hydration ($w_0 = 16$), the 2D lineshapes are preserved over a time range of 4 ps, as is also evident from the unchanged slopes of the center lines shown in blue. This fact demonstrates an essentially constant inhomogeneous broadening and, thus, a practically static disorder of the hydrated phospholipid.

For a detailed quantitative analysis, the measured 2D spectra were compared to lineshape calculations based on a Kubo formalism for the frequency-time correlation function (tcf). We find excellent agreement with the full data set for a tcf consisting of an initial 300 fs decay, necessary in order to account for the antidiagonal widths of the 2D envelopes, and a slow second component for which we estimate a lower limit of its decay time of 10 ps in order to account for the 'static' inhomogeneous broadening. The fast decay present also at low hydration reflects structural fluctuations of the phospholipid geometry and—at high w_0—restricted motions of water molecules. The slow contribution shows that frequency jumps from the breaking and reformation of water-phosphate hydrogen bonds are absent on a 10 ps time scale, suggesting a comparably rigid hydration pattern of the phosphate groups. This finding which is in line with theoretical molecular dynamics simulations of hydrated phospholipid membranes, also suggests a minor influence of structure fluctuations in the inner water nanopool on the hydration of the interface.

References

1. L. Saiz and M. L. Klein, "Structural properties of a highly polyunsaturated lipid bilayer from molecular dynamics simulations", Biophys. J. **81**, 204−218 (2001).
2. W. Zhao, D. E. Moilanen, E. E. Fenn, and M. D. Fayer, "Water at the surfaces of aligned phospholipid multibilayer model membranes probed with ultrafast vibrational spectroscopy. J. Am. Chem. Soc. **130**, 13927−13937 (2009).
3. N. E. Levinger, R. Costard, E. T. J. Nibbering, and T. Elsaesser, "Ultrafast energy migration pathways in self-assembled phospholipids interacting with confined water", J. Phys. Chem. A **115**, 11952−11959 (2011).

4. R. Costard, C. Greve, I. A. Heisler, and T. Elsaesser, "Ultrafast energy redistribution in local hydration shells of phospholipids: a two-dimensional infrared study", J. Phys. Chem. Lett. **3**, 3646−3651 (2012).
5. R. Costard, I. A. Heisler, and T. Elsaesser, "Structural dynamics of hydrated phospholipid surfaces probed by ultrafast 2D spectroscopy of phosphate vibrations", J. Phys. Chem. Lett. **5**, 506−511 (2014).

Femtosecond Time and Angle Resolved Photoemission Spectroscopy of Liquids

Yo-Ichi Yamamoto, Yoshi-Ichi Suzuki, Gaia Tomasello,
Takuya Horio, Shutaro Karashima, Roland Mitric
and Toshinori Suzuki

Abstract We report the time- and angle-resolved photoemission spectroscopy of a liquid laminar flow injected into vacuum.

Time-resolved photoemission spectroscopy (TRPES) enables unprecedented direct access to ultrafast electronic dynamics in molecules and materials. Angle-resolved PES provides further information on the dynamics, as the electronic character of a molecule or the dispersion relation of solid can be determined from the photoelectron angular distribution. In this contribution, we present the first application of time and angle-resolved photoemission spectroscopy (TARPES) to liquids [1].

We discharge a liquid laminar flow from a fused silica capillary with a 25 μm inner diameter into a time-of flight (TOF) photoelectron spectrometer at a flow rate of 0.5 mL/min. A femtosecond 226 nm pump pulse (5.49 eV) electronically excites solutes in the liquid at ca. 1 mm downstream from the nozzle, where the liquid temperature is estimated to be 275–280 K. Photoemission is induced by a femtosecond 260 nm probe pulse. The cross-correlation of the pump and probe pulses is ca. 120 fs. Photoelectrons emitted from the liquid surface are sampled into the TOF analyzer; the electron energy resolution is ca. 50 meV. A high-repetition-rate 100 kHz laser system is employed to compensate for the small detection solid angle of 9×10^{-4} steradians [2]. The linear polarization of the pump pulse is fixed

Y.-I. Yamamoto · Y.-I. Suzuki · T. Horio · S. Karashima · T. Suzuki (✉)
Department of Chemistry, Graduate School of Science, Kyoto University,
Kitashirakawa-Oiwakecho, Sakyo-Ku Kyoto 606-8502, Japan
e-mail: suzuki@kuchem.kyoto-u.ac.jp

Y.-I. Suzuki · T. Horio · T. Suzuki
RIKEN Center for Advanced Photonics, RIKEN, 2-1 Hirosawa,
Wako 351-0198, Japan

G. Tomasello · R. Mitric
Institut Für Physikalishce Und Theoretische Chemie, Am Hubland,
Universität Würzburg, Würzburg 97074, Germany

T. Horio · T. Suzuki
CREST, Japan Science and Technology Agency, Sanbancho,
Chiyoda-Ku, Tokyo 102-0075, Japan

© Springer International Publishing Switzerland 2015
K. Yamanouchi et al. (eds.), *Ultrafast Phenomena XIX*,
Springer Proceedings in Physics 162, DOI 10.1007/978-3-319-13242-6_74

perpendicular to the electron detection axis, and angle-resolved photoemission spectra are measured by rotating the linear polarization of the probe pulse from 0° to 90° with respect to the electron detection axis. The UV absorption spectrum of aqueous DABCO (1,4-diazabicyclo [2] octane) solution exhibits a gradually increasing intensity from 240 nm. The four lowest excited states of DABCO in water have dominant s and p Rydberg characters in agreement with those states in the gas phase although the Rydberg orbitals in water have delocalization over unoccupied orbitals of water molecules. We performed TARPES of aqueous 0.5 M DABCO solution.

Figure 1a presents the time-energy map of the observed photoelectron spectra: eBE is the difference between the probe photon energy (4.77 eV) and the observed photoelectron kinetic energy. The eBE rapidly increases within 1 ps and reaches 3.4 eV, which is the eBE of a hydrated electron [3–7]. The result clearly reveals an ultrafast charge transfer to solvent (CTTS) reaction from the excited state of DABCO to liquid water. Figure 1b shows the total photoelectron intensity profile obtained from Fig. 1a by integrating the distribution at each delay time over eBE.

Figure 2 presents TARPE spectra measured for the same CTTS reaction from DABCO to water. Each upper panel shows a photoelectron spectrum, $S(E, \theta)$, at different time delays from 100 fs to 3 ps, respectively. Rapid change of the photoelectron spectrum appears predominantly in the low eBE region, where strong photoemissioin anisotropy is observed. For closer examination of the anisotropic photoemission component, we calculated the difference spectrum defined by $\Delta S_{ani}(E, \theta) = S(E, \theta) - S(E, 90°)$ at each time delay, as shown in lower panels of Fig. 2. We have simulated the anisotropy of hydrated DABCO using continuum

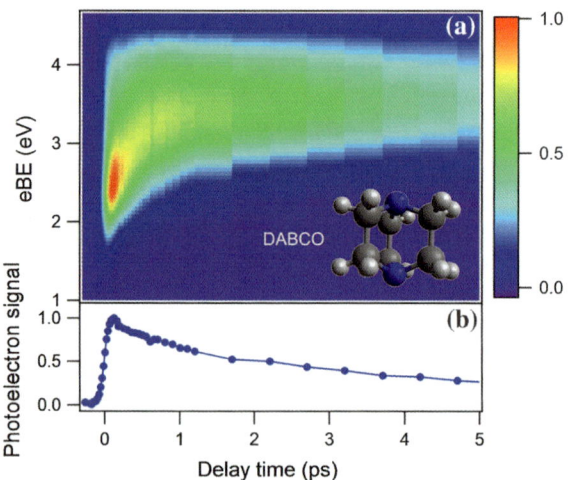

Fig. 1 **a** Two-dimensional false color map of photoelectron spectra measured for aqueous 0.1 M DABCO solutions at different pump-probe time delays. The pump and probe laser wavelengths are 226 and 260 nm, respectively. **b** The total electron intensity profile obtained from **a** by integrating the distribution at each delay time over eBE

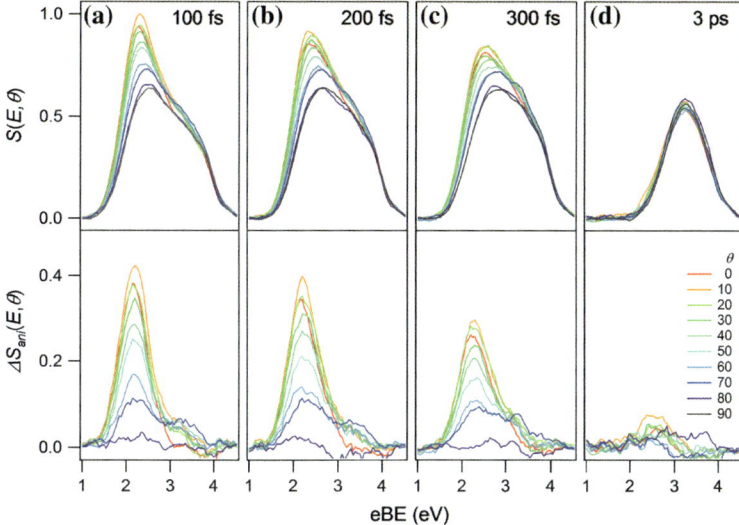

Fig. 2 Photoelectron spectra of aqueous 0.5MDABCO solution as a function of the polarization angle of the probe pulse θ with respect to the electron detection axis. The spectra were observed at different pump-probe time delays of **a** 100 fs, **b** 200 fs, **c** 300 fs and **d** 3 ps

multiple scattering Xα calculations; we assumed the 3 s Rydberg state of DABCO hydrated by 64 water molecules. The calculated photoemission distribution was isotropic. Thus, weak yet finite photoemission anisotropy observed for aqueous DABCO solution is ascribed to DABCO molecules segregated on the liquid surface. We performed soft X-ray photoemission spectroscopy of aqueous DABCO solution and confirmed that DABCO has indeed an enhanced molecular density on the liquid surface.

We observed no signature of the previously suggested band of a surface-bound electron at an eBE of 1.6 eV with a long lifetime [3]. If this eBE is accurate, it means that the surface-bound electron state is energetically higher than the excited state of DABCO. Thus, formation of the surface-bound electron is energetically unfavorable. Recent quantum chemical calculations suggest that the eBE of a surface-bound electron is 3.4 eV [8]. Further investigation is necessary for energetics and dynamics of a surface-bound electron.

Acknowledgments We thank N. Kurahashi for experimental assistance in soft X-ray photoemission spectroscopy. The synchrotron radiation experiments were performed at BL17SU of SPring-8 with the approval of RIKEN (proposal No. 20130075).

References

1. Y. Yamamoto, Y.-I. Suzuki, G. Tomasello, T. Horio, S. Karashima, R. Mitric, and T. Suzuki, Phys. Rev. Lett. 112, 187603 (2014).
2. H. Shen, S. Adachi, T. Horio, and T. Suzuki, Opt. Express 19, 22637 (2011).
3. K. R. Siefermann, Y. Liu, E. Lugovoy, O. Link, M. Faubel, U. Buck, B.Winter, and B. Abel, Nat. Chem. 2, 274 (2010).
4. Y. Tang, H. Shen, K. Sekiguchi, N. Kurahashi, T. Mizuno, Y.-I. Suzuki, and T. Suzuki, Phys. Chem. Chem. Phys. 12, 3653 (2010).
5. A. Lubcke, F. Buchner, N. Heine, I. V. Hertel, and T. Schultz, Phys. Chem. Chem. Phys. 12, 14629 (2010).
6. A. T. Shreve, T. A. Yen, and D. M. Neumark, Chem. Phys. Lett. 493, 216 (2010).
7. T. Horio, H. Shen, S. Adachi, and T. Suzuki, Chem. Phys. Lett. 535, 12 (2012).
8. F. Uhlig, O. Marsalek, and P. Jungwirth, J. Phys. Chem. Lett. 4, 338 (2013).

Ultrafast Vibrational Dynamics of Water at a Zwitterionic Lipid/Water Interface Revealed by Two-Dimensional Heterodyne-Detected Vibrational Sum Frequency Generation (2D HD-VSFG)

Ken-ichi Inoue, Prashant Chandra Singh, Satoshi Nihonyanagi, Shoichi Yamaguchi and Tahei Tahara

Abstract 2D HD-VSFG is applied to the study of ultrafast vibrational dynamics at a zwitterionic lipid/water interface for the first time. The 2D spectrum reveals spectral diffusion of three distinct water species existing at the interface.

1 Introduction

Biological membranes play an essential role in maintaining the cellular environment properly to enable a variety of biological processes. Therefore, a molecular-level understanding of structure and dynamics of the membrane/water interface is indispensable for obtaining physicochemical understanding of biological processes.

So far, lipid monolayers on the water surface have been extensively studied as model systems of the membrane/water interface. Vibrational spectra of the OH stretch region provide abundant information about water because the OH stretch frequency is sensitive to the environment around water. Thus, surface-sensitive vibrational sum frequency generation (VSFG) spectroscopy of the OH stretch can provide rich information about interfacial water. In particular, heterodyne-detected VSFG (HD-VSFG) enables us to measure not only the amplitude of the second-order nonlinear susceptibility ($\chi^{(2)}$), but also its phase, providing $\mathrm{Im}\chi^{(2)}$ spectra which can be directly compared to the absorption spectra ($\mathrm{Im}\chi^{(1)}$) in the bulk [1].

Previously, our group reported the steady-state HD-VSFG study of a zwitterionic lipid/water interface, which is a major constituent of biological membranes

K. Inoue (✉) · P.C. Singh · S. Nihonyanagi · S. Yamaguchi · T. Tahara
Molecular Spectroscopy Laboratory, Riken, 2-1 Hirosawa, Wako 351-0198, Japan
e-mail: ken-ichi.inoue@riken.jp

S. Nihonyanagi · S. Yamaguchi · T. Tahara
Ultrafast Spectroscopy Research Team, RIKEN Center for Advanced Photonics (RAP),
2-1 Hirosawa, Wako 351-0198, Japan

© Springer International Publishing Switzerland 2015
K. Yamanouchi et al. (eds.), *Ultrafast Phenomena XIX*,
Springer Proceedings in Physics 162, DOI 10.1007/978-3-319-13242-6_75

[2]. The analysis of the $Im\chi^{(2)}$ spectra showed that there are three distinct water species at the interface: the water in the vicinity of the negatively charged phosphate group, the water in the vicinity of the positively charged choline group, and the water existing in the hydrophobic region. In the present study, we applied two-dimensional (2D) HD-VSFG to the study of ultrafast dynamics of water at the zwitterionic lipid/water interface for the first time.

2 Experimental Methods

We used 1,2-dipalmitoyl-sn-glycero-3-phosphocholine (DPPC), shown in Fig. 1a, as a zwitterionic lipid. Water was isotopically diluted (H_2O:HOD:D_2O = 1:8:16) to eliminate the effects of the intra- and intermolecular couplings. HD-VSFG measurements of a Langmuir monolayer of DPPC at the water surface were performed at a temperature of 298 K and at a surface pressure of 35 ± 3 mN/m, which corresponds to the liquid condensed (LC) phase.

2D HD-VSFG is an extension of HD-VSFG to 2D spectroscopy, and it enables us to observe the time evolution of vibrationally excited states. The optical setup for 2D HD-VSFG measurements was described in detail previously [3]. Briefly, a narrow-band visible ω_1 pulse (center wavelength: 795 nm, bandwidth: 25 cm^{-1}, pulse width: 0.5 ps, S-polarized) and a broadband infrared ω_2 pulse (center

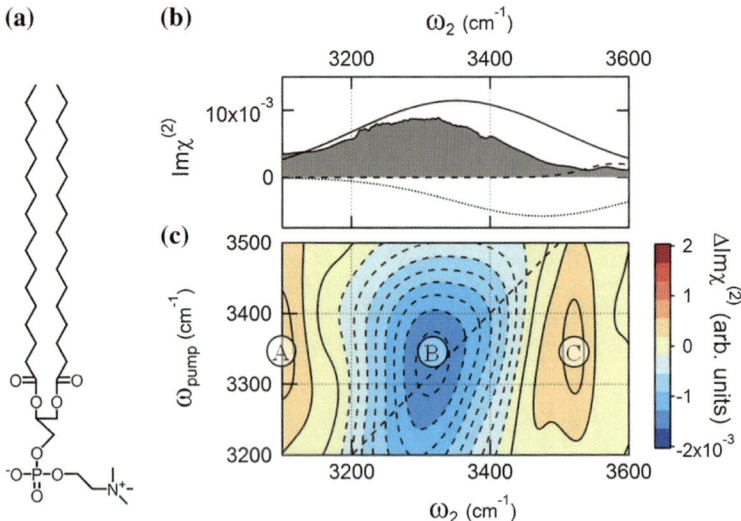

Fig. 1 **a** Chemical structure of DPPC. **b** (*filled*) Steady-state $Im\chi^{(2)}$ spectrum of DPPC/HOD interface measured by HD-VSFG. The steady-state spectrum was fitted with three Gaussians: (*solid line*) component 1 (*dotted line*) component 2 (*dashed line*) component 3. **c** 2D HD-VSFG spectrum of the DPPC/HOD interface at 0.0 ps delay

wavenumber: 3,350 cm^{-1}, bandwidth: 250 cm^{-1}, pulse width: 0.1 ps, P-polarized) were focused into a y-cut quartz crystal and then onto the lipid/water interface to generate sum frequency ($\omega_1 + \omega_2$, S-polarized). The former SFG was used as a local oscillator (LO) and passed through a glass plate (2 mm) to be delayed with respect to the latter SFG from the sample by 3.5 ps. The spectral interferograms were recorded by using a polychromator with a charge-coupled device camera. Time-resolved HD-VSFG measurements were carried out with pump ω_{pump} pulses (bandwidth: 160 cm^{-1}, pulse width: 0.2 ps, P-polarized) at 3200, 3300, 3400, and 3500 cm^{-1}. By linearly interpolating the observed spectra, 2D vibrational spectra of the zwitterionic lipid/water interface were obtained.

3 Results and Discussion

Figure 1b shows the steady-state Im$\chi^{(2)}$ spectrum (filled) and the fitting components (solid line, dotted line, and dashed line). As reported before, this spectrum was successfully decomposed into three components [2]. Components 1, 2, and 3 have been assigned to the water species associated with phosphate, choline, and the hydrophobic region of the lipid, respectively.

Figure 1c shows 2D HD-VSFG spectrum at 0.0 ps delay. In this spectrum, the following three bands are observed: a positive peak (denoted as A) on the low frequency side, a negative peak (B) in the center, and a positive peak (C) on the high frequency side. The peaks A and B are assignable to the hot band ($v = 2 \leftarrow v = 1$ transition) and bleaching of the $v = 1 \leftarrow v = 0$ transition of the OH stretch, respectively. The appearance of peak C is evidence that the single positive band in the steady-state spectrum is actually a superposition of negative and positive bands. Otherwise, the positive peak can never appear in the 2D spectrum in the high frequency side. We consider that peak C appears due to the following mechanism: with irradiation of the ω_{pump} pulses, spectral holes are simultaneously created in the positive component 1 and negative component 2 at the wavenumber of the pump pulse. Then, the center of mass of the bleaching of each component starts shifting toward its inherent peak frequency by spectral diffusion, so that a negative feature appears around the peak of the positive component 1, while a positive feature appears around the peak of the negative component 2. Therefore, the positive sign of peak C indicates that a negative component (i.e., the OH stretch band due to water associated with choline) is buried in the positive OH stretch band in the Im$\chi^{(2)}$ spectrum. Peak C disappears in 300 fs, as a result of energy transfer and/or interconversion among the three water species. The present 2D HD-VSFG experiments revealed ultrafast vibrational dynamics of distinct water species existing at the zwitterionic lipid/water interface.

References

1. S. Nihonyanagi, S. Yamaguchi, and T. Tahara, J. Chem. Phys. **130**, 204704 (2009).
2. J. A. Mondal, S. Nihonyanagi, S. Yamaguchi, and T. Tahara, J. Am. Chem. Soc. **134**, 7842-7850 (2012).
3. P. C. Singh, S. Nihonyanagi, S. Yamaguchi, and T. Tahara, J. Chem. Phys. **139**, 161101 (2013).

Ultrafast Vibrational Spectroscopy at Liquid Interfaces by Heterodyne-Detected Sum-Frequency Generation

Tahei Tahara

Abstract Ultrafast dynamics at liquid interfaces is still obscure. Femtosecond time-resolved heterodyne-detected vibrational sum-frequency generation spectroscopy now enables us to investigate vibrational/photochemical dynamics at liquid interfaces with the same clarity as in solution-phase ultrafast spectroscopy.

1 Introduction

Liquid interfaces are unique environments where a variety of important molecular processes take place. Therefore, it is highly desirable to elucidate steady-state and dynamic properties of interfaces at the molecular level. Nevertheless, our understanding of liquid interfaces is very limited, compared to the rich knowledge accumulated for molecules in solution. Vibrational sum-frequency generation (VSFG) spectroscopy is a second-order nonlinear spectroscopy that has intrinsic interface selectivity. VSFG spectroscopy has been extensively utilized for studying the steady-state properties of interfacial molecules, and a few pioneering time-resolved works have also been reported. However, ultrafast dynamics at liquid interfaces are still obscure.

In traditional VSFG measurements, sum-frequency signals generated at interfaces are directly detected. The interfacial vibrational spectra obtained with this homodyne detection are the spectra of the absolute square of the 2nd-order nonlinear susceptibilities ($|\chi^{(2)}|^2$). This absolute square nature of the signal causes a number of problems, e.g., low sensitivity and spectral deformation due to the

---

T. Tahara (✉)
Molecular Spectroscopy Laboratory, RIKEN, 2-1 Hirosawa, Wako, Saitama 351-0198, Japan
e-mail: tahei@riken.jp

T. Tahara
Ultrafast Spectroscopy Research Team, RIKEN Center for Advanced Photonics (RAP),
2-1 Hirosawa, Wako, Saitama 351-0198, Japan

© Springer International Publishing Switzerland 2015
K. Yamanouchi et al. (eds.), *Ultrafast Phenomena XIX*,
Springer Proceedings in Physics 162, DOI 10.1007/978-3-319-13242-6_76

interference between resonant peak(s) and nonresonant background. These issues of conventional VSFG have hindered extension to time-resolved measurements.

Recently, the drawbacks of conventional VSFG have been solved by heterodyne detection, which realizes direct measurements of the phase and amplitude of the sum-frequency signals [1–3]. Heterodyne-detected VSFG (HD-VSFG) provides $Im\chi^{(2)}$ spectra which can be directly compared to absorption spectra (i.e., $Im\chi^{(1)}$) in solution. In particular, broadband HD-VSFG, which was developed in our group [2, 3], can be readily extended to femtosecond time-resolved measurements by introducing pump pulses for photoexcitation. We have developed two types of femtosecond time-resolved HD-VSFG (TR-HD-VSFG) spectroscopy, i.e., infrared-excited TR-HD-VSFG and ultraviolet-excited TR-HD-VSFG.

2 Infrared-Excited TR-HD-VSFG and 2D HD-VSFG to Study Ultrafast Vibrational Dynamics at Liquid Interfaces

Water is the most important liquid, and its unique properties originate from the ability to form hydrogen bond networks. Vibrational dynamics of bulk water have intensively been studied by ultrafast time-resolved infrared and two-dimensional infrared (2D IR) spectroscopies. We developed infrared-excited TR-HD-VSFG and investigated the ultrafast vibrational dynamics of water at interfaces by measuring time-resolved $Im\chi^{(2)}$ spectra in the OH stretch region [4–6]. In this experiment, we first excite a part of the broad OH band of water by femtosecond infrared pulses with ~ 100 cm^{-1} bandwidth and then measured the pump-induced change of the $Im\chi^{(2)}$ spectra ($\Delta Im\chi^{(2)}$) after a certain delay time [4]. This infrared-excited TR-HD-VSFG have been extended to 2D spectroscopy (2D HD-VSFG) by measuring time-resolved $\Delta Im\chi^{(2)}$ spectra with different pump wavelengths [6].

Figure 1a shows the 2D HD-VSFG spectra in the OH stretch region obtained from a positively-charged surfactant/water interface [5]. We used isotopically diluted water to suppress the intra/intermolecular vibrational couplings, and the OH stretch band under this experimental condition predominantly arises from HOD species. The 2D HD-VSFG spectrum immediately after photoexcitaion is diagonally elongated, demonstrating inhomogeneity in the interfacial water. This elongation almost disappears at 300 fs, due to spectral diffusion. This measurement clearly revealed hole-burning and the subsequent spectral diffusion of interfacial water for the first time. We also carried out 2D HD-VSFG measurements for the OH stretch at the air/water (H$_2$O) interface [6]. We observed off-diagonal cross peaks among the three OH bands immediately after photoexcitation, indicating that the three distinct OH oscillators observed in the steady-state $Im\chi^{(2)}$ spectrum are strongly coupled [6].

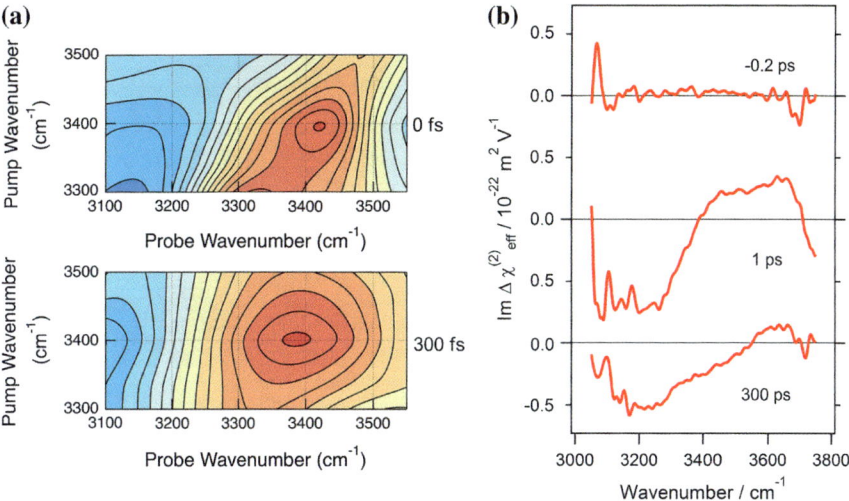

Fig. 1 **a** 2D HD-VSFG (infrared-excited TR-HD-VSFG) spectra of a positively charged water interfaces. [Reprinted with permission from J. Chem. Phys. **137**, 094706 (2012). Copyright 2012, AIP Publishing LLC.] **b** Ultraviolet-excited TR-HD-VSFG spectra at the air/indole aqueous solution interface

3 Ultraviolet-Excited TR-HD-VSFG to Study Ultrafast Photochemical Dynamics at Liquid Interfaces

For investigating the dynamics of chemical processes at the interface, we developed ultraviolet-excited TR-HD-VSFG, in which we induce a photochemical process by electronic excitation and observe the following dynamics by HD-VSFG. We applied this new method to study hydrated electrons at the water interface. Hydrated electrons are the most fundamental anion species, consisting only of electrons and surrounding water molecules. Although hydrated electrons have been extensively studied in the bulk aqueous solutions, even their existence is still controversial at the water interfaces.

Figure 1b shows time-resolved $\Delta \text{Im}\chi^{(2)}$ spectra at the air/indole aqueous solution interface obtained with ultraviolet excitation at 265 nm. With photoexcitation, indole molecules are photoionized, and electrons are generated. The time-resolved $\Delta \text{Im}\chi^{(2)}$ spectrum immediately after photoexcitation shows a transient signal with a negative peak around 3,200 cm^{-1} and a positive peak around 3,500 cm^{-1}. Then, the spectra exhibit substantial temporal changes, and a positive feature around 3,400 cm^{-1} vanishes in a few hundred picoseconds. The SVD analysis showed that the time-resolved $\Delta \text{Im}\chi^{(2)}$ spectra consist of two components. We assigned the positive component peaked around 3,400 cm^{-1} to the water interacting electrons at the interface. This assignment was confirmed by TR-HD-VSFG experiments at the air/water interface in which the electrons are generated by multi-photon ionization of water. The obtained data revealed the unique dynamics and solvation environment of the excess electrons at the aqueous interfaces [7].

References

1. N. Ji, V.Ostroverkhov, C. Y. Chen, Y. R. Shen, J. Am. Chem. **129**, 10056 (2007).
2. S. Yamaguchi, T. Tahara, J. Chem. Phys. **129**, 101102 (2008).
3. S. Nihonyanagi, S. Yamaguchi, T. Tahara, J. Chem. Phys. **130**, 204704 (2009).
4. S. Nihonyanagi, P. C. Singh, S. Yamaguchi, T. Tahara, Bull. Chem. Soc. Jpn. **85**, 758 (2012).
5. P. C. Singh, S. Nihonyanagi, S. Yamaguchi, T. Tahara, J. Chem. Phys. **137**, 094706 (2012).
6. P. C. Singh, S. Nihonyanagi, S. Yamaguchi, T. Tahara, J. Chem. Phys. **139**, 161101 (2013).
7. K. Matsuzaki, S. Nihonyanagi, S. Yamaguchi, T. Nagata, T. Tahara, in preparation.

Ultrafast Electron Solvation at the Room Temperature Ionic Liquid/Metal Interface

Alex J. Shearer, Benjamin W. Caplins, David E. Suich and Charles B. Harris

Abstract Ultrafast electron solvation was studied in thin films of the room temperature ionic liquid [Bmpyr]$^+$ [NTf$_2$]$^-$ on a Ag(111) substrate. Two-photon photoemission spectra reveal a solvation effect which increases from a 250 meV shift in under 400 fs for the monolayer to a 1 eV shift in over 100 ps for the trilayer. The state's binding energy relaxes along the same path for all coverages at a given temperature, suggesting that the solvation process is insensitive to film thickness. Time-dependent population analysis showed that the lifetime of solvation changes with coverage due to charge screening. In the monolayer coverage regime, the state has dispersive, delocalized character at early times and nondispersive, localized character after 200 fs.

1 Introduction

Due to their unique physical properties, room temperature ionic liquids (RTILs) are a novel class of compounds that have found widespread use as a substitute for traditional organic solvents. Though these materials have received an immense amount of attention and study in recent years, there remain areas where little is known about how RTILs behave. For example, it has been shown that electron solvation proceeds on multiple timescales in bulk RTILs [1], but ultrafast studies of these compounds on surfaces are limited in number.

Our group recently used ultrafast time- and angle-resolved two-photon photoemission (TPPE) to observe ultrafast electron solvation in a thin film of [Bmpyr]$^+$ [NTf$_2$]$^-$/Ag(111) [2]. The relaxation time and energy shift were strongly dependent

A.J. Shearer · B.W. Caplins · D.E. Suich · C.B. Harris (✉)
Department of Chemistry, University of California, Berkeley, CA 94720, USA
e-mail: cbharris@berkeley.edu

A.J. Shearer · B.W. Caplins · D.E. Suich · C.B. Harris
Chemical Sciences Division, Lawrence Berkeley National Laboratory,
Berkeley, CA 94720, USA

© Springer International Publishing Switzerland 2015
K. Yamanouchi et al. (eds.), *Ultrafast Phenomena XIX*,
Springer Proceedings in Physics 162, DOI 10.1007/978-3-319-13242-6_77

on sample temperature. The previous study characterized solvation in a thin RTIL film, but at that time no structural characterization was possible. Recent X-ray photoemisison spectroscopy (XPS) and scanning tunneling microscopy (STM) experiments have provided valuable insight into the structure of $[Bmpyr]^+ [NTf_2]^-$ films [3]. Using this new structural information, we build on our previous study by using temperature programmed desorption (TPD), as well as coverage-dependent TPPE, to assign accurate film thicknesses for the present study, allowing us to link ultrafast dynamics to the film's structure.

2 Results and Discussion

The result of a time-resolved measurement is shown in Fig. 1a in the form of a false color plot. The solvation response of the film can be clearly seen by the energetic shift of the single peak with increasing delay time. In the monolayer coverage regime, the injected electron state is observed to relax in energy by 250 meV in less than 400 fs. Thicker films show a similar solvation response, but the energetic relaxation increases to 700 meV for the bilayer and 1 eV for the trilayer. Figure 1b shows how the binding energy of the injected electron state changes over time for all coverages at 300 and 130 K. Interestingly, the relaxation of the state within the first 400 fs for all film coverages at high and low temperatures follows a similar curve, indicating that the initial steps occurring in the solvation process are alike regardless of coverage or temperature. This similarity in behavior across different coverage and temperature regimes is interesting considering that both our previous work and the recent STM work show that the structure of films is very different at 300 K versus 130 K [2, 3]. After 500 fs, the structural difference manifests itself as the relaxation behavior of 130 K films becomes markedly slower than that of the 300 K films.

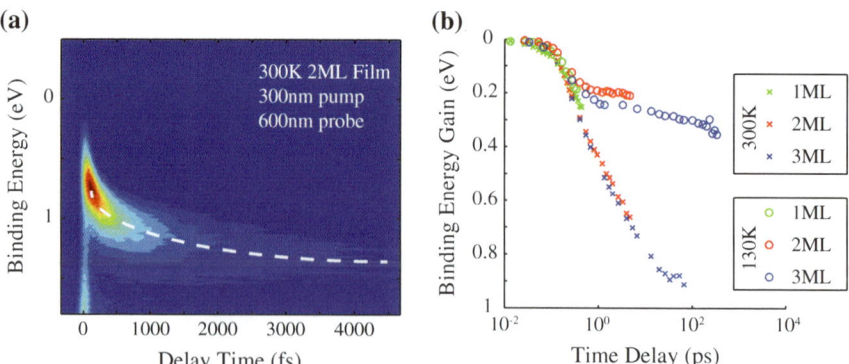

Fig. 1 **a** A representative contour plot of a time resolved TPPE experiment in a 2 ML film at 300 K. **b** Binding energy shift of the injected electron peak as a function of time at variable thickness and temperature

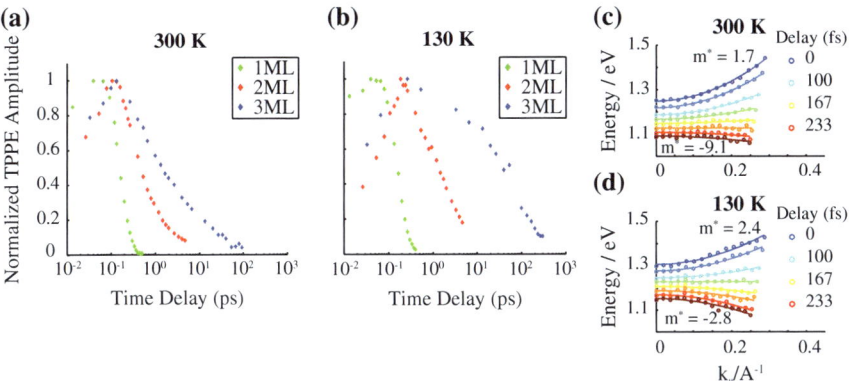

Fig. 2 **a** Population decay at 300 K. **b** Population decay at 130 K. **c** Time-dependent dispersions at 300 K. **d** Time- dependent dispersions at 130 K

The population of the state was monitored as a function of delay time for films of all coverages at 300 and 130 K, in order to investigate the injected electron's lifetime and its dependence on film structure. These data, presented in Fig. 2a, b, show that there is a multiple orders of magnitude increase in lifetime as the film thickness increases from 1 to 3 monolayers. This increase in lifetime indicates that the electron injected into the film is being screened by the RTIL ions, thicker films lead to more screening and therefore a decreased probability of tunneling back to the Ag(111) substrate.

Angle-resolved measurements were taken at different time delays to investigate how the spatial extent of the injected electron's state changes as solvation occurs. Angle-resolved spectra are taken by rotating the sample relative to the time-of-flight photoelectron detector, a dispersion relation for a given state can then be constructed and the curvature of this dispersion curve indicates whether the state is delocalized or localized in the surface plane. As shown in Fig. 2c, d, the spectra at initial time show a positively dispersing peak indicative of a delocalized state, while spectra at long time delays show a negatively dispersing peak indicative of a localized state [4].

Interestingly, the transition from delocalized to localized appears to be continuous, Fig. 2c, d show a continuous change in dispersion curvature as a function of time for both 300 and 130 K films. There are two possible explanations for our observation of a state transitioning from delocalized to localized in a continuous way. The first is that the injected electron is solvating in a dynamic process, as has been observed in other surface systems [5]. The second explanation is that the electron is being injected into a delocalized band, perhaps an image potential state or conduction band state, and then being transferred into a localized state within a few hundred femtoseconds. The latter situation can also lead to observation of a continuous change in dispersion curvature, as shown by previous work [4].

3 Conclusions

We have characterized the electron solvation and decay dynamics in three different coverages of the RTIL [Bmpyr]$^+$ [NTf$_2$]$^-$. The initial energetic relaxation was shown to be similar for films of all coverages and temperatures, structural differences only manifested themselves after 400 fs. Furthermore, extended lifetimes of these states showed that the RTIL is effectively screening the injected charge from the metal surface. Dispersion measurements have provided a first look at the type of processes that could be responsible for solvation in the <1 ps regime. These results are the first demonstration of coverage dependent ultrafast behavior at an RTIL/ metal interface.

References

1. H. Shirota, A. Funston, J. Wishart, E. Castner, J. Chem. Phys. **122**, 184512 (2005).
2. E. A. Muller, M. L. Strader, J. E. Johns, A. Yang, B. W. Caplins, A. J. Shearer, D. E. Suich, C. B. Harris, J. Am. Chem. Soc. **135**, 10646-10653 (2013).
3. F. Buchner, K. Forster-Tonigold, B. Uhl, D. Alwast, N. Wagner, H. Farkhondeh, A. Groß, R. J. Behm, ACS Nano **7**, 7773-7784 (2013).
4. U. Bovensiepen, C. Gahl, M. Wolf., J. Phys. Chem B. **107**, 8706-8715 (2003).
5. J. Johns, E. Muller, J. Frechet, C. B. Harris, J. Am. Chem. Soc. **132**, 15720 (2010).

Ultrafast Spectroscopy Reveals Bulk Heterojunction Morphology

Maxim S. Pshenichnikov, Almis Serbenta
and Paul H.M. van Loosdrecht

Abstract We propose a new technique to probe the nanoscale morphology in polymer-fullerene bulk heterojunctions by ultrafast spectroscopy. The method reveals characteristic sizes of fullerene clusters in an all-optical way which makes it directly applicable to functional photovoltaic devices.

1 Introduction

The morphology of the bulk heterojunction—the nanoscale texture of polymer and fullerene constituents—is one of the key ingredients for optimization of the efficiency of modern plastic solar cells [1]. The particular morphology emerges from self-organization of polymer and fullerene fractions into an interpenetrating network. This arrangement maximizes the interface area where optically excited excitons are separated into charges and provides percolated pathways of charge transport to the device electrodes. Morphology is a challenging property not only to control but even to characterize as this generally requires chemical selectivity combined with nanometer spatial resolution. Standard methods for studying nanostructures—such as AFM, STM, TEM, SAXS, etc.—either lack adequate spatial resolution or require special sample preparation (or both).

Here we show how ultrafast spectroscopy can be used to obtain information on morphology by making use of fullerene ($PC_{71}BM$) excitons [2]. Our approach (Fig. 1a) employs the substantial visible light absorption of $PC_{71}BM$ which is currently used in the overwhelming majority of novel organic photovoltaic blends. A pump pulse selectively creates an exciton in the $PC_{71}BM$ domain which after diffusion to the interface dissociates into a hole on the polymer and an electron on the fullerene. The hole modifies the electronic structure of the polymer leading to

M.S. Pshenichnikov (✉) · A. Serbenta · P.H.M. van Loosdrecht
Zernike Institute for Advanced Materials, University of Groningen, Nijenborgh, 4,
Groningen 9747, AG, The Netherlands
e-mail: M.S.Pchenitchnikov@RuG.nl

© Springer International Publishing Switzerland 2015
K. Yamanouchi et al. (eds.), *Ultrafast Phenomena XIX*,
Springer Proceedings in Physics 162, DOI 10.1007/978-3-319-13242-6_78

the so-called polaron absorption which is probed by a delayed IR pulse [3]. From the delay of the IR response and efficiency of exciton-to-hole generation, a characteristic size of the PC$_{71}$BM domains is retrieved via the ratio of short-time interfacial and long-time diffusion-mediated contributions. The key advantage of the proposed all-optical technique lies in its potential applicability to functional photovoltaic devices with no additional materials preparation requirements.

2 Experimental

As donor polymers we selected MDMO-PPV (known for extreme phase segregation), regiorandom (RRa-) P3HT (for excellent miscibility with PC$_{71}$BM) and regioregular (RRe-) P3HT (for formation of the crystalline structure). PC$_{71}$BM was selectively excited by choosing appropriate pump energies below the polymer absorption bandgap where the PC$_{71}$BM absorption dominates. Different weight ratios of constituents were used to decipher direct and diffusion-mediated hole transfer. The probe wavelength was chosen near the low-energy polaron peak at 3 μm.

3 Results

Interfacial and diffusion-mediated charge yields are summarized in Fig. 1c. The contribution of interfacial excitons decreases while the share of diffusion-mediated excitons increases. Knowing the fractions of interface and diffusion mediated excitons, we can readily estimate the typical size of PC$_{71}$BM clusters (Fig. 2a)

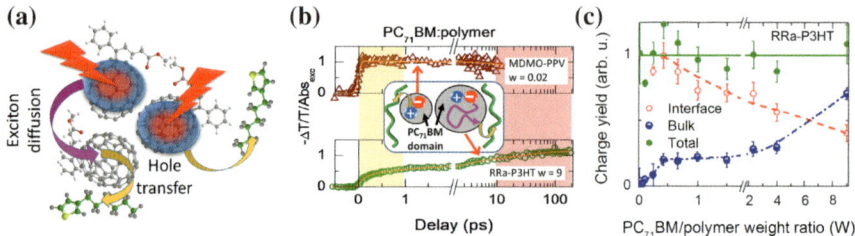

Fig. 1 a The concept of the proposed technique. *Lightning bolt* PC$_{71}$BM excitation, *magenta arrow* PC$_{71}$BM exciton diffusion, *yellow arrow* exciton dissociation via hole transfer to the polymer. **b** Representative (out of 30) transients for two polymers at different PC$_{71}$BM concentrations (indicated by PC$_{71}$BM/polymer weight ratio, *w*). *Symbols* experimental data, *solid curves* fits from Monte–Carlo simulations. *Yellow* and *red shaded areas* indicate characteristic times for interfacial and diffusion-mediated exciton dissociation, respectively. **c** Charge yield at different delays for interfacial (*red dots*), total (*green dots*), and bulk (*blue dots*) excitons for blends with different PC$_{71}$BM/RRa-P3HT weight ratios *w*

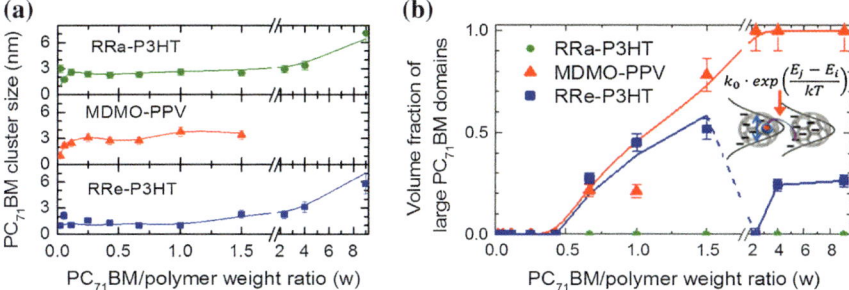

Fig. 2 **a** PC$_{71}$BM aggregate sizes at different PC$_{71}$BM/polymer ratios for the three investigated polymers. Closed symbols are obtained with the surface-to-volume model while *solid curves* resulted from Monte-Carlo simulations. **b** Volume fraction of large (>15 nm) PC$_{71}$BM domains which do not contribute to separated charges. *Inset* shows schematics of the exciton diffusion model. Equilibrium hopping rate is k$_0$ = 5 ps^{-1}

assuming them to be spherical, by a simple surface-to-volume argument. From this we find that the average PC$_{71}$BM cluster diameter varies from ~ 1 nm (i.e. a single PC$_{71}$BM molecule) to ~ 7 nm in blend. Hence, P3HT polymers exhibit reasonably good miscibility with optimal domain sizes for both polymer and fullerene exciton harvesting. In contrast, MDMO–PPV blends demonstrate explosive growth of the domain size at weight ratios higher than 1.5 resulting in a catastrophic decrease of PC$_{71}$BM exciton yield (Fig. 1b).

To obtain insight into the PC$_{71}$BM exciton dynamics, we performed Monte–Carlo simulations to fit to the experimental data. The fullerene cluster was modelled by 1 nm spherical molecules placed in a 3D hexagonal matrix. To account for energy disorder in the fullerene domains, each PC$_{71}$BM molecule was assigned a random energy from a Gaussian distribution with standard deviation 0.07 eV (Fig. 2b, inset). The exciton hopping rate between PC$_{71}$BM molecules was scaled with a Boltzmann factor [4]. The initial exciton was randomly placed in the PC$_{71}$BM domain, after which its diffusional trajectory was calculated until it either reached the interface or died after its 500 ps lifetime. The simulations were run over 100,000 realizations for each PC$_{71}$BM/polymer weight ratio.

The results of the simulations excellently reproduce the experimental data (Fig. 1b, solid lines). Figure 2a (solid lines) demonstrates that the simple surface-to-volume rationale used here is fully consistent with the Monte-Carlo simulations.

Summarizing, the main advantage of the proposed method lies in its capability to resolve nm-sized PC$_{71}$BM clusters "on-the-fly". Furthermore, it allows estimating the PC$_{71}$BM exciton dissociation yield which is highly relevant for initial charge generation in organic photovoltaics devices.

References

1. W. Chen, M.P. Nikiforov, S.B. Darling, Energy Environ. Sci., **5**, 8045 (2012).
2. A.A. Bakulin *et al.*, Adv. Funct. Mater., **20**, 1653 (2010).
3. A.A. Bakulin *et al.*, Science, **335**, 1340 (2012).
4. S. Westenhoff, I.A. Howard, R.H. Friend, Phys. Rev. Lett., **101**, 016102 (2008).

Toward Ultrafast In Situ X-ray Studies of Interfacial Photoelectrochemistry

S. Neppl, Y.-S. Liu, C.-H. Wu, A. Shavorskiy, I. Zegkinoglou,
T. Troy, D.S. Slaughter, M. Ahmed, A.S. Tremsin, J.-H. Guo,
P.-A. Glans, M. Salmeron, H. Bluhm and O. Gessner

Abstract Picosecond time-resolved in situ X-ray absorption and X-ray photoelectron spectroscopy techniques for atomic site-specific real-time studies of interfacial photoelectrochemistry are developed at the Advanced Light Source (ALS). First experiments monitor electronic dynamics in films of dye-sensitized nanocrystals and at hematite-electrolyte interfaces.

1 Introduction

Interfacial photoelectrochemistry is governed by the flow of charge, mass, and energy among molecules and between molecules and condensed phase substrates. By definition, the quest for a deeper understanding of interfacial photochemical processes is directly related to the capability to monitor non-equilibrium dynamics of complex systems in electronically and/or vibrationally excited states. Time-domain X-ray spectroscopy techniques offer new opportunities to unravel the fundamental

S. Neppl (✉) · O. Gessner
Ultrafast X-Ray Science Laboratory, University of California, Berkeley, CA 94720, USA
e-mail: SNeppl@lbl.gov

S. Neppl · A. Shavorskiy · I. Zegkinoglou · T. Troy · D.S. Slaughter · M. Ahmed · H. Bluhm ·
O. Gessner
Chemical Sciences Division, Lawrence Berkeley National Laboratory, University of
California, One Cyclotron Road, M/S 2-300, Berkeley, CA 94720, USA

Y.-S. Liu · J.-H. Guo · P.-A. Glans
Advanced Light Source, University of California, Berkeley, CA 94720, USA

C.-H. Wu · M. Salmeron
Materials Sciences Division, Lawrence Berkeley National Laboratory, University of
California, Berkeley, CA 94720, USA

C.-H. Wu
Department of Chemistry, University of California, Berkeley, CA 94720, USA

A.S. Tremsin
Space Sciences Laboratory, University of California, Berkeley, CA 94720, USA

© Springer International Publishing Switzerland 2015
K. Yamanouchi et al. (eds.), *Ultrafast Phenomena XIX*,
Springer Proceedings in Physics 162, DOI 10.1007/978-3-319-13242-6_79

electronic dynamics that underlie the chemical and electronic interactions in processes such as photoelectrochemical water splitting or photovoltaic power generation. The simultaneous element specificity and chemical sensitivity of X-ray transitions in combination with the time-structure of accelerator-based X-ray light sources provides real-time access to transient oxidation states and local bonding motives with atomic pinpoint accuracy.

Time-resolved X-ray photoelectron spectroscopy (TRXPS) and time-resolved X-ray absorption spectroscopy (TRXAS) provide complementary information on interfacial dynamics. TRXPS is particularly surface sensitive and, in combination with ambient pressure photoelectron spectroscopy (APPES) techniques, provides a detailed picture of the interaction of the first few molecular monolayers with a condensed phase substrate. Time-resolved X-ray absorption spectroscopy (TRXAS) is sensitive to both surface and bulk dynamics and, in particular, lends itself to in situ/*operando* applications in which the X-ray spectroscopic signatures of the sample are recorded under device-like operating conditions. Both TRXPS and TRXAS techniques for in situ studies of interfacial photoelectrochemistry are currently being implemented at the Advanced Light Source (ALS) at Lawrence Berkeley National Laboratory. Showcase applications include photoinduced electronic dynamics in films of N3 dye-sensitized ZnO nanocrystals and at hematite/electrolyte interfaces in working photoelectrochemical (PEC) cells.

2 Experiments and Results

The ALS based experiments are enabled by a combination of three fundamental components: (i) APPES and in situ XAS techniques, (ii) a mobile high-repetition rate, high-power picosecond laser system, (iii) time-stamping data acquisition techniques enabled by time-sensitive detectors. The laser system provides 10 ps long "pump" pulses that initiate interfacial processes, which are monitored by ~ 70 ps "probe" pulses from the ALS. The laser system is synchronized to the ALS bunch structure and is used at a variety of beamlines (BLs). At BL 11.0.2, an APPES enabled electron analyzer is interfaced with a time- and position-sensitive detector that marks every registered electron with a unique time stamp. Post-processing of the data provides picosecond TRXPS spectra extending over microsecond time ranges, all recorded simultaneously and making use of the entire X-ray fluence of the ALS in either multi-bunch or two-bunch operating mode [1, 2]. A recent proof-of-principle study at the Linac Coherent Light Source (LCLS) demonstrated the potential of femtosecond TRXPS to monitor interfacial electron dynamics at N3-sensitized films of ZnO nanocrystals during photoinduced charge injection [1, 3]. The ALS based experiments extend these studies to the picosecond to microsecond domain, giving access to the recombination dynamics during the dye recovery phase of the working cycle of dye-sensitized solar cells. TRXPS fingerprints of both injection and recombination dynamics can be discerned and are interpreted with the aid of ab initio constrained density functional theory (DFT) calculations [1–3].

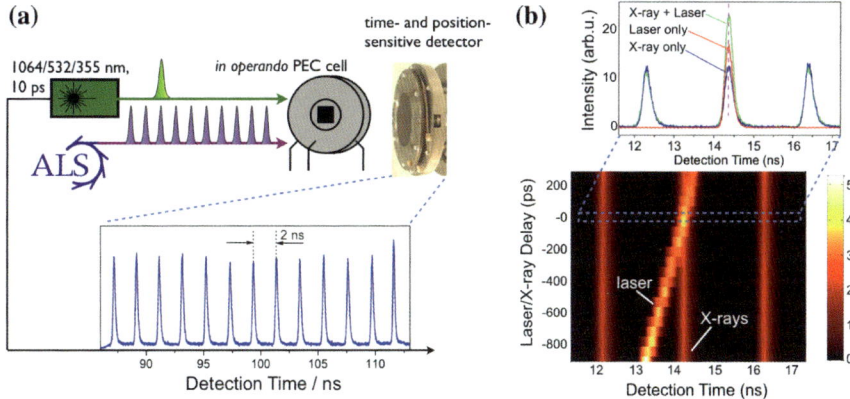

Fig. 1 a Setup for picosecond time-resolved in situ X-ray absorption spectroscopy of interfacial photoelectrochemistry. A high-repetition rate picosecond laser system is synchronized to the ALS bunch train. Laser and X-ray pulses are incident on an *in operando* photoelectrochemical (PEC) cell. The total X-ray fluorescence yield is recorded by a time- (and position-) sensitive detector, enabling the acquisition of picosecond time-resolved X-ray absorption spectra while simultaneously using multiple ALS X-ray pulses. **b** Demonstration of picosecond time-resolution by scanning the laser pump/X-ray probe delay

A schematic of the TRXAS experiment (BL 8.0.1 and BL 6.3.1) is shown in Fig. 1a. The laser system is operated at either 532 or 355 nm output wavelength, producing pump pulses with up to 8 MHz repetition rate and up to 75 µJ pulse energy at 100 kHz repetition rate. The repetition rate and pulse energy of the laser system can be set during operation, providing the crucial capability to efficiently tune the delicate balance of all operating parameters of photoactive interfaces. Both the laser and the ALS X-ray beams are focused into a PEC cell, designed for *in operando* X-ray absorption studies of photoelectrochemical processes [4]. The cell contains a semi-transparent ~ 30 nm thick hematite (α-Fe_2O_3) working electrode attached to a ~ 100 nm thick silicon nitride (Si_3N_4) window with a 10 nm thick Au film and a 2 nm Cr layer. The other side of the hematite film is exposed to a reservoir filled with an aqueous NaOH electrolyte. The electrical circuit of the PEC cell is completed by a Pt counter electrode and an Ag/AgCl reference electrode in the reservoir. Near-edge X-ray absorption fine structure (NEXAFS) spectra of the assembly are recorded by monitoring the total X-ray fluorescence yield emerging through the Si_3N_4 window as a function of the incident X-ray photon energy. A time- (and position-) sensitive fluorescence detector provides a unique time stamp for every detected fluorescence photon, enabling a clear distinction of fluorescence signals from all ALS bunches in either multi-bunch (Fig. 1a, bottom) or two-bunch operating mode. The picosecond time resolution of the setup is demonstrated in Fig. 1b. To record the false color map, the PEC was oriented at an angle at which the laser back reflection from the cell was incident on the fluorescence detector. The map shows the detected total light intensity as a function of time (horizontal axis) and the delay between the optical pump pulse and one of the X-ray probe pulses (vertical axis). The upper part of Fig. 1b illustrates the time-dependent signals at zero pump-probe delay when either only X-rays, or only the pump laser, or both X-ray and laser pulses are incident on the cell. After the timing calibration, the PEC cell is rotated such that the laser back reflection does not overlap with the TRXAS signal. In the next step, TRXAS spectra of working PEC cells will be used to monitor the interfacial hole dynamics that are critical for the efficiency of photo-electrochemical water splitting.

Acknowledgment This work was supported by the U.S. Department of Energy, Office of Basic Energy Sciences, Chemical Sciences, Geosciences and Biosciences Division, through Contract No. DE-AC02-05CH11231. O.G. was supported by the Department of Energy Office of Science Early Career Research Program.

References

1. A. Shavorskiy et al., "Time-resolved x-ray photoelectron spectroscopy techniques for real-time studies of interfacial charge transfer dynamics," AIP Conf. Proc. **1525**, 475 (2013).
2. S. Neppl et al., "Capturing interfacial photoelectrochemical dynamics with picosecond time-resolved X-ray photoelectron spectroscopy," Faraday Discuss. **171**, 219 (2014).
3. K. Siefermann et al., "Atomic Scale Perspective of Ultrafast Charge Transfer at a Dye-Semiconductor Interface", J. Phys. Chem. Lett., submitted **5**, 2753 (2014).
4. A. Braun, et al., "Direct Observation of Two Electron Holes in a Hematite Photoanode during Photoelectrochemical Water Splitting", J. Phys. Chem. C **116**, 16870 (2012).

Ultrafast Dynamics in Epitaxial Silicene on Ag(111)

E. Cinquanta, S.D. Conte, D. Chiappe, C. Grazianetti, M. Fanciulli, A. Molle, G. Cerullo, S. Stagira, F. Scotognella and C. Vozzi

Abstract Ultrafast transient reflectivity measurements were performed in epitaxial 4 x 4 silicene grown on Ag(111). Comparison with bulk silicon and silver response highlighted the occurrence of peculiar photo-physical mechanisms, suggesting a metallic-like behavior in silicene.

A 2D honeycomb crystalline form of silicon, the so called silicene, has recently attracted great attention as a novel graphene-like material [1]. Until now, free standing silicene has never been observed, because a 2D sp2 Si crystal is chemically unstable in ambient conditions [2]. Nevertheless, quasi-two dimensional honeycomb Si layers grown by molecular beam epitaxy can be stabilized by a supporting substrate, as shown in recent pioneering experiments [3]. To date, buckled silicene structures have been experimentally reported on substrates with metallic character, including Ag(111) [4, 5], Ir(111) [6] and ZrB2 [7]. Despite huge efforts devoted to the characterization of epitaxial silicene by means of STM/STS, LEED and ARPES, an unambiguous picture of its physical properties is still lacking.

Indeed the presence of degenerate surface superstructures together with a partial hybridization between Si and Ag atoms makes the valence band structure of the epitaxial silicene non-trivial, hence providing controversial interpretations [4, 8, 9]. Here we present a study of ultrafast transient reflectivity in 4×4 silicene superstructure grown on Ag(111) and capped with Al2O3 [10]. We compare the results with measurements performed in bulk silicon, on the silver substrate and on oxidized silicene. We observed a substantial difference in the silicene response with

E. Cinquanta · D. Chiappe · C. Grazianetti · M. Fanciulli · A. Molle
Laboratorio MDM, IMM-CNR, Agrate Brianza, Italy

S.D. Conte · G. Cerullo · C. Vozzi (✉)
Istituto Di Fotonica E Nanotecnologie IFN-CNR, Milan, Italy
e-mail: caterina.vozzi@cnr.it

M. Fanciulli
Dipartimento Di Scienza Dei Materiali, Università Degli Studi Di Milano-Bicocca, Milan, Italy

G. Cerullo · S. Stagira · F. Scotognella
Dipartimento Di Fisica, Politecnico Di Milano, Milan, Italy

© Springer International Publishing Switzerland 2015
K. Yamanouchi et al. (eds.), *Ultrafast Phenomena XIX*,
Springer Proceedings in Physics 162, DOI 10.1007/978-3-319-13242-6_80

respect to the other samples considered, confirming the presence of a peculiar electronic dynamics that suggests a metallic-like behavior similar to the one reported in graphene [11]. Figure 1 shows the STM image of the 4 × 4 silicene superstructure on Ag(111) as the one investigated in the experiments. The honeycomb nature of the silicene lattice has also been confirmed by multi-wavelength resonant Raman spectroscopy supported by ab initio calculations [12]. The ultrafast photo-physical properties of silicene were investigated by performing pump-probe measurements in reflection geometry with a temporal resolution of about 20 fs. We used as a probe the second harmonic of a visible optical parametric amplifier (OPA) [13] in order to achieve a pulse duration of approximately 20 fs in the probe region from 340 to 370 nm.

We excited the sample with a broadband visible pulse, spectrally peaked around 500 nm, obtained from another OPA. Both the two OPAs were driven by an amplified Ti:sapphire laser system (500 μ J, 150 fs, 1 kHz). After chirped mirror compression, the duration of the pump pulse was less than 15 fs. We measured the probe reflection of the sample with an optical multichannel analyzer working at the full repetition rate of the laser source. The acquisition of the pump-perturbed and pump-unperturbed probe spectra allowed extraction of the sample differential reflectivity $\Delta R/R$. The results are shown in Fig. 2 where the sample reflectivity is reported as a function of the pump-probe delay and probe wavelength. Panel (a) and (b) correspond to the Ag(111) substrate and oxidized silicene respectively. A very similar fast signal is observed around zero delay in both measurements. This could be attributed to the ultrafast response of silver. Indeed the oxidation process transforms the silicene in silicon oxide, which does not contribute to the transient reflectivity of the sample because of the strong increase of the optical gap due the oxidation process (9 eV). Panel (c) shows the reflectivity change in bulk silicon. One can see a decrease of reflectivity which takes place in the first 200 fs followed by a very slow recovery on the scale of tens of picoseconds; this is consistent with early studies on the ultrafast response of silicon [14].

Fig. 1 STM image of the silicene sample

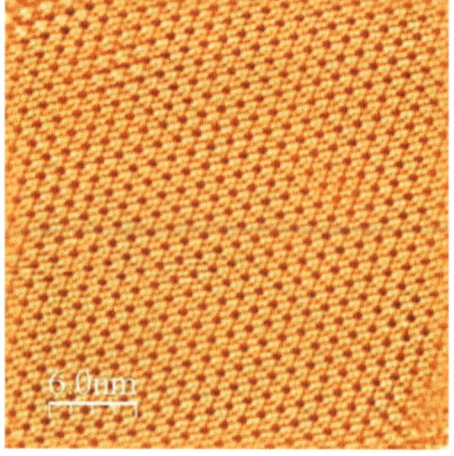

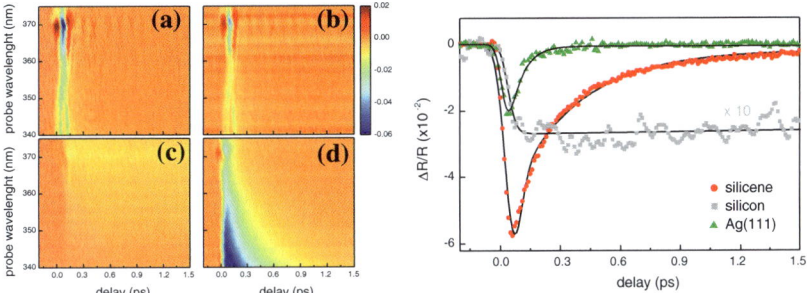

Fig. 2 *Left panel* transient reflectivity change as a function of pump-probe delay and probe wavelength in **a** Ag(111), **b** oxidized silicene on Ag(111), **c** bulk silicon and **d** silicene on Ag(111) with Al2O3 capping layer. *Right panel* transient reflectivity change as a function of pump-probe delay for the probe wavelength of 350 nm in silicene (*red dots*), bulk silicon (*grey squares*, signal multiplied by 10) and Ag(111) (*green triangles*). *Solid lines* are fits by multi-exponential decay convoluted with the instrument response function

The result of pump-probe measurement in silicene is reported in Fig. 2 panel (d). We observed a very intense feature in the $\Delta R/R$ spectrum followed by a fast recovery on the picosecond timescale. Figure 2e reports the transient reflectivity change corresponding to the probe wavelength of 350 nm for silicene (dots), bulk silicon (squares) and the silver substrate (triangles). The difference among the response of the samples is very clear. A multi-exponential fit is reported as a solid curve for each sample in Fig. 2e. The fit gives a recovery time of 65 fs for the silver substrate, that corresponds to the very fast electron relaxation occurring in metals. On the other hand, bulk silicon shows a slow recovery with a time constant of 35 ps. This can be attributed to the energy gap that hinders the electron-hole recombination in a semiconductor like silicon. In the case of silicene, the signal presents two components with time constant of 166 and 860 fs. These fast time constants suggest the metallic- like nature of this material. In conclusion, we reported ultrafast reflectivity measurements in 4 × 4 epitaxial silicene grown on Ag (111). The experimental results show peculiar characteristics which cannot be ascribed to the silver substrate and that are substantially different from the bulk silicon response. These results contribute to the understanding of silicene physical properties, suggesting a metallic-like behavior in the 4 × 4 superstructure.

Acknowledgments FP7 EU Projects 2D-Nanolattices (FET-OPEN grant number: 270749), Graphene Flagship (contract no. CNECT-ICT-604391).

References

1. S. Cahangirov et al., Phys. Rev. Lett. 102, 236804 (2009).
2. E. F. Sheka, Int. J. Quantum Chem. 113, 612 (2013).
3. J. Gao and J. Zhao, Sci. Rep. 2, 861 (2012).

4. P. Vogt et al., Phys. Rev. Lett. 108, 155501 (2012).
5. D. Chiappe et al., Adv. Mater. 24, 5088 (2012).
6. L. Meng et al., Nano Lett. 13, 685 (2013).
7. A. Fleurence et al., Phys. Rev. Lett. 108, 245501 (2012).
8. C.-L. Lin et al, Phys. Rev. Lett 110, 076801 (2013).
9. D. Tsoutsou et al., Appl. Phys. Lett. 103, 231604 (2013).
10. A. Molle et al, Adv. Funct. Mater. 23, 4340 (2013).
11. M. Breusing et al., Phys. Rev. Lett. 102, 086809 (2009).
12. E. Cinquanta et al, J. Phys. Chem C 117, 16179 (2013).
13. M. Beutler et al., Opt. Lett. 34, 710 (2009).
14. F. E. Doany and D. Grischkowsky, Appl. Phys. Lett. 52, 36 (1988).

Accessing Energy-Dependent Photoemission Delays in Solids

Matteo Lucchini, Luca Castiglioni, Reto Locher, Michael Greif,
Lukas Gallmann, Jürg Osterwalder, Matthias Hengsberger
and Ursula Keller

Abstract Our new detection scheme combines the RABBITT technique in solids with simultaneous measurements in a reference argon target. The experiment resolved attosecond delays in the photoemission from noble metal surfaces beyond simple ballistic transport.

1 Introduction

Studying attosecond delays involved in photoemission is of great interest for our understanding of fundamental processes in nature. Recent developments in the generation of extreme-ultraviolet (XUV) attosecond pulses (1 as = 1×10^{-18} s) provide powerful tools to access such dynamics that unfold on atomic time scales [1]. So far, two main techniques were employed to study photoemission delays in atoms: attosecond streaking [2] and RABBITT [3]. In the case of streaking, a single attosecond pulse is used to ionize the target in the presence of a delayed infrared (IR) reference pulse, while for RABBITT the excitation occurs via a train of attosecond pulses. In both cases information about the photoemission dynamics can be extracted from the delay dependence of the photoelectron spectrum [4, 5]. Despite the good results obtained with gaseous samples, the extension of attosecond techniques to condensed matter is not trivial. Phenomena absent in rare-gas samples like space charge, secondary electrons and above threshold photoemission (ATP)

M. Lucchini (✉) · R. Locher · L. Gallmann · U. Keller
Department of Physics, Institute for Quantum Electronics, ETH Zurich,
8093 Zurich, Switzerland
e-mail: mlucchini@phys.ethz.ch

L. Castiglioni · M. Greif · J. Osterwalder · M. Hengsberger
Department of Physics, University of Zurich, 8057 Zurich, Switzerland

L. Gallmann
Institute of Applied Physics, University of Bern, 3012 Bern, Switzerland

© Springer International Publishing Switzerland 2015
K. Yamanouchi et al. (eds.), *Ultrafast Phenomena XIX*,
Springer Proceedings in Physics 162, DOI 10.1007/978-3-319-13242-6_81

333

by the IR field, limited the applicability of existing methods to high excitation energies (~ 90 eV). Indeed, so far, only attosecond streaking has been successfully employed in time-resolved studies of photoemission from solids. In 2007 Cavalieri et al. [6] studied the delay between the electrons emitted from the 4f and conduction band in W(110) and found it to be 110 as. Five years later the same group showed that no delay exists in the relative photoemission between the 2p and valence band of Mg(0001) [7]. We present the first extension of the RABBITT technique to solid-state samples. Compared to streaking, RABBITT exhibits several advantages: (i) the required IR intensity is lower ($\sim 1 \times 10^{11}$ W/cm^2), thereby enabling its application even at lower XUV photon energies; (ii) the XUV generation scheme is simplified since no single attosecond pulse is required and the carrier-envelope phase of the driving pulse does not have to be stabilized; (iii) in contrast to streaking, this method yields the energy dependence of the photoemission delay in a single measurement.

2 Results

In order to be able to study photoemission from the noble metal surfaces Ag(111) and Au(111), we introduced a refocusing toroidal mirror in the existing attosecond beamline [8]. This peculiar refocusing geometry offers the possibility to perform simultaneous measurements with two different, spatially separated targets. In this way, a well-known reference sample can be chosen to calibrate the experiment in one interaction region. For this experiment, a surface-science end station equipped with a hemispherical electron analyzer was installed in the second interaction region. This combination enables measuring photoemission delays from a single initial state. We used few-cycle IR pulses with time duration between 7 and 12 fs to generate a comb of odd harmonics in a cell filled with argon. The two targets are ionized by the harmonics producing comb-like photoelectron spectra, which peak at the harmonic positions.

An example of simultaneously acquired RABBITT traces is shown in Fig. 1. The delayed IR pulse modulates the photoelectron spectrum. When the attosecond pulse train and the IR pulse overlap in time, satellites of the harmonic peaks (sidebands) appear in the recorded trace. The amplitude of these sidebands oscillates with twice the frequency of the IR beam. It can be shown that the phase of the photoelectron is recorded in the phase of the oscillatory signal. Thus, after subtraction of the calibration signal, one can access the surface specific photoemission phases displayed in Fig. 2. The energy dependence of the experimental phases is not monotonic. In particular the photoemission phases show a tendency to increase with energy. This is in contrast to the explanation of the previous results from the streaking experiments [6, 7] where transport was assumed to be the dominant mechanism.

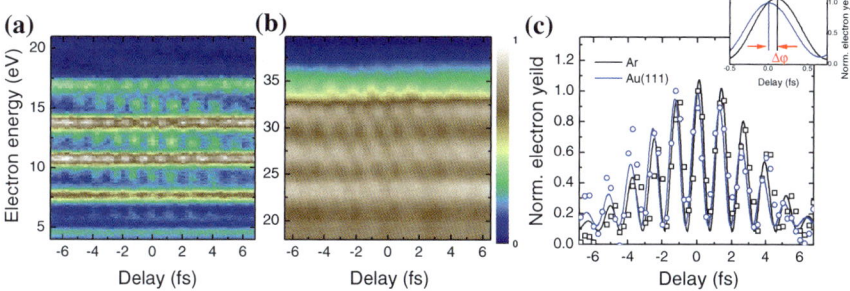

Fig. 1 Simultaneously recorded RABBITT traces for Ar *3p* and Au(111) *4d*-band. **a** and **b** Show the experimental photoelectron spectra as a function of delay between the attosecond pulse train and the IR pulse. The two RABBITT traces are offset with respect to each other in energy due to the different ionization potential/work function of the respective targets. **c** Shows the signal of sideband 18 extracted from **a** around 12.25 eV and **b** around 25.5 eV, *black squares* and *blue rounds*, respectively. The *solid lines* represent the fitting curve used to extract the experimental phases as is enlightened in the *inset*

Fig. 2 Experimental surface specific phases for four different sidebands (SB 16, 18, 20, 22) for Ag(111) *4d*-band and Au(111) *5d*-band, *black squares* and *green rounds*, respectively

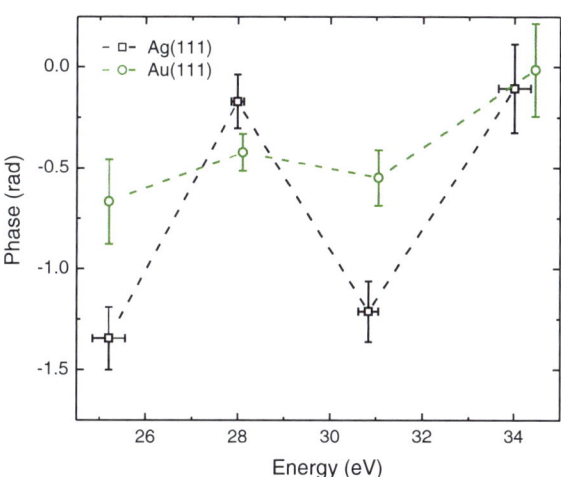

3 Conclusion and Outlook

In conclusion, we presented a new experimental approach that extends the RAB-BITT technique to condensed matter and accesses the surface specific response with the simultaneous acquisition of two traces. The experimental results revealed that a simple model dominated by transport is not suitable to describe the photoemission process in noble metals like Au(111) and Ag(111). It is worth to point out that the new detection scheme disposes of the requirement for a reference state in the second target and makes timing information for a single starting state accessible.

Furthermore, a different scheme for the harmonic generation could allow a better energy resolution of the measurement and yield access to photoemission dynamics at other photon energies.

References

1. Krausz F., Ivanov M., Rev. Mod. Phys. **81**, 163 (2009).
2. Itatani J. et al., Phys. Rev. Lett. **88**, 173903 (2002).
3. Paul P. M., Toma E. S., Breger B., Mullot G., Augé F., Balcou Ph., Muller H. G., Agostini P., Science **292**, 1689 (2001).
4. Schultze M. et al., Science **328**, 1658 (2010).
5. Klünder K. et al., Phys. Rev. Lett. **106**, 143002 (2011).
6. Cavalieri A. L. et al., Nature **449**, 1029 (2007).
7. Neppl S. et al, Phys. Rev. Lett. **109**, 087401 (2012).
8. Locher R. et al, Rev. of Sci. Instrum. **85** 013113 (2014).

Visualization of Ultrafast Electron Dynamics Using Time-Resolved Photoemission Electron Microscopy

K. Fukumoto, Y. Yamada, T. Matsuki, K. Onda, T. Noguchi,
R. Mizokuchi, S. Oda and S. Koshihara

Abstract We constructed a TR-PEEM which can directly image the photo-generated electron dynamics in semiconductor on nm and fs scales. Carrier transport properties relating to device performance, carrier lifetime, drift velocity and mobility, are investigated.

1 Introduction

The transport properties of charge carriers govern various characteristics of semiconductor devices, and thus, intensive studies have been devoted for the development of tools and techniques for probing of carrier dynamics. Recent developments of nanoscale semiconductor fabrication technology have made it possible to achieve quantum confinement of carrier movement. These achievements in semiconductor physics have also elevated the temporal and spatial resolution requirements of systems for the direct observation of carrier dynamics. Only a few methods, such as TR-PEEM [1, 2] and also TR-SEM, TR-STM can examine ultrafast carrier dynamics with high spatial resolution. Until today, however, there no reports investigating the electron dynamics on fs and nm scales in semiconductor, because it has been thought to be difficult to do it due to space charging and surface charging effects.

Here we introduce a three-dimensional method for the direct observation of electron motion in semiconductors, which is conducted in two spatial dimensions

K. Fukumoto (✉) · Y. Yamada · T. Matsuki · K. Onda · T. Noguchi · R. Mizokuchi · S. Oda · S. Koshihara
Tokyo Institute of Technology, 2-12-1 Ookayama, Meguro-Ku, Tokyo 152-8551, Japan
e-mail: fukumoto.k.ab@m.titech.ac.jp

K. Fukumoto · S. Koshihara
JST-CREST, 4-1-8 Honcho, Kawaguchi, Saitama 332-0012, Japan

K. Onda
JST-PRESTO, 4-1-8 Honcho, Kawaguchi, Saitama 332-0012, Japan

© Springer International Publishing Switzerland 2015
K. Yamanouchi et al. (eds.), *Ultrafast Phenomena XIX*,
Springer Proceedings in Physics 162, DOI 10.1007/978-3-319-13242-6_82

337

and the time domain by utilizing time-resolved photoelectron emission microscopy (TR-PEEM) conducted with femtosecond laser pulses. By utilizing a repetition-rate-variable fs laser system and carefully optimizing laser parameters, lower the photon density per pulse to avoid charging and higher the repetition rate to gain more signal, we could observe photo-excited carrier dynamics on semiconductor surfaces in ultrafast timescale.

2 Experimental

Figure 1 provides experimental scheme for dynamical carrier measurements on a semiconductor surface. The pump pulses of 2.4 eV excite electrons into the conduction band, and the probe pulses of 4.8 eV is used to project the spatial distribution of photo-generated electrons density to the PEEM screen where photoemission intensity reflects the carrier density with the spatial resolution of 40 nm. By controlling the temporal delay between the two pulses, the electron dynamics in the conduction band could be visualized. The temporal resolution is defined by the cross-correlation of the two pulses (230 fs).

3 Results

Using the constructed TR-PEEM, we observed the carrier relaxation and/or recombination on the timescale of sub-ps, and also observed the lateral motion of photo-generated electron bunch driven by the electric field gradient to estimate the drift velocity and the mobility [3].

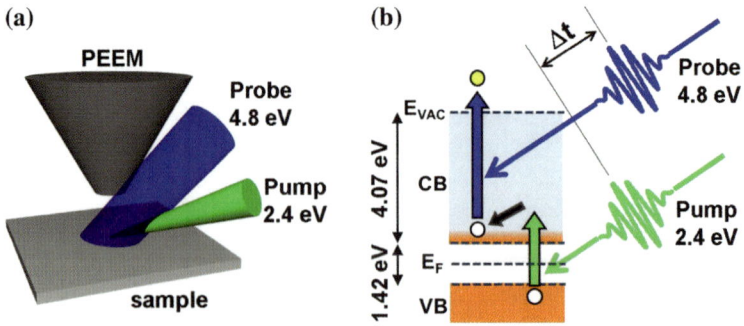

Fig. 1 Pump-probe setup for the observation of photogenerated carrier dynamics on GaAs

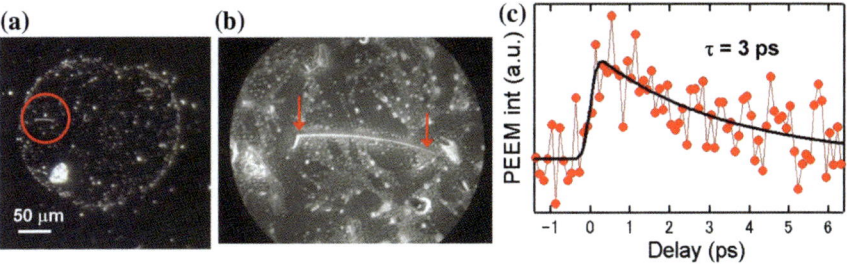

Fig. 2 **a** Optical microscopy image of Si nano-wire. **b** Enlarged view of the Si nano-wire using PEEM. **c** Carrier recombination in the Si nano-wire

3.1 Carrier Lifetime in a Si Nanowire

Here in Fig. 2 we show a demonstration of utilizing both high spatial and temporal resolutions as observing carrier lifetime in a single Si nano-wire (SiNW). Figure 2a is an optical microscopy image of SiNWs. One of them is highlighted by a red circle and an enlarged view is shown in Fig. 2b as a PEEM image. In Fig. 2c, integrated PEEM intensity along NW is plotted against the delay time. Numerical fitting gives a fast annihilation of the carrier density in the wire in an exponential manner.

3.2 Electron Drift Velocity and Mobility on a GaAs Surface [3]

This equipment can be also used to estimate electron drift velocity and mobility by directly imaging the motion of photo-generated electron bunch on a GaAs surface. Two metal electrodes are evaporated on to the surface and the electric field gradient (1.8 kV/cm) is applied to drive the electrons which are excited between the electrodes (Fig. 3a). A PEEM image at 0 ps is shown in Fig. 3b, in which the two electrodes are appeared at the top and bottom of the image, and the bright elliptical region at around the middle indicate the distribution of photo-generated electrons. Plots in Fig. 3c are the intensity profiles along the vertical direction in PEEM images obtained at different pump-probe delay times of 20, 40, 100, and 132 ps after the pumping. The lateral motion of the electron bunch to upwards is recognized by numerical fittings with a Gauss function to each plot. The peak position is plotted against delay time in Fig. 3d, and the slope by linear fit gives the drift velocity of 5.7×10^6 cm/s. By considering the mount of field gradient, 3,200 cm^2 V^{-1} s^{-1} of the mobility has been estimated, which is relatively smaller than the bulk value of 8,000 cm^2 V^{-1} s^{-1}. This slightly smaller value obtained here can be due to surface roughness, which work as carrier trapping center to lower the drift velocity.

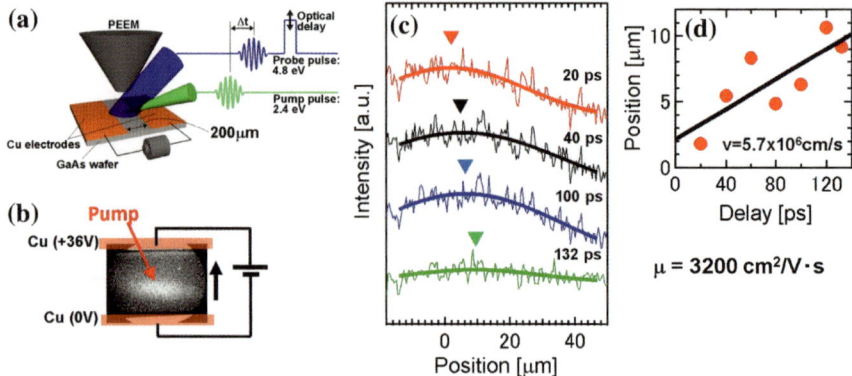

Fig. 3 **a** and **b** Experimental setup for the carrier motion imaging. **c** Intensity profiles along the vertical direction in TR-PEEM images obtained with different time delay. Fitting with Gauss function indicates the position of electron bunch. **d** Electron bunch versus delay

In summary, our newly developed TR-PEEM equipment using a repetition-rate-variable fs laser system can be used to evaluate some of the most important optical properties in semiconductor on nm and fs scales.

References

1. M. Aeschlimann, M. Bauer, D. Bayer, T. Brixner, S. Cunovic, F. Dimler, A. Fischer, W. Pfeiffer, M. Rohmer, C. Schneider, F. Steeb, C. Strueber, and D. Voronine, "Spatiotemporal control of nanooptical excitations" PNAS **107**, 5329-5333 (2010).
2. A. Kubo, K. Onda, H. Petek, Z. Sun, Y. S. Jung, and H. K. Kim, "Femtosecond Imaging of Surface Plasmon Dynamics in a Nanostructured Silver Film" Nanoletters **5**, 1123-1127 (2005).
3. K. Fukumoto, Y. Yamada, K. Onda, and S. Koshihara, "Direct imaging of electron recombination and transport on a semiconductor surface by femtosecond time-resolved photoemission electron microscopy" Appl. Phys. Lett. **104**, 53117-53121 (2014).

A New Regime of Nanoscale Thermal Transport: Collective Diffusion Counteracts Dissipation Inefficiency

Kathleen Hoogeboom-Pot, Jorge N. Hernandez-Charpak, Erik Anderson, Xiaokun Gu, Ronggui Yang, Henry Kapteyn, Margaret Murnane and Damiano Nardi

Abstract We uncover a new regime of nanoscale thermal transport that dominates when the separation between heat sources is small compared with the substrate's dominant phonon mean free paths. Surprisingly, the interplay between neighboring heat sources can facilitate efficient, diffusive-like heat dissipation.

1 Introduction

Understanding thermal transport from nanoscale heat sources is important for a fundamental description of energy flow in materials, as well as for thermal management in many technological applications including nanoelectronics, thermoelectric devices, nano-enhanced photovoltaics and nanoparticle-mediated thermal therapies. Recent work has shown the rate of heat dissipation from a heat source is reduced significantly below that predicted by Fourier's law for diffusive heat transport when the characteristic dimension of the source is smaller than the mean free path (MFP) of the dominant heat carriers (phonons in dielectric/semiconductor materials) [1, 2]. However, a complete fundamental description of nanoscale thermal transport is still elusive and current theory is limited by a lack of experimental validation.

Diffusive heat transport requires many collisions between heat carriers to establish a local thermal equilibrium and a continuous temperature gradient along which energy dissipates. However, when the dimension of a heat source is smaller

K. Hoogeboom-Pot · J.N. Hernandez-Charpak · H. Kapteyn · M. Murnane · D. Nardi (✉)
JILA and Department of Physics, University of Colorado, Boulder, CO 80309, USA
e-mail: damiano.nardi@jila.colorado.edu

E. Anderson
Center for X-Ray Optics, Lawrence Berkeley National Laboratory,
Berkeley, CA 94720, USA

X. Gu · R. Yang
Department of Mechanical Engineering, University of Colorado, Boulder, CO 80309, USA

© Springer International Publishing Switzerland 2015
K. Yamanouchi et al. (eds.), *Ultrafast Phenomena XIX*,
Springer Proceedings in Physics 162, DOI 10.1007/978-3-319-13242-6_83

than the phonon MFP, the diffusion equation is intrinsically invalid as phonons move ballistically without collisions, and the rate of nanoscale heat dissipation is significantly lower than the diffusive prediction. Furthermore, heat-carrying phonons in real materials have a wide distribution of MFPs, from several nanometers to hundreds of microns. For a given heat source size, phonons with MFPs shorter than the hot spot dimension remain fully diffusive and contribute to efficient heat dissipation and a high thermal conductivity (or equivalently, a low thermal resistivity). In contrast, phonons with long MFPs travel far from the heat source before scattering, with an effective thermal resistivity far larger than the diffusive prediction. Phonons with intermediate MFPs fall in between: heat transport is quasi-ballistic with varying degrees of reduced contributions to the conduction of heat away from the nanoscale source.

Most work to date explored the reduction in heat transport from functionally isolated micro- and nanoscale heat sources [1, 2]. Indeed, characterizing heat transport from nanostructures with varying size can be used to experimentally measure cumulative phonon MFP spectra of materials, with the proof-of-principle demonstrated for long-MFP (>1 μm) phonons in silicon [2].

2 A New Regime of Nanoscale Thermal Transport

In this work, we use tabletop extreme ultraviolet (EUV) high harmonic beams to investigate the different regimes of nanoscale thermal transport—purely diffusive and quasi-ballistic—and surprisingly uncover a new *collectively-diffusive* regime that occurs when the *separation* between nanoscale heat sources is smaller than the average phonon MFP. Quasi-ballistic transport dominates when the size of isolated nanoscale heat sources is smaller than dominant phonon MFPs as the long-MFP contributions to heat dissipation are suppressed relative to diffusive predictions. In the new collectively-diffusive regime, the separation between heat sources is small enough that long-MFP phonons can interact with phonons originating from a neighboring heat source as they would if both originated from the same, larger heat source, reintroducing their diffusive-like contributions to thermal transport. This can counteract the reduction in nanoscale heat dissipation to such an extent that heat transport recovers toward the diffusive limit, as shown in Fig. 1. In the limiting case, the spacing between heat sources vanishes and this regime approaches heat dissipation from a uniformly heated layer.

This work has two important implications. First, both size *and spacing* of heat sources are important for determining nanoscale heat dissipation, offering new ideas to mitigate scaling problems for thermal management in nanoelectronics [3]. Second, this new transport regime contains clear signatures of a material's phonon MFP spectrum, enabling detailed characterization of MFP-dependent thermal conductivity.

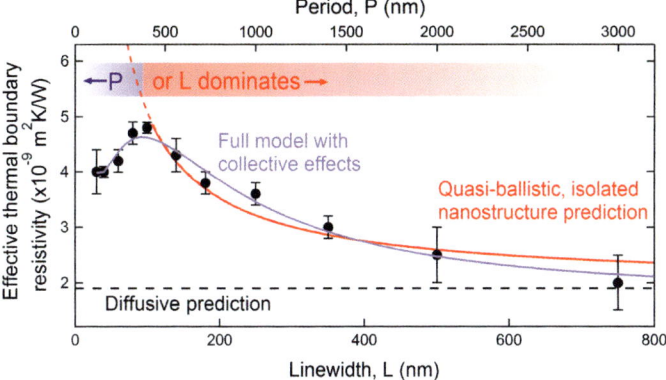

Fig. 1 Effective thermal boundary resistivity extracted from dynamic EUV diffraction. For each heat source linewidth L, on sapphire substrate, the resistivity increases with decreasing linewidths due to the suppression of the contribution to thermal conductivity of phonon modes with MFP larger than L. Decreasing the period P = 4L can reactivate modes with MFP larger than P, returning the effective resistivity towards the diffusive limit (*black dashed line*)

3 The Experiment

Arrays of nickel nanowires were fabricated by e-beam lithography to form periodic gratings on the surface of sapphire substrates. The linewidths L range from 750 to 30 nm, with period P = 4L and a rectangular profile height of ≈13.5 nm. The nanowires are heated by a 25 fs pump pulse centered at a wavelength of 800 nm. Laser excitation creates an array of nanoscale hot spots (lines) on the surface of a cold, transparent substrate. All nanowires are fabricated on the same substrate at the same time, for a constant intrinsic thermal boundary resistivity across all samples: any variation in heat dissipation efficiency as the hot spot size or spacing is varied can thus be attributed to different regimes of thermal transport. The thermal expansion and subsequent cooling of the gratings is probed using coherent EUV light centered at a wavelength of 29 nm, created by high harmonic up-conversion of an 800 nm Ti:Sapphire laser [4]. As EUV light diffracts from the periodic grating, the thermal expansion and relaxation of the nanowires can be extracted from the changes in the diffraction efficiency [5].

4 Characterizing Phonon Transport in Materials

We use this new phenomenon to extract the contribution to thermal transport from specific regions of the phonon MFP spectrum, opening up a new approach for thermal transport metrology and mean free path spectroscopy. This is because by varying both nanostructure size and separation, an effective phonon filter is introduced that suppresses specific MFP contributions to thermal conductivity, resulting in the trends pictured in Fig. 1. We compare our extracted phonon MFP spectra with predictions from first-principles calculations and find excellent agreement between experiment and theory.

This unique new capability for characterizing phonon transport in materials will enable for the first time the experimental characterization of MFP-dependent phonon thermal conductivity spectra down to MFP $\approx$ 10 nm for more complex nanostructured or metamaterials, where theoretical predictions are not yet possible.

Acknowledgments The authors gratefully acknowledge support from the US Department of Energy BES AMOS, the SRC, the National Science Foundation and NSSEFF.

References

1. M. Siemens *et al.* "Quasi-ballistic thermal transport from nanoscale interfaces observed using ultrafast coherent soft X-ray beams," Nature Mater. **9**, 26-30 (2010).
2. A. Minnich *et al.* "Thermal conductivity spectroscopy technique to measure phonon mean free paths," Phys. Rev. Lett. **107**, 095901 (2011).
3. S. King *et al.* "The three M's (materials, metrology, and modeling) together pave the path to future nanoelectronic technologies," Appl. Phys. Lett. Mater. **1**, 040701 (2013).
4. A. Rundquist *et al.* "Phase-matched generation of coherent soft X-rays," Science **280**, 1412 (1998).
5. D. Nardi *et al.* "Probing limits of acoustic nanometrology using coherent extreme ultraviolet light," Proc. SPIE **8681**, 86810N (2013).

Laser-Induced Plasma Dynamics Imaged by Femtosecond In-Line Holography

N. Rothe, C. Merschjann, C. Schuster, T. Fennel and S. Lochbrunner

Abstract The microplasma evolution in 30 nm thick Au-foils driven by 800 nm pump pulses is imaged via in-line holography using delayed 400 nm probe pulses. Time-resolved optical properties are extracted via numerical inversion of scattering images.

OCIS codes $300.0300 \cdot 300.6230 \cdot 300.6450$

Studying the dynamics of laser-induced solid-density plasmas is of key interest for understanding the response of condensed matter targets to intense laser radiation, e.g. for optimizing laser machining. Furthermore, corresponding experiments open a route to investigate the properties of warm dense matter. Here we describe a technique to analyze the spatio-temporal evolution of laser plasmas in thin metallic foils with high resolution by combining ultrafast pump-probe techniques with two-dimensional diffractive imaging. From the recorded diffraction pattern a lateral 2D-map of the complex transmittance is obtained by inverting the holographic phase problem. From the temporal evolution of the resulting 2D-optical parameter maps details of the ionization, heating and ablation dynamics realized in the microplasma will be extracted.

1 Experimental Setup and Numerical Approach

A dense laser plasma is generated by exciting a 30 nm thick gold foil with tightly focused pulses at 800 nm ($\tau_{pump} = 50\,\text{fs}$). The plasma evolution is probed by delayed 400 nm pulses in transmission and the resulting diffraction pattern of the probe beam is recorded by a CMOS camera. By compressing the probe pulses, a time resolution of less than 100 fs is realistic. The experimental setup is similar to

N. Rothe · C. Merschjann · C. Schuster · T. Fennel · S. Lochbrunner (✉)
Institute of Physics, University of Rostock, Universitätsplatz 3, 18055 Rostock, Germany
e-mail: stefan.lochbrunner@uni-rostock.de

© Springer International Publishing Switzerland 2015
K. Yamanouchi et al. (eds.), *Ultrafast Phenomena XIX*,
Springer Proceedings in Physics 162, DOI 10.1007/978-3-319-13242-6_84

those used by Widmann et al. and Liu et al. [1, 2]. The evolution of the plasma is investigated in the time range from 0 to 2000 ps. The recorded diffraction pattern can be mathematically treated as a superposition of reference and object light field, and the inverse phase problem can be solved using an iterative algorithm similar to that described in [3]. Here, we choose a Gaussian beam reflecting the undisturbed probe beam as reference, while the object field describes only the changes induced by the plasma object. Assumption of a support with known complex transmittance (Au-foil) facilitates an efficient and reliable fit result. The spatially resolved complex transmittance of the plasma object is deduced after sufficient convergence of the iterative algorithm [3].

2 Results and Discussion

The experimentally observed diffraction patterns exhibit structure changes between 0 and 2000 ps. Figure 1a shows a typical diffraction pattern at the maximum measured delay. Note that the reference intensity (Gaussian beam profile) has been subtracted in order to emphasize the diffraction structure. Changes in these structures indicate a temporal evolution of the plasma over time. The radial diffraction intensity as a function of time is shown in Fig. 1b. The inserted horizontal lines mark transitions between four distinguishable temporal regimes. The first regime resembling the shortest delay times shows a diffraction pattern which exhibits almost no changes, corresponding to the quasi-steady-state described in [1]. During the first picoseconds after the excitation only the electrons are heated while the lattice has still low energy and the plasma is too weak to produce any visible effects. Until 100 ps mainly the intensity of the zero-order maximum decreases and higher-order maxima appear. Here, the energy transfer to the lattice takes place until an equilibrium is reached [4]. After 100 ps up to the limit of measured delays

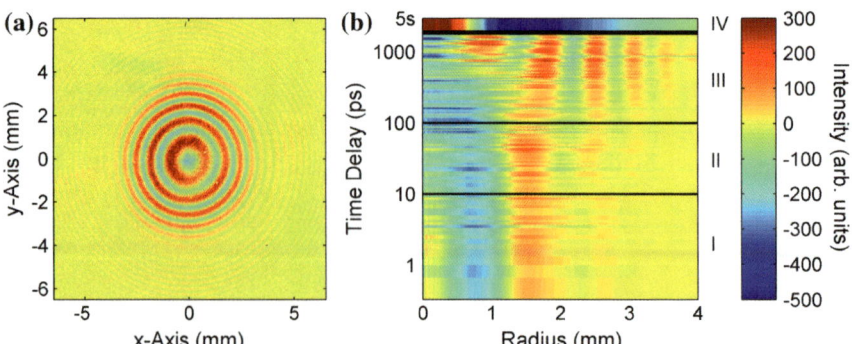

Fig. 1 a Pump-probe hologram after subtraction of the reference intensity (Gaussian beam) at a delay of 2000 ps; **b** Intensity-plot depending on the time delay between pump and probe pulse plotted against the diffraction pattern radius

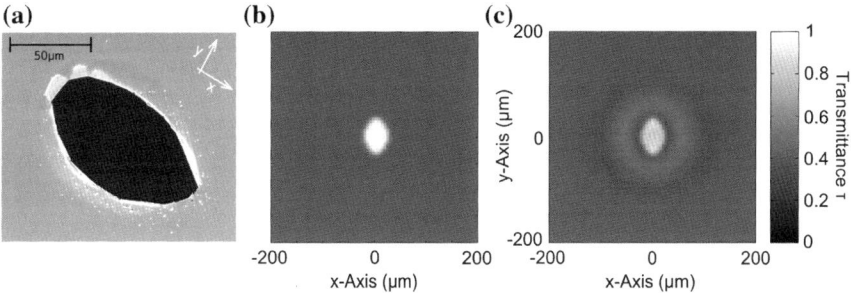

Fig. 2 a SEM picture of Au-foil after excitation; **b** Numerical test object based on the SEM; **c** Reconstruction of the simulated test object (hole in Au-foil)

substantial shifts and bifurcations are observed. The plasma plume is forming and expanding, while the ions and atoms escape out of the foil and start the ablation process [4]. The uppermost slide in Fig. 1b shows the steady-state situation several seconds after the pump pulse. It is evident from the differences in the pattern structure hat the ablation process is not finished within the observation window of 2000 ps.

The hole generated by the laser interaction, which is the object causing the steady-state hologram, is shown in the scanning electron microscope (SEM) picture of Fig. 2a. We have chosen these elliptical holes with half axes $r_y = 36$ μm, $r_y = 23$ μm as test objects for our numerical reconstruction algorithm (Fig. 2b). The complex transmittance of the support was calculated using the complex refractive index of Au ($\hat{n} = 1.658 + 1.956\,i$) [5] and a thickness of 30 nm. The object reconstructed from the hologram after 500 iterations is shown in Fig. 2c. Obviously the real part of the transmittance is reproduced in a convincing way, as compared to the original object. The ring-like structures in the reconstructed object are due to the twin-image unavoidable in in-line holography.

Acknowledgments The authors thank C. Peltz and L. Seiffert for technical support.

References

1. K. Widmann, T. Ao, M. E. Foord, D. F. Price, A. D. Ellis, P. T. Springer, and A. Ng, Phys. Rev. Lett. **92**, 125,002 (2004)
2. J. Liu, Z. Duan, Z. Zeng, X. Xie, Y. Deng, R. Li, Z. Xu, and S. L. Chin, Phys. Rev. E **72**, 026,412 (2005)
3. T. Latychevskaia and H.-W. Fink, Phys. Rev. Lett. **98**, 233,901 (2007)
4. X. Zhao and Y. C. Shin, Journal of Physics D: Applied Physics **45**, 105,201 (2012)
5. E. D. Palik, Handbook of Optical Constants of Solids (Academic Press, 1998)

Resonant Optical Kerr Response with Ultrashort Decay Time by Nonlocal Wave Coupling of Light and Excitons

Masayoshi Ichimiya, Takayuki Umakoshi, Hiroyuki Murata, Takashi Kinoshita, Hajime Ishihara and Masaaki Ashida

Abstract Resonant optical Kerr effects have been investigated in high-quality CuCl thin films. The peculiar spectral feature and ultrafast response below 200 fs due to a long-range coherent coupling between light and multinode-type excitons are observed.

An ultrafast optical response in semiconductor materials as well as a large non-linearity is essential to improve the performance of optical switching devices, while it has been considered to be difficult due to the long radiative decay time of excitons. In nanostructures, the optical responses are described by the long wave-length approximation (LWA), where the oscillator strength increases with the system size [1–3], while the increase has been believed to be saturated in the size region larger than the excitonic coherent length or the wavelength of light. On the other hand, the size-resonantly enhanced nonlinear optical response in a system with high crystalline quality is theoretically proposed, where the self-consistent interaction between the internal field and the induced polarization causes an enhancement of the response field at a particular size and photon energy [4].

Recently, we reported experiments of degenerate four-wave mixing (DFWM) in high-quality CuCl thin films [5]. We observed a strong coupling between light and a multinode-type exciton combined with the ultrafast radiative decay of 100 fs. The calculated result of the radiative corrections predicts large radiative width, corresponding to exceptionally short response time for excitons (below 10 fs in CuCl) [6]. Such a large radiative correction is caused by the long-range coupling between light and excitons over several wavelengths and is expected to appear in other

M. Ichimiya (✉)
Department of Electronic Systems Engineering, The University of Shiga Prefecture, Shiga 522-8533, Japan
e-mail: ichimiya.m@e.usp.ac.jp

M. Ichimiya · T. Umakoshi · H. Murata · M. Ashida
Department of Materials Engineering Science, Osaka University, Osaka 560-8531, Japan

T. Kinoshita · H. Ishihara
Department of Physics and Electronics, Osaka Prefecture University, Osaka 599-8531, Japan

© Springer International Publishing Switzerland 2015
K. Yamanouchi et al. (eds.), *Ultrafast Phenomena XIX*,
Springer Proceedings in Physics 162, DOI 10.1007/978-3-319-13242-6_85

optical phenomena. The optical Kerr effect is one of nonlinear optical effects and suitable for demonstrating the ultrafast switching action. In particular, the nonlinearity of optical Kerr effect is much enhanced by using light in the exciton resonance region. In the present work, we investigate the resonant optical Kerr effect of confined excitons using CuCl thin films with high crystalline quality and with thicknesses in the non-LWA regime.

1 Experimental Procedures

CuCl thin films were grown by our newly developed technique involving electron beam irradiation before the molecular beam epitaxy growth [7]. Grown films were mounted in a helium flow cryostat and cooled to below 10 K. Resonant optical Kerr spectra were measured with the second harmonic of a mode-locked Ti:sapphire laser with a pulse duration of 110 fs. The photon energy of the light was set around the exciton energy in CuCl, and the spectral width of 20 meV covered the whole exciton resonance region. Both pump and probe beams were linearly polarized. The polarization of the pump beam was set at 45° to that of the probe beam set to vertical. The horizontal component of the transmitted probe beam was detected as the Kerr signal by a monochromator equipped with a charge-coupled device. The spectral resolution was 0.1 meV. The film thickness and the phase decay constant of excitons (Γ) at the focused spot were derived by fitting to the reflection spectrum measured with the same geometry.

2 Results and Discussion

The blue line shown in Fig. 1a indicates a resonant optical Kerr spectrum of a high-quality CuCl thin film with a thickness of 224 nm. E_T and E_L show the transverse and longitudinal exciton energies of CuCl bulk crystal, respectively. The spectrum exhibits a peculiar structure with several peaks, which have never been observed in any Kerr spectrum. Figure 1b shows the film thickness dependence of the eigenenergies including the radiative shift in the coupled system of light and multinode-type excitons [6]. The energy at the intersection of the calculated curve and a horizontal dashed line corresponds to the eigenenergy at the film thickness of 224 nm. The eigenenergies precisely coincide with the energies of the peaks in the resonant optical Kerr spectrum. The DFWM spectrum measured using the same geometry as that of resonant optical Kerr effect is shown as red line in Fig. 1a for the comparison. The spectral feature with several peaks is in good agreement with the feature of Kerr spectrum. The photon energy at each peak corresponds to the eigenenergy including the radiative shift due to long-range coupling between light and exciton [5]. A similar agreement between the Kerr spectrum and the DFWM spectrum is observed for CuCl thin films with other thicknesses. These results

Fig. 1 **a** Observed Kerr and DFWM spectra in a CuCl thin film. **b** Film thickness dependences of calculated eigenenergies including the radiative shift and uncoupled excitonic modes [6]

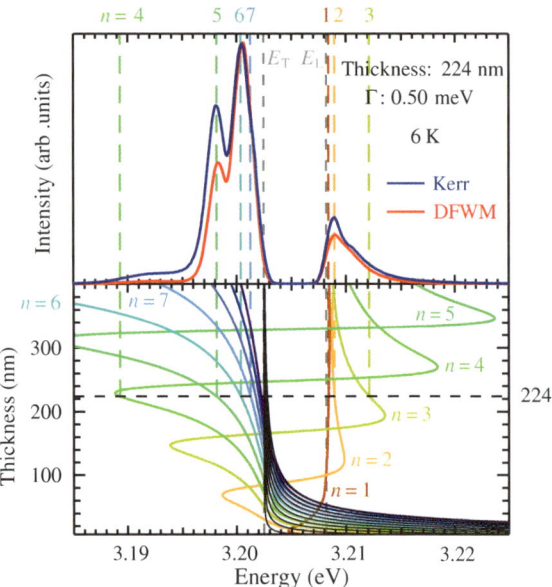

Fig. 2 Observed delay time dependence of the Kerr intensity in a CuCl thin film

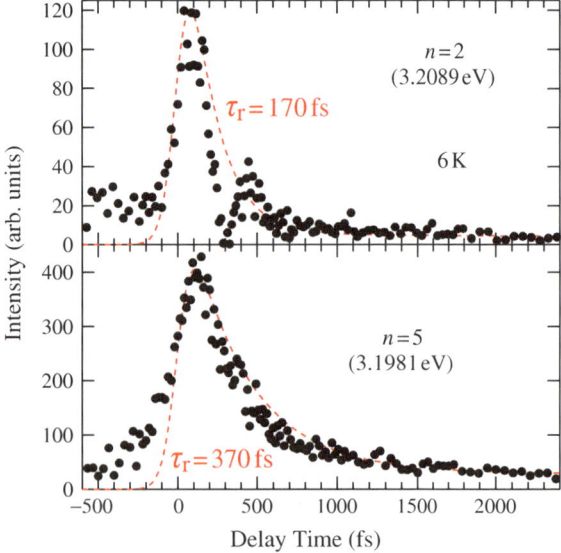

suggest that the shape of resonant optical Kerr spectrum closely reflects the radiative corrections similar to that of DFWM, where 100 fs-class response and signal at room temperature are observed [8]. Figure 2 shows the delay time dependence of the probe beam for the excitonic states of $n = 2, 5$. The dashed red line represents the exponential functions convoluted by the excitation laser profile, where the time constants of the calculated radiative decay time ($\tau_r = \hbar/2\gamma$) derived from the radiative width (γ) are substituted [8]. At the film thickness of 224 nm, γ of $n = 2, 5$ are 1.9 and 0.90 meV [6], and τ_r are 170 and 370 fs, respectively. The measured resonant optical Kerr response reflects the calculated radiative decay profile of exciton. These agreements between experimental results and the calculated mode structures suggest that the resonant optical Kerr response below 10 fs will be achieved for a CuCl thin film with high crystalline quality and appropriate film thickness.

Acknowledgments   The present work was supported by a Grant-in-Aid for Young Scientists (B) (24740210) and a Grant-in-Aid for Scientific Research (A) (60273611) from the Ministry of Education, Culture, Sports, Science and Technology of Japan.

References

1. E. Hanamura, Phys. Rev. B **37**, 1273 (1988).
2. A. Nakamura, H. Yamada, and T. Tokizaki, Phys. Rev. B **40**, 8585 (1989).
3. T. Itoh, M. Furumiya, T. Ikehara, and C. Gourdon, Solid State Commun. **73**, 271 (1990).
4. H. Ishihara, T. Amakata, and K. Cho, Phys. Rev. B **65**, 035305 (2001).
5. M. Ichimiya, M. Ashida, H. Yasuda, H. Ishihara, and T. Itoh, Phys. Rev. Lett. **103**, 257401 (2009).
6. H. Ishihara, J. Kishimoto, and K. Sugihara, J. Lumin. **108**, 343 (2004).
7. M. Ichimiya, L. Q. Phuong, M. Ashida, and T. Itoh, J. Cryst. Growth **378**, 372 (2013).
8. M. Ichimiya, K. Mochizuki, M. Ashida, H. Yasuda, H. Ishihara, and T. Itoh, Phys. Status Solidi B **248**, 456 (2011).

Single-Shot Real-Time Observation of Ultrafast Amorphization in $Ge_2Sb_2Te_5$ Thin Film

W. Oba, I. Katayama, Y. Minami, T. Saiki and J. Takeda

Abstract Ultrafast dynamics of photo-induced amorphization in $Ge_2Sb_2Te_5$ thin film has been studied using broadband single-shot real-time pump-probe imaging spectroscopy. We successfully observed the transient absorption changes accompanied with the ultrafast amorphization with a single-shot detection.

1 Introduction

$Ge_2Sb_2Te_5$ (GST) is one of the rapid phase-change materials, which has been applied to rewritable optical media such as DVD-RAM. The phase-change from the amorphous to the crystalline phase is induced by an irradiation of a cw-laser through transient temperature ramping [1]. The phase-change from the crystalline to the amorphous phase, on the other hand, is induced by an irradiation of a single femtosecond laser pulse through a non-thermal process [2]. A model of the amorphization has been recently proposed [3, 4]; the relatively weaker and longer Ge-Te bond is broken due to an intense photoexcitation, and subsequently, the amorphization occurs due to displacement of the Ge atoms from an octahedral to a tetrahedral arrangement with a subpicosecond time scale (the phonon period).

In order to detect the photo-induced amorphization in GST, a measurement of transient absorption changes by using pump-probe spectroscopy is required. However, the amorphization is irreversible, and therefore, it is difficult to reveal the details of the dynamics with the conventional pump-probe spectroscopy that

W. Oba · I. Katayama · Y. Minami · J. Takeda (✉)
Graduate School of Engineering, Yokohama National University,
Yokohama 240-8501, Japan
e-mail: jun@ynu.ac.jp

T. Saiki
Graduate School of Science and Technology, Keio University,
Yokohama 223-8522, Japan

K. Yamanouchi et al. (eds.), *Ultrafast Phenomena XIX*,
Springer Proceedings in Physics 162, DOI 10.1007/978-3-319-13242-6_86

requires many repetitions of pump-probe sequence. In the present work, we performed real-time measurements of the transient absorption changes in GST using single-shot real-time pump-probe imaging spectroscopy [5–7].

2 Experimental Method

A schematic configuration of the single-shot real-time pump-probe imaging spectroscopy is depicted in Fig. 1. The key aspect of this technique is the use of an echelon mirror, which is fabricated on a Ni block ($40 \times 40 \times 20$ mm^3) with 500 steps having 80 μm step-width and 10 μm step-height. As shown in Fig. 1, using the double four focusing configuration, the echelon mirror does not act as diffractive optics but generates spatially encoded delay times for the probe pulse. The probe pulse with spatially encoded delay times is focused onto a sample together with a pump pulse ($\sim 0.1 \times 15$ mm^2). After passing through the sample, the probe pulse is imaged onto an entrance slit of a spectrometer coupled with a two-dimensional (2D) charge coupled device (CCD) detector ($1{,}340 \times 1{,}300$ pixels). The vertical axis of the CCD detector corresponds to the delay time, whereas the horizontal axis corresponds to the detected wavelength. Using this method, we can obtain time-frequency 2D image of transient absorption changes on a single-shot basis with the signal-to-noise ratio of about 10.

The light source was a Ti:sapphire regenerative amplifier system with center wavelength of 800 nm and pulse duration of 100 fs. The probe pulse was focused into CaF$_2$ thin plate to generate a white-light continuum by self phase modulation. As a result, we could perform the single-shot mapping of the transient absorption changes in the range of 530–660 nm and the time window of 15 ps. The sample used in this study was a 10 nm thick GST film deposited on a glass substrate.

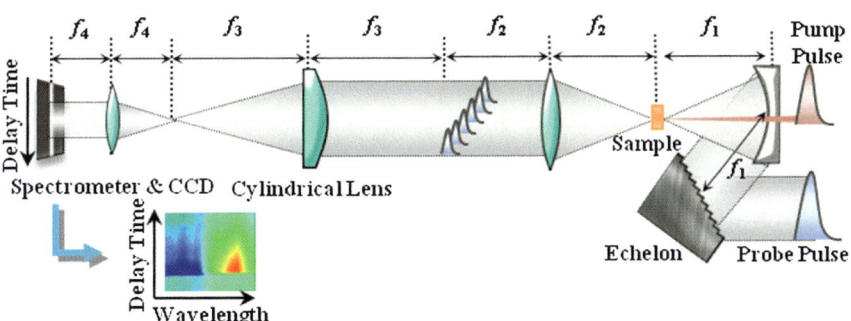

Fig. 1 Schematic configuration of single-shot real-time pump-probe imaging spectroscopy

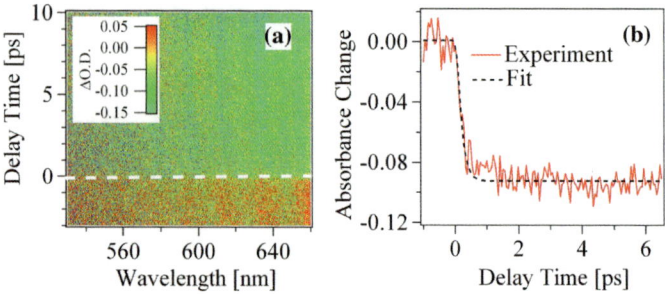

Fig. 2 **a** Time-frequency 2D image of the absorption change obtained in GST. **b** Time evolution of the absorbance change obtained by cutting off the 2D image at 650 nm. The *solid* and *dotted lines* are experimental data and the best fits to the data, respectively

3 Results and Discussion

Figure 2a shows the time-frequency 2D image of the absorption change in the crystalline GST film measured by a single-shot detection. Here, the excitation laser fluence was 26.3 mJ/cm^2 where the recording mark associated with amorphization is formed. As shown by a dotted line, the absorption rapidly decreases at the time origin and the change does not recover even after long delay time, suggesting that the irreversible phase-change to the amorphous phase takes place. Figure 2b shows the time evolution of the absorption change obtained by cutting off the 2D image at 650 nm. The fast rise time was evaluated by a fitting procedure convoluted by the system response function, and was estimated to be 150 fs as shown by a dotted curve. The estimated rise time of 150 fs is consistent with the half period of the A1 mode in the GeTe6 structure [8, 9]. This fact strongly indicates that the amorphization in GST is due to the rearrangement of Ge atoms from an octahedral structure to a tetrahedral structure.

We also observed the excitation power dependence of the absorption change, which shows saturation behavior without a critical threshold. On the other hand, the recording mark due to amorphization is engraved with a clear threshold laser fluence. This fact suggests that the stabilization of the amorphous phase might occur in the longer time scale. The details of the stability of the photo-induced amorphous phase is discussed elsewhere [10].

4 Summary

In conclusion, we successfully observed ultrafast amorphization in GST for the first time using single-shot real-time pump-probe imaging spectroscopy. The phase-change occurs within the inverse of the phonon frequency, suggesting that the amorphization is due to the rearrangement of Ge atoms.

Acknowledgments We would like to thank T. Shintani for providing the GST thin film sample.

References

1. N. Yamada *et al.*, J. Appl. Phys. **69**, 2849 (1991).
2. M. Konishi *et al.*, Appl. Opt. **49**, 3470 (2010).
3. A. V. Kolobov *et al.*, J. Phys. Condens. Mat. **19**, 455209 (2007).
4. X.-B. Li *et al.*, Phys. Rev. Lett. **107**, 015501 (2011).
5. I. Katayama, H. Sakaibara, and J. Takeda, Jpn. J. Appl. Phys. **50**, 102701 (2011).
6. H. Sakaibara, Y. Ikegaya, I. Katayama, and J. Takeda, Opt. Lett. **37**, 1118 (2012).
7. Y. Minami, H. Yamaki, I. Katayama, and J. Takeda, Appl. Phys. Express **7**, 022402 (2014).
8. J. Hernandez-Rueda *et al.*, Appl. Phys. Lett. **98**, 251906 (2011).
9. K. Makino, J. Tominaga, and M. Hase, Opt. Express **19**, 1260 (2011).
10. J. Takeda, W. Oba *et al.*, Appl. Phys. Lett. **104**, 261903 (2014).

Electrochemical Control of Coherent Phonon Generations in Single-Walled Metallic Carbon Nanotubes

Keisuke Maekawa, Kenji Sato, Yasuo Minami, Ikufumi Katayama,
Jun Takeda, Kazuhiro Yanagi and Masahiro Kitajima

Abstract Coherent phonons in single-walled metallic carbon nanotubes were measured under the application of a gate voltage through ionic liquid. We found that the frequencies, amplitudes and phases of the phonons strongly depend on the voltage.

1 Introduction

Carbon nanotubes (CNTs) consist of graphene sheets rolled up to form one-dimensional cylinder structure, which results in unusual electronic structures such as van-Hove singularities in the density of states. CNTs can either be semiconductors or metals, depending on the structure specified by the so-called chiral vector [1]. Metallic CNTs are of particular interest because they exhibit strong electron-phonon coupling due to Kohn anomaly, and thus the phonon properties shows drastic dependence on the Fermi energy and the non-equilibrium electron distribution [2–4]. However, the Fermi energy dependence on the dynamics of carriers and electron-phonon coupling has not been reported until now, partly because the control of the Fermi energy requires micro-sized devices such as FETs that are

K. Maekawa · K. Sato · Y. Minami · I. Katayama (✉) · J. Takeda · M. Kitajima
Department of Physics, Graduate School of Engineering,
Yokohama National University, Yokohama 240-8501, Japan
e-mail: katayama@ynu.ac.jp

K. Yanagi
Department of Physics, Tokyo Metropolitan University,
Hachioji 192-0397, Japan

M. Kitajima
LxRay Co. Ltd., Nishinomiya 663-8172, Japan

M. Kitajima
Department of Applied Physics, National Defense Academy,
Yokosuka 239-8686, Japan

© Springer International Publishing Switzerland 2015
K. Yamanouchi et al. (eds.), *Ultrafast Phenomena XIX*,
Springer Proceedings in Physics 162, DOI 10.1007/978-3-319-13242-6_87

difficult to apply to the ultrafast measurement. Here, we use a more convenient approach to investigate the Fermi energy dependence on the ultrafast carrier and phonon dynamics in CNTs, in which we apply a voltage through an ionic liquid and CNT electrodes [5]. The observed coherent phonon signals depend strongly on the applied voltage, which are inaccessible with the conventional Raman spectroscopy.

2 Experiments

Metallic single-walled CNTs (SWCNTs) with average diameters of 1.4 nm were obtained from purification of initial SWCNTs produced by an arc-discharge (Arc-SO, Meijo nano-carbon Co.). The purifications were performed by using density-gradient ultracentrifuge method using deoxycholate sodium salt for a surfactant in a dispersion step to obtain high-purity metallic SWCNTs in a manner similar to that in [6]. The thin films were formed on glass substrates by using a method reported in [5]. From the optical absorption spectra, purity of metallic SWCNT is confirmed to be greater than 95 %. We used an ionic liquid (N, N, N-trimethyl-N–propylammonium bis(trifluoromethanesulfonyl) imide (TMPA-TFSI), Kanto-Kagaku Co.) for electrolyte solution, which is known to exhibit large electrochemical window, high transparency, low volatility, and high environmental stability. A 7.5-fs Ti-sapphire laser (center wavelength of 800 nm, repetition rate of 80 MHz, and output power of 200 mW) was used for pump-probe transmission measurements. The output of the laser was separated into pump and probe beams. The pump pulse is delayed by an optical shaker with a scanning range of 15 ps, and focused on the sample together with the probe pulse. The probe pulse transmitted through the sample was delivered to a photodiode for detecting the probe signal. To balance the current of the photodiode for the probe signal, the reference signal was also detected. The difference between the probe and reference signals was amplified and collected by an analogue-to-digital converter [7].

3 Results and Discussion

Figure 1 shows the coherent phonons in metallic SWCNTs with different gate voltages. The amplitudes and phases of the observed coherent phonons in SWCNTs depend strongly on the gate voltage. The Fourier transformed spectra obtained from the time-domain data show specific high-frequency phonons in SWCNTs up to 100 THz: radial breathing mode (RBM), D-mode, G-mode, 2D-mode, and 2G-mode. We also note that the frequencies of these peaks slightly change upon the gate voltage, which are consistent with those of the micro-Raman measurement [3].

The decrease of the coherent phonon amplitude at the high gate voltage can be understood as the change of Raman scattering cross section due to the Pauli blocking. When the electronic states at the van Hove singularity are occupied by

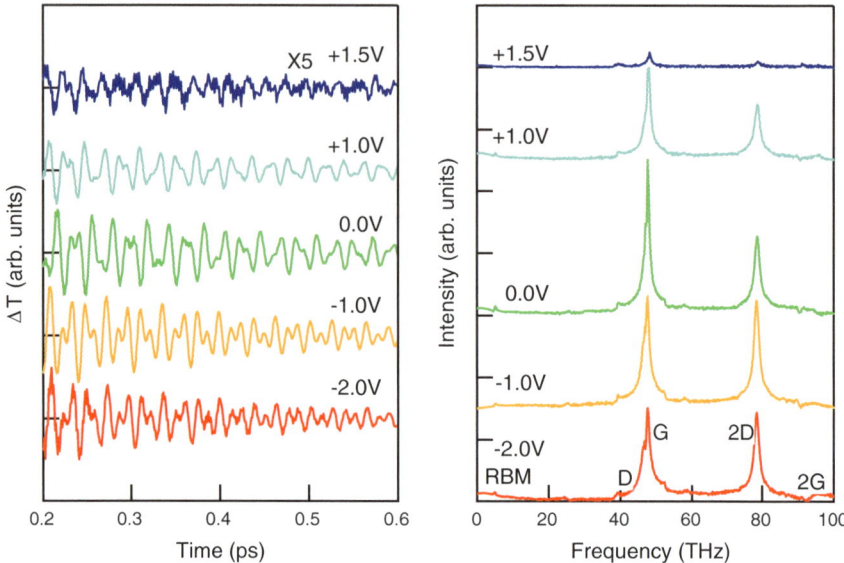

Fig. 1 Observed transmission change of high frequency coherent phonons in metallic SWCNTs at different gate voltages. The slow responses due to the transient absorptions of SWCNTs are subtracted by filtering out the low frequency components. The Fourier transformed spectra are shown in the *right*

applying the gate voltage, the resonant transition would be suppressed, and subsequently, gives the smaller coherent phonon signal. We also observed an effect of Kohn anomaly to the resonant frequencies and dephasing times of the G modes, in which the phonon softening and enhanced dephasing are observed at the Dirac point. Time-dependent frequency-shift of the G-mode was also observed by applying the windowed Fourier transformation on the time profiles. These results demonstrate that our method for measuring the Fermi energy dependence will open a new possibility to observe novel ultrafast phenomena in materials.

4 Conclusion

We observed the coherent phonons in metallic SWCNTs up to 100 THz under the application of a gate voltage through ionic liquid. The amplitudes, frequencies, and phases of the observed coherent phonons show strong dependence on the gate voltage, which can be explained by considering the Pauli blocking and Kohn anomaly. The results demonstrate that our method for applying the gate voltage is promising for investigating ultrafast dynamics of materials with different chemical potentials.

References

1. S. Reich, C. Thomsen, and J. Maultzsch, *Carbon Nanotubes: Basic Concepts and Physical Properties* (Wiley-VCH, Berlin, 2004).
2. J. Yan, Y. Zhang, P. Kim, and A. Pinczuk, "*Electric Field Effect Tuning of Electron-Phonon Coupling in Graphene*," Phys. Rev. Lett. **98**, 166802 (2007).
3. H. Farhat, H. Son, Ge. G. Samsonidze, S. Reich, M. S. Dresselhaus, and J. Kong, "*Phonon Softening in Indivisual Metallic Carbon Nanotubes due to the Kohn Anomaly*," Phys. Rev. Lett. **99**, 145506 (2007).
4. K. Ishioka, M. Hase, M. Kitajima, L. Wirtz, A. Rubio, and H. Petek, "*Ultrafast electron-phonon decoupling in graphite*," Phys. Rev. B **77**, 121402 (2008).
5. K. Yanagi, R. Moriya, Y. Yomogida, T. Takenobu, Y. Naitoh, T. Ishida, H. Kataura, K. Matsuda, and Y. Maniwa, "*Electrochromic Carbon Electrodes: Controllable Visible Color Changes in Metallic Single-Wall Carbon Nanotubes*," Adv. Mater. **23**, 2811 (2011).
6. S. Ghosh, S. M. Bachilo, and R. B. Weisman, "*Advanced Sorting of Single-walled Carbon Nanotubes by Nonlinear Density-gradient Ultracetrifugation*," Nat. Nanotech. **5**, 443 (2010).
7. I. Katayama, S. Koga, K. Shudo, J. Takeda, T. Shimada, A. Kubo, S. Hishita, D. Fujita, and Masahiro Kitajima, "*Ultrafast Dynamics of Surface-Enhanced Raman Scattering Due to Au Nanostructures*," Nano Lett. **11**, 2648 (2011).

Ultrafast Charge Photogeneration and Dynamics in Semiconducting Carbon Nanotubes

Giancarlo Soavi, Francesco Scotognella, Daniele Viola, Timo Hefner, Tobias Hertel, Guglielmo Lanzani and Giulio Cerullo

Abstract We show that charge-carriers are instantaneously photogenerated in semiconducting carbon nanotubes by identifying their spectral signature in transient absorption. We exploit carbon nanotubes as ideal systems for the study of charge-carriers dynamics in one dimension.

Single-walled carbon nanotubes (SWNTs) are excellent model systems for the study of photoexcitation dynamics in one-dimensional (1D) quantum confined systems. Theory predicts that Wannier-Mott excitons are the elementary photo-excitations in SWNTs, due to the strong Coulomb interaction caused by the weak screening [1]. Such excitons have peculiar 1D characteristics, such as extraordinary large binding energies, large size and 1D transport. Experimental observations, such as the measured binding energy, typically 0.1–1 eV [2], and the electron-hole correlation length, in the 1–10 nm domain [3] confirm theoretical predictions. The exciton model alone, however, fails to capture the whole dynamics following photoexcitation, and many other photoexcited species have crowded the complex scenario of SWNTs' optical response, ranging from triplets to bi-excitons and trions. Experiments based on photocurrent, transient absorption and THz spectroscopy also point out a non-negligible photogeneration of free charge-carriers. This is in contrast with the excitonic model and the reduced Sommerfeld factor that predicts that excitons should be the only species generated upon photoexcitation. Attempts to solve this discrepancy proposed non-linear processes, such as exciton-exciton annihilation,

G. Soavi (✉) · F. Scotognella · D. Viola · G. Cerullo
Dipartimento Di Fisica, Politecnico Di Milano, P.Zza Leonardo Da Vinci 32,
20133 Milan, Italy
e-mail: giancarlo.soavi@polimi.it

T. Hefner · T. Hertel
Institute for Physical and Theoretical Chemistry, Department of Chemistry and Pharmacy,
University of Wuerzburg, 97074 Würzburg, Germany

G. Lanzani
Center for Nano Science and Technology@PoliMi, Istituto Italiano Di Tecnologia,
Via Giovanni Pascoli, 70/3, 20133 Milan, Italy

© Springer International Publishing Switzerland 2015
K. Yamanouchi et al. (eds.), *Ultrafast Phenomena XIX*,
Springer Proceedings in Physics 162, DOI 10.1007/978-3-319-13242-6_88

360

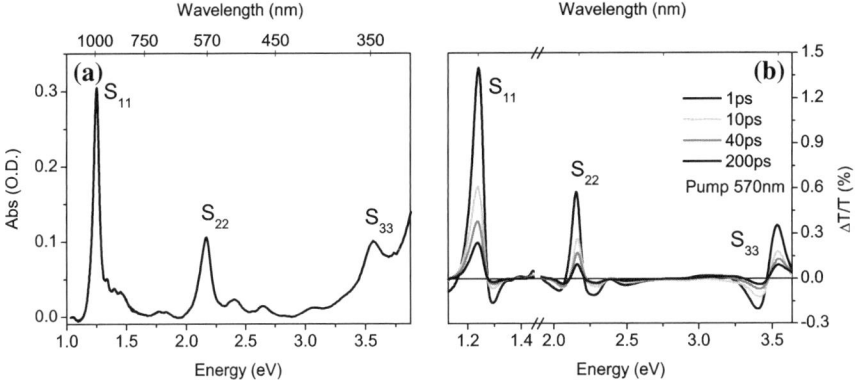

Fig. 1 **a** Ground state absorption spectrum of the (6,5) semiconducting carbon nanotubes used for the experiments. The sharp peaks are due to excitonic transitions, in particular S_{11} at approximately 1 μm, S_{22} at 570 nm and S_{33} at 350 nm. **b** $\Delta T/T$ spectra for the (6,5) SWNT sample at different pump-probe delays for 570 nm excitation wavelength

as a mechanism of charge-carrier photogeneration. However, there is solid experimental evidence that charge-carrier photogeneration is linear with the pump fluence.

We apply broadband ultrafast transient absorption spectroscopy to the semi-conducting (6,5) SWNT (Fig. 1a) and show that charge-carriers can be identified by their effect on excitonic resonances, in particular the large Stark shift that they induce on high-energy, easily polarizable excitons (S_{22} and S_{33}). Having identified the Stark shift as a good spectroscopic fingerprint for charge-carriers, we are able to study their dynamics in a nearly ideal 1D system. We find that a fraction of the absorbed photons generates geminate charge-carrier pairs within our temporal resolution ($\approx$ 50 fs), which then recombine on the picosecond timescale following the characteristic kinetic law of random walk in 1D.

Our analysis starts from the observation that the first derivative of the ground state absorption spectrum, i.e. a photoinduced red shift of the excitonic transition, can reproduce many of the features observed in the differential transmission ($\Delta T/T$) spectra of semiconducting SWNTs (Fig. 1b). This effect can be ascribed to different physical mechanisms, such as bi-excitons or trions formation, thermal effects or Stark effect. Here we unambiguously demonstrate that this derivative shape is indeed due to strong local electric field induced by photo-generated charge-carriers, which shifts the electronic transitions by Stark effect [4]. In particular, we show that the energy shift is stronger for excitons with lower binding energy, as for S_{33} with respect to S_{11}, as expected for the Stark effect (Fig. 1b). This suggests that the higher energy S_{33} exciton is ideally suited for directly probing charge-carriers in SWNTs and can be exploited to further analyze the charge photogeneration process and to study charge-carrier dynamics in one dimension. Our data show that the free charge-carriers are generated within 50 fs. The temporal evolution of the S_{33} pump-probe dynamics is very accurately reproduced by a power law $\sim t^{-0.5}$ (Fig. 2b).

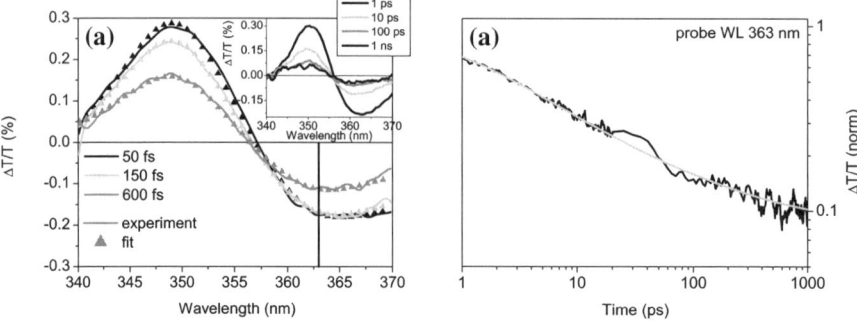

Fig. 2 **a** $\Delta T/T$ spectra at different pump-probe delays for the S_{33} exciton, following excitation at S_{11}, together with a fitting model based on the energy shift contribution from Stark effect. **b** Dynamics at 363 nm (*black line*) for excitation at S_{11} and fitting (*light gray line*) with $t^{-0.5}$ power law

A monomolecular power law decay is the predicted dynamics for geminate recombination of free particles after random walk in an infinite one-dimensional chain. A more detailed modelling of the geminate recombination process indicates that the initial distance between the geminate e-h pair is of the same order of magnitude of the exciton correlation length, thus suggesting that charge-carriers arise from instantaneous linear exciton dissociation.

Our results shed new light onto the charge photogeneration mechanism in SWNTs, suggesting that the nascent exciton dissociates spontaneously, perhaps in presence of extrinsic screening of the Coulomb attraction, possibly due to water or other ambient contamination [5].

Acknowledgments C.G. acknowledges support by the EC under Graphene Flagship (contract no. CNECT-ICT-604391). F.S., G.L. and T.H. acknowledge the ITN project 316633 "POCAONTAS".

References

1. T. Ando, "Excitons in Carbon Nanotubes", J. Phys. Soc. Jpn. 66, 1066–1073 (1997).
2. F. Wang, G. Duckovic, L. E. Brus and T. F. Heinz, "The optical resonances in carbon nanotubes arise from excitons", Science 308, 838–840 (2005).
3. L. Luer, S. Hoseinkhani, D. Polli, J. Cochet, T. Hertel and G. Lanzani, "Size and mobility of excitons in (6,5) carbon nanotubes", Nature Phys. 5, 54–58 (2009).
4. G. Soavi, F. Scotognella, D. Brida, T. Hefner, F. Spath, M. R. Antognazza, T. Hertel, G. Lanzani and G. Cerullo, "Ultrafast charge photogeneration in semiconducting carbon nanotubes", J. Phys. Chem. C. 117, 10849–10855 (2013).
5. P. G. Collins, K. Bradley, M. Ishigami and A. Zettl, "Extreme oxygen sensitivity of electronic properties of carbon nanotubes", Science 287, 1801–1804 (2000).

Thickness Dependent Hot-phonon Effects Observed by Femtosecond Mid-infrared Luminescence in Graphene

Tohru Suemoto, Tomohiro Kawasaki, Hiroshi Watanabe, Takushi Iimori and Fumio Komori

Abstract Femtosecond luminescence of graphene and graphite is studied from near- to mid-infrared regions. Remarkable reduction of lifetime at 0.3 eV is found in mono- and bi-layer graphenes, and interpreted as an effect of non-zero Fermi energy.

1 Introduction

Graphene receives considerable attention as a platform of high speed electronics such as field effective transistors, single electron transistors and optical sensors, because it has a unique symmetric linear dispersion for electron and hole, and high carrier mobilities. The dynamics of high energy electrons and coupled phonons have been investigated with various experimental methods, such as transient absorption [1], reflectance [2], anti-Stokes Raman scattering [3], photoelectrons [4, 5], and luminescence [6, 7]. In these reports, the response of photoexcited electrons (holes) ranging from 100 fs to several ps were found. The time constants strongly depends on the observation photon energy, excitation fluence, thickness (layer number n), and sample condition.

In spite of these efforts, the reported lifetimes are not consistent each other and especially the thickness dependence is controversial.

To understand the whole picture of photo-excited electron dynamics, it is important to observe the response in a wide energy range especially at low energies close to Dirac point. In this report, we observed luminescence in the energy region down to 0.3 eV, corresponding to 0.15 eV electron. The result shows a considerable thickness dependence of the carrier lifetime at low energies.

T. Suemoto (✉) · T. Kawasaki · H. Watanabe · T. Iimori · F. Komori
Institute for Solid State Physics, The University of Tokyo, 5-1-5, Kashiwanoha, Kashiwa-Shi, Chiba 277-8581, Japan
e-mail: suemoto@issp.u-tokyo.ac.jp

© Springer International Publishing Switzerland 2015
K. Yamanouchi et al. (eds.), *Ultrafast Phenomena XIX*,
Springer Proceedings in Physics 162, DOI 10.1007/978-3-319-13242-6_89

2 Experiment

We used $n = 1, 2, 6 \sim 8$ graphene sheets purchased from ACS MATERIAL®. The graphene sheets are transferred onto fused silica substrates. For a reference we measured also HOPG graphite ($n = \infty$). The sample was excited at 1.57 eV (790 nm), by 70 fs pulses at a repetition rate of 200 kHz, and the infrared luminescence signal was up-converted to visible light and analyzed by a double grating monochromator and detected with a photon counting system.

3 Results and Discussion

Photon energy dependences of the decay profiles in $n = 2$ graphene under relatively high excitation fluence are shown in Fig. 1a. In graphite, the lifetime becomes longer at lower photon energy (not shown) as reported in [8]. The $n = 2$ graphene has a similar tendency, whereas the lifetime at 0.3 eV is not so long as that in graphite. On the other hand, the lifetime at 1.2 eV is almost the same as that in graphite. Elongation of the lifetime at lower energy corresponds to slower decay of electron population at lower energy, that is described by the time dependent Fermi-Dirac distribution, assuming cooling of the electron system.

Luminescence signals from graphenes with different n are shown in Fig. 1b. At the lowest energy 0.3 eV, the n dependence is clearly seen, that is, the lifetime in mono-layer graphene is roughly one half of that in graphite, while that of $n = 6-8$ sample is almost the same as that of graphite. Decrease of the lifetime in thinner graphene has been reported by several groups and explained by assuming additional energy decay channels, such as the energy flow to substrate, or flexural mode,

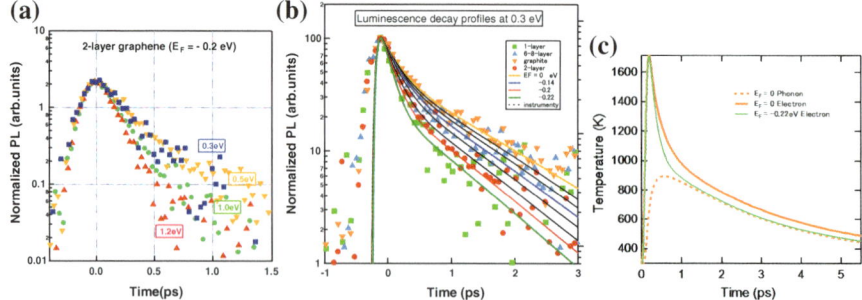

Fig. 1 a Decay profiles of luminescence intensity in bi-layer graphene at 0.3 (*square*), 0.5 (*inverted triangle*), 1.0 (*circle*), and 1.2 (*triangle*) eV. **b** Decay profiles at 0.3 eV for $n = 1$ (*square*), 2 (*circle*), $6 \sim 8$ (*triangle*) and ∞ (*inverted triangle*). The *solid curves* are the calculated results for $E_F = -0.22, -0.2, -0.18, -0.16, -0.14, -0.12, -0.1$ and 0 from the lower side. **c** Calculated electron (*thick solid*) and phonon (*dashed*) temperatures for $E_F = 0$ eV. The *thin solid curve* shows the electron temperature for $E_F = -0.22$ eV

which corresponds to bending motion of the thin film. In this report, we discuss the effect of the Fermi energy on the carrier dynamics. From a photoemission spectrum, we found that the Fermi energy of the $n = 1$ graphene supported by a silica plate is located at -0.2 eV from the Dirac point. On the other hand, the Fermi energy in graphite can be assumed to be zero, because deviation of the conduction band minimum (valence band maximum) from the Fermi energy is only 0.015 eV.

We try to understand these observations in terms of the two-temperature model [7], taking the Fermi energy into account. We assume that the energy of the electron system is transferred to the 0.2 eV optic phonon (G mode) as the first step and that the energy of the hot optic phonon is dissipated to the large heat bath within the graphene sheet.

The parameters are chosen to reproduce the decay curve of graphite with $E_F = 0$. The experimental curve is well reproduced in positive t, while the disagreement at $t < 0$ is mainly ascribed to the finite time resolution of the measurements. The electron (T_e) and phonon (T_{ph}) temperatures in this calculation are shown in Fig. 1c with thick solid and dashed curves, respectively. Around 1 ps, the cooling rate of T_e decreases due to hot phonon effect. This kink is reflected in the decay profile in Fig. 1b.

As the electron transfer from the graphene layer to the substrate is the cause of the negative E_F in mono-layer graphene attached to the substrate, E_F of the subsequent layers in thicker graphene sheets comes closer to zero. Then it should converge to zero at n = ∞. If we define "average E_F" for a luminescent volume, it is expected to be a monotonic increasing function of n, varying from a finite negative value to zero. Decay profiles for finite negative E_F are calculated with the fixed values for the rest of the parameters and shown in Fig. 1b by solid curves. It is found that the experimental curve for $n = 1$ is in agreement with the curve with $E_F = -0.22$, in consistence with the photoemission result. The decay profiles for $n = 2$ and 6 $\sim$ 8 are reproduced with $E_F = -0.2$ and -0.14 eV, respectively. The faster decay for $n = 1$ corresponds to the faster cooling of T_e shown as a thin solid curve in Fig. 1c.

An intuitive explanation is as follows. The phonon emission probability is proportional to the density of state (DOS) of the final electronic state. As the DOS becomes small near the Dirac point (bottleneck effect), the cooling rate decreases as the mean electron energy is comparable or smaller than the optical phonon energy, if E_F is close to zero. However, if E_F has a large value (either negative or positive), the DOS near E_F has always a large value and the phonon emission process does not suffer from the bottleneck effect, resulting in a fast cooling.

In conclusion, the layer number dependence of the carrier lifetime in a silica supported graphenes is successfully interpreted in terms of average Fermi energy without assuming an additional relaxation pass. This shows the crucial importance of the Fermi energy in understanding the ultrafast carrier relaxation process in graphenes.

References

1. B. Gao *et al.*, "Studies of Intrinsic Hot Phonon Dynamics in Suspended Graphene by Transient Absorption Microscopy," Nano Lettres **11**, 3184–3189 (2011).
2. P. J. Hale, S. M. Hornett, J. Moger, D. W. Horsell, and E. Hendry "Hot phonon decay in supported and suspended exfoliated graphene," Phys. Rev. B 83 121404-1-4 (2011).
3. Shiwei Wu *et al.* "Hot Phonon Dynamics in Graphene," Nano Letters **12**, 5495–5499 (2012).
4. J. C. Johannsen *et al.* "Direct View of Hot Carrier Dynamics in Graphene," Phys. Rev. Lett. **111**, 027403-1-5 (2013).
5. I. Gierz *et al.* "Snapshots of non-equilibrium Dirac carrier distributions in graphene," Nat. Materials, **12**, 1119–1124 (2013).
6. C. H.Lui, K. F.Mak, J.Shan, and T. F. Heinz, "Ultrafast Photoluminescence from Graphene," Phys. Rev. Lett. **105**, 127404-1-4 (2010).
7. T. Koyama *et al.* "Near-Infrared Photoluminescence in the Femtosecond Time Region in Monolayer Graphene on SiO₂," ACS NANO **7**, 2335–2343 (2013).
8. T. Suemoto, S. Sakaki, M. Nakajima, Y. Ishida, and S. Shin "Access to hole dynamics in graphite by femtosecond luminescence and photoemission spectroscopy," Phys. Rev. B **87**, 224302-1-5 (2013).

Part V
Chemistry—Liquid Phase

Discriminating Racemic from Achiral Solutions with Femtosecond Accumulative Spectroscopy

Andreas Steinbacher, Patrick Nuernberger and Tobias Brixner

Abstract We follow asymmetric photodissociation reactions of chiral substances with a sensitive polarimeter specifically designed for applications with femtosecond pulses. The accumulative detection scheme allows the discrimination of racemic and achiral solutions with high sensitivity.

1 Introduction

Steady-state circular dichroism (CD) or optical activity (OA) are common techniques to analyze chiral samples, even on the ultrafast timescale [1]. However, those techniques cannot discriminate a racemic sample from an achiral one. Here, we introduce the possibility to discriminate racemic from achiral solutions by optical means without prior spatial separation.

2 Measurement Principle and Experimental Results

In order to be able to perform such a discrimination one has to generate a chirality-sensitive signal. We chose to generate an enantiomeric excess (ee) via asymmetric photodecomposition [2] with circularly polarized fs pulses. Since the difference of the absorption coefficients for left (LC) or right circularly (RC) polarized light is very small, we employ an accumulative measurement scheme with a sensitive polarimeter. It is capable of detecting optical rotation (OR) changes $\Delta\alpha$ of only

A. Steinbacher · T. Brixner (✉)
Universität Würzburg, Am Hubland, 97074 Würzburg, Germany
e-mail: brixner@phys-chemie.uni-wuerzburg.de

P. Nuernberger
Ruhr-Universität Bochum, 44780 Bochum, Germany

© Springer International Publishing Switzerland 2015   369
K. Yamanouchi et al. (eds.), *Ultrafast Phenomena XIX*,
Springer Proceedings in Physics 162, DOI 10.1007/978-3-319-13242-6_90

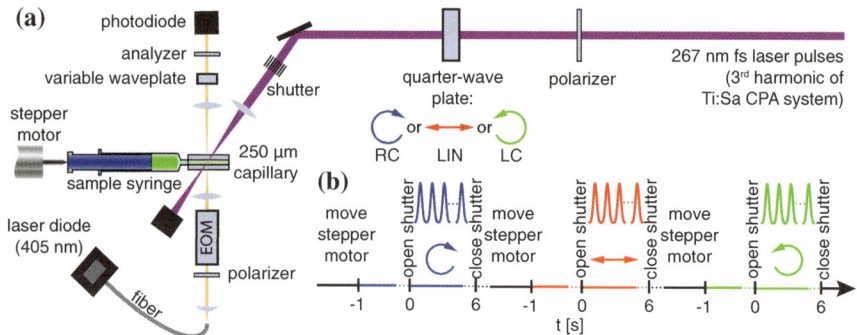

Fig. 1 **a** Pump (*purple*) and probe (*blue*) are spatially overlapped in the 250 μm capillary to record the optical rotation before and during illumination with fs laser pulses. The polarization state (LIN, LC, RC) of the pump beam is controlled via a polarizer and a quarter-wave plate. **b** The measurement cycle consists of three single accumulation steps with RC, LIN, and LC. Before the recording of the OR starts, fresh sample is pushed into the capillary in every case. After 1 s of recording the shutter is opened for 6 s

0.10 mdeg in a capillary with only 250 μm optical path length [3]. In our scheme, not only one pump pulse is used, but several subsequent fs laser pulses irradiate the same sample volume and initiate the photoreaction.

The setup and measurement scheme is depicted in Fig. 1a. The polarization of the 267 nm pump pulses is varied between LC, RC, and linear (LIN) light. For each polarization the following measurement procedure is conducted. The sample in the capillary is renewed and the acquisition of $\Delta\alpha$ is started. We measure 1 s to have a baseline for $\Delta\alpha$, then the pump shutter is opened at $t = 0$ and the sample is irradiated with UV laser pulses for 6 s, while $\Delta\alpha$ is continuously recorded (see Fig. 1b).

The UV fs laser pulses trigger an asymmetric photoreaction of a racemic mixture of 1,1'-binaphthyl-2,2'-diyl hydrogenphosphate, leading to achiral photoproducts. In Fig. 2a, the CD spectra of the pure enantiomers and the racemate dissolved in methanol ($\gamma = 13.5$ mg/ml) are depicted. The CD signal does not reveal whether an unknown solution is racemic or contains achiral molecules.

The result of an OR measurement using the scheme of Fig. 1 with the racemate in solution is shown in Fig. 2b. In the case of LIN pump pulses (red), no $\Delta\alpha$ signal is visible (right axis). In contrast, RC pulses (blue) lead to a positive change, while for LC (green) the sign flips, and for both (LC and RC) a clear extremum is visible. After UV irradiation the molecules form non-chiral fragments. Depending on the handedness of the polarization, more R- or S-enantiomers are gradually decomposed, leading to the generation of ee. Due to the small difference of the absorption coefficients of both enantiomers for LC or RC, a single fs laser pulse would not lead to a detectable signal. However, by accumulating the outcome of several subsequent photoreactions, this photoinduced ee can be magnified and hence observed.

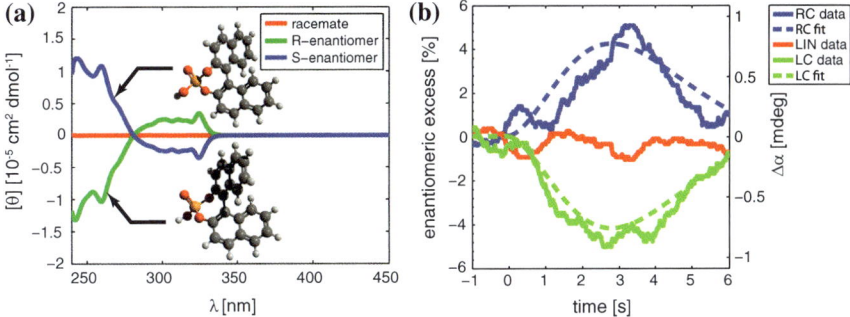

Fig. 2 **a** Steady-state CD spectra of the racemate (*red*) and for the pure enantiomers (*green, blue*). **b** Experimental outcome of an accumulative OR experiment with racemic solution. While LIN polarization (*red*) does not change the ee, LC (*green*) and RC (*blue*) polarized light leads to an enrichment of one or the other enantiomer. The *dashed lines* correspond to the fitting model

We model the data by solving the differential equations for the time-dependent change in the number of molecules for the R- and S-enantiomer with their corresponding cross sections after irradiation. Due to symmetry, the resulting function changes sign if the polarizations (LC ↔ RC) are exchanged. Fitting both curves together (dashed lines in Fig. 2b) reveal that the relative difference of the cross sections is only 0.15 %. In order to further quantify the induced ee, we perform experiments with just one enantiomer (not shown) and determine the maximal possible $\Delta\alpha$ to 19.18 mdeg. Assuming the ee scales linearly with the OR, we obtain the ee axis of Fig. 2b (left), revealing an ee of several percent at the optimal time.

3 Discussion and Outlook

For an exemplary racemic solution, we monitored the evolution of photoinduced ee, which reaches several percent at a certain illumination time and decreases again for continued exposure to UV light. This is a direct demonstration that the approach allows the discrimination of racemic and achiral solutions if the pump pulses generate stable achiral photoproducts. Furthermore, if the pump illumination is stopped at the right moment, this ee persists in the sample volume and might be exploited for other purposes. Among those, quantum control approaches could be beneficial, since the setup allows for fs time resolution [3] if pump pulse pairs or possibly shaped fs laser pulses are employed.

References

1. Meyer-Ilse, J., Akimov, D., and Dietzek, B.: Recent advances in ultrafast time-resolved chirality measurements: perspective and outlook. Laser Photon. Rev. 7, 495–505 (2013)
2. Inoue, Y. and Ramamurthy, V.: Chiral Photochemistry. CRC Press, Boca Raton (2004)
3. Steinbacher, A., Buback, J., Nuernberger, P., and Brixner, T.: Precise and rapid detection of optical activity for accumulative femtosecond spectroscopy. Opt. Express 20, 11838–11354 (2012)

Quantum Dynamics of Molecular Reactions Directed by Explicit Solvent Environment

Sebastian Thallmair, Julius Zauleck and Regina de Vivie-Riedle

Abstract We present the first method that combines molecular quantum dynamics of the solute with classical molecular dynamics of the solvent. Its mechanical impact on the ultrafast internal motions is decisive for the reaction outcome.

1 Ultrafast Chemical Reactions in Solution

It is well known that solvents can influence the outcome of chemical and biochemical reactions. Their electrostatic effects stabilize polar configurations along the reaction path. In addition dynamic solvent effects can emerge. They intervene on the ultrafast time scale and are in the focus of this study. A direct connection to the reaction process, involving either bond cleavage or formation or both, can be established. Besides the change in the molecular structure of the reactants also the volume and the shape of the solute's cavity within the solvent are affected. Therefore the dynamics of the solute is not governed by its internal degrees of freedom alone but in addition the position and orientation of the surrounding molecules can play an important role. In a given arrangement of solvent molecules specific internal motions of the solute can be hindered and areas of the molecular potential surface, dissimilar to the free molecule, are explored. The areas within reach change with the temporal arrangement of the solvent molecules. The average over all accessible arrangements defines the final yield. The dynamic solvent effects become especially important for ultrafast reactions with a substantial change of the shape of the reactants like in photochemical dissociations or isomerizations. Not only can internal motions be hindered, but additional channels can be opened.

S. Thallmair (✉) · J. Zauleck · R. de Vivie-Riedle
Department Chemie, Ludwig-Maximilians-Universität München,
Butenandtstraße 11, 81377 Munich, Germany
e-mail: sebastian.thallmair@cup.uni-muenchen.de

© Springer International Publishing Switzerland 2015
K. Yamanouchi et al. (eds.), *Ultrafast Phenomena XIX*,
Springer Proceedings in Physics 162, DOI 10.1007/978-3-319-13242-6_91

2 Photochemical Bond Cleavage in Acetonitrile

Our example is the photochemical bond cleavage of diphenylmethyl triphenyl-phosphonium ions ($Ph_2CH-PPh_3^+$) in acetonitrile. $Ph_2CH-PPh_3^+$ is a common precursor for the generation of Ph_2CH^+ cations which was experimentally demonstrated by ultrafast broadband transient absorption measurements [1]. Electrostatic effects play no role for the product formation as each of the two possible product channels include one charged and one neutral fragment and their relative energy gap remains unchanged [2]. Therefore $Ph_2CH-PPh_3^+$ is perfectly suited to test our new method for the dynamic solvent effect combining quantum dynamics on ab initio potential energy surfaces (PES) for the solute with classical molecular dynamics (MD) of solvent molecules. After a π-π* excitation of $Ph_2CH-PPh_3^+$ induced by an ultrashort UV laser pulse (275 nm, 35 fs FWHM), the system crosses a small barrier in the excited state S_1 [2]. Subsequently the leaving group PPh_3 is separated by C-P bond cleavage. Previous quantum dynamical simulations that included the dynamic solvent effect in an implicit formalism demonstrated its importance [3]. The solvent controls the reaction outcome. It guides the system through a conical intersection (CoIn), non accessible in the gas phase, to form the experimentally observed Ph_2CH^+ cations. Our new approach goes beyond the implicit solvent model, it includes the environment explicitly.

3 Solvent Potential Extracted from Classical Molecular Dynamics Simulations

The number of accessible arrangements of the solvent molecules surrounding a solute is very large. To average statistically over all possible solvent orientations, we perform classical MD simulations of the solute in a box of solvent molecules. Snapshots are taken at every 200 fs and the interaction in each solute-solvent arrangement is taken individually into account in form of a potential energy term $E_{solv}(r,\phi)$, which depends on the chosen internal reactive coordinates r, the C-P distance and the P-C-X angle ϕ (see Fig. 1, top right). This potential energy term consists of all contributions $E_{sf}(d_{sf}(r,\phi))$ between every solvent molecule s and the solute splitted into the fragments f of the leaving group PPh_3 and the Ph_2CH group. E_{sf} depends explicitly on the s-f distance $d_{sf}(r,\phi)$, which in turn is evaluated along the two-dimensional surface spanned by the reaction coordinates r and ϕ. Thus $E_{solv}(r,\phi)$ is given as:

$$E_{solv}(r, \phi) = \sum_{s=1}^{N_{solv}} \sum_{f=1}^{2} E_{sf}\left(d_{sf}(r,\phi)\right) \tag{1}$$

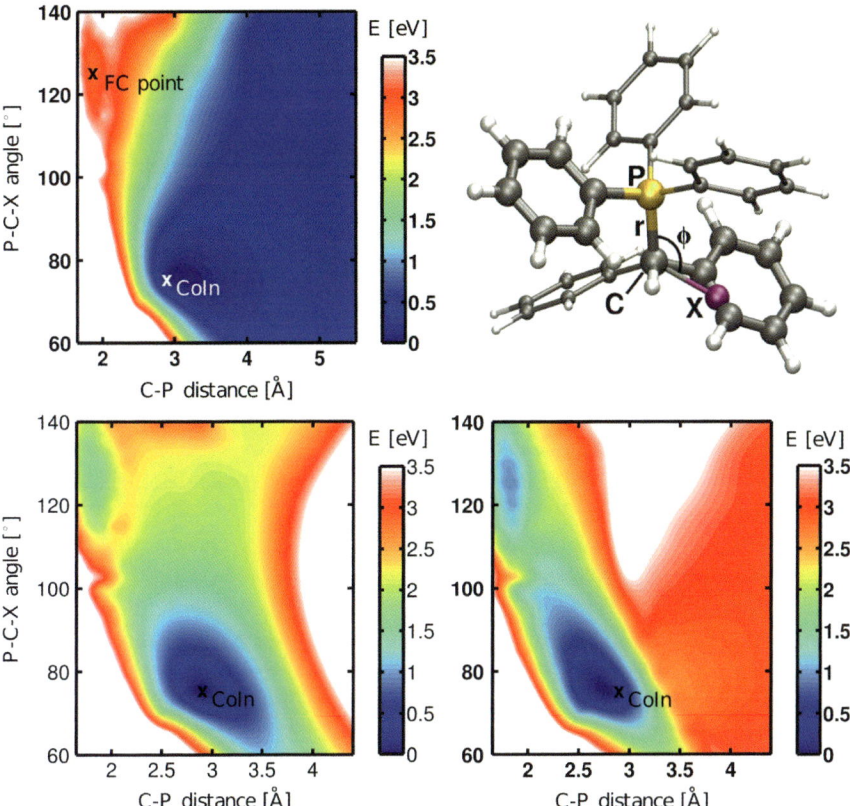

Fig. 1 Molecular PES of a diphenylmethyltriphenylphosphonium ion (Ph$_2$CH–PPh$_3^+$) calculated at the ONIOM level of theory (CASSCF(10,10)/M06-2X/6-31g(d)) for the two reactive coordinates r and ϕ (*top left*). Visualization of the two reactive coordinates in Ph$_2$CH–PPh$_3^+$ (*top right*). r denotes the distance between the C and the P atom; ϕ the angle between P, C and the dummy atom X. The thicker atoms are part of the high-level system of the ONIOM calculations; the thinner ones belong to the low-level system. Two exemplary total PES used for the quantum dynamics (*bottom*). The crucial influence of the solvent cage is in the area where r is larger than 3.5 Å. The solvent hinders the free dissociation along the r coordinate and guides the wave packet toward the conical intersection (CoIn). The Franck-Condon (FC) point lies around $r = 1.87$ Å and $\phi = 125°$, the CoIn where the experimentally observed Ph$_2$CH$^+$ cations are formed is located at $r = 2.9$ Å and $\phi = 75°$

The total PES $E_{tot}(r,\phi)$, used for the quantum dynamics, is the sum of $E_{solv}(r,\phi)$ and the molecular PES $E_{mol}^{ONIOM}(r,\phi)$ (Fig. 1, top left) calculated at the ONIOM level of theory (CASSCF(10,10)/M06-2X/6-31g(d)).

$$E_{tot}(r, \phi) = E_{mol}^{ONIOM}(r, \phi) + E_{solv}(r, \phi) \qquad (2)$$

Figure 1 (bottom) shows the total PES for two exemplary snapshots of a MD trajectory. The individual solvent configurations show up in the different height and shape of the potential barrier beyond C-P distances of 3.5 Å. The CoIn connecting the competing reaction channels of the higher lying radicals Ph$_2$CH$^•$ and the lower lying experimentally observed Ph$_2$CH$^+$ cations is located at $r = 2.9$ Å and $\phi = 75°$.

4 Quantum Dynamics in the Combined Potential of the Molecule and Its Surrounding

Figure 2 shows a dissociating wave packet in a selected total PES (right) about 170 fs after crossing the barrier close to the FC region together with its explicit solvent arrangement (left). Obviously the dissociation of the wave packet along the r coordinate is hindered by the solvent cage and the wave packet is reflected toward the CoIn.

This new ansatz reproduces the experimental findings and the results obtained with the implicit solvent formalism. Furthermore it takes into account the explicit arrangements of a given solvent and its effect on the PES, allows for the description of solvent mixtures and opens a way to include the feedback of the decelerated wave packet on solvent cage.

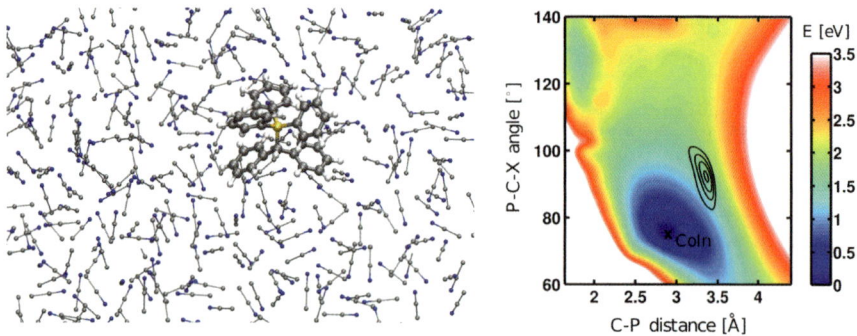

Fig. 2 Explicit solvent arrangement of an exemplary snapshot of the MD trajectory (*left*) and the dissociating wave packet (*black isolines*) in the corresponding total PES (*right*). The dissociation along the r coordinate is hindered by the potential barrier around $r \approx 3.5 - 4.0$ Å and the wave packet is guided toward the CoIn next to the global minimum

References

1. J. Ammer, C. F. Sailer, E. Riedle, and H. Mayr, "Photolytic generation of benzhydryl cations and radicals from quaternary phosphonium salts: How highly reactive carbocations survive their first nanoseconds", J. Am. Chem. Soc. **134**, 11481-11494 (2012).
2. S. Thallmair, B. P. Fingerhut, and R. de Vivie-Riedle, "Ground and excited state surfaces for the photochemical bond cleavage in phenylmethylphenylphosphonium ions", J. Phys. Chem. A **117**, 10626-10633 (2013).
3. S. Thallmair, M. Kowalewski, B. P. Fingerhut, C. F. Sailer, and R. de Vivie-Riedle, "Molecular wave packet dynamics decelerated by solvent environment: A theoretical approach" in Ultrafast Phenomena XVIII, M. Chergui, S. Cundiff, A. Taylor, R. de Vivie-Riedle, K. Yamanouchi, eds., EPJ Web of Conferences **41**, 05043-p.1–05043-p.3 (2013).

Excited-State Dynamics of Catalytically Active Transition Metal Complexes Studied by Transient Photofragmentation in Gas Phase and Transient Absorption in Solution

D. Imanbaew, Y. Nosenko, K. Chevalier, F. Rupp, C. Kerner,
F. Breher, W.R. Thiel, R. Diller and C. Riehn

Abstract Femtosecond photofragmentation (gas phase) and transient absorption (solution) revealed ultrafast electronic coupling (0.1–3 ps) and energy transfer (7–12 ps) in a Ru(II)-complex and ultrafast formation (~ 0.4 ps) of a long-lived triplet state in a Pd_3-complex.

1 Introduction

Transition metal-ligand complexes are important and selective catalysts or pre-catalysts for homogeneous reactions, among them transfer hydrogenation (addition of H_2) and carbon-carbon bond formation (C–C-coupling). Moreover, due to their synthetically tunable photophysical properties these complexes succeed in the fields of light-harvesting and photocatalysis. In order to elucidate the primary competing photoreactive pathways, we report here studies on ultrafast excited state dynamics and catalyst activation processes of new Ru(II)- and Pd_3-complexes.

D. Imanbaew · Y. Nosenko · C. Kerner · W.R. Thiel · C. Riehn (✉)
Department of Chemistry, Forschungszentrum OPTIMAS, TU Kaiserslautern,
Erwin-Schrödinger-Str.52, 67663 Kaiserslautern, Germany
e-mail: riehn@chemie.uni-kl.de

K. Chevalier · F. Rupp · R. Diller
Department of Physics, TU Kaiserslautern,
Erwin-Schrödinger-Str.46, 67663 Kaiserslautern, Germany

F. Breher
Institute of Inorganic Chemistry, Karlsruhe Institute of Technology (KIT),
Engesser Str.15, 76131 Karlsruhe, Germany

© Springer International Publishing Switzerland 2015
K. Yamanouchi et al. (eds.), *Ultrafast Phenomena XIX*,
Springer Proceedings in Physics 162, DOI 10.1007/978-3-319-13242-6_92

2 Results and Discussion

The first molecular target of interest consists of new ruthenium(II) complexes $[(\eta^6\text{-cymene})\text{RuCl(apypm)}]\text{PF}_6$ (apypm = 2-NR_2-4-(pyridine-2-yl)-pyrimidine, R = $CH_3(\mathbf{1})/H(\mathbf{2})$), which catalyze the transfer hydrogenation of arylalkyl ketones in the absence of a base upon thermal activation [1]. The ultrafast gas phase studies are based on pump-probe transient photofragmentation employing a newly designed setup consisting of a kHz 50 fs-amplified Ti:Sa-laser system equipped with two optical parametrical generator/amplifier units for independent wavelength tuning of pump and probe pulses and an electrospray ion trap mass spectrometer for ion selection, storage and mass analysis [2].

Femtosecond transient photofragmentation on mass selected ions ($\mathbf{1^+}$, $\mathbf{2^+}$) prove the formation of the activated catalyst species (HCl abstraction) after photoexcitation of the MLCT bands [2]. The excited state dynamics in the gas phase could be fitted to a biexponential decay (time constants <100 fs and 1–3 ps, Fig. 1b). In parallel, transient absorption spectroscopy in acetonitrile using femtosecond UV/Vis and IR probe laser pulses revealed additional deactivation processes on longer time scales ($\sim$7–12 ps, Fig. 1c). The formation of the active catalyst species after photoexcitation could not be observed in solution, emphasizing an efficient deactivation process and strong influence of the solvent-catalyst interaction [2].

The second molecular target is represented by a trinuclear Pd-complex ([$Pd_3\{Si(mt^{Me})_3\}_2$], $\mathbf{3}$, mt^{Me} = methimazole, Fig. 2a) [3], which acts as a C–C-coupling agent in a Suzuki-Miyaura type reaction.

Upon electrospray ionization of $\mathbf{3}$ from acetonitrile solution a cationic species $\mathbf{3^+}$ could be identified by its isotope pattern (1,054 m/z). The large intensity of this mass peak points towards a stable cationic structure. Its fragmentation is dominated by the loss of one mt^{Me} unit ($C_4H_5N_2S$, remaining ion 941 m/z, Fig. 2b, top)

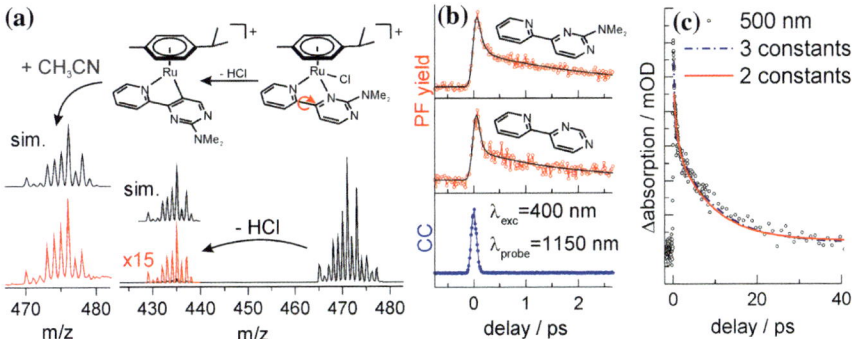

Fig. 1 **a** Proposed catalyst self-activation [1] and mass spectrum demonstrating the photoactivation process ($\mathbf{1^+}$, λ_{exc}: 400 nm) (*bottom*). Complexation of activated species by acetonitrile inside an ion trap (*left*). **b** Transient fragment ion intensity of $\mathbf{1^+}$ (*top*) and of a similar complex bearing a bipyridine ligand (*centre*) in gas phase; experimental cross correlation (125 fs, *bottom*). **c** Transient absorption of $\mathbf{1^+}$ in acetonitrile (λ_{exc}: 400 nm/λ_{probe}: 500 nm)

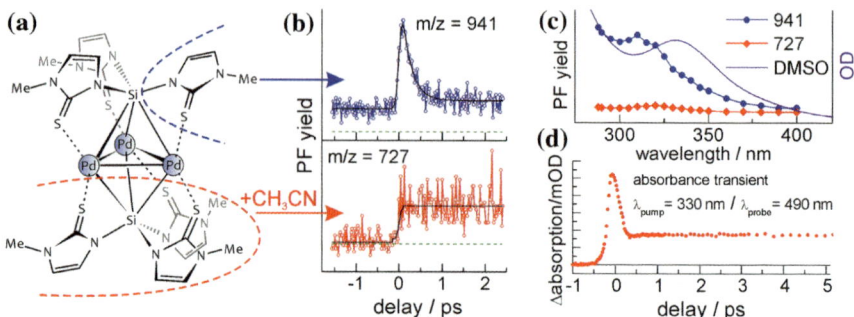

Fig. 2 **a** Structure of the Pd$_3$-complex. **b** Transient signals depending on the fragmentation channel (λ_{exc}: 320 nm/λ_{probe}: 720 nm). **c** UV absorption in DMSO versus photofragmentation spectra in gas phase. **d** Transient absorption in DMSO (λ_{exc}: 330 nm/λ_{probe}: 490 nm)

together with a second channel forming 727 m/z (loss of Si(mt^{Me})$_3$ plus addition of acetonitrile, Fig. 2b, bottom). Interestingly, the opening of the hemilabile bond between the sulfur donor of the methimazole unit and the Pd$_3$ core was considered crucial for its catalytic activity [3]. The photofragmentation spectrum (Fig. 2c) shows, besides a blue shift of 15 nm, strong similarity to the absorption spectrum of the neutral species in DMSO (dimethyl sulfoxide). The dynamics of the cationic species (320 nm/720 nm, Fig. 2b) reveal for 941 m/z a fast relaxation of the primary state with a lifetime of ca. 200 fs and an additional long-lived (>100 ps) component whereas for 727 m/z a direct formation of the long-lived (probably triplet) state was observed. These findings are remarkably close to the behavior of the neutral compound in solution (static spectrum, Fig. 2c). In DMSO after photoexcitation at 330 nm the observed absorption changes are dominated by an instantaneous bleach in the region of electronic ground state absorption and a concomitant fast ($\sim$400 fs) build-up of spectrally broad and long-lived (ns time-scale) excited state absorption at 370–750 nm, which is assigned to the lowest triplet state ($^3A_1$). No hints on ligand dissociation or other structural/electronic changes have been observed in solution [4].

3 Conclusions

The ultrafast dynamics of the Ru(II)- and Pd$_3$-complexes exhibit remarkably similar timescales in both, the gas phase and solution. Whereas the Ru(II)-complex displays a fast and efficient internal conversion into a vibrationally hot ground state, the Pd$_3$-complex proceeds, independent of charge status (neutral versus monocationic), towards a long-lived triplet state, that might be the starting point for further reactions. For both complexes, the investigation of gas phase fragmentation proved valuable in characterizing the possible catalytically active species.

Acknowledgments We acknowledge collaborations with: G. Niedner-Schatteburg (TUKL) and F. Menges (TUKL). Financial Support from the DFG (SFB/TRR88, "3MET"), Stift. Rh-Pfalz f. Innovation and research center OPTIMAS is gratefully acknowledged.

References

1. L. T. Ghoochany, S. Farsadpour, F. Menges, Y. Sun, G. Niedner-Schatteburg, W. R. Thiel, Eur. J. Inorg. Chem. **24**, 4305 (2013)
2. D. Imanbaew, Y. Nosenko, K. Chevalier, F. Rupp, C. Kerner, C. Riehn, R. Diller, W. R. Thiel, Excited state dynamics of a ruthenium(II) catalyst studied by transient photofragmentation in gas phase and transient absorption in solution, Chem. Phys. (2014). doi:10.1016/j.chemphys.2014.03.005
3. F. Armbruster, J. Meyer, A. Baldes, P. O. Burgos, I. Fernandez, F. Breher, Chem. Commun. **47**, 221 (2011)
4. Y. Schmitt, K. Chevalier, F. Rupp, M. Becherer, A. Grün, A. M. Rijs, F. Walz, F. Breher, R. Diller, M. Gerhards and W. Klopper, Phys. Chem. Chem. Phys. **16**, 8332 (2014)

Coherent Control of the Photodissociation of Triiodide in Solution Reveals New Pathways

Rui Xian, Valentyn I. Prokhorenko, Ryan L. Field and R.J. Dwayne Miller

Abstract We perform closed- and open-loop coherent control experiments of the photodissociation of triiodide (I_3^-) in ethanol using shaped UV pulses. In the regime where the relative diiodide (I_2^-) yield is excitation-independent, we observe a symmetrical second-order chirp dependence of the I_2^- yield, and that the I_2^- production can be controlled within 40 % by varying the second-order chirp of the pump pulse. We show that additional pathways involving higher-lying potential energy surfaces are viable means to explain such observations.

1 Introduction

Photodissociation in the solution phase is complicated by the involvement of solvent cages, leading to altered reaction and energy disposal pathways compared with equivalent gas phase experiments [1]. In the case of I_3^- in ethanol, certain gas phase dissociation products are not seen in the solution phase and doubts remain on the number of pathways involved in dissociation via excitation to the high-lying absorption band (peaks at 290 nm) [2, 3]. Here we perform coherent control and pump-probe experiments to examine the influence of the second-order chirp and the energy of excitation on the I_2^- yield. Our observations point to additional mechanisms involved in the photodissociation that haven't so far been discussed [2, 3].

R. Xian · V.I. Prokhorenko · R.L. Field · R.J. Dwayne Miller (✉)
Max Planck Institute for the Structure and Dynamics of Matter, Centre for Free Electron Laser Science, Bld. 99, Luruper Chaussee 149, 22761 Hamburg, Germany
e-mail: dwayne.miller@mpsd.mpg.de

R.L. Field · R.J. Dwayne Miller
Departments of Chemistry and Physics, University of Toronto, 80 St. George Street, Toronto, Ontario M5S 3H6, Canada

© Springer International Publishing Switzerland 2015
K. Yamanouchi et al. (eds.), *Ultrafast Phenomena XIX*,
Springer Proceedings in Physics 162, DOI 10.1007/978-3-319-13242-6_93

2 Experimental Methods

The sample, tetra-*n*-butylammonium triiodide, or TBAT (Sigma-Aldrich), was dissolved in UV-grade ethanol (Uvasol, Merck Millipore) and carried through a home-built liquid delivery system involving a 100 μm-path-length flow cell (Starna), a sample reservoir, and a peristaltic pump with Teflon tubing (Cole-Parmer Masterflex). The reservoir maintains the sample concentration (OD ~ 1 at 290 nm) during the course of an experiment due to partial reversibility of the reaction. A commercial Ti:sapphire ultrafast laser system (Coherent Elite USP) provides 35 fs-long pulses at 800 nm used for monitoring the photoproduct I_2^- (with an absorption band around 750 nm) and as a pump for a home-built two-stage NOPA, whose output is frequency-doubled to generate transform-limited pulses at 290 nm (~ 60 fs) for excitation of triiodide. For closed-loop coherent control experiments and controlling of the second-order chirp rate we used a home-built pulse shaper, based on an AOM in the 4f-geometry [4] and designed for mid-UV operation with a technical efficiency of ~ 50 % around 290 nm. In open-loop control experiments the chirp was varied from −10,000 to 10,000 fs^2, and the excitation energy from 100 to 425 nJ.

3 Results and Discussion

Optimization of the I_2^- yield using a closed-loop coherent control protocol at 450 nJ excitation energy (1.24×10^{16} photons/cm^2) results in an optimal pulse approaching a transform-limited (TL) pulse. Chirp scans show asymmetry in the absorbed energy whereas the I_2^- yield (proportional to the differential absorption ΔA monitored at 800 nm) is symmetric with respect to the sign of the chirp and is maximal for TL pulses at all excitation energies applied (Fig. 1a, b). The relative yield (RY), $\Delta A/E_{exc}$, is a quantity proportional to the quantum yield in one-photon photochemistry. A flat $\Delta A/E_{exc}$ − E_{exc} dependence implies linear dissociation (a linear dependence indicates the involvement of two-photon dissociation, and so on). Taking into account the scattering contribution to measured signals at low excitation levels, the RY of I_2^- experiences very slight growth with increasing excitation energy (Fig. 1c, upper panel), indicating that the whole range lies mostly within the linear regime of dissociation, thereby excluding the possibility of two-or-more-photon absorption in the sample. The transmission of the sample (at 290 nm) exhibits saturation at high excitation levels, and fitting to a 2-level model ([5], (3)) shows more discrepancies towards high excitation level (Fig. 1c, lower panel). These signatures suggest the opening-up of additional pathways involving excited state absorption of triiodide.

The pump-probe trace with TL pump pulse in Fig. 2b displays very similar features as in previous works [2, 3], an initial fast rise up to ~ 10 ps, dominated by photoproduct formation, then followed by a slow delay when recombination

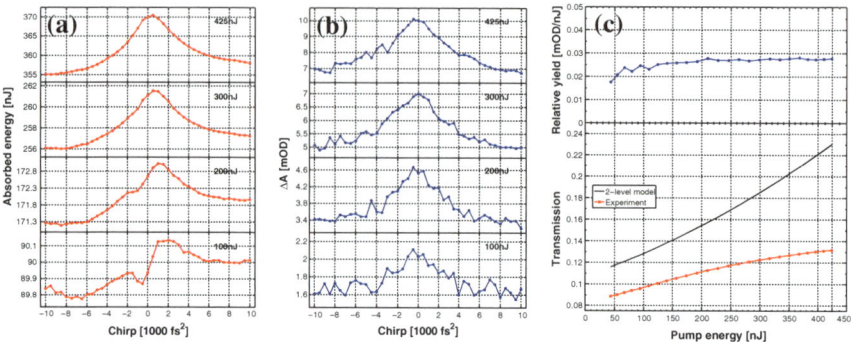

Fig. 1 Plots of the chirp dependence of **a** sample absorption and **b** ΔA of I_2^-, all taken with a constant pump-probe delay of 100 ps. The excitation energies used in **a** and **b** are marked on the *corner* of each subplot and as *ticks* in (**c**). **c** Excitation energy dependence of the relative I_2^- yield (*upper*) and the sample transmission (*lower*) in comparison with a 2-level model [5]

processes (geminate or non-geminate) outrun dissociation, which lasts until hundreds of picoseconds and further. It is demonstrated here that the TL pulse is more favorable to I_2^- formation than chirped pulses at all pump-probe delays and I_2^- formation is insensitive to the sign of the chirp, which reaffirms the chirp scan results. The possible processes involved are summarized in Fig. 2a. The cascaded one-photon absorption of the reactant could be more effective for short pulses, and in the case of additional dissociation from higher excited state(s) the overall I_2^- yield will be higher. Absorption of excitation photons by I_2^- during a long excitation pulse can be a parasitic effect that suppresses its absorption of the probe photon. The interplay between the two dissociation channels of I_3^- can eventually lead to the observed behavior of the I_2^- yield in our coherent control and chirp scan experiments.

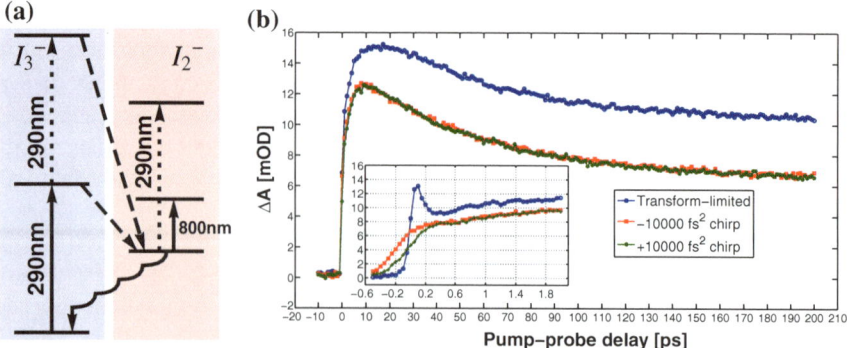

Fig. 2 **a** The proposed energy diagram of the $I_3^- - I_2^-$ system. The *arrows* represent the pump and probe (*both solid*), dissociation (*dashed*), re-absorption of the pump pulse (*dotted*), and recombination (*wavy*). **b** Pump-probe traces measured with different chirp rates with 450 nJ excitation

Spectrally-resolved studies are underway to pin down the principal mechanism as well as to investigate if other unknown species formed at high excitations influence pump-probe signal in the same spectral region, or a spectral shift of the I_2^- absorption band due to, e.g., dissociation of the solvent [5].

Acknowledgments This work was supported by the Max Planck Institute for the Structure and Dynamics of Matter (MPSD) and the Hamburg Center for Ultrafast Imaging (CUI).

References

1. C. G. Elles, F. F. Crim, Ann. Rev. Phys. Chem. 57, 273 (2006).
2. T. Kühne, R. Küster, P. Vöhringer, Chem. Phys. 233, 161 (1998).
3. E. Gershgoren, U. Banin, S. Ruhman, J. Phys. Chem. A 102, 9 (1998).
4. B. J. Pearson, T. C. Weinacht, Opt. Express 15, 4385 (2007).
5. V. I. Prokhorenko, A. Halpin, P. J. M. Johnson, R. J. D. Miller, L. S. Brown, J. Chem. Phys. 134 085105 (2011).

Multidimensional Photochemistry Model: Application to Aminobenzonitrile and Benzopyran

Aurelie Perveaux, Pedro J. Castro, Mar Reguero, Hans-Dieter Meyer, Fabien Gatti, Benjamin Lasorne and David Lauvergnat

Abstract A study of the photochemical reactivity of aminobenzonitrile and benzopyran is presented. The potential energy surfaces are investigated at the CASSCF level and full-dimensional models are developed to perform quantum dynamics calculations.

Over the last decades, progress in experimental techniques combined with theoretical simulations has given access to the analysis and control of the photochemical reactivity of large molecular systems with numerous technological applications. For example, it has been shown that the shape of the laser pulse that initiates the photoisomerisation of spiropyrans into merocyanines was determining for using them as optically-controlled molecular switches [1, 2]. Aminobenzonitrile-like molecules are another example where different fluorescence patterns are observed, depending on the solvent or the substituents. Such properties are crucial in the field of organic materials to understand and design fluorescent markers or photoswitches [3, 4].

One major challenge in theoretical chemistry concerns the study of photochemical processes of large molecules (several tens of modes), where Conical Intersections (CI) play a major role by transferring population between electronic state [5]. This requires the calculation of more than one electronic state and the non-adiabatic coupling between them, the development of models of potential energy

A. Perveaux · D. Lauvergnat (✉)
Laboratoire de Chimie Physique, Université Paris-Sud and CNRS, UMR 8000,
91405 Orsay, France
e-mail: david.lauvergnat@u-psud.fr

A. Perveaux · F. Gatti · B. Lasorne
CTMM, Institut Charles Gerhardt Montpellier, Montpellier Cedex 5, 34095 Montpellier,
France

P.J. Castro · M. Reguero
Departement de Química Física I Inorgànica, Pl. Imperial Tarraco 1, 43005 Tarragona, Spain

H.-D. Meyer
Theoretische Chemie, Physikalisch-Chemisches Institut, Ruprecht-Karls Universität,
Im Neuenheimer Feld 229, 69120 Heidelberg, Germany

© Springer International Publishing Switzerland 2015
K. Yamanouchi et al. (eds.), *Ultrafast Phenomena XIX*,
Springer Proceedings in Physics 162, DOI 10.1007/978-3-319-13242-6_94

386

surfaces, and the use of modern quantum dynamics simulation methods. Our strategy is summarised as follows.

(i) Exploring the potential energy surfaces and optimising specific points (e.g., CI, minima, and TS) with quantum chemistry calculations. In a first stage, these are run at the CASSCF level of theory with a polarised extended basis sets, and the solvent effect is described implicitly with the PCM model.

(ii) Generating the full-dimensional potential energy surfaces as analytical functions of the nuclear coordinates. More precisely, the diabatic surfaces and the electronic couplings are expressed as quadratic and linear expansions around reference geometries, respectively:

$$H_{ii}^{Diab}(Q) = e_{ii} + \frac{1}{2}\sum_{K}\sum_{L}\left(Q^K - Q_{ref_{ii}}^K\right)f_{ii}^{KL}\left(Q^L - Q_{ref_{ii}}^L\right) \qquad (1)$$

$$H_{ij}^{Diab}(Q) = \sum_{K}\lambda_{ij}^K\left(Q^K - Q_{ref_{ij}}^K\right) \qquad (2)$$

The energy (e_{ii}) and geometry (Q_{ref}) parameters of the model are extracted from the ab initio calculations. The linear parameters of the electronic couplings (λ_{ij}) are determined with the ab initio branching space vectors at the CI. The curvatures of the diabatic potentials (f_{ii}) are obtained through a self-consistent procedure using the second-order Jahn Teller effect.

(iii) Solving the time-dependent Schrödinger equation for the nuclei for all the degrees of freedom. This is achieved with the Multilayer Multiconfiguration Time-Dependent Hartree method (ML-MCTDH) [6].

This work is focused on two photoreactive processes:

1 Aminobenzonitrile (ABN, R = H) versus Dimethylaminobenzonitrile (DMABN, R = CH₃) Charge Transfer

The absorption of UV light excites the system to the second excited state (S_2), and CI's between S_2 and S_1 deactivate the system to S_1. Experimentally, one fluorescence band from a Local Excitation (LE) electronic state (S_1) is observed for both molecules in the gas phase. In polar solvents, DMABN presents a second fluorescence band from an Intramolecular Charge Transfer (ICT) electronic state (S_1) [4].

A *Cs* pathway is already known (Fig. 1, right panel) [7]. Nevertheless, we have found a new C_{2v} pathway (Fig. 1). Both are complementary and in competition during the photoinduced process.

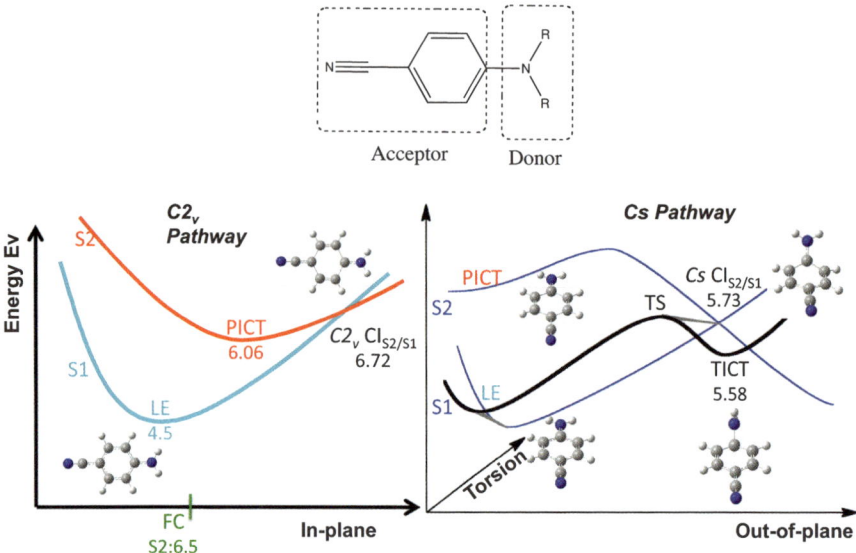

Fig. 1 Aminobenzonitrile photochemistry pathways. Energy in eV with respect to the ground state minimum

The *Cs* pathway is along the out-of-plane benzene deformation and linked to the formation of LE. The twisted (C-NH$_2$ torsion) ICT (TICT) minimum is reached when the wave packet is developed along the torsion mode once on the seam.

The *C$_{2v}$* pathway is along the in-plane quinoidal deformation. This motion drives the system to a sloped *C$_{2v}$* CI$_{S2/S1}$ leading to the formation of LE (Fig. 1, left panel). Similar results were obtained for DMABN in the gas phase. One should expect the *C$_{2v}$* pathway to be more efficient than the *Cs* one in the gas phase, thus explaining the absence of dual fluorescence for both molecules.

In a polar solvent, the ICT state will be strongly stabilised (Fig. 2 left panel). Therefore, the *C$_{2v}$* CI$_{S2/S1}$ is closer to the FC region than in the gas phase. To study the effect of the solvent during the photochemistry process, the full-dimensional (39 dimensions) potential energy surfaces of the *C$_{2v}$* pathway were calculated within the above procedure in the gas phase and in a polar solvent (Fig. 2 left panel); and quantum dynamics calculations were performed. The evolution of the diabatic populations (Fig. 2 right panel) confirms that the *C$_{2v}$* pathway becomes efficient in polar solvents.

Further studies, taking into account the effect of the polar solvent on the photoreactivity of DMABN are ongoing.

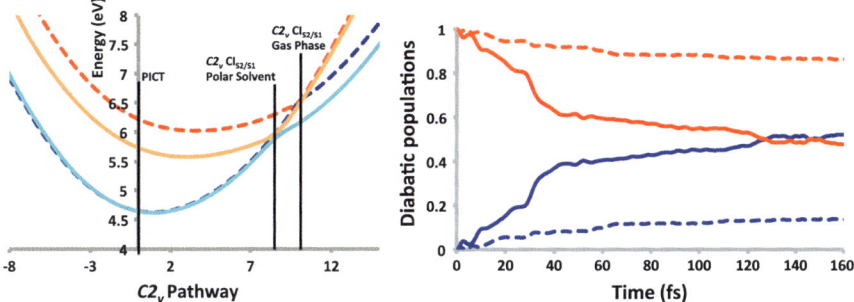

Fig. 2 Potentials and populations along the C_{2v} pathway. *Left* Comparison of the potential energy, with respect to the minimum in the ground state, in a polar solvent (*plain line*) and in the gas phase (*dashed line*). *Right* Time evolution of the diabatic populations. *Dashed line* gas phase. *Plain line* acetonitrile

2 Photoisomerisation of Benzopyran (the Spiropyran Chromophore)

The absorption of UV light excites the system to the S_1 state. Through vibrational relaxation, the system reaches a $CI_{S1/S0}$ where it deactivates to the ground state without fluorescence. Part of the wave packet goes back to the closed form (benzopyran), the other part goes to the open form (merocyanine). This photochemistry process is reversible and both isomers absorb different wavelengths. Thus, it is a photochromic system [8].

The quantum chemistry of benzopyran was the center of numerous studies and is nowadays well known. Therefore, we applied our multidimensional model, and a UV spectrum was obtained after a 100-fs dynamics. The theoretical absorption spectrum is in good agreement with the first band of experimental spectrum (Fig. 3). Intrinsic anharmonicity is critical to describe the photoreactivity and will be included in our future models.

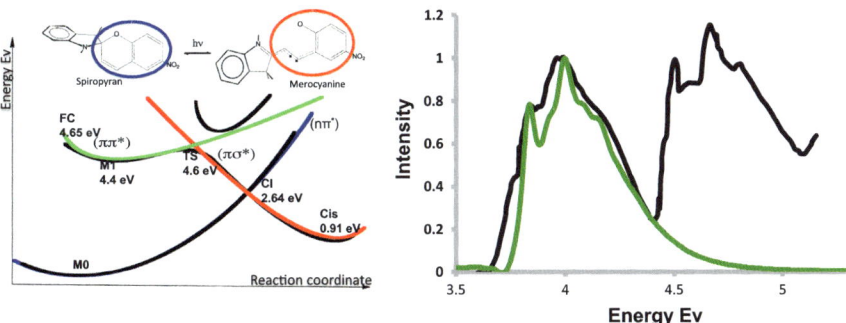

Fig. 3 *Left* scheme of the benzopyran potential energy surfaces along the reaction coordinate. *Black line* adiabatic potential; *color line* diabatic potential. *Right* benzopyran absorption spectrum. *Black* experimental; *green* model. Energies are in eV

References

1. S.-R. Keum, M.-S. Hur, P. M. Kazmaier, and E. Buncel. Canadian Journal of Chemistry 69, 12 (Dec. 1991), 1940–1947.
2. M. Saab, L. Joubert Doriol, B. lasorne, S. Guerin, and Fabien Gatti. Chem. Phys. 442, 93–102, (2014)
3. C. Bosshard, K. Sutter, P. Pretre, J. Hulliger, M. Florsheimer, P. Kaatz, and P. Gunter. Organic Nonlinear Optical Materials, Advances in Nonlinear Optics. Gordon and Breach Publishers, New York, 1995.
4. Z. R. Grabowski and K. Rotkiewicz. Chem. Rev **103**, 3899-4031 (2003).
5. G.A. Worth, L.S. Cederbaum. Chemical Physics Letters **338**, 219-223 (2001)
6. O. Vendrell and H-D Meyer. J. Chem. Phys. **134**, 044135 (2011)
7. I. Gómez, M. Reguero, M. Boggio-Pasqua, and M. A. Robb. J. Am. Chem. Soc **127**, 19 (2005)
8. L. Joubert-Doriol, B. Lasorne, D. Lauvergnat, H-D Meyer, F. Gatti. The Journal of Chemical Physics 140, 044301 (2014)

Tuning of Isomerization Rates
in Indigo-Based Photoswitches

E. Samoylova, B. Maerz, S. Wiedbrauk, S. Oesterling, A. Nenov,
H. Dube, R. de Vivie-Riedle and W. Zinth

Abstract Ultrafast excited-state dynamics in indigo-based photochromic compounds was studied with the transient absorption spectroscopy and ab initio calculations. We demonstrate an approach for adjusting excited state relaxation routes
and photoisomerization rates for applications where fast photoswitching is needed.

1 Introduction

Manipulation of chemical reactions in biomolecules represents a challenging task in
chemistry and is an important aspect for applications. Different classes of photochromic molecules were proposed as light-controllable triggers for incorporation in
biomolecules [1]. An ideal photochromic trigger molecule should fulfill several
requirements, such as separated absorption spectra for both photochromic isomers,
fast isomerization reactions, high thermo- and photostability as well as good fatigue
resistance for both isomers. Therefore, adjustment of photochemical and photo-
physical properties of the molecules of interest should be done. Commonly, those
properties can be tuned by changing external (e.g., solvent polarity) or internal
(e.g., chemical substitutions) conditions. Recently we have studied a wide range of
indigo-based photoswitches focusing our interest on the photoswitching rates [2, 3].

An example of isomerization in the hemithioindigo (HTI) photoswitch is shown
in Fig. 1.

The photoisomerization reaction takes place at the central double bond, where
the stilbene part and the thioindigo part form a donor-acceptor system. By attaching

E. Samoylova (✉) · B. Maerz · W. Zinth
Chair for Biomolecular Optics, Department of Physics, Ludwig-Maximilians-University,
Oettingenstr. 67, 80538 Munich, Germany
e-mail: elena.samoylova@physik.uni-muenchen.de

S. Wiedbrauk · S. Oesterling · A. Nenov · H. Dube · R. de Vivie-Riedle
Department of Chemistry, Ludwig-Maximilians-University, Butenadtstr. 5-13, 81377
Munich, Germany

© Springer International Publishing Switzerland 2015
K. Yamanouchi et al. (eds.), *Ultrafast Phenomena XIX*,
Springer Proceedings in Physics 162, DOI 10.1007/978-3-319-13242-6_95

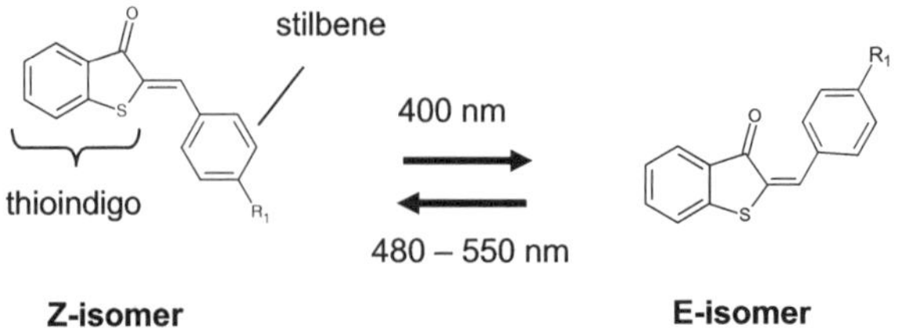

Fig. 1 Chemical structure of the Z- and E-isomers of HTI. The photoreaction between the two isomers is driven with visible light. R_1 indicates the attachment position for substituents

chemical groups with different electron-donating strength on the stilbene part and carrying out time-resolved experiments we demonstrated that the rates of the photoisomerization reaction can be successfully tuned. With the help of quantum chemical calculations a detailed understanding of the isomerization processes at the excited state was obtained.

2 Experimental

The time-resolved experiments were carried out on the Ti:Sapphire-based laser amplifier system (Spectra Physics) that delivers pulses at a central wavelength of 800 nm, with a repetition rate of 1 kHz and a pulse duration of 90 fs. The pump pulse is frequency doubled ($\lambda = 400$ nm) and/or converted using a home-built two-stage non-collinear optical parametric amplifier (NOPA). For the excitation of the HTI-derivatives presented here, the excitation (pump) wavelengths of 400, 480, 500, 550 nm were used with pulse duration down to 60 fs. The white-light pulses ($\lambda = 350$–700 nm), used as the probe, are generated by continuum generation from the 800 nm pulse in a rotating CaF_2 plate. The sample dissolved in dichloromethane is circulated with a peristaltic pump.

3 Results and Discussions

The hemithioindigo molecules, as a new class of photoswitchable compounds, are attractive for fundamental and applied studies due to their good isomerization efficiencies and absorption bands located in the visible spectral range for both isomers. The dynamics of the excited states was studied in a pump-probe experiment where changes in absorption were monitored in a time interval up to 3 ns with a time resolution of ~ 70 fs. Examples of transient absorption spectra for the Z- and E-isomers are shown in Fig. 2.

The positive signal in both sets of spectra is due to excited-state absorption (ESA) whereas the negative signal observed near 400 nm for the Z-isomer (Fig. 2a) and near 480 nm for the E-isomer (Fig. 2b) is a result of ground-state bleaching

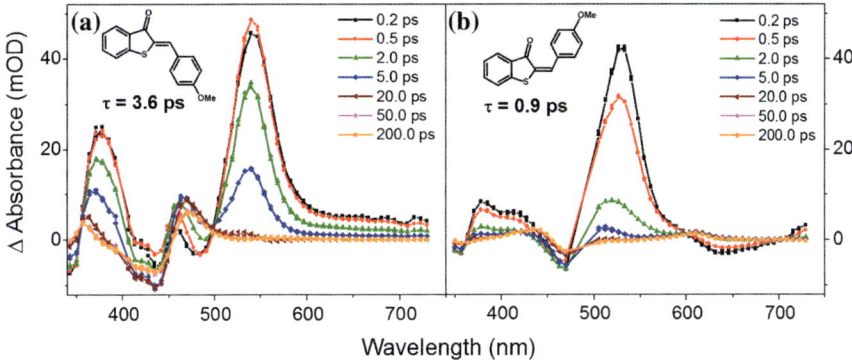

Fig. 2 Transient absorption spectra at fixed time delays for Z-isomer (**a**) and E-isomer (**b**)of the methoxy-substituted ($R_1 = OCH_3$) HTI in dichloromethane. The structure of the corresponding isomers and their excited state lifetimes extracted from the fit are given in the top

(GSB). The fast evolution of the absorption signals is observed within the first 10 picoseconds for both isomers and assigned to the relaxation of the excited states. The excited-state decay is followed by a slower process assigned to vibrational cooling in the hot ground state. The reaction is complete with a building of a sinusoidal absorption signal that remains constant within the studied time interval. Those signals are attributed to the formation of the photochromic product.

Comparison of the dynamics for all the derivatives, bearing single substitutions, showed that the increase in the donating strength of a substituent leads to an initial acceleration of the reaction rates as can be seen in the Table 1 (from 38 ps to 3.6 ps). The electron-donating strength is described by the Hammett parameter, σ^+, which is more negative for stronger electron-donating groups.

In the presence of very strong donors, the isomerization times for the Z/E reaction, however, decrease (10 ps ($\sigma^+ = -1.70$) and 29 ps ($\sigma^+ = -2.03$)). This phenomenon is attributed to a change in the excited state potential energy surfaces involved in the relaxation. It was established for HTI that the isomerisation proceeds via a barrier formed by a forbidden crossing between the locally-excited S_1 state and a charge-transfer S_2 state. Acceleration of the reaction is assigned to a decrease of the reaction barrier due to energetic lowering of the S_2 state solely. For very strong donors, both the energies of the S_1 and the S_2 states are lowered, leading to an increase of the barrier and therefore to the reduction of the reaction rates.

Table 1 Isomerisation times for the Z/E-reaction in selected HTI-derivatives with a single substitution

Isomerisation time, Z/E reaction	$R_1 = H$ ($\sigma^+ = 0$)	$R_1 = CH_3$ ($\sigma^+ = -0.31$)	$R_1 = OCH_3$ ($\sigma^+ = -0.78$)	$R_1 = N(CH_3)_2$ ($\sigma^+ = -1.70$)	$R_1 = C_6H_{12}N$ ($\sigma^+ = -2.03$)
τ_Z (ps)	38	13	3.6	10	29

From the experimental results supported by high-level ab initio calculations we conclude that the reaction rates can be tuned via the donor-acceptor character of the HTI molecule. We have shown that electron donating substituents may accelerate the reaction significantly. However, if the single, electron donating group becomes very strong, the potential energy surface is changed significantly and the original trend is inverted.

References

1. W. Szymanski, J. M. Beierle, H. A. V. Kistemaker, W. A. Welema, B. L. Feringa, "Reversible photocontrol of biological systems by the incorporation of molecular photoswitches," Chem. Rev. **113**, 6114-6178 (2013)
2. T. Cordes, T. Schadendorf, K. Rück-Braun, W. Zinth, „Chemical control of hemithioindigo-photoisomerization – substituents effects on different molecular parts," Chem. Phys. Lett. **455**, 197–201 (2008)
3. M. Dittmann, F. F. Graupner, B. Maerz, S. Oesterling, R. de Vivie-Riedle, W. Zinth, M. Engelhard, and W. Luettke "Photostability of 4,4'-dihydroxythioindigo, a mimetic of indigo," Angew. Chem. Int. Ed. **53**, 591-594 (2014)

Bimolecular Reactions on a Timescale Below 1 ps

Roland Wilcken and Eberhard Riedle

Abstract Access to the intrinsic reaction rate is gained by canceling out diffusion. The use of precursors on demand and reactive solvents allows the study of bimolecular reactions down to 200 fs. Even the molecular rotation is considerably slower and a preformed, favorable configuration is concluded.

1 Beating Diffusion to Get Access to the Intrinsic Rates of Bimolecular Reactions

Chemical reactions in solution are thought to be limited by the diffusive approach of the reaction partners. Even for reactions that occur on the first encounter, the observed reaction time cannot be faster than several nanoseconds [1]. This so-called diffusion limit masks the intrinsic reaction rate and allows no direct insight into the specific reaction mechanism at highest reaction rates. To overcome this limitation we use one reactant as the solvent (Fig. 1). Therefore it is already in close vicinity to the solute chosen to react with. The second condition to be met is the activation of one of the reaction partners on demand in order to be able to synchronize the reaction and the observation.

The reaction we study is a nucleophilic substitution reaction of benzhydryl compounds with leaving groups like chlorine. The precursor undergoes photolysis to produce a benzhydryl cation. In this way, the reaction can be triggered by light excitation to produce the cation on demand and it is free to react right away with a proper nucleophilic partner. Using the nucleophilic reaction partner as solvent cancels the need for a diffusional approach and allows for direct observation of the intrinsic reaction rate. The reaction time is now only dependent on the reactivity and on the orientation of the involved molecules. In our measurements we find time

R. Wilcken (✉) · E. Riedle
LS für BioMolekulare Optik, Ludwig-Maximilians-Universität München,
Oettingenstr. 67, 80538 Munich, Germany
e-mail: roland.wilcken@physik.uni-muenchen.de

© Springer International Publishing Switzerland 2015
K. Yamanouchi et al. (eds.), *Ultrafast Phenomena XIX*,
Springer Proceedings in Physics 162, DOI 10.1007/978-3-319-13242-6_96

395

(a)   (b)   (c)

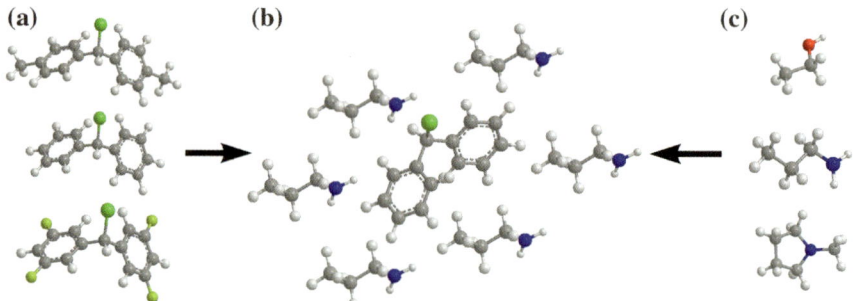

Fig. 1 **a** Benzhydryl chloride (B–Cl, *middle*) precursor with methyl (B-Me2-Cl, *top*) and fluorine substitution (B-F4-Cl, *bottom*). **b** Reaction mixture of precursor and nucleophilic solvent. **c** Strong nucleophiles: Ethanol, *n*-propylamine and *N*-methylpyrrolidine (from *top* to *bottom*)

constants down to 200 fs for the most reactive combinations. This is three orders of magnitude faster than the diffusion limit and faster than all bimolecular reactions observed up to now. Surprisingly, it is also at least one order of magnitude faster than the rotational relaxation times of the nucleophiles and at the timescale of molecular vibrations.

2 Tuning Reactivity to Speed up Reaction Times

We perform femtosecond pump-probe measurements with a 30 fs pump pulse at 270 nm and a supercontinuum probe pulse ranging from 290 to 720 nm. With this setup we can follow the reaction of the benzhydryl cations with a time resolution of about 50 fs. The benzhydryl compounds are well established reference electrophiles in nucleophilic substitution reactions. The photoexcitation of benzhydryl chloride leads to the cleavage of the carbon-chlorine bond. The bond can cleave either homolytically or heterolytically resulting in a benzhydryl cation and a chlorine anion or in the corresponding radical pair. A rapid removal of the chlorine in 125 fs leads to the cation [2]. The reaction of the cation with the nucleophile can be followed by the decay of its unique spectral signature around 440 nm [2, 3]. The decays of the cation signal shown in Fig. 2 demonstrate that we can follow extremely fast reactions that depend strongly on the individual pair of reactants. By chemical analysis we confirm that the decreasing cation population is indeed due to the bimolecular reaction.

We tune the reactivity of the cation by substitution on the phenyl rings. Substitution of electron withdrawing fluorine makes the cation much more reactive while an electron donating methyl substitution leads to longer reaction times. Additionally, we choose a variety of different alcohols and amines as nucleophiles to react with the benzhydryl cations (Fig. 1). For the alcohols we can adjust the reactivity mainly by the carbon chain length. For methanol we measure reaction

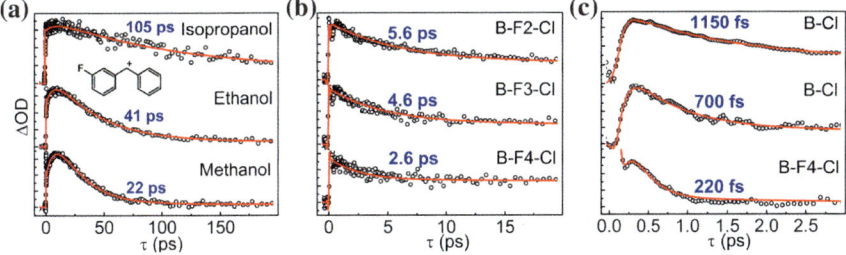

Fig. 2 Experimental results for the decay of the cation spectral signature at 440 nm. **a** Time traces and fits of the cation signal for alcohols with decreasing chain length. **b** Time constants for methanol reach the rotational limit. **c** For amines the reactions are even faster. Shown are examples of the cation trace for diaminopropane (*top*), N-methylpyrrolidine (*middle*) and n-propylamine (*bottom*)

Table 1 Time constants for the reaction of the cation. In the nomenclature of the precursor mainly the substitution of the used benzhydryl cations (B) is marked by Me (methyl) or F (fluorine). All time constants in ps

Cation	n-Propanol	Ethanol	Methanol	n-Propylamine	Pyrrolidine	N-Methyl-pyrrolidine
B-Me2	271	239	203	3.1	1.6	1.6
B-Me	146	103	76	2.5	1.2	1.1
B	89	56	32	1.6	0.9	0.7
B-F	62	41	22	0.8	0.6	0.5
B-F2	–	–	5.6	0.6	0.5	0.3
B-F3	–	–	4.6	0.3	0.4	0.3
B-F4	–	–	2.6	0.2	0.3	0.2

times from 200 ps down to 2.6 ps (Fig. 2a–b). This is on the edge of rotational relaxation times reported for small molecules. Even higher reaction rates are achieved with amines ranging from a few ps down to 200 fs (Fig. 2c).

For the amines we can reach higher reactivities by going from primary to tertiary amines. Remarkably, chemical variation loses its effect for the fastest reactions (Table 1). A typical rotational relaxation time for molecules of this size is 20 ps. This means the molecule cannot fully rotate before the reaction appears. Therefore the orientation of the molecules has to fit a certain angle and distance to make the extremely fast reaction possible. They have to be in the right orientation already before the light induced bond cleavage. Intermolecular interactions seem to play a crucial role here because precursor and nucleophiles both can form hydrogen bonds to bring the reactants into the favorable configuration. For the fastest reaction pairs the precursor absorption spectrum already starts to be altered.

The observation of the bimolecular reaction of a reactant produced on demand by photolysis and a surrounding partner allows the determination of the intrinsic reaction speed. It is found to be faster than even the rotational relaxation time. The reaction happens nearly immediately, likely on the first vibrational encounter and the highly reactive species already have a favorable orientation before photolysis.

References

1. J. Ammer, C. Nolte, H. Mayr, J. Am. Chem. Soc. **134**, 13902 (2012).
2. C.F. Sailer, N. Krebs, B. P. Fingerhut, R. de Vivie-Riedle, E. Riedle, to be submitted to J. Phys. Chem. Lett.
3. B. P. Fingerhut, C. F. Sailer, J. Ammer, E. Riedle, R. de Vivie-Riedle, J. Phys. Chem. A **116**, 11064 (2012).

Ultrafast Dynamics of a Bistable Intramolecular Proton Transfer Switch

Julia Bahrenburg, Michał F. Rode, Andrzej L. Sobolewski and Friedrich Temps

Abstract The stepwise formation of the proton transfer product of a bistable molecular switch was revealed by femtosecond fluorescence and absorption spectroscopy. The interpretation was supported by ab initio excited-state calculations.

1 Introduction

Excited-state intramolecular proton transfer (ESIPT) reactions belong to the fastest chemical reactions known [1]. Moreover, many molecules showing proton transfer after UV excitation exhibit record photostabilities. ESIPT molecules thus offer huge advantages in numerous fields, e.g. as photostabilizers in sunscreens for protection against solar UV light [2] or for the development of novel photochromic molecular switches [3].

Here, we report on the stepwise ultrafast formation of the proton transfer product of the bistable ESIPT switch N-(3-pyridinyl)-2-pyridinecarboxamide (NPPCA, Fig. 1) by femtosecond fluorescence and absorption spectroscopy. The experimental results are complemented by ab initio excited-state calculations at the MP2/cc-pVDZ, CC2/cc-pVDZ and ADC(2)/cc-pVDZ levels of theory.

J. Bahrenburg (✉) · F. Temps
Institute of Physical Chemistry, Christian-Albrechts-University Kiel,
Olshausenstr. 40, 24098 Kiel, Germany
e-mail: bahrenburg@phc.uni-kiel.de

F. Temps
e-mail: temps@phc.uni-kiel.de

M.F. Rode · A.L. Sobolewski
Institute of Physics Polish Academy of Science, al. Lotników 32/46,
02-668 Warsaw, Poland
e-mail: mrode@ifpan.edu.pl

A.L. Sobolewski
e-mail: sobola@ifpan.edu.pl

© Springer International Publishing Switzerland 2015
K. Yamanouchi et al. (eds.), *Ultrafast Phenomena XIX*,
Springer Proceedings in Physics 162, DOI 10.1007/978-3-319-13242-6_97

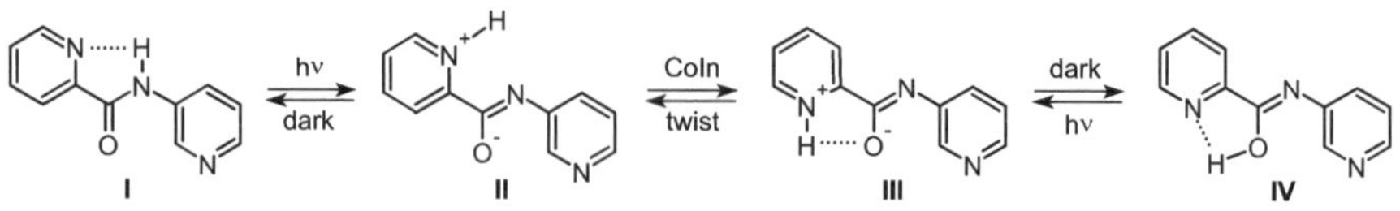

Fig. 1 Molecular structures and reaction scheme of the ESIPT switch NPPCA after UV excitation

2 Results

Measured fluorescence–time profiles, the two–dimensional (2D) transient absorption map and the transient absorption–time profiles at selected wavelengths after excitation of NPPCA in acetonitrile at $\lambda = 264$ nm are given in Fig. 2 a–e. The analysis of the fluorescence decay curves yielded lifetimes of $\tau_{fl,1} < 180$ fs as upper limit and $\tau_{fl,2} = 500 \pm 100$ fs mainly at longer emission wavelengths. The transient absorption map exhibits stimulated emission at $\lambda = 320 - 350$ nm and $450 - 550$ nm with lifetimes $\tau_1 = 100 \pm 10$ fs and $\tau_2 = 500 \pm 10$ fs, respectively, which correlate well with the fluorescence times. Additionally, we observe pronounced excited-state absorption (ESA) bands around $\lambda \approx 340, 470$ and 730 nm which feature a main decay time of $\tau_3 = 20 \pm 1$ ps next to the ultrashort τ_1 and τ_2 components.

3 Discussion

The experimental time constants determined by a global analysis were assigned to sequential dynamical transformations of the photo-induced molecules as illustrated in Fig. 2f. Accordingly, the ≈ 100 fs component is the lifetime of the initially excited Franck-Condon state of the keto isomer **I**. The applied electronic excitation induces the subsequent ultrafast proton transfer to the excited enol tautomer **II**. The observed 500 fs fluorescence lifetime has to be attributed to the barrierless transition of the enol **II** to the electronic ground state through the CoIn encountered at a twisted configuration of the pyridine moiety. The ESIPT from **I** to **II** should thus be completed in <500 fs. The main decay time of 20 ps seen in the absorption experiment must belong to states **II** and **III** in the electronic ground state and the proton transfer to the final product **IV**. As can be seen from the data in Fig. 2b, c, this 20 ps decay contribution features a 500 fs rise time precisely as expected by the observed 500 fs electronic deactivation time of the excited state. Last but not least, the formation of **IV** is confirmed by the weak positive permanent absorption ($\tau \gg 1$ ns) in the time profile at $\lambda = 247$ nm (Fig. 2c).

In conclusion, our experimental and computational results demonstrate the stepwise proton transfer reaction in the bistable photochromic ESIPT switch NPPCA. The ultrafast fluorescence decay within a time $\tau = 500$ fs, the delayed rise within

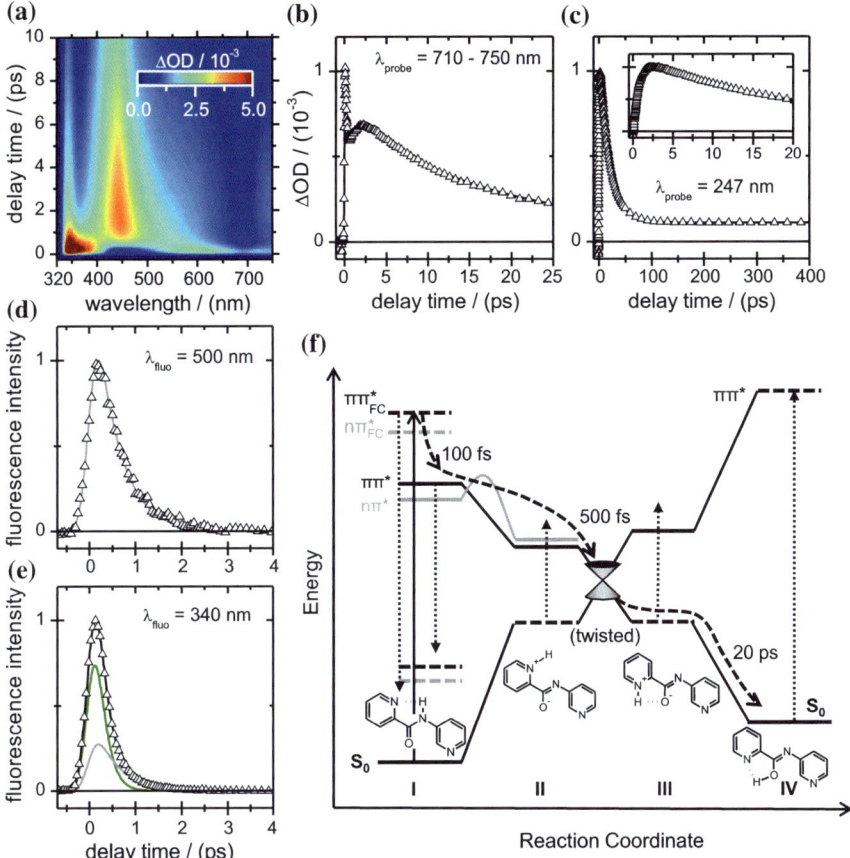

Fig. 2 **a** 2D transient absorption map. **b** Transient absorption–time profile at $\lambda_{probe} = 710 - 750$ nm. **c** Transient absorption–time profile at $\lambda_{probe} = 247$ nm. **d** Fluorescence–time profile at $\lambda_{fl} = 500$ nm. **e** Fluorescence–time profile at $\lambda_{fl} = 340$ nm. **f** Scheme of the photo-induced proton transfer in the investigate switch NPPCA obtained by combination of the experimental and computational [CC2 and ADC(2)] results. The assigned time scales belong to the distinctive reaction steps denoted by the *dashed arrows*. *Dotted arrows* indicate the observed transient emission and absorption bands

$\tau \approx 500$ fs of the absorption at $\lambda > 700$ nm and the persistent (permanent) absorption at $\lambda = 247$ nm provide strong evidence for the proposed ESIPT switching process.

JB and FT acknowledge the support of their work by the CRC 677. MFR and ALS have been supported by the National Science Centre of Poland and the ICM Warsaw.

References

1. A. Douhal, F. Lahmani and A. Zewail, " Proton-transfer reaction dynamics," Chem. Phys. **207**, 447–498 (1996).
2. T. Elsaesser and H. J. Bakker, ed., in *Ultrafast hydrogen bonding dynamics and proton transfer processes in the condensed phase* (Kluwer, Dordrecht, 2002).
3. L. Lapinski, M. J. Nowak, J. Nowacki, M. F. Rode and A. L. Sobolewski, "A bistable molecular switch driven by photoinduced hydrogen-atom transfer," ChemPhysChem **10**, 2290–2295 (2009).

Excited State Structural Dynamics Probed with Time-Resolved Sulfur K-Edge X-Ray Absorption Spectroscopy

Matthew Ross, Benjamin E. Van Kuiken, Mathew L. Strader, Amy Cordones-Hahn, Hana Cho, Robert W. Schoenlein, Tae Kyu Kim and Munira Khalil

Abstract Time-Resolved X-ray absorption spectroscopy at the sulfur K-edge (~ 2.4 keV) is used to monitor structural dynamics following excited state proton transfer in an organosulfur molecule. The timescales of electronic structural relaxation are solvent dependent.

1 Introduction

Many biological and chemical systems rely on the transfer of protons between bonding sites. In some cases this transfer occurs as excited state proton transfer (ESPT) after a molecule or protein has been promoted to an excited state [1]. Much of the work to study ESPT using optical methods utilizes broad, overlapping visible absorptions which lack spatial resolution. Ultrafast time-resolved X-ray absorption (XA) is unique in its ability to measure chemical dynamics in solution with atomic specificity by probing excitations from highly localized core-level electrons. In recent years, time-resolved XA spectroscopy has been demonstrated to study photoexcited structural dynamics in disordered media [2]. We, for the first time, present transient XA spectroscopy at the S K-edge. The molecule of interest is 2-thiopyridon (2TP, C5H5NS). After photo-excitation, the proton, bonded initially to nitrogen in the 2TP form, undergoes ESPT to bond with sulfur in the 2MP

M. Ross · B.E. Van Kuiken · M. Khalil (✉)
Department of Chemistry, University of Washington, Seattle, WA, USA
e-mail: mkhalil@chem.washington.edu

M.L. Strader
SLAC National Accelerator Laboratory, Menlo Park, CA, USA

A. Cordones-Hahn · H. Cho · R.W. Schoenlein
Chemical Sciences Division, Lawrence Berkeley National Laboratory, Berkeley, CA, USA

T.K. Kim
Department of Chemistry, Pusan National University, Geumjeog-gu, Busan
Republic of South Korea

© Springer International Publishing Switzerland 2015
K. Yamanouchi et al. (eds.), *Ultrafast Phenomena XIX*,
Springer Proceedings in Physics 162, DOI 10.1007/978-3-319-13242-6_98

(a)   **(b)**

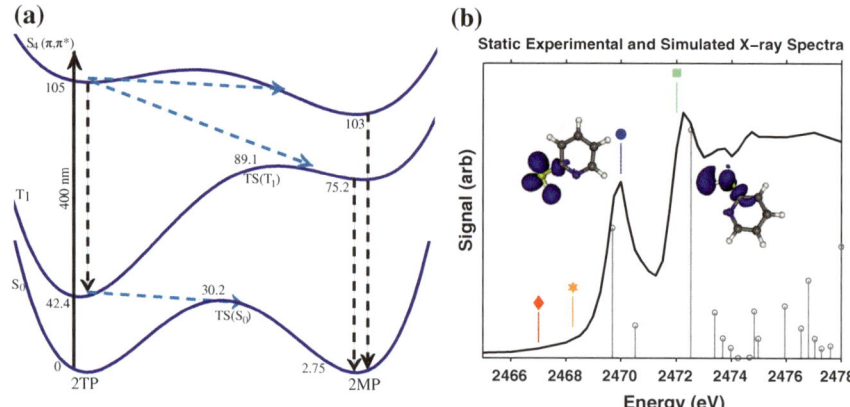

(2- mercaptopyridine) form (see Fig. 1a). By probing the S K-edge in this system, we are able to monitor the proton transfer process with site specificity and we measure how the solvent modulates the excited state dynamics. Our experiments open up the possibility of performing transient S K-edge experiments as a sensitive probe of the covalency of metal-sulfur bonds in small molecules and metalloproteins [2].

2 Transient Sulfur K-Edge Spectroscopy

Experiments on the sulfur K-edge of 2TP after photo excitation at 400 nm were performed at BL 6.0.1 at the advance Light Source using monochromated (~ 0.2 eV width) 70 ps camshaft-driven X-ray pulses with $\sim 10^4$ photons/pulse using a previously described set-up [3]. The transient XA experiments were performed in transmission mode with a 0.75 M solution of 2TP in acetonitrile in a 50 μm thick free flowing jet housed in a He atmosphere. In polar solutions the 2TP species, with the proton bonded to nitrogen dominates and is thought to be excited to the fourth excited singlet, $S_4(\pi, \pi^*)$ upon irradiation by near UV light [4]. Subsequent decay is thought to proceed through intermediate states to the first excited triplet, T_1. This is diagramed in Fig. 1a. Figure 1b shows the measured differential absorption spectra of the sulfur K-edge of 2TP in acetonitrile solution. The simulated spectra agree qualitatively with the observed experimental peak positions and amplitudes of near-edge transitions at the sulfur K- edge. We assign the peak at 2,470 eV to a 1 s to σ_1^* transition and the peak at 2,472 eV to a 1 s to σ_2^* transition in 2TP.

Figure 2a shows the differential pump probe spectra at S K-edge for several delay times. We observe bleaches at the strongest edge features seen in Fig. 1b, as well as new transient absorption features at lower energies. Figure 2b shows pump-probe time traces for several X-ray probe energies. The initial rise/decay for most of the energies can be fit to 70 ± 10 ps, which is consistent with the time resolution of synchrotron based X-ray pulses. The traces in Fig. 2b show longer dynamics on timescales of 0.8 ns for the 2,467 eV feature and 3.4–3.8 ns for the other spectral features. The mechanisms for these longer timescale dynamics are still under investigation; however, it is clear that there are multiple timescales accompanying ESPT in 2TP dissolved in acetonitrile.

Figure 2c compares the transient sulfur K-edge XA spectroscopy of 2TP dissolved in acetonitrile and water. As in the acetonitrile, the 2,470–2,472 eV ground state bleaches recover at the same rates, but slow by from 1.9 to 19 ns in water. The dynamics of the excited state feature at 2,468 eV also slows by an order of magnitude from 7 to 68 ns. These slower responses may be due to increased interaction with water compared with acetonitrile which could increase the time required for the proton to settle into the excited T_1 state of 2MP after proton transfer.

In order to verify initial states and better understand interpretation of the slow and fast rises presented in Fig. 2b, c, we preformed TD-DFT calculations. Figure 3a shows a comparison of the experimental absorption spectra and calculated absorption for the ground state 2TP compound. These data confirm that our system starts in the expected state. Both photoinduced x-ray absorptions occur at lower energy, which is also observed in several transient states possible in this system. The faster, lowest energy rise is consistent with the predicted absorption of the triplet 2TP state. This would indicate that the triplet forms quickly after initial photoabsorption and subsequent relaxation, faster than the 70 ps x-ray pulse. The slow peak, at 2,468 eV, however, is not so clearly assigned. As seen in Fig. 3b, two calculated states, a 2MP triplet and a radical are potential states, but more work is needed to positively identify the slow peak at 2,468 eV.

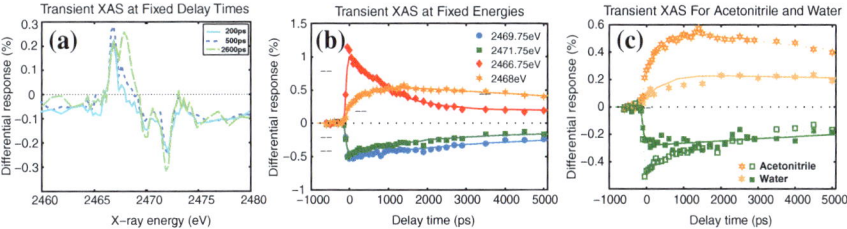

Fig. 2 **a, b** Pump probe traces for several X-ray energies demonstrating multiple time scales, with most initial responses too fast to resolve with 70 ps X-ray pulses. **c** Comparison of time traces for acetonirile (*open*) and water (*solid*)

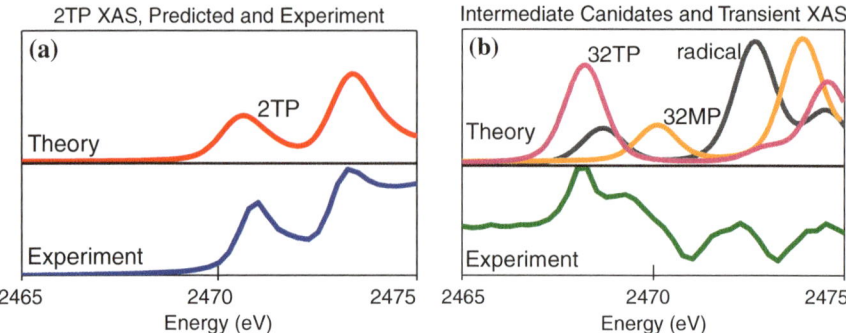

Fig. 3 a Comparison of measured static sulfur k-edge and closest match of predicted spectra, 2TP. **b** Comparison of measured transient sulfur K-edge spectra and three potential transient state predicted spectra

3 Summary

In summary, our data shows detailed changes to the electronic structure on the picosecond time-scale to an organosulfur compound engaged in ESPT using atomic site specific x-ray probes at the sulfur K-edge. We demonstrate the viability of transient S K-edge XA spectroscopy to probe proton transfer and other redox processes in solution with microscopic details.

References

1. R. Du, C. Liu, Y. Zhao, K. Pie, H. Wang, X. Zheng, M. Li, J. Xue, D.L. Phillips, "Resonance Raman Spectroscopic and Theoretical Investigation of the Excited State Proton Transfer Reaction Dynamics of 2-Thiopyridone," *J. Phys. Chem. B* 115, 8266-8277 (2011)
2. C.J. Mine; M. Chergui "Time-resolved X-ray absorption spectroscopy," *Spectroscopy Europe* 24, 16 (2012)
3. B. Van Kuiken, M. Valiev, S. Daifuku, C. Bannan, M. Strader, H. Cho, N. Huse, R.W. Schoenlein, N. Govind, M. Khalil "Simulating Ru L3- edge X-ray absorption spectroscopy with time-dependent density functional theory:model complexes and electron localization in mixed valence metal dimers," *J. Phys. Chem. A* 117 4444-54 (2013)
4. E. I. Solomon, B. Hedman, K. O. Hodgson, A. Dey, and R. K. Szilagyi, Coordination Chemistry Reviews **249,** 97 (2005).

Solvent Environment Revealed by Positively Chirped Pulses

Arkaprabha Konar, Vadim V. Lozovoy and Marcos Dantus

Abstract The spectroscopy of large organic molecules and biomolecules in solution has been investigated using various time-resolved and frequency-resolved techniques. Of particular interest is the early response of the molecule and the solvent, which is difficult to study due to the ambiguity in assigning and differentiating inter- and intra-molecular contributions to the electronic and vibrational populations and coherence. Our measurements compare the yield of fluorescence for two laser dyes IR144 and IR125 as a function of chirp. While negatively chirped pulses are insensitive to solvent viscosity, positively chirped pulses are found to be uniquely sensitive probes of solvent viscosity.

1 Introduction

Understanding molecular dynamics soon after photon absorption, taking into account the solvent environment surrounding the molecule, is central to predicting the course of chemical reactions and biophysical processes. The relevant timescales regarding photo-excitation in solution are determined by the inter- and intra-molecular interactions and their corresponding energy fluctuations, which occur in the 10–100 fs regimes [1, 2]. The multiple inter- and intra-molecular processes occurring during this time, convoluted by inhomogeneous broadening as well as the spectral and temporal response function of the experimental setup complicate assignment of the observed decay processes. Here we focus on the optical response from IR125 to IR144, which have been studied in solution as a function of temperature and solvent. The organic dye molecule IR125 undergoes non-polar

A. Konar · V.V. Lozovoy · M. Dantus
Department of Chemistry, Michigan State University, East Lansing, MI 48824, USA

M. Dantus (✉)
Department of Physics and Astronomy, Michigan State University, East Lansing,
MI 48824, USA
e-mail: dantus@msu.edu

© Springer International Publishing Switzerland 2015
K. Yamanouchi et al. (eds.), *Ultrafast Phenomena XIX*,
Springer Proceedings in Physics 162, DOI 10.1007/978-3-319-13242-6_99

solvation, while IR144 undergoes polar solvation given the reduction in its dipole moment upon excitation. The difference between both molecules is caused by the piperazine functional group in IR144 [3]. Three-pulse photon-echo peak shift (3PEPS) measurements revealed a solvent independent coherent response for times <0.1 ps, described as inertial, followed by a solvent dependent component changing from 4.04 to 1.36 ps in ethylene glycol at 297–397 K [4]. Our goal is to find spectroscopic probes [5] of solvation environment that are sensitive and easier to implement in a microscope, in particular we evaluate here chirped femtosecond pulses. These single-beam methods will be of paramount importance when investigating microenvironment effects on single molecules, due to the relative ease of the experimental implementation.

2 Experimental

Pulses from a femtosecond regeneratively amplified Ti: Sapphire laser producing 25 nm FWHM, (corresponding to 36 fs when transform limited) were used. Pulses were compressed and shaped using a pulse shaper (MIIPS-HD, Biophotonic Solutions Inc.) placed after the amplifier. A chirped-pulse scan consisted of recording molecular emissions as a function of φ'', the spectral chirp from negative to positive 20,000 fs^2, for $\varphi(\omega) = 0.5\varphi''(\omega - \omega_0)^2$. In a sense, a chirp scan can be interpreted as two-color time-resolved measurements for which early changes such as those occurring at 100 fs are observed near 1,000 fs^2.

3 Results and Discussion

Measurements of the dependence of integrated fluorescence intensity (detected at 90°) were recorded as a function of chirp. Results for both dyes dissolved in ethylene glycol at different temperatures are shown in Fig. 1. A secondary axis denoting the pulse duration of the chirped pulses has been provided in the figures to help elucidate the timing of the inter-and intra-molecular processes taking place during the chirp scans. The data is normalized on the asymptotic values of the chirp effect. The curves were normalized on the asymptotic values attained for negative chirp. Measurements for the two different dyes dissolved in ethylene glycol were performed at 278, 294 and 323 K and have been color coded as blue, black and red respectively. Negative chirp experiments can be thought of as being similar to pump-probe measurements, in which the bluer wavelength pump precedes the redder wavelength probe. This observation is consistent with pump-probe measurements comparing IR144 in methanol and ethylene glycol, and finding no difference. Positive chirp, on the other hand, yields different dynamics as a function of temperature. When exploring IR125 fluorescence, we find that colder solvent leads to enhanced fluorescence near 800 fs^2. This enhancement is not observed at higher

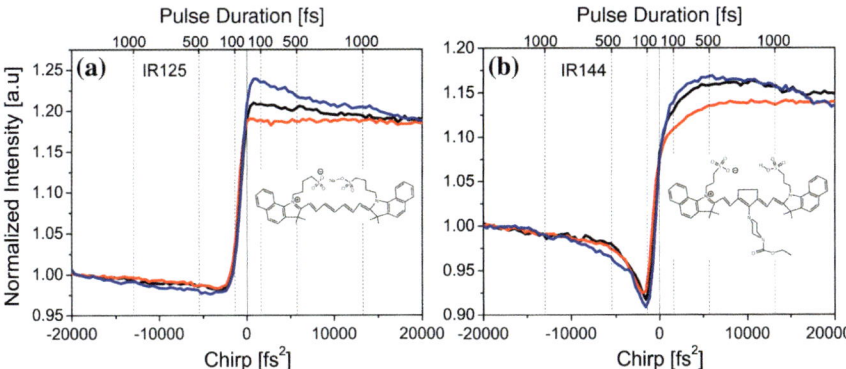

Fig. 1 Integrated fluorescence response to chirped pulses for **a** IR125 and **b** IR144 in ethylene glycol at three temperatures: 278 K (*blue*), 294 K (*black*), and 323 K (*red*)

temperatures. A similar fluorescence enhancement for colder solvent is observed for IR144, however, the maximum fluorescence is reached around 5,000 fs^2 which corresponds to 500 fs. The overall difference in the shape and decay rates between the two dyes can be attributed to molecular properties of the probe molecules. IR144 undergoes a change in dipole moment upon excitation and therefore undergoes polar solvation, which depends on solvent reorientation.

This accounts for the slower dynamics of IR144 in ethylene glycol. IR125 on the other hand undergoes non-polar solvation due to the absence of any significant change in dipole moment upon excitation. The viscoelastic model for non-polar solvation predicts a rapid viscosity independent inertial response and slower viscosity dependent diffusive dynamics following excitation. Our positive chirp findings for IR125 are consistent with the rapid viscosity independent inertial response (rise close to zero chirp) followed by the slower viscosity dependent diffusive dynamics observed as pulse duration increases. In contrast, the dipolar response of IR144 depends on solvent reorientation, a process that is viscosity dependent, and thus delaying the point where maximum fluorescence is observed. The ability of phase shaped laser pulses to probe the solvent environment is particularly exciting given the relative ease of these experiments compared to the much more complicated four wave mixing setups. We plan to take advantage of chirped pulses to probe solvent environment effects of probe molecules in interesting environments such as protein pockets, membranes and under single molecule conditions.

References

1. T. Joo et. al., "Electronic Dephasing Studies of Molecules in Solution at Room-Temperature by Femtosecond Degenerate 4-Wave-Mixing," Chem. Phys. **176,** 233-247 (1993).
2. R. Jiminez et. al., "Femtosecond Solvation Dynamics of Water," Nature **369,** 471-473 (1994).
3. A. C. Yu et. al., "Solvatochromism and Solvation Dynamics of Structurally Related Cyanine Dyes," J. Phys. Chem. A. **106,** 9407-9419 (2002).
4. S. A. Passino et. al., "Three-Pulse Echo Peak Shift Studies of Polar Solvation Dynamics," J. Phys. Chem. A. **101,** 725-731 (1997).
5. A. Konar et. al., "Solvation Stokes-Shift Dynamics Studied by Chirped Femtosecond Laser Pulses," J. Phys. Chem. Lett. **3,** 2458-2464 (2012).

Coherent Wavepacket Motion in Ultrafast Intermolecular Electron Transfer in Electron-Donating Solvent

Yusuke Yoneda, Shohei Nambu, Eisuke Takeuchi, Yutaka Nagasawa and Hiroshi Miyasaka

Abstract Coherent wavepacket motion in an ultrafast electron transfer (ET) system consisting of electron accepting solute, 5,12-bis(phenylethynyl)-naphthacene (BPN), and donating solvent, N,N-dimethylaniline (DMA), was investigated by means of femtosecond transient absorption spectroscopy and coherent wavepacket motions in the ground and in the excited states were observed.

1 Introduction

Photoinduced electron transfer (ET) is one of the most important fundamental chemical processes and its relation with the solvation dynamics has been studied extensively by time-resolved spectroscopies. In polar solvents, solvation is the main driving force because it is capable of stabilizing the charge or electric dipole

Y. Yoneda · S. Nambu · E. Takeuchi · Y. Nagasawa (✉) · H. Miyasaka
Graduate School of Engineering Science, Osaka University, Osaka 560-8531, Japan
e-mail: nagasawa@chem.es.osaka-u.ac.jp

Y. Yoneda
e-mail: yoneda@laser.chem.es.osaka-u.ac.jp

S. Nambu
e-mail: nanbu@laser.chem.es.osaka-u.ac.jp

E. Takeuchi
e-mail: takeuchi@laser.chem.es.osaka-u.ac.jp

H. Miyasaka
e-mail: miyasaka@chem.es.osaka-u.ac.jp

Y. Nagasawa · H. Miyasaka
Center for Quantum Science and Technology Under Extreme Conditions,
Osaka University, Osaka 560-8531, Japan

Y. Nagasawa
PRESTO, Japan Science and Technology Agency (JST), Saitama 332-0012, Japan

© Springer International Publishing Switzerland 2015
K. Yamanouchi et al. (eds.), *Ultrafast Phenomena XIX*,
Springer Proceedings in Physics 162, DOI 10.1007/978-3-319-13242-6_100

moment produced by ET reaction. On the other hand, intramolecular nuclear reorganization is also considered to be important for ultrafast ET, although its understanding is still limited and further investigation is necessary. In the Marcus inverted region and near the top of the bell-shaped energy gap dependence of ET, it becomes possible for the reactant to access the intramolecular vibrational levels of the product state and ultrafast ET can manifest without much solvation. Higher vibrational levels in the excited state of a reactant are also accessible by photo-excitation with excess energy. It should be possible to accelerate the reaction by selecting a particular vibrational level with the strongest electronic coupling with the product state.

By employing femtosecond ultrafast spectroscopies that can induce and monitor coherent intramolecular wavepacket motions, dynamical aspects of the molecular vibrations in ET systems can be investigated. Photoinduced ultrafast intermolecular ET with time constants in the order of hundreds of femtoseconds or shorter are reported for dyes in an electron donating solvent, N,N-dimethylaniline (DMA). We have monitored wavepacket motions accompanying the ET for DMA solution of 5,12-bis(phenylethynyl)-naphthacene (BPN) by means of femtosecond white-light supercontinuum transient absorption (TA) spectroscopy with an excitation wavelength centered at 560 nm and a pulse duration of ca. 23 fs. In this system, ground state and excited state oscillations can be readily separated by changing the monitoring wavelength.

2 Results and Discussion

The differential absorbance of BPN in DMA and in 1-chloronaphthalene (1-CN) are contour plotted against wavelength and time in Fig. 1. 1-CN is an inert solvent with polarity similar to DMA and often used for comparison. For BPN in 1-CN (Fig. 1a), a strong negative band appears at ~ 565 nm which is the mixture of ground state bleach and stimulated emission. The band is strongly modulated by wavepacket motion with frequency of ca. 310 cm^{-1} which can be the wavepacket motion in either the ground or the excited state. In Fig. 1b, time dependence of the second vibrational structure in the stimulated emission at ~ 620 nm is shown which is free of the ground state bleach. Thus, the oscillation in this band is safely ascribable to the excited state wavepacket motion with frequency of 310 cm^{-1}. The frequency of this oscillation is similar in both ground and in excited states.

In DMA (Fig. 1c), the intensity of the negative band centered at ~ 565 nm rapidly reduces because of the reduction of the stimulated emission by ET. The reduction of the stimulated emission is even more drastic at ~ 620 nm (Fig. 1d). The dephasing time of the oscillation modulating the intensity of the negative band is much faster than that in 1-CN while the dephasing time of the oscillation of the peak wavelength is almost the same. Because the excited state vibration of BPN in DMA disappears rapidly and only the ground state vibration will remain, it is

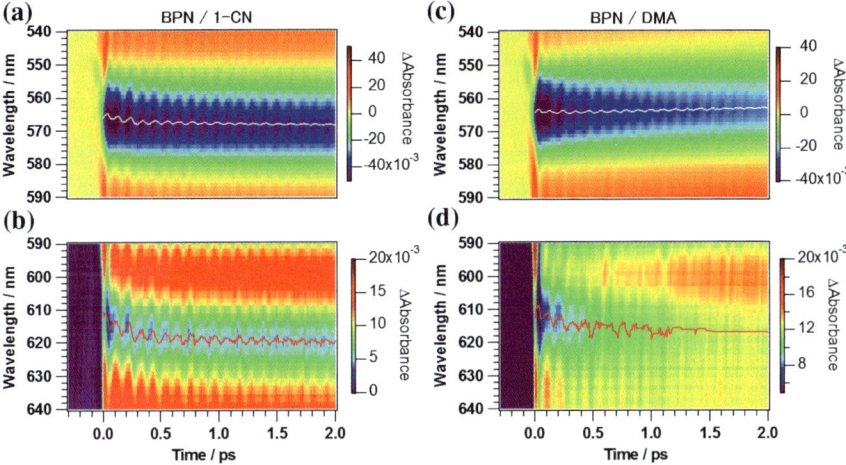

Fig. 1 Contour plot of TA spectrum of BPN in 1-CN (**a**, **b**) and DMA (**c**, **d**) excited at 560 nm. The presented wavelength ranges are 540–590 nm for (**a**) and (**c**) and 590–640 nm for (**b**) and (**d**). The peak wavelengths of the negative band are shown in white and red curves

indicated that the excited state wavepacket motion modulates the emission intensity more strongly than the peak wavelength.

Figure 2a shows time profiles of transient absorbance of BPN in 1-CN and in DMA. For the positive TA signal at 470 nm (upper panel), which is ascribed to the excited state of BPN, nearly no decay was observed in 1-CN, while ultrafast decay due to the ET was observed in DMA. The signal could be fitted by two decay components with time constants of 230 fs and 3.2 ps. The TA signals at 470 nm exhibit no wavepacket motion, while those at 630 nm exhibit strong oscillations. It can be also seen that the oscillation of BPN in DMA is significantly reduced by the ultrafast ET and a rise of the absorbance, which is ascribable to the production

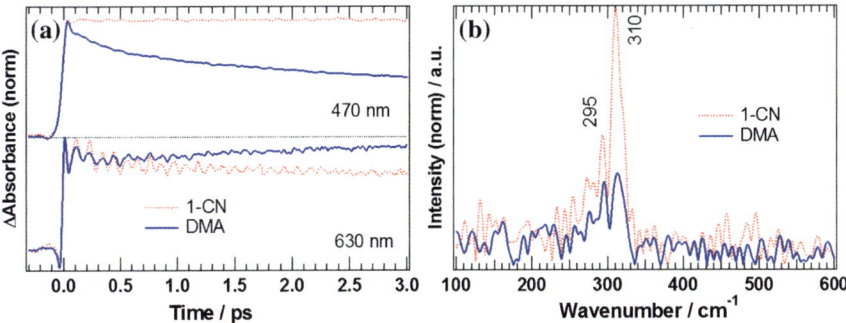

Fig. 2 **a** Time profile of transient absorption of BPN in 1-CN (*red dotted curves*) and DMA (*blue solid curves*) at 470 nm (*upper panel*) and 630 nm (*lower panel*). **b** The real parts of the Fourier transformed spectra of the oscillations at 630 nm

of the radical ion pair, can be also seen. Figure 2b shows the real parts of the Fourier transformed spectra of oscillations observed at 630 nm. The band at 310 cm^{-1} was the dominant mode in 1-CN with a weak shoulder at $\sim$295 cm^{-1}, and it was significantly reduced in DMA. Interestingly, the intensity of the shoulder at 295 cm^{-1} is comparably not much reduced. The vibrational dephasing time of the shoulder is originally shorter than that of the main band at 310 cm^{-1}, thus it is not much affected by the ultrafast ET.

3 Conclusion

Ultrafast solvent-solute ET of naphthacene derivative, BPN, in DMA solution occurred multi-exponentially with time constants of 230 fs and 3.2 ps. BPN in inert 1-CN solution exhibited strong low-frequency wavepacket motion in both ground and excited states with similar frequency of $\sim$310 cm^{-1}. The wavepacket motion in the excited state was significantly reduced by the ultrafast ET, which indicates that the oscillation is strongly coupled to the optical transition of the neutral BPN while it is not coupled to that of the produced radical ion pair.

Elementary Electron and Ion Dynamics in Ionized Liquid Water

Jialin Li, Zhaogang Nie, Yi Ying Zheng, Shuo Dong
and Zhi-Heng Loh

Abstract Polarization-resolved femtosecond coherence spectroscopy is used to observe the dealignment of the injected electron, hole orbital motion, solvent reorganization, and ballistic proton transport in ionized liquid water. The lifetime of the H_2O^+ cation is also determined.

1 Introduction

The ionization of liquid water is a universal phenomenon that accompanies the interaction of high-energy radiation with matter in aqueous environments [1]. The ensuing cascade of chemical reactions that involves ions, electrons, and radicals forms the basis of radiation chemistry and biology. The direct products of water ionization are the H_2O^+ radical cation and the injected electron, both of which are highly reactive species. H_2O^+ undergoes an ion-molecule reaction with a neighboring H_2O molecule to yield the hydronium cation and a hydroxyl radical. The timescale of this process is predicted to be ~ 100 fs, although experimental attempts at the determination of its lifetime have thus far been inconclusive [2]. In comparison, the spectroscopy and dynamics of the hydrated electron have been studied extensively [3]. However, the possible formation of an intermediary p state electron by the ionization of liquid water has yet to be verified [4].

Here, femtosecond coherence spectroscopy is combined with polarization anisotropy measurements to investigate the early-time electron and ion dynamics triggered by the multiphoton ionization (MPI) of liquid water [5]. The peak intensity of the 800-nm ionization pump pulse (2×10^{13} W/cm^2) yields a Keldysh parameter of $\gamma \approx 2$. Accordingly the MPI process can be characterized as being in the strong-field regime. The 800-nm probe pulse interrogates the $s \rightarrow p$ transitions

J. Li · Z. Nie · Y.Y. Zheng · S. Dong · Z.-H. Loh (✉)
School of Physical and Mathematical Sciences, Nanyang Technological University,
21 Nanyang Link, Nanyang S637371, Singapore
e-mail: zhiheng@ntu.edu.sg

© Springer International Publishing Switzerland 2015   415
K. Yamanouchi et al. (eds.), *Ultrafast Phenomena XIX*,
Springer Proceedings in Physics 162, DOI 10.1007/978-3-319-13242-6_101

of both the hydrated s electron and its presolvated precursor. The manner in which the intermolecular vibrational coherences of liquid water modulate the electron $s \rightarrow p$ absorption is analogous to the way in which a vibrational wave packet that is launched on a molecular potential energy surface modifies the energy and amplitude of its associated electronic transitions. By relying on the observation of coherences and polarization anisotropies that are associated with the various intermediates, our approach complements earlier methods based on optical pump–probe transient absorption (TA) spectroscopy.

2 Experimental Methods

The femtosecond coherence spectroscopy setup employs the 800-nm, 30-fs output from an amplified Ti:sapphire laser system as both pump and probe pulses. The sample target consists of a 40-μm-thick, wire-guided flowing jet of distilled water. The peak intensity of the ionization pump pulse is 2×10^{13} W/cm^2. Fluence dependence measurements reveal a 9-photon ionization pump process. The total energy deposition of 14.0 eV by the pump pulse is significantly above the vertical ionization potential of liquid water (11.16 eV) [6]. At this input energy, the large electron ejection length of 35 Å [7] allows access to the collective dynamics of $\sim 10^4$ water molecules.

3 Results and Discussion

Ionization of liquid water results in the observation of a pronounced differential absorption polarization anisotropy at early times, which vanishes for time delays that exceed ~ 350 fs (Fig. 1a). Analysis of the time-dependent polarization anisotropy within a framework that considers only the s ground and p_z excited states of the injected electron yields, in the case of H$_2$O (D$_2$O), values of 0.94 ± 0.05 (0.92 ± 0.04) for the initial fractional p_z population and 79 ± 5 fs $(101 \pm 6$ fs$)$ for the p_z lifetime. The creation of an aligned electron distribution in this work is facilitated by the use of a strong-field-ionizing pump pulse. While the initial fractional populations of the p_z electron generated in H$_2$O and D$_2$O are identical, the retrieved p state lifetimes are clearly dependent on H/D isotopic substitution. The factor of $\sim 1.3 \times$ longer lifetime in D$_2$O than H$_2$O is reminiscent of previous results obtained on the internal conversion dynamics of the photoexcited hydrated electron [8], which suggests the participation of solvent librational motion in the relaxation of the injected p electron.

Oscillatory features in the polarization anisotropy decay indicate a periodic reorientation of the probe transition dipole moment that is concomitant with the electronic relaxation of the p state electron. The FFT spectrogram reveals oscillation frequencies of 150 and 500 cm^{-1} at early times, neither of which exhibits a

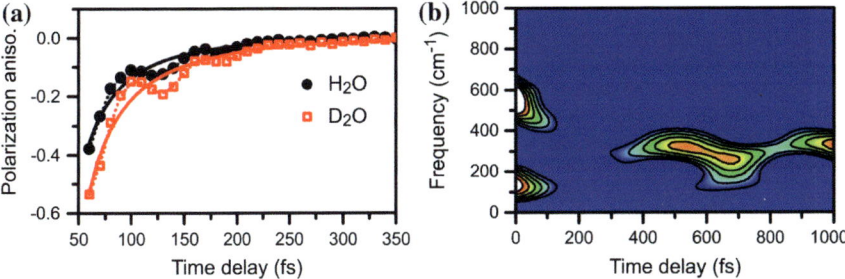

Fig. 1 **a** The polarization anisotropy shows that the *p* electron is produced by the MPI of liquid water. The symbols correspond to the experimental data, *dashed lines* serve as guides, and *solid lines* are fits to the theoretical model. **b** The time-frequency spectrogram for the anisotropic signal shows the existence of the 150 and 500 cm^{-1} frequency components at early times and the rise of the frequency component at 310 cm^{-1} at later time delays (>300 fs)

perceptible shift with H/D substitution (Fig. 1b). For both frequency components, time-domain analysis yields exponential damping times of 70 ± 12 and 64 ± 7 fs for H_2O and D_2O, respectively. The 150-cm^{-1} component that arises in both isotropic and anisotropic signals is assigned to a Raman-active intermolecular hindered translation along the hydrogen bond coordinate. The appearance of this mode is a manifestation of solvent reorganization that accompanies the ionization of water. The isotopic behavior of the 500-cm^{-1} component rules out solvent librational motion as its origin. Instead, this frequency is attributed to valence hole orbital motion in ionized liquid water that involves closely-spaced ion states with orthogonal hole densities.

At long time delays (>300 fs), a new oscillation feature characterized by a frequency of 310 cm^{-1} appears and persists beyond 1 ps (Fig. 1b). This component is assigned with the aid of ab initio calculations to the hindered translation between a hydronium (H_3O^+) species and its neighboring water molecules. The observation of this intermolecular mode, which is associated with excess proton transport [9], signifies long-lived ballistic proton transport in ionized liquid water. Because H_3O^+ is initially produced by the transfer of a proton from the H_2O^+ cation to a neighboring water molecule, the onset of ballistic proton transport is expected to correlate with the population decay of the H_2O^+ species. Fitting the growth of the 310-cm^{-1} component to pseudo-first-order kinetics yields a lifetime of 196 ± 5 fs for the H_2O^+ radical cation.

Acknowledgments This work is supported by an NTU start-up grant, the A*Star Science and Engineering Research Council (122-PSF-0011).

References

1. B. C. Garrett, et al., "Role of water in electron-initiated processes and radical chemistry: issues and scientific advances," Chem. Rev. **105,** 355–389 (2005).
2. O. Marsalek, C. G. Elles, P. A. Pieniazek, E. Pluhařová, J. VandeVondele, S. E. Bradforth, P. Jungwirth, "Chasing charge localization and chemical reactivity following photoionization of liquid water," J. Chem. Phys. **135,** 224510 (2011).
3. L. Turi, P. J. Rossky, "Theoretical studies of spectroscopy and dynamics of hydrated electrons," Chem. Rev. **112,** 5641–5674 (2012).
4. P. Kambhampati, D. H. Son, T. W. Kee, P. F. Barbara, "Solvated dynamics of the hydrated electron depends on its initial degree of electron delocalization," J. Phys. Chem. A **106**, 2374–2378 (2002).
5. J. Li, Z. Nie, Y. Y. Zheng, S. Dong, Z.-H. Loh, "Elementary electron and ion dynamics in ionized liquid water," J. Phys. Chem. Lett. **4**, 3698–3703 (2013).
6. B. Winter, R. Weber, W. Widdra, M. Dittmar, M. Faubel, I. V. Hertel, "Full Valence Band Photoemission from Liquid Water using EUV Synchrotron Radiation," J. Phys. Chem. A **108**, 2625–2632 (2004).
7. R. A. Crowell, D. M. Bartels, "Multiphoton Ionization of Liquid Water with 3.0–5.0 eV Photons," J. Phys. Chem. **100**, 17940–17949 (1996).
8. M. S. Pshenichnikov, A. Baltuška, D. A. Wiersma, "Hydrated electron population dynamics," Chem. Phys. Lett. **389**, 171–175 (2004).
9. D. Marx, M. E. Tuckerman, J. Hutter, M. Parrinello, "The nature of the hydrated excess proton in water," Nature **397**, 601–604 (1999).

Signatures of Conical Intersection Mediated Relaxation Dynamics in Time-Resolved Broadband Raman Detection

Benjamin P. Fingerhut, Konstantin E. Dorfman and Shaul Mukamel

Abstract Ab-Initio simulations of Raman signals reveal the excited state deactivation mechanism of uracil. The signals provide sub-molecular sensitivity of out-of-plane displacements during conical intersection mediated relaxation and properly describe the time-resolution of the techniques.

1 Ultrafast Relaxation Dynamics of RNA and DNA Bases

The strong UV absorption bands of DNA and RNA nucleobases lead to the population of bright valence excited states with $\pi\pi^*$ character. The nucleobases have been engineered by nature to be photostable with respect to UV irradiation. At the core of this self-protecting property are ultrafast excited-state deactivation mechanisms involving conical intersections where electronic energy is very rapidly converted into vibrational energy, allowing to minimize harmful photochemical processes that can eventually lead to DNA photolesions. Two relaxation paths have been proposed for pyrimidine bases, one being an ultrafast direct $\pi\pi^* \longrightarrow$ ground state (gs) channel, the other one involves relaxation into a dark $n_O\pi^*$ state. The access to the conical intersection seams of either channel determines the excited-state lifetime and deactivation mechanism of the pyrimidine bases.

Structural information about the required rearrangements of atoms can be derived directly from time- and frequency-resolved vibrational spectroscopy with Raman probes. Here we demonstrate that the high-frequency modes of uracil serve as sensitive fingerprints of the excited-state photoreaction and the associated non-adiabatic relaxation dynamics. Transient out-of-plane deformations of the π-ring

B.P. Fingerhut (✉)
Max-Born-Institut für Nichtlineare Optik und Kurzzeitspektroskopie, 12489 Berlin, Germany
e-mail: fingerhut@mbi-berlin.de

K.E. Dorfman · S. Mukamel
Department of Chemistry, University of California, Irvine, CA 92697-2025, USA

© Springer International Publishing Switzerland 2015
K. Yamanouchi et al. (eds.), *Ultrafast Phenomena XIX*,
Springer Proceedings in Physics 162, DOI 10.1007/978-3-319-13242-6_102

system can be monitored and mapped on characteristic dispersive features reflecting the local molecular changes of involved conical intersection structures.

2 Off-Resonant Raman Detection Schemes

We consider different off-resonant Raman techniques (i.e. the homodyne-detected frequency-resolved spontaneous Raman signal (FR-SPRS), heterodyne-detected time-resolved impulsive stimulated Raman signal (TR-ISRS), transient grating impulsive stimulated Raman signal (TG-ISRS), and femtosecond stimulated Raman signal (FSRS)) which provide information about the same four-point matter correlation functions:

$$F_i(t_1, t_2, t_3) = \left\langle VG^\dagger(t_1)\alpha G^\dagger(t_2)\alpha G(t_3)V^\dagger \right\rangle, \tag{1}$$

$$F_{ii}(t_1, t_2, t_3) = \left\langle VG^\dagger(t_1)\alpha G(t_2)\alpha G(t_3)V^\dagger \right\rangle. \tag{2}$$

They contain all relevant information about the matter, but with different detection windows. In FSRS, TR-ISRS, and TG-ISRS the symmetry between the both branches of the loop diagram is broken (see Fig. 1), which explains microscopically the origin of dispersive peak shapes. In FR-SPRS the signals can be recast as a modulus square of the transition amplitude and only absorptive lineshapes appear [1].

The different time-resolved Raman signals are simulated within a recently developed semi-classical simulation protocol [2, 3], which treats nuclear motions classically but fully retains the quantum character of electrons thus allowing to follow the complex dynamics over conical intersections on CASSCF (14/10) level of theory. The excited state instantaneous frequencies of spectator modes (Fig. 2a), modulated in the classical bath of remaining modes, are reconstructed in a mode tracking procedure which decouples the numerical effort from system size. The signal expressions contain a path integral over the excited state instantaneous frequencies and thus properly describe the time-resolution of the techniques.

3 Time-Resolved Raman Signatures of Conical Intersection Mediated Relaxation of Uracil

The simulated excited state dynamics of uracil is characterized by ultrafast population redistribution (after ~ 190 fs), where the bright S_2 state ($\pi\pi^*$ character) is depopulated and serves as a reservoir for both relaxation channels ($\pi\pi^* \rightarrow$ gs channel and dark $n_O\pi^*$ state population). The derived $\pi\pi^*$ lifetime is 516 fs (single exponential fit) which is in good agreement with the 530 fs time constant reported by Ullrich et al. [4]. The molecular changes of involved conical intersection structures

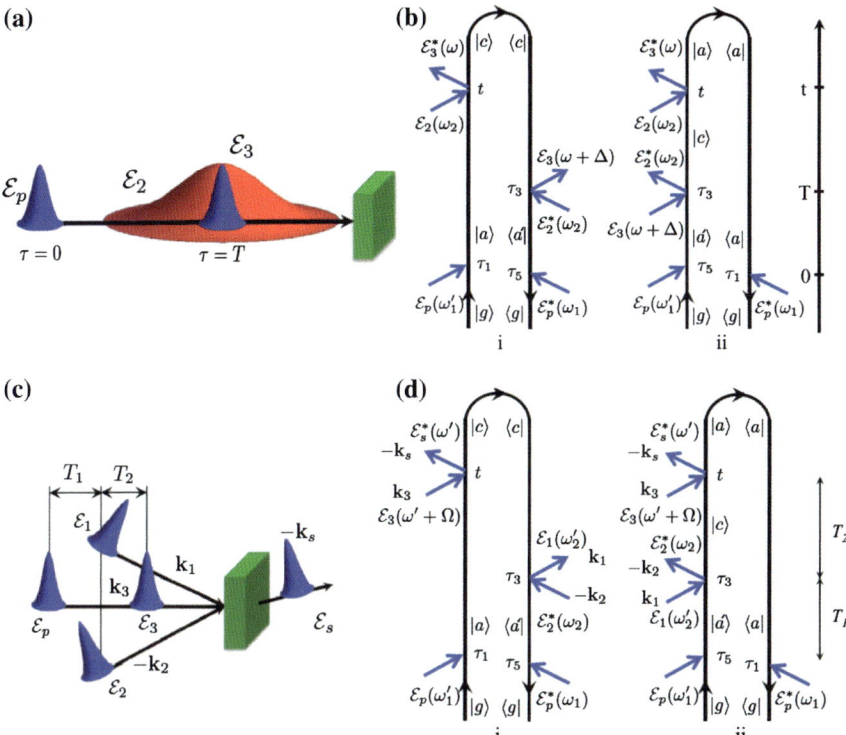

Fig. 1 Schematic layout and loop diagrams of the FSRS (**a, b**) and the TG-ISRS (**c, d**) technique: in FSRS pulse E_p initiates the vibrational excited state dynamics. The Raman probe sequence consists of a picosecond E_2 and a femtosecond pulse E_3, the signal is given by frequency-dispersed probe transmission $E_3(\omega)$. In TG-ISRS, two short pulses ($\mathbf{k_1}$ and $\mathbf{k_2}$) coincide after a delay T_1 and form an interference pattern with wave vector $\mathbf{k_1} - \mathbf{k_2}$. After a second delay period T_2, a third beam with wave vector $\mathbf{k_3}$ is scattered off the grating to generate a signal with wave vector $\mathbf{k_S} = \mathbf{k_1} - \mathbf{k_2} + \mathbf{k_3}$

are characterized by an out-of-plane deformation of the pyrimidine π system (see inlay in Fig. 2b) which instantaneously modifies the local potential of high-frequency C–H spectator modes (see Fig. 2a for the $\pi\pi^* \rightarrow$ gs channel).

Analysis of the joint time/frequency resolution of the Raman techniques is provided on the basis of the Δ-dispersed signal (Fig. 2b). Even though not an experimental observable, the Δ-dispersed signal contains all matter contributions to off-resonant Raman probes where the dispersion in the Δ-axis reveals the inherent matter chirp contributions that limit the temporal resolution of the techniques but determine the required probe pulse bandwidth.

Figure 2c depicts the FSRS signal of the direct $\pi\pi^* \rightarrow$ gs relaxation mechanism of uracil which is initially characterized by a single, system dynamics broadened resonance. Due to the non-adiabatic relaxation induced frequency shift the resonance evolves toward $\omega - \omega_3 = 3{,}100 \text{ cm}^{-1}$ with complex dispersive line shape

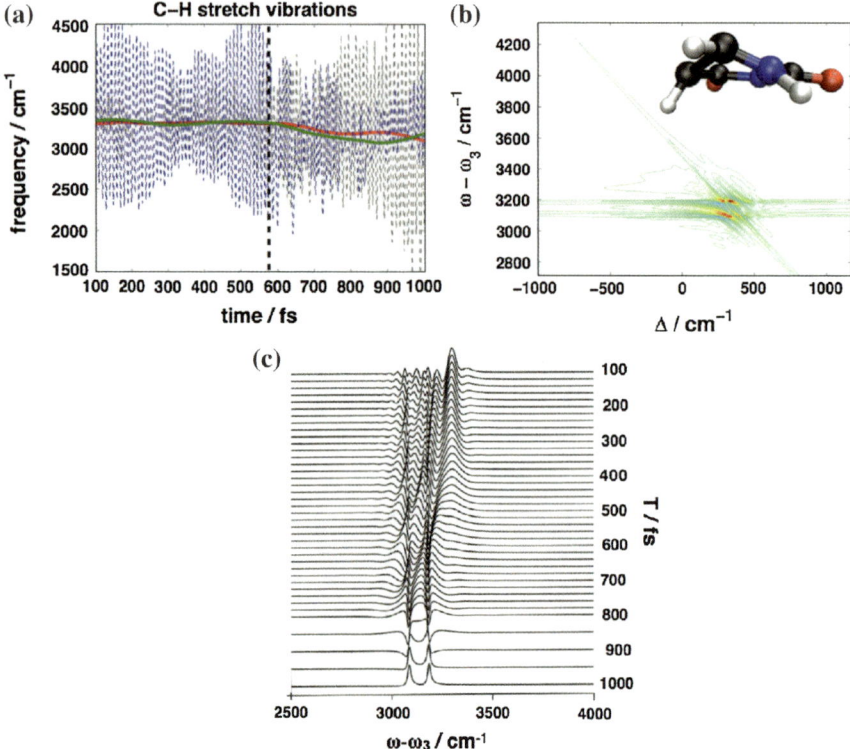

Fig. 2 **a** Instantaneous frequencies, **b** Δ-dispersed signal for T = 586 fs and **c** FSRS signal of C–H stretch vibrations of the direct $\pi\pi^* \rightarrow$ gs relaxation mechanism of uracil. Dispersive features in FSRS detection are induced by passage through conical intersection structures shown as inlay. The required bandwidth is represented in the Δ-axis of **b**

(T > 550 fs). As the delay time T and the frequency resolution of the detection axis $\omega - \omega_3$ are independent experimental knobs, the FSRS signal at time T does not represent a snapshot of the system dynamics. Resonances appear broadened due to the matter dynamics induced chirp contribution which modulates line-shapes and induces the dispersive features in the FSRS signals. These reflect the partial re-hybridization pattern of the displaced carbon atoms and also survive ensemble averaging. TG-ISRS and TR-ISRS allow the manipulation of both, the pump and probe pulses. In the limit of ultrashort pump TG-ISRS/TR-ISRS provide the same resolution as FSRS.

Acknowledgments We gratefully acknowledge support from the National Science Foundation (Grant No. CHE 1361516), and the Chemical Sciences, Geosciences and Biosciences Division, Office of Basic Energy Sciences, Office of Science, and (U.S.) Department of Energy (DOE). B.P.F. gratefully acknowledges support from the Alexander-von-Humboldt Foundation through the Feodor-Lynen program.

References

1. K. E. Dorfman, B. P. Fingerhut, and S. Mukamel, "Time-resolved broadband Raman spectroscopies: A unified six-wave-mixing representation", J. Chem. Phys. **139**, 124113 (2013).
2. K. E. Dorfman, B. P. Fingerhut and S. Mukamel, "Broadband infrared and Raman probes of excited-state vibrational molecular dynamics: simulation protocols based on loop diagrams", PCCP **15**, 12348-12359 (2013).
3. B. P. Fingerhut, K. E. Dorfman and S. Mukamel, "Probing the Conical Intersection Dynamics of the RNA Base Uracil by UV-Pump Stimulated-Raman-Probe Signals; Ab Initio Simulations", J. Chem. Theory Comput. **10**, 1172–1188 (2014).
4. S. Ullrich, T. Schultz, M. Z. Zgierski and A. Stolow, "Electronic relaxation dynamics in DNA and RNA bases studied by time-resolved photoelectron spectroscopy Phase-coherent generation of tunable visible femtosecond pulses", PCCP **6**, 2796–2801 (2004).

VIPER 2D-IR: Novel Pulse Sequence to Track Exchange Beyond the Vibrational Lifetime

Luuk J.G.W. van Wilderen, Andreas T. Messmer
and Jens Bredenbeck

Abstract We present a new IR/UV-VIS pulse sequence that uses an IR pulse to pick a molecule within a mixture, in order to monitor its photochemistry. The benefits of this sequence over commonly used ones discussed.

1 Mixed IR/Vis 2D Spectroscopies

The rich possibilities of 2D-IR spectroscopy can be even further enhanced by incorporating Vis or UV pulses in the IR pulse sequence. Several mixed ultrafast IR/Vis pulse sequences have been developed so far, each having specific applications. Transient 2D-IR spectroscopy (TRIR) takes 2D-IR spectra of populations that are photoactivated by a Vis pulse preceding the 2D-IR pulses [1–4]. In non-equilibrium 2D-IR exchange spectroscopy (2D-IR EXSY, also termed triggered exchange), a resonant UV/Vis pulse is applied during the population time of the 2D-IR experiment. In this way cross peaks are generated that correlate the vibrations of the transient species with the vibrations of the starting state [5, 6]. Another pulse scheme is designed to probe surfaces via sum-frequency generation (SFG 2D-IR) between the 2D-IR signal and a non-resonant Vis or NIR pulse [7, 8]. The latter is a 4th order experiment, while the others are 5th order experiments.

Here we present a novel 5th order IR-UV/Vis-IR pulse scheme designed to measure molecular exchange processes beyond the vibrational lifetime. Moreover, subensemble-specific photochemistry can be performed in a mixture of molecular species. The scheme is called VIPER 2D-IR (Vibrationally Promoted Electronic Resonance). An important feature of VIPER, in contrast to triggered exchange spectroscopy which also uses an IR-UV/Vis-IR scheme, is that the pulse for electronic excitation is off-resonant. The resulting implications are explained below. While the SFG method also includes a non-resonant Vis or NIR pulse, the interaction

L.J.G.W. van Wilderen · A.T. Messmer · J. Bredenbeck (✉)
Institute for Biophysics, Goethe University, Max-von-Laue-Str. 1, 60438 Frankfurt, Germany
e-mail: bredenbeck@biophysik.uni-frankfurt.de

© Springer International Publishing Switzerland 2015

424

K. Yamanouchi et al. (eds.), *Ultrafast Phenomena XIX*,
Springer Proceedings in Physics 162, DOI 10.1007/978-3-319-13242-6_103

is a different one: in VIPER the UV/Vis pulse is used to transfer populations to the excited state, while in SFG it is used to up-convert to generated signal.

2 Viper 2D-IR

In the visible spectral region it is often difficult to distinguish between multiple molecular species due to spectral overlap. IR spectroscopy is usually much more selective. However, the measurement of exchange between different species by 2D-IR EXSY is limited by the vibrational lifetime, which is typically a few picoseconds. In contrast, visible pump-probe spectroscopy probes excitations which exhibit lifetimes that can be several orders of magnitude longer than those in infrared spectroscopy. In VIPER 2D-IR, we use the 'selectivity power' of infrared spectroscopy and the long lifetime of electronically excited states to monitor chemical exchange processes in the infrared that occur on time scales that are much longer than the vibrational lifetime of the investigated species [9].

To achieve this goal, VIPER 2D-IR (Fig. 1a) makes use of an IR pulse that vibrationally tags a molecule, followed by an off-resonant UV/Vis pulse and an IR probe pulse. Only molecules that are vibrationally excited will be transferred to the electronically excited state, because the IR pulse shifts the molecule into resonance with the UV/Vis pulse. This way a long lived 2D-IR signal is generated that allows monitoring chemical exchange until the initial ground state is repopulated. The population in the electronically excited state can also perform exchange or undergo photochemistry. Exchange can thus be investigated both in the electronic ground and excited states. A strong vibronic coupling between the initially excited vibration and the electronic transition is not essential for the generation of a VIPER signal, because a modulation of the visible cross-section by intramolecular vibrational redistribution (IVR) into Frank-Condon-active modes may serve the same purpose [10]. This expands the applicability of the VIPER pulse sequence.

The VIPER pulse sequence allows important applications, which include probing chemical exchange processes which occur on time scales that are slower than the vibrational lifetime (T_1), or monitoring the photochemistry of one particular molecular species within a mixture of species, even when they are in fast equilibrium. In the case of molecules that are modified or destroyed (e.g. a photolabile compound) by the visible laser pulse, a very long-lived or even a permanent ground state bleach allows to monitor exchange from picoseconds to extremely long times. Here we demonstrate the applicability of this scheme to monitor chemical exchange of two molecular species (hydrogen bonded vs free) on a time scale that is much longer than the vibrational lifetime, and show that it is possible to select one of these species within the mixture for electronic excitation.

Our model system is the laser dye coumarin 6 (C6) which, upon addition of methanol, forms a dynamic equilibrium of methanol-associated and free coumarin molecules in solution [9]. The two species have distinct IR absorption bands (TRIR; Fig. 1b) which are assigned to the free and associated carbonyl of C6, i.e. the band

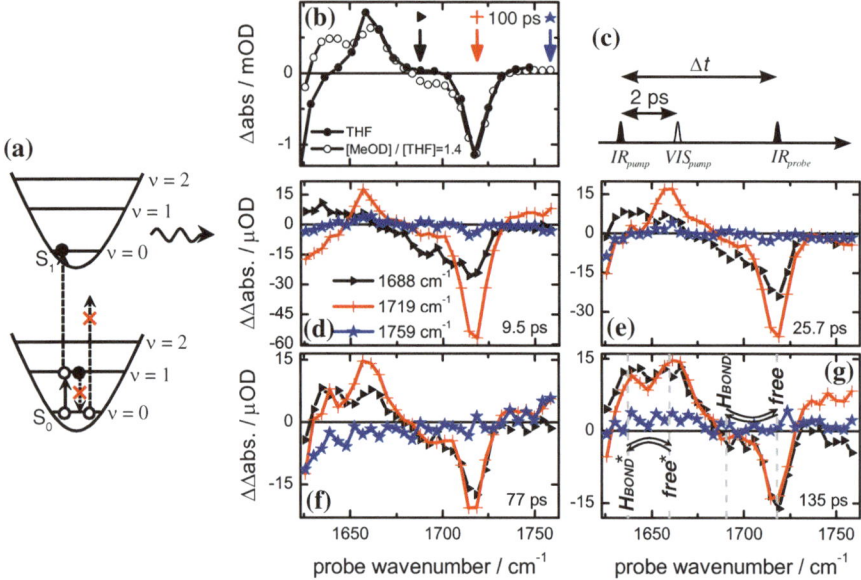

Fig. 1 Principle of VIPER 2D-IR and its application. **a** In VIPER 2D-IR, the off-resonant Vis_pump (*long-dashed arrow*) is applied after the IR_pump (*black arrow*) and removes population from $v = 1$ (not from $v = 0$, *long-dashed arrow* with *red cross*), thereby preventing ground-state recovery (*short-dashed arrow* with *red cross*) and generating a persistent 2D-IR signal. The population on S_1 can relax with T_{el} or exhibit photochemistry (*curved arrow*). VIPER 2D-IR EXSY beyond T_1. **b** TRIR spectra for C6 with and without MeOD in the C=O region. **d–g** Cross-sections of the VIPER 2D-IR spectra of the sample with MeOD, pumped at the wavenumbers corresponding to the colored *arrows* in *panel B*. Note that signals at wavenumbers other than the IR pump correspond to cross-peaks. The delay time (Δt in **c**) is depicted in (**d–g**), which share the same tick increment of 15mOD. **b** and **d–g** have symbols to improve visibility if printed in *black-white*. Copyright © 2013 WILEY-VCH Verlag GmbH & Co. KGaA, Weinheim

demonstrates a down-shift upon hydrogen bonding to methanol (for clarity all species have been labeled in Fig. 1g). By using VIPER (Fig. 1c, using a 14 cm^{-1} wide IR_pump pulse and a non-resonant Vis_pump at 500 nm), we see that at 9.5 ps delay the VIPER spectra are very different (compare the black and the red spectra), best seen in the regions where the excited states of the associated and free species absorb (at 1,639 and 1,657 cm^{-1}, respectively). This means that we observe transient IR spectra of a distinct species within a mixture of species, by selecting it via the IR pulse. In the normal TRIR spectrum (Fig. 1b) both species are excited. In contrast, regardless of the species we excite with the IR pump pulse (denoted by the different colored arrows in Fig. 1b), the 'end' spectrum we obtain at 135 ps is identical (still being different at 77 ps delay). This means that chemical exchange has occurred on a time scale much slower than the vibrational lifetime of 1.2 ps. Thus, we can follow a chemical exchange process with 2D-IR spectroscopy on a time scale that is far beyond the vibrational lifetime.

References

1. J. Bredenbeck, J. Helbing, R. Behrendt, C. Renner, L. Moroder, J. Wachtveitl, P. Hamm, J. Phys. Chem. B, **107**, 8654-8660 (2003).
2. J. Bredenbeck, J. Helbing and P. Hamm, Phys. Rev. Lett. **95**, 083201 (2005).
3. W. Xiong, J. E. Laaser, P. Paoprasert, R. A. Franking, R. J. Hamers, P. Gopalan, M.T. Zanni, J. Am. Chem. Soc., **131**, 50, 18040–18041 (2009).
4. C. R. Baiz, R. McCanne, M. J. Nee, K. J. Kubarych, J. Phys. Chem. A, **113**, 31, 8907–8916 (2009).
5. J. Bredenbeck, J. Helbing, P. Hamm, J. Am. Chem. Soc., **126**, 990-991 (2004).
6. J. Bredenbeck, J. Helbing, K. Nienhaus, G. U. Nienhaus and P. Hamm, Proc. Natl. Acad. Sci. **104**, 14243-14248 (2007).
7. J. Bredenbeck, A. Ghosh, M. Smits, M. Bonn, J. Am. Chem. Soc., **130**, 7, 2152-2153 (2008).
8. W. Xiong, J. E. Laaser, R. D. Mehlenbacher, M. T. Zanni, Proc. Natl. Acad. Sci. **108**, 20902-20907 (2011).
9. L. J. G. W. van Wilderen, A. T. Messmer, J. Bredenbeck, Angew. Chem. Int. Ed., **53**, 2667-2672 (2014).
10. C. M. Cheatum, M. M. Heckscher, D. Bingemann, F. F. Crim, J. Chem. Phys., **115**, 7086–7093 (2001).

Sagnac Interferometer for Two-Dimensional Spectroscopy in the Pump-Probe Geometry

Samuel D. Park, Trevor L. Courtney, Dmitry Baranov, Byungmoon Cho and David M. Jonas

Abstract An intrinsically phase-stable Sagnac interferometer is introduced for enhanced sensitivity detection in partially collinear two-dimensional spectroscopy. The sensitivity and phase accuracy of the apparatus are demonstrated on the dye IR-26 in the short-wave IR.

Two-dimensional (2D) Fourier transform (FT) spectra show how a nonlinear signal field, as a function of radiated frequency, depends on an excitation frequency, revealing coupling between excitations [1]. The pump-probe geometry offers intrinsic phasing, but the intense last pulse and weak nonlinear signal copropagate, which can make their interference more difficult to detect [2]. The new method presented here implements a Sagnac interferometer to increase the signal sensitivity by decreasing the local oscillator (LO) amplitude. A Sagnac has been used previously for optical background suppression in pump-probe spectroscopies [3]. Despite its simplicity, implementing the Sagnac for 2D experiments has presented a challenge because 2D spectra are sensitive to phase. We have recently demonstrated a Sagnac for enhanced sensitivity in detecting absorptive 2D spectra in the short-wave IR [4], and will discuss conditions needed to maintain the necessary π phase shift here. This work presents a partially collinear 2D spectrometer with a Sagnac (Fig. 1a), which creates a nearly background-free signal and selectively detects the absorptive 2D spectrum. The validity of the interferometer has been tested on IR-26 dye (Fig. 1b) and a new Germanium beam splitter has been characterized and may offer further improvements at longer wavelengths. Employing 2D spectroscopy in the short-wave IR enables access to low-energy electronic processes and next-generation photovoltaics, where sensitivity is at a premium.

Pulses from a 1 kHz Ti:Sapphire regenerative amplifier pump a single-pass, short-wave IR noncollinear optical parametric amplifier with a PPSLT crystal [5].

S.D. Park · T.L. Courtney · D. Baranov · B. Cho · D.M. Jonas (✉)
Department of Chemistry and Biochemistry, University of Colorado, 215 UCB,
Boulder, CO 80309, USA
e-mail: david.jonas@colorado.edu

© Springer International Publishing Switzerland 2015
K. Yamanouchi et al. (eds.), *Ultrafast Phenomena XIX*,
Springer Proceedings in Physics 162, DOI 10.1007/978-3-319-13242-6_104

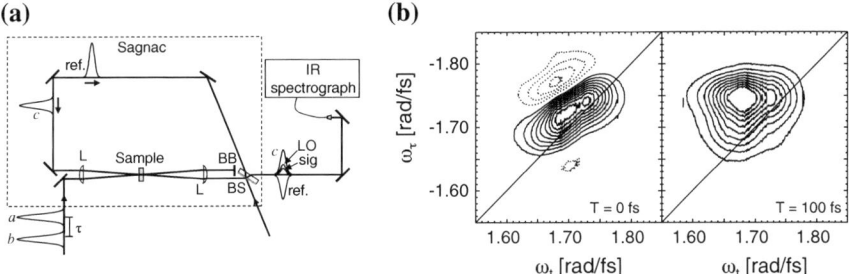

Fig. 1 a Brewster's angle Sagnac interferometer apparatus. The *dark* output of the Sagnac interferometer is used, where pulse *c* and the reference combine to result in an attenuated local oscillator used for interference with the 2D signal. **b** 2D spectra of IR26 dye with 10 % contours. *Solid* and *dashed contours* denote positive and negative amplitude, respectively. 2D spectra at T = 100 fs show the rapid loss of correlation between excitation and detection frequencies observed as the 2D spectra approach an uncorrelated product *line* shape at long T

The tunable pulses are compressed with a deformable mirror using SHG feedback in a genetic algorithm to pulse durations of 10–30 fs. After the compressor, the beam is spatially filtered with a 150-μm pinhole. All spectral IR detection uses single-mode fiber coupling to a 0.15-m Czerny-Turner spectrometer and a 1,024 × 1 pixel InGaAs array.

Pump pulse pairs are generated by an actively delay stabilized Mach-Zehnder interferometer [6] with inconel-coated beam splitters set at Brewster's angle to prevent multiple surface reflections. The probe path splits into counterpropagating probe and reference pulses in a Brewster's angle Sagnac interferometer (Fig. 1a). The pumps and probe intersect at the sample within the interferometer ∼1.5 ns after the reference excites the sample. When the beams are recombined at the thin-film gold-coated beam splitter, the attenuated probe (counterclockwise-propagating beam) and reference (clockwise-propagating beam) pulses exit the dark output of the interferometer nearly out of phase with each other, destructively interfering to generate an attenuated local oscillator. The Sagnac has an odd number of mirrors and a telescope. The telescope increases the nonlinear signal and introduces an additional inversion.

The Sagnac interferometer beam splitter requires careful attention to assure a π phase shift between dark outputs of the probe and reference while avoiding dispersion. The Brewster's angle beam splitter (Fig. 1a) has an ∼8-nm thin film of gold deposited on a 1-mm thick BK7 substrate. The refractive index, $\hat{n} = n + ik$, of amorphous gold has $k \approx 25 \times n$ [7] to assure a nearly π phase shift (170–171°, compared to ∼30° with inconel) between dark output pulses. For gold, k/n increases with increasing wavelength in the short-wave IR. Destructive interference in the dark output suppresses the in-phase component of the reference pulse, and increases the phase error of the LO as the LO is attenuated, yielding an LO phase error of 15°. LO phase correction in ω_t using 2D Kramers-Kronig relations amounts to less than 5 % rms.

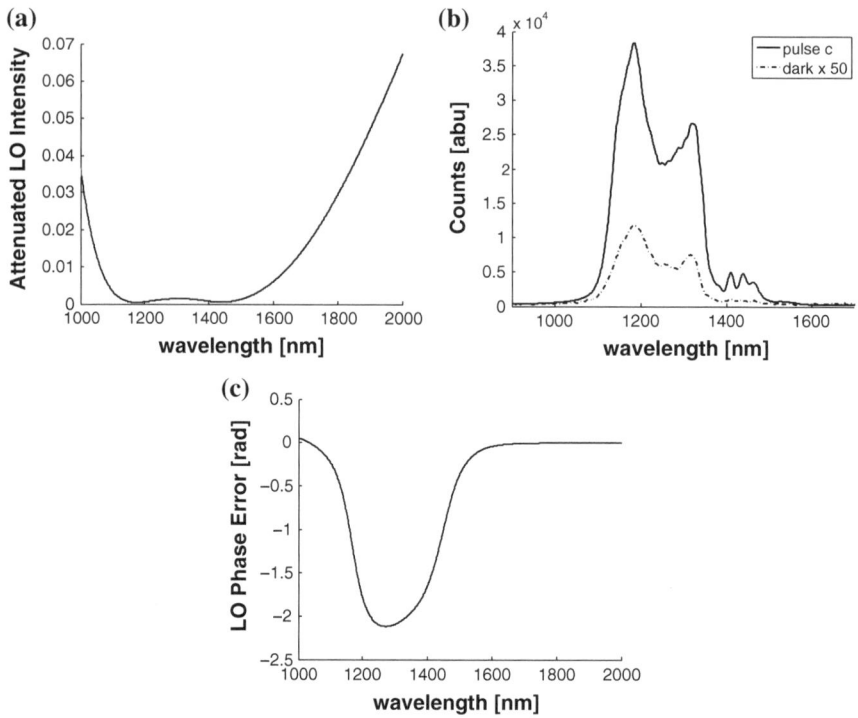

Fig. 2 **a** Calculated LO intensity normalized to original pulse c intensity. The LO intensity can be further controlled by varying the Germanium thickness on the beam splitter. **b** Experimental data for ∼ 80 nm Germanium beam splitter. Near complete destructive interference over the 1,100–1,600 nm range. **c** Calculated LO phase error resulting from near complete destructive interference

The resulting real 2D correlation spectra of IR-26 (Fig. 1b) show a diagonally elongated positive peak at early waiting times (T = 0), which reflects strong correlation between excitation and detection frequencies. The slight shift above the diagonal and the off-diagonal, negative region are indicative of vibrational and solvent frequency memory. By T = 100 fs, nearly all correlation between excitation and detection frequencies is lost; the peak is approaching a product line shape.

If the beam splitter coating index is not purely real or imaginary, large attenuations of the LO will amplify the phase error of the attenuated LO. The phase of the attenuated LO is given by $arg(\hat{t}\hat{t} + \hat{r}\hat{r}')$, where $\hat{t}$, $\hat{r}$, and $\hat{r}'$ are complex-valued Fresnel coefficients [8] for transmission, reflection, and back-surface reflection of the beamsplitter coating. Figure 2 shows calculated and experimental results from an 80-nm Germanium beam splitter. The nearly π phase difference between beams in the dark output results in a more complete destructive interference. However, the phase difference reaches ∼ 3.19 rad near 1,350 nm, which causes large LO phase errors over the 1,100–1,600 nm wavelength range (Fig. 2c); this suggests that a Germanium coated beam splitter may be more useful at wavelengths longer than 1,600 nm, where the imaginary part of the refractive index drops by over an order of magnitude [9].

Acknowledgments This material is based upon work supported by the National Science Foundation under Grant No. CHE-1112365. We thank D. Brida and G. Cerullo for advice on the NOPA.

References

1. D. M. Jonas, Annu. Rev. Phys. Chem. **54**, 425 (2003).
2. L. P. DeFlores, R. A. Nicodemus, A. Tokmakoff, Opt. Lett. **32**, 2966 (2007).
3. R. Trebino, C. C. Hayden, Opt. Lett. **16**, 493 (1991).
4. T. L. Courtney, S. D. Park, R. J. Hill, B. Cho, D. M. Jonas, Opt. Lett. **39**, 513 (2014).
5. D. Brida *et al.*, Opt. Express **17**, 12510 (2009).
6. M. K. Yetzbacher *et al.*, J. Opt. Soc. Am. B **27**, 1104 (2010).
7. E. D. Palik, *Handbook of Optical Constants of Solids* (Academic, 1985).
8. J. R. Reitz, F. J. Milford, R. W. Christy, *Foundations of Electromagnetic Theory* (Academic, 1992).
9. H. H. Li, J. Phys. Chem. Ref. Data **9**, 3 (1980).

Broadband Electronic Two-Dimensional Spectroscopy in the Deep UV

Valentyn I. Prokhorenko, Alessandra Picchiotti, Samansa Maneshi and R.J. Dwayne Miller

Abstract We developed an all-reflective fully-noncollinear setup for two-dimensional electronic spectroscopy in the broadband UV (2DUV) with high phase stability ($\Lambda/150$) and applied it to UV-chromophores dissolved in ethanol using 8-fs UV-pulses, generated in the 245–300 nm range. We are able to resolve 2D-spectra in an $\sim 6,000$ cm^{-1} spectral window.

1 Introduction

Ultraviolet electronic two-dimensional spectroscopy of solution phase chromophores and DNA requires broadband transform-limited UV-pulses in the 240–300 nm range. This requirement is dictated by the spectral width of their absorption spectra which normally have a FWHM of ~ 30 nm or more ($\geq 3,000$ cm^{-1}). Such UV-pulses can be generated from the broadband VIS-spectra using achromatic frequency doubling (ASHG) combined with a pulse shaper for reducing the phase distortions [1]. However, managing and manipulating these ultrashort pulses in the UV-domain, especially in the 240–300 nm spectral range, requires the use of explicitly reflective optics since any refractive optical element will heavily disturb their phase profiles and will thus significantly distort 2D-spectra. Using reflective diffractive optics (DO) [2] allows one to overcome these difficulties and creates the possibility for the design of a four-beam interferometer with the high phase stability necessary for acquiring accurate 2D-spectra in a fully-noncollinear geometry ("background-free" heterodyne measurement). In order to avoid parasitic phase distortions in the UV-pulses, we developed a fully reflective photon-echo-based 2DUV setup capable of measuring 2D-spectra in the UV spectral range of 230–350 nm (spectral window $\sim 15,000$ cm^{-1}).

V.I. Prokhorenko (✉) · A. Picchiotti · S. Maneshi · R.J. Dwayne Miller
Max Planck Institute for Structure and Dynamics of Matter,
Luruper Chaussee 149, 22761 Hamburg, Germany
e-mail: valentyn.prokhorenko@mpsd.mpg.de

© Springer International Publishing Switzerland 2015
K. Yamanouchi et al. (eds.), *Ultrafast Phenomena XIX*,
Springer Proceedings in Physics 162, DOI 10.1007/978-3-319-13242-6_105

432

2 Experimental Setup

Figure 1 (left) shows a layout of developed all-reflective 2DUV setup based on the reflective DO (crossed grating, custom designed by Holoeye), which diffracts the incoming beam into 4 first-orders with an efficiency of 60 %. Generically, it is similar to the 2D-setup developed earlier [3]; the main difference is the use of a reflective DO and roof mirrors instead of transmissive optics and retroreflectors (to minimize the number of reflections and thus transfer more energy to the sample). The construction is similar to the design published in [4]; however, uncoupling of translation stages MTS1&2 (XMS-50, Newport) is achieved by an additional pass of beam #2 through MTS1 but in the opposite direction, allowing an independent scan of "waiting" delay T and "dephasing" delay τ which leads to an increase of phase stability and overall robustness of this four-beam interferometer. Figure 1 (right) demonstrates the phase deviations monitored within 20 min (for treatment details, see [3]) and the achieved phase stability of Λ/150 is even better than for the VIS 2D-setup published in [3]. Note that the acquisition of a single 2D-spectrum with a good SNR takes typically 5 min.

For heterodyne detection of the generated photon echo signal we used either a 200-μm thick neutral density filter with an OD = 2, placed into the local oscillator beam (4), or one from the bottom mirrors in the assembly of MTS2, which was replaced by the fused silica substrate (reflection 0.8 %) and shifted upward in order to delay the local oscillator beam by 605 fs.

The 8-fs UV-pulses (measured in situ) have been generated using ASHG (the design is similar to the one published in [1]) of broadband VIS-pulses, generated by two-cascaded home-built non-collinear optical parametric amplifiers (NOPAs) pumped by a Coherent Elite USP laser system. The NOPA system delivers light pulses with a spectral width of 490–600 nm and energy of 8 μJ, while the energy of the UV-pulses was 650–700 nJ at the output of ASHG and ∼ 14–15 nJ per beam at the sample position. In order to minimize light scattering and pulse broadening, we

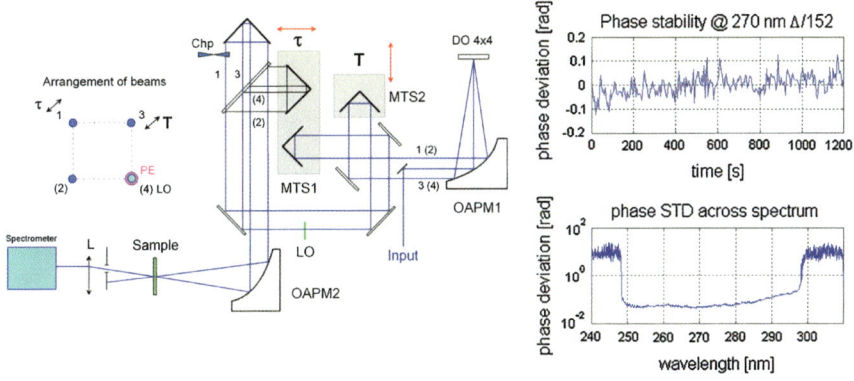

Fig. 1 Schematic layout of 2DUV setup (*right*) and phase stability, monitored within 20 min (*left*)

developed a high stable wire-guided jet for the circulation of samples, dissolved in ethanol. The thickness of the jet in the beam position was maintained at ~ 250 μm (can be varied by changing the speed of the jet); the OD of the samples used was 0.3 at the maximum of the absorbance.

3 Results and Discussion

We investigated several UV-chromophores dissolved in ethanol. As an example, Fig. 2 shows representative 2D-spectra of *para*-Terphenyl (PTP) at waiting times of T = 150, 300 and 1,000 fs, respectively, measured in a 34,000–41,000 cm^{-1} spectral range. At shorter T, the contribution to the photon-echo signal from ethanol increases significantly and obscures the measurement of 2D-spectra (at T = 0, it is ~ 200 times larger than the PTP-signal).

At all waiting times, the 2D-spectra exhibit homogeneous broadening for PTP in ethanol; however, there are some dynamics and structure in these spectra viewed at T = 150 fs, that disappears with increasing T. We can thus speculate that this fine structure reflects underlying vibrational transitions in the PTP-molecule not resolved in the absorption spectrum.

Another example is shown in Fig. 3 for the 2D spectrum of pyrene at a waiting time T = 540 fs. The absorption spectrum of pyrene already displays structure with well resolved peaks associated with the different vibrational levels in the S_3 electronic state (note that excitation at ~ 250–270 nm corresponds to the $S_0 \rightarrow S_3$ electronic transition in pyrene); however, this structure is much better resolved in the 2D spectra where, besides the main transitions located on the diagonal, many off-diagonal peaks, corresponding to the cross peaks, are also well resolved. It is interesting that these cross peaks are also clearly resolved in the modulus plot of the 2D spectrum (Fig. 3, right).

It should be noted that in the present study, due to the very short UV-pulses, the achieved spectral window ($\sim 6,000$ cm^{-1}) is more than one order of magnitude larger than was reported previously for this UV spectral range [4–6].

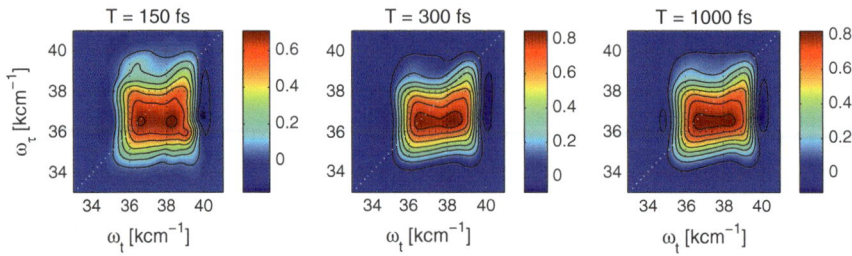

Fig. 2 Absorptive parts of 2D-spectra at different waiting times T (as indicated) for PTP/ethanol

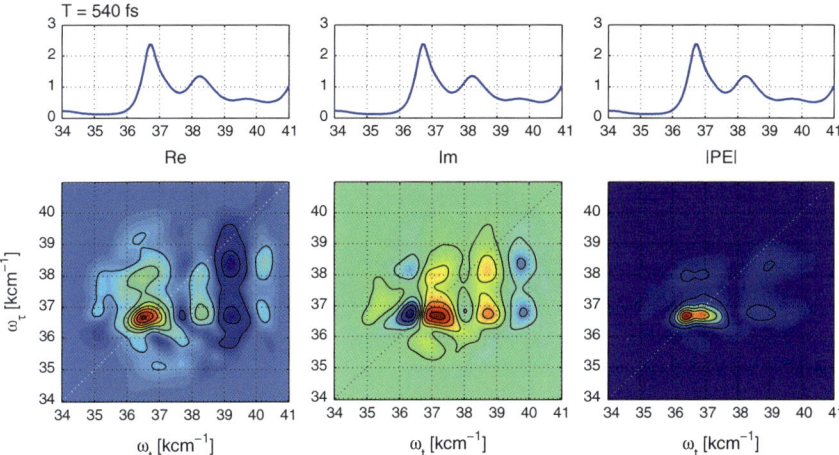

Fig. 3 2D spectrum of pyrene at T = 540 fs (*left*—real part, *center*—imaginary part, *right*—it modulus). The absorption spectrum is shown for comparison (on *top* of spectra)

References

1. P. Baum, S. Lochbrunner, E. Riedle, Opt. Lett. **29**, 1686 (2004).
2. M.L. Cowan, B.D. Bruner, N. Huse, J.R. Dwyer, B. Chung, E.T.J. Nibbering, T. Elsaeesser, R.J.D. Miller, Nature **434**, 199 (2005).
3. V.I. Prokhorenko, A. Halpin, R.J.D. Miller, Opt. Express **17**, 9764 (2009).
4. U. Selig, C-F. Schleussner, M. Foerster, F. Langhojer, P. Nuernberg, T. Brixner, Opt. Lett. **35**, 4178 (2010).
5. C-H. Tseng, P. Sandor, M. Kotur, T.C. Weinacht, S. Matiska, J. Phys. Chem. A **116**, 2654 (2012).
6. B.A. West, J.M. Womick, A.M. Moran, J. Phys. Chem. A **115**, 8630 (2011).

A Non time Ordered Pulse Scanning Protocol for Multidimensional Spectroscopy with Entangled Light

Konstantin E. Dorfman, Frank Schlawin and Shaul Mukamel

Abstract Quantum light can induce correlations in photo excited molecules and probe them with unusual spectral and temporal resolution. A new non-time-ordered pulse delay scanning protocol in multidimensional signals reveals resonances not accessible by standard techniques. This protocol allows to understand how entanglement of the light field is imprinted into entanglement of matter.

1 Introduction: Spectroscopy with Entangled Light

Quantum spectroscopy utilizes the quantum nature of light to reveal matter properties not available with classical light [1]. Entangled photons offer several advantages. First, the signals scale to lower order in the incoming intensity [2]. The pump-probe signal e.g. scales linearly rather than quadratically. This allows to perform nonlinear spectroscopy with lower intensity, limiting damage in e.g. imaging applications. Second, time-and-frequency entanglement allows to obtain higher temporal and spectral resolutions. The temporal resolution Δt depends on the entanglement time T, determined by the length of the nonlinear crystal, while the spectral resolution $\Delta \omega$ is determined by the pump envelope. These are independent control variables, not Fourier conjugates, and are thus not bound by the uncertainty $\Delta \omega \Delta t > 1$. The nonlinear response of the system is governed by a matter correlation function of the dipole operator windowed by a corresponding correlation function of the electric field. For instance the two photon absorption (TPA) signal given by a

K.E. Dorfman (✉) · F. Schlawin · S. Mukamel
Department of Chemistry, University of California, Irvine, CA 92697-2025, USA
e-mail: kdorfman@uci.edu

S. Mukamel
e-mail: smukamel@uci.edu

F. Schlawin
Institute of Physics Albert-Ludwigs University of Freiburg, Hermann-Herder-Str. 3, 79104 Freiburg, Germany

© Springer International Publishing Switzerland 2015
K. Yamanouchi et al. (eds.), *Ultrafast Phenomena XIX*,
Springer Proceedings in Physics 162, DOI 10.1007/978-3-319-13242-6_106

436

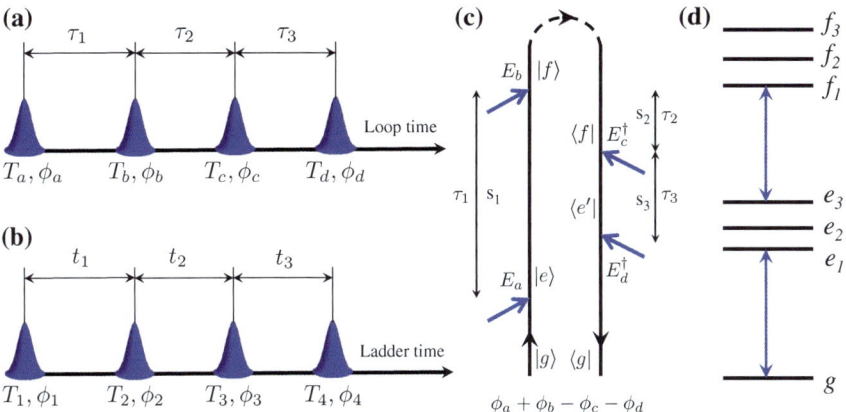

Fig. 1 The pulse sequence for unrestricted LOP (**a**), LAP (**b**). Loop diagrams for the TPA process with indicated loop delays for the phase cycling selected the signal with $e^{i(\phi_a+\phi_b-\phi_c-\phi_d)}$ (**c**). Level scheme for the molecular timer (**d**)

population of a double-excited state ρ_{ff} in an aggregate (Fig. 1d) can be read off the loop diagram in Fig. 1c

$$S(\Gamma) = \frac{1}{\hbar^4} \int_{-\infty}^{\infty} dr_a \int_{-\infty}^{\infty} dr_b \int_{-\infty}^{\infty} dr_c \int_{-\infty}^{\infty} dr_d \left\langle E_d^\dagger(r_d) E_c^\dagger(r_c) E_b(r_b) E_a(r_a) \right\rangle$$
$$\times \left\langle \mathcal{J} V(r_d) V r_c) V^\dagger(r_b) V^\dagger(r_a) \right\rangle. \tag{1}$$

Depending on the state of the light, the four-point correlation function in (1) may couple different interaction times between the bra- and/or the ket-. It would therefore be useful to define a delay scanning protocol for multidimensional signals that utilizes these properties.

2 LOP—A New Non time Ordered Delay Scanning Protocol

Multidimensional optical signals are commonly recorded by varying the delays between time ordered pulses $t_j, j = 1, 2, \ldots$ (see Fig. 1b). Spectra are displayed vs the Fourier conjugates $\tilde{\Omega}_j$ to these variables which describe the evolution of the density matrix and are represented by ladder diagrams. We denote it as the ladder delay scanning protocol (LAP). Here we propose a new non-time-ordered loop delay scanning protocol (LOP) obtained by following the time evolution of the wavefunction and described by loop diagrams [3]. We demonstrate that this protocol allows to observe different types of resonances and reveal information about intraband de-phasing not readily available by the LAP. The TPA signal (1) is given by the single loop diagram in Fig. 1c. a, b, c, d denote the pulse sequence ordered along the loop (not in real time); a represents "first" on the loop etc. To realize the LOP experimentally, the indices $a–d$ are assigned as follows: first by phase cycling we select a signal with phase $\phi_a + \phi_b - \phi_c - \phi_d$. The two pulses with positive phase detection are thus denoted a, b and with negative phase—c, d. In the a, b pair pulse a comes first. In the c, d pair pulse d comes first. The time variables in Fig. 1c are $\tau_1 = T_b - T_a$, $\tau_2 = T_c - T_b$, $\tau_3 = T_c - T_d$, where $T_j, j = a, b, c, d$ corresponds to the central time of the j-th pulse, Ω_j are frequency conjugates to τ_j. With this choice τ_1 and τ_3 are positive whereas τ_2 can be either positive or negative. The loop delay variables s_j in Fig. 1 are centered around $|\tau_j|$. The LOP protocol can be realized experimentally using a pulse shaper [4] or e.g. Franson interferometer [5].

3 Probing Intraband Dephasing Rates of Excitons in Aggregates

We consider a model aggregate with intraband dephasing rate $\gamma_{ee'}$ between excitonic states using LOP as shown in Fig. 2. Figure 2a shows the two photon fluorescence signal for a classical light with narrow intraband dephasing $\gamma_{ee'} = 1$ meV. It gives a diagonal cross peak $e = e'$ and one pair of weak side peaks parallel to the main diagonal at $(e, e') = (e_2, e_3)$ Fig. 2d shows that the signal obtained using

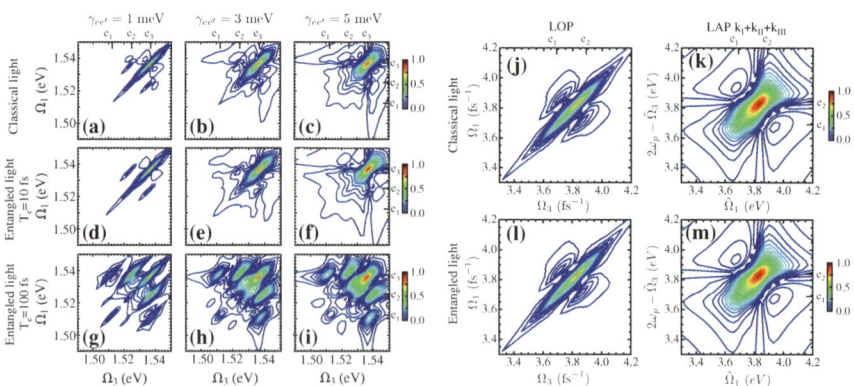

Fig. 2 *Left side* of the plot: TPA signal using LOP for a model of molecular trimer using classical light—*top row*, entangled light with entanglement time $T_e = 10$ fs—*middle row* and $T_e = 100$ fs—*bottom row*. Intraband dephasing $\gamma_{ee'} = 1$ meV—*left column*, 3 meV—*middle column* and 5 meV—*right column*. *Right side* of the plot is comparison between LOP and LAP protocols for a molecular dimer. LOP signal calculated using classical light (**j**), and entangled light (**l**). Corresponding LAP signal shown in *panels k and m*, respectively

entangled photons with short entanglement time $T_e = 10$ fs is similar to the classical signal in panel (a). For $T_e = 100$ fs in Fig. 2g we observe two additional strong side cross peak pairs with $(e, e') = (e_1, e_3)$ and $(e, e') = (e_1, e_2)$. The weak peak at $(e, e') = (e_2, e_3)$ is significantly enhanced as well. Thus, the LOP display (Ω_1, Ω_3) allows for effective determining of the intraband dephasing for distinct pair of e and e' states even if intraband dephasing is broad. The two right columns of the Fig. 2 compare the signals for LOP and LAP protocols. The LOP spectra for classical and entangled light are given in panels j and l. The corresponding LAP spectra are shown in the panels k and m. Entanglement makes no difference in this parameter regime (the two rows are virtually identical). However the signals of two scanning protocols are very different. The LOP signals are narrow and clearly resolve the e_1 and e_2 states whereas the corresponding LAP signals are broad and featureless.

Acknowledgements We gratefully acknowledge support from the National Science Foundation (Grant No. CHE 1361516), and the Chemical Sciences, Geosciences and Biosciences Division, Office of Basic Energy Sciences, Office of Science, and (U.S.) Department of Energy (DOE).

References

1. O. Roslyak, C.A. Marx, S. Mukamel, Phys. Rev. A **79**, 033832 (2009).
2. D.I. Lee, T. Goodson, J. Phys. Chem. B **110**, 25582 (2006).
3. K.E. Dorfman, S. Mukamel, New J. Phys. **16**, 033013 (2014).
4. F. Zäh, M. Halder, T. Feurer, Opt. Express **16**, 16452 (2008).
5. M.G. Raymer, A.H. Marcus, J.R. Widom, D.L.P. Vitullo, J. Phys. Chem. B **117**, 15559 (2013).

Ultrafast Interaction of Dark and Bright Electronic States in Open-Chain Carotenoids Investigated by Pump-DFWM

T. Miki, Tiago Buckup, M. Marek, R.J. Cogdell and Marcus Motzkus

Abstract Coupling between dark and bright electronic states in carotenoids was observed in the ultrafast evolution of vibrational coherence and in the non-oscillatory signal of pump-DFWM. Coupling efficiency depends on the number of conjugated double bonds.

1 Introduction

The ultrafast femtochemistry of carotenoids is governed by the interaction between electronic excited states. The two important functions of carotenoids, namely light harvesting and photo-protection, have been explained by the relaxation dynamics within a few hundred femtoseconds from the lowest optically allowed excited state S_2 to the optically dark state S_1 [1]. Extending this picture, some additional dark states (generally called S_x) and their interaction with S_2 state have been also suggested to play a major role in the ultrafast deactivation of carotenoids and their properties [2]. Here, we investigate the interaction between such dark and bright electronic excited states of open chain carotenoids, particularly its dependence on the number of conjugated double bonds (N). We focus on the ultrafast wave packet motion on the potential surface, which is modified by the interaction between bright and dark electronic states. In this regard, pump-degenerate four-wave mixing

T. Miki · T. Buckup (✉) · M. Marek · M. Motzkus (✉)
Physikalisch-Chemisches Institut, Ruprecht-Karls-Universität Heidelberg,
D-69120 Heidelberg, Germany
e-mail: tiago.buckup@pci.uni-heidelberg.de

M. Motzkus
e-mail: marcus.motzkus@pci.uni-heidelberg.de

T. Miki
e-mail: tiago.buckup@pci.uni-heidelberg.de

R.J. Cogdell
Institute of Biomedicine and Life Science, University of Glasgow,
Glasgow, Lanark G12 8QQ, Scotland

© Springer International Publishing Switzerland 2015
K. Yamanouchi et al. (eds.), *Ultrafast Phenomena XIX*,
Springer Proceedings in Physics 162, DOI 10.1007/978-3-319-13242-6_107

(pump-DFWM) is applied to a series of carotenoids with different number of conjugated double bonds $N = 9$, 10, 11 and 13 (neurosporene, spheroidene, lycopene and spirilloxanthin, respectively).

2 Experimental

Pump-DFWM measurements were carried out by using the experimental setup which has been described in detail previously [3]. All pulses were generated using non-collinear optical parametric amplifiers (nc-OPA). The Initial Pump pulse (IP) (18 fs) was resonant with the S_0-S_2 absorption. All three DFWM pulses were originated from a second nc-OPA (15 fs). The DFWM spectrum was resonant only with the excited state absorption (ESA). The IP pulse was sent to a delay line (delay T) while two DFWM pulses were further delayed via piezo stages (τ_{12} and τ_{23}). DFWM beams were configured in a folded boxcar geometry producing a spatially separate, background-free signal. Recrystallized carotenoids were dissolved in THF and circulated in a 450 μm flow cell during the measurements.

3 Results and Discussion

The pump-DFWM transient 2D signals of the four carotenoids show similar features (Fig. 1, top). Strong non-oscillatory signal at late T-delays ($\sim$ T > 200 fs) dephases fast and its dephase time constant scales with the lifetime of S_1-state.

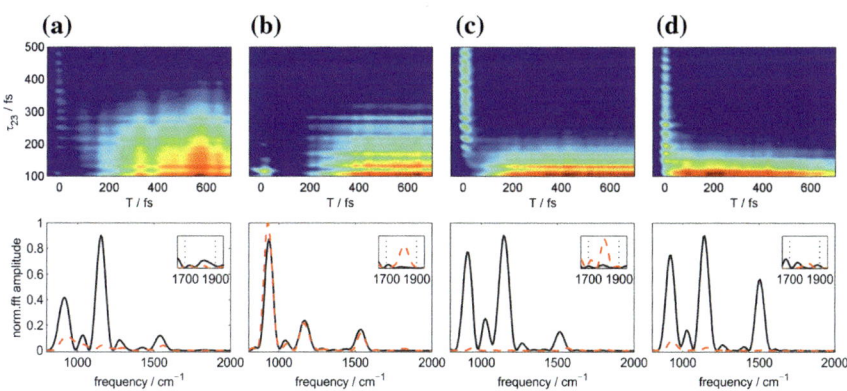

Fig. 1 2D-Plots of the pump-DFWM transients of the four different carotenoids (*top*). FFT power spectra of coherent signals at T = 10 fs (*black solid line*) and at 700 fs (*red broken line*) by frequency axis (*bottom*). **a** Neurosporene ($N = 9$), **b** spheroidene ($N = 10$), **c** lycopene ($N = 11$), **d** spirilloxanthin ($N = 13$). Detection wavelengths were chosen to maximize contribution from C–C and C=C modes

Further non-oscillatory contribution at early T-delays ($\sim T < 50$ fs) is related to the formation of a hot-S_0 signal via stimulated emission pumping (SEP) directly after excitation of the S_2 with the IP. This signal raises slowly along τ_{23} delay for all carotenoids except for spheroidene ($N = 10$). In spheroidene, such a SEP signal can be detected already at very early τ_{23} delay times. Numerical simulation of the whole 2D signal of pump-DFWM shows the SEP pathway does not occur directly from the S_2 state for spheroidene, but from an additional excited state (S_x), which is populated within 10–20 fs after leaving the Franck-Condon region [4, 5].

The 2D signals also display a strong oscillatory contribution (Fig. 1, bottom). After FFT, several vibrational modes can be clearly identified at 915 cm^{-1} (solvent), 1,030 cm^{-1} (methyl rocking mode), 1,150 cm^{-1} (C–C stretching modes) as well as 1,550 cm^{-1} (C=C stretching mode). Here, the amplitude ratio between C–C and C=C modes is similar for all measurements except, again, for spheroidene. Spheroidene's spectrum additionally differs from the other ones by showing asymmetric satellites, which are visible only at early T delays and vanish completely for later T. The satellites stem from interferences between the coherences of the SEP and the ESA signal, what further corroborates the existence of an additional dark (S_x) state being populated at early T delays in spheroidene [4, 5].

The evolution of the vibrational frequency in the T-delay reveals additional interactions between dark and bright states. Such an evolution of C–C (1,150 cm^{-1}) and C=C (1,545 cm^{-1}) stretching modes is shown in Fig. 2. While for longer carotenoids ($N = 11$ and 13), the vibrational frequency increases as the system follows the S_2-S_1 relaxation, for shorter carotenoids ($N = 9$ and 10), the frequency decreases. In general, in an anharmonic potential the vibrational frequency increases as the system relaxes towards lower levels. Our results indicate that short carotenoids display strongly distorted potentials when compared to longer carotenoids. The magnitude of this anharmonicity can be explained by the efficient coupling between S_2 and a dark S_x state (in this case a B_u type), which stabilizes the higher energy levels of the C–C and C=C vibrational modes at early delay times

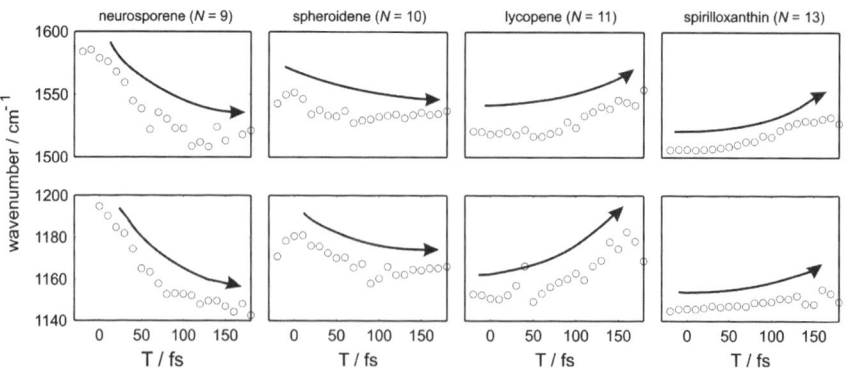

Fig. 2 The vibrational frequency evolution of C–C and C=C modes

(T < 50 fs). For longer carotenoids, this dark state is energetically shifted with respect to the S_2, decreasing the interaction and leading to a different degree of anharmonicity in the involved electronic excited states.

References

1. T. Polívka, V. Sundström, Chem. Rev., **104**, 20211 (2004)
2. E. Ostroumov et al., Phys. Rev. Lett., **103**, 108302 (2009)
3. J. Hauer et al., J. Phys. Chem. A, **111**, 10517 (2007)
4. M. S. Marek et al., J. Phys. Chem., **139**, 074202 (2013)
5. T. Buckup and M. Motzkus Annual Rev. Phys. Chem., **65**, 39 (2014)

Following the Excited State Dynamics of β-Apo-8′-Carotenal with Two-Dimensional Electronic-Vibrational Spectroscopy

Thomas A.A. Oliver, Nicholas H.C. Lewis and Graham R. Fleming

Abstract Two-dimensional electronic-vibrational spectroscopy is used to study the excited state relaxation of β-apo-8′-carotenal in acetonitrile solution. This new multidimensional spectroscopy technique is unique in its ability to directly follow the electronic and nuclear degrees of freedom simultaneously.

1 Introduction

The precise molecular mechanisms underlying the non-radiative decay pathways of the first optically bright S_2 state of carotenoids are complex and poorly understood, despite a vast number of studies. The main points of contention are the number, nature and role of lower lying electronic states that couple to the S_1 state, and role that nuclear motion plays driving the non-radiative transfer. It remains unclear if the reaction co-ordinate involves direct S_2–S_1 relaxation, or whether intermediate states such as $S*$, $S^{\ddagger}$ etc. are involved [1]. The rapid depopulation of the S_2 state is thought to be mediated by a conical intersection [2, 3]; a region of the potential energy surface where it is unreasonable to assume that the electronic and vibrational degrees of freedom are decoupled. Using a new experimental technique, two-dimensional electronic-vibrational (2DEV) spectroscopy, we investigate the excited state dynamics of a model carbonyl containing carotenoid, β-apo-8′-carotenal (displayed in Fig. 1a), correlating the evolution of the electronic and nuclear degrees of freedom simultaneously.

T.A.A. Oliver · N.H.C. Lewis · G.R. Fleming (✉)
Department of Chemistry, University of California, Berkeley, CA 94720, USA
e-mail: grfleming@lbl.gov

T.A.A. Oliver · N.H.C. Lewis · G.R. Fleming
Physical Biosciences Division, Lawrence Berkeley National Laboratory,
Berkeley, CA 94720, USA

© Springer International Publishing Switzerland 2015
K. Yamanouchi et al. (eds.), *Ultrafast Phenomena XIX*,
Springer Proceedings in Physics 162, DOI 10.1007/978-3-319-13242-6_108

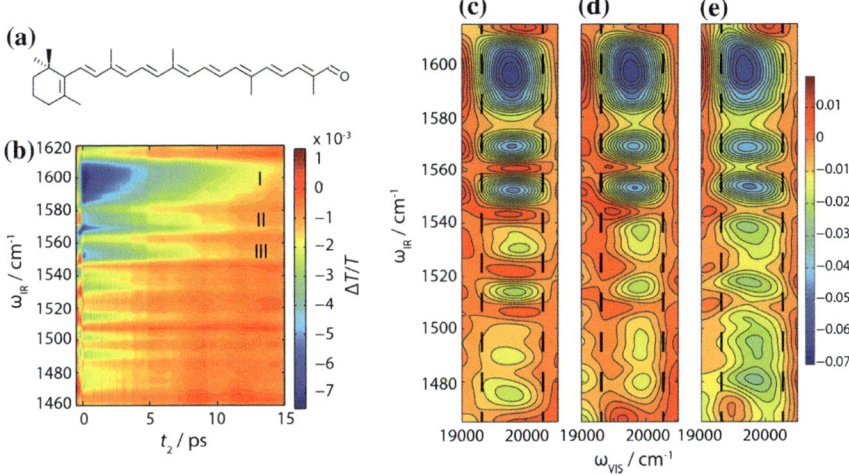

Fig. 1 a Skeletal structure of β-apo-8'-carotenal **b** 1D pump-probe spectrum, transient species I–III are assigned to the C=C symmetric, C=C anti-symmetric and C=C with ring breathing motions, respectively **c–e** 2DEV spectra for given values of t_2. *Dashed lines* indicate the initial bounds of peaks at $t_2 = 0$ fs. Data were collected under the magic angle polarization condition

2 Experimental

2DEV measurements were performed using a new experimental apparatus [4]. A commercial 1 kHz regenerative amplifier pumped a homebuilt non-collinear optical parametric amplifier (NOPA) and mid-IR optical parametric amplifier. The output of the NOPA was compressed (20 fs, centered at 515, 26 nm FWHM) with an acousto-optic programmable dispersive filter (AOPDF, Dazzler, Fastlite). The AOPDF was also utilized to create a collinear pair of pump pulses, k_1 and k_2, generating the coherence time delay, t_1, with control over the relative carrier envelope phase, ϕ_{12}. The total power of these pulses at the sample position was attenuated to 250 nJ. At waiting time, t_2, the probe mid-IR pulse, k_3 (~90 fs, ~80 nJ) interrogate evolution on the ground or excited states and k_{sig} is subsequently emitted at echo time, t_3. In the pump-probe geometry, k_{sig} is phase-locked with respect to the k_3 pulse, obviating any need to phase the resulting 2D spectra. The 2D spectrum can be retrieved by phase-cycling the pump-pulse pair, as previously demonstrated with visible and infrared multidimensional spectroscopies [5, 6] For a given waiting time, t_2, a 2DEV spectrum was collected by frequency dispersing k_{sig} (and k_3) onto an 64-element HgCdTe detector (Infrared Systems Development) as t_1 and ϕ_{12} were incremented. The instability of the mid-IR laser pulses was corrected by normalizing data with respect to a second mid-IR reference pulse that passed through the sample, but was not spatially overlapped with the pump pulse pair. A 1×3 phase cycling scheme was used to acquire data, from which a $t_1 - \omega_3$ matrix was generated for the respective re-phasing and non-rephasing pathways. These data were subsequently apodized, zero-padded and Fourier-transformed along the t_1 axis to generate a $\omega_1 - \omega_3$ 2DEV spectrum. The pump pulse pair were resonant with the $S_2 \leftarrow S_0$ absorption of β-apo-8'-carotenal. 2DEV spectra were recorded at room temperature in deuterated acetonitrile for many t_2 times. The β-apo-8'-carotenal sample had an optical density of 0.3 at the peak of the laser spectrum in a 250 μm path length flow cell.

3 Results and Discussion

Figure 1b displays the frequency resolved one-dimensional visible pump, mid-IR probe spectrum of β-apo-8'-carotenal in acetonitrile solution. This probe region was chosen to monitor the evolution of the C=C and mixed C=C/C–C back-bone vibrations. The spectrum is dominated by negative features: indicative of vibrational evolution on excited potential energy surface(s). The vibrational assignments are labeled in Fig. 1b from [7]. The vibrations on the excited potential energy surface are pertinent because the $S_2 \rightarrow S_1$ relaxation mechanism is proposed to involve a change in bond-order of the C=C/C–C backbone [2]. The kinetics of the three main vibrational features in Fig. 1b all decay with the same time constants and in good agreement with the S_2 and S_1 lifetimes (<200 fs and 9 ps, respectively). The frequency-dispersed pump-probe spectrum the central frequency for these bands does not shift, indicative of no change in bond order upon $S_2 \rightarrow S_1$ relaxation.

Real total 2DEV spectra are shown in Fig. 1c–e for $t_2 = 0$, 125 and 450 fs. Within the first few hundred femtoseconds, the center of the electronic lineshape shifts to lower frequencies (~ 100 cm^{-1}) within the convolution of our laser bandwidth and the absorption spectrum, indicative of a ballistic exit from the S_2 state. We interpret the subtle change in lineshape from square like ($t_2 = 0$ fs) to more elliptical in the first few hundreds of femtoseconds (see Fig. 1c–e) as a loss of correlation between the initially excited low frequency modes that comprise the electronic lineshape and the high frequency IR active mode probed.

4 Conclusions

We performed the first 2DEV measurements of a carotenoid (β-apo-8'-carotenal) in solution. We observed a small evolution along the electronic axis within the first few hundreds of femtoseconds. The absence of any change in fundamental frequencies of the C=C/C–C back-bone vibrations is contrary to prior expectations [2], and requires high level theoretical studies in order to understand this observation. Further analysis of the lineshapes in 2DEV spectra [4] is required to ascertain how to interpret the correlation function between the low-frequency modes and the probed mid-IR active vibration. Future 2DEV studies will seek to probe the dynamics of the carbonyl mode and its role in the S_2 relaxation.

Acknowledgments This work was supported by NSF under contract number NSF CHE-1012168 and the Director, Office of Science, Office of Basic Energy Sciences, of the USA Department of Energy under contract DE-AC02-05CH11231 and the Division of Chemical Sciences, Geosciences and Biosciences Division, of Basic Energy Sciences through grant DE-AC03-76F000098 (at LBNL and UC Berkeley).

References

1. T. Polívka and V. Sundström, Chem. Phys. Lett. 477, 1–11 (2009).
2. M. Garavelli et al. Faraday Discuss. 110, 51–70 (1998).
3. P. Kukura et al. J. Phys. Chem. A 108, 5921–5925 (2004).
4. T. A. A. Oliver et al. Proc. Natl. Acad. Sci. USA 111, 10061–10066 (2014).
5. J. A. Myers et al. Opt. Express 16, 17420–17428 (2008).
6. S. H. Shim et al. Proc. Natl. Acad. Sci. USA 104, 14197–14202 (2007).
7. Y. Pang et al. J. Phys. Chem. B 113, 13086–13095 (2009).

Survival of Nuclear Coherences for a Series of Internal Conversions in Free Base Tetraphenylporphyrin

S.Y. Kim, S. Kim and T. Joo

Abstract We observe that the stepwise internal conversions from B to Q_y and to Q_x states in free base tetraphenylporphyrin generate coherent nuclear wave packets in both Q_x and Q_y states. Theory and experiment show that these wave-packet motions involve out-of-plane vibrations of the porphyrin ring that are strongly coupled to the internal conversions.

1 Introduction

Porphyrins, including free base tetraphenylporphyrin (H_2TPP), are one of the most important group of molecules in nature and have been under intense investigation [1–3]. Unlike the steady-state spectroscopy, however, the ultrafast relaxation dynamics from B to Q_y and to Q_x states in H_2TPP has not been clearly revealed yet, although Baskin et al. reported the femtosecond time-resolved experimental results [1]. Because the excited-state dynamics of porphyrins often occurs on a femtosecond time scale, the nuclear coordinates responsible for the internal conversions as well as the precise internal conversion rates are not clear.

Coherent nuclear wave packets in the excited states can be launched by an internal conversion process occurring faster than the vibrational period. Such a reaction-driven wave-packet motion provides a valuable information on the reaction coordinates and electron-nuclear couplings along the relevant coordinates. These wave-packet motions can be measured most unambiguously by time-resolved fluorescence (TRF) with a resolution higher than the periods of vibrations [4].

S.Y. Kim · T. Joo (✉)
Department of Chemistry, Pohang University of Science and Technology (POSTECH),
Pohang 790-784, Korea
e-mail: thjoo@postech.ac.kr

S. Kim
Department of Chemistry, Korea Advanced Institute of Science and Technology (KAIST),
Daejeon 305-701, Korea

© Springer International Publishing Switzerland 2015   448
K. Yamanouchi et al. (eds.), *Ultrafast Phenomena XIX*,
Springer Proceedings in Physics 162, DOI 10.1007/978-3-319-13242-6_109

In this work, we present the ultrafast internal conversion dynamics by TRF, which gives the signal from the excited state exclusively. The TRF of H_2TPP in toluene was measured by fluorescence up-conversion with 50 fs time resolution, which was high enough to resolve the wave-packet oscillations up to 500 cm^{-1}.

2 Results and Discussion

Figure 1b shows the TRF spectra of H_2TPP over the Q emissions after the B excitation. Immediately after the photoexcitation, the TRF spectra show two peaks at 550 and 610 nm, which are absent in the steady-state spectra (Fig. 1a). The observed peaks can be assigned to the $Q_y(0, 0)$ and $Q_y(0, 1)$ bands, and both bands rise and decay within 100 fs in the same manner. Figure 2 shows the TRF of H_2TPP measured at different Q bands after the B excitation. The TRF in the Q_y band (610 nm, Fig. 2a) rises by 60 fs, which can be assigned to the B $\rightarrow$ Q_y internal conversion, and rapidly decays by 80 fs. On the other hand, the TRF in the Q_x band (650 nm, Fig. 2b) shows a delayed rise of 80 fs, which is the same as the fastest decay of the Q_y emission. These temporal features are well-matched with the TRF spectra, and they denote that the $Q_y \rightarrow Q_x$ internal conversion occurs in 80 fs.

The most notable feature is the distinct oscillations in the TRF signals at each Q band. These long-lived nuclear coherences in H_2TPP have been observed for the first time. In the Q_y emission, a unique low-frequency oscillation at 33 cm^{-1} was observed, while a higher frequency at 193 cm^{-1} was dominant in the Q_x emission. To identify these distinct wave-packet dynamics, quantum mechanical calculations were performed. Geometry optimizations and frequency calculations were carried out at the B3LYP/6-311G level for the first excited state (Q_x) and B3LYP/6-31G level for the second excited state (Q_y) by using the Gaussian 09 package. In an analogy to the Franck-Condon transition, vibrational modes with large displacements between the reactant and product states will be excited preferentially. We can define a general Huang-Rhys factor coupled to the electronic transition, and calculate the amplitudes of vibrational modes comprising the wave packets in each Q state by calculating the projections of the structural displacements onto the normal modes of the product state.

We have carried out the normal-mode-projection calculations for two different cases: ground $\rightarrow Q_x$ and ground $\rightarrow Q_y$. For both calculations, v_{22} is dominant and matches well with the 193 cm^{-1} mode observed in the Q_x emission. This mode corresponds to the out-of-plane motion of the entire porphyrin ring with some rocking/scissoring motion of the phenyl groups. This result is quite reasonable because the wave packets generated in the Q_y state quickly move to the Q_x state along the $Q_y \rightarrow Q_x$ internal conversion. Hence we can assign the calculated v_{22} to the 193 cm^{-1} mode observed in the TRF, which has a half period of 86 fs that is nearly the same as the internal conversion time of 80 fs. We propose that this out-of-plane vibrational mode can be regarded as the reaction coordinate of $Q_y \rightarrow Q_x$ in H_2TPP leading to the ultrafast internal conversion.

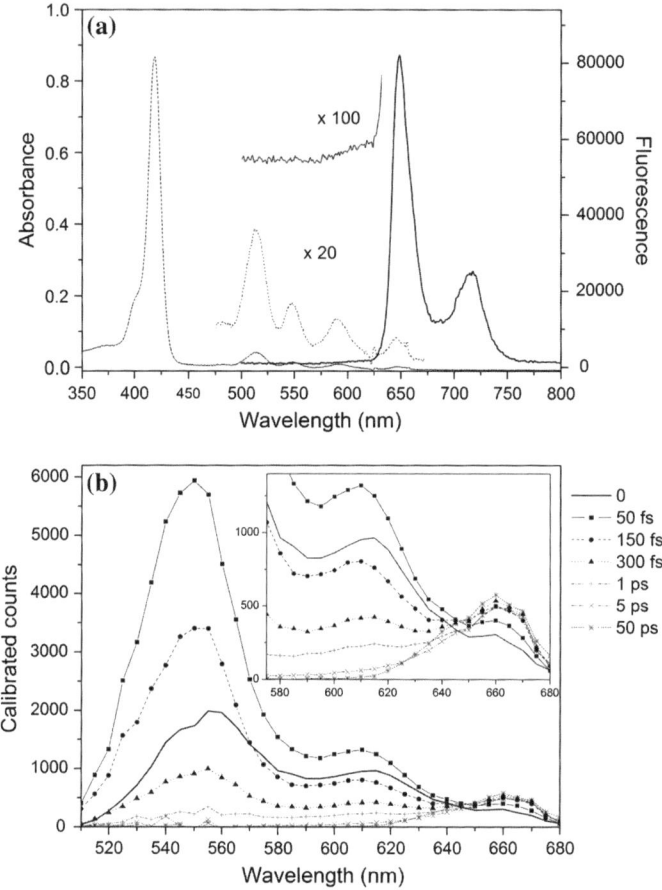

Fig. 1 a Steady-state absorption and emission spectra of H_2TPP in toluene. Only $Q_x(0, 0)$ and $Q_x(0, 1)$ bands appear in the emission spectrum after the B excitation, indicating the ultrafast internal conversions. **b** TRF spectra of H_2TPP in toluene. Excitation wavelength was 410 nm and time resolution for the TRF spectra measurement was 80 fs

Although most Q_y population transfers to the Q_x state, a small part of the population remains in the Q_y state, and shows the low-frequency oscillation in the TRF of Q_y emission. Interestingly, only the low-frequency mode remains in the Q_y state, although the internal conversion is ultrafast. For the second case (ground $\rightarrow Q_y$), there are about 10 vibrational modes in the low-frequency region, but only one mode at 37 cm^{-1} (v_5) shows nonzero displacement. Thus, the 33 cm^{-1} mode observed in the Q_y emission can be assigned to v_5. However, the 33 cm^{-1} mode may not be directly related to the B $\rightarrow Q_y$ internal conversion, unlike the 193 cm^{-1} mode. These vibronic coherences in porphyrins may be related to the long-lived electronic coherences in light-harvesting complex and requires more investigation.

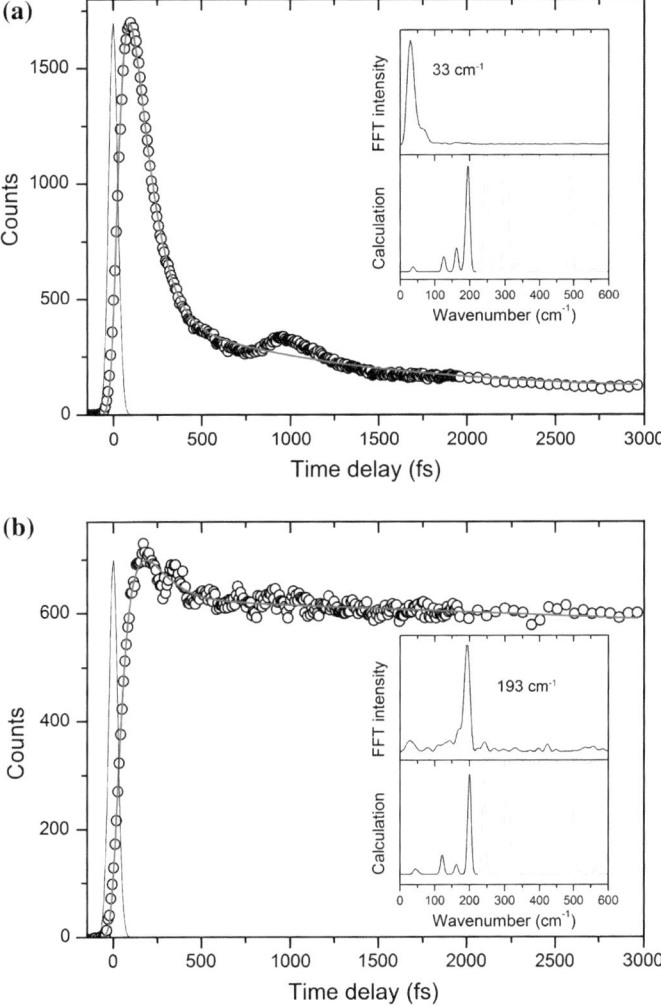

Fig. 2 TRFs of H_2TPP in toluene detected at **a** 610 nm and **b** 650 nm after the B excitation at 410 nm. *Insets* show the frequency spectra obtained by Fourier transform with the calculated frequency spectra

References

1. J. S. Baskin, H. Z. Yu, A. H. Zewail, J. Phys. Chem. A **106**, 9837 (2002).
2. K. Y. Yeon, D. Jeong, S. K. Kim, Chem. Comm. **46**, 5572 (2010).
3. B. Bialkowski, Y. Stepanenko, M. Nejbauer, C. Radzewicz, J. Waluk, J. Photochem. Photobiol. A: Chemistry **234**, 100 (2012).
4. C. H. Kim, T. Joo, Phys. Chem. Chem. Phys. **11**, 10266 (2009).

Distinctive Spectral Features of Exciton and Excimer States in the Ultrafast Electronic Deactivation of the Adenine Dinucleotide

Mayra C. Stuhldreier, Katharina Röttger and Friedrich Temps

Abstract We report the observation by transient absorption spectroscopy of distinctive spectro-temporal signatures of delocalized exciton versus relaxed, weakly bound excimer states in the ultrafast electronic deactivation after UV photoexcitation of the adenine dinucleotide.

1 Introduction

Extraordinarily efficient and ultrafast electronic deactivation processes in the natural nucleobases are widely considered as key factors determining the relative UV photostability of the DNA. Highly surprisingly, however, studies of single- and double-stranded DNA oligonucleotides revealed that the electronic lifetimes in these large base assemblies may be three to four orders of magnitude longer than in the monomers [1–3]. This huge discrepancy initiated intense debate, whether the long-lived excited states in the oligomers are due to dipole coupled, delocalized excitons extending over several bases [4] or whether they originate from localized excimers consisting of two neighboring, weakly bound π-stacked bases on the same strand with possible (partial) charge transfer (CT) character [5]. Here, we report the observation of distinct spectro-temporal signatures in transient excited-state absorption (ESA) spectra of the adenine dinucleotide $d(A)_2$ (Fig. 1a), which can be assigned to excitonic and to excimer-like excited states, respectively, shedding clear light on this important conundrum.

M.C. Stuhldreier · K. Röttger · F. Temps (✉)
Institute of Physical Chemistry, Christian-Albrechts-University Kiel, Olshausenstr. 40, 24098 Kiel, Germany
e-mail: temps@phc.uni-kiel.de

© Springer International Publishing Switzerland 2015
K. Yamanouchi et al. (eds.), *Ultrafast Phenomena XIX*,
Springer Proceedings in Physics 162, DOI 10.1007/978-3-319-13242-6_110

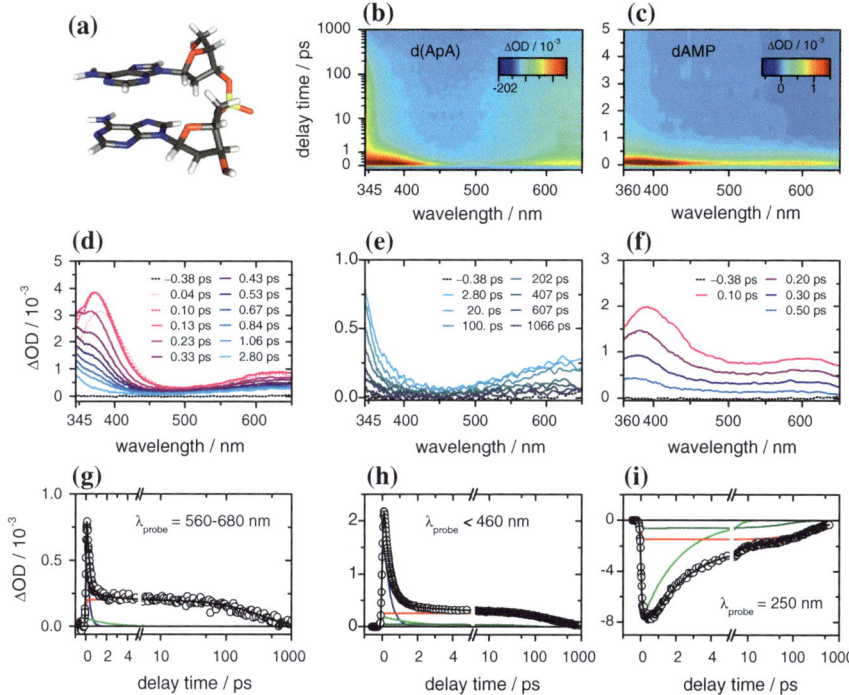

Fig. 1 **a** Structure of the d(A)$_2$ dinucleotide (B-DNA conformation). **b–c** 2-D transient absorption maps after 260 nm excitation of d(A)$_2$ and dAMP, respectively. **d–f** Excited-state absorption spectra of d(A)$_2$ (*panels d and e*) and dAMP (*panel f*) at selected delay times. **g–i** Temporal evolutions of the d(A)$_2$ ESA bands and recovery of the ground state bleach signal together with the respective least-squares fit curves

2 Experimental Results

The ultrafast electronic relaxation dynamics in the d(A)$_2$ dinucleotide were investigated following photoexcitation at λ = 260 nm in phosphate buffered aqueous solution (pH 7.0) by femtosecond UV/vis absorption spectroscopy with supercontinuum white-light and single-color deep-UV probe pulses [6]. The recorded two-dimensional (2-D) transient absorption map is displayed in Fig. 1b, the map for the mononucleotide dAMP is given for comparison in Fig. 1c. The vast differences between the two moieties are highlighted by the plots of the transient absorption spectra at selected delay times after excitation in Fig. 1d–e for the dimer versus Fig. 1f for the monomer. In the first place, the long-lived ESA of the dinucleotide peaking at λ < 345 nm is absent in the spectrum of the mononucleotide. Most importantly, however, the ESA spectra of d(A)$_2$ showcase not only substantial spectral evolution as function of time after excitation, but also partially resolved spectral structure, which is not visible in the considerably broader, unspecific transient spectra for dAMP. In particular, the broad initial ESA of the dinucleotide in the range 345–450 nm consisting of two overlapping, almost superimposed bands (Fig. 1d) collapses to the much longer-lived ($\tau \approx 400$ ps), and spectrally much narrower, single band at $\lambda < 345$ nm (Fig. 1d–e). This striking transition happens on an ultrashort time scale, $\Delta t \lesssim 500$ fs. Attempts of a spectro-temporal band analysis of Fig. 1d even suggest first relaxation steps taking place within ≤ 100 fs. Additionally, the blue-shift of the ESA is accompanied by a corresponding red-shift in the fluorescence spectra of the excited dinucleotide molecules (not shown).

3 Discussion and Conclusions

The observed spectral evolution in Fig. 1d and the temporal dynamics shown in Fig. 1g–i reveal the ultrafast relaxation in the initially excited excitonically coupled ladder of states via intraband scattering within $\Delta t \lesssim 100$–500 fs followed by subsequent larger-amplitude rearrangement of the nucleobase moieties within a few ps from the original B-DNA configuration to an energetically relaxed and therefore much longer-lived ($\tau \approx 400$ ps) excimer with presumable structure [7, 8] between near face-to-face or significantly distorted. With ongoing improvements of experimental time resolution, critical tests of the excited-state structures and dynamics proposed by quantum chemical calculations (e.g. [7, 8] and references therein) thus come into sight, which should resolve the challenging controversy surrounding the electronic dynamics in DNA oligonucleotides. Towards these ends, $d(A)_2$ constitutes an ideal model system for single-stranded DNA.

Acknowledgments The financial support of this work by the DFG is gratefully acknowledged.

References

1. C. E. Crespo-Hernández, B. Cohen, and B. Kohler, Base stacking controls excited-state dynamics in A·T DNA, Nature **436**, 1141–1144 (2005).
2. N. K. Schwalb and F. Temps, Base sequence and higher-order structure induce the complex excited-state dynamics in DNA, Science **322**, 243–245 (2008).
3. C. T. Middleton, K. de La Harpe, C. Su, Y. K. Law, C. E. Crespo-Hernández, and B. Kohler, DNA excited-state dynamics: from single bases to the double helix, Annu. Rev. Phys. Chem. **60**, 217–239 (2009).
4. D. Markovitsi, F. Talbot, T. Gustavsson, D. Onidas, E. Lazzarotto, and S. Marguet, Complexity of excited-state dynamics in DNA, Nature **441**, E7 (2006).
5. C. E. Crespo-Hernández, B. Cohen, and B. Kohler. Complexity of excited-state dynamics in DNA, Nature **441**, E8 (2006).
6. K. Röttger, R. Siewertsen, and F. Temps, Ultrafast electronic deactivation dynamics of the rare natural nucleobase hypoxanthine, Chem. Phys. Lett. **536**, 140–146 (2012).
7. F. Plasser and H. Lischka, Electronic excitation and structural relaxation of the adenine dinucleotide in gas phase and solution, Photochem. Photobiol. Sci. **12**, 1440–1452 (2013).
8. V. A. Spata and S. Matsika, Bonded excimer formation in π-stacked 9-methyladenine dimers, J. Phys. Chem. A **117**, 8718–8728 (2013).

Influence of Intramolecular Hydrogen Bonding on the Photodynamics of 2-(1-Ethynylpyrene)-Adenosine (PyA)

P. Trojanowski, C. Grünewald, F.F. Graupner, M. Braun, A.J. Reuss, J.W. Engels and J. Wachtveitl

Abstract We report on the influence of intramolecular hydrogen bonding between the 2'OH group of ribose and the N3 of adenine in 2-(1-ethynylpyrene)-adenosine (PyA) on the ultrafast dynamics, by comparing PyA with its deoxy derivate (PydA).

OSIC Codes (160.1435) biomaterials · (300.6500) spectroscopy, time-resolved · (260.5130) photochemistry

1 Introduction

Pyrene-modified molecules are commonly used as versatile fluorescent probes for studies of biomolecular systems like RNA, proteins and membranes [1–3]. We designed the 2-(1-ethynylpyrene)-adenosine (PyA, Fig. 1a) as a fluorescent probe for RNA dynamics [4–8]. Its high fluorescence quantum yield and sensitivity to environmental changes, e.g. its flanking bases [7], makes it an excellent probe for hybridization studies and for conformational changes, e.g. due to ligand binding.

Femtosecond transient-absorption spectroscopy, time-correlated single-photon counting (TCSPC) and streak-camera measurements revealed quite complex ultrafast dynamics [5, 6, 8]. These dynamics are influenced strongly by intramo-

P. Trojanowski · M. Braun · A.J. Reuss · J. Wachtveitl (✉)
Institute of Physical and Theoretical Chemistry, Johann Wolfgang Goethe-University,
Max-von-Laue-Str. 7, 60438 Frankfurt a. M., Germany
e-mail: wveitl@theochem.uni-frankfurt.de

C. Grünewald · J.W. Engels
Institute of Organic Chemistry and Chemical Biology,
Johann Wolfgang Goethe-University, Max-von-Laue-Str. 7,
60438 Frankfurt a. M., Germany

F.F. Graupner
Faculty of Physics, Center for Integrative Protein Science,
Ludwig-Maximilians-University Munich, Oettingenstrasse 67,
80538 Munich, Germany

© Springer International Publishing Switzerland 2015
K. Yamanouchi et al. (eds.), *Ultrafast Phenomena XIX*,
Springer Proceedings in Physics 162, DOI 10.1007/978-3-319-13242-6_111

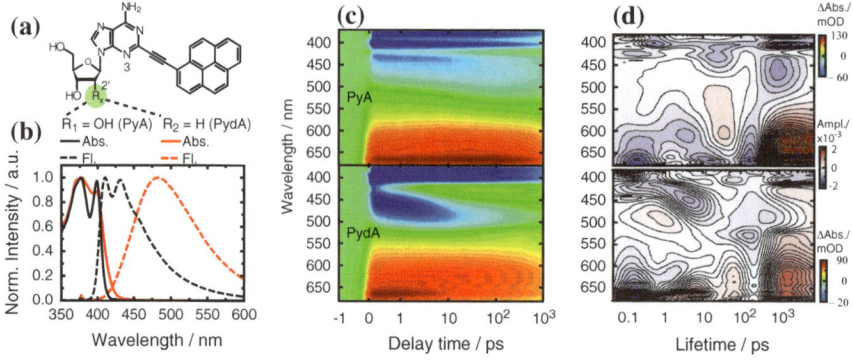

Fig. 1 PyA and PydA (**a**), steady-state absorption and fluorescence (λ_{exc} = 376 and 378 nm, respectively) spectra in DMSO (**b**) and corresponding transient absorption spectra (λ_{exc} = 388 nm) in DMSO (**c**), along with the LDA maps (**d**)

lecular hydrogen bonding [8]. Herein, we present our results on the influence of the 2′OH group on the photo-physical processes of pyrene modified adenosine (PyA) by comparison to the pyrene modified deoxyadenosine (PydA, Fig. 1b, c), which is the corresponding label for DNA.

2 Materials and Methods

2-(1-ethynylpyrene)-adenosine (PyA) and -deoxyadenosine (PydA) (Fig. 1a) were synthesized using methodologies described for the synthesis of protected PyA [4].

The time-resolved transient-absorption (TA) measurements (Fig. 1c) were performed on a home-built pump-probe setup [9]. The samples (150–200 μm) were excited with frequency-doubled (388 nm) pulses (150 nJ, 200 fs) of a CLARK CPA 2001 (Clark-MXR, Dexter, MI). TA measurements were corrected for solvent background and coherent artifacts and subjected to a lifetime density analysis (LDA, Fig. 1d). The pre-exponential amplitudes are determined in a sum of a large number (>50, typically ∼100) of exponential functions with fixed lifetimes. The LDA is, unlike global lifetime-analysis, model independent and able to describe non-exponential or stretched exponential kinetics [10].

3 Results

PyA and PydA show a complex time-resolved behavior in DMSO (Fig. 1c). Four time constants are needed to describe both PyA [5] and PydA dynamics (Fig. 1d). The fastest component (τ_1) is in the sub picosecond time range and caused by IVR of the S_1 state (excited state absorption (ESA1), 667 nm).

The second time constant (τ_2) describes the population of a second conformational substate S_{1x} (ESA2, ~ 625 nm) with a higher dipole moment than the S_1 state. The time constant for the population of the S_{1x} state (τ_2) for PyA and PydA is fairly similar with 3.7 and 2 ps, respectively. Both observed states are strongly emitting. Therefore, both contribute to the stimulated emission (SE) between ~ 400 and ~ 525 nm. The corresponding SE1 and SE2 bands are directly connected to the respective ESA1 and ESA2 signals. SE1 and ESA1 are assigned to a locally excited state S_1, while SE2 and ESA2 are assigned to a substate on the potential energy surface of the S_1 state (S_{1x}) [8]. The S_1 state is highly populated in case of PyA. In case of PydA, a quick loss of population is visible, which leads to dominantly populated S_{1x} state. The τ_3 component is mainly caused by the stabilization of the S_{1x} state [8], which is much slower for PyA (~ 40 ps) than for PydA (<10 ps). The population of the S_{1x} state occurs in the same time range as this spectral shift.

The ground state is repopulated from both states (S_1 and S_{1x}) with τ_4. The exact values were determined via streak-camera measurements (data not shown) and are 2.1 ns for PyA and 2.6 ns for PydA.

4 Discussion

Substitution of the 2′OH group of the ribose with hydrogen favors the population of the S_{1x} state with respect to the S_1 state, which is due to a lower energy barrier between these two states. Additionally, the spectral shift of the S_{1x} state is strongly accelerated due to the missing intramolecular hydrogen bond and the polar character of the S_{1x} state. Steady-state measurements indicate that the spectral shift increases gradually with increasing solvent permittivity (data not shown). In combination with the fact that the 2′OH group is too far away to interact directly with the conjugated electron system of pyrene, we can assume that hydrogen bonding of the 2′OH group with the adenine is responsible for the higher S_1 population in case of PyA. The only accessible position is the N3 of the adenine, due to its proximity to the 2′OH group (Fig. 1a). The result of the hydrogen bonding is a reduced energy barrier for the $S_1 \rightarrow S_{1x}$ transition, due to an altered electron density on the N3. In combination with the overall faster spectral shift for PydA (Fig. 1c), this indicates that PydA has a more polar S_{1x} state than PyA and, in addition, that PyA is sterically more hindered than PydA.

References

1. C. Armbruster, M. Knapp, K. Rechthaler, R. Schamschule, A. B. J. Parusel, G. Köhler, and W. Wehrmann, J. Photochem. Photobiol. Chem. 125, 29–38 (1999)
2. H. Maeda, T. Maeda, K. Mizuno, K. Fujimoto, H. Shimizu, and M. Inouye, Chem. - Eur. J. 12, 824–831 (2006)

3. K. K. Karlsen, A. Pasternak, T. B. Jensen, and J. Wengel, ChemBioChem 13, 590–601 (2012)
4. C. Grünewald, T. Kwon, N. Piton, U. Förster, J. Wachtveitl, and J. W. Engels, Bioorg. Med. Chem. 16, 19–26 (2008)
5. U. Förster, N. Gildenhoff, C. Grünewald, J. W. Engels, and J. Wachtveitl, J. Lumin. 129, 1454–1458 (2009)
6. U. Förster, C. Grünewald, J. W. Engels, and J. Wachtveitl, J. Phys. Chem. B 114, 11638–11645 (2010)
7. U. Förster, K. Lommel, D. Sauter, C. Grünewald, J. W. Engels, and J. Wachtveitl, ChemBioChem 11, 664–672 (2010)
8. P. Trojanowski, J. Plötner, C. Grünewald, F. F. Graupner, C. Slavov, A. J. Reuss, M. Braun, J. W. Engels and J. Wachtveitl, Phys. Chem. Chem. Phys. 16, 13875–13888 (2014)
9. L. Dworak, V. V. Matylitsky, and J. Wachtveitl, ChemPhysChem 10, 384–391 (2009)
10. A. A. Istratov and O. F. Vyvenko, Rev. Sci. Instrum. 70, 1233–1257 (1999)

S_2 to S_1 Relaxation Dynamics in Perylene Bisimide Dye Aggregates and Monomers

Steffen Wolter, Franziska Fennel, Marco Schröter, Jan Schulze, Frank Würthner, Oliver Kühn and Stefan Lochbrunner

Abstract The ultrafast relaxation from the S_2 to the S_1 state in perylene bisimides is investigated by femtosecond absorption spectroscopy. The relaxation takes place on a timescale of 150 fs and decelerates slightly upon aggregation.

OCIS Codes 320.7150 · 320.2250 · 160.4890

1 Introduction

Exciton-Exciton annihilation (EEA) has a strong impact on the exciton dynamics of supramolecular structures and is a useful tool to monitor exciton transfer [1]. In an EEA event two excitons encounter and merge on one chromophore leading to a highly excited electronic state S_n. So far, EEA models assume that this state is very short living and the relaxation back to the single exciton band takes a negligible amount of time. Nevertheless, this assumption has to be proven as the relaxation from the S_n state is complex and multiple states and transitions might be involved. In this work, we address the relaxation from the S_2 state to the S_1 state of a fourfold in the bay area tert-buthyl-phenoxy-substituted perylene bisimide (PBI, see Fig. 1c), which are popular building blocks for supramolecular structures [2], and study the influence of aggregation on this process. The aggregation is controlled by means of the solvent. PBI forms stable J aggregates in the nonpolar solvent methylcyclo-hexane (MCH) [3] while aggregation is prevented in dichloromethane (DCM). For monomers the different solvents do not effect the relaxation times from the S_2 to the S_1 state as it was tested for a non aggregating similar dye (data not shown).

S. Wolter · F. Fennel · M. Schröter · J. Schulze · O. Kühn · S. Lochbrunner (✉)
Institute of Physics, University of Rostock, Universitätsplatz 3, 18055 Rostock, Germany
e-mail: stefan.lochbrunner@uni-rostock.de

F. Würthner
Institut für Organische Chemie and Center for Nanosystems Chemistry,
Universität Würzburg, 97074 Würzburg, Germany

© Springer International Publishing Switzerland 2015
K. Yamanouchi et al. (eds.), *Ultrafast Phenomena XIX*,
Springer Proceedings in Physics 162, DOI 10.1007/978-3-319-13242-6_112

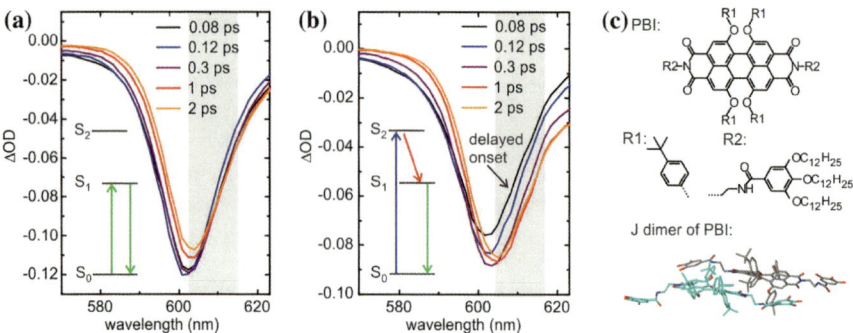

Fig. 1 Transient spectra of PBI aggregates in MCH after excitation to the S_1 state (**a**) and to the S_2 state (**b**). In the case of S_2 excitation a delayed onset of the stimulated emission from the S_1 level is observed. (**c**) Molecular structures of PBI and J dimer representing the aggregates

2 Experimental Approach

Transient absorption measurements are performed by exciting the sample with a tunable pump pulse delivered by a noncollinear optical parametric amplifier and probing it by a CaF_2 white light continuum. The white light is compressed by a prism sequence to obtain a high time resolution. We selectively excite either the S_1 state (Fig. 1a) by means of a 545 nm pump pulse or the S_2 state (Fig. 1b) at 440 nm to observe the impact of the S_2–S_1 relaxation. In the case of S_1 excitation of the aggregates (Fig. 1a) the dynamics in the first 2 ps mainly reflect intramolecular vibrational redistribution (IVR), manifested by a slight shift of ground state bleach and stimulated emission. In the case of S_2 excitation (Fig. 1b) the ground state bleach in the spectral region of the S_1–S_0 absorption (below 605 nm) is again present from time zero on and exhibits only minor changes with time. In contrast, above 605 nm an additional signal increase within a few hundred femtoseconds is observed. In this spectral region the stimulated emission from the S_1 state contributes strongly to the transient signal. It only appears after the population of the S_2 state has relaxed to the S_1 state, most probably via a conical intersection. Therefore, its delayed rise gives the relaxation time from the S_2 to the S_1 state.

3 Analysis of the Dynamics

In Fig. 2a time traces in the spectral region of the ground state bleach and the stimulated emission from the S_1 state for S_2 excitation are displayed for the aggregate and the monomer. To extract the time constants of the S_2–S_1 relaxation the traces are fitted with an exponential decay and the signal onset with an error function. For a fair comparison of the relaxation behavior in aggregates and monomers one has to take into account that the width and exact spectral position of

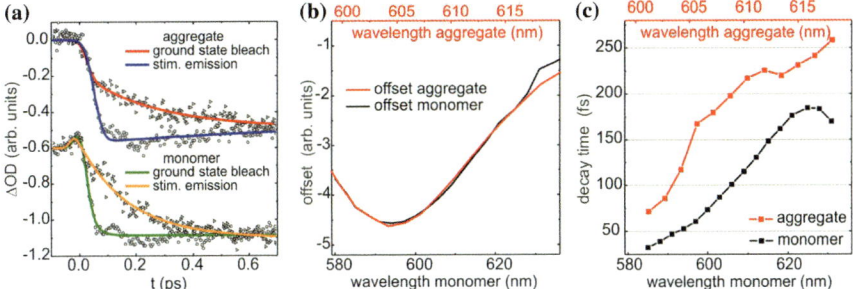

Fig. 2 a Time traces for PBI aggregates and monomers after S_2 excitation at characteristic wavelengths in the region of the ground state bleach and the stimulated emission of the S_1 state. **b** Transient spectra at a delay time of 750 fs. The *top* and *bottom* x-axis belong to aggregate (*red*) and monomer (*black*) respectively and are scaled in order to match the band positions and widths of the dominant transient bands. **c** Time constants of the fast signal component for aggregate (*red*) and monomer (*black*). The wavelength axes are scaled according to (**b**)

the bleach and stimulated emission depend somewhat on the aggregation state. To this end, the wavelength axes are scaled in such a way that the dominant transient bands of both, aggregate and monomer, are on top of each other after the fast processes are over (see Fig. 2b). Figure 2c shows a comparison of the extracted time constants which exhibit a wavelength dependence prohibiting a global fit. We attribute this wavelength dependence to the superposition of two fast processes, namely S_2–S_1 relaxation and IVR. In the spectral region of the ground state bleach the decay dynamics are dominated by IVR processes. In the region around 612 nm, the transient absorption signal is to a large extent due to stimulated emission and thus, the observed decay should mainly reflect S_2–S_1 relaxation. In both cases this process takes about 150–200 fs. However, for the aggregated compound the observed time constants are systematically longer than for the monomer and aggregation slightly slows down the depopulation of the S_2 state. Interestingly, the stationary absorption spectra show a small increase of the energy gap between the S_2 and S_1 state upon aggregation indicating that, in accordance with the energy gap rule, a larger energetic separation is associated with a slower relaxation rate. Aggregation seems to influence the relaxation of higher electronic states and thereby also the speed of the deactivation step in EEA.

References

1. S. Wolter, J. Aizezers, F. Fennel, M. Seidel, F. Würthner, O. Kühn, and S. Lochbrunner, New J. Phys., **14**, 105027, (2012).
2. F. Würthner, Pure Appl. Chem., **78**, 2341–2349 (2006).
3. F. Fennel, S. Wolter, Z. Xie, P.-A. Plötz, O. Kühn, F. Würthner, and S. Lochbrunner, J. Am. Chem. Soc., **135**, 18722-18725 (2013).

2D IR Spectroscopy with Phase-Locked Pulse Pairs from a Birefringent Delay Line

J. Réhault, M. Maiuri, D. Brida, C. Manzoni, Jan Helbing
and G. Cerullo

Abstract We demonstrate the functioning of a new scheme for two-dimensional IR spectroscopy in the partially collinear pump-probe geometry. With birefringent wedges, we can generate and delay a pair of pump pulses locked in phase which was already demonstrated by Brida et al. (Opt Lett 37:3027, 2012). For a proof-of-principle demonstration we use lithium niobate, which allows operation up to 5 μm.

Two-dimensional infrared spectroscopy (2D IR) has developed into important technique to describe molecular structure, measure the couplings between vibrational modes in molecules and track structural changes on the ultrafast timescale [1, 2].

Compared to standard pump-probe spectroscopy, the additional requirement of 2D spectroscopy is to resolve the frequency of the pump pulses. This is done in the time domain by scanning a delay t_1 ("coherence time") between two pump pulses and Fourier transforming the probe data against t_1 to obtain the pump frequency axis of the 2D spectra. The accurate control of delay t_1, within a fraction of the optical cycle is the key to obtaining proper 2D spectra, and this is the main technical difficulty of 2D spectroscopy in the time domain.

With TWINS (Translating-Wedge-Based Identical Pulses eNcoding System), we solved in a simple, compact and cheap way the problem of realizing an interferometrically stable collinear delay line. TWINS uses the birefringence in an optical material to set an arbitrary delay on two orthogonally polarization pulses by continuously varying the material thickness L (see Fig. 1).

J. Réhault (✉) · M. Maiuri · C. Manzoni · G. Cerullo
IFN-CNR, Dipartimento di Fisica, Politecnico di Milano,
P.zza Leonardo da Vinci 32, 20133 Milan, Italy
e-mail: julien.rehault@polimi.it

D. Brida
Department of Physics and Center for Applied Photonics,
University of Konstanz, Universitätsstraße 10, 78464 Constance, Germany

J. Helbing
Department of Chemistry, University of Zurich,
Winterthurerstrasse 190, 8057 Zurich, Switzerland

© Springer International Publishing Switzerland 2015
K. Yamanouchi et al. (eds.), *Ultrafast Phenomena XIX*,
Springer Proceedings in Physics 162, DOI 10.1007/978-3-319-13242-6_113

462

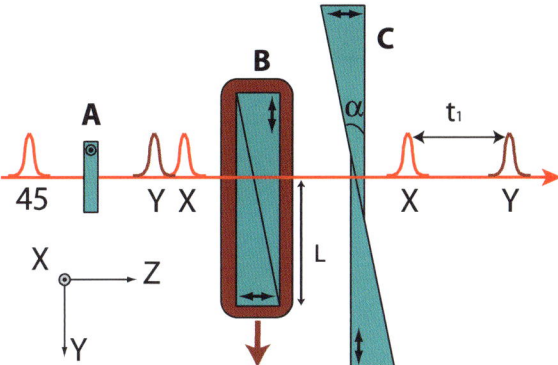

Fig. 1 Sequence of birefringent plates for the generation of interferometrically stable collinear pulse replicas of an input pulse polarized at 45°. *Thick arrows* indicate the optical axes. The delay t_1 is adjusted by changing L

The important advantage of the TWINS configuration is that it naturally produces a pair of perpendicularly polarized pump pulses. They can be projected to the same polarization with an additional polarizer in the pump beam for conventional 2D spectroscopy.

For demonstration, we show in Fig. 2a set of 2D IR spectra of a solution of 5 % of HOD in H_2O pumped at 4 µm wavelength and for different values of the population time t_2. As birefringent material, we used $LiNbO_3$, which is transparent at the pump frequencies. The OD stretch vibration gives rise to an intense and inhomogeneously broadened absorption band near 2,500 cm^{-1}. We observe a positive signal due to the 0–1 transition (bleach and stimulated emission, blue) and, at lower frequencies, a negative signal due to the 1–2 transition (photo-induced absorption, red). The tilt of the 2D IR signal is a measure of the correlation between pump and probe vibrational frequencies. Consistent with the literature [3, 4] this correlation is lost on a picosecond timescale, and this dynamics can be related to the fluctuations of the hydrogen bond network of water.

Without the output polarizer, the TWINS setup delivers a pair of pump pulses with very well-defined perpendicular polarizations. These pulses can be used directly for 2D spectroscopy with improved signal to noise when placing an additional polarizer into the probe beam behind the sample [5, 6]. Figure 2b shows the 2D IR measurement of the OD stretch spectrum with perpendicularly polarized pump pulses. The probe beam at the sample could be made approximately 10 times stronger than in the experiment with all parallel polarizations shown in Fig. 2a. As a result, the signal to noise ratio at short waiting times is much better. Since only the anisotropic signal is amplified, isotropic background due to heat is suppressed. However, the signal decays very quickly with the loss of anisotropy due to fast reorientation (<1 ps) of the O-D transition dipole [4].

In conclusion, we have introduced a method for performing 2D IR spectroscopy based on a birefringent delay line. The TWINS apparatus is simple and compact

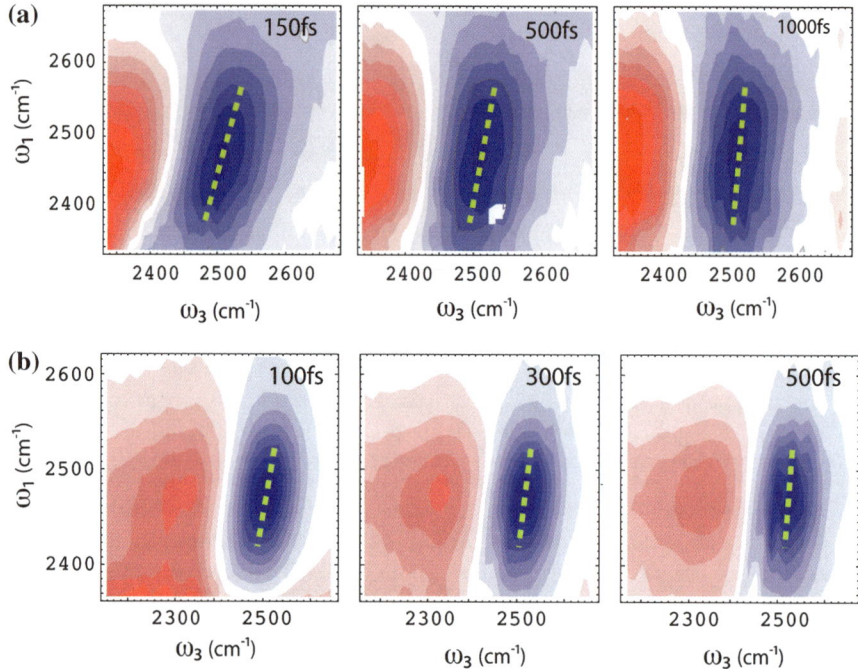

Fig. 2 2D IR spectra of the OD stretch in H_2O for the population times (t_2) specified in the *upper right corner* of each spectrum. **a** All parallel measurement (XXXX) **b** polarization measurements (YXXY). Spectral diffusion, which is revealed by the tilt of the bleaching signals highlighted by the *white dotted line*, decays in the sub-picosecond time scale

and allows generating phase-locked pulses with extremely high phase stability. TWINS can be introduced easily in the pump beam of a pump-probe set-up, and is by far the simplest method to produce phaselocked pulse pairs with perpendicular polarization. It is therefore particularly well suited for anisotropy measurements and signal amplification using polarizers. This can also be exploited to determine angles between transition dipole moments with improved accuracy [5] and to eliminate diagonal peaks by subtracting signals recorded with polarization orders YXYX and XYYX.

The current limitations of the set-up are due to the small birefringence and limited spectral transmission of $LiNbO_3$. Our goal is to use the TWINS device to generate phase-locked pulse pairs at arbitrary IR wavelengths. An excellent candidate material for applications of TWINS to the mid-IR appears to be crystalline mercurous chloride (Hg_2Cl_2), also known as calomel. Table 1 compares $LiNbO_3$ to Hg_2Cl_2. We are now developing strategies to be able to use smaller crystals and improve the throughput of the sequence, which will permit to utilize calomel for 2D IR spectroscopy with birefringent wedges.

Table 1 Comparison of Lithium Niobate and Calomel refractive indices and transparency

Material	Ordinary index	Birefringence $(n_e - n_o)$	GVM (fs/mm)	Range (μm)
$LiNbO_3$	2.1142	−0.0583	310	0.4–5.2
Hg_2Cl_2	1.8976	0.549	1,847	0.4–20

Acknowledgments The work was supported by the European Research Council Advanced Grant STRATUS (ERC-2011-AdG No. 291198). JR thanks the Swiss National fund for financial support (Grant No. 200020 129938).

References

1. D. Brida, C. Manzoni and G. Cerullo, "Phase-locked pulses for two-dimensional spectroscopy by a birefringent delay line" Opt. Lett. 37, 3027 (2012).
2. Peter Hamm and Martin T. Zanni. "Concepts and Methods of 2D Infrared Spectroscopy." Cambridge University Press, 2011.
3. F. Perakis and P. Hamm. "Two-dimensional infrared spectroscopy of supercooled water." J. Phys. Chem. B, 115(18):5289–5293, (2011).
4. T. Steinel, J.B. Asbury, J. Zheng and M.D. Fayer, "Watching Hydrogen Bonds Break: A Transient Absorption Study of Water." J. Phys. Chem. A 108, 10957–10964 (2004).
5. J. Réhault and J. Helbing. "Angle determination and scattering suppression in polarization-enhanced two dimensional infrared spectroscopy in the pump-probe geometry." Opt. Express, 20(19):21665–21677, (2012).
6. J. Réhault, V. Zanirato, M. Olivucci, and J. Helbing. "Linear dichroism amplification: Adapting a long-known technique for ultrasensitive femtosecond IR spectroscopy." J. Chem. Phys., 134 (12):124516–124516–10, (2011).

Hydrogen Bond Dynamics in Alcohols Studied by 2D IR Spectroscopy

Keisuke Shinokita, Ana V. Cunha, Thomas L.C. Jansen
and Maxim S. Pshenichnikov

Abstract Ultrafast hydrogen-bond dynamics in alcohols are studied by 2D IR spectroscopy and combined molecular dynamics—quantum mechanical simulations on the OH stretching mode. Fast memory loss in ~ 100 fs are attributed to intact hydrogen-bond fluctuations. Stable (at the experimental timescale) hydrogen bond structures lead to substantial inhomogeneity at longer times.

1 Introduction

Water dynamics have been put in the research spotlight thanks to tremendous advances in 2D IR spectroscopy and combined molecular dynamics (MD)—quantum chemistry calculations in the past decade. In particular, the important process of hydrogen-bond (HB) dynamics has been studied, and fascinating aspects of inter- and intramolecular vibrational couplings and vibrational exciton formation have been identified. Alcohols present another example of HB liquids in many important aspects. First, in alcohols the number of donated and accepting HB per molecule is different. This not only disrupts the 3D HB network observed in water, but also leads to long-lived structures. Second, alcohol molecules have complex HB dynamics due to their amphiphilic nature [1–3]. Third, the length of the alcohol molecules can be easily tailored by changing the alkyl chain. Furthermore, by going from primary to secondary or tertiary alcohols the shape can be varied.

Here we use 2D IR spectroscopy on the OH stretching mode to reveal HB dynamics in liquid alcohols. Experimental results are supported by combined MD—quantum mechanics calculations to elucidate the molecular picture of HB dynamics. We show that the dynamics occur at two prime time scales. The fast

K. Shinokita · A.V. Cunha · T.L.C. Jansen · M.S. Pshenichnikov (✉)
Zernike Institute for Advanced Materials, University of Groningen,
Nijenborgh 4, 9747 AG Groningen, The Netherlands
e-mail: Maxim.Pchenitchnikov@RuG.nl

© Springer International Publishing Switzerland 2015   466
K. Yamanouchi et al. (eds.), *Ultrafast Phenomena XIX*,
Springer Proceedings in Physics 162, DOI 10.1007/978-3-319-13242-6_114

~ 100 fs component is due to OH librations. The slow component longer than the time span of our experiments of ~ 2 ps arise from coexistence of different hydrogen-bonded structures.

2 Experiment

For 2D measurements, we used a pump-probe geometry with two collinear pump pulses and one probe pulse, all centered at 3,400 cm^{-1} and of ~ 75 fs duration. A sample, 6 % solution of MeOH in MeOD or EtOH in EtOD, was placed in a 50 μm-thickness free-standing sapphire jet which resulted in OD ~ 0.5 at the maximum of OH-stretch absorption. The circulatory pump system was purged with nitrogen to minimize water absorption from the air. All experiments were performed at room temperature.

3 Theory

MD simulations were performed using the OPLS/AA force field [4] for the alcohols using 300 molecules for methanol and 200 molecules for the other alcohols. After initial equilibration, a 50 ps trajectory with snapshots stored at 10 fs intervals was obtained. The vibrational hamiltonian was generated for the spectral simulations using a modified electrostatic mapping for the OH stretch vibration [5, 7]. The anharmonicity was fixed at 200 cm^{-1}, and the harmonic rule was applied for the |1>→|2> transition dipoles. The linear absorption and two-dimensional infrared spectra were obtained using the NISE simulation code [6] with coherence times of 320 fs and sampling at 1 ps intervals along the 50 ps trajectories.

4 Results and Discussion

As an example, Fig. 1a shows the linear absorption spectrum of the OH stretching mode of methanol with a peak position at 3,350 cm^{-1} and FWHM of ~ 210 cm^{-1}. The frequency-independent population lifetime deduced from the frequency-resolved pump-probe spectrum (Fig. 1b) amounts to ~ 600 fs which is consistent with [2] (albeit slightly faster). This value is lower than that for HDO:D$_2$O (~ 700 fs) which hints to CH stretching modes as possible acceptors. Nonetheless, the pump-probe data on MeOH diluted in acetonitrile showed a much longer, ~ 5 ps lifetime demonstrating that the intermolecular degrees of freedom are also necessary for OH-stretch energy dissipation. Development of the thermal response due to local heating of the solvent is also apparent at long (>1 ps) times.

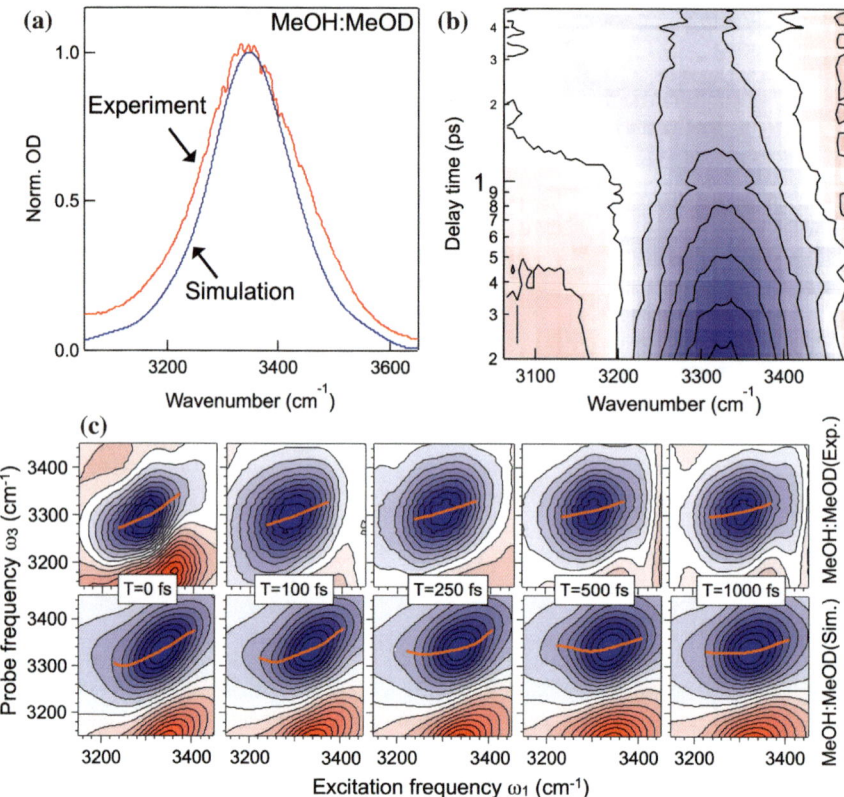

Fig. 1 a The experimental (*red*) and simulated (*blue*) linear absorption spectra of hydroxyl stretch in MeOH:MeOD. **b** Frequency-resolved pump-probe signal. The *blue* and *red colors* show bleaching/stimulated emission and induced absorption, respectively. **c** Normalized experimental (*top panels*) and simulated (*bottom panels*) absorptive 2D spectra at different waiting times. *Orange curves* connect the maxima of the ω_1 cuts to present the central line slopes (CLS). All simulated spectra are *red*-shifted by 130 cm^{-1} to match the experimental peak position

Figure 1c presents 2D IR spectra of methanol at different waiting times. At short times, the spectrum is elongated along the diagonal, which indicates a highly correlated response. With time, the memory for initial excitation frequency still exists, keeping the 2D signal slightly tilted. To quantify the phase memory, the center line slope (CLS) analysis was applied (Fig. 1c, orange lines), with results summarized in Fig. 2a (red and black dots). The retrieved correlation function of methanol and ethanol initially decays at a time scale of ~100 fs followed by a longer tail, in contrast with the HDO:D$_2$O case (Fig. 2a, blue dots).

The calculated 1D and 2D spectra (Fig. 1a, c) present a reasonable agreement with the experiment. The CLS (Fig. 2a) in both methanol and ethanol exhibit fast memory loss at short times and the subsequent long tail. Altogether this suggests that the used force field and mapping overestimate initial frequency fluctuations

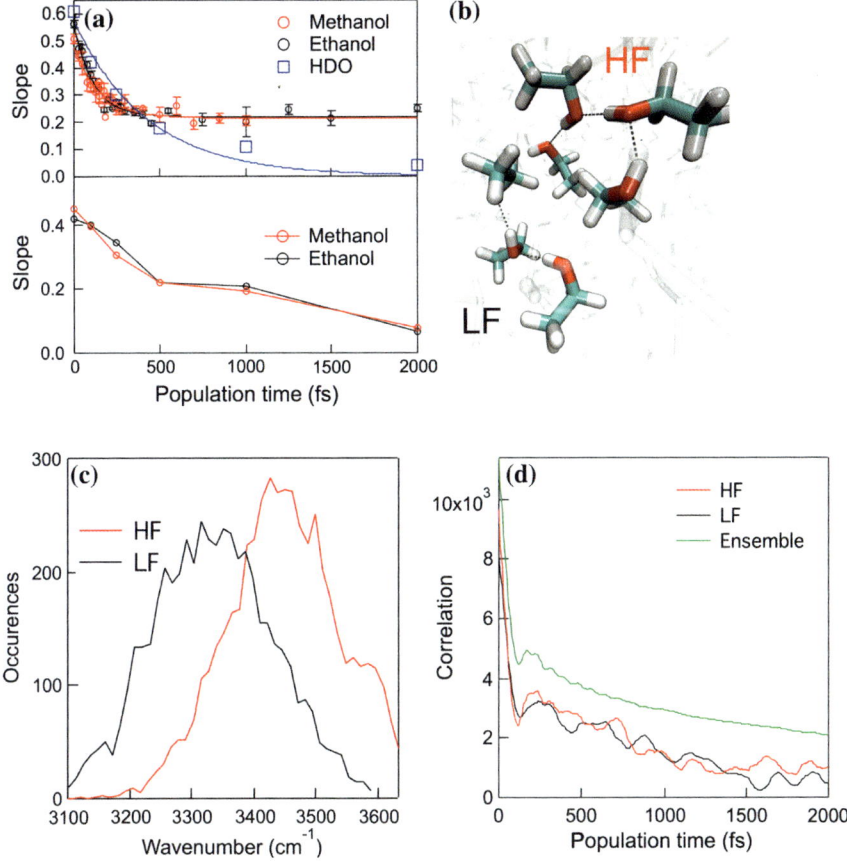

Fig. 2 **a** CLS values obtained from experiment (*upper panel*) and simulations (*bottom panel*). **b** Snapshots of two characteristic HB configurations from MD simulations. *Top* the high frequency configuration where the central ethanol molecule only donates one HB. *Bottom* the low frequency configuration where the central ethanol donates one HB and accepts one HB. **c** Frequency distributions for the two HB configurations during 50 ps. The histograms are *red*-shifted by 130 cm⁻¹ to match Fig. 1. **d** Frequency-frequency correlation functions for the two HB configuration (*black* and *red curves*) and for the whole ensemble of 200 molecules (*green*)

caused by the HB stretching motion and librations. On the other hand, the longer time dynamics (>2 ps) reflects the stability of HB structure in alcohols.

Figure 2b illustrates two examples of the low- and high-frequency HB clusters of ethanol. As shown in Fig. 2c, the two configurations have different frequency distributions with peak at ∼3,320 and ∼3,460 cm⁻¹, respectively. This suggests relatively slow (>50 ps) exchange between these clusters. Nonetheless, the faster exchanging configurations are also abundant. Figure 2d shows calculated correlation for the two configurations and the whole ensemble of 200 ethanol molecules. The correlation functions of the two configurations decay to almost zero in 2 ps.

However, there is a non-zero frequency offset in the case of the whole ensemble, most probably due to non-exchanged by 2 ps frequency-shifted structures.

In conclusion, we find that the excited-state lifetime in deuterated methanol and ethanol is shorter than for deuterated water. Similarly to water, fast timescales of ~ 100 fs of memory loss are attributed to HB fluctuations. The longer timescale is noticeably longer in alcohols, presumably due to slowly interchanging clusters with different HB configurations.

References

1. J.B. Asbury *et al.*, Chem. Phys. Lett. 374, 362–371 (2003).
2. S. Yamaguchi *et al.*, Rev. Laser Eng. 36, 1024–1027 (2008).
3. K. Kwac and E. Geva, J. Phys. Chem. B. 117, 7737–7749 (2013).
4. W.L Jorgensen *et al.*, J. Am. Chem. Soc. 118 (45): 11225–11236 (1996).
5. B. Auer *et al.*, PNAS, 104, 14215–14220 (2007).
6. C. Liang, T.L.C. Jansen, J. Chem. Theory Comput. 8, 1706–1713 (2012).
7. C. P. van der Vegte *et al.*, J. Phys. Chem. B, 118, 6256-6264 (2014).

Hydrogen Bond Enhancement of Fermi Resonances Explored with Ultrafast IR Two-Colour Pump-Probe and 2D-IR Spectroscopy

Christian Greve, Rene Costard, Henk Fidder and Erik T.J. Nibbering

Abstract Ultrafast polarisation-resolved 2D-IR mapping the fundamental and first overtone N-H stretching manifolds, and two-colour IR pump-probe experiments following transient population dynamics characterize a key role of a Fermi resonance with the NH_2-bending in aniline-dimethylsulfoxide complexes.

Probing hydrogen stretching modes is a well-established means to explore hydrogen bonds. Couplings of these hydrogen stretching modes to fluctuating solvent degrees of freedom are reflected in the line broadening of hydrogen stretching transitions, which can be grasped by ultrafast 2D-IR spectroscopy. Determination of vibrational couplings with other intramolecular degrees of freedom provides a detailed insight into the potential energy surfaces of hydrogen-bonded systems. These vibrational couplings are reflected in vibrational energy flow pathways, which can be deciphered with ultrafast vibrational pump-probe spectroscopy. Recently, we presented a combined NMR, FT-IR, 2D-IR, and DFT study of adenosine-thymidine (A-T) base pairs in chloroform solution [1]. Based on our findings we concluded that a Fermi resonance of the N-H stretching modes of adenosine with the NH_2-bending degree of freedom results in an enhancement of the NH_2-bending first overtone transition, making it as strong as the N-H stretching transitions.

To further substantiate this interesting feature of this characteristic structural motif of hydrogen-bonded amino-groups, we have chosen the model system of aniline-d_5 (An) hydrogen bonded with one or two dimethylsulfoxide (DMSO) molecules in nonpolar CCl_4 (Fig. 1a). Using spectral decomposition of concentration-dependent linear and first overtone spectra of An and DMSO, it is possible to determine the equilibrium constants between An, An$\cdots$DMSO and An$\cdots$(DMSO)$_2$, and with that extract the spectra of the different species (Fig. 1b, c). Whereas An monomer has its symmetric and asymmetric N-H stretching transitions at 3,395.5 and 3,480.5 cm^{-1}, respectively, hydrogen bonding results in frequency

C. Greve · R. Costard · H. Fidder · E.T.J. Nibbering (✉)
Max Born Institut für Nichtlineare Optik und Kurzzeitspektroskopie,
Max Born Strasse 2A, D-12489 Berlin, Germany
e-mail: nibberin@mbi-berlin.de

© Springer International Publishing Switzerland 2015   471
K. Yamanouchi et al. (eds.), *Ultrafast Phenomena XIX*,
Springer Proceedings in Physics 162, DOI 10.1007/978-3-319-13242-6_115

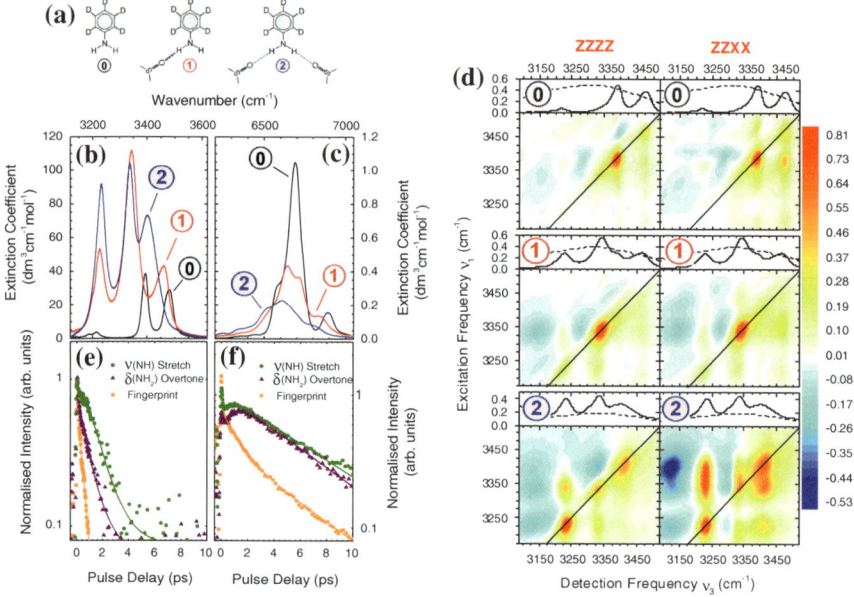

Fig. 1 **a** Chemical structures of An, An···DMSO and An···(DMSO)$_2$; **b**, **c** linear decomposed spectra of fundamental and first overtone transitions, color code according to chemical structures; **d** 2D-IR spectra for (ZZZZ) and (ZZXX) polarization geometries, linear spectra shown above the 2D-IR spectra; **e**, **f** transient pump-probe signals of the NH$_2$-bending and C=C ring stretching modes of An···(DMSO)$_2$, respectively, for different excitation conditions

downshifting of the N-H stretching transitions for An···DMSO and An···(DMSO)$_2$. The significant increase in absorption cross section of the fundamental N-H stretching manifold, and the decrease of oscillator strength of the associated first overtone N-H stretching manifold, are apparent in Fig. 1b, c. Although the oscillator strength increase of the N-H stretching modes in the region 3,300–3,500 cm^{-1} is impressive, it nevertheless pales in comparison to the immense intensity increase of the overtone transitions near 3,200 cm^{-1}.

Figure 1d shows the 2D-IR spectra of An, An···DMSO and An···(DMSO)$_2$, for the parallel (ZZZZ) and perpendicular (ZZXX) polarisation geometries, measured with a heterodyne-detection 2D-IR photon echo set-up using the four-wave mixing geometry with diffractive optics. The magnitudes of positive and negative cross peaks compared to the diagonal peaks is on the same order of magnitude. As a result, and—more importantly—in contrast to the strongly overlapping peaks in the linear first overtone spectrum, it is possible to fully map out the vibrational energy levels of the fundamental and first overtone N-H stretching manifolds [2]. The reason for this is the connectivity of positive and negative contributions in a femtosecond 2D-IR spectrum that enables a full determination of Liouville space pathways taken by single-quantum transitions caused by the applied femtosecond

IR pulses. The polarisation-resolved 2D-IR spectra support the relative orientations of the various IR transitions. Having established a mapping of the possible IR transitions between the ground, the fundamental and first overtone N-H stretching manifolds, a quantitative analysis of the fundamental, and first overtone N-H stretching manifolds can be achieved using a model vibrational Hamiltonian in a hybrid basis (local N-H stretching modes, normal bending mode; basis set notation $|v(NH)_1, v(NH)_2, \delta\rangle$. A proper description for all species is found using a coupling $J = -51 \text{ cm}^{-1}$ between the two local N-H stretching modes and a Fermi coupling between the NH_2-bending and the N-H stretching modes of $V = -37 \text{ cm}^{-1}$. The diagonal anharmonicity of the local N-H stretching mode is slightly larger for An⋯DMSO, $\Delta_{HB} = 88 \text{ cm}^{-1}$, than for An⋯$(DMSO)_2$, where it amounts to $\Delta_{HB} = 80 \text{ cm}^{-1}$, signifying a slightly stronger hydrogen bond in single hydrogen-bonded An⋯DMSO than the individual hydrogen bonds in the double-hydrogen bonded An⋯$(DMSO)_2$. From this analysis using the hybrid basis set we can make clear why a symmetrized N-H stretching representation $|v(NH)_S, v(NH)_{AS}\rangle$ is more appropriate for An and An⋯$(DMSO)_2$, and that a local mode representation $|v(NH)_f, v(NH)_b\rangle$ makes more sense for An⋯DMSO. In addition, the nature of the eigenstates changes for the first overtone manifold as well due to a different degree of interplay of intermode couplings and diagonal anharmonicities. A similar observation was found for adenosine monomer [3].

Fermi resonances with bending overtones have been postulated to play a key role in the vibrational energy redistribution of hydrogen stretching oscillators, ranging from C-H stretching modes in organic molecules [4] to the O-H stretching modes in liquid water [5]. Using single and two-colour pump-probe measurements we have investigated the population kinetics of the N-H stretching manifold as well as the C=C ring mode at $1{,}573 \text{ cm}^{-1}$, and the NH_2-bending band located in the range of $1{,}617–1{,}641 \text{ cm}^{-1}$. The N-H stretching lifetime is 5.5 ps in free An, and shortens to 0.6–0.9 ps in An⋯$(DMSO)_2$. Such a drastic lifetime shortening is not found for the NH_2-bending mode fundamental, which decreases from 0.6 ps in An to 0.44 ps in An⋯$(DMSO)_2$. Interestingly, both the C=C ring mode and the NH_2-bending mode show a delayed rise in red-shifted transient absorption when the pump pulse excites the Fermi resonance enhanced NH_2-bending overtone, which becomes even more accentuated when the initially excited modes are the N-H stretching transitions. These findings strongly suggest a dominant vibrational energy flow pathway scenario from the fundamental N-H stretching via Fermi resonances involving the NH_2-bending mode to the fundamental NH_2-bending and the C=C ring stretching states.

References

1. C. Greve, N. K. Preketes, H. Fidder, R. Costard, B. Koeppe, I. A. Heisler, S. Mukamel, F. Temps, E. T. J. Nibbering and T. Elsaesser, J. Phys. Chem. A 117, 594 (2013).
2. C. Greve, E. T. J. Nibbering and H. Fidder, J. Phys. Chem. B 117, 15843 (2013).

3. C. Greve, N. K. Preketes, R. Costard, B. Koeppe, H. Fidder, E. T. J. Nibbering, F. Temps, S. Mukamel and T. Elsaesser, J. Phys. Chem. A **116**, 7636 (2012).
4. D. D. Dlott, Chem. Phys. **266**, 149 (2001).
5. S. Ashihara, N. Huse, A. Espagne, E. T. J. Nibbering and T. Elsaesser, J. Phys. Chem. A 111, 743 (2007).

Observation of the Dark State in Ruthenium Complexes Using Femtosecond Infrared Vibrational Spectroscopy

Ken Onda, Tatsuhiko Mukuta, Sei'ichi Tanaka, Kei Murata and Akiko Inagaki

Abstract Comprehensive analyses of the excited states of prototypical ruthenium complexes using time-resolved infrared vibrational spectroscopy reveal a peak that can be assigned to the dark ^{3}MC (metal centered) state, which plays an important role for their photofunctions.

1 Introduction

Ruthenium complexes are used in various photofunctional materials, such as photocatalysts, artificial photosynthesis, and organic solar cells, because they absorb visible light efficiently and serve as the reaction centers. Extensive studies have established the fundamental processes involved in the absorption of a photon and subsequent dynamics in the excited state of a molecule [1, 2]. Irradiation of a metal complex using visible light excites an electron, leading to a singlet metal-to-ligand charge transfer (^{1}MLCT) state, and subsequently the intersystem crosses to the ^{3}MLCT state within 100 fs. With a lifetime over 100 ns, this metastable ^{3}MLCT plays a key role in photofunctional materials by transferring the excited electron to a

K. Onda (✉) · T. Mukuta · S. Tanaka
Graduate School of Science and Engineering, Tokyo Institute of Technology, 4259
Nagatsuta, Midori-ku, Yokohama 226-8503, Japan
e-mail: onda.k.aa@m.titech.ac.jp

K. Onda · A. Inagaki
PRESTO, Japan Science and Technology Agency (JST), 4-1-8 Honcho,
Kawaguchi, Saitama 332-0012, Japan

K. Murata
Chemical Resources Laboratory, Tokyo Institute of Technology, 4259 Nagatsuta,
Midori-ku, Yokohama 226-8503, Japan

A. Inagaki
Graduate School of Science and Engineering, Tokyo Metropolitan University,
1-1 Minami-Osawa, Hachioji, Tokyo 192-0397, Japan

© Springer International Publishing Switzerland 2015
K. Yamanouchi et al. (eds.), *Ultrafast Phenomena XIX*,
Springer Proceedings in Physics 162, DOI 10.1007/978-3-319-13242-6_116

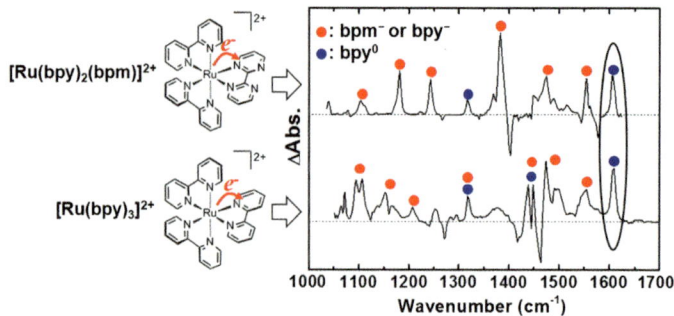

Fig. 1 Structures of the complexes investigated and their time-resolved IR spectra at 500 ps. *Red* and *blue solid circles* represent peaks attributed to vibrations localized on bpm⁻ (or bpy⁻) and bpy⁰, respectively

reactive state or to another molecule. A critical process to be considered during photoinduced electron transfer is the electron back transfer from the ^{3}MLCT to ^{3}MC (metal centered) state because the excited electron in ^{3}MC, the state favorable for most photochemical reactions, weakens the bonds between the metal center and ligands. However, because the ^{3}MC does not luminesce (dark state), its behavior and lifetimes have not been observed through direct means. In this study, we report the identification of a vibrational peak in the fingerprint region (1,000–1,700 cm^{-1}) that can be attributed to the ^{3}MC state of the prototypical ruthenium complexes, [Ru(bpy)$_3$]$^{2+}$ and [Ru(bpy)$_2$(bpm)]$^{2+}$ (bpy = 2,2′-bipyridine, bpm = 2,2′-bipyrimidine; Fig. 1). Furthermore, we demonstrate that the progress of a photochemical reaction can be monitored by observing the behavior of this vibrational peak.

2 Experimental

Time-resolved infrared (TR-IR) vibrational spectra were acquired using the pump-probe method using a femtosecond Ti:sapphire chirped pulse amplifier (CPA; wavelength = 800 nm, pulse duration = 120 fs, repetition rate = 1 kHz). A tunable mid-infrared pulse was generated by optical parametric amplification (OPA) and difference frequency generation (DFG) from the output of the CPA. The bandwidth of the IR pulse was 150 cm^{-1}. The 400 nm pulse was obtained by doubling the output of the CPA. To acquire the transient IR absorption spectrum, the probe pulse was passed through the IR cell and subsequently dispersed by a 19-cm polychromator, before being recorded with a 64-channel MCT detector. The spectral resolution was approximately 3 cm^{-1} and the detectable absorbance change (Δabs) was approximately 5 × 10^{-5}.

3 Results and Discussion

Whereas the ruthenium complexes used in this study are used in various applications, the vibrational peaks in the excited state of these complexes have not been comprehensively characterized. Therefore, we first measured the transient absorption spectra of $[Ru(bpy)_3]^{2+}$ and $[Ru(bpy)_2(bpm)]^{2+}$ and assigned all the observed peaks to the normal vibrational modes using deuterium substitution and quantum chemical calculations [3]. Figure 1 shows the spectra recorded 500 ps after photoexcitation. Here, in the lowest vibrational ^{3}MLCT excited state, the excited electron is localized on the ligand, bpy$^-$ or bpm$^-$ [2]. As shown in Fig. 1, several transient absorption peaks are assigned to the vibrational modes of the charge-localized ligands, bpy$^-$ or bpm$^-$. However, at least two peaks that can be assigned to the neutral ligand bpy^0 are observed, indicating that the transient moment of these modes are largely changed by oxidation (loss of electron) of the center metal without a change in the charge of the ligands. Intriguingly, one of these peaks (at approximately 1,600 cm^{-1}; indicated by an oval in Fig. 1) varies with the delay time. Figure 2 shows the expanded view of the TR-IR spectra around 1,600 cm^{-1}. The center of the absorption peak on the left is clearly blueshifted with increase in delay time. Typically, a shift in the wavenumber of a vibrational band indicates variation in charge and/or structure. To investigate this further, we varied the excitation wavelength because ^{3}MC states are thought to lie approximately 3,000 cm^{-1} above those of ^{3}MLCT [1, 2]. We observe that when the excitation wavelength is >480 nm, the observed peak of interest at approximately 1,600 cm^{-1} disappears, indicating that this peak can be attributed to a potential lying >2,000 cm^{-1} above the lowest vibrational ^{3}MLCT state. The disappearance of the peak is not surprising because it is assumed that excitation with longer wavelength does not populate the ^{3}MC state (Fig. 2). If the band at approximately 1,600 cm^{-1} is

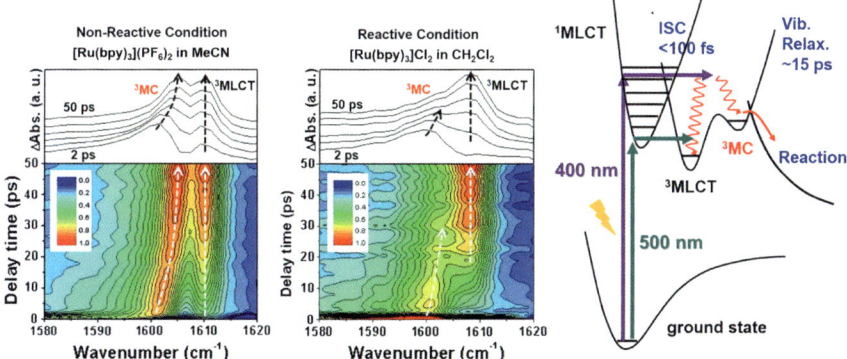

Fig. 2 Temporal evolution of the peaks at approximately 1,600 cm^{-1} under nonreactive and reactive conditions (*left*) and a representative scheme depicting the excited states of $[Ru(bpy)_3]^{2+}$ (*right*)

assigned to ^{3}MC, it is likely to be influenced by a photochemical reaction because such a reaction is believed to occur via the ^{3}MC state. Therefore, we have compared the temporal variations in the peak under nonreactive and reactive conditions. The reaction employed here involves the replacement of one of the bpy ligands by chloride (Cl) in the presence of the solvent CH_2Cl_2 [4]. As clearly evident from Fig. 2, the progress in the substitution reaction leads to the disappearance of the peak at 1,600 cm^{-1} in approximately 20 ps, indicating the likelihood that the reaction takes place via the ^{3}MC state. Conversely, the peak at approximately 1,610 cm^{-1} (peak at the right in Fig. 2) can be assigned to ^{3}MLCT because no peak shift is observed. This and other experimental results lead us to the conclude that the peak observed at approximately 1,600 cm^{-1} can be assigned to ^{3}MC, i.e., the dark state, which has not yet been observed despite its importance in photochemistry and photophysics.

References

1. D.W. Thompson, et al. "[Ru(bpy)$_3$]$^{2+}$ and Other Remarkable Metal-to-Ligand Charge Transfer (MLCT) Excited States" *Pure Appl. Chem.* **85**, 1257 (2013).
2. S. Campagna, et al. "Photochemistry and Photophysics of Coordination Compounds: Ruthenium" *Top. Curr. Chem.* **280**, 117 (2007).
3. T. Mukuta, et al. "Infrared Vibrational Spectroscopy of [Ru(bpy)$_2$(bpm)]$^{2+}$ and [Ru(bpy)$_3$]$^{2+}$ in the Excited Triplet State" *Inorg. Chem.* **53**, 2481 (2014).
4. B.D. Durham, et al. "Photochemistry of Ru(bpy)$_3^{2+}$" *J. Am. Chem. Soc.* **104**, 4803 (1982).

Vibrational Dynamics of Nitrosyl Stretch of Ru Complex in Aqueous Solution Studied by Two-Dimensional Infrared Spectroscopy

Kaoru Ohta, Kyoko Aikawa and Keisuke Tominaga

Abstract Vibrational dynamics of nitrosyl stretching mode of $[RuCl_5(NO)]^{2-}$ in water was studied by two-dimensional infrared spectroscopy. Observed temperature dependence of the correlation function of vibrational frequency fluctuation clearly shows that the time scale of the hydration dynamics was mostly determined by collective dynamics of hydrogen bonding network of water.

1 Introduction

Generally, molecular systems in the condensed phase exhibit a complex structural fluctuation so that it is very difficult to obtain detailed information on the solute-solvent interactions on a very fast time scale. In particular, it is well known that hydrogen bonding dynamics in aqueous solution takes place on a few picosecond time scale which may facilitate proton transfer and ion transport. Furthermore, the hydrogen bonding interactions are substantially influenced by the presence of charged species. Understanding of such interactions requires detailed knowledge of dynamical properties of the solvent molecules at the microscopic level. It is demonstrated that small ionic molecules such as N_3^- and SCN^- are excellent vibrational probes for studying the local interaction with the surrounding environment. Recently we studied the vibrational dynamics of the anti-symmetric stretching mode of N_3^- and SCN^- in various polar solvents by using three-pulse infrared (IR) photon echo methods [1, 2]. Our results showed that there exists a clear difference

K. Ohta (✉) · K. Tominaga
Molecular Photoscience Research Center, Kobe University,
1-1 Rokkodai, Nada, Kobe 657-8501, Japan
e-mail: kohta@kobe-u.ac.jp

K. Aikawa · K. Tominaga
Graduate School of Science, Kobe University, 1-1 Rokkodai,
Nada, Kobe 657-8501, Japan

© Springer International Publishing Switzerland 2015
K. Yamanouchi et al. (eds.), *Ultrafast Phenomena XIX*,
Springer Proceedings in Physics 162, DOI 10.1007/978-3-319-13242-6_117

479

of the vibrational dynamics between protic and aprotic solvents. Hydrogen bonding interaction plays an important role in the structural fluctuations of the solution.

To understand the microscopic origin of vibrational dynamics in detail, we chose the nitrosyl stretching mode of $[RuCl_5(NO)]^{2-}$ as a vibrational probe. Here we investigated the temperature dependence of the vibrational frequency fluctuation of the nitrosyl stretching mode in aqueous solution and compared with our previous studies of different ionic probe molecule. Temperature dependence of the vibrational dynamics of the solute in solution provides the systematic information about the coupling between the vibrational mode of the solute and the bath degrees of the system.

2 Results and Discussion

To obtain information on the frequency-frequency correlation functions of vibrational transitions, we used two-dimensional IR spectroscopy. The experimental setup used for the measurements has been described in detail elsewhere [1]. Briefly, a mid-IR pulse at around $1,900 \ cm^{-1}$ was generated by home-built optical parametric amplifier and difference frequency generator. 2D IR spectra were measured with a collinear pulse-pair pump and probe geometry. From the FT-IR spectra of the nitrosyl stretching mode of $[RuCl_5(NO)]^{2-}$, the peaks of the absorption spectrum are located at around $1,882 \ cm^{-1}$ in D_2O and H_2O. We measured the vibrational population relaxation for this mode by ultrafast IR pump-probe spectroscopy. We found that vibrational relaxation takes place on 31 ps in D_2O and 7.7 ps in H_2O, respectively. Similar fast relaxations in H_2O were observed for the other vibrational modes such as anti-symmetric stretching modes of N_3^- and SCN^- [2]. This is because the vibrational band at around $1,900 \ cm^{-1}$ couples strongly with a combination band of the bending and librational modes of H_2O. Spectral overlap of the solvent vibrational modes enhances the vibrational energy transfer from solute to solvent.

Figure 1a, b display 2D-IR spectra of the nitrosyl stretching mode of $[RuCl_5(NO)]^{2-}$ in D_2O taken at 283 and 313 K, respectively. At small values of population times, the 2D IR line shape is elongated along diagonal axis which reflects a positive correlation between the excitation (ω_{pump}) and detection (ω_{probe}) frequencies. At longer population times, the shape becomes more circular, reflecting the loss of the correlation of transition frequency. To determine the frequency-frequency correlation function from 2D IR spectra, we used the center line slope (CLS) method developed by Fayer and coworkers [3]. Figure 1c shows the decay profiles of CLS at different temperatures. At 293 K, the decay time constant of CLS is about 1.4 ps. This process reflects the hydrogen bonding dynamics under thermal equilibrium condition. The obtained time constant is very similar to that observed in the other ionic vibrational probes. This means that time scales of the vibrational frequency fluctuation are not simply determined by the interaction between solute and nearest solvent molecules. Collective dynamics of

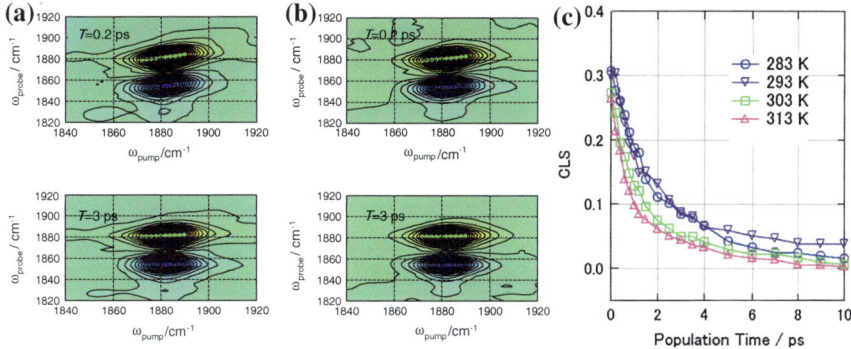

Fig. 1 **a** 2D IR spectra measured at 283 K with the population times of 0.2 and 3 ps. **b** 2D IR spectra measured at 313 K. *Green lines* and *dots* are the center line slopes. **c** Temporal profiles of CLS at temperatures of 283, 293, 303 and 313 K

the solvent molecules around solute plays a major role in determining the time scales of the vibrational frequency fluctuations in aqueous solutions.

With decreasing temperature, the decay of CLS becomes slower as clearly seen from Fig. 1c. In our previous studies, we investigated the temperature dependence of the vibrational frequency fluctuation of the anti-symmetric stretching mode of N_3^- in D_2O by three-pulse IR photon echo experiments [4]. The decay component of frequency-frequency correlation function varies from 1.4 to 1.1 ps in the temperature range from 283 to 353 K. The observed temperature dependence is very similar to that of current results. Furthermore, the dynamics of hydrogen bonding network in water has been intensively studied by ultrafast IR spectroscopy in conjunction with molecular dynamics simulations. In these studies, the vibrational frequency fluctuation of OH (OD) stretching mode was monitored for a dilute solution of HOD in D_2O (H_2O). Temperature dependent studies of such systems suggested that the mechanisms of vibrational frequency fluctuation, reorientational relaxation of OD stretch of HOD in H_2O, and hydrogen bond rearrangement are strongly correlated [5].

Together with these observations, we can conclude that picosecond vibrational dephasing dynamics of small ionic probes in aqueous solutions is driven by the electric field fluctuations originating from collective reorganization of hydrogen bonding network. Since the dynamic range of the 2D IR measurement is limited by the vibrational lifetime of the chromophore, the nitrosyl stretching mode of $[RuCl_5(NO)]^{2-}$ is a good vibrational probe for watching the local fluctuations of solvation structures of water in various environments on a longer time scale.

References

1. K. Ohta, J. Tayama, and K. Tominaga, "Ultrafast vibrational dynamics of SCN⁻ and N_3^- in polar solvents studied by nonlinear infrared spectroscopy," Phys. Chem. Chem. Phys. 14, 10455-10465 (2012).
2. K. Ohta, J. Tayama, S. Saito and K. Tominaga, "Vibrational frequency fluctuation of ions in aqueous solutions studies by three-pulse infrared photon echo method," Acc. Chem. Res. 45, 1982-1991 (2012).
3. K. Kwak, S. Park, I. J. Finkelstein, and M. D. Fayer, "Frequency-frequency correlation functions and apodization in two-dimensional infrared vibrational echo spectroscopy: A new approach," J. Chem. Phys. 127, 124503 (2007).
4. J. Tayama, A. Ishihara, M. Banno, K. Ohta, S. Saito, and K. Tominaga, "Temperature dependence of vibrational frequency fluctuation of N_3^- in D_2O," J. Chem. Phys. 133, 014505 (2010).
5. R. A. Nicodemus, S. A. Corcelli, J. L. Skinner, and A. Tokmakoff, "Collective hydrogen bond reorganization in water studied with temperature-dependent ultrafast infrared spectroscopy," J. Phys. Chem. B 115, 5604-5616 (2011).

Ultrafast IR Spectroscopy of O-H Stretching Modes in 2-Naphthol-Acetonitrile Photoacid-Base Complexes

Brian T. Psciuk, Mirabelle Prémont-Schwartz, Benjamin Koeppe,
Sharon Keinan, Dequan Xiao, Victor S. Batista
and Erik T.J. Nibbering

Abstract The O-H stretching mode is a direct hydrogen-bond probe. In a combined femtosecond IR spectroscopic and quantum chemical approach, we demonstrate how this local marker directly reflects charge distribution changes induced in photo excited photoacid-base complexes.

Photoacids are typically organic aromatic alcohols or protonated amines that exhibit a strong change in acidity (proton donating capability) upon electronic excitation. For example, the prototypical photoacid 2-naphthol (2N) undergoes a dramatic change of pK_a upon $S_1 \leftarrow S_0$ photoexcitation, shifting from $pK_a = 9$ in the S_0 state to $pK_a = 3$ in the S_1 state. The pK_a values can be used to predict ultrafast proton transfer rates to accepting bases under aqueous conditions by using Marcus-type relationships that correlate free energy changes and reactivity according to the Förster cycle [1]. This has been supported by measurements of proton-to-solvent transfer rates and data on steady-state electronic absorption and emission spectra. Such considerations have also led to the understanding that an additional electronic excited-state level crossing is involved for 1-naphthol (1N), only recently probed to

B.T. Psciuk · D. Xiao · V.S. Batista (✉)
Department of Chemistry, Yale University, 225 Prospect Street,
New Haven, CT 06520, USA
e-mail: victor.batista@yale.edu

M. Prémont-Schwartz · B. Koeppe · E.T.J. Nibbering (✉)
Max Born Institut für Nichtlineare Optik und Kurzzeitspektroskopie,
Max Born Strasse 2A, 12489 Berlin, Germany
e-mail: nibberin@mbi-berlin.de

S. Keinan
Department of Chemistry, Ben-Gurion University of the Negev, P.O. Box 653,
84105 Be'er Sheva, Israel

D. Xiao
Department of Chemistry and Chemical Engineering, University of New Haven,
300 Boston Post Road, West Haven, CT 06516, USA

© Springer International Publishing Switzerland 2015
K. Yamanouchi et al. (eds.), *Ultrafast Phenomena XIX*,
Springer Proceedings in Physics 162, DOI 10.1007/978-3-319-13242-6_118

occur with a 60 fs time constant [2]. To further characterize the electronic excited states of photoacid-base complexes, we report here the ultrafast probing of O-H stretching modes. The correlation between the O-H stretching frequency and the H-bond distance is well established [3]. Stronger H-bonds tend to have shorter H-bond distances and red-shifted frequencies of the O-H stretching oscillator. However, it has also been well understood that the differences in frequency values is caused by electrostatic interactions imposed by the surrounding solvent molecules, leading to spectral line broadening as exemplified by the O-H stretching band of water under ambient conditions.

Here, we focus on the study of O-H stretching modes of photoacid molecules in hydrogen-bonded complexes by analyzing the 2N-CH$_3$CN model system. We combine ultrafast UV/IR pump-probe spectroscopy with ab initio quantum chemistry calculations to parameterize the Pullin model of solvent-induced vibrational frequency shifts [4]. Calculations were performed with GAUSSIAN09, using density functional theory at the B3LYP/TZVP and ωB97XD/TZVP levels of theory and excited stated calculations at the TD-DFT level with implicit solvation using IEF-PCM. The 2N-CH$_3$CN complex exhibits a moderate solvent-induced vibrational frequency shift for the O-H stretching mode when changing the surrounding solvent from nonpolar cyclohexane to polar acetonitrile. With O-H stretching frequencies well-above 3,000 cm^{-1}, the observed O-H stretching bands reflect intramolecular charge distributions of the photoacid-base complexes as well as couplings with the surrounding solvent of a given polarity, while vibrational couplings to fingerprint combination/overtone bands as well as anharmonic couplings with H-bond modulation modes can be neglected. Figure 1a, b shows the steady-state and transient IR spectra of the 2N-CH$_3$CN complex in various solvents after UV photoexcitation at 330 nm. Upon promotion to the S$_1$-state, the O-H stretching frequency downshifts primarily due to the underlying electronic excitation while the solvent rearrangements in the (sub)picosecond time scale have only a minor additional contribution [4]. The frequency shift exhibits a linear dependence on $F_0 = (2\varepsilon_0 - 2)/(2\varepsilon_0 + 1)$, where ε_0 is the static dielectric constant of the solvent (Fig. 1c). This is in line with our previous findings of uncomplexed 2N and 1N, which validates treating the 2N-CH$_3$CN complexes as a point dipole immersed in a spherical Onsager cavity according to Pullin's derivation of the solvent-induced vibrational frequency shifts.

To obtain microscopic insights into the observed frequency changes of 2N-CH$_3$CN, we compare our results with observed O-H stretching frequency shifts for a large collection of phenol-derivatives as well as a 1N-derivative (Fig. 1c). We plot the observed frequency shifts as a function of F_0 and we derive the intercept as its hypothetical gas phase value ($\varepsilon = 1$; $F_0 = 0$). The solute-solvent coupling sensed by the O-H stretching mode is represented by the slope of the observed frequency shift. In this way, we have characterized the different aromatic alcohol complexes in a slope vs. intercept plot (Fig. 1d) showing that the complexes can be characterized as regular acids.

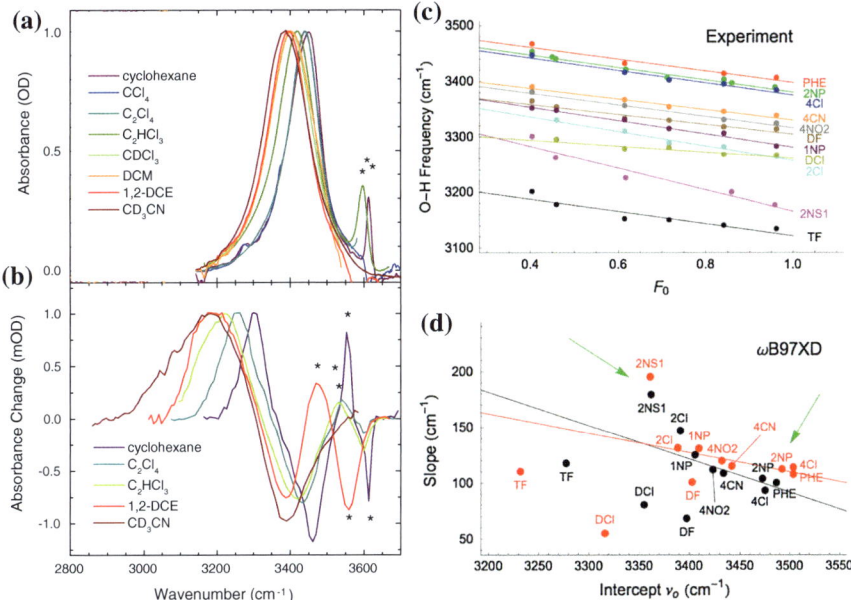

Fig. 1 a Steady-state IR spectra of 2 N-CH₃CN complexes in various solvents (*asterisks* denote uncomplexed 2N); **b** transient IR spectra after UV photoexcitation at 330 nm; **c** the O-H frequency shift exhibits a linear dependence on $F_0 = (2\varepsilon_0 - 2)/(2\varepsilon_0 + 1)$, where ε_0 is the static dielectric constant of the solvent, which validates treating the 2N-CH₃CN complexes (*dashed curves*) according to Pullin's model of the solvent-induced vibrational frequency shifts: *PHE* phenol, *4Cl* 4-chlorophenol; *2NP* 2-naphthol (S_0), *4CN* 4-cyanophenol, *4NO2* 4-nitrophenol, *1NP* 4-nitro-1-naphthol, *DF* 2,6-difluorophenol; *2Cl* 2-chloro-4-nitrophenol, *2NS1* 2-naphthol (S_1), *DCl* 2,6-dichoro-4-nitrophenol, *TF* 2,3,5,6-tetrafluoro-4-nitrophenol; **d** values for the intercept and slope of various photoacid-base complexes derived from experimental *curves* in (**c**) are shown in *red* and compared with calculated values presented in *black*. The 2N-CH₃CN points are indicated with *green arrows*

Based on our findings, we conclude that a free-energy correlation with the vibrational spectroscopic observable, namely the O-H stretching marker mode of the 2N-CH₃CN complexes can be made. As such O-H stretching frequency shifts follow changes in acidity (as characterized by the pK_a-value). The underlying mechanisms of charge distribution changes govern how the solute-solvent couplings can have an impact on the O-H stretching potential, leading to solvent-induced O-H stretching frequency shifts. These mechanisms can be quantified using (time-dependent) density functional theory at the B3LYP/TZVP and ωB97XD/TZVP levels of theory using solvation continuum models.

References

1. M. Prémont-Schwarz, T. Barak, D. Pines, E. T. J. Nibbering and E. Pines, J. Phys. Chem. B 117, 4594 (2013).
2. F. Messina, M. Prémont-Schwarz, O. Braem, D. Xiao, V. S. Batista, E. T. J. Nibbering and M. Chergui, Angew. Chem. Int. Ed. 52, 6871 (2013).
3. E. T. J. Nibbering and T. Elsaesser, Chem. Rev. 104, 1887 (2004).
4. D. Xiao, M. Prémont-Schwarz, E. T. J. Nibbering and V. S. Batista, J. Phys. Chem. A 116, 2775-2790 (2012).

Vibrational Dynamics of the CN Stretching in the Electronically Excited State by UV and Visible-Pump and Infrared-Probe Spectroscopy

Sho Hiraoka, Kaoru Ohta and Keisuke Tominaga

Abstract We have carried out UV and visible-pump infrared-probe measurements on a CN-containing coumarin in a protic solvent. The time-dependent changes of the infrared spectra are measured on a picoseconds time scale, likely because of vibrational cooling.

1 Introduction

Understanding of mechanisms of photochemical reactions in solution is one of the important problems in current chemistry. After photoexcitation to the electronically excited state of a solute molecule in solution, various relaxation processes occur such as vibrational energy relaxation or solvation dynamics. The fate of the excited state is strongly influenced by these relaxation processes, therefore, it is very important to understand these relaxation processes in detail. Furthermore, hydrogen-bonding liquids such as water and alcohol form characteristic network structures, which affect various properties of a solute molecule such as the electronic state and vibrational structures. In this work, we focus on the vibrational dynamics of a solute in the electronically excited state in alcohol solutions. The vibrational mode we investigate is the CN stretching mode of 3-cyano-7-hydroxy-4-methyl coumarin (C183m, Fig. 1). Generally, coumarin dyes change their permanent dipole moments largely by photoexcitation, causing change of the dielectric interaction between the dyes and surrounding solvents. We have carried out UV and visible-pump and infrared (IR)-probe spectroscopic measurements on this mode to examine influence of the solvent interactions on the vibrational dynamics of the solute.

S. Hiraoka · K. Tominaga (✉)
Graduate of Science, Kobe University, Rokkodai, Nada, Kobe 657-8501, Japan
e-mail: tominaga@kobe-u.ac.jp

K. Ohta · K. Tominaga
Molecular Photoscience Research Center, Kobe University, Rokkodai, Nada,
Kobe 657-8501, Japan

© Springer International Publishing Switzerland 2015
K. Yamanouchi et al. (eds.), *Ultrafast Phenomena XIX*,
Springer Proceedings in Physics 162, DOI 10.1007/978-3-319-13242-6_119

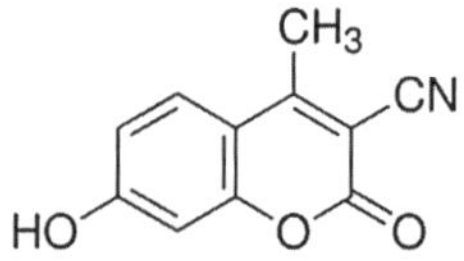

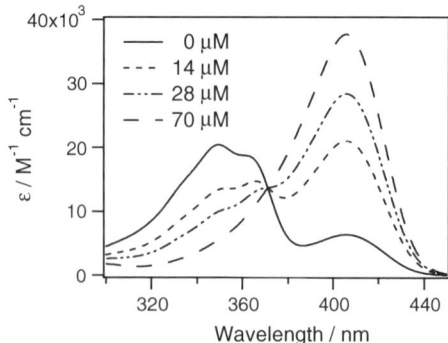

Fig. 1 Molecular structure of C183m

2 Results and Discussion

For UV and visible-pump infrared-probe measurements, the pump pulse was generated by frequency-doubling or tripling of the output from a Ti:sapphire regenerative amplifier. The tunable IR pulse was generated by difference frequency mixing of the signal and idler from the optical parametric amplifier. The absorption changes were measured with an MCT array detector. The experiments were conducted using 100-μm-thick liquid sample in a static cell. Figure 2 shows the absorption spectrum of C183m in methanol in the UV-visible region. In this solvent, C183m exists as either a neutral form or an anionic form. We added a solution of CH_3ONa in methanol to the C183m solution so that a single species, the anion, exists in the sample. Figure 2 illustrates acid-base reactions of C183m in methanol. All the experiments are made for this anionic C183m.

Figure 3 shows the temperature dependence of the IR spectra of the CN stretching mode of C183m in methanol. At room temperature (293 K), the spectral lineshape was reproduced well by a single Gaussian function. It was found that the band peak shows a red shift and its shape becomes asymmetric with temperature increase. It is suggested that there are more than one component in this band, probably hydrogen-bonded and non-hydrogen-bonded CN stretching mode, and the free CN stretching band component increases with increasing temperature.

The transient IR spectra of the CN stretching band are shown in Fig. 4. There are a bleach centered at $2,225~cm^{-1}$ and a transient absorption at $2,180~cm^{-1}$, and we fit

Fig. 2 UV-visible absorption spectra of C183m in methanol. A solution of CH_3ONa in methanol is added. The concentrations of CH_3ONa are shown in the *inset*

Fig. 3 Temperature-
dependent IR spectra of the
CN stretching mode of
C183m in methanol

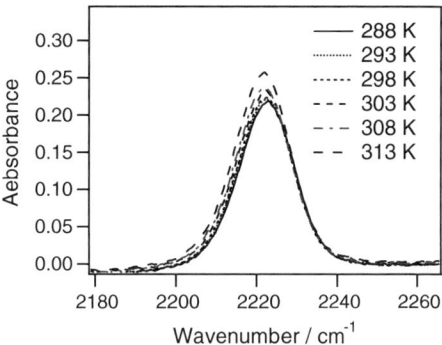

Fig. 4 Transient absorption
spectra of C183m in methanol
at different delay times

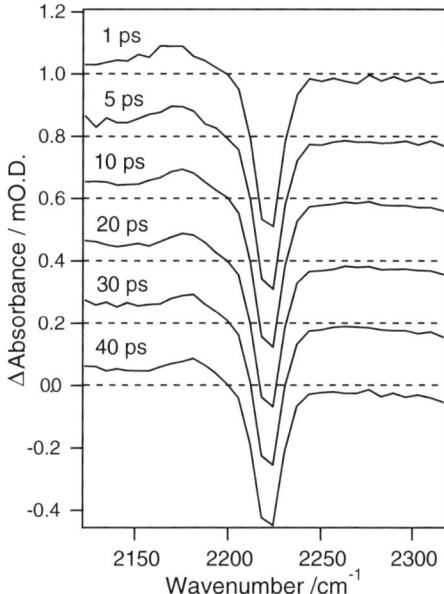

the transient spectra with a sum of two Gaussian functions to determine the peak wavenumber of the transient absorption component. The peak of the CN stretching mode shows a blue shift with delay time, and the time-dependence of this blue shift is reproduced well by a single exponential with a time constant of about 10 ps (Fig. 5). One of the possible mechanisms for this blue shift is vibrational cooling. Vibrational cooling resulted from anharmonic couplings between the high-frequency mode and some low-frequency modes. Just after the photoexcitation the local temperature around the solute is higher than the room temperature because of the dissipation of the excess energy, producing population in the higher vibrational states of the low-frequency modes. This vibrational cooling has been observed in several transient vibrational spectroscopic studies [1, 2]. Another possible mechanism for the blue shift is solvation dynamics [3, 4]. If the photoexcitation

Fig. 5 The time evolution of the peak wavenumber of the CN stretching band of C183m in S_1 measured by visible-pump and IR-probe spectroscopy

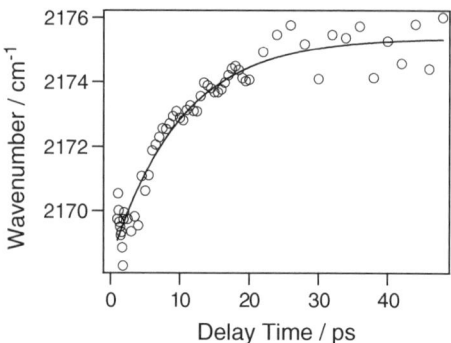

accompanies a large change of the dipole moment of the solute, dielectric response from the polar solvent causes the stabilization of the electronic and vibrational states of the solute. Depending on the directions of the permanent dipole moments of the S_0 and S_1 states and the charge distribution in the normal modes, the peak wavenumber shifts either the blue side or red side during the solvation dynamics.

We further carried out UV-pump and IR-probe measurement, where the third harmonics at 266 nm was used as a pump pulse. In this case the excess energy is about 12,450 cm^{-1}, which is much larger than that of the second harmonic excitation (1,420 cm^{-1}). The transient absorption spectrum at $t = 0$ ps is further blue-shifted compared to the second harmonic excitation case, and the time evolution of the blue shift can be described by a similar time constant to that of the visible-pump case. These results suggest that the blue shift is caused by vibrational cooling rather than solvation dynamics, since the total shift due to vibrational cooling depends on the excess energy whereas that due to solvation dynamics does not depend on it so much.

3 Conclusion

We investigated vibrational dynamics of the CN stretching of the S_1 state for C183m in methanol by sub-picosecond UV and visible-pump and IR-probe spectroscopy. The peak of the CN stretching band shows a blue shift in time, which is probably due to vibrational cooling.

References

1. P. Hamm, S. M. Ohline, and W. Zinth, "Vibrational Cooling after Ultrafast Photoisomerization of Azobenzene Measured by Femtosecond Infrared Spectroscopy," J. Chem. Phys. 106, 519-529 (1997).
2. K. Iwata and H. Hamaguchi, "Microscopic Mechanism of Solute-Solvent Energy Dissipation Probed by Picosecond Time-Resolved Raman Spectroscopy," J. Phys. Chem. A 101, 632-637 (1997).

3. J. B. Asbury, Y. Wang, and T. Lian, "Time-Dependent Vibration Stokes Shift during Solvation: Experiment and Theory," Bull. Chem. Soc. Jpn. 75, 973-983 (2002).
4. D. Xiao, M. Prémont-Schwarz, E. T. J. Nibbering, and V. S. Batista, "Ultrafast Vibrational Frequency Shifts Induced by Electronic Excitations: Naphthols in Low Dielectric Media," J. Phys. Chem. A 116 2775–2790 (2012).

Structural Motifs of Liquid Acetic Acid from Ultrafast CARS Spectroscopy

Matthias Lütgens, Frank Friedriszik and Stefan Lochbrunner

Abstract The carbonyl vibration of acetic acid is analyzed by spontaneous and ultrafast coherent anti-Stokes Raman spectroscopy. The complex band is decomposed into four contributions from different structural motifs and the cyclic dimer signature is extracted.

OCIS Codes 300.0300 · 300.6230 · 300.6450

1 Introduction

Molecular structures in gaseous and crystalline phases are typically well characterized since in the first case the molecules are isolated and in the second case diffraction methods provide detailed information. In liquids the situation is more complicated due to disorder and the possible coexistence of different local structural motifs. Hydrogen bonds (HBs) play a decisive role for the molecular network of many liquids. HBs are local and directed interactions which lead to the formation of complexes. Especially, if several HB donor and acceptor sites are present different structural motifs can appear. Acetic acid (AA) exhibits two acceptor sites, the carbonyl and the hydroxyl oxygen, and four donor sites, the hydroxyl and methyl hydrogen atoms. The structure of liquid AA is still under debate. In detail, linear chain oligomers and cyclic dimers are discussed to be the predominate molecular aggregates. Since the carbonyl stretch vibration is involved in hydrogen bonding it is a good probe for the local structure of AA.

In this contribution we report a comparison of spontaneous Raman measurements with time and frequency resolved ultrafast coherent anti-Stokes Raman spectroscopy (CARS). Spontaneous Raman probes the static time averaged CO vibration and dephasing as well as static vibrational distributions are encoded in the line width of the Raman resonance. In CARS the pump and Stokes pulse stimulate a coherent

M. Lütgens · F. Friedriszik · S. Lochbrunner (✉)
Institute of Physics, University of Rostock, Universitätsplatz 3, 18055 Rostock, Germany
e-mail: stefan.lochbrunner@uni-rostock.de

K. Yamanouchi et al. (eds.), *Ultrafast Phenomena XIX*,
Springer Proceedings in Physics 162, DOI 10.1007/978-3-319-13242-6_120

excitation of a vibration. A time delayed probe pulse with a duration of ca. 1 ps scatters off this driven vibration. Mapping the coherence decay with the narrowband probe pulse allows the determination of dephasing times with an intrinsic spectral resolution defined by the spectral width of the probe [1]. Under these conditions strongly dephasing contributions fade out very fast and slowly dephasing vibrational modes persists longer and can be observed separated from other contributions.

2 CO Vibration of Acetic Acid

The spontaneous Raman spectrum of CO reveals three clear features (Fig. 1a). Nakabayashi et al. (1999) assigned all components to vibrations originating from a linear chain aggregate [2]. This assignment is not intuitive at first glance since the most stable structure formed by two monomers is the cyclic dimer and not the dimer from which the linear chain can be constructed. E.g. D'Amico et al. (2012) used for analyzing the CO vibrational band an additional mode which is attributed to the cyclic dimer [3]. From the spontaneous Raman spectrum no clear statement can be made with respect to the number of modes, since a reconstruction with three as well as four modes mirrors the essential shape of the spectrum. Probing the carbonyl vibration with time resolved CARS gives additional insight into the interpretation of this band since it discriminates the vibrational contributions in the time domain. Sub-50 fs Stokes pulses at 570 nm and 0.9 ps long pump and probe pulses at 510 nm are focused into a 2 mm thick sample cell applying a phase matched boxcars geometry and the resulting CARS signal is spectrally dispersed detected by a CCD camera. After the cross correlation two distinct modes are observed in the CARS spectra (Fig. 1c). The higher frequency mode corresponds according to its center frequency as well as the dephasing time (Fig. 1d) to mode III in the linear

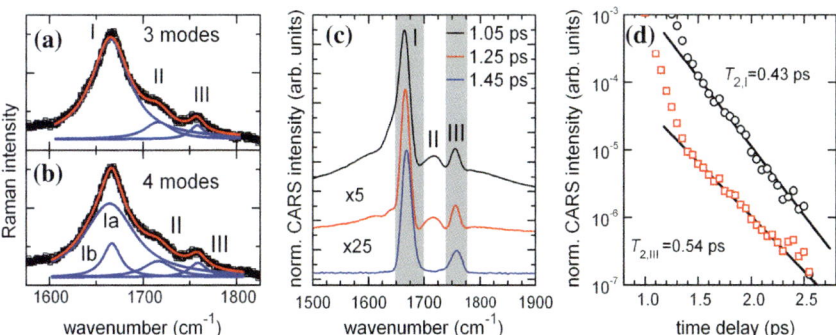

Fig. 1 Raman and CARS measurements of the carbonyl stretch vibration of neat acetic acid. Decomposition of the Raman spectrum by three and four contributions are shown in (**a**) and (**b**), respectively. **c** Selected CARS spectra showing the passage from the cross correlation region to time delays after temporal pulse overlap. The decays of the two remaining slow dephasing signals are fitted with mono exponentials (**d**)

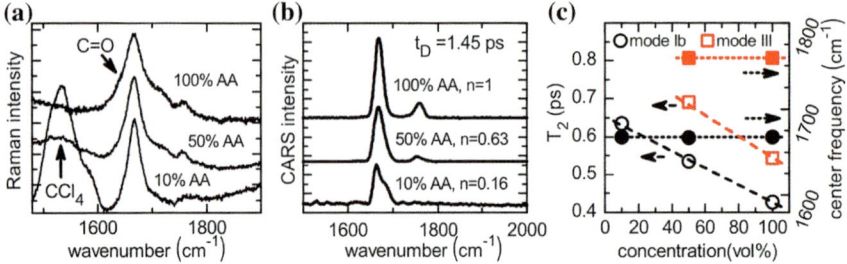

Fig. 2 **a** Raman and **b** CARS spectra after cross correlation of the carbonyl stretch region of acetic acid—CCl₄ mixtures, **c** dephasing constants T_2 and center frequencies of the slowly dephasing CO stretch vibrations extracted from CARS measurements

Raman spectrum. The lower frequency contribution is located within the broad band I but the dephasing constant determined from the decay of the CARS signal corresponds to a line width only half as large as the Raman line width. This leads to the conclusion that band I has to be interpreted as a superposition of two independent contributions most likely originating from different aggregation structures.

The assignment of the modes can be clarified by concentration dependent measurements. The Raman spectrum of AA in the unpolar solvent CCl₄ shows a systematic change with decreasing AA content towards a spectrum with a single remaining, rather narrow band located at approx. 1,668 cm^{-1} (Fig. 2a). In unpolar organic solvents at low concentrations AA forms cyclic dimers [4]. They exhibit two equally strong HBs and for symmetry reasons a single Raman active mode in the relevant part of the spectrum to which the band at 1,668 cm^{-1} is assigned. In CARS measurements of diluted AA also only one strong contribution can be observed located at the same frequency and with a dephasing time of 0.63 ps (Fig. 2b). Figure 2c shows the concentration dependent frequencies and dephasing constants found by CARS for different concentrations. Since the center frequency does not vary and the dephasing time shows a linear behavior with respect to the concentration the mode has to be assigned to the same vibration in all discussed samples, i.e. to the cyclic dimer. This in turn means that in neat AA the cyclic dimer mode can be clearly separated from other contributions by ultrafast CARS with time delayed probe pulses.

References

1. M. Lütgens, S. Chatzipapadopoulos, and S. Lochbrunner, Opt. Express **20**, 6478–6487 (2012); M. Lütgens, S. Chatzipapadopoulos, F. Friedriszik, and S. Lochbrunner, J. Raman Spectrosc. **45**, 359–368 (2014)
2. T. Nakabayashi, K. Kosugi, and N. Nishi, J. Phys. Chem. A **103**, 8595–8603 (1999)
3. F. D'Amico, F. Bencivenga, A. Gessini, E. Principi, R. Cucini, and C. Masciovecchio, J. Phys. Chem. B **116**, 13219–13227 (2012)
4. Y. Fujii, H. Yamada and M. Mizuta, J. Phys. Chem. **92**, 6768–6772 (1988)

Ultrafast Time-Domain Raman Study to Visualize Large-Amplitude Distortions in Copper Complexes

Satoshi Takeuchi, Munetaka Iwamura and Tahei Tahara

Abstract Time-resolved impulsive-Raman with narrowband photoexcitation was utilized to study structural dynamics of bis-diimine copper complex in solution. A copper-ligand symmetric stretch band showed up with frequency oscillation, demonstrating its anharmonic coupling with large-amplitude distortional motions.

1 Introduction

Transition metal complexes play important roles in both fundamental science and numbers of applications because of their variety of photochemical properties. In particular, a photoinduced metal-to-ligand charge transfer (MLCT) state of Cu(I) complexes attracts much interests as a prototype system showing ultrafast "Jahn-Teller" distortions. As shown in Fig. 1, a representative bis-diimine copper complex forms a perpendicular structure in the S_0 state, where the two ligands are aligned perpendicularly to each other. Once excited to the S_1 state, on the other hand, it is considered to distort toward the square-planar structure, which is most typical for Cu(II) complexes, by decreasing the dihedral angle between the two ligands. In fact, a nanosecond X-ray absorption study confirmed such a "flattened" structure in the T_1 state formed in later times [1]. Our recent femtosecond emission [2] and absorption [3] measurements have further shown that the initial

S. Takeuchi (✉) · T. Tahara
Molecular Spectroscopy Laboratory, RIKEN, 2-1 Hirosawa,
Wako 351-0198, Japan
e-mail: stake@riken.jp

S. Takeuchi · T. Tahara
Ultrafast Spectroscopy Research Team, RIKEN Center for Advanced
Photonics (RAP), 2-1 Hirosawa, Wako 351-0198, Japan

M. Iwamura
Graduate School of Science and Engineering, University of Toyama,
3190 Gofuku, Toyama 930-8555, Japan

© Springer International Publishing Switzerland 2015
K. Yamanouchi et al. (eds.), *Ultrafast Phenomena XIX*,
Springer Proceedings in Physics 162, DOI 10.1007/978-3-319-13242-6_121

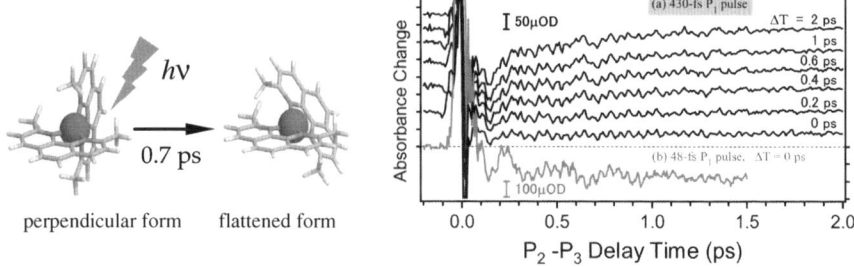

Fig. 1 TR-ISRS experiment of [Cudmphen₂]⁺ in dichloromethane. *Left* Photoinduced flattening distortion in the MLCT state. *Right* TR-ISRS signals at selected ΔT times obtained with **a** 430-fs and **b** 48-fs actinic pump (P₁) pulses

perpendicular S_1 state has a well-defined vibrational structure and that it exhibits emission spectral changes with a time constant of 0.7 ps. However, we still have limited structural knowledge on the early stage of this fundamental phenomenon, i.e., when and how such a large-amplitude distortion actually takes place. Therefore, it is highly desirable to clarify the initial structural events with firm enough vibrational evidences.

Motivated by this idea, we investigated the ultrafast structural dynamics of this copper complex in the MLCT state by time-resolved impulsive stimulated Raman spectroscopy (TR-ISRS) [4, 5] that we developed and upgraded to date. With careful bandwidth control in actinic excitation, we successfully observed intrinsic time evolution of the Raman vibrational structure, which is free from unwanted fifth-order contributions. It further enabled us to recognize frequency modulation of a key vibration, leading to a visualization of the vibrational anharmonic coupling in the excited state.

2 Experimental

The TR-ISRS measurements were carried out using a combination of actinic pump (P₁), Raman pump (P₂), and Raman probe (P₃) pulses. Briefly, the fundamental output of a 1-kHz Ti:sapphire amplifier (840 nm) was equally divided into two parts. The first part was frequency-doubled in a BBO crystal, and the second harmonic pulse was used as the P₁ pulse for photoexcitation after spectral filtering by a grating-slit setup. The second part of the amplifier output was used to drive a home-built one-stage noncollinear optical parametric amplifier (NOPA). The NOPA output was tuned to a wavelength range of 600–700 nm, and it was compressed in time down to 12 fs using a prism pair and a grating pair simultaneously. This ultrashort pulse was divided into two, and they were used as the P₂ and P₃ pulses for the ISRS measurements. All the three pulses were focused together into a 0.3-mm-thick flow cell, where the sample solution was circulated. The TR-ISRS signal, i.e., the P_2-induced absorbance change monitored by the P_3 pulses, was evaluated by chopping every other P_2 pulse. A dichloromethane solution (8.8 mM) of $[Cu(dmphen)_2]PF_6$ (dmphen = 2,9-dimethyl-1,10-phenanthroline) was prepared and used in the TR-ISRS measurements.

3 Results and Discussion

In our TR-ISRS measurement, the copper complex was initially excited to the S_2 state by the P_1 pulse at 420 nm, and then it undergoes a rapid internal conversion to the S_1 state. To track the structural dynamics, we introduced the ultrashort P_2 pulse at a ΔT delay and generated the vibrational coherence in the S_1 state. The resultant nuclear wavepacket motion was monitored by the P_3 pulse as oscillatory features in the P_2-induced absorbance change. It was found that the oscillatory feature substantially differs, depending on the duration of the P_1 pulse (48 and 430 fs), as shown in Fig. 1. This difference arises, since the P_1-induced vibrational coherence is still remaining in the molecule when the P_2 pulse reaches the sample at ΔT, i.e., the fifth-order contribution that does not reflect the vibrational structure at ΔT is significant. To remove this unwanted contribution, we carefully narrowed the spectral bandwidth of the P_1 pulse, and confirmed that the vibrational coherence is no longer generated by the 430-fs P_1 pulse having a bandwidth as narrow as 63 cm^{-1}. As shown in Fig. 1, the TR-ISRS data taken under this condition, which is free from the fifth-order processes, showed clear oscillations, reflecting the intrinsic vibrational structure of the S_1 state. Therefore, their Fourier transform gives "instantaneous" Raman spectrum at each ΔT delay (Fig. 2). Most importantly, the copper-ligand symmetric stretch at 125 cm^{-1} is not observed around $\Delta T = 0$ ps, but it gradually shows up within a few picoseconds. The time scale of this Raman spectral change is fully consistent with the 0.7-ps dynamics that we observed in both emission and absorption [2, 3]. Therefore, the present Raman result provides

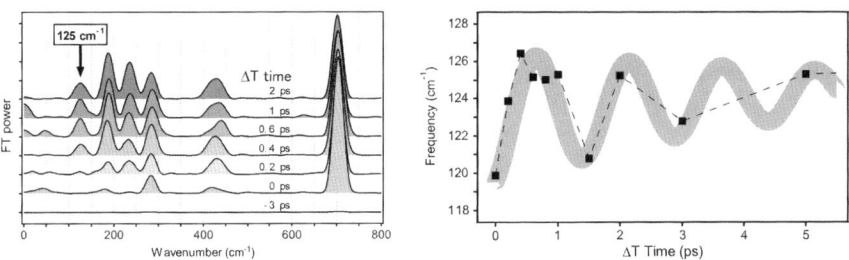

Fig. 2 Fourier transform analysis of the TR-ISRS signal. *Left* FT power spectra of the oscillatory component at selected ΔT times. *Right* Plot of the peak frequency of the 125-cm^{-1} mode against the ΔT time, showing an oscillation feature

a vibrational evidence for the ultrafast distortion occurring on the femtosecond time scale.

A close look at the Raman spectra suggested that the peak frequency of the symmetric stretch band changes with the ΔT time. In fact, a plot of the peak frequency versus ΔT time (Fig. 2) seems to show a slow oscillation with a period of ca. 1.5 ps. Interestingly, this period coincides well with that of the flattening mode (~ 22 cm^{-1}), in which the dihedral angle between the two ligands oscillates with time. This observation implies that the symmetric stretch mode is anharmonically coupled with the flattening mode. Our TDDFT calculation also supported this idea, and showed that the symmetric stretch frequency varies, depending on the dihedral angle. The present observation of the frequency change with ΔT time provides a new scheme to visualize the vibrational anharmonic coupling in the excited state.

References

1. L. X. Chen, G. B. Shaw, L. Novozhilova, T. Liu, G. Jennings, K. Attenkofer, G. J. Meyer, P. Coppens, J. Am. Chem. Soc. 125, 7022 (2003).
2. M. Iwamura, S. Takeuchi, T. Tahara, J. Am. Chem. Soc. 129, 5248 (2007).
3. M. Iwamura, H. Watanabe, K. Ishii, S. Takeuchi, T. Tahara, J. Am. Chem. Soc. 133, 7728 (2011).
4. S. Fujiyoshi, S. Takeuchi, T. Tahara, J. Phys. Chem A, 107, 494 (2003).
5. S. Takeuchi, S. Ruhman, T. Tsuneda, M. Chiba, T. Taketsugu, T. Tahara, Science 322, 1073 (2008).

Investigation of Vibrational Dynamics by Femtosecond Time-Resolved CARS

Yuanqin Xia, Yang Zhao, Sheng Zhang, Ping He, Zhiwei Dong, Deying Chen and Zhonghua Zhang

Abstract We report the femtosecond time-resolved CARS in BBO crystal, ethanol, cresyl violet 670 and pyrromethene 650 using the various degrees of freedom such as the timing, polarization and wavelengths of the laser pulses.

OCIS Codes (320.7150) Ultrafast spectroscopy · (300.6230) Spectroscopy, coherent anti-Stokes Raman scattering

Femtosecond time-resolved coherent anti-Stokes Raman spectroscopy (CARS) has emerged as an attractive method for studying the vibrational dynamics of Raman modes in time-domain. Multiple Raman modes can be coherently excited simultaneously due to the spectrally broad femtosecond laser pulses. The restriction of the CARS resolution is especially severe in the case of biological macromolecule, the separation and recognition of individual Raman modes becomes rather challenging. As three laser pulses are used for the generation of the CARS signal, many degrees of freedom can be varied [1]. By controlling the timing, polarization, and wavelengths of the laser pulses, the vibrational dynamics of different Raman modes are coherently excited and probed by CARS.

The experimental setup and methodology of femtosecond time-resolved CARS presented here has been described in detail elsewhere [2]. A commercial regenerative amplifier and two optical parametric amplifiers provide the three laser beams. The generation of CARS signal requires temporal and spatial overlap of the beams in the samples. The delay time between the pump and Stokes pulses defines T_{12}. The delay time between the probe and pump pulses defines T_{13}. This folded BOXCARS beam geometry ensures that the CARS signal propagates in a direction

Y. Xia (✉) · Y. Zhao · Z. Dong · D. Chen · Z. Zhang
National Key Laboratory of Science and Technology on Tunable Laser,
Harbin Institute of Technology, Harbin 150080, China
e-mail: xiayuanqin@hit.edu.cn

S. Zhang
Department of Physics, Harbin Institute of Technology, Harbin 150001, China

P. He
College of Foundation Science, Harbin University of Commerce, Harbin 150028, China

© Springer International Publishing Switzerland 2015   499
K. Yamanouchi et al. (eds.), *Ultrafast Phenomena XIX*,
Springer Proceedings in Physics 162, DOI 10.1007/978-3-319-13242-6_122

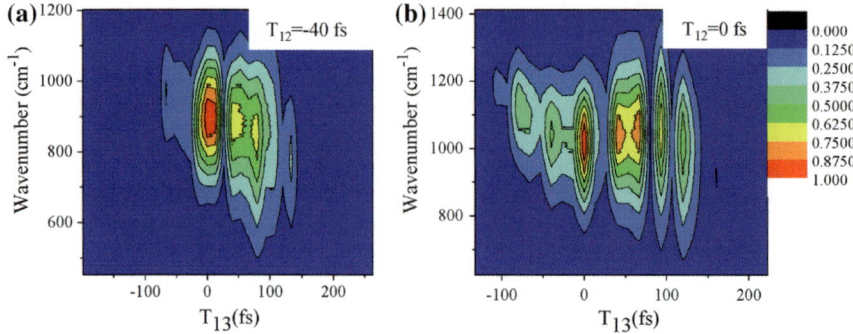

Fig. 1 The contour plots of spectrally dispersed transient CARS intensity of the ethanol solutions for pump wavelengths $\lambda_{pr} = \lambda_{pu} = 547$ nm and a Stokes wavelength $\lambda_{st} = 580$ nm

($k_{CARS} = k_{pr} - k_{st} + k_{pu}$). The CARS can be used in different applications by changing the delay time of laser pulses. Several Raman modes can be excited selectively by changing T_{12} due to the bandwidth and chirp of Stokes and pump pulse. The stretching vibration of C-C bonds and C-O bonds in ethanol are also selective excited by changing T_{12}. Figure 1 shows the contour plots of spectrally dispersed transient CARS intensity of the ethanol solutions, when $T_{12} = -40, 0$ fs (a, b), a pump of wavelength $\lambda_{pu} = \lambda_{pr} = 547$ nm and a Stokes wavelength $\lambda_{st} = 580$ nm. In Fig. 1a, b, the anti-Stokes signal has relative larger intensities centered on the detection wavenumbers of approximately 860 and 1,000 cm^{-1}. If T_{13} was set to zero, the signal can be got by changing T_{12}. The duration of pump and Stokes beams can be got by fitting the signals [2].

The ratio of intensity between vibrational modes can be changed by the polarization of pump pulses or crystal geometries. Zero degree is the best angle for second harmonic generation in BBO crystal. Figure 2a, c shows the time-integrated CARS intensity of BBO crystal with the different geometrical configurations (the angle between the crystal axis and the polarization) for a pump of wavelength $\lambda_{pu} = \lambda_{pr} = 714$ nm and a Stokes wavelength $\lambda_{st} = 800$ nm.

The fast Fourier transform (FFT) power spectra of the time-resolved CARS signal from Fig. 2a, c is exhibited in Fig. 2b, d, respectively. The frequency difference between the modes is 200 ± 2 and 185 ± 8 cm^{-1}. This corresponds to the

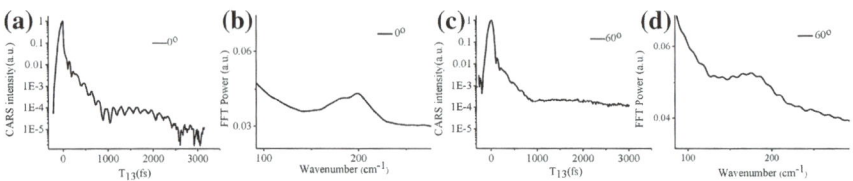

Fig. 2 The CARS of BBO crystal under different geometrical configurations (714 nm pump, 800 nm stokes)

Raman modes of B-O stretching vibrations of extra-ring B-O bonds (A′, 1,570 cm^{-1}) and extra-ring B-O' bonds (E′, 1,408 cm^{-1}). Furthermore, signal noise ratio is smaller with the increase of angle. From Fig. 2 we can see that geometrical configuration will influence the ratio of intensity between vibrational modes. The reason is that polarizability tensor influences vibration modes. The polarization of pump pulses will also influence the intensity of the vibrational mode [2].

The signal-to-noise ratio of CARS can be increased by controlling the wavelength of laser pulse. For the measurements on dye the wavelengths of the pump and probe beams were therefore tuned to the absorption band of the dye. In order to generate the vibrational coherence, the difference frequencies of the pump and Stokes laser pulses must be tuned to match the Raman resonance. Recently He et al. [3, 4] have achieved the selective excitation of the vibrational modes in cresyl violet 670 (CV670) and pyrromethene 650 (PM650) dye in diluted solutions (5×10^{-5} and 1×10^{-4} mol/L) by time-resolved CARS.

Figure 3a shows dependence of femtosecond CARS signal on probe pulse delay for PM650-ethanol solution and CV670-ethanol solution. The red line shows result in CV670-ethanol solution obtained for $\lambda_{pu} = \lambda_{pr} = 628$ nm and $\lambda_{st} = 671$ nm. The black line shows result in PM650-ethanol solution obtained for $\lambda_{pu} = \lambda_{pr} = 615$ nm and $\lambda_{st} = 655$ nm.

The FFT power spectra from Fig. 3a are exhibited in Fig. 3b, respectively. These power spectra confirm that the oscillations in the transient signal result from a coherent superposition of vibrational eigenstates with an energy spacing centered around 48 and 81 cm^{-1}. The frequency difference 48 cm^{-1} may be link to the Raman modes between $v'' = 0$ and $v' = 1$ bands in the ground electronic state in PM650 dye molecules. The vibrational dynamics in CV670 may be linked to the ring-breathing modes and ring-deformation modes of the six-folded ring.

The vibrational dynamics between the excited Raman transitions in organic molecules can be detected and investigated by simply changing wavelengths of pump or Stokes pulses. The three vibrational modes of ethanol are around 2973, 2927, and 2878 cm^{-1}. We observe a strong Raman vibrational mode with a

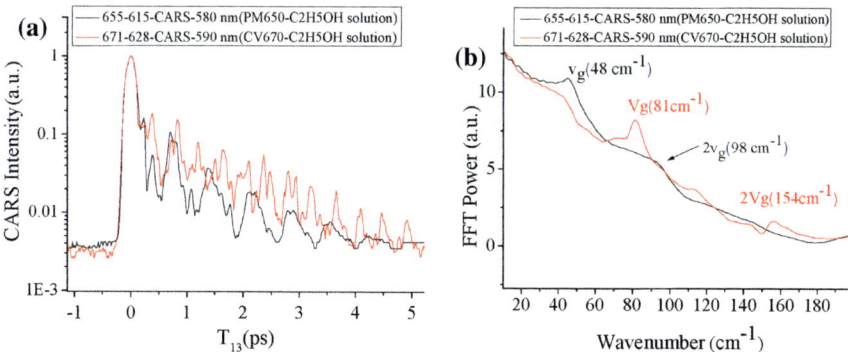

Fig. 3 **a** Dependence of femtosecond CARS signals on T_{13} for CV670 and PM650 in ethanol solution; **b** the FFT power spectra of the CARS signal from (**a**)

wavenumber difference of about 46 and 49 cm^{-1} in ethanol solution by varying wavelengths of the laser pulses.

In conclusion, the ultrafast vibrational dynamics in BBO crystal, ethanol, and dye were investigated by femtosecond time-resolved CARS. The vibrational modes can be changed by geometrical configuration and the polarization of pump pulses, varying the pulse sequence and wavelengths of the laser pulses.

Acknowledgments This work is supported by the National Natural Science Foundation of China (Grant Nos. 11174068, 61275157, 11004042), the Research Fund for the Doctoral Program of Higher Education (RFDP20102302120022), CERS (Cers-1-43) and the HIT. NSRIF (HIT. NSRIF.2009011).

References

1. A. Materny, T. Chen, M. Schmitt, T. Siebert, A. Vierheilig, V. Engel, and W. Kiefer, "Wave packet dynamics in different electronic states investigated by femtosecond time-resolved four-wave-mixing spectroscopy," Applied Physics B: Lasers and Optics **71**, 299-317 (2000)
2. Y. Xia, Y. Zhao, Z. Wang, S. Zhang, Z. Dong, D. Chen, and Z. Zhang, "Investigation of vibrational characteristics in BBO crystals by femtosecond CARS," Optics & Laser Technology **44**, 2049-2052 (2012).
3. D. Chen, P. He, R. Fan, Y. Xia, X. Yu, J. Wang, and Y. Jiang, "Femtosecond time-resolved ERE-CARS of CV670 dye in solutions," The Journal of Physical Chemistry C **116**, 5881-5886 (2012).
4. P. He, R. Fan, D. Chen, X. Li, Y. Xia, X. Yu, J. Wang, and Y. Jiang, "Femtosecond time-resolved ERE-CARS of PM650," Opt Commun **285**, 3284-3288 (2012).

Two-Dimensional Fourier Transform Infrared-Visible and Infrared-Raman Spectroscopies

Trevor L. Courtney, Zachary W. Fox, Karla M. Slenkamp, Michael S. Lynch and Munira Khalil

Abstract Femtosecond nonlinear spectroscopies using new IR and visible pulse sequences are demonstrated, including 2D IR-visible spectroscopy to study vibrational-electronic couplings and 2D IR-Raman spectroscopy to study anharmonic inter- and intramolecular vibrational couplings.

We report multidimensional spectroscopies to probe electronic and vibrational couplings and vibrational anharmonic inter- and intramolecular couplings to understand the roles played by coupled vibrations in ultrafast charge-transfer processes. An IR vibrational pump or collinear pump pulse pair for 2D Fourier transform (FT) spectroscopy [1, 2] excites a sample, which is then probed by either a visible (Vis) pulse or Raman pulse sequence (Fig. 1a, b). Our Raman detection scheme is analogous to femtosecond stimulated Raman spectroscopy (FSRS) [3] but with IR excitation to study structural dynamics on the electronic ground state. Resulting 2D IR-Raman (5th-order) spectra enable the measurement of cross terms between the IR dipole moment and the polarizability of a particular Raman mode in a molecule. Fully resonant 2D IR-Vis (3rd-order) spectra should measure IR and optical dipole moment cross terms. In the present work, we demonstrate the recovery of these 3rd- and 5th-order signals in neat benzonitrile.

The output from a 1 kHz Ti:Sapphire regenerative amplifier (800 nm, 35 fs, 2.7 W) is used to generate all pulses in this experiment (Fig. 1a). Mid-IR pump pulses ($v = 2{,}100$ cm^{-1}, 310 cm^{-1} bandwidth, 65 fs duration) are created via near-IR optical parametric amplification and subsequent difference frequency mixing. A Mach-Zehnder interferometer in the IR path generates pulse pairs (0.45 µJ/pulse) from the train of single IR pump pulses with a computer-controlled delay, τ_1, which transforms to ω_1 in 2D FT experiments. A computer-controlled stage defines the delay, τ_2, between the final IR pulse and the Raman continuum probe, generated in sapphire from 800 nm light. Prism compression and transmission through a 750-nm short-pass filter yield 35 fs pulses for anti-Stokes FSRS. Tunable ($\lambda = 800 \pm 15$ nm), narrowband (≤ 10 cm^{-1}), 3 ps Raman pump pulses (~ 0.5 µJ/pulse) from a grating

T.L. Courtney · Z.W. Fox · K.M. Slenkamp · M.S. Lynch · M. Khalil (✉)
Department of Chemistry, University of Washington, Seattle, Washington 98195, USA
e-mail: mkhalil@uw.edu

© Springer International Publishing Switzerland 2015
K. Yamanouchi et al. (eds.), *Ultrafast Phenomena XIX*,
Springer Proceedings in Physics 162, DOI 10.1007/978-3-319-13242-6_123

503

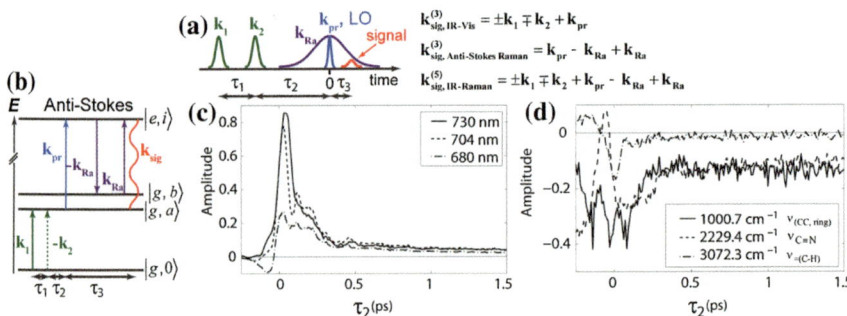

Fig. 1 Overview of IR-visible and IR-Raman spectroscopies. **a** Collinear IR pumps, k_1 and k_2, are separated by the vibrational coherence time, τ_1; after a population time, τ_2, the noncollinear visible probe and picosecond near-IR Raman pump impinge on the sample, generating a signal at τ_3. **b** Energy level diagram for vibrational excitation with ω_a from ground state and a stimulated Raman process involving an electronic excited state. **c** IR-Vis pump-probe spectra for three probe wavelengths versus τ_2. **d** IR-Raman signal versus τ_2 at three benzonitrile Raman peak frequencies

filter [3] are fixed at delay ~ 0 fs with respect to the probe for maximum Raman signal. Lenses focus all three vertically polarized beams into the 250 µm sample. The probe serves as the local oscillator (LO) and copropagates with nonlinear signal fields; a double-chopping scheme [4] isolates each nonlinear signal (Fig. 1) for single-shot detection with a CCD spectrograph.

These experiments allow us to collect three nonlinear signal fields simultaneously: 3rd-order IR-Vis, 5th-order IR-Raman, and 3rd-order ground-state stimulated anti-Stokes Raman (shifts from 12,500 cm^{-1}). Single IR pump scans of τ_2 yield 1D spectra to measure dynamics of the IR-Vis (Fig. 1c) and IR-Raman (Fig. 1d) signals. From a symmetric scan of two collinear IR pump pulses about $\tau_1 = 0$, we generate a 2D FT spectrum for a fixed τ_2 delay. Here, the 3rd-order (Fig. 2a) or 5th-order (Fig. 2b) signal is located at the ω_1 excitation frequency of the specific IR-active mode responsible for the change in visible probe transmission or Raman signal strength. Linearity of 3rd- and 5th-order signals with concentration was verified with stepwise benzonitrile/carbon tetrachloride mixtures to rule out contamination from 3rd-order cascades.

Since benzonitrile has no resonant transition in the probe frequency range, the IR-Vis signal follows the probe spectrum in ω_3 and lacks other spectral features. However, each probe frequency slice of the 2D spectrum maps the IR pump in ω_1 with absorption at benzonitrile vibrational modes (Fig. 2a): $\nu_{CN} \approx 2,230$ cm^{-1}, 1,850–2,000 cm^{-1} combination bands, and a continuum baseline. Longer-lived vibrational coherences of frequencies within the pump bandwidth [5] appear in Figs. 1c and 2a at ~ 450 cm^{-1} (benzonitrile) and 322 cm^{-1} (CaF$_2$ window). By extension of Fig. 1b, low-frequency peaks could arise from vibrational coherences between two substate populations, accessed via ω_a and $\omega_{a'}$ IR pump frequencies, respectively. In the 5th-order experiments, the same benzonitrile modes as in the 3rd-order spectra are resonantly excited with the IR pump. A broad signal appears

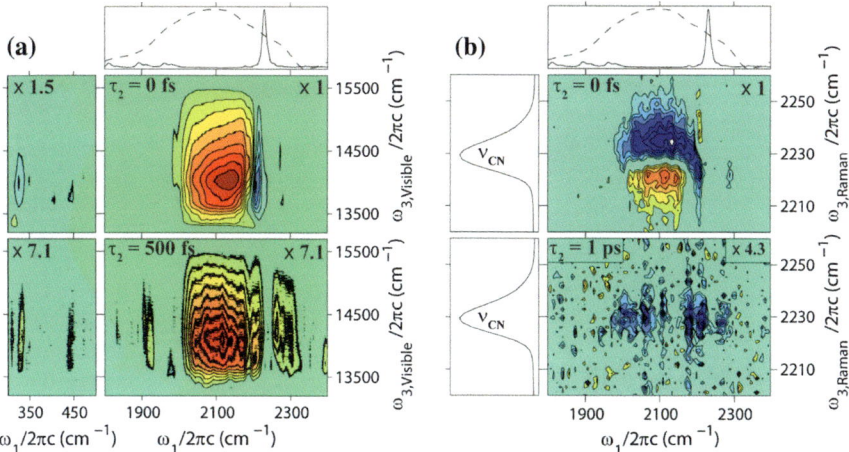

Fig. 2 2D IR-Vis and 2D IR-Raman spectra of benzonitrile, each with FTIR (*solid line*) and IR pump (*dashed line*) spectra in *top panel*. **a** 3rd-order 2D IR-Vis spectra and **b** 5th-order 2D IR-Raman spectra in the CN region (ground-state stimulated Raman spectra in *left panels*)

near $\tau_2 = 0$ from the IR-Vis signal at all benzonitrile Raman peaks followed by an exponential decay of the 5th-order signal with τ_2 (Fig. 1d). With increasing τ_2, the 5th-order signal at the $\nu_{CN} = 2229.4$ cm^{-1} Raman peak is more clearly coupled to the IR-active modes (Fig. 2b).

We have demonstrated new FT multidimensional spectroscopies using IR and visible pulse trains with sub-100 fs time resolution. Analysis of 2D IR-Raman and 2D IR-Vis benzonitrile spectra indicate that these experiments are well-suited to studies of vibrational energy transfer pathways in charge transfer complexes [4]. The ability to probe vibronic couplings with high time and frequency resolutions should make these ultrafast spectroscopies invaluable to studies of correlated electronic and nuclear motions in chemistry, biology, and material science.

Acknowledgments This work was supported by the Office of Basic Energy Sciences of the U.S. Department of Energy (Grant No. DE-SC0002190), the National Science Foundation (Grant No. CHE 0847790), and the David and Lucille Packard Fellowship for Science and Engineering.

References

1. S. M. Gallagher Faeder, and D. M. Jonas, J. Phys. Chem. A, **103**, 10489, 1999.
2. M. Khalil, N. Demirdoven, and A. Tokmakoff, J. Phys. Chem. A, **107**, 5258, 2003.
3. D. W. McCamant, P. Kukura, S. Yoon, and R. A. Mathies, Rev. Sci. Instrum., **75**, 4971, 2004.
4. M. S. Lynch, K. M. Slenkamp, M. Cheng, and M. Khalil, J. Phys. Chem. A, **116**, 7023, 2012.
5. K. Ishii, S. Takeuchi, and T. Tahara, J. Chem. Phys., **131**, 044512, 2009.

Part VI
Biological Systems

Ultrafast Intersystem Crossing in SO₂ and Nucleobases

Sebastian Mai, Martin Richter, Philipp Marquetand
and Leticia González's

Abstract Mixed quantum-classical dynamics simulations show that intersystem crossing between singlet and triplet states in SO_2 and in nucleobases takes place on an ultrafast timescale (few 100 fs), directly competing with internal conversion.

1 Introduction

Transitions involving a change in the spin state of a molecule are formally forbidden in a non-relativistic theoretical frame and are only mediated by spin-orbit couplings (SOCs). Including SOCs in the description of the excited-state dynamics allows investigating the so-called intersystem crossing (ISC) process [1]. Through ISC a singlet state can be transformed non-radiatively into a triplet state, or in general a transition between two states of different multiplicity can occur. ISC is promoted by large SOC and small energy gaps [1]. Accordingly, in molecules containing heavy atoms, such as transition metals, where a high density of states and large SOC are present, ultrafast ISC has been observed both experimentally and theoretically, see e.g. [2–4]. However, because SOCs are typically small for organic molecules, ISC is assumed to be one of the slowest forms of relaxation in photochemistry and photophysics, taking place on the order of picoseconds (ps) to microseconds [5]. Only recently, for a number of molecules including light atoms it has been shown that ISC can indeed occur on a femtosecond (fs) timescale and compete with internal conversion (IC) [6].

The study of photoinduced reactions can be tackled theoretically in two ways, which are complementary to each other. Solving the time-independent Schrödinger equation allows for optimizing critical points on the excited-state potential energy surfaces and proposing pathways that connect these points. A time-dependent approach provides the relevant reaction pathways following the dynamics of the

S. Mai · M. Richter · P. Marquetand · L. González's (✉)
Institute of Theoretical Chemistry, University of Vienna, Währinger Strasse 17,
1090 Vienna, Austria
e-mail: leticia.gonzalez@univie.ac.at

© Springer International Publishing Switzerland 2015
K. Yamanouchi et al. (eds.), *Ultrafast Phenomena XIX*,
Springer Proceedings in Physics 162, DOI 10.1007/978-3-319-13242-6_124

system in real time. An advantage of dynamical simulations is that they also allow to obtain the time constants and quantum yields associated to these pathways.

In this work we study the excited-state dynamics of SO_2 and the photophysics of DNA/RNA nucleobases using ab initio molecular dynamics in the form of surface-hopping, according to Tully's fewest switches criterion [7] within the recently developed SHARC variant [8] (Surface Hopping including ARbitrary Couplings), which permits to describe simultaneously ISC and internal conversion (IC) on the same footing. In both examples, it is illustrated that despite the lack of heavy atoms, ultrafast ISC takes place directly competing with IC.

2 Methodology

At least two bases to describe electronic states are distinguished within SHARC. One is the basis of the eigenfunctions of the molecular Coulomb Hamiltonian (MCH)—this is the basis employed in most electronic structure codes. In this representation, the interaction between states of the same multiplicity is described by non-adiabatic couplings, while states of different multiplicity are coupled by off-diagonal elements in the Hamiltonian matrix. Alternatively, another basis, in which the Hamiltonian is diagonalized and its eigenfunctions are spin-mixed states coupled only by non-adiabatic couplings, is possible. The transformation between the two bases is governed by a unitary matrix $\mathbf{U}$,

$$\mathbf{H}^{\mathrm{diag}} = \mathbf{U}^{\dagger}\mathbf{H}^{\mathrm{MCH}}\mathbf{U} \qquad \text{and} \qquad \mathbf{c}^{\mathrm{diag}} = \mathbf{U}^{\dagger}\mathbf{c}^{\mathrm{MCH}}, \tag{1}$$

which is also used to transform the gradients and non-adiabatic couplings.

The latter basis is the one employed for surface hopping within the SHARC methodology [8, 9], as it presents a number of advantages. First, it includes the effect of the SOC on the shape of the potential energy surfaces, which is known as the Zeeman effect. Second, in this basis all couplings are localized at regions of the potential energy surfaces where the states are close in energy. This implies that trajectories only hop in these localized regions—in the spirit of the fewest-switches approach [7]. Third, it correctly accounts for rotational invariance between the multiplet components and thus allows to easily include all of these components. [10].

3 Results

In the following, the photophysics of two different molecular systems will be discussed: SO_2 and the DNA/RNA nucleobases, the latter especially focusing on the two main tautomers of cytosine.

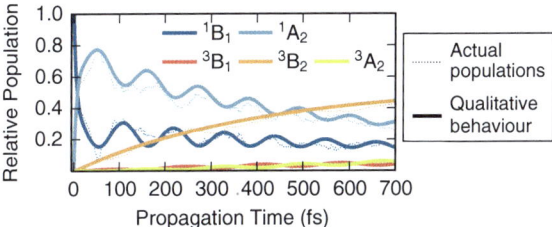

Fig. 1 Time evolution of the excited-state populations in SO₂. The main processes are periodical IC between $^1B_1$ and $^1A_2$ and ISC towards $^3B_2$

3.1 Dynamics of SO₂

In SO₂ a complicated vibrational structure, arising from IC between the $^1A_2$ and $^1B_1$ singlet states, has been observed in the first allowed band of the absorption spectrum. Another band of very weak intensity exists at energies slightly lower than the first allowed band, which shows an appreciable Zeeman effect [11] and hence is proposed to arise from ISC. Modern quantum chemistry indicates the presence of three triplet states ($^3A_2$, $^3B_1$, $^3B_2$) in this energetic region of the spectrum, but neither experiments nor static ab initio calculations could elucidate the relative importance of these states.

The results of SHARC dynamics simulations [12] based on on-the-fly MRCI wavefunctions are shown in Fig. 1. Starting from the spectroscopically bright $^1B_1$ state, oscillatory population transfer by IC between $^1B_1$ and $^1A_2$ occurs. Additionally, the $^3B_2$ triplet state is populated (mainly from the $^1A_2$ state) on a 410 fs timescale, showing that this state is strongly participating in the excited-state dynamics. Already after 700 fs, more than 40 % of the population has crossed to the $^3B_2$ state. On the contrary, the triplet states $^3A_2$ and $^3B_1$ do not show a large participation in the dynamics. These findings agree well with recent experimental studies based on the TRPEPICO (time-resolved photo-electron photo-ion coincidence spectroscopy) method [13] as well as with recent quantum dynamical simulations [14].

3.2 Dynamics of DNA Nucleobases

A number of experiments in gas phase have been performed in order to understand the excited-state dynamics of isolated nucleobases. Cytosine is a particularly complex case since it involves several stable tautomers in the gas phase and the presence of multiple deactivation pathways [15]. As Fig. 2 shows, the time evolution of the excited-state populations of the keto and enol tautomers are

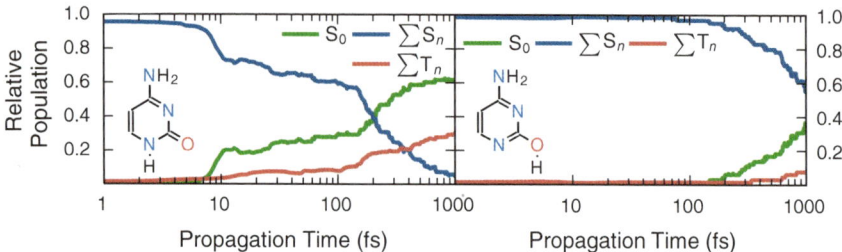

Fig. 2 Populations of the electronic states in cytosine tautomers (*left* keto, *right* enol). The keto form shows fast ground state relaxation and ISC, while the enol shows only much slower IC relaxation

significantly different [9]. In the keto tautomer, already after 10 fs, a notable fraction of the population is transferred to the electronic ground state and after 1 ps more than 60 % of the population is non-radiatively relaxed to the electronic ground state. Additionally, almost 30 % of the population undergoes ISC to the triplet manifold during this time. On the contrary, in the enol tautomer, ground state relaxation is much slower. After the first ps, still about 60 % of the population remains in the excited singlet states. Moreover, the enol form shows almost no ISC to the triplet manifold. According to our calculations, the experimentally observed multiexponential decay of cytosine [16] could be then explained by the presence of both the keto and enol tautomers with their significantly different relaxation timescales. The analysis of our trajectories [9] reveals that the keto tautomer relaxes to the ground state mainly through a three-state conical intersection, which is planar and thus quickly accessible from the Franck-Condon region (hence explaining that relaxation to the ground state starts already after 10 fs). In the enol tautomer, ground state relaxation involves conical intersections with highly distorted ring structures, necessitating large-scale motion of the ring atoms. Additionally, it was found that ISC in the keto tautomer is fast, compared to the enol tautomer, due to the presence of the carbonyl group that increases SOC substantially.

Analogous SHARC simulations on thymine and uracil [17] reveal that ISC also takes places on an ultrafast time scale. Especially in thymine, ISC is important since it is proposed to lead to thymine-thymine dimerization, one of the most prevalent types of DNA photodamage.

4 Conclusion

This contribution shows that the SHARC methodology is an efficient tool to study the excited-state dynamics of molecules in the presence of non-adiabatic and spin-orbit couplings, providing significant insight into elementary processes occurring on a molecular level during the first few ps.

References

1. C. M. Marian, "Spin-orbit coupling and intersystem crossing in molecules," WIREs Comput. Mol. Sci. **2**, 187 (2012).
2. A. Cannizzo, F. van Mourik, W. Gawelda, G. Zgrablic, C. Bressler, and M. Chergui, "Broadband femtosecond fluorescence spectroscopy of $[Ru(bpy)_3]^{2+}$," Angew. Chem. **118**, 3246 (2006).
3. I. Tavernelli, B. F. Curchod, and U. Rothlisberger, "Nonadiabatic molecular dynamics with solvent effects: A LR-TDDFT QM/MM study of ruthenium (II) tris (bipyridine) in water," Chem. Phys. **391**, 101 (2011).
4. L. Freitag and L. González, "Theoretical spectroscopy and photodynamics of a ruthenium nitrosyl complex," Inorg. Chem. **53**, 6415 (2014).
5. J. Cadet and P. Vigny, "The photochemistry of nucleic acids," in "Bioorganic Photochemistry 1: Photochemistry and the Nucleic Acids,", H. Morrison, ed. (Wiley-Interscience, 1990).
6. T. J. Penfold, R. Spesyvtsev, O. M. Kirkby, R. S. Minns, D. S. N. Parker, H. H. Fielding, and G. A. Worth, "Quantum dynamics study of the competing ultrafast intersystem crossing and internal conversion in the 'channel 3' region of benzene," J. Chem. Phys. **137**, 204310 (2012).
7. J. C. Tully, "Molecular dynamics with electronic transitions," J. Chem. Phys. **93**, 1061 (1990).
8. M. Richter, P. Marquetand, J. González-Vázquez, I. Sola, and L. González, "SHARC: Ab initio molecular dynamics with surface hopping in the adiabatic representation including arbitrary couplings," J. Chem. Theory Comput. **7**, 1253 (2011).
9. S. Mai, P. Marquetand, M. Richter, J. González-Vázquez, and L. González, "Singlet and triplet excited-state dynamics study of the keto and enol tautomers of cytosine," ChemPhysChem **14**, 2920 (2013).
10. G. Granucci, M. Persico, and G. Spighi, "Surface hopping trajectory simulations with spin-orbit and dynamical couplings," J. Chem. Phys. **137**, 22A501 (2012).
11. A. E. Douglas, "The Zeeman effect in the spectra of polyatomic molecules," Can. J. Phys. **36**, 147 (1958).
12. S. Mai, P. Marquetand, and L. González, "Non-adiabatic and intersystem crossings dynamics in SO_2: II. The role of triplet states in the bound state dynamics studied by surface-hopping simulations," J. Chem. Phys. **140**, 204302 (2014).
13. I. Wilkinson, A. E. Boguslavskiy, J. Mikosch, D. M. Villeneuve, H.-J. Wörner, M. Spanner, S. Patchkovskii, and A. Stolow, "Non-adiabatic and intersystem crossing dynamics in SO_2 I: Bound state relaxation studied by time-resolved photoelectron photoion coincidence spectroscopy," J. Chem. Phys. **140** 204301 (2014).
14. C. Lévêque, R. Taeb, and H. Köppel, "Communication: Theoretical prediction of the importance of the $^3B_2$ state in the dynamics of sulfur dioxide," J. Chem. Phys. **140**, 091101 (2014).
15. J. González-Vázquez and L. González, "A time-dependent picture of the ultrafast deactivation of keto-cytosine including three-state conical intersections," ChemPhysChem **11**, 3617 (2010).
16. J.-W. Ho, H.-C. Yen, W.-K. Chou, C.-N. Weng, L.-H. Cheng, H.-Q. Shi, S.-H. Lai, and P.-Y. Cheng, "Disentangling intrinsic ultrafast excited-state dynamics of cytosine tautomers," J. Phys. Chem. A **115**, 8406 (2011).
17. M. Richter, S. Mai, P. Marquetand, L. González: Phys. Chem. Chem. Phys. 16, 24423 (2014).

Detection of the G(–H)• Radical in the Electronic Deactivation of the G–C Watson-Crick Base Pair

Katharina Röttger and Friedrich Temps

Abstract Transient absorption spectroscopy of the G–C base pair revealed the formation of the G(–H)• radical with 3 ps lifetime in the electronic deactivation. This radical is the key intermediate in an electron-coupled proton transfer.

1 Introduction

The absence of resolved vibronic structure in the electronic spectra of the guanosine–cytidine Watson–Crick (WC) base pair in the gas phase has been explained by an extraordinarily short lifetime of the excited $\pi\pi^*$ state due to an ultrafast G-to-C charge-transfer (CT) transition followed by a rapid proton transfer along the central (G)N–H$\cdots$N(C) hydrogen bond [1, 2] and subsequent return to the electronic ground state through a conical intersection (CoIn). The existence of this process in G–C in solution, however, has remained highly controversial [3, 4]. Here, we report the observation of a short-lived, distinctive, new excited-state absorption (ESA) band after UV excitation of G–C in $CHCl_3$ with pronounced absorption maximum at $\lambda \approx 390$ nm, which is not present in the excited-state spectra of G or C alone. By comparison with the known spectrum of the G(–H)• radical [5], this characteristic 390 nm feature provides clear evidence for the occurrence of the proposed electron-driven G-to-C proton transfer in photoexcited G–C.

K. Röttger (✉) · F. Temps
Institute of Physical Chemistry, Christian-Albrechts-University Kiel,
Olshausenstr. 40, 24098 Kiel, Germany
e-mail: roettger@phc.uni-kiel.de

© Springer International Publishing Switzerland 2015
K. Yamanouchi et al. (eds.), *Ultrafast Phenomena XIX*,
Springer Proceedings in Physics 162, DOI 10.1007/978-3-319-13242-6_125

2 Experimental Details

Guanosine–cytidine heterodimers with WC structure are formed as decidedly predominant hydrogen-bonded assemblies of the two nucleosides in solution in dry CHCl$_3$ [6]. The ultrafast dynamics in the base pair were thus studied after excitation at $\lambda = 260$ nm by femtosecond transient UV/vis absorption spectroscopy [7] of G resp. C in separate solutions and G + C under the same conditions in an equimolar mixture. To investigate the effect of increasing dimer concentration, each sample was measured at concentrations corresponding to optical densities of OD = 0.2 and 1.0. The recorded two-dimensional spectro-temporal maps of transient absorption versus probe wavelength and time after excitation were then scaled according to concentration and relative fractions of absorbed UV pump light by the appropriate actinometric factors.

3 Results and Discussion

The two-dimensional absorption maps after excitation of G and C separately are given in Fig. 1a, b. No concentration dependence of the spectro-temporal shapes was observable for the samples in the investigated concentration range. The resulting synthetic superposition of the transient absorptions of G + C depicted in Fig. 1c is practically identical with the map for pure G, because the ESA by C is very weak. A moderately strong transient absorption maximum is visible around $\lambda \approx 340$ nm, followed by a weaker unstructured band at $\lambda \geq 400$ nm. In strong

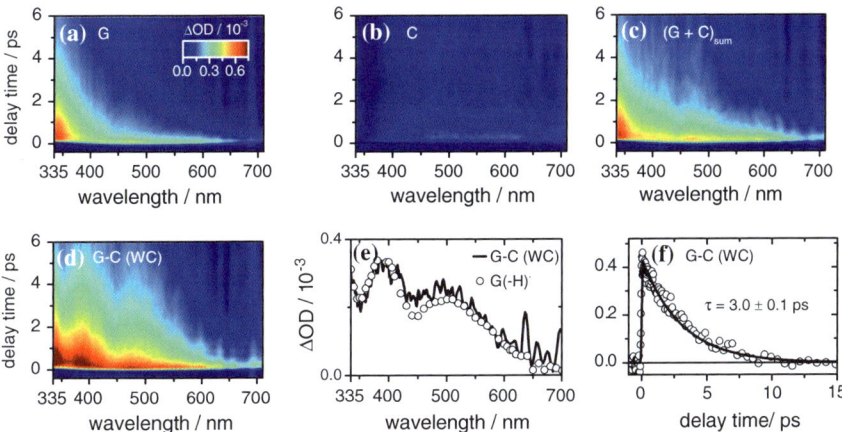

Fig. 1 Transient absorption maps after 260 nm excitation of **a** G, **b** C, **c** G + C as synthetic superposition, and **d** the measured equimolar mixture of G and C (93 % G–C) in CHCl$_3$. Each absorption map is shown with the same intensity scale. **e** Transient difference spectrum at $\Delta t = 2$ ps (*black line*) and normalized spectrum of G(–H)$^\bullet$ (*circles*, taken from [5]). **f** Time profile of the difference signal at $\lambda = 400$ nm together with its least-squares fit

contrast, the measured transient absorption map belonging to the equimolar mixture of G and C in Fig. 1d showcases an intense additional, new ESA band with distinctive absorption maximum at $\lambda \approx 390$ nm next to less pronounced bands around 340 and 480 nm. At the experimental G and C concentrations of $c_0 = 5.6 \times 10^{-3}$ M used, the G–C WC dimer accounts for ≈ 93 % of the dissolved molecules in the mixed solution. The "new" transient absorption in Fig. 1d around 390 nm therefore reflects a characteristic transition that is unique to G–C, but absent in G or C. For further analysis, Fig. 1e depicts a plot of the difference spectrum resulting from the G–C measurement (Fig. 1d) minus the synthetic sum spectrum of unbound G + C (Fig. 1c) at a selected delay time of $\Delta t = 2$ ps after excitation. Included in this plot is the known UV/vis absorption spectrum of the G(−H)$^{\bullet}$ radical of Candeias and Steenken [5]. The striking resemblance of the two spectra strongly suggests the formation of the G(−H)$^{\bullet}$ radical as short-lived transient during the electronic deactivation of the G–C WC pair. Evidently, a sizable fraction of the G(−H)$^{\bullet}$–C(+H)$^{\bullet}$ biradical intermediate formed via the unique electron-driven proton transfer deactivation route in G–C [2] gets trapped for a short time dynamically or in a shallow potential energy well. The analysis of the temporal evolution of the difference spectrum afforded a lifetime of $\tau = 3.0 \pm 0.1$ ps (cf. Fig. 1f). Transient absorption by the C(+H)$^{\bullet}$ counterpart appears to be weak, similar to the absorption of electronically excited C itself.

In conclusion, the close resemblance of the characteristic, G–C-specific transient absorption spectrum after electronic excitation with the spectrum of the G(−H)$^{\bullet}$ radical provides strong support for the deactivation of the photoexcited base pairs via the proposed electron-driven proton transfer pathway. In this light, the reported ultrashort (≈ 300 fs) fluorescence lifetime of G–C [3] may be attributed more likely to the transition from the initially excited state to the "optically dark" CT state and/or formation of the G(−H)$^{\bullet}$–C(+H)$^{\bullet}$ biradical intermediate rather than to the return of the excited dimer through the CoIn to the electronic ground state.

Acknowledgments The support of this work by the CRC 677 "Function by Switching" is gratefully acknowledged.

References

1. A. Abo-Riziq, L. Grace, E. Nir, M. Kabelac, P. Hobza, M. S. de Vries, PNAS **102**, 20–23 (2005).
2. A. L. Sobolewski, W. Domcke, C. Hättig, PNAS **102**, 17903–17906 (2005).
3. N. K. Schwalb, F. Temps, J. Am. Chem. Soc. **129**, 9272–9273 (2007).
4. L. Biemann, S. A. Kovalenko, K. Kleinermanns, R. Mahrwald, M. Markert, R. Improta, J. Am. Chem. Soc. **133**, 1966419667 (2011).
5. L. P. Candeias, S. Steenken, J. Am. Chem. Soc. **111**, 1094–1099 (1989).
6. M. Yang, L. Szyc, K. Röttger, H. Fidder, E. T. J. Nibbering, T. Elsaesser, F. Temps, J. Phys. Chem. B **115**, 5484–5492 (2011).
7. K. Röttger, R. Siewertsen, F. Temps, Chem. Phys. Lett. **536**, 140–146 (2012).

Ultrafast Photoisomerization of Chiral Biomimetic Molecular Switches

M. Gueye, S. Haacke, S. Fusi, M. Olivucci, E. Gindensperger and J. Léonard

Abstract Transient absorption spectroscopy on chiral biomimetic molecular switches reveals a critical influence of methyl substitutions on the photoreaction speed and on the observation of vibrational coherence in both isomerization directions.

1 Introduction

Biomimetic molecular switches based on the N-alkylated indanylidene-pyrroline (NAIP) structure have been designed to mimic the ultrafast, vibrationally coherent cis-trans photoisomerization observed in rhodopsin (Rho) [1]. In solution these molecules undergo an ultrafast photoisomerization in which a significant amount of mechanical energy is delivered into a few specific vibrational modes such that vibrational wavepackets are observed during the photoreaction [2–4].

Newly-synthesized chiral (Ch) NAIP compounds (see Fig. 1a, b) are available in which a H atom replaces one of the two methyl (Me) groups carried by the $C_{2'}$ atom in the previous, non-chiral NAIP's. We apply broadband transient absorption spectroscopy to study the photoisomerization dynamics of a racemic mixture of the chiral Ch-MeO-NAIP and Ch-dMe-MEO-NAIP in methanol. In the latter, the Me

M. Gueye · S. Haacke · J. Léonard (✉)
Institut de Physique et Chimie Des Matériaux de Strasbourg & Labex NIE,
CNRS - Université de Strasbourg, Strasbourg, France
e-mail: Jeremie.Leonard@ipcms.unistra.fr

S. Fusi · M. Olivucci
Dipartimento Di Chimica, Università Degli Studi Di Siena, Siena, Italy

M. Olivucci
Chemistry Department, Bowling Green State University, Bowling Green, USA

E. Gindensperger
Laboratoire de Chimie Quantique, Institut de Chimie, CNRS - Université de Strasbourg,
Strasbourg, France

© Springer International Publishing Switzerland 2015   517
K. Yamanouchi et al. (eds.), *Ultrafast Phenomena XIX*,
Springer Proceedings in Physics 162, DOI 10.1007/978-3-319-13242-6_126

group on C_5 in the pyrrolinium moiety is also replaced by a H atom. We observe that the presence of a Me group on the sp2 hybridized C_5 carbon atom is required (but not sufficient) to observe a faster, vibrationally coherent photoreaction. However, unlike in a recent investigation of methyl substitutions of the protonated Schiff base retinal in solution [5] there is no clear correlation between photoreaction speed and yield.

2 Results

In the chiral Ch-dMe-MeO-NAIP (Fig. 1d) in the E isomer, the Excited State Absorption (ESA) spectrum almost perfectly overlaps the Ground State Bleach (GSB) which is thus non apparent at short time delays. Instead, for the E isomer of the chiral Ch-MeO-NAIP (Fig. 1c), the overlap between ESA and GSB is partial. The spectroscopic signature of the Photoproduct Absorption (PA) emerges after ~ 0.5 ps in the range 400–450 nm. For Ch-dMe-MeO-NAIP the ESA and PA are spectrally separated, while in the case of the chiral MeO-NAIP the PA appears in the continuity of the ESA. After ~ 30 ps, a quasi-static, vibrationally-relaxed

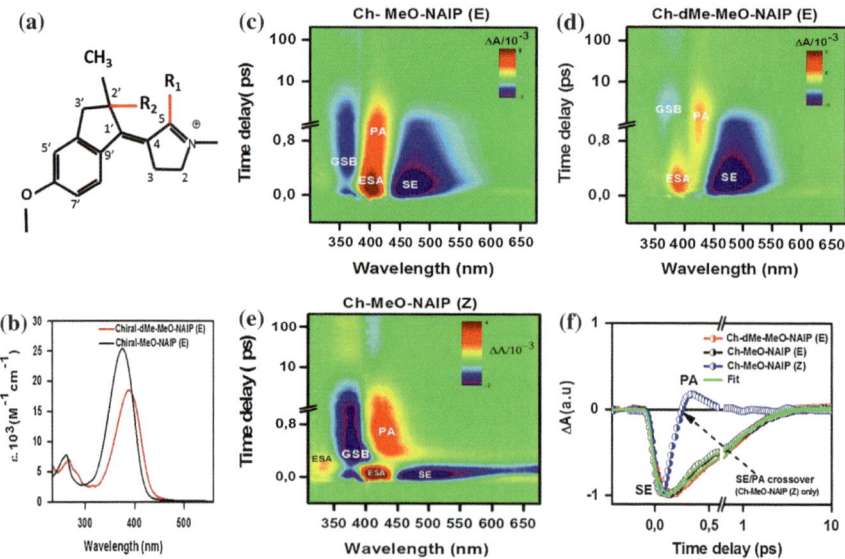

Fig. 1 **a** Chemical structure of the molecular switches: $R_2 = H$ = chiral; $R_2 = Me$ = non chiral; $R_1 = CH_3$ or H (demethylated). **b** Steady-state absorption spectra of the E Ch-dMe-MeO-NAIP (*red*), and E Ch-MeO-NAIP (*black*) in MeOH. **c**, **d** and **e** UV-VIS transient absorption data for E Ch-MeO-NAIP, E Ch-dMe-MeO-NAIP and Z Ch-MeO-NAIP in methanol, respectively. **f** Selected transient absorption kinetics showing the SE decay at 500 nm, followed (Z Ch-MeO-NAIP) or not (the other two) by the impulsive PA signature

spectrum is observed for both molecules which overlaps with the difference between the pure Z and E spectra (not shown). Quantitative analysis reveals bi-exponential decay of the SE signal in both cases, with time constants of ~ 0.6 ps and ~ 1.6 ps and relative amplitudes 57 and 43 % respectively for the Chiral dMe-MeO-NAIP and of ~ 0.4 ps (50 %) and 1.3 ps (50 %) for the Chiral-MeO-NAIP.

For the Z isomers, while the photoreaction dynamics is essentially identical to that of the E isomer in the case of Ch-dMe-MeO-NAIP (not shown), Z Ch-MeO-NAIP displays a qualitatively very different dynamics (Fig. 1e) exhibiting the signatures of vibrational reactive motion as already observed in the non chiral NAIP's [2, 3], including a far red-shifting SE impulsively followed by the far-red-detuned early PA signature. The rapid cross over from SE to PA after ~ 260 fs in the red part of the spectrum is attributed to the signature of a vibrational wavepacket travelling through the CI (see Fig. 1f).

3 Discussion

For the chiral (here) and non chiral [6] NAIP molecules which do not carry a Me substitute on the sp2 hybridized C_5 atom, all kinetic traces may be adjusted by exponential decay functions as would be expected for an incoherent photoreaction modeled by rate equations. The SE decay is relatively slow (≥ 300 fs) and bi-exponential, and no PA signature follows in the red part of the spectral observation window. Instead when C_5 carries a Me group, a vibrationally coherent photoreaction is observed in the non chiral NAIP's as well as in Ch-MeO-NAIP and photoreaction speeds are faster (<300 fs). Interestingly, in the latter molecule, this behavior is observed only in the Z isomer, while the E isomer undergoes an incoherent-like photoreaction as in the C_5-non-methylated compounds. Thus, the presence of this Me group is not sufficient to observe vibrational coherence signatures. No correlation is observed between the occurrence of vibrational coherence and the photoisomerization quantum yields.

Besides, the methylation of the sp3 hybridized $C_{2'}$ may have a minor effect on the photoreaction speed and yield since chiral compounds seem to isomerize slightly slower and with a slightly weaker quantum yield than their non chiral versions, independently on whether or not vibrational coherence is observed. While C_5 methylation affects the π-electron system, $C_{2'}$ methylation controls steric effects. Modeling may evidence the influence of steric hindrance by comparing ground state structures of chiral and achiral compounds. In addition, it may help at unraveling the origin of the different behavior observed in both isomers of the Ch-MeO-NAIP.

References

1. Lumento, F., et al., *Quantum Chemical Modeling and Preparation of a Biomimetic Photochemical Switch.* Angew. Chem., Int. Ed. **46,** p. 414-420 (2007).
2. Briand, J., et al., *Coherent ultrafast torsional motion and isomerization of a biomimetic dipolar photoswitch.* PCCP **12,** p. 3178-3187 (2010).
3. Léonard, J., et al., *Mechanistic Origin of the Vibrational Coherence Accompanying the Photoreaction of Biomimetic Molecular Switches.* Chem Eur J. **18,** p. 15296-15304 (2012).
4. Léonard, J., et al. *Isomer-dependent vibrational coherence in ultrafast photoisomerization.* New J. Phys. **15,** p. 105022 (2013).
5. Bassolino,et al., *Synthetic Control of Retinal Photochemistry and Photophysics in Solution.* J. Am. Chem. Soc. **136,** 2650−2658 (2014).
6. Dunkelberger, et al., *Photoisomerization and Relaxation Dynamics of a Structurally Modified Biomimetic Photoswitch.* J. Phys. Chem. A. **116** p. 3527-3533 (2012).

Snapshots of Sub-picosecond Dynamics in Heme-proteins Captured by Femtosecond Stimulated Raman Scattering

C. Ferrante, E. Pontecorvo, G. Batignani and T. Scopigno

Abstract The reaction pathway in photoexcited hemeproteins (ligand dissociation, energy redistribution and structural dynamics) has been unraveled by Femtosecond Stimulated Raman Scattering. The possible existence of short living intermediates as opposed to vibrational relaxation is discussed.

1 Background and Significance

Photoinduced dynamics in heme proteins embed several concurring processes: bond breaking events, excited state dynamics, vibrational energy redistribution (within the protein and to the solvent) and conformational changes. The way such different aspects syncretize on a coherent picture of the reaction pathway is a matter of open debate. Transient absorption studies provide valuable information, subject to contrasting interpretations. Specifically, the existence of short living intermediates on the way to the ground state has been hypothesized, as opposed to the direct decay from the reactive energy surface down to a cold ground state. Vibrational spectroscopy, on the other hand, is endowed with structural sensitivity, but suffers from resolution limitations. Time Resolved Resonance Raman (TR3), in particular, is a powerful technique to study protein dynamics, whose time resolution has been improved over the years from microseconds to a few picoseconds. If a sharp spectral resolution (<15 cm^{-1}) is to be maintained, however, no further improvement of the time resolution (<1 ps) is obtainable due to the Fourier Transform limit.

C. Ferrante · E. Pontecorvo · G. Batignani · T. Scopigno (✉)
Physics Department, University "Sapienza", Rome, Italy
e-mail: tullio.scopigno@phys.uniroma1.it

© Springer International Publishing Switzerland 2015   521
K. Yamanouchi et al. (eds.), *Ultrafast Phenomena XIX*,
Springer Proceedings in Physics 162, DOI 10.1007/978-3-319-13242-6_127

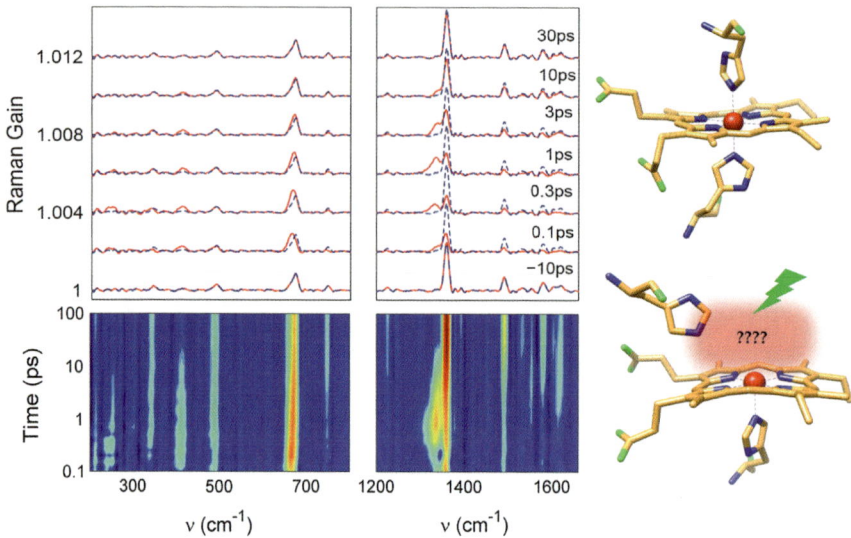

Fig. 1 FSRS data of photoexcited 6-coordinated deoxy Neuroglobin. The high spectral resolution combined to sub picosecond time precision allows revealing the sub-ps vibrational dynamics following photoexcitation process

2 Methods

Femtosecond Stimulated Raman Spectroscopy (FSRS) is a powerful method for studying ultrafast reaction dynamics, in which the simultaneous presence of a narrowband picosecond pulse (the Raman pulse) and a broadband femtosecond continuum (the probe pulse) stimulates vibrational Raman transitions over a wide frequency range [1, 2]. It represents a smart way to "circumvent" the aforementioned time-energy transform limit, allowing for simultaneously high temporal precision and spectral resolution. Using dispersed detection of the probe beam, indeed, spectral resolution is fundamentally limited by the vibrational dephasing time only. Nonetheless, the time resolution is only determined by the duration of the pulse initiating the macroscopic polarization in the sample and, of course, by the photochemical pump. This "disentanglement" of time and energy resolution allows reaching values as low as ~ 30 fs/10 cm^{-1}.

We recently combined the principles of FSRS and TR3, developing a Femtosecond Stimulated Resonance Raman (FSRRS) setup with broadly tunable Raman pulse optimized to exploit resonance enhancement in diverse biomolecules [3, 4].

3 Results

We applied Femtosecond Stimulated Resonance Raman Scattering (FSRS) to study photoinduced dynamics in Myoglobin and Neuroglobin, representative cases of five and six coordinated hemes, respectively. Taking advantage of unrestricted frequency resolution (10 cm^{-1}) and time precision (30 fs), combined with broad tunability of the Raman excitation, we have been able to unveil the details of the reaction pathway. These include the early stages following ligand dissociation, ascertaining the possible existence of short living intermediates, the subsequent energy redistribution among different vibrational channels, and the underlying structural rearrangements.

FSRS data suggest that the photoexcited hemeprotein, once the ligand bond breaking is accomplished, evolves to a hot ground state photoproduct. Distinct signatures of anharmonicities are observed with structural sensitivity, evolving during the cooling process (Fig. 1). In the specific case of the six coordinated deoxy Neuroglobin, we address the iron-histidine stretching to ascertain the possibility of a bond breaking event on either the proximal or distal side of the iron atom. Taken all together, these observations allow drawing a clear scenario for the photoinduced dynamics of heme proteins, rationalizing previous transient absorption results subject to contrasting interpretations.

References

1. D. W. McCamant, P. Kukura, S. Yoon, and R. A. Mathies, "Femtosecond broadband stimulated Raman spectroscopy: apparatus and methods," Rev. Sci. Instrum. 75, 4971-4980 (2004).
2. M. Yoshizawa and M. Kurosawa, "Femtosecond time-resolved Raman spectroscopy using stimulated Raman scattering", Phys. Rev. A 61, 013808 (1999)
3. E. Pontecorvo, S.M. Kapetanaki, M. Badioli, D. Brida, M. Marangoni, G. Cerullo and T. Scopigno, "Femtosecond Stimulated Raman Spectrometer in the 320-520 nm range.", Optics Express, 19, 1107 (2011).
4. E. Pontecorvo, C. Ferrante, C.G. Elles and T. Scopigno. "Spectrally tailored narrowband pulses for femtosecond stimulated Raman spectroscopy in the range 330 nm-750 nm", Optics Express, 21, 6866-6872 (2013)

Vibrational Dynamics in Photoactive Yellow Protein Revealed by Mid-IR Pump/Visible Probe Spectroscopy

Ryosuke Nakamura and Norio Hamada

Abstract Vibrational dynamics in photoactive yellow protein is studied by mid-IR pump/visible probe spectroscopy. Vibrational relaxation with time constants of 0.2–0.3 and 5–7 ps in the chromophore are obtained upon excitation of the vibrational modes of the chromophore. On the other hand, when the pump pulse is resonant to the C=O stretching mode of the protein backbone, an additional rise component with a 4.3 ps time constant is observed, which indicates that the vibrational energy flows from the protein backbone to the chromophore.

1 Introduction

Photoactive yellow protein (PYP) is a 125-residue, 14 kDa photoreceptor protein isolated from *Ectothiorhodospira halophile*. The chromophore of PYP is a 4-hydroxycinnamic acid which is covalently bound to Cys69 through a thioester linkage. After photoexcitation, PYP undergoes a photocycle with a number of intermediate states, which involves *trans-cis* isomerization of the chromophore, rearrangement of the hydrogen-bonding network surrounding the chromophore, and large structural changes of the protein. PYP returns to its initial ground state in a few hundred milliseconds [1]. This unique photoreaction is achieved by an efficient vibrational coupling between the chromophore and the surrounding protein. FTIR difference spectroscopy has revealed that the C=O stretching mode of the protein backbone of PYP is downshifted in response to the photo-induced structural change of the chromophore [2]. In the case of heme protein, time-resolved Raman study has found that the excess energy within the heme is transferred to modes of the protein in approximately 5 ps [3]. In this paper, vibrational energy flow from the protein to the chromophore in photoactive yellow protein is studied by mid-IR pump/visible probe spectroscopy. Vibrational modes of the chromophore or the surrounding

R. Nakamura (✉) · N. Hamada
Science & Technology Entrepreneurship Laboratory, Osaka University,
Suita 565-0871, Japan
e-mail: r.nakamura@uic.osaka-u.ac.jp

© Springer International Publishing Switzerland 2015
K. Yamanouchi et al. (eds.), *Ultrafast Phenomena XIX*,
Springer Proceedings in Physics 162, DOI 10.1007/978-3-319-13242-6_128

524

protein are excited with a mid-IR pump pulse, and the following vibrational dynamics within the chromophore is selectively probed with a visible probe pulse which is resonant to the electronic transition of the chromophore.

2 Experimental

90 % of the output from a Ti:sapphire regenerative amplifier (800 nm, 120 fs, 900 mW, 1 kHz) was introduced into an optical parametric amplifier (OPA), while a portion of the remaining output was used to generate a broadband probe pulse with a CaF_2 plate. A mid-IR pump pulse was generated by difference frequency generation in a type-I $AgGaS_2$ crystal of the signal and idler pulses from OPA. The spectral width and the pulse energy of the pump pulse used in this study were 150 cm^{-1} and 1.5 µJ, respectively. The probe pulse after the sample was dispersed onto a linear image sensor with a spectrometer. The output signals were digitized and collected at the repetition rate of the laser system (1 kHz). The mid-IR pump beam was modulated at 500 Hz by a mechanical chopper, which was frequency locked to the laser pulse train.

We used a PYP sample held in a moving cell with two BaF_2 windows (50 µm of sample thickness) with an optical density of 0.5 OD at 445 nm in heavy water solution.

3 Results and Discussion

A typical pump-probe spectrum of PYP is shown in Fig. 1a. The stationary absorption spectrum is also shown by a broken line. The negative signal is due to the ground state bleach. On the other hand, the positive signal is originated from the transient absorption (TA) that is the transition from a vibrational excited state in the electronic ground state to the electronic excited state of the chromophore. Figure 1b shows time profiles of TA with pump pulses at 1340, 1420, 1500, and 1660 cm^{-1}, respectively. The time profiles pumped at 1340, 1420, and 1500 cm^{-1} are similar to each other. They have a rise component at 0.2–0.3 ps and a slow decay component at 5–7 ps. We also observed that the ground state absorption of the chromophore was immediately bleached after pumping. Therefore, these dynamics are ascribed to vibrational relaxation within the chromophore after excitation of the vibrational modes of the chromophore. On the other hand, an additional rise component with a time constant of 4.3 ps is observed when the pump pulse energy is 1660 cm^{-1}, which is resonant to the C=O stretching mode of the protein backbone. We have not observed a corresponding decay component in the dynamics of the chromophore. This suggests that the energy is transferred from the outside of the chromophore. We conclude that the 4.3 ps rise component is ascribed to the energy flow from the protein backbone to the chromophore.

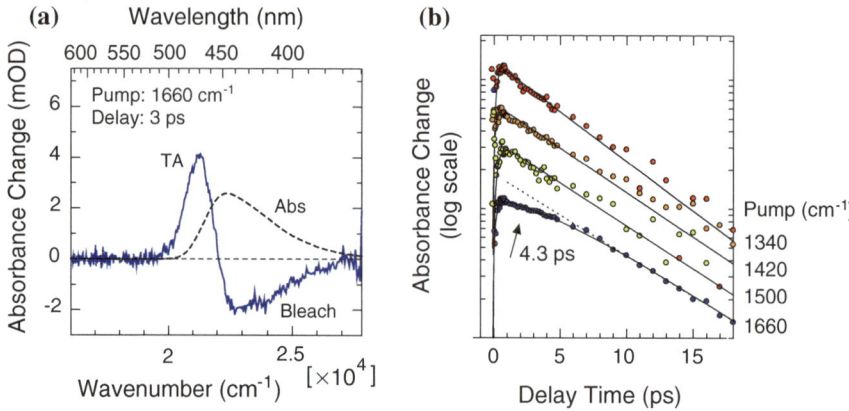

Fig. 1 **a** Pump-probe spectrum at 3.0 ps with 1660 cm^{-1} pump pulse. Positive signal is transient absorption (TA) while negative signal is ground state bleach. Stationary absorption spectrum is also shown (*broken*). **b** Time profiles of TA (470 nm) obtained with pump pulses at 1340, 1420, 1500, and 1660 cm^{-1} respectively

It should be noted that the spectrum and the decay time of TA are quite similar to those of 'ground state intermediate (GSI)' reported so far [4, 5], which is formed from the electronic excited state on the different pathway from the photocycle. The visible pump/mid-IR probe spectroscopy [5] proposed that GSI is likely to be a *cis* isomer, not 'hot' ground state (the vibrational excited state of the initial electronic ground state). Although our experiment does not give insight into the conformation of the chromophore, it does not affect the conclusion about the energy flow from the protein backbone to the chromophore.

4 Conclusion

Mid-IR pump/visible probe spectroscopy is effective to reveal vibrational dynamics in the chromophore and also the energy flow into the chromophore embedded in a protein environment. Mid-IR pump pulses at 1340, 1420 and 1500 cm^{-1} excite vibrational modes of the chromophore in PYP. Vibrational relaxation with time constants of 0.2–0.3 and 5–7 ps are observed. When the mid-IR pulse at 1660 cm^{-1} is used, we observed that the energy flow from the protein backbone to the chromophore occurs with a time constant of 4.3 ps.

References

1. K.J. Hellingwerf, J. Hendriks, and T. Gensch, "Photoactive yellow protein, a new type of photoreceptor protein: Will this "Yellow Lab" bring us where we want to go?", J. Phys. Chem. A **107**, 1082 (2003).
2. W.D. Hoff, A. Xie, I.H.M. van Stokkum, X. Tang, J. Gural, A.R. Kroon, and K.J. Hellingwerf, "Global conformational changes upon receptor stimulation in photoactive yellow protein", Biochemistry **38**, 1009 (1999).
3. R.Lingle, X. Xu, H. Zhu, S.C.Yu, and J.B. Hopkins, "Picosecond Raman study of energy flow in a photoexcited heme protein", J. Phys. Chem. **95**, 9320 (1991).
4. D.S. Larsen, I.H.M. van Stokkum, M. Vengris, M.A. van der Horst, F.L. de Weerd, K. J. Hellingwerf, and R. van Grondelle, "Incoherent manipulation of the photoactive yellow protein photocycle with dispersed pump-dump-probe spectroscopy" Biophys. J. **87**, 1858 (2004).
5. L.J.G.W. van Wilderen, M.A. van der Horst, I.H.M. van Stokkum, K.J. Hellingwerf, R. van Grondelle, M.L. Groot, "Ultrafast infrared spectroscopy reveals a key step for successful entry into the photocycle for photoactive yellow protein", PNAS **103**, 15050 (2006).

Probing Ultrafast Structural Dynamics of Photoactive Yellow Protein with Femtosecond Time-Domain Raman Spectroscopy

Hikaru Kuramochi, Satoshi Takeuchi, Kento Yonezawa, Hironari Kamikubo, Mikio Kataoka and Tahei Tahara

Abstract Ultrafast dynamics of photoactive yellow protein was investigated by time-resolved impulsive stimulated-Raman spectroscopy. Time-Domain vibrational data revealed rapid change of the hydrogen-bonding structure in the excited state and vibrational structure of the first ground-state intermediate.

1 Introduction

Photoactive yellow protein (PYP) is a water-soluble, cytosolic protein, which was discovered in a halophilic purple phototrophic bacterium, *Halorhodospira halophila* (Fig. 1). PYP is widely considered to function as a blue-light photoreceptor for negative phototactic response of this organism. The function of PYP is realized by a photocycle, which is triggered by the photo-induced trans-to-cis isomerization of the chromophore, p-coumaric acid (pCA). The isomerization reaction has been considered to occur on the femto-to-picosecond time scale, and it gives rise to the formation of the first ground-state intermediate called the I_0 state. The I_0 state eventually transforms into the pR state on the ns time scale, followed by a transition to the long-lived, putative signaling state pB. Since the discovery of PYP, a great deal of effort has been made to elucidate the photocycle and to characterize intermediates, using various experimental techniques, and these studies provided detailed insights into the structure of each intermediate, that is, the conformation of the chromophore, the hydrogen bonding between the chromophore and surrounding

H. Kuramochi (✉) · S. Takeuchi · T. Tahara
Molecular Spectroscopy Laboratory, RIKEN, 2-1 Hirosawa, Wako 351-0198, Japan
e-mail: h.kuramochi@riken.jp

S. Takeuchi · T. Tahara
Ultrafast Spectroscopy Research Team, RIKEN Center for Advanced Photonics (RAP),
2-1 Hirosawa, Wako 351-0198, Japan

K. Yonezawa · H. Kamikubo · M. Kataoka
Graduate School of Materials Science, Nara Institute of Science and Technology,
8916-5 Takayama, Ikoma 630-0192, Japan

© Springer International Publishing Switzerland 2015
K. Yamanouchi et al. (eds.), *Ultrafast Phenomena XIX*,
Springer Proceedings in Physics 162, DOI 10.1007/978-3-319-13242-6_129

528

amino acid residues, and so forth [1]. However, most of these experimental approaches have only provided information about the events that take place on time scales longer than 100 ps. Therefore, the structural dynamics of PYP on the femto- to picosecond region have been still veiled largely, regardless of its fundamental importance in activating the function.

Aiming to shed new light on the ultrafast primary photoreaction process of PYP, we used time-resolved impulsive stimulated Raman spectroscopy (TR-ISRS [2, 3]) to track the structural evolution of PYP from the excited state down to the I_0 state on the femto-to-picosecond time scale.

2 Experimental

PYP was dissolved in 10-mM Tris-HCl buffer at pH 7. Time-resolved impulsive stimulated Raman spectroscopy (TR-ISRS) was carried out using a setup based on noncollinear optical parametric amplifiers (NOPA) that were driven by the output of the Ti:sapphire regenerative amplifier (1 mJ, 1 kHz, 780 nm). The output of the first NOPA was used as the actinic excitation pulse (P1, 450 nm, 250 fs). The output of the second NOPA (500−700 nm, 6.5 fs) was divided into two, and they were used as the impulsive excitation pulse (P2) to induce coherent nuclear wavepacket motion in the excited state through resonant impulsive stimulated Raman process and the probe pulse (P3) to monitor the transient absorbance change, respectively.

3 Results and Discussion

In the left panel of Fig. 2, oscillatory components of TR-ISRS signals measured at various P1-P2 delay times (ΔT) are shown after the subtraction of slowly-varying population components from the raw data. The oscillatory features are due to the

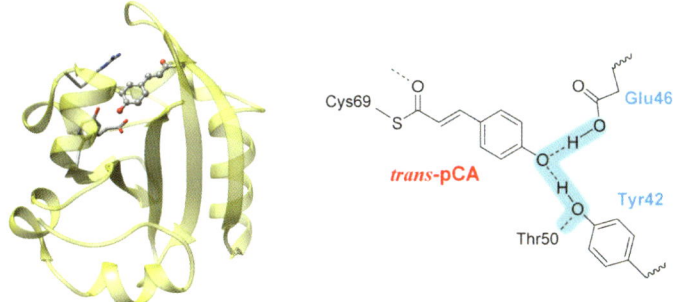

Fig. 1 (*Left*) crystallographic structure of PYP. (*Right*) chemical structure around the chromo-phore cavity

nuclear wavepacket motion on the excited state. Fourier transform analysis of these oscillatory components yields frequency-domain vibrational spectra of excited-state PYP at respective delay times (ΔT) as shown in the right panel of Fig. 2. The observed vibrational spectrum of the excited state exhibits prominent vibrational features at 1,160 cm^{-1} (in-plane bending vibration of C_{ph}-H and O-H of the phenolic part of pCA), 751 cm^{-1} (C_{ph}-O and C_{ph}-C_{et} stretch), and 536 cm^{-1} (phenolic-ring deformation). Furthermore, several noticeable bands were observed in the lower-frequency region (100−400 cm^{-1}). Remarkable feature in the obtained time-resolved vibrational spectra in Fig. 2 is threefold. First, although the decay rate constants of the higher-frequency bands are consistent with those of the population decay of the excited state ($\sim$2 ps), the 135-cm^{-1} band was found to decay within 1 ps. Second, the 1,160-cm^{-1} band showed a gradual upshift by $\sim$4 cm^{-1} towards 5 ps. Notably, the spectral signatures of these bands have been proposed as a sensitive marker of the hydrogen-bonding (HB) structure around pCA. Therefore, the observed dynamics of these bands indicate the rapid change of the HB structure in the excited state upon photoexcitation of the chromophore. Third, a growth of the 633-cm^{-1} band can be clearly recognized in the time-resolved spectra in Fig. 2, which could be attributable to the formation of the product I_0 state. The obtained spectrum of the I_0 state exhibits pronounced bands around 630 cm^{-1}, with several weaker bands in the other frequency region. Importantly, the obtained vibrational spectrum of the I_0 state significantly differs from those of the pG or pR state, whose chromophore configuration is planar trans and cis, respectively. This indicates the

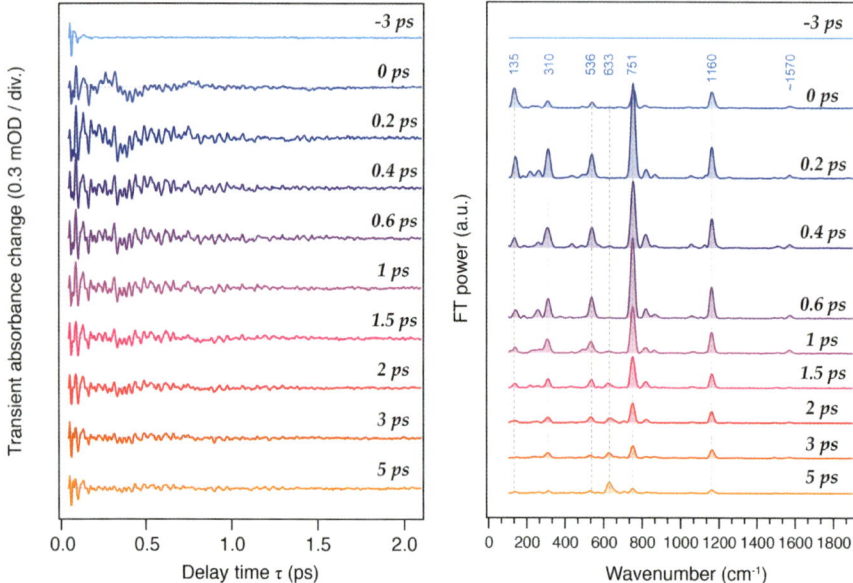

Fig. 2 (*Left*) oscillatory components of TR-ISRS signals obtained at various delay time ΔT. (*Right*) FT power spectra of the oscillatory components

distinct structure of the I_0 state. All these experimental data reveal the first comprehensive overview of the primary structural dynamics of PYP that includes the ultrafast change of the HB structure followed by the isomerization of the chromophore.

References

1. K. J. Hellingwerf, J. Hendriks and T. Gensch, J. Phys. Chem. A **107**, 1082 (2003).
2. S. Fujiyoshi, S. Takeuchi and T. Tahara, J. Phys. Chem. A **107**, 494 (2003).
3. S. Takeuchi, S. Ruhman, T. Tsuneda, M. Chiba, T. Taketsugu and T. Tahara, Science **322**, 1073 (2008).

Vibrational Energy Flow in Hemeproteins

Yasuhisa Mizutani, Naoki Fujii, Mitsuhiro Miyamoto, Misao Mizuno
and Haruto Ishikawa

Abstract We demonstrate that time-resolved anti-Stokes ultraviolet resonance Raman spectroscopy is a powerful tool for studying the vibrational energy flow in proteins with a spatial resolution of a single amino acid residue.

1 Introduction

Flow of excess energy from a reacting molecule is one of the key issues for understanding how chemical reactions take place in the condensed phases, such as liquid and protein. Excess energy is often deposited in many degrees of freedom right after photoreactions or internal conversions. Many experiments have been performed to study the dissipation processes of excess energy after photoexciting the chromophores. Particularly, hemeproteins have been extensively studied because the heme exhibits ultrafast internal conversion (<100 fs) and, hence, large amount of excess energy is deposited by photoexcitation. The cooling processes of the heme [1] and the heating of solvent molecules [2] have been well characterized by ultrafast spectroscopy. However, the energy flow within protein moiety has not been directly observed. In this study, we succeeded in observing the vibrational energy flow in photoexcited hemeproteins by using anti-Stokes ultraviolet resonance Raman (UVRR) spectroscopy. UVRR spectroscopy can selectively monitor Raman bands of aromatic amino acid residues, such as tryptophan (Trp), tyrosine (Tyr), and phenylalanine (Phe) [3].

Y. Mizutani (✉) · N. Fujii · M. Miyamoto · M. Mizuno · H. Ishikawa
Department of Chemistry, Graduate School of Science, Osaka University,
1-1 Machikaneyama, Toyonaka, Osaka 560-0043, Japan
e-mail: mztn@chem.sci.osaka-u.ac.jp

© Springer International Publishing Switzerland 2015
K. Yamanouchi et al. (eds.), *Ultrafast Phenomena XIX*,
Springer Proceedings in Physics 162, DOI 10.1007/978-3-319-13242-6_130

2 Vibrational Energy Flow in Cytochrome *c* [4]

Vibrational energy flow in ferric cytochrome *c* has been examined by time-resolved anti-Stokes UVRR measurements. We chose cytochrome *c* because it has a single tryptophan residue (Trp59 in bovine cytochrome *c*) near the heme group. The ferric form of cytochrome *c* is considered to be photoinert, and hence, effective deposition of excess energy is possible through the photoexcitation of the heme unit. The time-resolved anti-Stokes UVRR spectra of Trp59 disclosed the kinetics of the energy flow from the heme group and the energy release of the residue in cytochrome *c*. It was found from temporal changes of the anti-Stokes UVRR intensities that the energy flow from the heme to Trp59 and the energy release from Trp59 took place with the time constants of $1-3$ and 8 ps, respectively. These data are consistent with the time constants for vibrational relaxation of the heme [1] and heating of water [2] reported for hemeproteins. Comparison of the data generated upon excitation to the two different electronic excited states showed that the kinetics of the energy flow were not affected by the amount of excess energy deposited to the heme group.

3 Vibrational Energy Flow in Myoglobin

We investigated distance dependence of energy flow from the heme to discuss the energy transport mechanism in protein moiety, measuring time-resolved anti-Stokes UVRR spectra of myoglobin mutants upon the excitation of heme. In the mutants, two original Trp residues were replaced with Phe and Tyr residues, to prepare Trp-free myoglobin mutants. Then, a Trp residue was introduced at a desired position. Thus, we prepared several mutants in which a Trp residue located at different distance from the heme. For all mutants we studied, anti-Stokes bands attributed to the Trp residue at 760 (W18 band) and 1,012 cm^{-1} (W16 band) were observed. These bands appeared in a few picoseconds after the photoexcitation and diminished in tens of picoseconds. The increase and decrease of band intensities can be ascribed to energy transfer from the heme and energy release to the surrounding residues, respectively. In the time-resolved spectra, the anti-Stokes band intensities due to the transient species became lower as the distance between the Trp residue and the heme was longer. This is consistent with the simple idea that spatial density of the excess energy decreases as the energy diffuses. The time evolution of anti-Stokes intensities were compared among the myoglobin mutants. The intensity rise of the anti-Stokes band became slower as the distance between the Trp residue and the heme was longer. This means that it takes longer time for the excess energy to arrive at the position farer from the heme. We analyzed the data to see if a classical heat transport model reproduces the data. The model was not able to reproduce whole set of the observed anti-Stokes data, suggesting that the molecular-level description is necessary to account for the energy transport in the protein.

4 Advantages of the Present Technique for Studies on Intermolecular Vibrational Energy Transfer in Condensed Phases

Despite extensive experimental and theoretical investigations, intermolecular vibrational energy transfer between two different polyatomic molecules in solution has received very little attention due to experimental difficulties. In hemeprotein, the distance and relative orientation between heme and the amino acid residues are well characterized by X-ray crystallography. Because of the extremely short non-radiative lifetime, excess energy as great as 25,000 cm^{-1} can be deposited into the heme by photoexcitation via the Soret transition. Moreover, we can investigate distance dependence of energy flow from the heme by introducing the probe residue at the desired position by site-directed mutagenesis. Accordingly, studies using the present technique based on hemeproteins will provide us new insights for understanding the mechanism of vibrational energy transfer in condensed phases.

Acknowledgements This work was supported by a Grant-in-Aid for Scientific Research in the Priority Area "Molecular Science for Supra Functional Systems" (Grant No. 19056013) to Y.M. from the Ministry of Education, Science, Sports and Culture of Japan, a Grant-in-Aid for Scientific Research on Innovative Areas "Soft Molecular Systems" (Grant No. 25104006) to Y.M. from the Ministry of Education, Science, Sports and Culture of Japan, and a Grant-in-Aid for Scientific Research (B) (Grant No. 20350007) to Y.M. from the Japan Society for the Promotion of Science.

References

1. Mizutani, Y., Kitagawa, T.: Direct observation of cooling of heme upon photodissociation of carbonmonoxy myoglobin. Science **278**, 443–446 (1997).
2. Lian, T., Locke, B., Kholodenko, Y., Hochstrasser, R. M.: Energy flow from solute to solvent probed by femtosecond IR spectroscopy: malachite green and heme protein solutions. J. Phys. Chem. **98**, 11648–11656 (1994).
3. Asher, S. A.: UV resonance Raman studies of molecular structure and dynamics: applications in physical and biophysical chemistry. Ann. Rev. Phys. Chem. **39**, 537–588 (1988).
4. Fujii, N., Mizuno, M., Mizutani, Y.: Direct observation of vibrational energy flow in cytochrome *c*. J. Phys. Chem. B **115**, 13057–13064 (2011).

Towards Direct Measurement of Ultrafast Vibrational Energy Flow in Proteins

Henrike M. Müller-Werkmeister, Martin Essig, Patrick Durkin,
Nediljko Budisa and Jens Bredenbeck

Abstract Vibrational energy transfer (VET) within a molecule can be investigated in great detail with ultrafast IR spectroscopy. We report on progress towards mapping of VET pathways in proteins using unnatural amino acids as site-specific probes.

A long-standing question in biophysics is about the existence of networks of coupled side chains in protein domains which are believed to function as pathways for energy flow and allosteric communication [1]. The measurement of vibrational energy flow between arbitrary sites in a protein is up to today not possible in experiments. Using UAAs as site-specific vibrational probes might allow for the direct tracking of vibrational energy transfer between sites of interest and thus for the direct mapping of energy transfer pathways in proteins.

Vibrational energy transfer (VET) is a widely studied phenomenon in ultrafast spectroscopy. By 2D-IR spectroscopy or pump-probe spectroscopy it is possible to follow the energy flow between functional groups, e.g. in small molecules [2] or peptides [3]. The dynamics of the vibrational bands provides important information about the system under investigation. The transfer times, for example, reflect the spatial proximity of functional groups [4]. The gained information can be of great importance for different applications, given that signals can be assigned to a certain functional group in the studied systems, which might be difficult in case of spectral

H.M. Müller-Werkmeister · M. Essig · J. Bredenbeck (✉)
Institute of Biophysics, University of Frankfurt,
60438 Max-Von-Laue-Str. 1, Frankfurt, Germany
e-mail: bredenbeck@biophysik.uni-frankfurt.de

H.M. Müller-Werkmeister
Departments of Chemistry & Physics, University of Toronto,
80 St. George Street, Toronto, ON M5S 3H6, Canada

P. Durkin · N. Budisa
Department of Chemistry, Technical University Berlin,
Biocatalysis Group, Müller-Breslau-Str. 10, 10623 Berlin, Germany

© Springer International Publishing Switzerland 2015   535
K. Yamanouchi et al. (eds.), *Ultrafast Phenomena XIX*,
Springer Proceedings in Physics 162, DOI 10.1007/978-3-319-13242-6_131

congestion and especially for larger molecules such as proteins. A recent approach
to overcome the limitation by spectral congestion in vibrational spectra of proteins
is the use of unnatural amino acids (UAAs) with functional groups, which absorb
well separated from protein vibrations, as site-specific vibrational probes [5]. They
can be incorporated during protein expression and allow for labeling side chains at
positions of interest [6].

To test the usability of azide-containing UAAs in proteins we studied azidophenylalanine (N$_3$Phe) by two-color 2D-IR spectroscopy. Even for this seemingly small
system the assignment of the absorption bands (Fig. 1b) to functional groups of the
molecule (Fig. 1a) is not straight forward. DFT calculations are regularly used to
assist with assignment but are contradictory for this molecule. Only by using the
additional information from the time-dependent VET induced cross peaks (transients
shown in Fig. 1c) an unambiguous assignment of all absorption bands to functional
groups is possible (shown in Fig. 1a, atoms involved in the vibrations of different
functional groups are color coded; color intensities are proportional to the kinetic
energy of the atom, as derived by DFT calculations) [2]. Following excitation of the
azide group (blue), the other modes respond at different times, depending on the
distance to the azide. This allows e.g. to distinguish the ring modes and the amide-II
mode, which are in close spectral proximity. The results demonstrate the concept of
mapping energy flow in a molecule by the tracking of vibrational energy transfer
using localized oscillators.

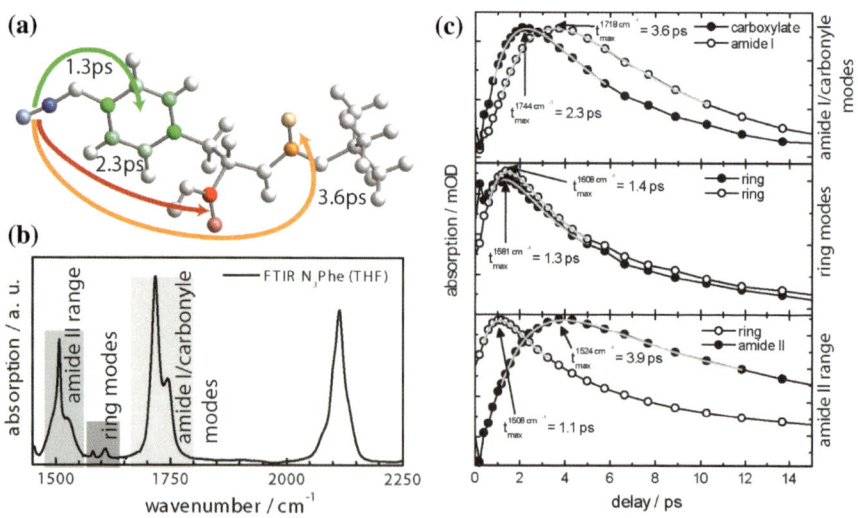

Fig. 1 Time dependence of cross-peaks in 2D-IR spectra originating from vibrational energy
transfer allows unambiguous assignment of FTIR spectra. **a** Structure of the unnatural amino acid
azidophenylalanine (N$_3$Phe) with VET after excitation of the azide group indicated by *arrows*, the
intensity in the colour coding reflects the kinetic energy per atom for the shown vibrations as
derived by DFT calculations. **b** FTIR absorption spectra of N$_3$Phe in tetrahydrofuran (THF).
c Transients for vibrational modes in amide I region (*upper panel*), ring mode region (*middle
panel*), amide II region (*lower panel*). Figures adapted from [2]

For the application of such experiments in proteins to investigate the above mentioned energy transfer pathways further requirements have to be met. To track VET in more complex systems explicit donor- and acceptor moieties need to be introduced, both absorbing separated from the natively occurring modes. Additionally they need to be optically orthogonal and have decent oscillator strength.

We report here on a donor-acceptor pair specially designed to match those requirements [7]. Azulene has been successfully used before in studies of IVR or VET in smaller systems [8]. It's a famous example of the violation of Kasha's rule, meaning that upon visible excitation to S_1 at 600 nm the molecule undergoes internal conversion instead of fluorescence, thereby dumping the energy in the vibrational modes of the electronic ground state within ~ 1 ps. This is the ideal chromophore to serve as a donor for tracking VET in proteins, because the visible excitation allows for 10 times higher energy deposition than IR excitation.

Azulenylalanine (AzuA) is the corresponding UAA we have tested. As an acceptor we use the azide-containing UAA azidohomoalanine, which is compatible with AzuA in terms of independent incorporation into proteins. Figure 2a shows the structures of the VET donor and acceptor.

Different model peptides with this VET pair have been investigated: A pentamer containing additional amino acids (Tyr, Asn, Gly) was tested to demonstrate the distance-dependency and feasibility of the proposed approach for systems with the size matching the diameter of small proteins (2 nm, Fig. 2b). A dimer of the two UAAs was tested under different conditions to prove the usability in aqueous solution, an important prerequisite for applications in proteins. The transients for those systems are shown in Fig. 2c. The transfer time is not only correlated to the distance between the donor and acceptor (comparison between AzuAha and Az-NYAhaG in d6-DMSO) but also affected by the solvent with a much faster peak time observed in H2O than in d6-DMSO. We furthermore demonstrated the application of the donor-acceptor pair in a small protein domain. In summary, the proposed VET pair might not only be useful for the anticipated study of VET pathways in proteins, but is as well a versatile tool to study the principles of vibrational energy transfer in greater detail.

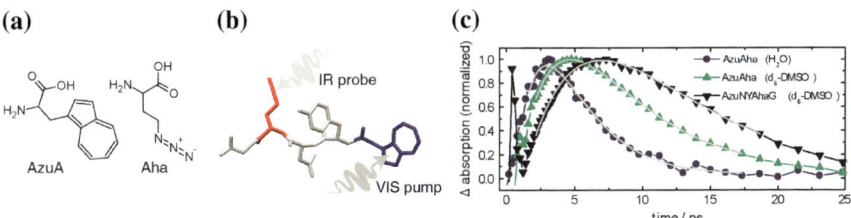

Fig. 2 A donor-acceptor pair of unnatural amino acids (UAAs) is used to track VET in model peptides. **a** Structures of the UAAs azulenylalanine (AzuA), used as donor and zidohomoalanine (Aha), used as acceptor. **b** Sketch of the performed experiment in the model peptide AzuNYAhaG. **c** Transients for the azide stretch vibration in different systems and different solvents (see legend)

References

1. David M. Leitner and John E. Straub, *Proteins: Energy, Heat and Signal Flow (Computations in Chemistry)*. (CRC Press, 2009).
2. H. M. Müller-Werkmeister, Y.-L. Li, E.-B. W. Lerch, D. Bigourd, J. Bredenbeck, Angew. Chem. Int. Ed. **52,** 6214-6217 (2013).
3. V. Botan, E. H. G. Backus, R. Pfister, A. Moretto, M. Crisma, C. Toniolo, P. H. Nguyen, G. Stock, P. Hamm, Proc. Natl. Acad. Sci. U. S. A. **104**, 12749–12754 (2007).
4. N. I. Rubtsova, I. V. Rubtsov, Chem. Phys. **422**, 16–21 (2013).
5. H. Kim, M. Cho, Chem. Rev. **113**, 5817-5847 (2013).
6. M. G.Hoesl, N. Budisa, Angew. Chem. Int. Ed. **50,** 2896–2902 (2011).
7. H. M. Müller-Werkmeister, J. Bredenbeck, Phys. Chem. Chem. Phys. **16**, 3261-3266 (2014).
8. D. Schwarzer, C. Hanisch, P. Kutne, J. Troe, J. Phys. Chem. A **106,** 8019–8028 (2002).

Time-Resolved Impulsive Raman Study of Excited State Structures of Green Fluorescent Protein

Tomotsumi Fujisawa, Hikaru Kuramochi, Satoshi Takeuchi and Tahei Tahara

Abstract Structural dynamics of green fluorescent protein was studied by femtosecond time-resolved impulsive Raman spectroscopy. The excited-state vibrational spectra of the protein with three different chromophore forms were obtained and they revealed their structural differences and excited-state deprotonation.

1 Introduction

Green fluorescent protein (GFP, from *Aequorea victoria*) is the most famous fluorescent protein that is widely used in bio-imaging application. The photophysics of GFP attracts much attention to understand how the bright green fluorescence is emitted. So far, spectroscopic measurements revealed that the intrinsic chromophore, *p*-hydroxybenzylideneimidazolinone, is in the equilibrium between protonated state (A state) and deprotonated state (B state) inside GFP. The A state exhibits a major absorption band at ~ 398 nm while the B state gives another minor band at ~ 475 nm (Fig. 1, left).

When photoexcited, the A state undergoes excited state proton transfer (ESPT) and is converted to the highly emissive deprotonated form of the chromophore, which is origin of the green fluorescence of GFP. This deprotonated state after ESPT contains the same deprotonated chromophore as the B state, but it is called the I state because the fluorescence after ESPT exhibit a slightly different spectrum from that of the B state (Fig. 1). This difference has been attributed to the distinct chromophore pockets of the I and B states. Although the photochemistry of GFP and involvement

T. Fujisawa · H. Kuramochi · S. Takeuchi · T. Tahara (✉)
Molecular Spectroscopy Lab, RIKEN, 2-1 Hirosawa, Wako, Saitama 351-0198, Japan
e-mail: tahei@riken.jp

S. Takeuchi · T. Tahara
Ultrafast Spectroscopy Research Team, RIKEN Center for Advanced Photonics (RAP),
2-1 Hirosawa, Wako, Saitama 351-0198, Japan

© Springer International Publishing Switzerland 2015
K. Yamanouchi et al. (eds.), *Ultrafast Phenomena XIX*,
Springer Proceedings in Physics 162, DOI 10.1007/978-3-319-13242-6_132

539

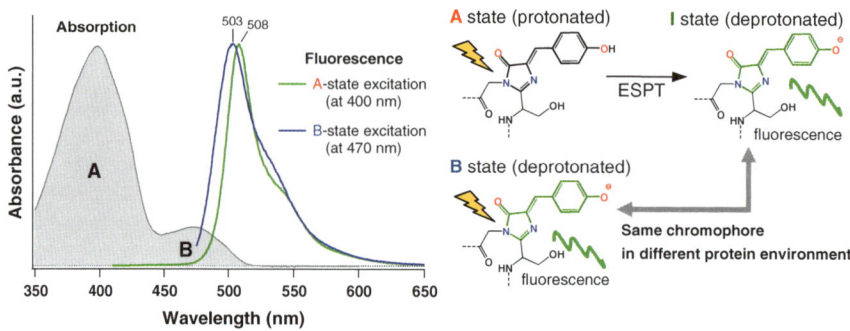

Fig. 1 Three states of GFP (*right*) and their absorption/fluorescence spectra (*left*)

of multiple excited states are the bases of the widely-utilized GFP fluorescence, the structures of the A, B, and I states, especially those of their excited states (A^{ex}, B^{ex}, and I^{ex} states), have not been clearly characterized yet by experiments. In this study, we used time-resolved impulsive stimulated Raman spectroscopy (TR-ISRS) [1] to investigate excited-state structures of GFP.

2 Experimental Method

TR-ISRS measures excited-state vibrational modes in time domain by using three pulses P1, P2 and P3 (Fig. 2a). The first P1 pulse of ~ 150 fs duration prepares excited state population. The wavelength of P1 pulse is set at 390 nm (or 470 nm) to excite the A state (or the B state). The second P2 pulse interacts with the excited state at a certain delay time ΔT and induces the vibrational coherence in the excited state. Then, the induced vibrational coherence is probed by the third P3 pulse as a function of P2-P3 interval (τ). The P2 and P3 pulses cover the spectral range of 550–650 nm and are compressed in time down to ~ 10 fs. The signal from the excited state is almost selectively observed by tuning the wavelength of the P2 and P3 pulses to be resonant with the stimulated emission of the excited state.

3 Results and Discussion

Figure 2b shows the TR-ISRS signals obtained after the A-state excitation at selected P1-P2 delays (ΔT), together with that obtained after B-state excitation for comparison. The TR-ISRS signal, i.e. the absorbance change induced by the P2 pulse, consists of the component arising from the excited-state population change and the oscillatory component due to impulsively excited Raman-active vibrations.

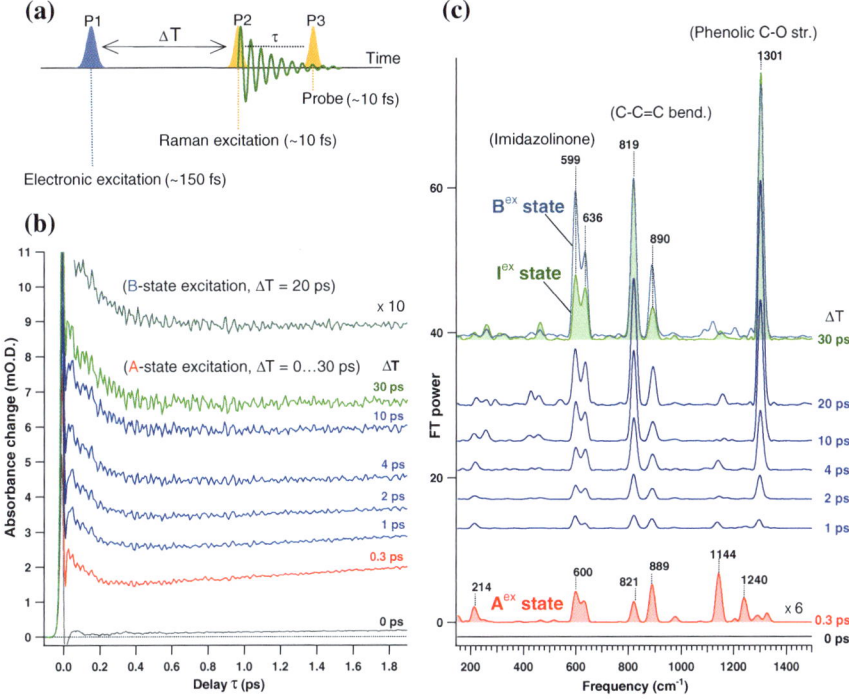

Fig. 2 a Experimental scheme of TR-ISRS. **b** TR-ISRS time-domain signal after exciting A state in comparison to the signal after exciting B state. **c** TR-ISRS spectra obtained by Fourier transform analysis. I^{ex} state spectrum at 30 ps is compared with B^{ex} state spectrum on the *top*

Fourier transform analysis of the oscillatory component yields the time-resolved vibrational spectra of the excited states, as is shown in Fig. 2c. The spectrum clearly changes with ΔT. Immediately after photoexcitation, the spectrum shows vibrational bands at 214, 600, 821, 889, 1144, 1240 cm^{-1} (red, at $\Delta T = 0.3$ ps). As ΔT increases, the vibrational band at 1,240 cm^{-1} gradually disappears and a new band at 1,301 cm^{-1} grows significantly within 30 ps. The rise of the 1,301 cm^{-1} band (time constant: 6.6 ps) is concurrent with ESPT (time constant: ∼6 ps), which indicates that the spectral evolution reflects the structural transition from A^{ex} state to I^{ex} state.

The major difference of the A^{ex} and I^{ex} spectra is the frequency difference of their phenolic C-O stretching modes. The ∼60 cm^{-1} upshift on going from the A^{ex} to I^{ex} states (1,240 cm^{-1} → 1,301 cm^{-1}) is fully consistent with the deprotonation in the ESPT process. Our experiment further showed that the C-O stretching frequency of the I^{ex} state is equal to that of the B^{ex} state which was measured by direct excitation of the B state. Although X-ray crystallography of GFP mutants proposed that B and I states have quite different hydrogen bond lengths between Thr203 and C-O group of the chromophore [2], this argument is not supported because the C-O stretching frequency is the same between the B^{ex} and I^{ex} states. Rather, the significant difference between the I^{ex} and B^{ex} states is found for the relative intensities of imidazolinone ring vibrations at 599 and 636 cm^{-1}, which suggests that a distortion of imidazolinone ring is involved in the structural differences of the I^{ex} and B^{ex} states.

References

1. S. Takeuchi, S. Ruhman, T. Tsuneda, M. Chiba, T. Taketsugu, T. Tahara, *Science* 322, 1073 (2008).
2. K. Brejc, T. K. Sixma, P. A. Kitts, S. R. Kain, R. Y. Tsien, M. Ormo, S. J. Remington, *Proc. Natl. Acad. Sci. USA* 94, 2306 (1997).

Nonlinear Fourier-Transform Spectroscopy Using Ultrabroadband Femtosecond Pulses for the Measurement of Photobleaching of Fluorescent Proteins

Akira Suda, Hiroshi Takahashi and Keisuke Toda

Abstract We investigate the mechanism of photobleaching of fluorescent proteins using nonlinear Fourier-transform spectroscopy with ultrabroadband femtosecond pulses. Photobleaching of fluorescent molecules occurs through one-photon excited-state absorption following two-photon excitation from the ground state.

1 Introduction

Fluorescent protein is widely used as a tool in biological researches such as multi-photon fluorescent imaging [1]. However, photobleaching (PB) of fluorescent molecules is a crucial issue for correctly observing biological samples and for interpreting their molecular function. Although a number of experiments have been carried out to explore the mechanism of PB, little is known about PB in multi-photon processes.

We apply nonlinear Fourier-transform spectroscopy to the measurement of the action spectrum for the PB of fluorescent proteins using an ultrabroadband pulse with a spectrum ranging from 650 to 1,200 nm. We show that sequential three-photon excitation process, excited-state absorption (ESA) following two-photon excitation (TPE) of fluorescent molecules, is one of the routes to the PB.

A. Suda (✉) · H. Takahashi · K. Toda
Department of Physics, Tokyo University of Science, 2641 Yamazaki, Noda,
Chiba 278-8510, Japan
e-mail: asuda@rs.tus.ac.jp

© Springer International Publishing Switzerland 2015   543
K. Yamanouchi et al. (eds.), *Ultrafast Phenomena XIX*,
Springer Proceedings in Physics 162, DOI 10.1007/978-3-319-13242-6_133

2 Nonlinear Fourier-Transform Spectroscopy

The principle of nonlinear Fourier-transform spectroscopy based on interferometric autocorrelation (IAC) measurements is shown as follows [2]. The second-order interferometric autocorrelation signal $S^{(2)}(\tau)$ is given by

$$S^{(2)}(\tau) = \int_{-\infty}^{\infty} \left| \int_{-\infty}^{\infty} R^{(2)}(t-t_1)E(t_1,\tau)^2 dt_1 \right|^2 dt, \qquad (1)$$

where τ, $R^{(2)}(t)$ and $E(t,\tau)$ are the delay time between two broadband pulses, the response function of the TPE process and the interferometric field, given by $E(t,\tau) = [A(t) + A(t-\tau)\exp(-i\omega_0\tau)]\exp(i\omega_0 t)$, respectively. To obtain the spectral information $\tilde{R}^{(2)}(\Omega)$, that is, the Fourier transform of $R^{(2)}(t)$, the power spectrum $\tilde{S}^{(2)}(\Omega)$ is calculated using the Fourier transform of $\tilde{S}^{(2)}(\tau)$ Then, the power spectrum at frequencies around $2\omega_0$ can be written as

$$\tilde{S}_{2\omega_0}^{(2)}(\Omega) = \left| \tilde{R}^{(2)}(\Omega) \right|^2 \left| \tilde{A}^{(2)}(\Omega - 2\omega_0) \right|^2, \qquad (2)$$

where $\left| \tilde{A}^{(2)}(\Omega - 2\omega_0) \right|^2$ is the second harmonic power spectrum obtained from a reference sample and $\left| \tilde{R}^{(2)}(\Omega) \right|^2$ is the TPE spectrum of a resonant sample. Therefore, the TPE spectrum can be acquired by dividing the second harmonic power spectrum of the resonant sample by that of the reference sample.

Assuming that PB occurs through sequential three-photon excitation process, one-photon ESA following TPE from the ground state of the fluorescent molecules, the corresponding IAC signal $S_{PB}^{(3)}(\tau)$ can simply be expressed as

$$S_{PB}^{(3)}(\tau) = S_{TPE}^{(2)}(\tau) \cdot S_{ESA}^{(1)}(\tau) \qquad (3)$$

where $S_{TPE}^{(2)}(\tau)$ and $S_{ESA}^{(1)}(\tau)$ are the IAC signals for TPE from the ground state and one-photon ESA, respectively. Since $S_{PB}^{(3)}(\tau)$ and $S_{TPE}^{(2)}(\tau)$ are obtained from the measurements of TPE with and without PB, $S_{ESA}^{(1)}(\tau)$ can be obtained by dividing $S_{PB}^{(3)}(\tau)$ by $S_{TPE}^{(2)}(\tau)$. Finally, the ESA spectrum is calculated from the Fourier-transform of $S_{ESA}^{(1)}(\tau)$.

3 Experimental

We used a Ti:sapphire mode-locked laser with a spectrum ranging from 670 to 1,100 nm as an excitation light source. We precompensated for dispersion of the broadband laser pulses caused by the optical elements, including the microscope objective lens, using a pair of chirped mirrors and a liquid-crystal spatial light modulator placed in the Fourier plane in a 4f pulse shaper configuration. The prechirped pulses were sent to a Michelson interferometer. The output from the interferometer was focused within a fluorescent sample through the microscope objective lens. The sample used was enhanced Green Fluorescent Protein (eGFP). In order to prepare unbleached fluorescent molecules for each delay time, the sample set in a glass container, was translated using a three-dimensional electrical stage. The exposure time for each delay time was 200 ms.

Figure 1a and b show the IAC traces for PB and TPE of eGFP, respectively. The IAC trace for PB was obtained from the reduction in the fluorescence signal due to PB during the 200 ms irradiation for each delay time. Dividing the IAC trace for PB by that for TPE, we have a trace shown in Fig. 1c, in which the peak to the background ratio is 2:1 indicating that it is one-photon ESA process. Figure 1d shows the result of the ESA spectrum for eGFP obtained by Fourier transforming the IAC trace to the spectral domain and by dividing it by the reference fundamental spectrum, together with the TPE spectrum.

The ESA spectrum has a peak at around 700 nm. In a separate experiment, we found that this absorption corresponds to the $T_1 \rightarrow T_2$ transition after intersystem crossing.

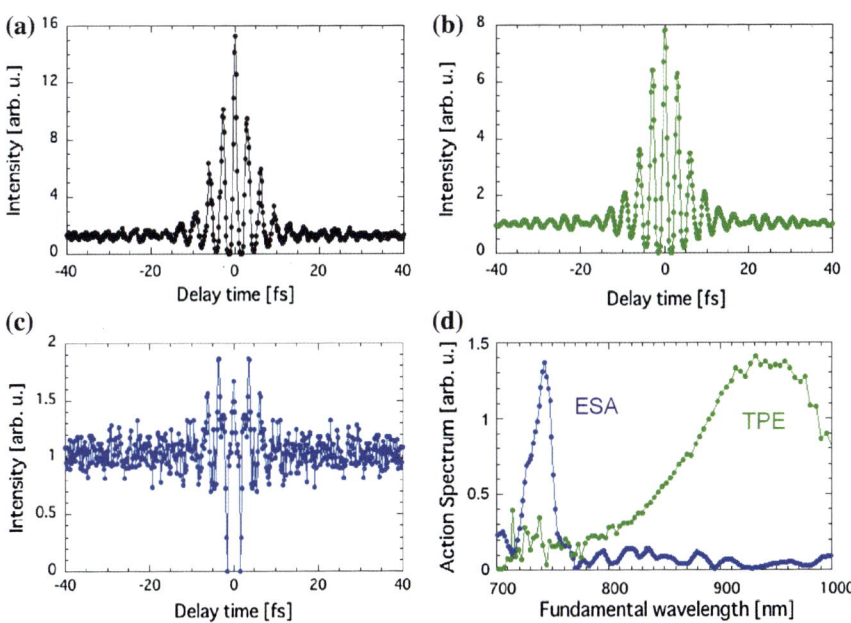

Fig. 1 IAC traces for **a** PB, **b** TPE, and **c** ESA of eGFP, and **d** The corresponding TPE and ESA spectra

4 Conclusions

The action spectrum for the PB of a fluorescent protein was observed by using nonlinear Fourier-transform spectroscopy. We found that sequential three-photon excitation process, one-photon ESA following two-photon excitation from the ground state, is one of the routes to the PB.

Acknowledgements We would like to thank Drs. K. Isobe, K. Midorikawa, and A. Miyawaki of RIKEN, and Prof. F. Kannari of Keio Univ. for their support and cooperation during the work. This work was supported in part by the Grants-in-Aid for Scientific Research (No. 24560052) and the Strategic Research Foundation Grant-aided Project for Private Universities, both from the MEXT, Japan.

References

1. W. R. Zipfel, R. M. Williams and W. W. Webb, *Nat. Biotec.* 21, 1369-1377 (2003).
2. H. Hashimoto, K. Isobe, A. Suda, F. Kannari, H. Kawano, H. Mizuno, A. Miyawaki, and K. Midorikawa, *Appl. Opt.* 49, 3323-3329 (2010).

Femtosecond Vibrational Spectroscopic Study on Photoexcitation Dynamics of DNO-Bound Myoglobin

Taegon Lee, Seongchul Park and Manho Lim

Abstract Time-resolved vibrational spectra of DNO-bound myoglobin showed instantaneous bleach that decays on a picosecond time scale, suggesting that most of the photoexcited MbDNO undergoes picosecond geminate rebinding of DNO to Mb after its immediate deligation.

OCIS codes (300.6340) General · (350.5130) General [8-pt. type. For codes, see www.opticsinfobase.org/submit/ocis.]

1 Introduction

Nitric oxide (NO) plays various important physiological roles. Recently, nitroxyl (HNO), a reduced form of NO, has been proposed to be an immediate precursor of NO and an intermediate in the denitrification process [1]. HNO was also found to play various biologically important roles [2]. In particular, the reactivity of HNO with various hemes was proposed to be physiologically important [1].

Although HNO exists as a dimer in aqueous solution that decomposes to N_2O and H_2O rapidly, it reacts with the ferrous atom of myoglobin (Mb) to form a very stable HNO-bound ferrous Mb (MbHNO). Thus MbHNO has been intensively studied using various spectroscopic method [3], which elucidated the structures of HNO moiety and the porphyrin ring in MbHNO reasonably well. However, the binding dynamics of HNO to heme proteins has hardly been investigated. Only a recent experiment showed that HNO with Mb was generated when it was excited by a visible pulse and about 90 % of the photodeligated ligand geminately rebinds in 70 µs. The reported geminate rebinding (GR) rate of HNO to Mb is remarkably slow compared with the picosecond GR of NO to Mb [4]. Here, we report

T. Lee · S. Park · M. Lim (✉)
Department of Chemistry and Chemistry Institute for Functional Materials,
Pusan National University, Busan 609-735, Korea
e-mail: mhlim@pusan.ac.kr

© Springer International Publishing Switzerland 2015
K. Yamanouchi et al. (eds.), *Ultrafast Phenomena XIX*,
Springer Proceedings in Physics 162, DOI 10.1007/978-3-319-13242-6_134

547

femtosecond vibrational spectra of photoexcited MbDNO in D_2O solution at 294 K with a 575-nm pulse that evolves on the picosecond time scale. The picosecond evolution of the bands was used to estimate the GR kinetics of DNO to Mb.

2 Experimental Methods

Femtosecond vibrational spectrometer was described elsewhere [4]. Briefly, a 575 nm visible pump pulse with 3 µJ of energy and a 1,280–1,450 cm^{-1} mid-IR probe pulse were generated by two optical parametric amplifiers, pumped by a Ti:S amplifier at a repetition rate of 1 kHz. The probe pulse was dispersed by the 120 l/mm grating of a 320 mm monochromator and detected with a 64-element-N_2(l)-cooled HgCdTe array detector. Instrument response function was about 0.3 ps.

A 15 mM metMb solution was prepared using 200 mM carbonate buffer in D_2O solution (pD = 9.6). The solution was bubbled with N_2 gas to remove oxygen. After preparing $NaNO_2$ in pD 9.6 buffer and $NaBH_4$ in 1 N NaOD, 20 equivalents of $NaNO_2$ and 60 equivalents of $NaBH_4$ were added to the metMb solution in sequence under N_2 to produce MbDNO [3]. Once MbDNO was generated, the solution was immediately diluted into 200 mM phosphate buffer with pD = 7.4 and concentrated using an Amicon Ultra-15 filter. The concentrated sample was loaded into a sample cell with a 30 µm-thick Teflon spacer. Temperature of the sample cell was maintained at 294 ± 1.

3 Results and Discussion

Protein samples have strong absorption bands in our spectral range of interest (amide II at 1,450–1,580 cm^{-1} and amide III at 1,200–1,350 cm^{-1}). Therefore, the N–O stretching (v_{N-O}) mode of MbDNO is expected to be buried between two strong amide bands. To effectively assign the vibrational absorption bands related to the v_{N-O} mode, time-resolved spectra of both MbDNO and MbD^{15}NO were collected. To remove conformational band not related to the NO band, time-resolved spectra of MbCO were also obtained and subtracted from transient spectra of MbDNO.

Figure 1 shows the treated spectra of MbDNO and MbD^{15}NO. The treated spectra show features that are minimally affected by thermal and conformational relaxation of the photoexcited protein except the spectral features related to the N atom. When the two treated spectra of MbDNO and MbD^{15}NO are compared, most of the features shift, except an absorption band near 1,338 cm^{-1}. The non-shifting band at 1,338-cm^{-1} likely arises from the band change of the histidine residues, which is not fully compensated by subtracting the MbCO spectrum. Clearly, DNO-related bands are negative features (bleaches) near 1,399, 1,385, and 1,368 cm^{-1} in

Fig. 1 Representative treated spectra of MbDNO (*red*) and MbD^{15}O (*blue*) (see text)

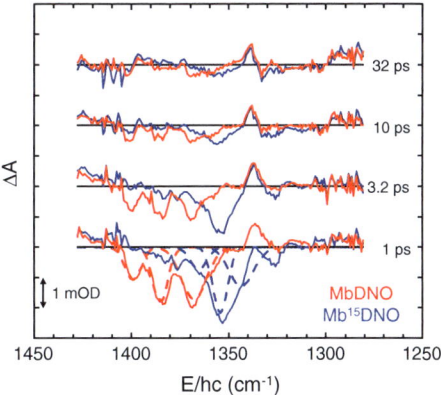

the treated spectra of MbDNO and a broad bleach near 1,353 cm^{-1} in the treated spectra of MbD^{15}NO.

The bleach in the treated spectra, showing instantaneous bleach that decays on a picosecond time scale, was described by three vibrational bands, each modeled with a Gaussian function. The bleach in the treated spectra of MbDNO was composed of three bands at 1,399, 1,385, and 1,368 cm^{-1} with relative intensities of 0.7, 1, and 1.6, respectively. In general, when the spectrum of an isotope-labeled molecule is compared with that of the unlabeled molecule, the vibrational band related to the isotope-labeled atom shifts according to the reduced mass change of the isotope. Thus, the bleach in the treated spectra of MbD^{15}NO was also described by three bands. The fitted bands for MbD^{15}NO peaked at 1,366, 1,354, and 1,343 cm^{-1} with relative intensities of 0.4, 1, and 0.8, respectively. The decay kinetics of three bands turned out to be about the same and thus, we treated the decay kinetics of three bands globally in the final fitting. The decay kinetics was well described by 0.92 exp $(-t/4$ ps$)$ + 0.08.

The fitted band positions imply that the bands at 1,399, 1,385, and 1,368 cm^{-1} shift by 33, 31, and 25 cm^{-1} upon the ^{15}N isotope substitution in MbDNO. The band at 1,385 cm^{-1} that shifts to 1,354 cm^{-1} upon ^{15}N labeling is consistent with the v_{N-O} mode of HNO in MbHNO observed by Raman spectroscopy [5]. The position of the v_{N-O} mode in MbDNO was calculated to be almost the same as that of MbHNO [6]. The remaining two bands at 1,399 and 1,368 cm^{-1} have not been observed before but they are clearly related to the vibrational band involving the N atom in MbDNO.

Based on the transient spectra of MbD^{15}NO as well as vibrational wavenumber calculation using the *ab initio* calculation of the vibrational frequency on a DNO-bound model heme, three bands were attributed to Fermi interaction between the strong v_{N-O} mode and weak overtone and combination modes involving the N atom. The picosecond decay of the bleach likely arises from either photodeligation of DNO from MbDNO and ps GR of DNO to Mb or rapid thermal relaxation subsequent to rapid electronic relaxation of the photoexcited MbDNO, without

DNO deligation. The same appearance and decay in the conformational bands near 1,410 cm^{-1} in MbCO and MbDNO indicates that there is an instantaneous photodeligation of DNO and, thus, the ps decay in the ν_{N-O} band in MbDNO likely arises from GR of DNO to Mb. GR of DNO to Mb appears to be the fastest rebinding in heme proteins, which can be attributed to the reactivity and the bonding structure of the ligand.

References

1. B. A. Averill, "Dissimilatory nitrite and nitric oxide reductases," *Chem. Rev.* **96**, 2951-2964 (1996).
2. K. M. Miranda, "The chemistry of nitroxyl (HNO) and implications in biology," *Coord. Chem. Rev.* **249**, 433-455 (2005).
3. M. R. Kumar, J. M. Fukuto, K. M. Miranda, P. J. Farmer, "Reactions of HNO with heme proteins: New routes to HNO-heme complexes and insight into physiological effects," *Inorg. Chem.* **49**, (14), 6283-6292 (2010).
4. S. Kim, M. Lim, "Protein conformation-controlled rebinding barrier of NO and its binding trajectories in myoglobin and hemoglobin at room temperature," *J. Phys. Chem. B* **116**, 5819-5830 (2012).
5. C. E. Immoos, F. Sulc, P. J. Farmer, K. Czarnecki, D. F. Bocian, A. Levina, J. B. Aitken, R. S. Armstrong, P. A. Lay, "Bonding in HNO-myoglobin as characterized by X-ray absorption and resonance Raman spectroscopies," *J. Am. Chem. Soc.* **127**, 814-815 (2005).
6. D. P. Linder, K. R. Rodgers, "Structural, electronic, and vibrational characterization of Fe-HNO porphyrinates by density functional theory," *Inorg. Chem.* **44**, 8259-8264 (2005).

Part VII
Charge and Energy Transfer—Photovoltaic and Light Harvesting

Interpreting Oscillations in Numerically Exact Simulations of 2D Electronic Spectra

Daniele M. Monahan, Lukas V. Whaley-Mayda, Akihito Ishizaki and Graham R. Fleming

Abstract 2D electronic spectroscopy signals are simulated with accurate hierarchy method treatment of an electronic heterodimer coupled to a bath and local vibrations. We examine the effect of vibrations on the correspondence between exciton and population dynamics.

1 Introduction

The mechanisms of excitation energy transfer (EET) in biological systems depend on the interplay between pigment-pigment and pigment-bath interactions [1]. Strong electronic coupling between pigments creates delocalized excitons, helping to avoid trapping at local energy minima. Fluctuations in coupled pigment or protein nuclear modes act to destroy the site superpositions, localizing the excitation and halting reversible coherent energy transfer. Underdamped intramolecular vibrations also play a role, for example by allowing transfer via resonances in the vibronic energy landscape [2, 3]. In this work, we simulate 2D electronic spectra (2D-ES) of a heterodimer in a quantum environment, to study the correspondence between exciton and spatial dynamics in the intermediate coupling regime.

The lifetime of a superposition of excitons, generated by broadband laser pulses in 2D-ES, measures a system's support for delocalization despite environmental

D.M. Monahan · L.V. Whaley-Mayda · G.R. Fleming (✉)
Department of Chemistry, University of California, Berkeley CA 94720, USA
e-mail: grfleming@lbl.gov

D.M. Monahan · L.V. Whaley-Mayda · G.R. Fleming
Physical Biosciences Division, Lawrence Berkeley National Laboratory,
Berkeley CA 94720, USA

A. Ishizaki
Institute for Molecular Science, National Institutes of Natural Sciences,
Okazaki 444-8585, Japan

© Springer International Publishing Switzerland 2015
K. Yamanouchi et al. (eds.), *Ultrafast Phenomena XIX*,
Springer Proceedings in Physics 162, DOI 10.1007/978-3-319-13242-6_135

fluctuations only if the superposition is a mixture of states with different locations. In molecules with substantial electronic-vibrational coupling, two spectroscopically resolvable excitons may be largely confined on a single pigment [4, 5] (i.e. a vibrationally hot state and a cold state). Population transfer between such localized excitons does not measure the transport of energy through space; similarly, the lifetime of localized coherences cannot probe the localization timescale. We focus on this effect of intramolecular vibrations on our ability to study EET via 2D-ES, and on the EET mechanism. We pay special attention to the case of a vibration that is resonant with the electronic energy gap and the effect of the resulting vibronic resonances.

2 Model

We simulate the spectroscopic response of an electronic heterodimer with site energies 200 cm^{-1} apart and electronic coupling $J = 50$ cm^{-1}. Each site is coupled to an independent bath of overdamped harmonic oscillators, plus an underdamped vibration that dephases over 2 ps. The bath reorganization energy and correlation time, which describe the magnitude and rapidity of the bath's response to electronic excitation, are 35 cm^{-1} and 50 fs, respectively. The vibrational frequency is chosen near resonance with the gap between electronic eigenstates (e.g. $\omega = 223$ cm^{-1} for $J = 50$ cm^{-1}); the Huang-Rhys parameter, S, describes the strength of the vibrational-electronic coupling. We vary the Huang-Rhys factor, adjusting the bath reorganization energy, λ, to hold the total reorganization energy constant. The dynamics including the dissipation are accurately treated using the hierarchy equations of motion (HEOM) method of Ishizaki and Fleming [6]. We calculate the spectroscopic response in the form of 4-point time correlation functions and Fourier transform over the t_1 and t_3 periods to give simulated 2D-ES spectra as in Fig. 1a.

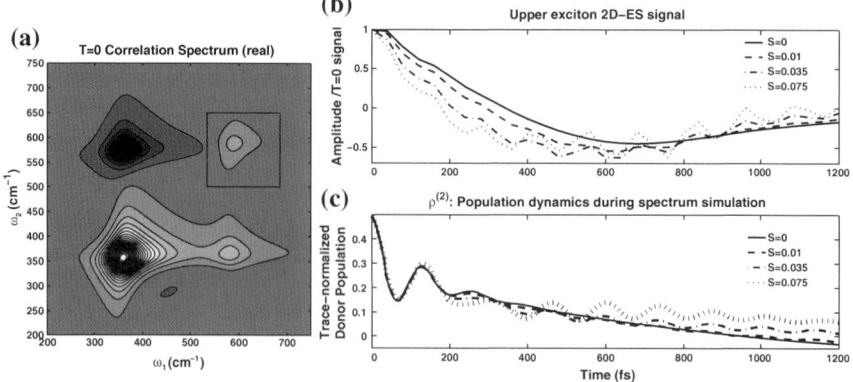

Fig. 1 a 2D-ES spectrum of electronic heterodimer simulated at 77 K, for $t_2 = 0$ fs. **b** 2D-ES signal integrated over the boxed area (*upper* diagonal peak), as a function of t_2 for given values of S. **c** diagonal element of $\rho^{(2)}$ corresponding to donor populations in the same simulations

In attempt to compare dynamics measured by these spectra with site population dynamics during the waiting time, we record diagonal elements of

$$\rho^{(2)}(t_2) = \int\limits_0^\infty d\omega_1 \int\limits_{-\infty}^\infty dt_1 e^{-i\omega_1 t_1} G(t_2)[G(t_1)[\rho(0),\mu],\mu]$$

where $G(t_i)$ describes time propagation calculated with the HEOM in the period t_i.

3 Results and Discussion

Our preliminary work on pure population dynamics demonstrates that the bath greatly damps mixed vibrational-electronic coherences, compared to dynamics without dissipation. This implies significant bath-induced energy gap fluctuations are characteristic of the parameter regime described, and that resonance between two vibronic excitons will be fragile. Thus we argue that the accurate quantum-mechanical treatment of dissipation is required to produce the spectroscopic phenomena typical in the intermediate coupling regime. We note that we have not included the low-temperature corrections in the HEOM, leading to breakdown of positivity in the equilibrium populations (though not affecting oscillatory behavior). We accordingly take care to examine only the initial decay rates in Fig. 1.

In the simulated 2D-ES, we note long-lived oscillations that depend on the size of the vibronic coupling *and* are accompanied by oscillatory transfer between pigments (as measured by $\rho^{(2)}$). The long-lived oscillations are significant for simulations with $S > 0.01$. The $\rho^{(2)}$ site dynamics are significantly damped after the first picosecond, but the 2D-ES oscillations persist. Figure 1b shows that increasing S raises the initial rate for exciton relaxation, we have seen a similar effect in site dynamics initialized with pure donor population. We do not observe such a rate enhancement in the initial decay of $\rho^{(2)}$ depicted in Fig. 1c. Future studies will aim to understand this observation.

4 Conclusions

This work illustrates that for most systems with vibrational modes with energy close to the excitonic energy gap, spectral oscillations at late times are the result of localized (vibrational) dynamics, and the lifetime of oscillations in spectra alone cannot indicate the timescale for localization. The environment destroys the mixed-character vibronic coherence (dominating the site dynamics at intermediate times) on a faster timescale than the vibrational dephasing. Under the influence of vibrations, the time for localization is longer than the pure electronic dephasing time, which may be due to vibronic resonances between pigments.

Acknowledgments This work was supported by the Director, Office of Science, Office of Basic Energy Sciences, of the USA Department of Energy under contract DE-AC02-05CH11231 and the Division of Chemical Sciences, Geosciences and Biosciences Division, of Basic Energy Sciences through grant DE-AC03-76F000098 (at LBNL and UC Berkeley), and by Grants-in-Aid for Scientific Research (Grant Number 25708003) from the Japan Society for the Promotion of Science. D.M. thanks the NSF Graduate Research Fellowship Program.

References

1. A. Ishizaki and G. R. Fleming, *Annu. Rev. Condens. Matter Phys.*, 3, 333–361, 2012.
2. A. Kolli *et al., J. Chem. Phys.*, 137, 174109, 2012.
3. J. M. Womick and A. M. Moran, *J. Phys. Chem. B*, 115, 1347–1356, 2011.
4. V. Tiwari, *et al., Proc. Natl. Acad. Sci. U.S.A.*, 110, 1203–1208, 2013.
5. Y.-C. Cheng and G. R. Fleming, *J Phys Chem A*, 112, 4254–4260, 2008.
6. A. Ishizaki and G. R. Fleming, *J. Chem. Phys.*, 130, 234111, 2009.

Coherent Ultrafast Charge Transfer in an Organic Photovoltaic Blend

Antonietta De Sio, Sarah M. Falke, Carlo A. Rozzi, Daniele Brida, Margherita Maiuri, Michele Amato, Ephraim Sommer, Angel Rubio, Giulio Cerullo, Elisa Molinari and Christoph Lienau

Abstract Combining high-time resolution pump-probe spectroscopy and time-dependent density functional theory calculations, we show that coherent vibronic coupling is of key importance in triggering charge transfer in a technologically relevant organic photovoltaic blend.

A. De Sio (✉) · S.M. Falke · E. Sommer · C. Lienau
Institut für Physik, Carl von Ossietzky Universität, 26129 Oldenburg, Germany
e-mail: antonietta.de.sio@uni-oldenburg.de

A. De Sio · S.M. Falke · E. Sommer · C. Lienau
Center of Interface Science, Carl von Ossietzky Universität,
26129 Oldenburg, Germany

C.A. Rozzi · E. Molinari
Istituto Nanoscienze—CNR, Centro S3, via Campi 213a, 41125 Modena, Italy

D. Brida · M. Maiuri · G. Cerullo
IFN-CNR, Dipartimento di Fisica, Politecnico di Milano, 20133 Milan, Italy

D. Brida
Department of Physics and Center for Applied Photonics,
University of Konstanz, 78457 Konstanz, Germany

M. Amato
Institut d'Électronique Fondamentale, UMR8622, CNRS,
Universitè Paris-Sud, 91405 Orsay, France

A. Rubio
Nano-Bio Spectroscopy Group and ETSF Scientific Development Centre,
Dpto. Física de Materiales, Universidad del País Vasco,
Centro de Física de Materiales CSIC-UPV/EHU-MPC and DIPC,
Av Tolosa 72, 20018 San Sebastián, Spain

A. Rubio
Fritz-Haber-Institut der Max-Planck-Gesellschaft, 14195 Berlin, Germany

© Springer International Publishing Switzerland 2015
K. Yamanouchi et al. (eds.), *Ultrafast Phenomena XIX*,
Springer Proceedings in Physics 162, DOI 10.1007/978-3-319-13242-6_136

1 Introduction

Polymer:fullerene blends are reference material systems for organic solar cells [1, 2]. Light-induced charge transfer from the photoexcited polymer donor to the fullerene acceptor represents the key process in organic photovoltaic (OPV) devices [3, 4]. It is now accepted that much of the initial, non-diffusive charge separation occurs on an ultrafast, sub-100 fs time scale. However, still very little is known about the initial quantum dynamics of this process in technologically relevant OPV materials. Here we study the ultrafast optical response of a model OPV material system, the conjugated polymer *poly-3-hexylthiophene* (P3HT) and its blend with the fullerene derivative *[6, 6]-phenyl-C61 butyric acid methyl ester* (PCBM), combining high-time resolution pump-probe spectroscopy and time-dependent density functional theory (TDDFT) simulations. Our experimental and theoretical results provide strong evidence that the driving mechanism in the primary steps of the current photogeneration is a quantum-correlated wavelike motion of electrons and nuclei on a timescale of few tens of femtoseconds [5, 6].

2 Results

We have performed ultrafast spectroscopic studies on P3HT:PCBM films, as well as on the single components, using a two-color pump-probe spectrometer providing independently tunable pulses [7]. The overall time resolution of the setup is better than 15 fs. Pump pulses centered at 540 nm resonantly excite the polymer, whereas the transient absorption is monitored by broadband probe pulses in the blue-to-green wavelength region. For both the pristine P3HT and the P3HT:PCBM blend (Fig. 1a), the transient absorption map shows pronounced temporal oscillations throughout the entire range of probe wavelengths [5]. Fourier transform analysis (Fig. 1b, c) reveals that such oscillations in the 500–520 nm probe wavelength range correspond to the symmetric C=C stretching mode of the polymer at 1450 cm^{-1} in both samples [5]. For shorter probe wavelengths, we detect additional oscillatory components in the blend sample. In particular, we observe vibrations at 1470 cm^{-1} and at 1289 cm^{-1}. Both these frequencies correspond to vibrational modes of the fullerene C_{60} [8]. Control experiments on PCBM films [5] rule out direct excitation of the acceptor as a cause for the observed fullerene oscillations. Thus we assign the positive differential transmission signal observed in the 460–470 nm wavelength region (Fig. 1a) in the P3HT:PCBM sample as a stimulated emission from the fullerene. The observation of pronounced temporal oscillations in this spectral region with a period of about 23 fs is hard to explain in the framework

E. Molinari
Dipartimento di Scienze Fisiche, Matematiche e Informatiche,
Università di Modena e Reggio Emilia, via Campi 213a, 41125 Modena, Italy

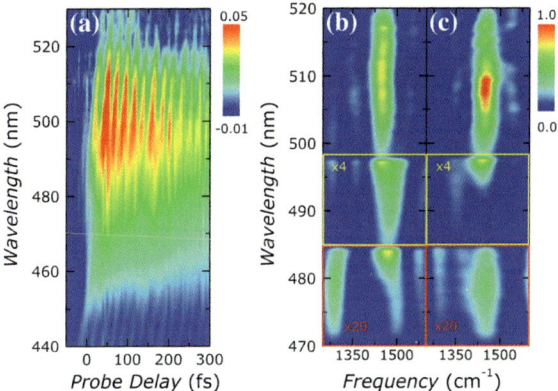

Fig. 1 Differential transmission map (**a**) of the P3HT:PCBM blend after impulsive excitation of the polymer with sub-10 fs pump pulses centered at 540 nm. Fourier transform maps of the P3HT: PCBM blend (**b**) and of a pristine P3HT (**c**) film. Adapted from [5]

of an incoherent charge transfer model. Instead these results provide evidence for a coherent charge transfer mediated by strong vibronic coupling between polymer and fullerene.

TDDFT simulations were performed on the simplest possible model of the blend [5]. Periodic boundary conditions were imposed in order to mimic the experimental configuration. After impulsive photoexcitation of the donor moiety, we observe temporal oscillations of the charge transfer probability to the fullerene with a period of about 25 fs (Fig. 2a). This approximately matches the vibrational period observed in the experiments. The time-dependent Kohn-Sham eigenvalues, representing the donor and acceptor lowest unoccupied molecular orbitals (LUMOs), undergo pronounced oscillations in anti-phase with each other (Fig. 2b, solid). At each crossing of the two LUMOs, the charge density is free to move through the

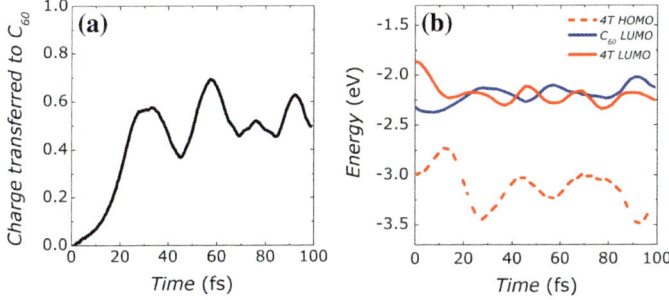

Fig. 2 Simulated charge transfer dynamics (**a**) in the simplest possible model of the blend. Time-dependent Kohn-Sham eigenvalues (**b**). The *red lines* refer to the 4T HOMO (*dashed*) and 4T LUMO (*solid*) levels, whereas the *green lines* to the C_{60} LUMO. Adapted from [5]

system. The charge flow is suppressed when the levels are energetically detuned. This explains the coherent charge oscillations (Fig. 2a). TDDFT simulations thus predict that vibronic coupling is necessary for charge transfer to occur [5].

3 Conclusions

We show that vibronic coupling between electronic and nuclear degrees of freedom promotes delocalization of the photoexcited electronic wave packet across the interface. Both the electronic density and the nuclei display correlated oscillations on the same time scales, which are essential for ultrafast charge transfer from the donor to the acceptor. The observation of coherent electron-nuclear motion in this system is strong evidence for the dominant role of quantum coherences in the early stages of the charge transfer dynamics in this class of OPV materials.

References

1. G. Yu, J. Gao, J. C. Hummelen, F. Wudl, and A. J. Heeger, Science 270, 1789-1791 (1995).
2. W. Ma, C. Yang, X. Gong, K. Lee, and A. J. Heeger, Adv. Funct. Mater. **15**, 1617-1622 (2005).
3. C. J. Brabec, G. Zerza, G. Cerullo, S. De Silvestri, S. Luzzati, J. C. Hummelen, and S. Sariciftci, Chem. Phys. Lett. **340**, 232-236 (2001).
4. J. Guo, H. Ohkita, H. Benten, and S. Ito, J. Am. Chem. Soc. **132**, 6154-6164 (2010).
5. S. M. Falke, C. A. Rozzi, D. Brida, M. Maiuri, M. Amato, E. Sommer, A. De Sio, A. Rubio, G. Cerullo, E. Molinari, and C. Lienau, Science **344**, 1001-1005 (2014).
6. C. A. Rozzi, S. M. Falke, N. Spallanzani, A. Rubio, E. Molinari, D. Brida, M. Maiuri, G. Cerullo, H. Schramm, J. Christoffers, and C. Lienau, Nat. Commun. **4**, 1602 (2013).
7. C. Manzoni, D. Polli, and G. Cerullo, Rev. Sci. Instrum. **77**, 023103 (2006).
8. S. Falke, P. Eravuchira, A. Materny, and C. Lienau, J. Raman Spectr. **42**, 1897-1900 (2011).

Ultrafast Energy and Charge Transfer Processes in a Flexible Molecular Triad Designed for Organic Photovoltaics

T. Roland, L. Liu, E. Heyer, A. Ruff, S. Ludwigs, R. Ziessel and S. Haacke

Abstract A detailed spectro-temporal analysis of the ultrafast transient absorption and fluorescence signals allows deciphering multiple energy and charge transfer processes in a light-harvesting molecular triad designed as photo-sensitizing unit featuring a novel BODIPY compound.

1 Introduction

Replacing the standard electron-donating P3HT polymer in organic solar cells made of blends with the acceptor PCBM is a very active field in organic material chemistry, with the specific aim to harvest photons in the red or near-IR part of the spectrum. BODIPY (boron-dipyrromethene) is a widely studied dye with extinction coefficients larger than 100,000 $M^{-1}cm^{-1}$, which offers a versatile chemical platform for the tuning of the optical properties. A solar cell based on a thienyl-BODIPY/PCBM blend was recently reported with a remarkable 4.7 % power conversion efficiency [1]. Here, we study a BODIPY dye conjugated with styryl and polyoxyethylene groups that shift the absorption spectrum into the 600–680 nm range. Together with a diketopyrrolopyrrole (DPP) and a triphyenylamine (TPA) unit, a molecular triad (depicted in Fig. 1a) is formed, whose extinction coefficient

T. Roland · L. Liu · S. Haacke (✉)
Institut de Physique et Chimie des Matériaux de Strasbourg,
Strasbourg University—CNRS, Strasbourg, France
e-mail: stefan.haacke@ipcms.unistra.fr

E. Heyer · R. Ziessel
Laboratoire de Chimie Organique et Spectroscopies Avances (ICPEES-LCOSA),
Strasbourg University—CNRS, Strasbourg, France
e-mail: ziessel@unistra.fr

A. Ruff · S. Ludwigs
Institute of Polymer Chemistry, University of Stuttgart, Stuttgart, Germany

© Springer International Publishing Switzerland 2015
K. Yamanouchi et al. (eds.), *Ultrafast Phenomena XIX*,
Springer Proceedings in Physics 162, DOI 10.1007/978-3-319-13242-6_137

561

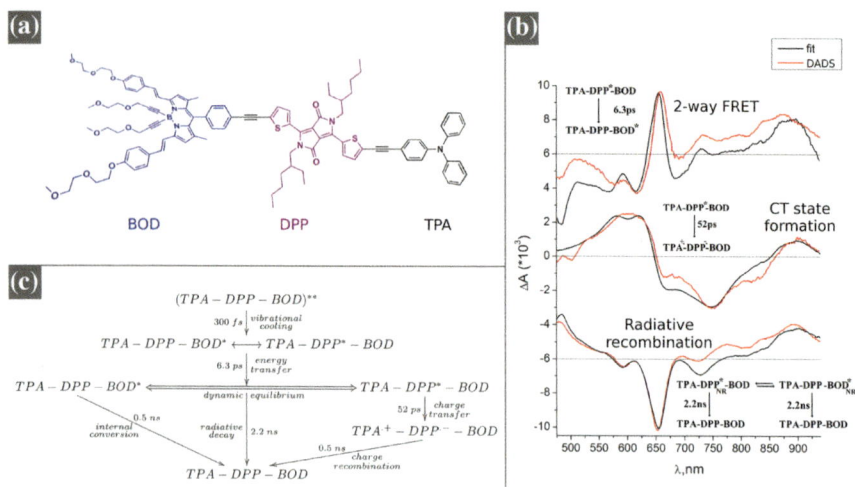

Fig. 1 **a** Sketch of the molecular triad. **b** Selected DAS and their fit. **c** Reaction scheme deduced from the DADS analysis

fully spans from 300 to 700 nm. The aim of our combined fluorescence and transient absorption (TA) studies is to pinpoint the ultrafast light-induced processes in terms of nature, time scales and efficiency.

2 Results

The static absorption of the triad matches well the algebraic sum of both TPA-DPP-TPA and BOD minus that of TPA at wavelengths longer than 500 nm, indicating small electronic coupling between DPP and BOD.

The spectral signature of radical anions and cations were characterised by spectrovoltametry on the isolated BOD, TPA-DPP-TPA and TPA compounds and on the full triad. The TPA radical cation has a broad peak centered at 730 nm, the BOD radical anion shows a main peak centered at 595 nm and a smaller peak at 455 nm, the DPP radical anion peak is centered at 753 nm (data not shown).

Upon excitation at 320 nm, the time-resolved fluorescence (measured by a streak camera with 10 ps resolution) displays a mono-exponential decay for the separated components TPA, TPA-DPP-TPA and BOD-TPA, of 1.5, 2.2 and 4.5 ns, respectively. In contrast, the triad shows a three-exponential decay. A detailed analysis of the decay-associated spectra (DAS) shows that approximately 70 % of excited DPP* quenches in 63 ps, 80 % of BOD* in 0.5 ns and the remainder of both species fluoresce with a lifetime close to the isolated unquenched reference molecules. This multi-exponential decay is a clear indication for structural heterogeneity of the triads, leading to these quenched and unquenched sub-populations.

Transient absorption (TA) spectroscopy was first performed on the separated compounds, giving access to the characteristic differential spectra of the excited species DPP^* and BOD^*.

A global fit with five lifetimes allows to cast the femtosecond TA data into a small set of DAS (3 are shown in Fig. 1b, red). These can be reconstituted by linear combinations of the excited state and radical ion-associated difference spectra of all three molecular components (Fig. 1b, black). This approach reveals that after a fast vibrational cooling, an out-of-equilibrium mixture of $TPA-DPP-BOD^*$ and $TPA-DPP^*-BOD$ is observed, that reaches its equilibrium within 6.3 ps through energy transfer processes. From this state, three paths can be defined: one that leads in 52 ps to the formation of a (TPA^+/DPP^-) intramolecular charge transfer state that recombines within 0.5 ns. The 0.5 ns BOD* quenching does not lead to a detectable photoproduct and is thus assigned to internal conversion [2, 3]. The last decay path is attributed to the decay of the equilibrated mixture of DPP^* and BOD^*, in 2.2 ns. This dynamic is reported in the reaction scheme (Fig. 1c).

3 Conclusions

Despite multiple conformations and reaction pathways the flexible TPA-DPP-BOD triad shows very efficient photo-sensitizing properties featuring sub-10 ps energy transfer to BOD and ~ 50 ps formation of a CT state. When blended with PCBM, efficient charge transfer from BOD and DPP is expected, as observed for other BOD [2] and DPP [4] derivatives. CT formation would then also quench the species involved in the longer-lived non-reactive pathways (0.5 ns) reported here for the isolated triad. Further work is in progress, and a detailed publication is under completion [5].

References

1. Bura T. et al, "High-performance solution-processed solar cells and ambipolar behavior in organic field-effect transistors with thienyl-BODIPY scaffoldings". J. Am. Chem. Soc. **134**, 17404–17407 (2012).
2. R. Ziessel, et al., "Selective triplet-state formation during charge recombination in a fullerene/Bodipy molecular dyad (Bodipy = borondipyrromethene)". Chem. Eur. J. **15**, 7382–7393 (2009).
3. R. P. Sabatini, et al., "Intersystem Crossing in Halogenated Bodipy Chromophores Used for Solar Hydrogen Production". J. Phys. Chem. Lett. **2**, 223–227 (2011).
4. B. P. Karsten, et al., "Charge separation and (triplet) recombination in diketopyrrolopyrrole-fullerene triads". Photochem. & Photobiol. Sci. **9**, 1055–1065 (2010).
5. T. Roland, et al., "A detailed analysis of multiple photo-reaction kinetics in a molecular TPA/DPP/BODIPY triad with overlapping spectra by ultrafast spectroscopy', J. Phys. Chem. C, **118**, 24290–24301 (2014), http://dx.doi.org/10.1021/jp507474r

Ultrafast Electron and Hole Dynamics in Novel Conjugated Star-Shaped Molecules

Oleg V. Kozlov, Yuriy N. Luponosov, Sergei A. Ponomarenko, Dmitry Yu. Paraschuk, Yoann Olivier, Jérôme Cornil, Nina Kausch-Busies and Maxim S. Pshenichnikov

Abstract Charge dynamics in organic photovoltaic blends based on novel star-shaped molecules are studied by ultrafast visible-IR spectroscopy. Pathways of intra-and intermolecular electron and hole transfer and their recombination are identified and discussed.

1 Introduction

Bulk heterojunction organic solar cells (OSCs) based on small molecules (SMs) have recently attracted much attention as a promising alternative to conventional polymer-based OSCs, with the latest sunlight-to-power efficiency of ~ 10 % [1]. SMs-based OSCs combine advantages of polymer OSCs with benefits of small molecules like excellent batch-to-batch reproducibility, well-defined molecular structure and molecular weight, easy mass-scale production, etc. However, because

O.V. Kozlov · M.S. Pshenichnikov (✉)
Zernike Institute for Advanced Materials, University of Groningen,
Groningen, The Netherlands
e-mail: m.s.pchenitchnikov@rug.nl

O.V. Kozlov · D.Yu.Paraschuk
International Laser Center and Faculty of Physics, Moscow State University,
Moscow, Russian Federation

Y.N. Luponosov · S.A. Ponomarenko
Institute of Synthetic Polymeric Materials of the Russian Academy of Science,
Moscow, Russian Federation

Y. Olivier · J. Cornil
Service de Chimie des Materiaux Nouveaux, Université de Mons, Mons, Belgium

N. Kausch-Busies
Conductive Polymers Division, Heraeus Precious Metals GmbH & Co. KG,
Leverkusen, Germany

© Springer International Publishing Switzerland 2015
K. Yamanouchi et al. (eds.), *Ultrafast Phenomena XIX*,
Springer Proceedings in Physics 162, DOI 10.1007/978-3-319-13242-6_138

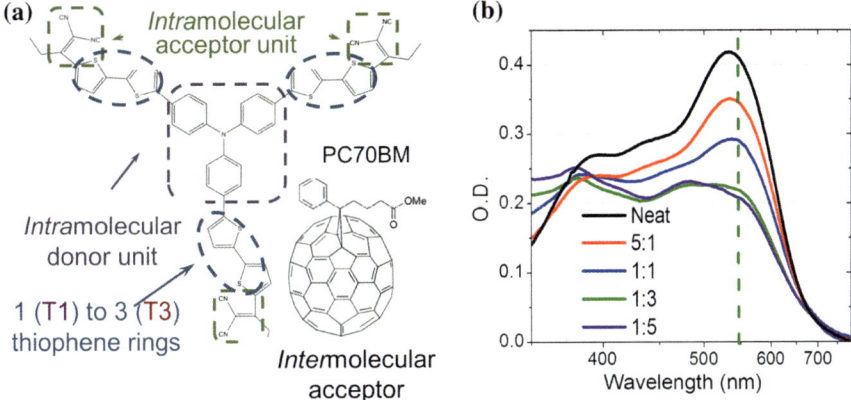

Fig. 1 **a** Structure of the star-shaped molecules [2] and the fullerene acceptor. **b** Representative absorption spectra of T2-based blends with different donor:acceptor weight content. The wavelength of the pump is shown by *vertical dashed line*

of the novelty of such OSCs, the ultrafast photophysics of initial photogeneration of charges and their subsequent recombination is barely known.

In this contribution, charge generation and recombination processes are studied in photovoltaic blends based on a series of novel star-shaped small molecules [2] as donors and PC70BM as acceptor (Fig. 1a). The conjugation length of the star arms is systematically varied by including one, two, or three thiophene rings. Charge dynamics are investigated by ultrafast photoinduced absorption (PIA) spectroscopy, using a visible pump to mimic sun photons and broadly-tunable IR probe to monitor concentration of hole polarons at the SM's conjugated system.

2 Experimental and Results

Blend films were prepared by spin-casting from solutions on a glass substrate. For the pump–probe experiments, the wavelength of the pump pulse was set close to the maximal absorption of the blends at 550 nm (Fig. 1b). Figure 2a shows polaron spectra [3] of the molecules in the IR region. A red-shift and narrowing of the spectra are observed with elongation of the conjugated arm. We choose the probe wavelength near the polaron absorption maxima and consider the PIA response to be proportional to the concentration of photogenerated charges.

Figure 2b shows the PIA transients for the T1:PC70BM blends. In all samples, a fast build-up of the PIA response is observed within the apparatus temporal resolution indicating almost instantaneous charge separation. In the pristine films, PIA decays biexponentially, which suggests charge recombination via two independent channels. The contribution of the fast ($\sim$100 ps) component depends strongly on the PC70BM content and vanishes completely at high fullerene concentrations.

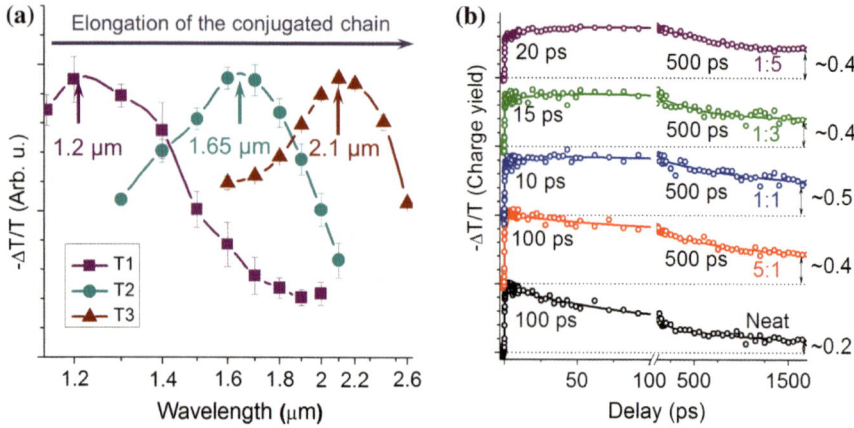

Fig. 2 **a** Normalized polaron spectra of thin films at 3 ps pump-probe delay. **b** PIA isotropic transients for films of T1:PC70BM blends with different donor:acceptor concentrations (indicated). Transient amplitudes are normalized by film absorption for direct comparison of the amount of generated charges per absorbed photon. Experimental data and biexponential global fits are shown by *open circles* and *solid lines*, respectively; the relaxation times are indicated next to the transients. The share of long-lived charges is indicated to the right of the transients. Similar transients were obtained for T2 and T3 molecules

We assign this decay to *inter*molecular ("interstar") recombination of charges initially separated onto neighboring molecules. The slow, ∼500 ps decay component is also observed in diluted solutions where the intermolecular interactions are negligible. This component is assigned to *intra*molecular recombination of the charges separated within the same molecule [4], similarly to the solution case.

In the blends with high PC70BM content (more than 50 %), the initial fast decay is replaced by the ingrowing component with a characteristic time of ∼3–20 ps (Fig. 2b) which amplitude and time increase with PC70BM concentration. Taking into consideration substantial PC70BM absorption, we attribute this growth to exciton migration within PC70BM domains with its subsequent dissociation at the PC70BM:SM interface, i.e. to the hole transfer process. Nevertheless, the maximal amplitude of the PIA decreases with increase of the PC70BM content. This means that not all PC70BM excitons reach the interface which points towards existing of large PC70BM domains with sizes exceeding the exciton diffusion length of ∼10 nm.

The process of charge generation and recombination are identified as follows. In pristine films, there are two recombination pathways after charge creation: the *inter*star one where the charges at two different star-shaped molecules are involved, and the *intra*star one where the charges are generated and recombined within the same molecule. When PC70BM is added to the blends, the electron is transferred to PC70BM thereby making the interstar recombination channel inefficient. At the same time, the new hole transfer channel opens due to direct excitation of PC70BM molecules; both processes result in the long-lived separated charges.

The share of long-lived charges for different PC70BM concentrations is indicated in Fig. 2b. A > 50 % survival probability of charges demonstrates that the star-shaped donors do provide efficient charge separation in blends with PC70BM which makes them prospectively promising donor materials in OSCs. The ways for further efficiency increase include modification of the donor structure to suppress intermolecular charge trapping, and optimization of the blend morphology by decreasing the PC70BM clusters size. We believe that all these result in an increase of ~5 % efficiency already achieved in the 2T:PC70BM blends [2].

OVK acknowledge support by "Aurora—Towards Modern and Innovative Higher Education" program, and by RFBR, research project №14-02-31632.

References

1. Y. Liu *et al.*, Sci. Rep. **3** (2013).
2. J. Min *et al.*, Advanced Energy Materials **4**, 1301234 (2014).
3. X. Wei *et al.*, Physical Review B **53**, 2187 (1996).
4. E. Ripaud *et al.*, Journal of Physical Chemistry B **115**, 9379 (2011).

Photoinduced Charge Transfer Occurs Naturally in DNA

D.B. Bucher, B.M. Pilles, T. Carell and W. Zinth

Abstract We show by femtosecond IR-spectroscopy that excited states in oligonucleotides decay with high yields by charge transfer to delocalized charged radicals. For the 6-4 lesion, charge transfer protects the DNA from Dewar formation.

1 Introduction

Charge transfer in DNA has been extensively investigated in the past decade in special DNA constructs [1]. Mainly motivated by DNA electronics or by trying to understand oxidative damage, the molecular mechanism was investigated by photoinduced charge injection [2]. In many experiments, excitation of artificial DNA bases or intercalating chromophores initiated the charge transport, which was followed by time resolved UV/Vis-spectroscopy.

Recently, we could directly show by UV-pump/IR-probe spectroscopy that photoinduced charge transfer is also an inherent process in natural DNA strands [3]. We observe long-living charge separated states in high yields in oligonucleotides after photo-excitation. This is in strong contrast to the situation found in single nucleotides, where the excited state decays ultrafast to the ground state, thus impeding photochemical damage formation. Base stacking in the biologically important oligonucleotides opens a new decay channel, which generates charge separated states. The charge is delocalized along stacked domains and recombines on the 100 ps time scale. The formation of reactive charged radicals along DNA

D.B. Bucher · B.M. Pilles · W. Zinth (✉)
BioMolecular Optics and Center for Integrated Protein Science,
Ludwig-Maximilians-Universität München, Oettingenstr 67, 80538 Munich, Germany
e-mail: wolfgang.zinth@physik.uni-muenchen.de

D.B. Bucher · T. Carell
Center for Integrated Protein Science at the Department of Chemistry,
Ludwig-Maximilians-Universität München, Butenandtstr 5-13, 81377 Munich, Germany

© Springer International Publishing Switzerland 2015
K. Yamanouchi et al. (eds.), *Ultrafast Phenomena XIX*,
Springer Proceedings in Physics 162, DOI 10.1007/978-3-319-13242-6_139

strands might induce chemical reactions—such as damage formation or repair—currently not considered in DNA photochemistry. Charge transfer is not only observed in intact DNA strands. Excitation of the 6-4 chromophore, a photo lesion between two adjacent pyrimidine bases, normally leads with high yields to the Dewar-lesion, a secondary photoproduct in DNA [4]. Our time-resolved experiments show that Dewar formation is efficiently quenched in stranded DNA by charge transfer with a neighboring base.

In all of these ultrafast experiments, we took advantage of the fingerprint spectra in the mid-IR which enables us not only to distinguish between different nucleobases, but also to identify the charge transfer states.

2 Results and Discussion

The investigated oligonucleotides contain three nucleobases, 2′-deoxyuridine (U), 2′-deoxyadenosine (A), 5-methyl-2′ deoxycytidine (mC). The red shifted absorbance of mC in comparison to A and U allows selective excitation with UV-light at 295 nm. In the IR, the contribution of each base in the excited state can be deduced from characteristic marker bands at 1,625 cm^{-1} (A), 1,655 cm^{-1} (U) and 1,667 cm^{-1} (mC). This approach is illustrated by using the trinucleotide mCAU (Fig. 1a). Although mC is selectively excited, the U and even the A bands show a

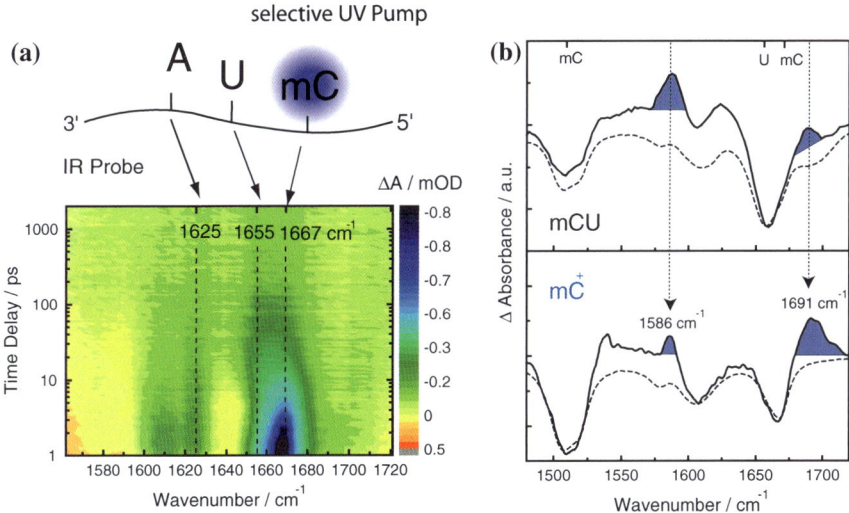

Fig. 1 **a** Selective excitation of mC in the trinucleotide mCAU leads to a bleach of all three bands (see *marker bands*) which recovers on a 100 ps time scale. **b** Decay spectra of the 100 ps living states of the dinucleotide mCU shows characteristic positive *marker bands*, which can be assigned to the mC-cation (*bottom*). The *dashed lines* correspond to the inverted ground state absorbance spectra of mCU and mC

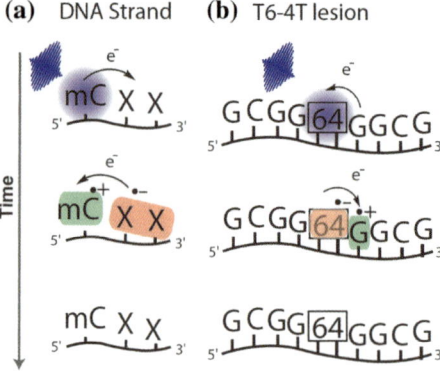

Fig. 2 **a** Model for the excited state decay in DNA oligonucleotides (X is a natural nucleobase), the charge distribution depends on the sequence. **b** Model for the quenching of the excited state of the 6-4 chromophore by a charge transfer processes

ground state bleach, which recovers on a timescale of 100 ps (Fig. 1a, bottom). This process is not observable in a mixture of the single nucleobases and is thus a direct consequence of the base stacking in the oligonucleotides.

In simple dinucleotides we find direct evidence of the molecular nature of these states: The decay associated spectra of mCU dinucleotide show marker bands which can be assigned to the radical cation spectrum of mC$^{•+}$ (Fig. 1b) [5]. Investigation of several different dinucleotides shows that light excitation leads to charge separation, which is directed by the redox potential of the involved nucleobases. The negative charge moves to the nucleobase with a higher redox potential and the charges recombine on a time scale of 100 ps. This charge-separated state is not only located on two neighboring bases but on all three bases.

It can be seen in Fig. 1a that all three bases of mCAU bleach instantaneously. Therefore a charge transfer via a hopping mechanism can be excluded since it would occur on a much longer time scale. Obviously, the charge is delocalized over a well stacked domain along the DNA strand. We thus propose the following model (Fig. 2a): Excitation of a nucleobase in a DNA strand with stacked bases leads ultrafast to the charge separated state. The direction of charge transfer is governed by the redox potential of the nucleobases and is thus encoded in the DNA sequence. These ionic states decay by charge recombination on the 100 ps time scale reforming the neutral ground-state.

We observe a similar process after excitation of the 6-4 chromophore in single and double stranded DNA. The 6-4 lesion has a characteristic absorbance band in the UV-A regime at 325 nm which allows selective excitation. Upon the decay of the excited state we find contributions from a guanine, which is in direct neighborhood of the 6-4 lesion. Apparently the excited state of the 6-4 lesion is quenched by charge transfer. This process reduces the quantum yield of Dewar formation and thus protects the DNA (Fig. 2b).

3 Conclusion

With the help of time resolved IR-spectroscopy we were able to show that charge separation in DNA plays an important role in natural systems and is a major decay channel of photoexcited DNA oligonucleotides. The identification of radical cation and anionic species after light absorption of DNA adds a new dimension to DNA photochemistry, which has been known until now only from artificial systems.

References

1. J. C. Genereux and J. K.Barton, "Mechanisms for DNA Charge Transport", *Chemical reviews* **110**, 1642-1662 (2010).
2. H. A.Wagenknecht, "Principles and Mechanisms of Photoinduced Charge Injection, Transport, and Trapping in DNA" in Charge transfer in DNA (Wiley-VCH, Weinheim, 2005), Chap. 1.
3. D. B. Bucher, B. M. Pilles, T. Carell and W. Zinth, "Charge separation and charge delocalization identified in long-living states of photoexcited DNA", *PNAS* **111**, 4369-4374 (2014).
4. K. Haiser, B. P. Fingerhut, K. Heil, A. Glas, T. T. Herzog, B. M. Pilles, W. J. Schreier, W. Zinth, R. de Vivie-Riedle and T. Carell, "Mechanism of UV-Induced Formation of Dewar Lesions in DNA", *Angew. Chem. Int. Ed.* **51**, 408-411 (2012).
5. D. B. Bucher, B. M. Pilles, T. Pfaffeneder, T. Carell and W. Zinth, "Fingerprinting DNA Oxidation Processes: IR Characterization of the 5-Methyl-2'-Deoxycytidine Radical Cation", *ChemPhysChem* **15**, 420-423, (2014).

A Regulation of Energy Flow in Purple Bacterial Photosynthetic Antennas

D. Kosumi, S. Maruta, R. Fujii, M. Sugisaki, S. Takaichi,
R.J. Cogdell and H. Hashimoto

Abstract Ultrafast energy transfer dynamics in photosynthetic antennas were investigated by femtosecond pump-probe measurements. Photo-excited carotenoids with short-polyene chains efficiently transfer energy to bacteriochlorophylls, while the energy rapidly dissipates to carotenoids in antennas containing longer carotenoids.

1 Introduction

Carotenoids (Cars) are naturally occurring photosynthetic pigments that play the two important roles of light-harvesting (LH) and photo-protection in photosynthesis of plants, algae, and bacteria [1]. Cars absorb sunlight in the blue-green regions of the spectrum and efficiently transfer it to nearby bacteriochlorophyll a (Bchl a). The S_0 ground state of Cars has A_g^- symmetry assuming that their linear polyene backbone has C_{2h} point group symmetry. The lowest singlet excited state, S_1

D. Kosumi (✉) · R. Fujii · H. Hashimoto
The Osaka City University Advanced Research Institute for Natural
Science and Technology (OCARINA), 3-3-138 Sugimoto, Sumiyoshi-ku,
Osaka 558-8585, Japan
e-mail: kosumi@ocarina.osaka-cu.ac.jp

S. Maruta · M. Sugisaki · H. Hashimoto
Department of Physics, Graduate School of Science, Osaka City University,
3-3-138 Sugimoto, Sumiyoshi-ku, Osaka 558-8585, Japan

R. Fujii
JST/PRESTO, 4-1-8 Hon-chou, Kawaguchi, Saitama 332-0012, Japan

S. Takaichi
Department of Biology, Nippon Medical School, 297, Kosugi-cho 2,
Nakahara, Kawasaki 211-0063, Japan

R.J. Cogdell
Glasgow Biomedical Research Centre, University of Glasgow,
126 University Place, G12 8QQ, Scotland

© Springer International Publishing Switzerland 2015
K. Yamanouchi et al. (eds.), *Ultrafast Phenomena XIX*,
Springer Proceedings in Physics 162, DOI 10.1007/978-3-319-13242-6_140

$(2^1A_g^-)$, is optically forbidden, therefore the S_2 $(1^1B_u^+)$ state is the lowest optically allowed state. The ultrafast spectroscopic measurements on purple bacterial LH complexes proposed that the singlet-singlet excitation energy transfer (EET) from Car to Bchl a involves two pathways of $S_2 \rightarrow Q_x$ and $S_1 \rightarrow Q_y$. The overall efficiency of Car $\rightarrow$ Bchl a EET has been reported to be 30–100 % depending strongly on number of conjugated double bonds n of the Cars involved [1]. Recently, we reported a new EET pathway of Bchl a $Q_x \rightarrow$ Car S_1 in LH1 from *Rhoddospirillum (Rsp.) rubrum* S1 involving the Car, spirilloxanthin (Spx) with $n = 13$ [2]. The reverse EET pathway of Bchl a $Q_x \rightarrow$ Car S_1 in LH2 has been also confirmed by broadband two-dimensional electronic spectroscopy [3]. The $Q_x \rightarrow S_1$ EET is rather efficient (40 %) in LH1 from *Rsp. rubrum* S1, whereas this channel is almost closed in LH2 from *Rhodobacter (Rba.) sphaeroides* 2.4.1, containing spheroidene (Sph) with $n = 10$ [4]. These results imply that the EET of $Q_x \rightarrow S_1$ depends on n. In this study, the reverse EET dynamics in LH2 containing Car with different n were investigated by femtosecond pump-probe spectroscopy.

2 Experimental

Femtosecond pump-probe measurements were carried out using a mode-locked Ti: Sapphire laser system. Excitation pulses were obtained from an optical parametric amplifier. A white continuum probe pulse was generated using a 5.0 mm sapphire plate. The instrument response function was determined to be less than 100 fs. The isolated LH2 complexes were dispersed in 20 mm Tris-HCl buffer with detergents to avoid aggregation. The samples were circulated in a 1.0 mm optical path length flow cell.

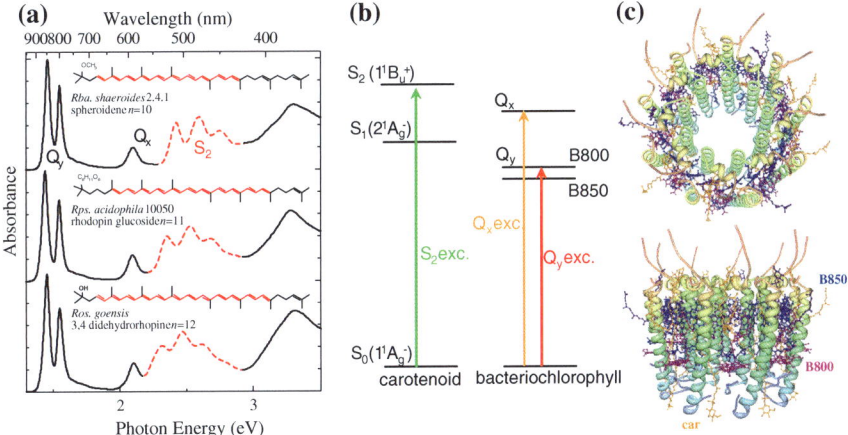

Fig. 1 **a** Steady-state absorption spectra of LH2 complexes. Insets show the chemical structures of carotenoids contained in antenna complexes. **b** Energy diagrams of carotenoid and bacteriochlorophyll a. **c** The crystal structure of LH2 from *Rps. acidophila* 10050

3 Results and Discussion

Figure 1 shows the steady-state absorption spectra of three different LH2 complexes. The absorption bands ranging from 2.1 to 3.0 eV are originated from the S_0 to S_2 transition of Cars, depending on n. Bchl a has three distinction absorption bands at 3.3 eV (Soret), 1.9 eV (Q_x), and 1.4 eV (Q_y). Two absorption bands of Q_y originates from the two different Bchl a molecules assemblies of LH2 (called as B800 and B850 shown in Fig. 1c) [1]. Figure 2a–c represent the photo-induced absorption (PIA) spectra, taken at 0.5 ps upon excitation either to S_2, Q_x, or Q_y. The PIA spectra of all LH2 complexes excited into S_2 contain the S_1-S_n and T_1-T_n transient absorption and ground state bleaching signals of Cars. In the PIA spectra

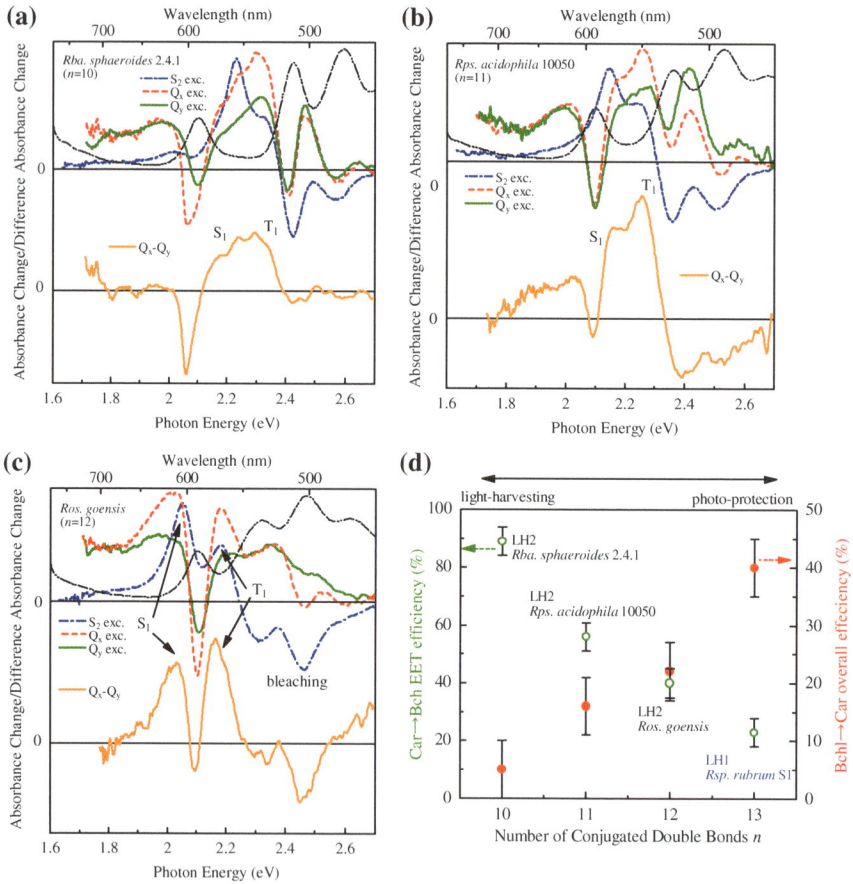

Fig. 2 a–c PIA spectra of the LH2 complexes upon excitation either to S_2, Q_x, or Q_y, taken at 0.5 ps. Lower panels represent difference spectra between Q_x and Q_y excitations. **d** EET efficiencies of Car $\rightarrow$ Bchl and Bchl $\rightarrow$ Car as a function of n

Table 1 The obtained EET rates of each pathway in the LH2 complexes

	$S_2 \rightarrow Q_x$	$S_1 \rightarrow Q_y$	$Q_x \rightarrow S_1$
Rba. sphaeroides 2.4.1 (LH2)	$(166 \text{ fs})^{-1}$	$(2.1 \text{ ps})^{-1}$	–
Rps. acidophila 10050 (LH2)	$(130 \text{ fs})^{-1}$	$(18 \text{ ps})^{-1}$	$(262 \text{ fs})^{-1}$
Ros. goensis (LH2)	$(216 \text{ fs})^{-1}$	$(40 \text{ ps})^{-1}$	$(177 \text{ fs})^{-1}$
Rsp. rubrum S1 (LH1)[a]	$(111 \text{ fs})^{-1}$	–	$(75 \text{ fs})^{-1}$

[a] The EET rates in LH1 from *Rsp. ruburum* S1 was reported in [2]

after excitation into Q_x of Bchl *a*, the S_1 and T_1 transient absorption and the ground state bleaching of Cars were also observed, owing to the reverse $Q_x \rightarrow S_1$ energy transfer [2, 5]. To clarify the transient signals due to Cars in the PIA spectra, the difference spectra between the Q_y and Q_x excitations were examined. The Q_y state of Bchl *a* is lower lying than S_1 of Cars (Fig. 1b), thus the $Q_y \rightarrow S_1$ EET is negligible. In the difference spectrum (lower panels of Fig. 2a–c), the transient signals due to Cars, the S_1 and T_1 transient absorption and the ground state bleaching, obviously depend strongly on *n*. Table 1 and Fig. 2d summarized the EET rates and efficiencies of each pathway in the LH2 complexes determined by the analysis of the kinetic traces of the transient signals. The LH2 complexes containing the short carotenoid (*n* = 10) efficiently transfer the absorbed energy to Bchl *a*, and then Car $\rightarrow$ Bchl *a* EET efficiencies decrease with *n*. On the other hand, the reverse Bchl $\rightarrow$ Car EET is negligible in LH2 containing the short carotenoid (*n* = 10), while the reverse EET become efficient (40 %) with *n*. In conclusion, the ultrafast EET dynamics between Cars and Bchl *a* in the LH2 complexes have been investigated by femtosecond pump-probe measurements. The results suggest that LH2 containing the short carotenoid has an efficient light-harvesting function, while that the long polyene carotenoids bound to LH complexes regulate an excess energy of Bchl *a*.

References

1. T. Polivka; V. Sundstrom *Chem. Rev.* **104**, 2021-2071, (2004).
2. D. Kosumi; S. Maruta; T. Horibe; R. Fujii; M. Sugisaki; R. J. Cogdell; H. Hashimoto *Angew. Chem., Int. Ed.* **50**, 1097-1100, (2011).
3. E. E. Ostroumov; R. M. Mulvaney; R. J. Cogdell; G. D. Scholes *Science* **340**, 52-56, (2013).
4. S. Maruta; D. Kosumi; T. Horibe; R. Fujii; M. Sugisaki; R. J. Cogdell; H. Hashimoto *Phys. Status Solidi B* **248**, 403-407, (2011).
5. D. Kosumi; S. Maruta; T. Horibe; Y. Nagaoka; R. Fujii; M. Sugisaki; R. J. Cogdell; H. Hashimoto *J. Chem. Phys.* **137**, 064505-064510, (2012).

Elucidation and Control of Ultrafast Intramolecular Charge Transfer Dynamics of Marine Photosynthetic Pigments

D. Kosumi, T. Kajikawa, K. Yano, S. Okumura, M. Sugisaki, K. Sakaguchi, S. Katsumura and H. Hashimoto

Abstract Ultrafast Intramolecular Charge Transfer (ICT) state dynamics of fucoxanthin have been investigated by femtosecond pump-probe measurements. A modification of conjugated polyene chain length of fucoxanthin enabled us to clarify and control an ICT character.

1 Introduction

Carotenoids (Cars) are essential pigments for light-harvesting in natural photosynthesis. They absorb light in the blue-green region of the spectrum and transfer to chlorophyll (Chl) rapidly and efficiently in photosynthetic antennas [1]. The excited states of Cars are described by a three-electronic level system [2]. The optically allowed S_2 $(1^1B_u^+)$ state is generated by a one-photon transition from the S_0 ground state $(1^1A_g^-)$, and rapidly decays to the lower-lying dark S_1 $(2^1A_g^-)$ state with ~ 100 fs. A nonradiative relaxation from S_1 to S_0 is comparatively slower

D. Kosumi (✉) · H. Hashimoto
The Osaka City University Advanced Research Institute for Natural Science and Technology (OCARINA), 3-3-138 Sugimoto, Sumiyoshi-ku, Osaka 558-8585, Japan
e-mail: kosumi@ocarina.osaka-cu.ac.jp

T. Kajikawa · K. Yano · S. Okumura · S. Katsumura
Department of Chemistry, School of Science and Technology, Kwansei Gakuin University, Gakuen, Sanda Hyogo 669-1337, Japan

M. Sugisaki · H. Hashimoto
Department of Physics, Graduate School of Science, Osaka City University, 3-3-138 Sugimoto, Sumiyoshi-ku, Osaka 558-8585, Japan

K. Sakaguchi
Department of Chemistry, Graduate School of Science, Osaka City University, 3-3-138 Sugimoto, Sumiyoshi-ku, Osaka 558-8585, Japan

© Springer International Publishing Switzerland 2015
K. Yamanouchi et al. (eds.), *Ultrafast Phenomena XIX*,
Springer Proceedings in Physics 162, DOI 10.1007/978-3-319-13242-6_141

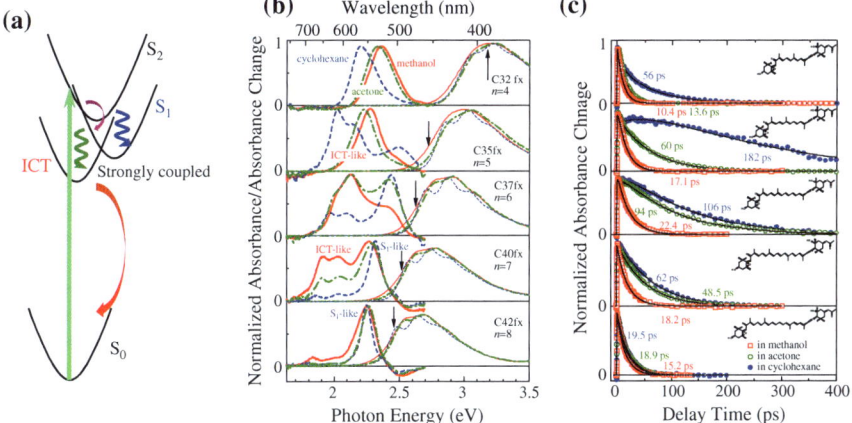

Fig. 1 **a** A schematic representation of energy diagram of fx. **b** Steady-state and PIA spectra at 3.0 ps of the fx homologs. *Arrows* indicate excitation energies. **c** Kinetic traces of the S_1/ICT state. Insets show the chemical structures of the fx homologs

(ps–ns), while the S_1-S_0 transition is optically forbidden [2]. The optically forbidden character (S_1) of Cars appears to not function as an effective energy donor. In fact, energy transfer from the Car S_1 state to Chl is often inefficient [1]. On the other hand, Cars containing a carbonyl group in their polyene backbone generate the intramolecular charge transfer (ICT) state below S_1 in polar environments, forming the strongly coupled S_1/ICT state as shown in Fig. 1a [3]. Consequently, the S_0-S_1/ICT transition dipole moment of carbonyl Cars increases due to their charge transfer character, and efficient energy transfer from carbonyl Cars to Chl via S_1/ICT is realized in photosynthetic antennas [4, 5]. The present study focused on the ICT character of the carbonyl Cars fucoxanthin (fx) from marine algae that could be correlated to its conjugated polyene chain lengths. Femtosecond pump-probe spectroscopic measurements with 100 fs time-resolution were performed on the fx homologues with varying numbers of conjugated double bonds n (4–8).

2 Experimental

Femtosecond pump-probe spectroscopic measurements were carried out using a mode-locked Ti:Sapphire laser system [6]. Excitation pulses were obtained from an optical parametric amplifier or by the second harmonic generation. A white continuum probe pulse was generated using a 5.0 mm sapphire plate. The instrument response function was determined to be ~ 100 fs. Sample preparations of the fx homologues with $n = 4$–8 are described elsewhere [7]. Samples were dissolved in three different solvents: methanol, acetone and cyclohexane.

3 Results and Discussion

Figure 1b represents the steady-state and photo-induced absorption (PIA) spectra of the fx homologues. The steady-state absorption bands, corresponding to the S_0-S_2 transitions, red-shifted with an increase in n [2]. In the PIA spectra, the native fx (C40fx) in cyclohexane shows a monotonic S_1-S_n transient absorption [1], while an additional band just below the S_1 transient absorption band, assigned to the ICT-like band [3], increased with solvent polarity. In the longest C42fx, the only S_1-like transient absorption band appeared even in polar solvents, due to insufficient stabilization of the ICT state. In contrast, both of the S_1 and ICT-like transient absorption bands were observed clearly in the shorter C37fx in all solvents; the transient absorption bands of C32fx and C35fx became monomeric in polar solvents. The monomeric absorption bands of C32fx and C35fx observed in polar solvents were assigned to the ICT-like transient absorption. In addition, the ICT-like transient absorption band predominated and the stimulated emission from S_1/ICT was observed in the shortest C32fx, even in cyclohexane.

Figure 1c shows the kinetic traces of the S_1/ICT transient absorption of the fx homologues. The solid lines represent the best-fit curves for the rise and decay phases convoluted with the instrumental response function. In all fx homologs, the S_1/ICT lifetimes decreased with an increase in solvent polarity, while the S_1/ICT lifetimes did not exhibit a monotonic dependence on n, i.e., they had the longest lifetimes with $n = 6$ in polar solvents or with $n = 5$ in a nonpolar solvent. The S_1 nonradiative relaxation rates of Cars without a carbonyl group are well explained by the energy gap law [1]. Figure 2 plots the nonradiative relaxation rates of S_1/ICT as a function of $1/(2n + 1)$ as well as those for S_1 of β-carotene homologues ($n = 7$–15). The solid line in Fig. 2 denotes the best-fit curve using the energy gap law for β-carotene homologues [8]. The S_1/ICT decay rates of fx homologues should also obey the energy gap law if the S_1/ICT decay cannot be enhanced by the coupling of the ICT state. However, the S_1/ICT decay rates clearly deviate from the energy gap law with a decrease in n, suggesting that the S_1/ICT decay of the shorter fx homologs

Fig. 2 The nonradiative relaxation of S_1/ICT for fx homologs and S_1 for β-carotene homologues as a function of $1/(2n + 1)$ or n

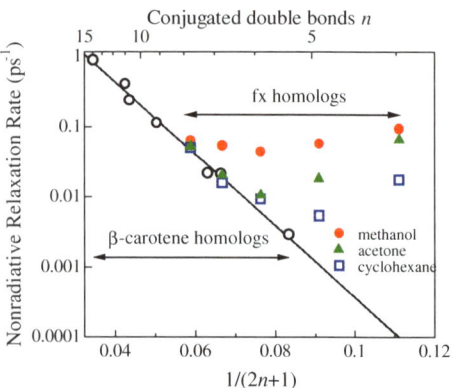

can be enhanced by the strong coupling of the S_1 and ICT states, even in a nonpolar solvent. The results suggest that the ICT property of fx becomes stronger with a decrease of n and that the shortest C32fx exhibits a strong ICT character even in a nonpolar solvent. In conclusion, the conjugation length dependence of the S_1/ICT state dynamics of the fx homologues demonstrates that the ICT property of fx can be systematically controlled by a modification of its conjugated polyene length [9].

References

1. T. Polívka; V. Sundström *Chem. Rev.* **104**, 2021-2071, (2004).
2. *Linear Polyene Electronic Structure and Potential Surfaces*; B. S. Hudson; B. E. Kohler; K. Schulten, Eds.; Academic Press: New York, 1982; Vol. 6.
3. J. A. Bautista; R. E. Connors; B. B. Raju; R. G. Hiller; F. P. Sharples; D. Gosztola; M. R. Wasielewski; H. A. Frank *J. Phys. Chem. B* **103**, 8751-8758, (1999).
4. D. Zigmantas; R. G. Hiller; V. Sundström; T. Polívka *Proc. Natl. Acad. Sci. USA* **99**, 16760-16765, (2002).
5. D. Kosumi; M. Kita; R. Fujii; M. Sugisaki; N. Oka; Y. Takaesu; T. Taira; M. Iha; H. Hashimoto *J. Phys. Chem. Lett.* **3**, 2659-2664, (2012).
6. D. Kosumi; T. Kusumoto; R. Fujii; M. Sugisaki; Y. Iinuma; N. Oka; Y. Takaesu; T. Taira; M. Iha; H. A. Frank; H. Hashimoto *Chem. Phys. Lett.* **483**, 95-100, (2009).
7. T. Kajikawa; S. Okumura; T. Iwashita; D. Kosumi; H. Hashimoto; S. Katsumura *Org. Lett.* **14**, 808-811, (2012).
8. D. Kosumi; M. Fujiwara; R. Fujii; R. J. Cogdell; H. Hashimoto; M. Yoshizawa *J. Chem. Phys.* **130**, 214506, (2009).
9. D. Kosumi; T. Kajikawa; S. Okumura; M. Sugisaki; K. Sakaguchi; S. Katsumura; H. Hashimoto *J. Phys. Chem. Lett.* **5**, 792-797, (2014).

The Primary Photosynthetic Energy Conversion in Bacterial Reaction Centers—Stepwise Electron Transfer and the Effect of Elevated Exposure Levels

Pablo Nahuel Dominguez, Matthias Himmelstoss, Jeff Michelmann, Florian Lehner, Alastair Gardiner, Richard Cogdell and Wolfgang Zinth

Abstract The primary reaction in photosynthetic reaction centers from Rhodobacter sphaeroides is investigated for different experimental conditions. Agreement with stepwise electron transfer via a reduced bacteriochlorophyll was observed at low excitation rates.

1 The Primary Reactions of Photosynthetic Reaction Centers

Photosynthetic energy conversion in bacterial reaction centers (RC) can be taken as a prototype reaction not only for other photosynthetic systems but also for artificial solar energy conversion devices. In this context a sound understanding of the primary reaction steps is of major importance. The observation of these processes requires excitation at the appropriate wavelength, weak intensities to avoid overexposure and a very sensitive detection over a wide range of probing wavelengths. In the 1990s, a simple stepwise reaction model (1), based on non-adiabatic Marcus-type electron transfer (ET), was proposed to explain the molecular processes and to understand optimization of photosynthesis. Here, an electron is transferred from the originally excited special pair P via the accessory bacteriochlorophyll B_A and the

P.N. Dominguez · M. Himmelstoss · J. Michelmann · F. Lehner · W. Zinth (✉)
BioMolekulare Optik and Center of Integrated Protein Science,
CIPSM Ludwig-Maximilians-Universität München,
Oettingenstr 67, 80538 Munich, Germany
e-mail: wolfgang.zinth@physik.uni-muenchen.de

A. Gardiner · R. Cogdell
Institute of Molecular Cell and Systems Biology, University of Glasgow,
126 University Place, Glasgow G12 8TA, Scotland

© Springer International Publishing Switzerland 2015
K. Yamanouchi et al. (eds.), *Ultrafast Phenomena XIX*,
Springer Proceedings in Physics 162, DOI 10.1007/978-3-319-13242-6_142

bacteriopheophytin H_A to the quinone Q_A (for the spatial arrangement of the chromophores in the RC see [1, 2]).

$$P \xrightarrow{h\nu} P^* \xrightarrow{3.5ps} P^+B_A^- \xrightarrow{0.9ps} P^+H_A^- \xrightarrow{200ps} P^+Q_A^- \tag{1}$$

Detailed simulations using evolutionary optimization have shown that this simple reaction scheme is able to explain the high quantum yield $\eta > 95\%$ and the good energy efficiency of the primary photosynthetic reaction [3]. More recently, new experimental investigations have been published which partially deviate from original publications and suggest alternative reaction principles [4, 5]. In this context we present here the results of new experiments recorded with a high signal to noise ratio and use improved data analysis algorithms. We find (i) a pronounced influence of exposure and repetition rate on the dynamics of the RC and that (ii) the data recorded at low exposure levels are in excellent agreement with the stepwise ET presented in (1).

2 Techniques

Pump-probe experiments were performed as described in [6]. The excitation rate was adjusted by choppers between 50 and 500 Hz. Stirring combined with rapid transversal motion of the sample cuvette allowed to replace the irradiated volume and to ascertain a defined exposure for the whole sample volume (0.3 ml). A special method was used to identify the relevant signal components in a Singular Value Decomposition (SVD) based procedure and to quantify the components related to the ET reaction (see supporting information in [6]).

3 Results

Experiments at low excitation intensities, with optimized sample exchange and low exposure levels (established by the low repetition rate of 50 Hz) yielded data which could be well fitted by a sum of exponential functions with time constants of ca. 1.2, 3.5, 220 ps and a long-lasting component. For details see [6]. The SVD-analysis shows that the time traces related to significant singular values (SV) are fitted well by this exponential 4-component model. Conversely, a 3-component model does not fit the significant time traces with sufficient quality. Figure 1a, b show the time traces of the fourth and fifth SV together with a multi-exponential fit for the 3-component (broken) and the 4-component model (solid). The comparison demonstrates that a fourth kinetic of ca. 1 ps is required to describe the primary ET-reaction. In order to obtain additional molecular information on the related processes we performed different analysis procedure. Spectral signatures for the

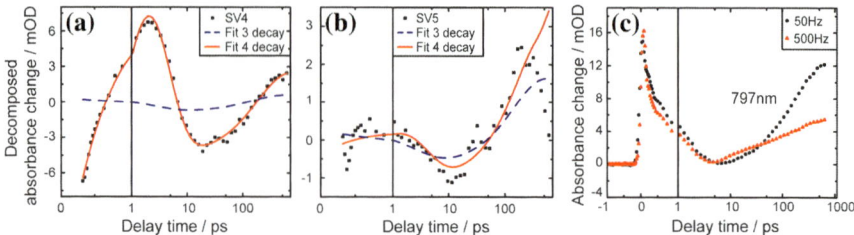

Fig. 1 Decomposed transient signal for measurements with 50 Hz rep. rate corresponding to the 4th SV (**a**) and 5th SV (**b**). Transient absorption signal at 797 nm measured with different repetition rates (**c**). Displayed data has parallel polarization. The time scale is linear until 1 ps and logarithmic afterwards

intermediate states indicated in (1) can be computed from the fit amplitudes using the corresponding rate equation model. The rate constants are assumed in the order 3.5, 1.2 and 220 ps. On the contrary, if the two first reaction rates are interchanged, difficulties arise to consistently describe the time dependent emission recorded in a previous publication [7]. Finally, a diverse investigation of the spectral signatures can be performed with target analysis [8]. In this case we obtain again nice agreement with the reaction scheme (1). Both procedures clearly show that the intermediate state populated upon the decay of the excited special pair P* contains the reduced bacteriochlorophyll B_A.

Experiments performed at higher repetition rates, with reduced sample exchange speed and with chemically treated RC-solutions, show only qualitatively similar results. At high excitation rate discrepancies are evident in the spectral range around 797 nm (Fig. 1c). The amplitude of the ca. 1 ps component is decreased and an additional kinetic appears with a time constant in the 7 ps range.

In conclusion, the photosynthetic reaction centers display primary reactions with a pronounced change in the transfer dynamics, depending on exposure and repetition rate. Nevertheless, data recorded at low exposure levels are in excellent agreement with the stepwise ET model described in the early 1990s (see 1).

References

1. W. Zinth and J. Wachtveitl, "The First Picoseconds in Bacterial Photosynthesis-Ultrafast Electron Transfer for the Efficient Conversion of Light Energy", Chem. Phys. Chem. **6**, 871–880 (2005).
2. W. Holzapfel, U. Finkele, W. Kaiser, D. Oesterhelt, H. Scheer, H. U. Stilz and W. Zinth, "Initial electron-transfer in the reaction center from Rhodobacter sphaeroides", PNAS **87**, 5168-5172 (1990).
3. B. P. Fingerhut, W. Zinth and R. de Vivie-Riedle, "Design criteria for optimal photosynthetic energy conversion", Chem. Phys. Lett. **466**, 209-213 (2008).

4. Y. Kakitani, A. Hou, Y. Miyasako, Y. Koyama and H. Nagae, "Rates of the initial two steps of electron transfer in reaction centers from Rhodobacter sphaeroides as determined by singular-value decomposition followed by global fitting", Chem. Phys. Lett. **492**, 142-149 (2010).
5. J. Zhu, I. H. M. van Stokkum, L. Paparelli, M. R. Jones and M. L. Groot, "Early Bacteriopheophytin Reduction in Charge Separation in Reaction Centers of Rhodobacter sphaeroides", Biophysical Journal **104**, 2493–2502 (2013).
6. P. N. Dominguez, M. Himmelstoss, J. Michelmann, F. T. Lehner, A. T. Gardiner, R. J. Cogdell and W. Zinth, "Primary reactions in photosynthetic reaction centers of Rhodobacter sphaeroides–Time constants of the initial electron transfer", Chem. Phys. Lett. **601**, 103–109 (2014).
7. P. Hamm, K. A. Gray, D. Oesterhelt, R. Feick, H. Scheer and W. Zinth, "Subpicosecond emission studies of bacterial reaction centers", BBA **1142**, 99-105 (1993).
8. Y. Koyama, Y. Kakitani and H. Nagae, "Mechanisms of Cis-Trans Isomerization around the Carbon–Carbon Double Bonds via the Triplet State", in *cis-trans Isomerization in Biochemistry* (Wiley-VCH Verlag GmbH & Co. KGaA, Weinheim, Germany, 2006), pp. 15-51.

Resonant Stimulated X-Ray Raman Spectroscopy of Molecule Following Core Ionization

Yu Zhang, Jason D. Biggs, Weijie Hua and Shaul Mukamel

Abstract We investigate computationally the valence electronic excitations of the amino acid glycine following a sudden nitrogen core ionization induced by an attosecond X-ray pump pulse. The molecule left in a superposition of cationic excited states is then probed by a second broadband X-ray pulse together with a narrowband pulse tuned to the carbon K-edge. Transient X-ray absorption and attosecond stimulated X-ray Raman signals are simulated and related to the evolution of the valence excited state wavepacket.

1 Introduction

A sudden ionization by removal of a core electron, which is faster than the electron correlation time, can bring molecules to a superposition of large number of cationic states. A typical response time for spectator electrons to a sudden removal of an electron is about 50 as [1, 2]. The sudden-ionizing pulse should therefore be shorter than this time. With the development of attosecond pulses [3], sudden ionization is becoming feasible.

After an electron has been ejected from the core orbital, the molecule is prepared in a nonstationary state [4] which can be monitored by attosecond stimulated X-ray Raman spectroscopy (ASRS). Core excitations are spatially selective because core transitions for different types of atoms are spectrally well removed from each other [5]. In an X-ray Raman process [6–8], only those valence excitations with transition density in the region near the selected atom will be active.

In vibrational spectroscopy, femtosecond stimulated Raman spectroscopy (FSRS) [9, 10] has been widely used to monitor nuclear dynamics. This technique uses a combination of narrow and broadband pulses to collect the Raman signal in a single shot. ASRS extends FSRS to the X-ray regime. We further use resonant

Y. Zhang · J.D. Biggs · W. Hua · S. Mukamel (✉)
Department of Chemistry, University of California, Irvine, CA 92697, USA
e-mail: smukamel@uci.edu

© Springer International Publishing Switzerland 2015
K. Yamanouchi et al. (eds.), *Ultrafast Phenomena XIX*,
Springer Proceedings in Physics 162, DOI 10.1007/978-3-319-13242-6_143

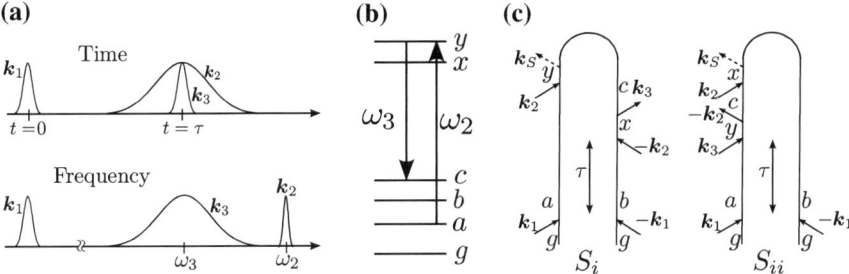

Fig. 1 **a** Pulse sequence for the narrow pump/broad probe stimulated Raman experiment. **b** Level scheme. **c** The two contributing loop diagrams

rather than off-resonant pulses. In an ASRS experiment, two probe pulses hit the sample simultaneously. The first is narrowband (long in duration) and resonant with a core transition, and the second probe is broadband (short in duration), and red-shifted from the first, as shown in Fig. 1. The signal is given by the spectrally dispersed transmission of the broadband pulse. We neglect nuclear motion in these calculations. The short lifetime of the initial ionized state ($\Gamma_{ab} = 0.09\,\text{eV} = \hbar/7.3\,\text{fs}$ [5]) means that the entire signal is collected on a timescale that is very fast in comparison to nuclear motion.

2 Results and Discussion

In Fig. 2 we show the electron density difference in glycine, defined as the time-dependent density of the cation minus the ground state electronic density, for five different times following core ionization of nitrogen. The large number of states with nonnegligible amplitudes in the superposition indicates a collective electronic motion. The negative region surrounding the nitrogen atom represents the removed electron. Positive electron density regions surrounding the nitrogen correspond to relaxation of the system in the presence of the hole. The fact that this relaxation is not negligible at short time delays is attributed to the truncation error of the eigenstate basis used to project the initial state. A truncated basis must be used since the Raman signal scales as the number of valence-states cubed. Considerable changes in the electronic density in the region around the two carbon atoms is seen for interpulse delays $\tau = 110.1$ and 317.2 as. For delays of 216.8 and 424.7 as the density changes around the carbon atoms is diminished. This matches the changes in intensity seen in the time-dependent ASRS signal in the right column of Fig. 2. Large differences in the density around the carbon atoms corresponds to larger carbon K-edge Raman signal, and small differences likewise lead to smaller signals.

It should be emphasized that even though the time delay between the ionizing pulse and the broadband Raman probe is well defined, the resulting signal may not be viewed as an instantaneous snapshot of the Raman spectrum, but is averaged

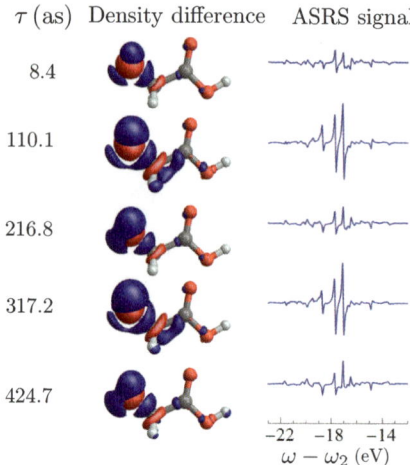

τ (as)   Density difference   ASRS signal

8.4

110.1

216.8

317.2

424.7

-22 -18 -14
$\omega - \omega_2$ (eV)

Fig. 2 Density difference for the state following ionization, defined as the time-dependent density minus the ground-state density for the cationic wavepacket, at time delays which correspond with the maxima and minima for the Raman peak at $\omega - \omega_2 = -17.2$ eV alongside the corresponding ASRS signals. *Red* isosurfaces show negative density differences, and *blue* surfaces show positive difference. The nitrogen atom is hidden by the *red* surface, corresponding to the suddenly created core hole

over the dephasing time of the final state. Correspondence between the time-dependent ASRS signal and changes in the electronic density are seen for short times following ionization. However, the ASRS signal depends upon the multi-electron dynamics, while the density is a single-electron property and not as sensitive to interference effects, so correspondence at later times is not as obvious.

Acknowledgments The support of the Chemical Sciences, Geosciences and Biosciences Division, Office of Basic Energy Sciences, Office of Science, U.S. Department of Energy and the National Science Foundation (Grant CHE-1361516) is gratefully acknowledged.

References

1. J. Breidbach, L.S. Cederbaum, Phys. Rev. Lett. **94**(3), 033901 (2005)
2. A.I. Kuleff, L.S. Cederbaum, Phys. Rev. Lett. **98**(8), 083201 (2007)
3. K.T. Kim, D.M. Villeneuve, P.B. Corkum, Nat. Photonics **8**(3), 187 (2014)
4. E. Goulielmakis, Z.H. Loh, A. Wirth, R. Santra, N. Rohringer, V.S. Yakovlev, S. Zherebtsov, T. Pfeifer, A.M. Azzeer, M.F. Kling, S.R. Leone, F. Krausz, Nature **466**(7307), 739 (2010)
5. G. Zschornack, *Handbook of X-Ray Data* (Springer-Verlag Berlin Heidelberg, 2007)
6. L.J. Ament, M. van Veenendaal, T.P. Devereaux, J.P. Hill, J. van den Brink, Rev. Mod. Phys. **83**(2), 705 (2011)
7. A. Kotani, S. Shin, Rev. Mod. Phys. **73**(1), 203 (2001)
8. S. Mukamel, D. Healion, Y. Zhang, J.D. Biggs, Annu. Rev. Phys. Chem. **64**, 101 (2013)
9. P. Kukura, D.W. McCamant, R.A. Mathies, Annu. Rev. Phys. Chem. **58**, 461 (2007)
10. J.M. Rhinehart, J.R. Challa, D.W. McCamant, J. Phys. Chem. B **116**(35), 10522 (2012)

Light Harvesting Dynamics in Gloeobacter Rhodopsin (GR)

E. Siva Subramaniam Iyer, Itay Gdor, Tamar Eliash, Mordechai Sheves and Sanford Ruhman

Abstract GR is directly shown by ultrafast pump-probe measurements to bind the carotenoid Salinixanthin, which acts as an efficient light harvesting antenna. Along with Xanthorhodopsin, This proves light harvesting to be a prevalent strategy in retinal proteins.

1 Introduction

Retinal Proteins (RP) are thought of as single chromophore molecular machines where retinal functions both for light harvesting (LH) and processing. Recent studies reveal that, as in chlorophyll based photosynthetic complexes, the microbial proton pump Xanthorhodopsin (XR) includes an associated carotenoid antenna which efficiently harvests blue-green light. In addition, conservation of the carotenoid binding site in a variety of newly discovered microbial RPs suggests that this strategy for more efficient LH is not a curiosity but a general trend. To prove this GR, expressed from a sequence in *Gloeobacter violaceus*, was shown capable of binding salinixanthin (SX), and by fluorescence measurements suggested to act as an antenna assisted proton pump. Here we complete this investigation by directly measuring transient absorption spectra with ultrafast time resolution and photoselective probing, following excitation either of the carotenoid antenna, or the retinal moiety. Results show not only that GR acts as an antenna assisted proton pump, but that it does so with higher energy transfer efficiency than that in XR.

E. Siva Subramaniam Iyer · I. Gdor · S. Ruhman (✉)
Institute of Chemistry, Hebrew University, Jerusalem, Israel
e-mail: sandy@mail.huji.ac.il

E. Siva Subramaniam Iyer
e-mail: Siva.iyer@mail.huji.ac.il

T. Eliash · M. Sheves
Department of Organic Chemistry, Weizmann Institute Rehovot, Rehovot, Israel

© Springer International Publishing Switzerland 2015   587
K. Yamanouchi et al. (eds.), *Ultrafast Phenomena XIX*,
Springer Proceedings in Physics 162, DOI 10.1007/978-3-319-13242-6_144

2 Experimental Methods

GR was extracted from *G. violaceus* and expressed in *E. coli* using previously published methods and was reconstituted with Salinixanthin [1]. To ascertain the fraction of energy transfer from carotenoid to retinal, the latter was reduced in order to inhibit energy transfer. The energy transfer process was studied using visible pump and visible and near IR probe spectroscopy. The pump pulses were generated by a TOPAS (light conversion) and white light probe generated in 2 mm of Sapphire was read out on multichannel detection systems using near IR probing with InGaAs array spectrograph [2, 3].

3 Results and Discussion

Without an associated carotenoid, GR exhibits photochemical dynamics reminiscent of retinal proteins involving bi-exponential internal conversion and all trans to 13 cis isomerization of retinal with 500 fs and 2.5 ps decay constants. As in other microbial retinal proteins, the slower of these timescales obscures the "J" to "K" transition documented in BR and assigned to vibrational relaxation of the primary isomerized photoproduct [4]. The absorption band of GR peaks at 550 nm, partially overlapping the S_0-S_2 absorption band of SX. The S_0-S_1 absorption of the carotenoid occurs at 1,150 nm, making S_2 the only state energetic enough to act as donor to the retinal. Accordingly, to determine the efficiency of energy transfer from XR to retinal, the S_2 lifetime was measured in GR reconstituted with SX (GRS), and in a GRS where the retinal conjugation length has been shortened by double bond reduction (RGR). This shifts the retinal absorption to the near UV, and prohibits energy transfer with minimal disruption of protein structure.

Figure 1a presents the transient absorption spectra following SX excitation for both samples in the form of evolution associated difference spectra (EADS) obtained by fitting to a sequential kinetic model. The short lived initial EADS is assigned to S_2. The lifetime assigned to this state in GRS increases from $\sim$83 to 133 fs upon retinal reduction, as expected if it blocks an effective route of energy transfer. Again the S2 EADS shifts in lifetime from 83 to 133 upon retinal reduction. This is graphically depicted in Fig. 1c, for both GRS and RGR. In this graph the kinetic traces at 1,150 nm are shown for RGR and GRS. The elementary kinetic considerations lead to:

$$\phi_{ET} = \frac{k_{ET}}{k_{ET} + k_{IC}} = \left(\frac{1}{83\,fs} - \frac{1}{133\,fs} \right) \Big/ \frac{100}{83\,fs} = 38\,\%$$

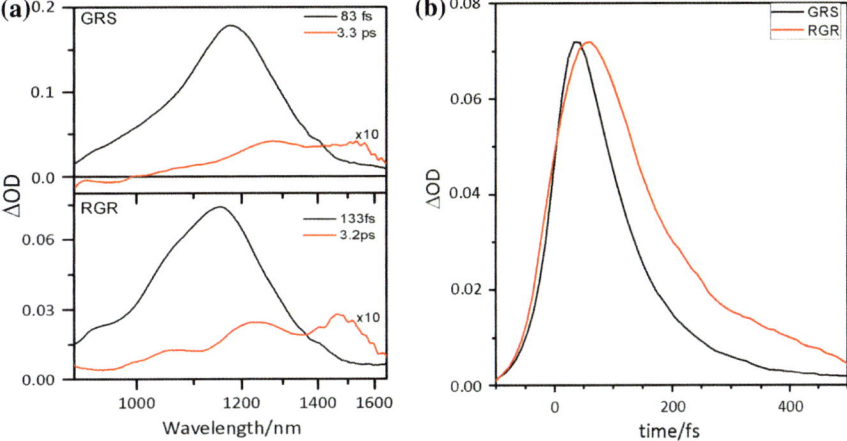

Fig. 1 EADS of GR-salinixanthin conjugate with retinal (GRS) and reduced retinal (RGR) using IR probe measurements. The spectrum of the long lived S_1 state of salinixanthin in **a** magnified by 10 times for clarity. **b** Kinetic traces of GRS and RGR at 1,150 nm

with $k_{ET} = 4.4 \times 10^{-3}$ fs^{-1}, even higher than the ~ 30 % ET efficiency recorded for XR. ϕ_{ET} is the efficiency for energy transfer, and k_{ET} and k_{IC} represent the rates of energy transfer and of internal conversion respectively.

4 Conclusion

The major conclusion of all these measurements is that efficient energy transfer takes place from carotenoid to the retinal in GR. From the lifetime of the S_2 state the quantum efficiency of energy transfer is evaluated to be 0.38, even higher than that in XR, and a rate of energy transfer is evaluated to be 4.4×10^{-3} fs^{-1}. This conclusion was also corroborated by measuring rhodopsin photocycle intermedates populations following excitation of the carotenoid leading to similar estimates for energy transfer efficiency. The importance of this finding is a realization that carotenoid light harvesting, prevalent in "green" photosynthetic complexes, is a substantial strategy in microbial RPs as well. This is likely not limited to the two proteins discussed here alone, since many proteorhodopsins for instance also have protein sequences which can anchor carotenoid antennae [5]. Finer details of our experiments point to additional similarities of XR and GRS dynamics, including an unexplained reduction of polarization ratio of the carotenoid absorption upon internal conversion of the SX from S_2 to S_1. The reduced ratio indicates the ultrafast development of an angle between the transition dipoles of the S_1 and S_2 during the radiationless relaxation. Future work will be required to fully reveal not only the prevalence of carotenoid light harvesting in other RPs, but to identify the organisms which harbor them and the degree of their expression in those hosts.

References

1. S.P. Balashov, E.S. Imasheva, V.A.B. Boichenko, J. Anton, J.M. Wang, J.K. Lanyi, *Science*, **309**, 2061–2064, (2005)
2. I. Gdor, J. Zhu, B. Loevsky, E. Smolensky, N. Friedman, M. Sheves, S. Ruhman, *Phys. Chem. Chem. Phys.* **13**, 3782-3787 (2011)
3. J. Zhu, I. Gdor, E. Smolensky, N. Friedman, M. Sheves, S. Ruhman, *J. Phys. Chem. B*, **114**, 3038–3045 (2010)
4. J.K. Lanyi, *Biochim. Biophys. Acta,* **2006**, *1757*, 1012.
5. DeLong, E. F.; Béjà, O. *PLoS Biology* **2010**, *8*, e1000359.

Disentangling Electronic and Vibrational Coherence in the Phycocyanin-645 Light-Harvesting Complex

Jeffrey A. Davis, Gethin H. Richards, Krystyna E. Wilk and Paul M.G. Curmi

Abstract We selectively excite coherence pathways in the light-harvesting complex PC645 and with wavelength and polarization control identify contributions from both electronic and vibrational coherences. Insight into the interactions between excited electronic and vibrational states follows.

1 Introduction

Energy transfer between chromophores in photosynthesis proceeds with near unity quantum efficiency. Understanding the precise mechanisms of these processes is made difficult by the complexity of the electronic structure and interactions with different vibrational modes. Two-dimensional spectroscopy has helped resolve some of the ambiguities and identified quantum effects that may be important for highly efficient energy transfer [1, 2]. Many questions remain, however, including whether the coherences observed are electronic and/or vibrational in nature and what role they play. We utilise a two-colour four-wave mixing experiment with control of the wavelength and polarization to selectively excite specific coherence pathways and identify the nature of these coherences in the PC645 light-harvesting complex from cryptophyte algae.

Previous work has reported coherences at several different frequencies in the vicinity of the blue band in the absorption spectrum [2, 3]. The electronic states in this vicinity are predicted to be the excitonic states, labeled DBV+ and DBV−, resulting from the strongly coupled DBV chromophores. The precise spectral location of these excitonic states, and indeed the other electronic states in the complex, is unclear,

J.A. Davis (✉) · G.H. Richards
Centre for Quantum and Optical Science, Swinburne University of Technology,
John St, Hawthorn, Melbourne 3122, Australia
e-mail: JDavis@swin.edu.au

K.E. Wilk · P.M.G. Curmi
School of Physics and Centre for Applied Medical Research, St Vincents Hospital,
The University of New South Wales, Sydney, NSW 2052, Australia

© Springer International Publishing Switzerland 2015   591
K. Yamanouchi et al. (eds.), *Ultrafast Phenomena XIX*,
Springer Proceedings in Physics 162, DOI 10.1007/978-3-319-13242-6_145

which increases the difficulty in attributing coherences to electronic, vibrational or vibronic coherent superpositions. In this work we focus on this blue spectral region to better understand the excited state structure and the nature of coherences observed.

2 Results and Discussion

The two-color four-wave mixing experiment can selectively excite coherence pathways when the wavelengths of first two pulses are different. The subsequent evolution of this coherence is probed by the third pulse arriving a variable time later [4, 5]. With the pulses set to 2.179 and 2.066 eV, close to the DBV absorption bands, previous work has reported extended coherence signals lasting beyond 1 ps [4]. In the present work we varied the wavelength of the first two pulses in this vicinity and reveal a ladder of discrete states separated by 24 meV that are coherently coupled and produce coherence signals that persist for several hundred femtoseconds [5]. We attribute this ladder of states to a vibrational mode with energy 23 meV. This is consistent with a vibrational mode identified in previous Raman experiments in similar bili-proteins [6].

An additional coherence with energy difference of 100 meV is also consistently present and is the longest and strongest signal we measure. Figure 1a shows this coherence for the case when the excitation energies were 2.179 and 2.091 eV. It can also be seen in this figure that there are other signal contributions in this vicinity with different coherence energies and coherence times.

To help separate these different contributions we performed an additional experiment where the polarizations of the excitation pulses and signal were oriented in a configuration ($0°$, $90°$, $-45°$, $45°$). In this configuration signal pathways where the first two pulses and the second two pulses interact with transitions dipoles that are parallel are eliminated [5]. This polarization scheme is thus predicted to eliminate purely vibrational coherences. Figure 1b shows the result of these experiments. The long-lived 100 meV coherence is removed, but there remain clear signal contributions at ~ 86 and ~ 108 meV. It is evident then that these coherences involve states with non-parallel transition dipoles, while the 100 meV coherence involves states with parallel transition dipoles and is likely due to an additional vibrational mode. Indeed, previous work [6] on similar complexes has shown a vibrational mode at around this energy. The coherence at 86 meV is attributed to electronic coherence between the two DBV excitonic states, while the 108 meV coherence is expected to involve these same electronic states, but an additional excitation of the 24 meV vibrational mode identified earlier [5].

Based on these two electronic states and two vibrational modes we are able to determine an energy level scheme that explains all of our results in this spectral range, as well as the results observed by other groups exploring coherences in PC654. Furthermore, the 86 meV coherence also closely matches a coherence seen by Turner et al. [3] in PC645 that they suggested may be electronic based on comparisons of rephasing and non-rephasing data.

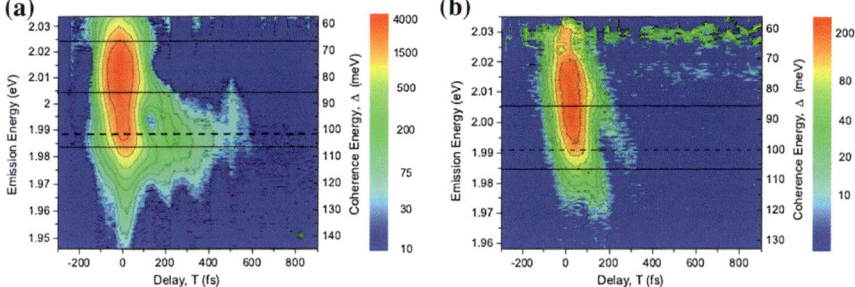

Fig. 1 The measured signals for pulse energies (E1, E1, E2, E3) = (2.179, 2.091, 2.091 eV) with **a** all pulses polarized parallel (0,0,0,0) and **b** the cross-polarized scheme, (0; 90; −45; 45), as described in the text

One additional factor that is raised by these experiments is the extent to which electronic and vibrational degrees of freedom can be separated. It would appear from these results and previous works that they are strongly coupled and so it may be more appropriate to think of these states as vibronic rather than electronic states. In which case, the question of whether the coherences observed are electronic of vibrational in nature becomes misleading and unhelpful. Rather, perhaps the question we should be asking, and which we begin to explore with these results, is how the interactions between the electronic and vibrational degrees of freedom modify the excited state landscape and potentially enhance energy transfer in light harvesting complexes.

3 Conclusions

The ability to selectively excite coherence pathways has allowed us to separate and identify both electronic and vibrational coherences and observe their evolution. In the present work this has allowed us to determine a self-consistent energy level-scheme that describes all reported observations of coherences in PC645. This is based on two electronic states separated by 84 meV, which we attribute to the DBV+ and DBV− states, and two vibrational modes with energy 23 and 100 meV. With the approach used we are able to reveal more spectral details than broadband experiments and have greatly enhanced the understanding of the excited state structure.

References

1. G. S. Engel, T. R. Calhoun, E. L. Read, T. Ahn, T. Mančal, Y-C. Cheng, R. E. Blankenship and G.R. Fleming, Nature **446**, 782 (2007)
2. E. Collini, C.Y. Wong, K. E. Wilk, P. M. G. Curmi, P. Brumer and G. D. Scholes, Nature, **463**, 644-647 (2010)

3. D.B. Turner, R. Dinshaw, K.K. Lee, M.S. Belsley, K.E. Wilk, P.M.G. Curmi, and G.D. Scholes, Phys. Chem. Chem. Phys 14, 4857 (2012)
4. G. H. Richards, K. E. Wilk, P. M. G. Curmi, H. M. Quiney, and J. A. Davis, J. Phys. Chem. Lett., 3, 272–277 (2012)
5. G. H. Richards, K. E. Wilk, P. M. G. Curmi, and J. A.Davis, J. Phys. Chem.Lett., 5, 43-49 (2014)
6. J. M. Womick, B. A. West, N. F. Scherer, A. M. Moran, J. Phys B 45, 154016 (2012).

Ultrafast Energy Flow and Equilibration Dynamics in Photosynthetic Light-Harvesting Complexes

Margherita Maiuri, Larry Lüer, Sarah Henry, Anne-Marie Carey,
Richard J. Cogdell, Giulio Cerullo and Dario Polli

Abstract We disentangle various energy transfer pathways in the bacterio-chlorophyll excitation cascade from LH2 to LH1 in *Chromatium vinosum* grown under high-light or low-light illumination using tunable narrowband selective excitation and broadband infrared probing.

Purple bacteria are excellent model organisms to study the basic light-harvesting (LH) mechanisms in Nature because of their simplicity and the availability of high-resolution X-ray crystallographic structures [1]. Their photosynthetic unit exhibits a quasi-2D architecture made up of circular LH pigment-protein complexes: the core LH1-reaction center (RC) complex is surrounded by several peripheral LH2 complexes. Through a cascading effect, the absorbed energy is efficiently transferred from high to low photon-energy complexes towards the RC.

Here we investigate the energy transfer (ET) pathways in the photosynthetic membranes of *Chromatium vinosum*, grown either under high-light (HL) or low-light (LL) conditions, using pump-probe spectroscopy in the near-infrared (NIR) region [2]. In these bacteria, illumination intensity during growth strongly affects the type of LH2 complexes synthesized and their optical spectra. We designed a tailored pump-probe apparatus for this study [2]. To provide selective excitation of each BChl species we generated narrowband ($\approx$5 nm) pump pulses, tunable from 1.57 to 1.30 eV photon energy, by an optical parametric amplifier. Broadband probe pulses in the NIR spectral region were generated via supercontinuum in a sapphire plate and detected by a single-shot low-noise spectrometer.

M. Maiuri · G. Cerullo · D. Polli (✉)
Dipartimento di Fisica, Politecnico di Milano, P.zza L. Da Vinci 32, 20133 Milan, Italy
e-mail: dario.polli@polimi.it

L. Lüer
Department of Nanoscience, Madrid Institute for Advanced Studies,
28049 Cantoblanco, Spain

S. Henry · A.-M. Carey · R.J. Cogdell
Institute for Molecular, Cell and Systems Biology, University of Glasgow,
Glasgow G12 8TA, UK

© Springer International Publishing Switzerland 2015
K. Yamanouchi et al. (eds.), *Ultrafast Phenomena XIX*,
Springer Proceedings in Physics 162, DOI 10.1007/978-3-319-13242-6_146

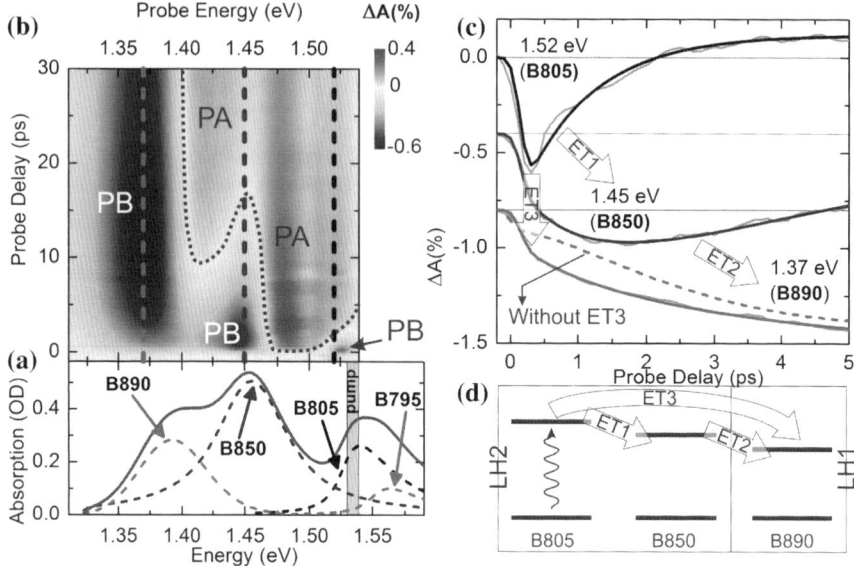

Fig. 1 **a** Ground-state absorption spectrum of THL40 and its decomposition into BChl bands.
b Pump-probe map of THL40 upon 1.53 eV excitation. **c** Measured (*thin lines*) and fitted (*thick
lines*) pump-probe dynamics at selected probe energies. *Dashed line* is a simulated curve
considering only ET1 and ET2 channels active. **d** Level scheme indicating the ET processes

Figure 1a shows the absorption spectrum of the THL40, grown under HL illumination. We observe the presence of three different Bacterio-Chlorophylls (BChls)
in the LH2 complexes, named B795, B805 and B850, while the LH1 only contains
B890 BChls. Figure 1b shows the measured transient absorption (ΔA) map for the
THL40 sample as a function of probe energy and delay after excitation at 1.53 eV,
thus predominantly pumping the B805 BChls.

The pump-probe map can be decomposed into three contributions from the
B805, B850 and B890 BChls, each providing a transient spectrum made of a
negative exciton photobleaching (PB) signal and a positive photo-induced
absorption (PA) signal from the exciton to the bi-exciton state, slightly blue-shifted
in energy with respect to the corresponding PB. Around time zero the predominant
contribution is from B805. Within LH2 complexes a first ET process (ET1) occurs
from B805 to B850, responsible for the rise of the signal at 1.45 eV completed
within ≈2 ps. This is also clear looking at the pump-probe traces at selected probe
energies plotted in Fig. 1c. Subsequent ET from the B850 in LH2 towards the B890
in LH1 (called ET2) occurs in a few ps and is responsible for the delayed formation
of the signal at 1.37 eV. The B890 relaxes back to the ground state in hundreds of
ps (not shown here). We performed global analysis of the data (see the fits as thick
solid lines in Fig. 1c): the comparison of the extracted decay-associated spectra
clearly confirmed the existence of the ET3 mechanism. Looking into the early
dynamics of Fig. 1c, we note two further very important details: (i) The transient

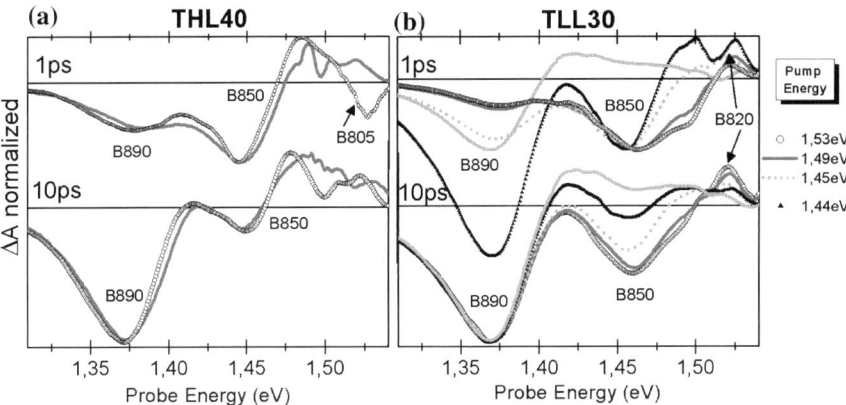

Fig. 2 Transient absorption spectra for THL40 (**a**) and TLL30 (**b**) samples at various pump photon energies at selected delays: 1 ps (*top panels*, normalized for the B850 signal) and 10 ps (*bottom panels*, normalized for the B890 signal)

signals at 1.45 and 1.37 eV show a partial instantaneous formation (within the pump pulse) due to the fact that at 1.53 eV pump photon energy there is also a partial excitation of the B850 and B890 BChls (see Fig. 1a). (ii) After this instantaneous formation, the signal at 1.37 eV, which purely monitors the dynamics of the B890 moiety, further rises with a slope which is steepest at early delays (i.e. the slope reduces in time and reaches a plateau around 15 ps delay). This demonstrates that a direct, one-step B805 (in LH2) → B890 (in LH1) ET is also active in the THL40 sample (named ET3), parallel to the sequential B800 → B850 → B890 two-step ET1 and ET2 processes. If this was not the case, the dynamics would look different: see for comparison the simulated time trace (dashed line in Fig. 1c), showing an inflection point at early delays. A scheme of the energy levels and ET processes is sketched in Fig. 1d.

We performed a thorough study of the energy flow and equilibration among LH2 and LH1 as a function of pump photon energy and illumination condition during growth. ΔA spectra at selected delays for various excitation energies are reported for the THL40 sample (Fig. 2a) and for the TLL30 sample (Fig. 2b), which was grown under LL illumination and contains an extra BChl in the LH2, the B820. In THL40 (Fig. 2a), the spectral shapes are weakly dependent on the pump energy. At 1 ps delay we can see the sharp PB bands of B805 in THL40, peaked at 1.53 eV, which is not present in TLL30 due to the overlap with the B820 band. In TLL30 (Fig. 2b) on the contrary we observe a strong dependence on pump excitation: (i) at B850 (1.45 eV) and B890 (1.40 eV) no signature from B820 is found at 1 ps and 10 ps delays, suggesting that no significant back transfer from B850 nor B890 occurs towards B820. (ii) Upon B805 excitation at 1.53 eV, we observe a very strong B820 feature at 1 ps delay, indicating efficient B805 → B820 ET. (iii) In all the transient spectra with a B820 signature, the B820/B850 ratio at 10 ps delay is small but not vanishing: the low B820 → B850 ET rate interestingly suggests that the B820 and B850 BChls reside in different LH2 complexes.

Our study sheds new light onto the selective advantage for the *Chromatium vinosum* bacterium to synthetize different LH2 complexes under HL/LL growth conditions.

References

1. R.J. Cogdell, A. Gall, J. Köhler, "The architecture and function of the light-harvesting apparatus of purple bacteria: From single molecule to in vivo membranes" Q. Rev. Biophys. **39**, 227-324 (2006).
2. L. Lüer *et al.*, "Tracking energy transfer between light harvesting complex 2 and 1 in photosynthetic membranes grown under high and low illumination" Proc. Natl. Acad. Sci. USA **109**, 1473-1478 (2012).

Primary Process in Light-Harvesting Complex Studied by Pump-Repump-Probe Spectroscopy

K. Sobue, K. Abe, S. Sakai, M. Nango, H. Hashimoto and M. Yoshizawa

Abstract Dark excited states of carotenoid in LH1 complex have been investigated by measuring recovery dynamics following the repump. The S* state in LH1 is different from the S_1 state but is similar to the T state.

1 Introduction

Carotenoids (Cars) have light-harvesting (LH) and photoprotecting functions in photosynthesis [1, 2]. In the LH process of bacterial photosynthesis, light energy is absorbed by Car and transferred to nearby bacteriochlorophyll (BChl) as shown in Fig. 1. The S_2 state in Car is the lowest optically allowed singlet excited state capturing the light energy. The photoexcited S_2 state relaxes to the S_1 state in femtosecond time scale, but excitation energy transfer (EET) to BChl occurs efficiently competing with the ultrafast relaxations. The recently assigned S* state has considerable importance in LH complexes [3, 4]. The S* state in LH1 is a precursor of the triplet (T) state [5] and has similar Raman signal with the T state [6]. However, essential properties of the $S*_{LH1}$ state have not been well-understood.

K. Sobue · K. Abe · M. Yoshizawa (✉)
Department of Physics, Graduate School of Science, Tohoku University,
6-3 Aramaki-Aza-Aoba, Aoba-Ku, Sendai 980-8578, Japan
e-mail: m-yoshizawa@m.tohoku.ac.jp

S. Sakai
Department of Life and Materials Engineering, Nagoya Institute of Technology,
Nagoya 466-8555, Japan

M. Nango · H. Hashimoto
OCARINA, Osaka City University, Osaka 558-8585, Japan

H. Hashimoto
Department of Physics, Graduate School of Science, Osaka City University,
Osaka 558-8585, Japan

© Springer International Publishing Switzerland 2015   599
K. Yamanouchi et al. (eds.), *Ultrafast Phenomena XIX*,
Springer Proceedings in Physics 162, DOI 10.1007/978-3-319-13242-6_147

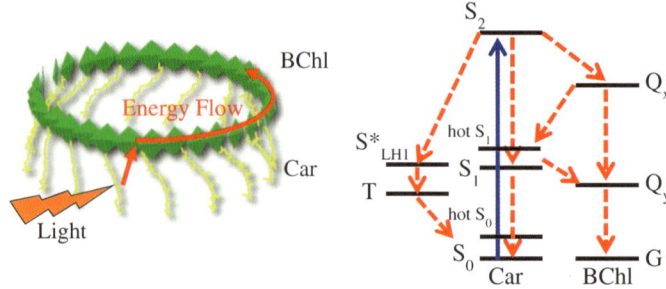

Fig. 1 Schematic picture and energy level diagram of LH1 complex

In this study, the S_1, S^*_{LH1}, and T states in reconstituted LH1 complex have been investigated by pump-repump-probe spectroscopy. Relations and properties of the excited states are discussed by comparing dynamics after selective re-excitation of the excited states.

2 Experimental

The reconstituted LH1(Sph) was prepared from *Rhodospirillum rubrum* G9+ with purified spheroidene [7]. Native LH1 contains spirilloxanthin as a major carotenoid but spheroidene was used in this study, because the S^*_{LH1} signal is more pronounced in LH1(Sph) than in LH1 with spirilloxanthin. The solution of LH1(Sph) was dispersed in a poly-vinyl alcohol film on a glass plate. During the laser spectroscopic measurements, the sample was translated to avoid sample degradation and accumulation of any potential photoproducts.

The femtosecond pump-repump-probe spectroscopy setup was based on an amplified mode-locked Ti:Sapphire laser system. Parts of the amplified pulse were used to drive two independent optical parametric amplifiers generating the first pump and the second repump. Wavelength and delay time of the repump pulse were controlled for selective re-excitation of the excited states generated by the first pump pulse. Additional absorbance change ($\Delta\Delta A$) induced by the repump pulse was observed by white continuum probe.

3 Results and Discussion

Figure 2 shows absorbance change (ΔA) spectra of LH1(Sph) following the first S_2 pump (500 nm, 100 fs). The signal at a delay time of 2 ps has two peaks at 525 nm and 550 nm. They are assigned to the S^*_{LH1} and S_1 states, respectively. A small dip at 590 nm is bleaching of the Q_x state of BChl. The S_1 state relaxes with a time

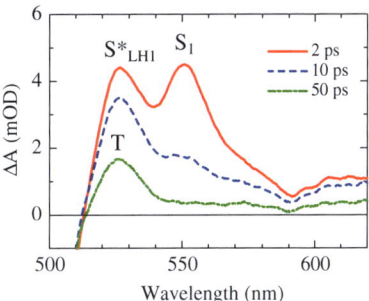

Fig. 2 Absorbance change in LH1(Sph) following the S_2 pump

constant of 5.3 ps. The S^*_{LH1} signal decreases with a time constant of 16 ps, but a long-lived component remains much longer than 100 ps at 525 nm. It is assigned to the T state.

The 560 nm repump pulse at a delay time of 2 ps was used to re-excite the S_1 state. The $\Delta\Delta A$ spectrum at 0.2 ps after the repump has a negative peak at 550 nm as shown in Fig. 3. It is assigned to bleaching of the S_1 transient absorption. The S_1 bleaching recovers with a time constant of 0.3 ps. The signal assigned to the S^*_{LH1} state does not affected by the S_1 repump.

The $\Delta\Delta A$ spectrum following the S^*_{LH1} repump (525 nm, 2 ps) has a negative peak at 525 nm and broad tail around 550 nm. The 525 nm signal is assigned to bleaching of the S^*_{LH1} state. It recovers with a time constant of 0.14 ps. On the other hand, the broad tail doesn't have the fast recovery. The broad tail is assigned to bleaching of the hot S_0 state, because the transient absorption of the S_2 state generated from the hot S_0 state is observed. The 525 nm repump excites both the S^*_{LH1} and the hot S_0 states. The independent $\Delta\Delta A$ signals between the 560 nm repump and the 525 nm repump show that the highly excited states from the S_1, S^*_{LH1}, and hot S_0 states are independent.

The 525 nm repump at a delay of 50 ps after the S_2 pump re-excites the T state. The $\Delta\Delta A$ signal is similar to that of the S^*_{LH1} repump except lack of the broad tail. It is consistent with the assignment of the broad tail to the hot S_0 state. The

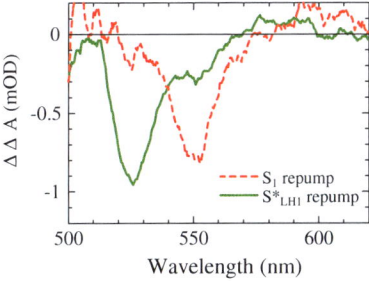

Fig. 3 Absorbance changes at 0.2 ps following the S_1 repump (560 nm, 2 ps) and the S^*_{LH1} repump (525 nm, 2 ps)

bleaching at 525 nm recovers with a time constant of 0.15 ps. The 525 nm bleaching and its recovery time suggest that the re-excited state from the T state is equal to that from the S^*_{LH1} state. However, the S^*_{LH1} state has been distinguished from the T state by the transient changes of the photoinduced absorption and Raman signal [6]. The transient state so-called the S^*_{LH1} state should be assigned to the vibrational excited triplet (hot T) state.

LH1(Sph) has more effective S^*_{LH1} formation and EET from the S_2 state to the Q_x state in BChl than LH1 with spirilloxanthin. The ultrafast formation of the triplet excited state may be due to interaction with BChl.

4 Conclusion

The transient absorbance change previously assigned to the S* state of carotenoid in LH1 has two components. The broad repump signal with slow recovery is assigned to the hot S_0 state. The sharp repump signal at 525 nm with fast recovery is tentatively assigned to the hot T state.

References

1. R.J. Cogdell, A. Gall, and J. Köhler,, Quart. Rev. Biophys. **39**, 227-324 (2006).
2. T.Polívka and V.Sundström, Chem. Rev. **104**, 2021-2071 (2004).
3. E. Papagiannakis, J.T.M. Kennis, I.H.M. van Stokkum, R.J. Cogdell, and R. van Grondelle, Proc. Natl. Acad. Sci. USA **99**, 6017-6022, (2002).
4. H. Cong, D.M. Niedzwiedzki, G.N. Gibson, A.M. LaFountain, R.M. Kelsh, A.T. Gardiner, R. J. Cogdell, and H.A. Frank, J. Phys. Chem. B **112**, 10689-10703 (2008).
5. R. Nakamura, K. Nakagawa, M. Nango, H. Hashimoto, and M. Yoshizawa, J. Phys. Chem. B **115**, 3233-3239 (2011).
6. O. Yoshimatsu, K. Abe, S. Sakai, T. Horibe, R. Fujii, M. Nango, H. Hashimoto, and M. Yoshizawa, Ultrafast Phenomena XVIII (EDP Sciences 2013), p.08007.
7. K. Nakagawa, S. Suzuki, R. Fujii, A.T. Gardiner, R.J. Cogdell, M. Nango, and H. Hashimoto, J. Phys. Chem. B **112**, 9467-9475 (2008).

Ultrabroadband Two-Dimensional Spectroscopy by a Birefringent Delay Line

J. Réhault, A. Oriana, M. Maiuri, D. Brida, D. Polli, C. Manzoni
and G. Cerullo

Abstract We introduce a passive birefringent delay line for the generation of collinear, interferometrically locked ultrashort pulse pairs. Their delay is controlled with attosecond precision and stability $<\lambda/360$, enabling two-dimensional electronic spectroscopy from UV to infrared.

Two-dimensional electronic spectroscopy (2DES) allows fundamentally new insights into the structure and dynamics of multi-chromophore systems, by measuring how the electronic states of chromophoric units within a complex molecular architecture interact with one another and transfer electronic excitations [1]. The main technical challenge of 2DES is the requirement of interferometric stability between pulse pairs, which need to be phase-locked to within a small fraction of their carrier wavelengths. This becomes more and more difficult when moving to shorter wavelengths, particularly the UV range, where most of the molecules of biochemical interest absorb [2].

So far two schemes have been successfully used to implement 2DES in the visible range: the heterodyne detected three-pulse photon echo [3] and the partially collinear pump-probe geometry [4]. The latter employs two phase-locked collinear pump pulses and a non-collinear probe pulse which is dispersed on a spectrometer, thus working in a self-heterodyning mode. Advantages of this configuration are its simplicity, its applicability to pump-probe setups and the fact that it naturally retrieves absorptive spectra. Phase-locked collinear pulses have been generated by actively stabilized interferometers or by pulse shapers. The former solution is very difficult to implement with short wavelengths, while the latter considerably increases the complexity of the setup and provides limited spectral acceptance.

J. Réhault · A. Oriana · M. Maiuri · D. Polli · C. Manzoni · G. Cerullo (✉)
IFN-CNR, Dipartimento di Fisica, Politecnico di Milano,
Piazza Leonardo da Vinci 32, 20133 Milano, Italy
e-mail: giulio.cerullo@polimi.it

D. Brida
Department of Physics and Center for Applied Photonics, University of Konstanz,
Universitätsstraße 10, 78464 Konstanz, Germany

© Springer International Publishing Switzerland 2015
K. Yamanouchi et al. (eds.), *Ultrafast Phenomena XIX*,
Springer Proceedings in Physics 162, DOI 10.1007/978-3-319-13242-6_148

603

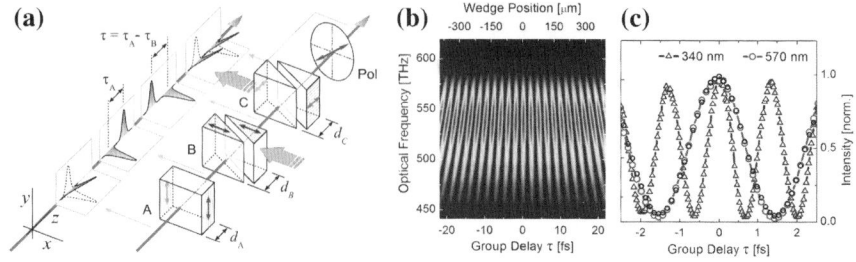

Fig. 1 **a** Scheme of the TWINS setup for the generation of phase-locked pulses; **b** spectrally-resolved linear autocorrelation of the visible NOPA pulses (center wavelength at 570 nm); **c** *large and small circles* detail of two correlation measurements for a visible pulse acquired at 30 min distance; *blue triangles* detail of a correlation trace of a UV pulse centered at 340 nm

In this work we introduce the Translating-Wedge-based Identical pulses eNcoding System (TWINS) as a compact optical device generating two phase-locked collinear replicas of an input pulse with a passive approach [5] that ensures simplicity and broadband operation. The device concept is inspired by the Babinet–Soleil compensator, but adapted to ultrabroadband pulse operation. With the proper choice of materials, TWINS can be implemented in a broad range of wavelengths, spanning from the ultraviolet to the mid-IR, and can support sub-10-fs pulses.

The TWINS device (see Fig. 1a) consists of three building blocks made of a birefringent material. Block A is a parallel faces plate with the optical axis aligned along the y direction and with fixed thickness d_A; block B consists of two wedges with the optical axis along the x direction and with an overall thickness d_B which can be varied by the insertion of one of the wedges. In this way we can control with very high precision the relative delay between the two orthogonally polarized pulse replicas, which is $\tau = (d_A - d_B)\delta_{eo}$, where $\delta_{eo} = 1/v_{ge} - 1/v_{go}$ is the group velocity mismatch between ordinary and extraordinary polarizations. Block C is a pair of wedges, cut at the same angle as B, with optical axis aligned along the propagation direction: it allows to keep constant the arrival time of one of the two pulses and to compensate, to the first order, the dispersion introduced by the variable thickness of block B. After the wedge sequence, a polarizer projects the two delayed and orthogonally polarized pulse replicas to a common linear polarization.

We constructed a TWINS device based on α-barium borate (α-BBO) as bire-fringent material and tested it with broadband visible pulses generated by a Non-collinear Optical Parametric Amplifier (NOPA). The wedges are mounted on a translation stage with 0.1 μm positioning accuracy; this allows controlling the delay down to ≈3.6 as, with a large demultiplication of the wedges translation. The two collinear pulse replicas produced by the device are sent to a spectrometer and their relative delay is characterized by spectral interferometry. By recording a sequence of interferograms for a fixed position of the wedges, for an observation time of 30 min, we can extract rms fluctuations of the relative phases of ≈17 mrad, cor-responding to delay fluctuations 1/360th of the optical cycle (≈5 as at 600-nm),

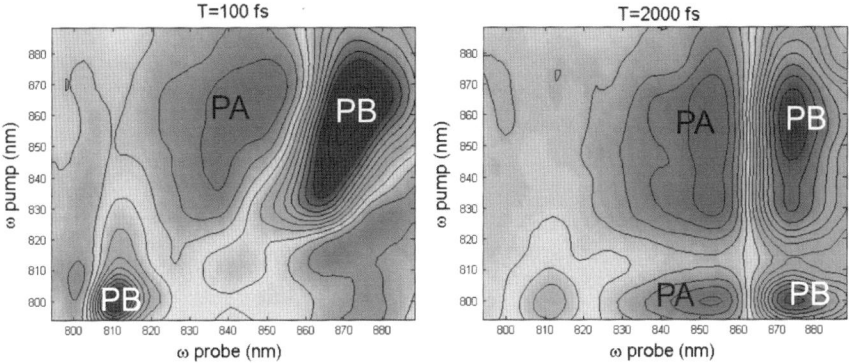

Fig. 2 2DES spectra of the LH2 complex of *Rps. Acidophila* obtained with the TWINS setup and 13-fs pulses at 850 nm. *PB* photobleaching; *PA* photoinduced absorption

mainly limited by laser intensity noise. The excellent phase-locking stability of the device is ascribed to the common optical path followed by the two pulse replicas.

Figure 1b shows a 2D map reporting a sequence of pulse pair interferograms for different values of the wedges insertion, i.e. of the delay τ. The map is a spectrally resolved linear autocorrelation of a pulse at 570 nm with symmetry axis around τ = 0. Small and large circles in panel (c) report portions of two autocorrelation traces acquired at 30 min distance; these data show that our device is able to produce a noise-free linear correlation trace in the visible spectral range, which would be particularly demanding in interferometer-based correlators. In addition, nearly perfect reproducibility demonstrates the extremely high accuracy and long-term stability of the pulse pair generator. The device has also been tested with UV pulses with 340 nm carrier wavelength, and it exhibits accuracy comparable to the one in the visible spectral range (triangles in Fig. 1c).

The TWINS device can be inserted into the pump arm of a high-time-resolution pump-probe apparatus to turn it into a 2DES setup. As an example, Fig. 2 reports 2DES spectra of the LH2 complex of *Rhodopseudomonas Acidophila* purple bacterium, excited by broadband 13-fs pulses produced by an IR NOPA, covering the absorption of B800 and B850 bacteriochlorophylls. One clearly observes energy transfer from B800 to B850 on the picosecond timescale by the build-up of the cross-peak.

We believe that TWINS has the advantages of simplicity, broadband operation and extreme delay precision which make it especially promising for extension of 2DES spectroscopy to the UV range.

Acknowledgments The work was supported by the European Research Council Advanced Grant STRATUS (ERC-2011-AdG No. 291198). JR thanks the Swiss National fund for financial support (Grant No. 200020 129938).

References

1. S. Mukamel, "Multidimensional femtosecond correlation spectroscopies of electronic and vibrational excitations," Annu. Rev. Phys. Chem. 51, 691 (2000).
2. J. Jiang and S. Mukamel, "Two Dimensional Ultraviolet (2DUV) Spectroscopic Tools for Identifying Fibrillation Propensity of Protein Residue Sequences", Angew. Chem. Int. Ed. 49, 9666 (2010).
3. T. Brixner, I.V. Stiopkin, and G. R. Fleming, "Tunable two-dimensional femtosecond spectroscopy," Opt. Lett. 29, 884 (2004).
4. E. M. Grumstrup, S. H. Shim, M. A. Montgomery, N. H. Damrauer, and M. T. Zanni, "Facile collection of two-dimensional electronic spectra using femtosecond pulse-shaping technology," Opt. Express 15, 16681 (2007).
5. D. Brida, C. Manzoni and G. Cerullo, "Phase-locked pulses for two-dimensional spectroscopy by a birefringent delay line" Opt. Lett. 37, 3027 (2012).

Part VIII
THz Generation and Application

Filling the Entire Terahertz Frequency Gap by Single-Cycle MV/Cm Pulses

C. Vicario, B. Monoszlai, F. Ardana-Lamas and C.P. Hauri

Abstract We demonstrate highly efficient Terahertz production and absolute phase control in the hardly accessible THz frequency gap (1–15 THz) by optical rectification in organic crystals leading to single-cycle field oscillations beyond 150 MV/m and 0.5 Tesla.

OCIS codes 140.3070

1 Introduction

Intense radiation in the Terahertz frequency gap (0.1–15 THz) has been challenging in the past. Single-cycle Terahertz transients at high fields are a novel and adequate tool to investigate fundamental properties of condensed matter and ultimately to initiate coherently collective motions directly by light [1]. While intense pulses at mid-infrared frequencies between 20–100 THz have been recently demonstrated [2], the production of single-cycle fields >1 MV/cm in the Terahertz gap remains a challenge. Here we demonstrate single-cycle pulses generated in the organic crystals DAST, OH1 and DSTMS with unprecedented high fields (up to 1.5 MV/cm and 0.5 Tesla) covering the THz gap [3–5]. The direct control of the absolute phase is also presented, which has shown to be an important technological step towards high-field THz science [6].

2 Experimental Results

A TW Ti:Sa amplifier system at 100 Hz drives an optical parametric amplifier (OPA) which delivers pulses with 60 fs FWHM duration and energy up 3 mJ at wavelengths between 1.35 and 1.5 μm.

C. Vicario (✉) · B. Monoszlai · F. Ardana-Lamas · C.P. Hauri
SwissFEL, Paul Scherrer Institute, 5232 Villigen-PSI, Switzerland
e-mail: carlo.vicario@psi.ch

F. Ardana-Lamas · C.P. Hauri
Ecole Polytechnique Federale de Lausanne, 1015 Lausanne, Switzerland

© Springer International Publishing Switzerland 2015
K. Yamanouchi et al. (eds.), *Ultrafast Phenomena XIX*,
Springer Proceedings in Physics 162, DOI 10.1007/978-3-319-13242-6_149

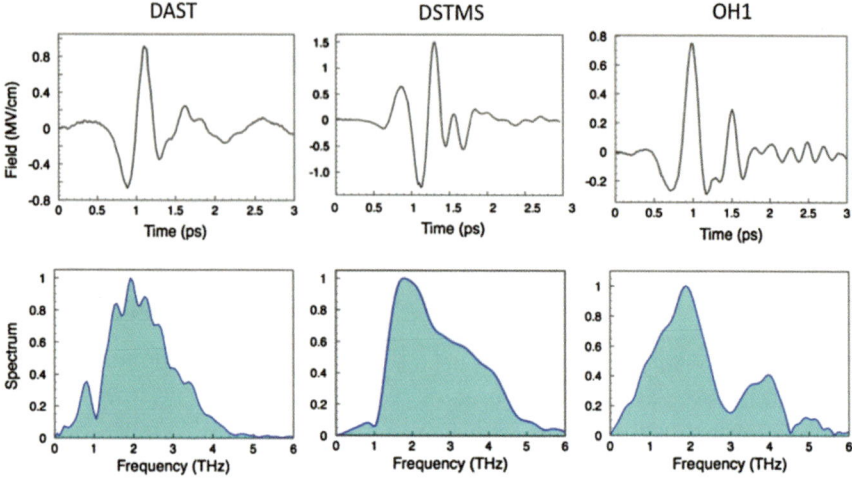

Fig. 1 High-field THz field generated in DAST, OH1, DSTMS (*upper part*) and corresponding multi-octave spectra (*lower part*)

This source is used to pump the organic crystal in order to provoke intense Terahertz emission by optical rectification. The single-cycle Terahertz pulses are emitted collinearly with the pump pulse in a close-to-collimated manner. Different to other prominent THz schemes based on optical rectification in Lithium Niobate, for example, our scheme provides highest conversion efficiency (up to 2 %) without cumbersome pulse front tilting and imaging and generates single-cycle THz transient paired to broadband spectra (Fig. 1).

Our simple setup allows for excellent focusing capabilities and THz spot size close to the diffraction limit. Equally important for high-field experiment is the control and the manipulation of the absolute phase, as shown in Fig. 2. The absolute phase control of ultra broadband THz pulses is non-trivial and has been achieved by combining dispersion properties of different materials.

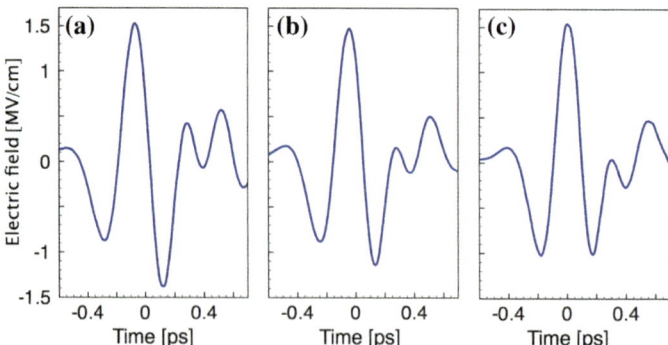

Fig. 2 Direct manipulation of the THz pulse absolute phase by dispersion management

Fig. 3 Multi-octave spectrum filling the THz gap generated in thin OH1

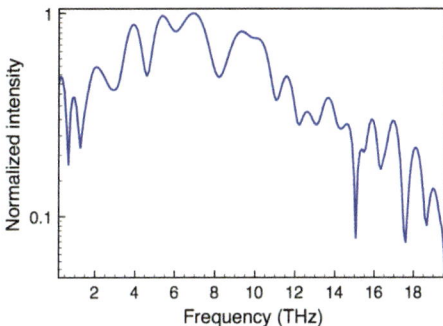

Finally we demonstrate THz spectra filling the THz gap generated in 200 μm thin OH1 crystals. The spectrum covers the full THz gap and continues further up to 20 THz (Fig. 3). The source offers the capability to produce highly asymmetric electric field shapes.

In conclusion we demonstrate record-high field strength in the hard-to-access Terahertz frequency range between 1–15 THz. The emitted spectrum covers more than 7 octaves and give rise to single cycle pulses with field strength beyond the MV/cm barrier. The availability of such pulses, and the technology to control their absolute phase, is going to pave the way towards nonlinear Terahertz photonics.

References

1. M. Tonouchi, "Cutting-edge terahertz technology," Nature Photon. 1, 97 (2007).
2. F. Jungiger, A. Sell, O. Schubert, B. Mayer, D. Brida, M. Marangoni, G. Cerullo, A. Leitenstorfer, R. Huber, "Single-cycle multiterahertz transients with peak fields above 10 MV/cm," Opt. Lett. 35, 2645 (2010).
3. C.P. Hauri, C. Ruchert, C. Vicario and F. Ardana Lamas, "Strong-field single-cycle THz pulses generated in an organic crystal", Appl. Phys. Lett. **99**, 161116 (2011).
4. C. Ruchert et al. "Scaling Sub-mm single-cycle transients towards MV/cm fields via optical rectification in organic crystal OH1," Opt. Lett. 37, 899 (2012).
5. C. Ruchert et al. Phys. Rev. Lett. 110, 123902 (2013).
6. C. Vicario et al. Nature Photon. 7, 720 (2013).

Terahertz Imaging with Optical Resolution by Femtosecond Laser Filament in Air

Jiayu Zhao, Lanjun Guo and Weiwei Liu

Abstract We introduce a sub-wavelength resolution THz imaging technique which uses the THz radiation generated by a femtosecond laser filament in air as the probe, based on the fact that the femtosecond laser filament forms a waveguide for the THz wave in air. The diameter of the THz beam, which propagates inside the filament, varies from 20 to 50 μm, which is significantly smaller than the wavelength of the THz wave. Using this highly spatially confined THz beam as the probe, THz imaging with resolution as high as 20 μm ($\sim \lambda/38$ at 0.4 THz) is promising.

1 Introduction

Resolution enhancement of terahertz (THz) imaging is one of central concerns in the THz science and technology research. Because of the used long wavelength (λ_1 $_{THz}$ = 300 μm), THz imaging's resolution is generally in the scale of millimetre. This constitutes a major obstacle for the application of THz imaging in bio-medical diagnosis and semiconductor device inspection [1–3]. Here, we report on a novel sub-wavelength THz imaging method via the femtosecond laser filament in air [4].

2 Experimental Results

2.1 THz Sub-wavelength Imaging by Femtosecond Laser Filament

Figure 1 schematically shows the experimental setup. A two-color (1 kHz, 50 fs, 1 mJ/pulse, 800 + 400 nm) laser filament was created at the focus of the lens (f = 30 cm). The generated THz pulse was detected by a standard electric-optic

J. Zhao · L. Guo · W. Liu (✉)
Institute of Modern Optics, Nankai University, Key Laboratory of Optical Information
Science and Technology, Ministry of Education, Tianjin 300071, China
e-mail: liuweiwei@nankai.edu.cn

© Springer International Publishing Switzerland 2015
K. Yamanouchi et al. (eds.), *Ultrafast Phenomena XIX*,
Springer Proceedings in Physics 162, DOI 10.1007/978-3-319-13242-6_150

Fig. 1 Experimental setup

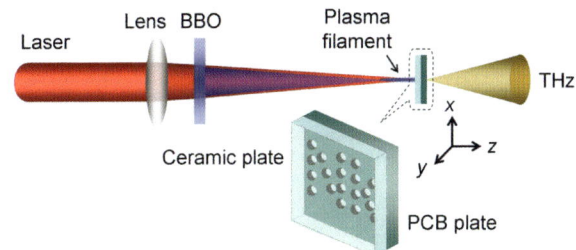

sampling (EOS) setup [5]. Sub-wavelength resolution THz imaging was carried out by inserting a Printed-Circuit-Board plate with multiple through-holes in the middle of the filament. A ceramic plate was placed in front of and in touch with the PCB plate. The ceramic plate terminates the filament and excludes the laser damage of the PCB plate. The THz power transmission of the ceramic plate is about 50 %. On the other hand, the PCB plate has poor transmission of THz wave. Mainly the THz energy passing through the holes could be detected by EOS setup. The THz image of the holes on the PCB plate was taken by moving the two plates together in the $x - y$ plane (z axis is defined as the laser propagation direction). The step sizes were 100 and 100 μm along x and y axes, respectively.

Figure 2a illustrates the image of the multiple holes under optical microscope (resolution: 5 μm). The diameter of each hole is about 600 μm. The holes form two characters of "NK", abbreviation for NanKai. The corresponding scanning THz image is displayed in Fig. 2b. Comparing Fig. 2a, b, no significant blurring effect could be noticed. Reminding ourselves that the peak wavelength of the THz pulse generated in our experiment is about 750 μm, which is even larger than the size of the holes on the PCB plate. However, Fig. 2 indicates that the minimum resolvable structure by THz imaging is less than 100 μm. For example, according to the optical image (Fig. 2a), three hole pitches pointed by arrow A, B and C have characterized widths of about 60, 80 and 75 μm, respectively. And they can be clearly resolved by the THz imaging (Fig. 2b). Hence, the resolution of the obtained THz image by our method is much smaller than the THz pulse wavelength.

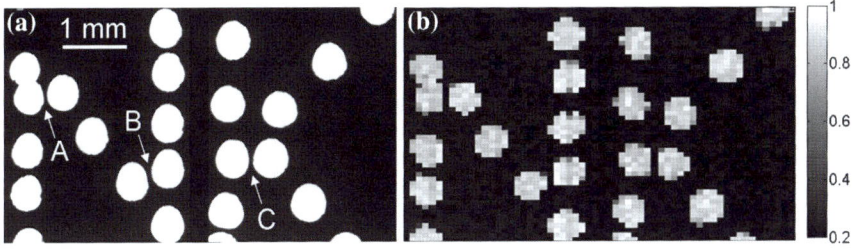

Fig. 2 Comparing optical microscope image (**a**) and THz image (**b**)

Fig. 3 THz beam diameter
d as a function of z

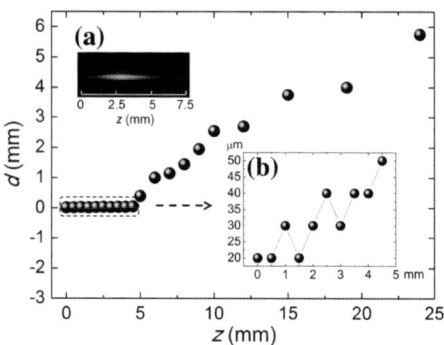

2.2 Resolution of THz Imaging (Diameter of THz Beam Emission from Femtosecond Laser Filament)

The knife-edge method [6] was applied to measure the THz beam diameter at different positions z, which determines the THz imaging resolution in our experiment. Note that $z = 0$ corresponds to the starting position where significant THz signal was able to be detected. The obtained THz beam diameters d as a function of z is depicted in Fig. 3. What Fig. 3 impresses us most is that the THz pulse energy is spatially constrained inside a space which is much smaller than the wavelength. The jump of the curve in Fig. 3 taking place between $z = 5$ mm and $z = 6$ mm gives a hint that the THz wave is in fact strongly guided from $z = 0$ mm to $z = 5$ mm. Coincidentally, this region superposes with the zone where significant plasma, i.e. a filament, was produced during the filamentation (see Fig. 3a). We have reason to believe that a THz waveguide is created inside the filament. It is this waveguide that strongly confines the THz energy into a region with only a few ten microns in diameter. And this phenomenon makes a novel sub-wavelength THz imaging technique feasible as we have demonstrated.

3 Numerical Simulation

In order to confirm that creation of a THz waveguide by the filamentation, we have also calculated the THz wave eigenmodes by the full-vector finite-element method (FEM) with the commercial software COMSOL Multiphysics [7]. The THz doublet degenerated modes (diameters <100 μm) localized in the filament area are found in our simulation, as shown in Fig. 4.

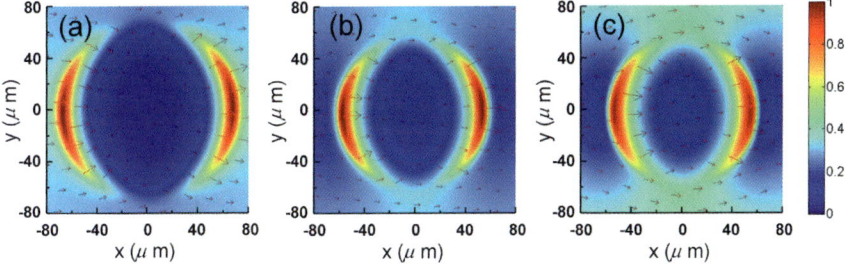

Fig. 4 Three typical THz eigenmodes around the filament area at **a** 0.2 THz, **b** 0.4 THz and **c** 0.6 THz with z-component of the Poynting vector (*color*) and electric vector (*arrows*)

References

1. M. Tonouchi, Nat. Photonics **1**, 97-105 (2007).
2. X. C. Zhang, Phys. Med. Biol. **47**, 3667 (2002).
3. R. Inoue *et al.* Jpn. J. Appl. Phys. **45**, L824-L826 (2006).
4. C. D Amico *et al.* Phys. Rev. Lett. **98**, 235002 (2007).
5. Y. Zhang *et al.* Opt. Express **16**, 15483-15488 (2008).
6. M. Peccianti *et al.* IEEE J. Sel. Top. Quantum Elect. **19**, 8401211 (2013).
7. J. Zhao *et al.* CLEO: QELS_Fundamental Science. OSA, FTu3D. 3. (2014).

Ultrafast Optical Modulation of Efficiently-Generated Terahertz-Wave in Charge Ordered Organic Ferroelectrics

Hirotake Itoh, Keisuke Itoh, Kazuki Goto, Junichi Ichimura, Yota Naito, Kaoru Yamamoto, Kyuya Yakushi, Hideo Kishida and Shinichiro Iwai

Abstract Terahertz-wave generation in organic ferroelectrics α-(ET)$_2$I$_3$ is over 70 times more efficient than prototypical ZnTe. Ultrafast (<0.1 ps) and sensitive ($\sim$40 %) photoresponse of the terahertz wave results from strongly-correlated electrons therein.

1 Introduction

Toward bright terahertz (THz)-wave sources and their broad applications, ferroelectrics have been mandatory. Among them, the layered organic salt α-(ET)$_2$I$_3$ (ET: bis(ethylenedithio)-tetrathiafulvalene) illustrates the potentiality of the charge ordering (CO) formed by strong Coulomb repulsion [1]. Therein, spontaneous electric polarization P is peculiarly driven by the CO resulting in large nonlinear optical susceptibility [2, 3]. Moreover, the strongly-correlated electrons host ultrafast (<100 fs) 'melting' of the CO upon photoexcitation, and consequent insulator-to-metal transition, or quenching of P [2–5]. Such characteristics, called as electronic ferroelectricity [6], should be promising for developing unprecedented THz-sources with brightness and ultrafast controllability [3].

H. Itoh (✉) · K. Itoh · K. Goto · J. Ichimura · Y. Naito · S. Iwai
Department of Physics, Tohoku University, Sendai 980-8578, Japan
e-mail: hiroitoh@m.tohoku.ac.jp

H. Itoh · H. Kishida · S. Iwai
JST, CREST, Sendai 980-8578, Japan

K. Yamamoto
Department of Applied Physics, Okayama University of Science, Okayama 700-0005, Japan

K. Yakushi
Toyota Physical and Chemical Research Institute, Nagakute 480-1192, Japan

H. Kishida
Department of Applied Physics, Nagoya University, Nagoya 464-8603, Japan

© Springer International Publishing Switzerland 2015
K. Yamanouchi et al. (eds.), *Ultrafast Phenomena XIX*,
Springer Proceedings in Physics 162, DOI 10.1007/978-3-319-13242-6_151

α-(ET)$_2$I$_3$ is in the metallic phase above the CO transition temperature T_{CO} = 135 K. Below T_{CO}, charge disproportionation among ET molecules sets in to break spatial inversion symmetry (Fig. 1a). Consequently, $\boldsymbol{P}$ as induced by the CO, but not by the structural deformation, shows up and hence it alternates the sign upon inversion of the CO pattern [2].

Here we show the THz-wave generation characteristics for a-(ET)$_2$I$_3$ [3]. The THz-wave generation via the optical rectification process is over 70 times more efficient than prototypical ZnTe. We also observed ultrafast (<0.1 ps) and sensitive (~ 40 %) response of the THz-wave generation upon photoexcitation.

2 Experiments

A single crystal of a-(ET)$_2$I$_3$ (ab-plane, typical size of $2 \times 2 \times 0.05$ mm^3) was irradiated by a femtosecond laser pulse (1.55 eV, 25 fs, 1 kHz) with a spot of 3 mm in diameter as the fundamental light (parallel to the a-axis of the sample) for the THz-wave generation. The generated THz-wave ($\parallel a$) was detected via electro-optic sampling using ZnTe. A pump pulse (0.89 eV, $\parallel b$) was generated in an optical parametric amplifier with time resolution of 150 fs.

3 Results and Discussions

Figure 1b shows temperature dependence of the THz-wave generation in α-(ET)$_2$I$_3$, plotting electric field amplitude E_{THz} after Fourier transform of the observed temporal waveforms. It is clear that E_{THz} onsets at ferroelectric transition temperature T_{CO}, revealing that the THz-wave originates from the CO-induced $\boldsymbol{P}$.

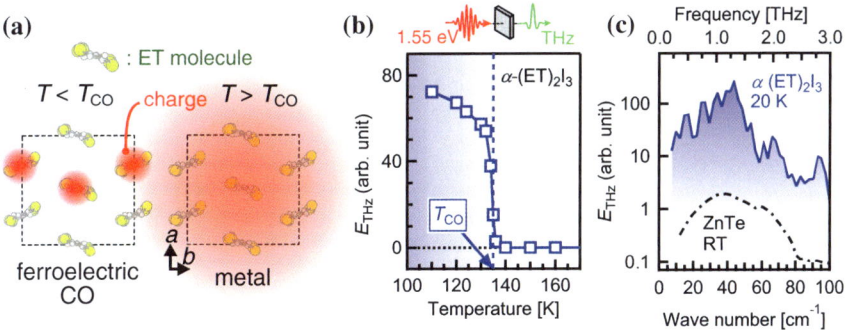

Fig. 1 **a** Schematics of arrangements of ET molecules and (ordered) charges in metallic and CO phases. **b** Temperature dependence of electric field amplitude E_{THz} (17−60 cm^{-1}) of the THz wave generated from α-(ET)$_2$I$_3$. **c** E_{THz} spectra for α-(ET)$_2$I$_3$ and ZnTe measured with a fundamental light fluence of 0.3 mJ/cm^2. The E_{THz} for ZnTe is normalized by penetration depth of α-(ET)$_2$I$_3$ at fundamental light energy

Figure 1c shows the E_{THz} spectrum observed at 20 K. The spectrum has the bandwidths of up to 70 cm^{-1} below optical gap (~ 0.1 eV [7]), along with ripples as a result of combined effects involving dispersion of (non)linear electric suscepti-bilities, absorption structures, and back reflection at the sample surface. It should be noted that E_{THz} was comparably large to that of (110)-oriented ZnTe (1 mm thick) observed with the same experimental setup. For comparison, we show the spectrum for ZnTe normalized by the penetration depth of α-(ET)$_2$I$_3$ at fundamental light energy (5 μm). As obviously shown, the THz-wave generation is highly efficient in α-(ET)$_2$I$_3$.

Considering Fresnel loss, absorption, and the aforementioned penetration depth, nonlinear optical susceptibility $\chi^{(2)}$(1 THz; $\omega - \omega$) for α-(ET)$_2$I$_3$ has been estimated to be 1×10^{-8} (m/V) by using ZnTe as a benchmark. Such a large value, which is more than 70 times larger than that of ZnTe, should be attributable to the polari-zation of electronic origin, or the CO, which is absent in ZnTe [3].

Another intriguing property of the electronic ferroelectricity is the ultrafast response upon photoexcitation. We have performed time-resolved THz-wave-gen-eration measurements (Fig. 2a), using the pump light (0.89 eV) above the charge transfer band [4]. Figure 2b shows the transient spectra of E_{THz} upon photoexcitation at 124 K. While a double-peak-like lineshape might result from several origins as mentioned above, the amplitude works as a direct measure of P (Fig. 1b). At the delay time $\tau = 0.1$ ps after the pump, E_{THz} showed a large decrease of approximately 40 %. The observed P quenching is not due to mere thermalization, but to the delocalization of correlated electrons upon photoexcitation and concomitant insulator(CO)-to-metal transition as reported previously [2–5]. Consequently, the photoinduced response of the THz-wave generation is instantaneous and sensitive revealing the capability of ultrafast photoswitching. In other words, the THz-wave generation works as a powerful probe for such transient P.

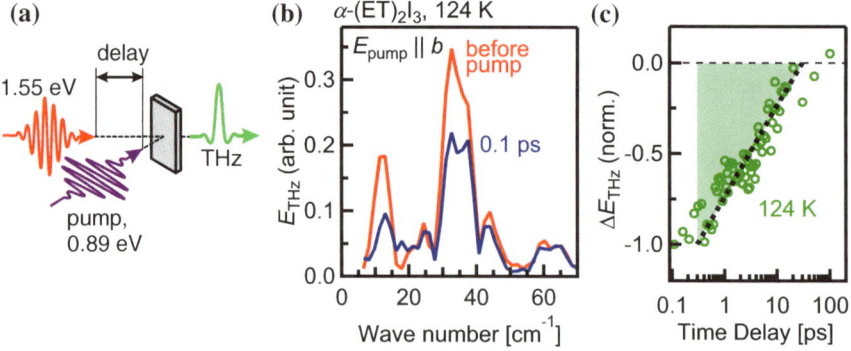

Fig. 2 a Schematics of the experimental setup for the time-resolved THz-wave-generation measurement. **b** Transient spectra of E_{THz} upon photoexcitation (0.3 mJ/cm^2) at 124 K. **c** Time evolution of E_{THz} (10−13 cm^{-1}) at 124 K. The *broken line* is a guide for the eyes

In Fig. 2c we show the time evolution of the ΔE_{THz} observed at 124 K. As shown, the recovery rate is approximately 10 ps, which is slower than those observed at lower temperatures [3]. This represents the diminishing stability of the ferroelectric CO in the vicinity of the transition temperature T_{CO}, which presumably corresponds to the stabilization of the photoinduced metal.

References

1. H. Seo, J. Phys. Soc. Jpn. **69**, 805 (2000).
2. K. Yamamoto *et al.*, J. Phys. Soc. Jpn. **77**, 074709 (2008); Appl. Phys. Lett. **96**, 122901 (2010).
3. H. Itoh *et al.*, Appl. Phys. Lett. **104**, 173302 (2014).
4. S. Iwai *et al.*, Phys. Rev. Lett. **98**, 097402 (2007).
5. Y. Kawakami *et al.*, Phys. Rev. Lett. **105**, 246402 (2010).
6. S. Ishihara, J. Phys. Soc. Jpn. **79**, 011010 (2010).
7. Y. Yue *et al.*, Phys. Rev. B **82**, 075134 (2010).

Ultrafast Terahertz Response of Lithium Niobate in the Nonperturbative Regime

Carmine Somma, Klaus Reimann, Christos Flytzanis,
Michael Woerner and Thomas Elsaesser

Abstract The response of a LiNbO$_3$ crystal to THz pulses in the nonperturbative regime is studied by two-dimensional spectroscopy. Phase-resolved detection allows for separating the THz bulk photovoltaic effect from other nonlinear contributions.

1 Introduction

Laser-driven sources for terahertz (THz) transients with very high electric field amplitudes allow for investigating the nonlinear response of condensed matter under conditions of a nonperturbative light-matter interaction where the coupling of electrons to the THz field represents their predominant coupling. Experiments in this regime have led to new insight into quantum coherent charge transport in semiconductors and carbon-based materials such as graphene [1]. Here, we study the nonlinear THz response of the well-known nonlinear material LiNbO$_3$ under nonperturbative conditions and separate the different contributions to the overall nonlinear response with the help of two-dimensional (2D) THz spectroscopy. Our data give the first evidence of a field-induced bulk photovoltaic effect in this material. While the conventional bulk photovoltaic effect manifests itself by the occurrence of a shift current (SC) of electrons generated by above-bandgap photoexcitation [2], the strong THz field applied in our experiments enables Zener tunneling of electrons from the valence to the conduction band [3].

C. Somma (✉) · K. Reimann · M. Woerner · T. Elsaesser
Max-Born-Institut, 12489 Berlin, Germany
e-mail: somma@mbi-berlin.de

C. Flytzanis
Laboratoire Pierre Aigrain, École Normale Supérieure, 75231 Paris, France

© Springer International Publishing Switzerland 2015
K. Yamanouchi et al. (eds.), *Ultrafast Phenomena XIX*,
Springer Proceedings in Physics 162, DOI 10.1007/978-3-319-13242-6_152

2 Experiment and Results

We perform collinear two-dimensional (2D) THz spectroscopy on an undoped single-domain $LiNbO_3$ crystal of 50 μm thickness. The ferroelectric c axis of the crystal is parallel to the electric field of the THz pulses. In our experiment, we use two strong phase-locked THz pulses A and B with a center frequency $v_0 = 2$ THz, generated by optical rectification of 800 nm pulses from a Ti:sapphire oscillator-amplifier system in two GaSe crystals. The electric-field transients transmitted through the sample are measured by electro-optical sampling in a thin ZnTe crystal as a function of both the real time t and the delay τ between the two THz pulses. With two choppers synchronized to the pulse repetition rate, three transients are determined, one when both pulses are present, and two for the single pulses. From these transients, the nonlinear electric field emitted by the sample is the difference $E_{NL}(t, \tau) = E_{AB}(t, \tau) - E_A(t, \tau) - E_B(t)$ (for further details see [4]).

As shown in Fig. 1b, the $LiNbO_3$ crystal emits a strong nonlinear electric field. The nonlinear signal $E_{NL}(t, \tau)$ reaches the highest value, about 20 kV/cm, and has the opposite direction of the total electric field $E_{AB}(t, \tau)$ (shown in Fig. 1a), when the two THz pulses overlap, i.e., for $t = 0$ and $\tau = 0$. For a sample of a thickness smaller than the THz wavelength, the nonlinear emitted terahertz field $E_{NL}(t, \tau)$ is proportional to the nonlinear current $J_{NL}(t, \tau)$ induced by the incident THz field.

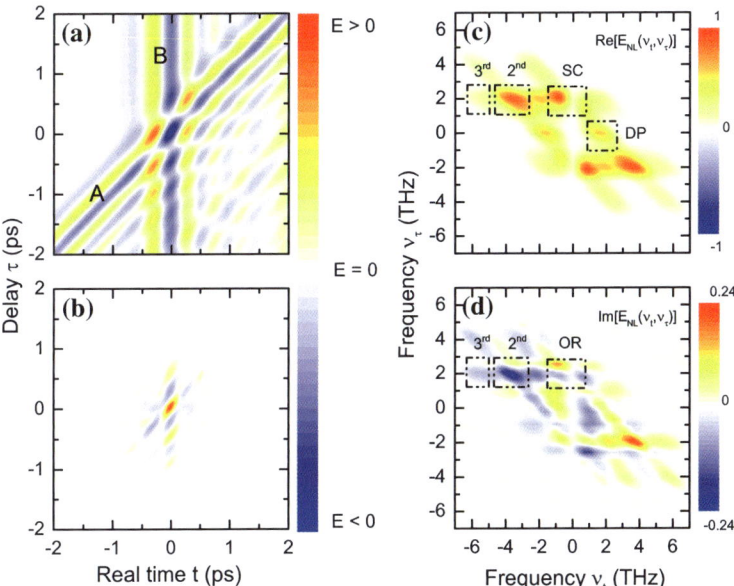

Fig. 1 a Electric field $E_{AB}(t, \tau)$ transmitted through the sample. **b** Nonlinear THz signal $E_{NL}(t, \tau)$ emitted by the sample. **c, d** The real and the imaginary parts of the spectrum of the nonlinear signal $E_{NL}(t, \tau)$ show the signatures of the shift current (SC), the optical rectification (OR), the second (2nd) and the third (3rd) harmonics, and the pump depletion signal (DP)

A 2D Fourier Transform (2DFT) of $E_{NL}(t, \tau)$ along the two time variables t and τ allows to separate the different terms of the nonlinear signal in the 2D frequency space spanned by v_t, the detection frequency, and by v_τ, the excitation frequency. Since electro-optic sampling yields the complete phase information, the 2DFT determines unambiguously real and imaginary parts of the nonlinear response. As shown in Fig. 1c, d, the 2DFT signal $E_{NL}(v_t, v_\tau)$ has contributions corresponding to second (2nd, at $v_t = \pm 2v_0$) and third (3rd, at $v_t = \pm 3v_0$) harmonic generation. Additionally, there are contributions at frequencies close to $v_t \approx 0$, labeled as shift current (SC) and as optical rectification (OR), and contributions at the fundamental THz frequency $v_t = \pm v_0$. The latter contribution (DP) is caused by the conversion of the fundamental into other frequencies.

3 Discussion

LiNbO$_3$ displays a substantial second-order nonlinearity, allowing for nonlinear frequency conversion. Our THz frequencies (2 THz) are way below the band gap of LiNbO$_3$ and, thus, one expects purely real nonlinear susceptibilities and only an imaginary part of the emitted nonlinear field. However, the 2D spectra in Fig. 1c, d display a real part even larger than the imaginary part, a behavior well beyond nonresonant nonlinear optics. The results give evidence of a bulk photovoltaic effect [5], which underlies the observed real part and is connected with the generation of free carriers in the crystal. Despite the low frequencies of the THz radiation, its very strong electric field induces Zener tunneling of electrons from the valence into the conduction band, i.e., it generates a real shift current (SC) of electrons giving rise to the observed new frequency components. The frequency spectrum of the SC reflects the dynamics of electron motion along the c axis of the LiNbO$_3$ crystal and contains components around zero frequency as well as harmonics up to—in principle—arbitrary order.

References

1. P. Bowlan, E. Martinez-Moreno, K. Reimann, T. Elsaesser, and M. Woerner, "Ultrafast terahertz response of multi-layer graphene in the nonperturbative regime," Phys. Rev. B **89**, 041 408(R) (2014)
2. M. Glass, D. von der Linde, and T. J. Negran, "High-voltage bulk photovoltaic effect and the photorefractive process in LiNbO3," Appl. Phys. Lett. **25**, 233–235 (1974)
3. W. Kuehn, P. Gaal, K. Reimann, M. Woerner, T. Elsaesser, and R. Hey, "THz-induced interband tunneling of electrons in GaAs," Phys. Rev. B **82**, 075 204 (2010)
4. M. Woerner, W. Kuehn, P. Bowlan, K. Reimann, and T. Elsaesser, "Ultrafast two-dimensional terahertz spectroscopy of elementary excitations in solids," New J. Phys. **15**, 025 039 (2013)
5. C. Somma, K. Reimann, C. Flytzanis, T. Elsaesser, M. Woener, Phys. Rev. Lett. **112**, 146602 (2014)

Inherent Resistivity of Graphene to Strong THz Fields

Dmitry Turchinovich, Zoltán Mics, Søren Jensen, Khaled Parvez,
Ivan Ivanov, Klaas-Jan Tielrooij, Frank H.L. Koppens,
Xinliang Feng, Klaus Müllen and Mischa Bonn

Abstract The THz conductivity of graphene at high driving THz fields was characterized using nonlinear ultrafast THz spectroscopy. We found that efficient carrier heating by strong THz signals leads to increased effective carrier scattering time. However, counter-intuitively, the heating also results in reduced high-frequency conductivity.

1 Introduction

The charge carriers in graphene behave like massless Dirac particles, which results in remarkably high dc conductivity of the material. Due to the ability to control the conductivity via the field effect [1], graphene was found to be highly promising for ultra-high-speed electronics applications such as nanoscale THz-rate transistors [2]. In such devices due to the short gate length 70–250 nm [2] the free charge carriers are driven by electric fields stronger than 100 kV/cm. We apply high-field THz spectroscopy as a contact-free probe to study the high-frequency conductivity of graphene at such high driving fields, i.e. the conductivity under conditions similar to a typical THz transistor. We find that the high-frequency conductivity of graphene decreases when the free carriers are driven by electric fields higher than 10 kV/cm. This is the result of efficient carrier heating via interaction with a strong high-frequency electric signal.

D. Turchinovich (✉) · Z. Mics · S. Jensen · K. Parvez · I. Ivanov · X. Feng ·
K. Müllen · M. Bonn
Max Planck Institute for Polymer Research, Ackermannweg 10, 55128 Mainz, Germany
e-mail: turchino@mpip-mainz.mpg.de

K.-J. Tielrooij · F.H.L. Koppens
The Institute of Photonic Sciences, Mediterranean Technology Park, 08860 Castelldefels,
Barcelona, Spain

© Springer International Publishing Switzerland 2015
K. Yamanouchi et al. (eds.), *Ultrafast Phenomena XIX*,
Springer Proceedings in Physics 162, DOI 10.1007/978-3-319-13242-6_153

623

2 Measurement

Our graphene sample was grown by chemical vapor deposition [3] and subsequently was transferred onto a fused silica substrate. The substrate causes a p-doping of graphene. The Fermi level of graphene was determined by linear THz spectroscopy [4] to be ~ 70 meV with respect to the Dirac point. We generated ultrashort THz pulses in the frequency range of 0.4–1.2 THz and with peak field strength up to 120 kV/cm in a LiNbO$_3$ crystal, which was pumped by ultrashort laser pulses with tilted pulse front [5]. Using these pulses as probes in a THz time-domain transmission spectroscopic arrangement [6, 7] we have measured the peak-field dependent frequency-resolved conductivity of graphene.

3 Results and Discussion

We find that at low fields the carrier conductivity can be well described by the free-carrier Drude model. As the peak driving field increases, the conductivity of graphene decreases. Moreover, this suppression of the conductivity is more noticeable at high frequencies (see Fig. 1).

The free charge carriers are efficiently heated up by the strong THz pulses. This heating causes the shift of the Fermi-level such that the concentration of the carriers is preserved, while the thermal energy of the carriers is increased. The change of the shape of the carrier distribution function results in the change of the conductive properties of graphene.

The conductive properties of the carriers are described by the energy-dependent scattering time $\tau(E)$. It has been shown that in graphene the free charge carriers scatter dominantly on scatterers with Coulomb long-range potential. For these

Fig. 1 The measured THz conductivity spectra at variable peak field strength (*symbols*). The measured data are fitted by a model accounting for the carrier heating (*lines*)

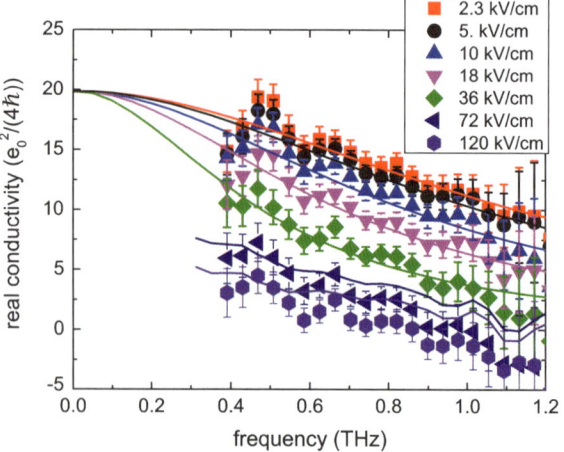

scatterers the energy-dependent scattering time is proportional to the energy of the carrier $\tau(E) \propto E$ [8]. Consequently, the dc conductivity of the carriers in our experiment depends only on the carrier concentration [9, 10]. This behaviour of the scatterers completely describes the results of our experiments:

- The concentration of the carriers does not change during the experiment, therefore the dc conductivity is constant at variable peak field strengths.
- Due to the heating by the strong THz fields the excess energy of the carriers increases, leading to an increase in the effective carrier scattering time.

The constant dc conductivity and the increasing effective carrier scattering time completely describe the graphene conductivity spectra for peak THz fields below 40 kV/cm (see Fig. 1). For higher fields, the high-field conductivity is described well using split step time-domain simulations, taking into account both heating of carriers by THz field, and the opposite process of cooling via phonon emission [11] (see Fig. 1). This dynamic model describes significant and ultrafast changes in the instantaneous temperature of carrier distribution in graphene during the interaction with the THz field, which in turn defines its dynamical conductivity.

4 Conclusion

In conclusion, we investigated the nonlinear conductivity of graphene using high-field THz time domain spectroscopy. We found that the high-frequency conductivity of graphene is suppressed at high driving fields, as result of free carrier heating in THz fields which leads to the *increase* in carrier scattering time.

References

1. K.S. Novoselov, A.K. Geim, S.V. Morozov, D. Jiang, Y. Zhang, S.V. Dubonos, I.V. Grigorieva, A.A. Firsov, Science **306**, 666 (2004).
2. F. Schwierz, Nature Nanotechnology **5**, 487 (2010).
3. X. Li, W. Cai, J. An, S. Kim, J. Nah, D. Yang, R. Piner, A. Velamakanni, I. Jung, E. Tutuc, S. K. Banerjee, L. Colombo, R.S. Ruoff, Science **324**, 1312 (2009).
4. G. Jnawali, Y. Rao, H. Yan, T.F. Heinz, Nano Lett. **13**, 524 (2013).
5. K.L. Yeh, M.C. Hoffmann, J. Hebling, K.A. Nelson, Applied Physics Letters **90**, 171121 (2007).
6. M.C. Hoffmann, B.S. Monozon, D. Livshits, E.U. Rafailov, D. Turchinovich, App. Phys. Lett. **97**, 231108 (2010).
7. D. Turchinovich, J.M. Hvam, and M.C. Hoffmann, Phys. Rev. B **85**, 201304 (2012).
8. T. Ando, J. Phys. Soc. Jpn. **75**, 074716 (2006).
9. S. Das Sarma, S. Adam, E. Rossi, Rev. Mod. Phys. **83**, 407 (2011).
10. K.J. Tielrooij, J.C.W. Song, S.A. Jensen, A. Centeno, A. Pesquera, A. Zurutuza Elorza, M. Bonn, L.S. Levitov, F.H.L. Koppens, Nature Phys. **9**, 248 (2013).
11. I. Gierz, J.C. Petersen, M. Mitrano, C. Cacho, I.C.E. Turcu, E. Springate, A. Stöhr, A. Köhler, U. Starke, A. Cavalleri, Nature Mat. **12**, 1119 (2013).

Nonlinear Carrier Responses in Gold Thin Films Induced by Intense Terahertz Waves

Yasuo Minami, Thang Duy Dao, Tadaaki Nagao, Jun Takeda, Masahiro Kitajima and Ikufumi Katayama

Abstract Terahertz transmittances of gold thin-films with thicknesses ranging from 1.4 to 5.8 nm were investigated. As terahertz field becomes intense, the transmittance decreases, suggesting the decrease of the damping constant.

1 Introduction

Gold (Au) has been extensively studied in the fields of microelectronics and bio-sensing, since many attractive electromagnetic properties of Au nanostructures were recently reported; the effective sheet conductivity of Au thin film dramatically decreases with the thickness less than 2 nm [1]. This suggests that the nano-scale size and morphology of Au nanostructures play key roles on the electronic properties. In the present study, we investigate the carrier dynamics in Au thin film with different thicknesses using intense terahertz electric field [2].

Y. Minami (✉) · J. Takeda · M. Kitajima · I. Katayama
Department of Physics, Graduate School of Engineering, Yokohama National University, Yokohama 240-8501, Japan
e-mail: minamiyasuo@ynu.ac.jp

T.D. Dao · T. Nagao · M. Kitajima
International Center for Materials Nanoarchitectonics, National Institute for Materials Science, Tsukuba 305-0044, Japan

T.D. Dao · T. Nagao · M. Kitajima
CREST, Japan Science and Technology Agency, Kawaguchi 332-0012, Japan

M. Kitajima
LxRay Co., Ltd, Nishinomiya 663-8172, Japan

M. Kitajima
Department of Applied Physics, National Defense Academy, Yokosuka 239-8686, Japan

© Springer International Publishing Switzerland 2015
K. Yamanouchi et al. (eds.), *Ultrafast Phenomena XIX*,
Springer Proceedings in Physics 162, DOI 10.1007/978-3-319-13242-6_154

2 Experiments

Au thin film was evaporated on the high resistivity Si substrate with Si (111) —7 × 7 surface in a chamber with the ultrahigh vacuum of $\sim 10^{-7}$ Pa at room temperature. The crystallization process was confirmed using reflection high-energy electron diffraction, and the polycrystalline Au films were successfully prepared. The thicknesses of the obtained Au films were 1.4, 1.6, 1.9, 3.0, 4.8, and 5.8 nm. The details of the preparation process were reported previously [3, 4]. Figure 1 shows the SEM image of the specimen with 1.9 nm thickness, which is fully covered by Au. The non-porous specimen enabled us to investigate the terahertz responses of Au thin films precisely.

A Ti: Sapphire amplifier system (repetition rate: 1 kHz, pulse duration: 130 fs, center wavelength: 800 nm, pulse energy: ~ 1.6 mJ/pulse) was employed to generate intense terahertz waves and to detect them. In generating terahertz fields, tilted laser pulses were irradiated onto a $LiNbO_3$ via Cherencov-type phase matching process [5]. The generated terahertz waves were incident on the specimen. Then, the transmitted waves were forwarded to a 0.4 mm-thick GaP crystal using off-axis parabolic mirrors and finally observed by the electro-optic (EO) sampling. The maximum-terahertz field was ~ 280 kV/cm. The intensity was tuned by the wire grid polarizers set in front of the specimen. Transmittance of the specimen was obtained by normalizing the transmitted spectrum of the specimen (Au on Si) by that of the reference (Si).

3 Results and Discussion

Figure 2 shows the Au film thickness dependence of the transmittance with different terahertz fields. The transmittance becomes lower with the intense terahertz field illumination. To clarify the reason, Drude analysis was carried out on the complex dielectric dispersion obtained from the terahertz transmittance (see, Fig. 3). The relationship between the observed complex transmittance $\tilde{T}$ and the dielectric constant at each angular frequency ω under the thin film approximation is expressed as

Fig. 1 SEM image of Au thin film surface with 1.9 nm thickness

50 nm

Fig. 2 Transmittance of
terahertz waves as a function
of the Au film thickness with
several terahertz intensities

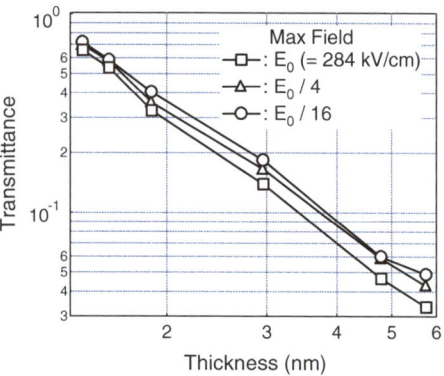

Fig. 3 Complex dielectric
constant of Au film with the
thickness of 1.9 nm. *Solid
curves* indicate the best-fit to
the experimental data
obtained by Drude analysis

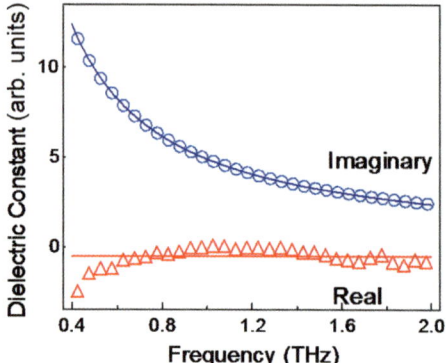

$$\tilde{T} = \frac{(n_{Si} + 1)}{n_{Si} + 1 - i\omega\tilde{\varepsilon}d/c}. \tag{1}$$

Drude dispersion is written as $\tilde{\varepsilon} = \varepsilon_\infty - \omega_p^2/[\omega(\omega + i\gamma)]$, where ε_∞ is the dielectric constant at high frequency limit, ω_p is the plasma frequency of free carriers, γ is the damping constant, c is the velocity of light, n_{Si} is the refractive index of Si substrate, and d is the thickness of the specimen. The damping constant represents the rate of the carrier-carrier scattering and/or carrier-boundary scattering.

In the weak terahertz field region, the plasma frequencies of Au films are similar to those obtained in Siegel et al. [6]. Assuming that the plasma frequency is constant, decrease of the damping constant with increasing electric field strength could induce the observed transmittance decrease. Actually, from the analysis, the damping constant is strongly decreased by the intense terahertz field.

4 Conclusions

We investigated the transmittance of the Au thin films with thickness ranging from 1.4 to 5.8 nm induced by intense terahertz electric fields. The transmittance became lower by the irradiation of intense terahertz fields. The Drude analysis clearly shows that the damping constant is suppressed by the intense terahertz fields.

References

1. J. J. Tu, C. C. Homes, and M. Strongin, Phys. Rev. Lett. **90**, 017402 (2003).
2. Y. Minami, J. Takeda, T. D. Dao, T. Nagao, M. Kitajima, and I. Katayama, (to be published in Appl. Phys. Lett.)
3. T. Nagao, J. T. Sadowski, M. Saito, S. Yaginuma, Y. Fujikawa, T. Kogure, T. Ohno, Y. Hasegawa, S. Hasegawa, and T. Sakurai, Phys. Rev. Lett. **93**, 105501 (2004).
4. S. Yaginuma, K. Nagaoka, T. Nagao, G. Bihlmayer, Y. M. Koroteev, E. V. Chulkov, and T. Nakayama, J. Phys. Soc. Jpn. **77**, 014701 (2008).
5. J. Hebling, G. Almási, I. Z. Kozma, and J. Kuhl, Opt. Express **10**, 1161 (2002).
6. J. Siegel, O. Lyutakov, V. Rybka, Z. Kolská, and V. Švorčík, Nanoscale Res. Lett. **6**, 96 (2011).

THz-Controlled Photoelectron Emission from Nanotips

L. Wimmer, G. Herink, K.E. Echternkamp, S.V. Yalunin, D.R. Solli, M. Gulde and C. Ropers

Abstract We introduce terahertz gating and streaking of photoelectron emission at a single nanostructure. The THz-near-field enhancement allows for far-reaching electron trajectory control, including phase-resolved streaking by the momentary THz field and propagation-induced spectral reshaping.

1 Introduction

The high localization of nanostructure field-enhancements leads to unique photoelectron dynamics at high intensities and long optical wavelengths [1]. Suggestions to use such dynamics in streaking-type experiments have been theoretically discussed in detail [2, 3]. Typical streaking experiments in attosecond spectroscopy employ a diffraction-limited focus, which imprints the vector potential of the streaking pulse onto the photoelectron kinetic energy [4]. If photoelectrons leave the field-enhanced region of a nanostructure within a fraction of an optical half-cycle, their kinetic energy is determined by the instantaneous electrical near-field rather than the time-integrated field. Specifically, long wavelengths facilitate sub-cycle electron dynamics. Our streaking-type experiment, implemented in the THz range, allows not only for a direct characterization of the spatio-temporal near-field, but also to widely tailor electron pulses in terms of the emitted charge and the kinetic energy distributions [5]. This THz-field-driven electron pulse control bears potential for optimizing electron pulses in ultrafast electron microscopy and diffraction experiments.

L. Wimmer (✉) · G. Herink · K.E. Echternkamp · S.V. Yalunin · D.R. Solli · M. Gulde · C. Ropers
4. Physical Institute, University of Göttingen, 37077 Göttingen, Germany
e-mail: wimmer@ph4.physik.uni-goettingen.de

© Springer International Publishing Switzerland 2015
K. Yamanouchi et al. (eds.), *Ultrafast Phenomena XIX*,
Springer Proceedings in Physics 162, DOI 10.1007/978-3-319-13242-6_155

630

2 Experimental Methods

In a two-color streaking experiment, few-cycle THz transients and 50-fs near-infrared (NIR) pulses are focused onto a single metal nanotip, as shown in Fig. 1a. The NIR-emitted electrons propagate through the enhanced THz-near-field and are accelerated additionally in a static field, which is caused by a moderate bias voltage applied to the tip. The photocurrent and the electron kinetic energy are detected by a detector assembly comprising a microchannel-plate and a retarding voltage grid. The delay between both pulses is scanned to map the impact of the THz field on the photoelectrons. In-situ characterization of the THz transient with electro-optic sampling enables a direct comparison of the incident THz field with its impact on the electron spectra.

3 Results and Discussion

An experimental streaking spectrogram is shown in Fig. 1d. The electron energy is strongly modulated by the enhanced THz-induced near-field (peak field—2 MV/cm). The electron acceleration by the THz-pulse not only affects the kinetic energy distribution, but also the detected photocurrent (Fig. 1c). Depending on the relative delay

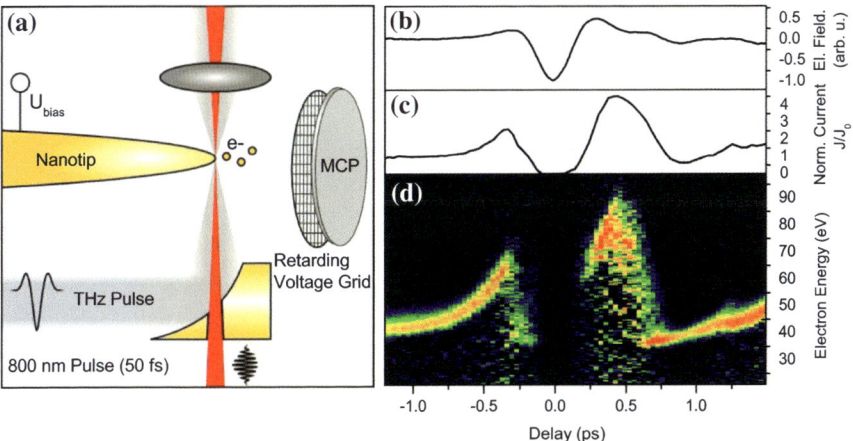

Fig. 1 a Schematic of the experimental setup. Single-cycle THz transients and near-infrared femtosecond pulses are collinearly focused onto a metal nanotip at variable delay. The photoelectrons are detected by a MCP detector combined with a retarding voltage electron spectrometer. **b** Incident electric field of the THz pulse detected by electro-optic sampling (peak field strength ∼100 kV/cm). **c** Normalized photo-current as a function of delay. **d** Streaking spectrogram for a sharp tip (bias voltage −40 V, radius of curvature 20 nm): photoelectron spectrogram formed by recording energy spectra as function of relative pulse delay between the NIR and THz pulses, the THz-transient enhances or suppresses the photocurrent due to an instantaneous THz modulation of the work function, i.e., via the Schottky effect.

The phase of the transient, detected by electro-optic sampling (Fig. 1b), is imprinted onto the photocurrent trace and the kinetic energy spectrogram. We find only a very minor phase-shift between the electro-optic field maxima and the maxima in kinetic energy, which is characteristic for the field-driven interaction between the THz-pulse and the photoelectrons at sharp nanostructures. In particular, this in-phase behavior results from the extremely short, sub-cycle interaction time of the photoelectrons with the THz near-field [5].

At longer interaction times, e.g. for lower THz-field strengths or spatially less confined fields, propagation effects arise that reshape the initial kinetic energy distributions and allow for strong spectral modulations. A temporally increasing streaking field transfers more energy to those electrons with low initial energy than to faster electrons, which leave the field enhanced region more rapidly. The same considerations explain the spectral broadening for temporally decreasing fields. This mechanism occurs in the transition region to field-driven dynamics.

The streaking measurements can be reproduced in detail by simulations based on the propagation of the NIR-induced photoelectrons in a temporally and spatially varying THz near-field and a static bias potential. In these simulations, initial energy distributions are included as measured in the absence of the THz field.

In conclusion, we demonstrate nanostructure gating and streaking experiments utilizing the high field localization of THz transients at nanostructures that give direct access to the phase-resolved near-field. Further control over the propagation dynamics can be achieved by modifications in the THz generation conditions and a tailoring of the static bias voltage. This opens up new prospects for nanoscopic electron pulse shaping to be used in ultrafast electron diffraction or microscopy.

References

1. G. Herink et al., Nature 483, 190 (2012).
2. M. I. Stockman et al., Nat, Photonics 1, 539 (2007).
3. F. Süßmann and M. F. Kling, Phys. Rev. B 84, 121406(4) (2011).
4. E. Goulielmakis et al., Science 317, 769 (2007).
5. L. Wimmer et al., Nature Physics 10, 432 (2014).

Nonlinear Carrier Dynamics in Semi-metal Bismuth Induced by Intense Terahertz Field

Kotaro Araki, Yasuo Minami, Thang Duy Dao, Tadaaki Nagao,
Jun Takeda, Masahiro Kitajima and Ikufumi Katayama

Abstract We investigated nonlinear carrier response of semi-metal bismuth under intense terahertz pulse illumination. By applying the intense terahertz field, the transmittance increases more than 10 %, indicating an increase of the effective mass.

1 Introduction

Semimetal bismuth (Bi) has been investigated over the years, and many interesting properties such as very small effective mass, high Hall coefficient, and non-parabolic band dispersion have been observed [1]. In addition, semimetal-to-semiconductor transition (SMSC) and structural phase transformation have been reported for the ultrathin Bi films below 30 and 1–2 nm, respectively [2, 3]. Therefore, the optical properties of Bi ultrathin film have been attracting attention. Recently, we have experimentally revealed the optical response of the surface state of Bi ultra-

K. Araki · Y. Minami · J. Takeda · M. Kitajima · I. Katayama (✉)
Department of Physics, Faculty of Engineering, Yokohama National University,
Yokohama 240-8501, Japan
e-mail: katayama@ynu.ac.jp

K. Araki
e-mail: araki-kotaro-kz@ynu.jp

T.D. Dao · T. Nagao · M. Kitajima
International Center for Materials Nanoarchitectonics, National Institute for Materials
Science, Tsukuba 305-0044, Japan

T.D. Dao · T. Nagao · M. Kitajima
CREST, Japan Science and Technology Agency, Kawaguchi 332-0012, Japan

M. Kitajima
LxRay Co., Ltd, Nishinomiya 663-8172, Japan

M. Kitajima
Department of Applied Physics, National Defense Academy, Yokosuka 239-8686, Japan

© Springer International Publishing Switzerland 2015
K. Yamanouchi et al. (eds.), *Ultrafast Phenomena XIX*,
Springer Proceedings in Physics 162, DOI 10.1007/978-3-319-13242-6_156

thin films by using broadband terahertz time domain spectroscopy (THz-TDS) [4]. In the present work, we have found an anomalous transmittance increase in the transmission spectrum in Bi film induced by intense THz pulses, revealing the nonlinear carrier response coming from the non-parabolic band structure of Bi.

2 Experiment

Single-crystalline Bi (001) film was deposited at room temperature on Si (111)-7×7 surface after the cleaning by DC resistive heating at 1,500 K in an ultrahigh vacuum chamber. After the deposition, the sample was annealed at 350 K to obtain atomically flat film. The thickness of Bi film was 40 nm. Details of the method were previously described elsewhere [5].

The experimental setup of THz-TDS is shown in Fig. 1a. The light source is a Ti-Sapphire regenerative amplifier with center wavelength of 800 nm, pulse duration of 130 fs, repetition of 1 kHz, and pulse energy of 1.7 mJ. The wave front-tilted laser pulse satisfying the non-collinear phase matching condition of optical rectification in $MgO:LiNbO_3$ generates an intense single cycle terahertz pulse [6]. The maximum field of the terahertz pulse is more than 160 kV/cm. The terahertz pulse is focused onto Bi film by off-axis parabolic mirrors, and then the transmitted terahertz pulse is detected by electro-optic (EO) sampling method using a 0.4-mm-thick GaP (110) crystal. All the measurements are performed at room temperature and under a nitrogen atmosphere to remove the influence of water vapor. Wire grid polarizers are used to adjust the intensity of the generated terahertz wave.

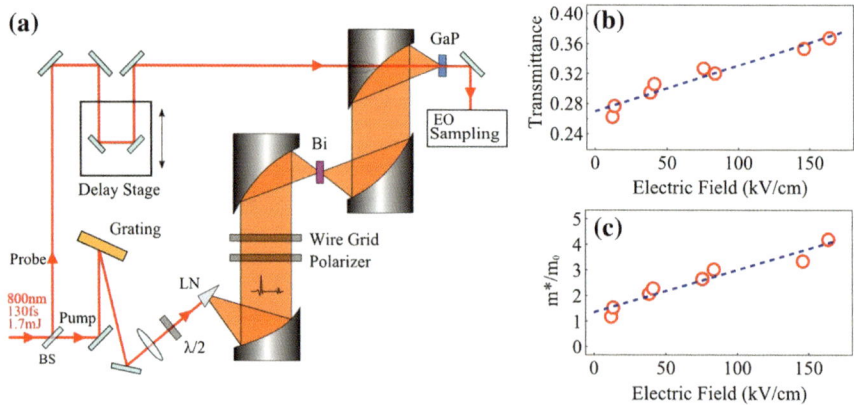

Fig. 1 **a** Schematic view of the experimental setup. An intense terahertz pulse is generated by the tilted pulse front excitation for $MgO:LiNbO_3$ (LN) and detected by EO sampling using 0.4-mm-thick GaP crystal. **b** Transmittance intensity as a function of the maximum electric field intensity. *Open circle* is experimental data, and a *dashed line* is the guide for eyes. **c** Maximum electric field intensity dependence of the effective mass of carriers normalized by the effective mass at the low-field limit (<1 kV/cm)

3 Result and Discussion

Figure 1b shows the maximum terahertz field dependence of the transmittance of Bi. When increasing the terahertz field up to 160 kV/cm, transmittance increased by 10 % or more compared with the low-field limit. This result shows the anomalous nonlinearity of the carrier response in Bi. In order to explain this behavior, we analyzed the transmitted spectra using a Drude model. In this model, the complex dielectric constant is described as

$$\tilde{\varepsilon}(\omega) = \varepsilon_\infty - \frac{\omega_p^2}{\omega(\omega + i\gamma)}, \tag{1}$$

where ε_∞ is dielectric constant at high frequency limit, ω_p is the plasma frequency, and γ is the damping constant, respectively. Here, the plasma frequency is expressed as

$$\omega_p = \sqrt{\frac{ne^2}{\varepsilon_0 m^*}}, \tag{2}$$

where n is the carrier density, m^* is the effective mass of the carriers, e is the elementary charge of electrons, and ε_0 is the dielectric constant in a vacuum. Assuming that the carrier density is constant, the effective mass can be estimated from (1) and (2). Figure 1c shows the field intensity dependence of the effective mass normalized by that at low electric field limit. As shown in Fig. 1c, the effective mass dramatically increases more than 300 % under the intense terahertz field illumination. This result indicates that the electrons at L-point of Bi are intensively moved to a nonlinear region. Because the band structure of electrons at L-point is similar to that of Dirac electrons, the electrons can be easily accelerated along the linear dispersion, leading to the anomalous nonlinear response.

4 Conclusion

We have investigated the electric field strength dependence of the terahertz transmittance of Bi film. Based on the Drude model analysis, we found the anomalous nonlinear response of electrons whose effective mass dramatically increases. The result comes from characteristics of electrons with non-parabolic band dispersion in Bi, which is analogous to that of Dirac electrons.

References

1. Y. Liu and R. E. Allen, *Phys. Rev. B* **52**, 1566 (1995).
2. C. A. Hoffman, J. R. Meyer, and F. J. Bartoli, *Phys. Rev. B* **48**, 11431 (1993).
3. T. Nagao, J. T. Sadowski, M. Saito, S. Yaginuma, Y. Fujikawa, T. Kogure, T. Ohno, Y. Hasegawa, S. Hasegawa, and T. Sakurai, *Phys. Rev. Lett.* **93**, 105501 (2004).
4. K. Yokota, J. Takeda, C. Dang, G. Han, D. N. McCarthy, T. Nagao, M. Hishita, M. Kitajima, and I. Katayama, *Appl. Phys. Lett.* **100**, 251605 (2012).
5. S. Yaginuma, K. Nagaoka, T. Nagao, G. Bihlmayer, Y. M. Koroteev, E. V. Chulkov, and T. Nakayama, *J. Phys. Soc. Jpn.* **77**, 014701 (2008).
6. J. Hebling, G. Almási, I. Z. Kozma, and J. Kuhl, *Opt. Express* **10**, 1161 (2002).

Ultrafast Insulator-Metal Transition in VO$_2$ Driven by Intense Multi-THz Pulses

A. Grupp, B. Mayer, C. Schmidt, J. Oelmann, R.E. Marvel,
R.F. Haglund Jr., A. Leitenstorfer and A. Pashkin

Abstract We demonstrate sub-100 fs metallization of VO$_2$ induced by few-cycle electric transients at frequencies around 25 THz. Interband tunneling is identified as an instantaneous excitation mechanism.

Vanadium dioxide (VO$_2$) is a prime example of a transition metal oxide exhibiting a sharp insulator-metal transition at 340 K (67 °C). This critical temperature around ambient conditions enables potential applications in optics and high-speed electronics. The delicate balance of competing interactions results in a high sensitivity to external perturbations. Starting from the dielectric phase, the metallic state can be induced by temperature, pressure, photoexcitation or electric fields.

Renewed interest towards VO$_2$ was triggered by the discovery of a photoinduced ultrafast insulator-metal transition (IMT) [1]. The essential role of lattice dynamics in the photoinduced transition was demonstrated by means of pump-probe spectroscopy [2, 3]. Nevertheless, control of the electronic state of VO$_2$ by applied electric bias is widely believed to be governed by resistive heating, dramatically limiting the switching speed [4].

Intense terahertz (THz) pulses offer an attractive possibility to transiently apply extremely high electric fields without irreversible destruction of the structure. Thus, an excitation by THz transients may offer an efficient way for the ultrafast non-thermal metallization of VO$_2$. A first demonstration of the field-induced switching in a VO$_2$ metamaterial using intense THz pulses has been reported recently [5]. The observed switching times of 8 ps indicate a thermal character of the transition driven by few-THz transients.

Here, we demonstrate for the first time an insulator-metal transition in VO$_2$ films induced by high-field multi-THz waveforms on a sub-100 fs timescale. Our sample is a polycrystalline film of a thickness of 200 nm grown by pulsed laser deposition

A. Grupp · B. Mayer · C. Schmidt · J. Oelmann · A. Leitenstorfer · A. Pashkin (✉)
Department of Physics and Center for Applied Photonics, University of Konstanz, 78457
Konstanz, Germany
e-mail: o.pashkin@hzdr.de

R.E. Marvel · R.F. Haglund Jr.
Department of Physics and Astronomy, Vanderbilt University, Nashville, TN 37235, USA

© Springer International Publishing Switzerland 2015
K. Yamanouchi et al. (eds.), *Ultrafast Phenomena XIX*,
Springer Proceedings in Physics 162, DOI 10.1007/978-3-319-13242-6_157

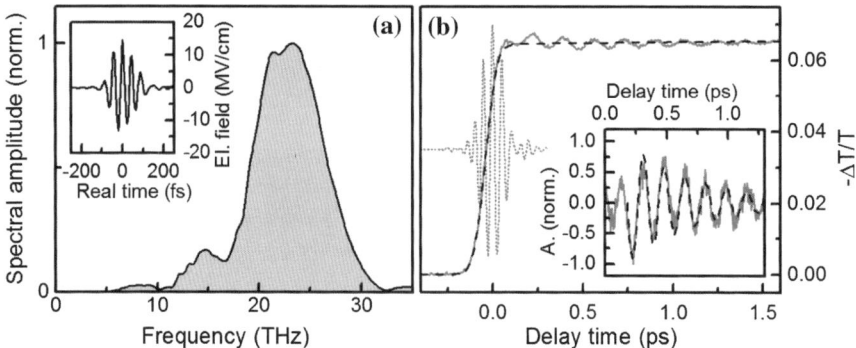

Fig. 1 a Amplitude spectrum and electric field profile (*inset*) of the multi-THz pump pulse;
b Relative transmission change ΔT/T of the 200-nm-thick VO₂ film (*grey line*) induced by the
multi-THz transient (*dotted line*) with incident fluence of 8 mJ/cm² measured at room temperature.
Dashed line in the main panel corresponds to a non-oscillating fit of the pump-probe signal. *Inset*
Remaining coherent oscillation fitted by a damped harmonic oscillation (*dashed line*) with a
frequency of 5.93 THz

on a CVD diamond substrate. It exhibits a transition temperature of 340 K. We
drive the IMT by broadband and phase-stable multi-THz transients with extremely
high peak electric fields of up to 17 MV/cm, generated via difference frequency
mixing [6]. A typical waveform and its corresponding amplitude spectrum centered
at 24 THz is shown in Fig. 1a. The electronic state of the sample is monitored by
probing the change of transmission of 8-fs-short near-infrared pulses with a central
wavelength of 1.2 μm.

The ultrafast switching dynamics measured by a multi-THz pump/near-infrared
probe experiment at room temperature is shown in Fig. 1b. The relative transmission change ΔT/T is depicted as a function of delay time between pump and probe
pulses (grey line) and compared to the driving multi-THz waveform (dotted line).
An ultrafast change of the transmission corresponds to the optically driven
switching into the metallic state, succeeded by a relatively slow relaxation on longer
timescales. Under a strong excitation the insulating state recovers on a thermal
timescale longer than 1 ns (not shown). A fit of the measured data (dashed line in
Fig. 1b) reveals a step-like decrease of transmission on a timescale limited by the
duration of the excitation pulse. Clearly, the ultrafast phase transition also includes
structural aspects happening on a longer time scale. The lattice dynamics manifests
itself by the coherent oscillation at a frequency of 5.9 THz (see Fig. 1b). The same
signature has been observed in ultrafast switching experiments under near-infrared
excitation. It is attributed to the coherent wave packet motion of the vanadium
dimers [2].

Finally, in a second experiment, a dependence of the IMT on the THz peak
electric field is systematically studied. Figure 2a demonstrates the time-resolved
relative transmission change of the probe pulse for incident peak electric fields
ranging from 3 to 15 MV/cm. Remarkably, the qualitative dynamics of the

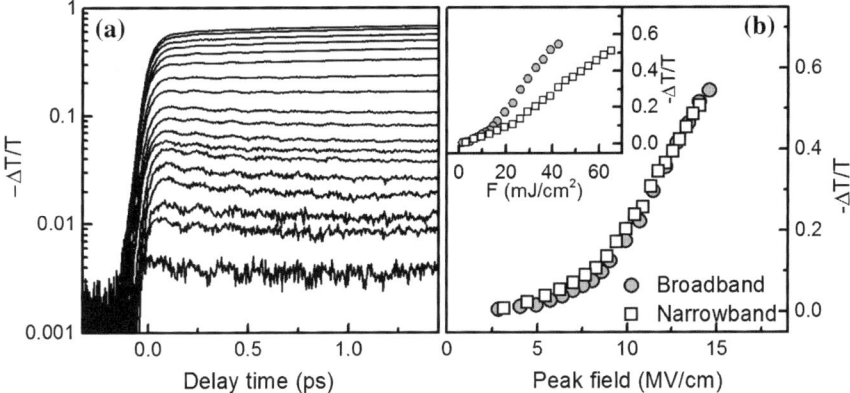

Fig. 2 **a** Relative transmission change $\Delta T/T$ measured at peak electric fields from 3 to 15 MV/cm;
b Maximal $|\Delta T/T|$ as a function of the peak electric field for broadband and narrowband multi-THz
excitation. *Inset* Same data plotted as function of the excitation fluence F

switching process is almost independent of the driving peak field. However, the
amplitude of the pump-probe signal shows a strongly nonlinear dependence on
the THz peak field. This behavior is emphasized in Fig. 2b where the maximal
transmission change is plotted as a function of the peak electric field for excitation
with broadband (circles) and narrowband (squares) multi-THz pulses. Obviously,
the broadband pulses show a more efficient excitation when plotted as a function of
fluence (see inset Fig. 2b). However, as a function of peak electric field both
measurements are almost coincident, indicating that the observed phase transition is
a *field-driven* phenomenon, and the switching is not governed by the number of
incident THz photons. In comparison to previous studies with near-infrared pump
pulses we do not observe a distinct threshold behavior when probing in the near-
infrared [2, 3]. Interestingly, switching the system into a long-lived metallic state is
very efficient despite the off-resonant excitation. The mechanism responsible for
this field-induced IMT is nonadiabatic interband tunneling leading to an inherent
separation of charge carriers. In this way the counter-acting effect of exciton self-
trapping occurring for resonant near-infrared excitation is bypassed in the field-
driven IMT.

In conclusion, we have observed sub-100 fs switching timescales and coherent
lattice dynamics in VO$_2$, strongly supporting a non-thermal and initially electronic
scenario of the field-induced insulator-metal transition. The degree of metallization
is determined by electric field strength and can be well-described by a tunneling
mechanism.

References

1. M. F. Becker et al., "Femtosecond laser excitation of the semiconductor-metal phase transition in VO$_2$", Appl. Phys. Lett. **65**, 1507–1509 (1994).
2. C. Kübler et al., "Coherent Structural Dynamics and Electronic Correlations during an Ultrafast Insulator-to-Metal Phase Transition in VO$_2$", Phys. Rev. Lett. **99**, 116401 (2007).
3. S. Wall et al., "Ultrafast changes in lattice symmetry probed by coherent phonons", Nat. Commun. **3**, 721 (2012).
4. A. Zimmers et al., "Role of Thermal Heating on the Voltage Induced Insulator-Metal Transition in VO$_2$", Phys. Rev. Lett. **110**, 056601 (2013).
5. M. Liu et al., "Terahertz-field-induced insulator-to-metal transition in vanadium dioxide metamaterial", Nature **487**, 345-348 (2012).
6. F. Junginger et al., "Single-cycle multi-THz transients with peak fields above 10 MV/cm", Opt. Lett. **35**, 2645-2647 (2010).

Coherent Ultrafast Magnetization Dynamics Non-resonantly Induced in Cobalt by an Intense Terahertz Transient

C. Vicario, F. Ardana-Lamas, P.M. Derlet, B. Tudu, J. Luning and C.P. Hauri

Abstract We demonstrate non-resonant magnetization dynamics in the ferromagnetic cobalt thin film induced by a record high-field Terahertz pulse. The magnetization dynamics are coherent and exactly follow the THz carrier oscillations.

OCIS Codes 140.3070

1 Introduction

Since the 90 s femtosecond near-IR laser pulses (with central wavelength $\lambda = 800$ nm) have been used to induce fast de-magnetization in ferromagnetic thin films [1]. In those experiments optical pulses modify the magnetization by heating the spin system via ultrafast electronic excitation. The associated cooling dynamics resulted orders of magnitude slower than the ultrafast demagnetization process and evolved over hundreds of picoseconds. The incoherent nature of laser-induced heating in the ferromagnetic film precluded any possibility of imprinting the phase information of the stimulus onto the magnetization dynamics. Furthermore the thermal relaxations are slow and do not allow potentially for precise ultrafast switching of magnetic domain on the femtosecond time-scale.

Here we demonstrate a new class of laser-matter interaction thanks to the advent of non-ionizing, high-field Terahertz (THz) pulses. The unique high field THz pulses [2] induce non-*resonant* femtosecond magnetization dynamics. These

C. Vicario (✉) · F. Ardana-Lamas · C.P. Hauri
SwissFEL, Paul Scherrer Institute, 5232, Villigen, Switzerland
e-mail: carlo.vicario@psi.ch

P.M. Derlet
Paul Scherrer Institute, Condensed Matter Theory Group, 5232, Villigen, Switzerland

B. Tudu · J. Luning
Université Pierre et Marie Curie, LCPMR, UMR CNRS 7614, 75005 Paris, France

C.P. Hauri
Ecole Polytechnique Federale de Lausanne, 1015, Lausanne, Switzerland

© Springer International Publishing Switzerland 2015
K. Yamanouchi et al. (eds.), *Ultrafast Phenomena XIX*,
Springer Proceedings in Physics 162, DOI 10.1007/978-3-319-13242-6_158

641

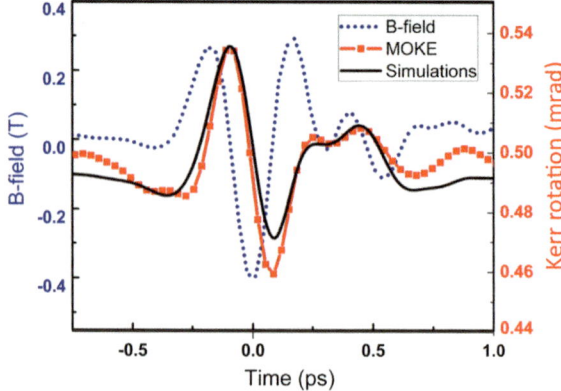

Fig. 1 Magnetization dynamics measured by MOKE (*red line*) are initiated by intense Terahertz magnetic field (*dotted line*). After the THz pulse left the magnetization dynamics are terminated corroborating the non-thermal excitation of this new interaction mechanism [3]

ultrafast magnetization is phase-locked to the THz carrier and thus follows exactly the ultra-strong, phase-stable laser magnetic THz field. In this novel interaction regime the laser's phase and field magnitude characteristics are directly imprinted onto the magnetization response, in absence of any resonant modes. This becomes possible since the off-resonant phase-locking mechanism presented here, injects only minor entropy into the system.

2 Experimental Results

For our experiment we used the magnetic component of a single-cycle Terahertz transient with a field strength of 0.4 T (Fig. 1, dotted line). The strong Terahertz transient is generated by optical rectification of a powerful mid-infrared pulse in an organic crystal. Upon illuminating the ferromagnetic sample (cobalt thin film), the Terahertz field induces an ultrafast magnetic response which follows the THz stimulus field almost instantaneously (Fig. 1. red line). The magnetization dynamics were measured by recording the magneto-optical Kerr effect (MOKE).

Shown in Fig. 2 is the spectrum of the THz and the measured magnetic response represented by the MOKE frequency spectrum (red). It manifests the occurrence of all the THz frequency constituents as observed for the THz stimulus spectrum (dashed blue line). This is the first evidence that ultrastrong Terahertz radiation is capable to initiate non-resonant magnetization dynamics. The experimental results are excellently reproduced by the Landau-Lifschitz-Gilbert (LLG) formalism where the magnetization evolution is described by:

$$d\vec{M}/dt = -\gamma(\vec{M} \times \vec{B}) - \gamma\alpha(\vec{M} \times (\vec{M} \times \vec{B}))/|\vec{M}|.$$

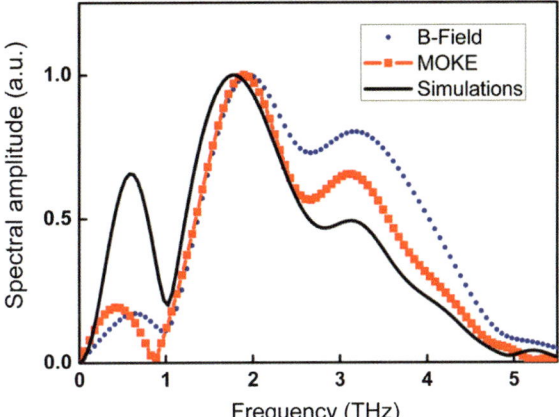

Fig. 2 Spectral intensities of the THz stimulus (*dashed line*) and the corresponding magnetic response in the Co thin film (*red line*). The calculated magnetization dynamics are excellently described by Landau-Lifshitz-Gilbert equation both in the temporal and spectral domain

As corroborated by LLG model, the observed out-of-plane magnetic response is entirely determined by the precessional term and the THz field contribution. The observed inherently phase-locked interaction between the THz stimulus and magnetization occurs on a femtosecond timescale which has not been expected to be that fast. The experiment gives for the first time evidence that the LLG equation is not only valid for DC magnetic fields but also for rapidly oscillating stimulus on the femtosecond time-scale.

In conclusion we demonstrate that the advent of THz field transient of >0.4 T opens a new class of laser-matter interaction. Previously inaccessible sub-cycle magnetization dynamics are visualized in a ferromagnetic thin film thanks to the non-ionizing THz field. The measured coherent magnetization response is phase-locked to the stimulus and occurs on a femtosecond timescale. The key-tool is the intense single-cycle Terahertz pulses recently developed by our group. The opto-magnetic interaction is remarkably well described by the phenomenological LLG approach which is usually valid only for almost DC field. The presented new concept of phase-stable non-ionizing high field stimulus opens the door to light-induced phase-locked magnetization control in view of the realization of ultrafast switching of magnetic domains.

References

1. Beaurepaire et al. Phys. Rev. Lett. 76,4250 (1996)
2. C. Vicario et. al. Phys. Rev. Lett 110, 123902 (2013)
3. C. Vicario et al. Nature Photon. 7, 720 (2013)

Beating of Terahertz Pulse Induced Spin Precession in ErFeO$_3$

Keita Yamaguchi, Takayuki Kurihara, Hiroshi Watanabe, Makoto Nakajima, Takeo Kato and Tohru Suemoto

Abstract Terahertz pulse induced spin precession in ErFeO$_3$ was observed via the Faraday rotation of the near infrared probe pulse. Unreported splitting of the magnetic resonance was discovered and mechanism explaining this splitting is proposed.

1 Introduction

Recent developments on the terahertz (THz) wave technology triggered intensive THz spectroscopic research on various materials. One such newly developing topic is the ultrafast excitation of spins [1–4], where the THz magnetic field instantaneously tilts the spins out of equilibrium. Here, we report ultrafast spin dynamics observed with high power THz pump—near infrared probe experiment in an orthoferrite ErFeO$_3$. The measurement revealed long-lived spin precession with a beating, which, to the best of our knowledge, has never been reported in the past. Orthoferrites have four Fe^{3+} spin sublattices that order antiferromagnetically below about 600 K. Owing to Dzyaloshinskii-Moriya (DM) interaction, the angle between the nearest neighbor Fe^{3+} spins deviates about 1 degree from the antiparallel orientation. This deviation gives rise to a weak macroscopic magnetization. In the temperature region studied here (10–80 K), the easy axis for each sublattice spins is along the c axis and the macroscopic magnetization is along the a axis. Here, precession of the weak macroscopic magnetic moment known as the quasiferromagnetic (F) mode resonance of the orthoferrite was studied.

K. Yamaguchi · T. Kurihara · H. Watanabe · T. Kato · T. Suemoto
Institute for Solid State Physics, The University of Tokyo, Kashiwa, Chiba 277-8581, Japan

M. Nakajima
Institute of Laser Engineering, Osaka University, Suita, Osaka 565-0871, Japan

K. Yamaguchi (✉)
Yokohama Research Laboratory, Hitachi Ltd., Yokohama 244-0817, Japan
e-mail: ykeith126@gmail.com

© Springer International Publishing Switzerland 2015
K. Yamanouchi et al. (eds.), *Ultrafast Phenomena XIX*,
Springer Proceedings in Physics 162, DOI 10.1007/978-3-319-13242-6_159

2 Result

The THz induced spin precession in a single crystal $ErFeO_3$ (100 μm thick, (001) surface) was observed through time dependent Faraday rotation of the transmitting near infrared probe pulse. The Faraday rotation reflects the magnetization dynamics in the thickness direction of the sample. Therefore, the oscillation of the magnetization in the c axis direction is observed. Since the spin dynamics are probed by ϕ10 μm diameter near infrared pulses, spatial resolution is expected to be higher than the THz time domain spectroscopy (TDS). The THz pulse used for the spin excitation was generated from $LiNbO_3$ crystal with tilted pulse front technique.

Figure 1a, b shows the time dependent THz induced Faraday rotation obtained with this sample. The polarization of the incident THz pulse was $H_{THz} \parallel b$ axis, which excites the F mode precession. As Fig. 1b shows, the spin precession is clearly observed. Unlike in the previous reports [1, 2], the precession lasts for a long time, and a beating of the oscillation with a period over 100 ps is also observed. The Fourier spectra of the oscillation (Fig. 1c) reveal that the signals consist of two separate sharp peaks. The magnitude of these splitting corresponds to the period of the beatings. However, as far as we are concerned, such splitting of the F mode resonant frequency has never been reported in the past.

These discrepancies with the previous researches may be ascribed to the difference in the probed sample area size. In our previous THz-TDS experiments

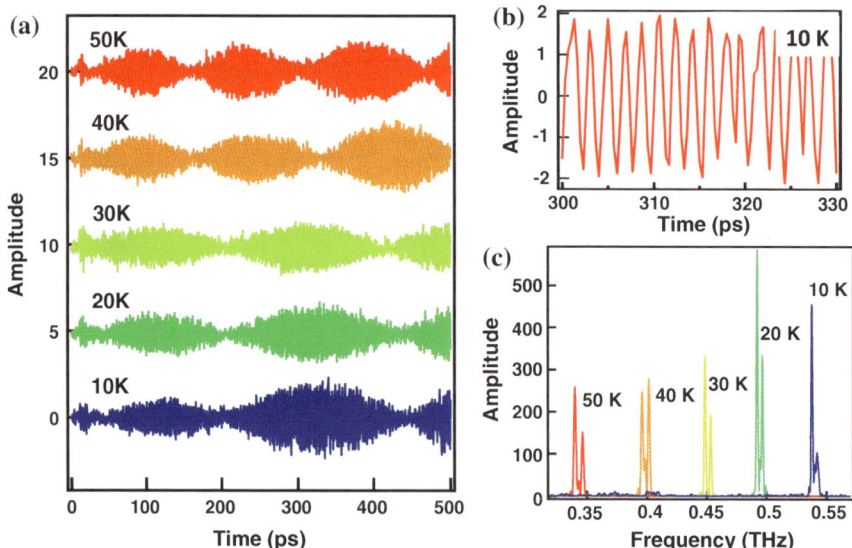

Fig. 1 a Time dependent THz induced Faraday rotation signal obtained with $ErFeO_3$ (001) single crystal. **b** Magnification of the time dependent signal at 10 K. **c** Fourier spectra of the oscillations shown in (**a**)

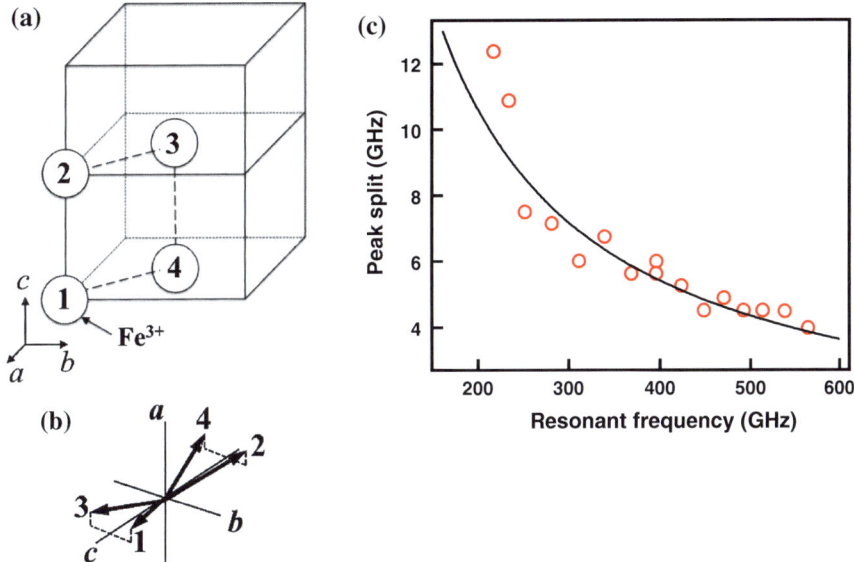

Fig. 2 **a** Crystallographic structure of $ErFeO_3$ and the position of the four Fe^{3+} sublattices. **b** The Fe^{3+} spin configuration in the temperature region below 90 K. **c** The splitting of the F mode resonance observed at each temperature. The *solid line* shows the calculated splitting with $D'_z = 1.3 \, cm^{-1}$

[1, 2], comparatively large area (around 1 mm in diameter) of the sample was observed, and such fine structures of the spectrum may have been averaged out by spatial inhomogeneity. In the case of the present experiment, owing to the high spatial resolution, such inhomogeneous effects may be resolved. This explains the long precession lifetime and the observation of the F mode resonance splitting.

Although there are several known mechanisms (standing spin wave or splitting due to the domain walls) that causes the magnetic resonance mode to split, the magnitude of the splitting does not match with that observed in the experiments. Here, instead of these known mechanisms, we speculated the DM interaction to be the cause of the splitting. Owing to the symmetry of the crystal, the DM interaction works differently for different nearest neighbor spin pairs [5, 6]. For example, spin 1 in Fig. 2a, b, interacts with the nearest neighbor spins 2 and 4 differently. Normally, such differences are ignored for deriving the resonant frequency [7]. We modeled the system as a three-spin system considering 1–2 pair and 1–4 pair DM interactions, and it was confirmed that D'_z, the z component of the DM vector, working on 1–4 (and 2–3) pair causes the resonant frequency to split. The splitting is described by $\Delta\omega = \sqrt{\omega^2 + D'^2_z} - \omega$, where ω is the resonant frequency of the 1–2 pair. Using this model, the observed splitting can be reproduced reasonably with $D'_z = 1.3 \, cm^{-1}$ deduced from [7] (Fig. 2c). It should be noted that this calculation becomes invalid when all 4 sublattices are handled in the equation of motion at once, and the splitting cannot be derived. Thus, more adequate splitting model must be established. However, our experimental and theoretical results show that the magnitude of the splitting is comparable to the DM interaction, and implies its involvement in the occurrence of the splitting.

3 Conclusion

THz pulse induced F mode spin precession was observed via visible Faraday rotation in $ErFeO_3$ and extremely long-lived precession with an unknown splitting of the resonant frequency was observed. We confirmed that magnitude of the splitting is comparable to the magnitude of the DM interaction, and speculate that this interaction involves in the splitting.

References

1. K. Yamaguchi, M. Nakajima, and T. Suemoto "Coherent control of spin precession motion with impulsive magnetic fields of half-cycle terahertz radiation," Phys. Rev. Lett 105, 237201-1 – 237201-4 (2010).
2. K. Yamaguchi, T. Kurihara, Y. Minami, M. Nakajima, and T. Suemoto "Terahertz Time-Domain Observation of Spin Reorientation in Orthoferrite $ErFeO_3$ through Magnetic Free Induction Decay," Phys. Rev. Lett 110, 137204-1 – 137204-5 (2013).
3. M. Nakajima, A. Namai, S. Ohkoshi, and T. Suemoto "Ultrafast time domain demonstration of bulk magnetization precession at zero magnetic field ferromagnetic resonance induced by terahertz magnetic field," Opt. Exp. 18, 18260-18268 (2010).
4. T. Kampfrath, et al., "Coherent terahertz control of antiferromagnetic spin waves," Nature Photon. 5, 31, 2011.
5. M. Marezio, J.P. Remeika, and P.D. Dernier, "The Crystal Chemistry of the Rare Earth Orthoferrites," Acta. Crystallogr. B 26, 2008-2022 (1970).
6. T. Moriya, "Weak Ferromagnetism" in *Magnetism I*, Eds. G.T. Rado, and H. Suhl (Academic Press, New York, 1963).
7. N. Koshizuka, and K. Hayashi, "Raman Scattering from Magnon Excitations in $RFeO_3$," J. Phys. Soc. Jpn. 57,4418-4428 (1988).

Resonant Antiferromagnetic Spin Wave Excitation by Terahertz Magnetic Near-Field with Split Ring Resonator

Y. Mukai, H. Hirori, T. Yamamoto, H. Kageyama and K. Tanaka

Abstract A spin wave of $HoFeO_3$ was excited by a terahertz magnetic near-field of a split ring resonator. The quantitative analysis shows that the spin wave was excited by the resonantly enhanced magnetic field.

1 Introduction

Ultrafast control of spins in solids has attracted considerable attention from researchers because of its importance in fundamental physics and for technological applications such as spintronics and spin-based information processing. The direct, magnetic field–induced, ultrafast excitation and control of spin waves is most advantageous for these purposes. Recent terahertz (THz) pulse generation and detection techniques allow us to manipulate spins coherently and to elucidate spin dynamics [1]. Split ring resonator (SRR) enables us to achieve field enhancement and subwavelength field localization [2], which provide a good platform to study magnetic phenomena in condensed matter and also a route to developing electronically controlled hybrid spintronics and dynamic magnetooptic devices such as isolators. However, the coupling between magnetic excitation and SRR resonant

Y. Mukai (✉) · K. Tanaka
Department of Physics, Graduate School of Science, Kyoto University,
Kyoto, Sakyo-Ku 606-8502, Japan
e-mail: mukai@scphys.kyoto-u.ac.jp

H. Hirori · H. Kageyama · K. Tanaka
Institute for Integrated Cell-Material Sciences (WPI-ICeMS), Kyoto University,
Kyoto, Sakyo-Ku 606-8501, Japan

H. Hirori · K. Tanaka
CREST, Japan Science and Technology Agency, Saitama, Kawaguchi 606-0012, Japan

T. Yamamoto · H. Kageyama
Department of Energy and Hydrocarbon Chemistry, Graduate School of Engineering,
Kyoto University, Kyoto, Nishikyo-Ku 615-8510, Japan

© Springer International Publishing Switzerland 2015
K. Yamanouchi et al. (eds.), *Ultrafast Phenomena XIX*,
Springer Proceedings in Physics 162, DOI 10.1007/978-3-319-13242-6_160

649

mode has remained elusive because of complex magnetic field distribution in near-field regime. Here, we study the dynamics of antiferromagnetic spin waves in a HoFeO$_3$ crystal excited with THz electromagnetic pulses with SRR by near-field Faraday microscopy.

2 Experimental Setup

Magnetization dynamics induced by the THz near-field of split ring resonators SRRs was studied using the time-resolved THz pump near-infrared (NIR) Faraday measurement setup shown in Fig. 1a, b [3]. As shown in the Fig. 1a, a planar array of SRRs was fabricated on the surface of c-cut single crystal HoFeO$_3$. The incident THz electric field causes a current to flow circularly at the LC resonant frequency of the SRR, thereby inducing a magnetic near-field perpendicular to the surface (c-axis) [2]. Figure 1c, d shows the spatial distribution of the magnetic field component H_z calculated from the THz electric field measured with an electro-optic sampling method and the THz magnetic field waveform in the time domain at the same position indicated by the circle in Fig. 1c.

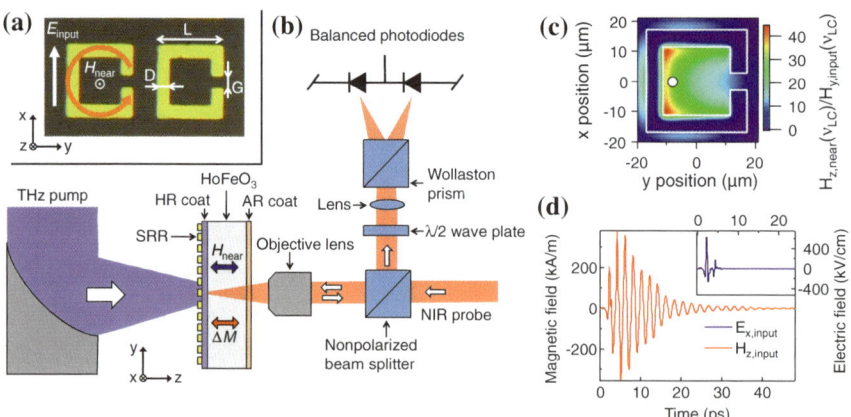

Fig. 1 **a** The SRR fabricated on the ab-plane of HoFeO$_3$. **b** Schematic setup of THz pump-NIR Faraday rotation measurement. The experimental coordinate system is such that the x, y, and z axes are parallel to the crystal axes (**a**, **b**, **c**). The length of the ring L was 34 μm, the width D was 6 μm, and the split gap G was 6 μm. **c** Finite-difference time-domain method (FDTD) simulation of the magnetic field near the SRR. Two-dimensional distribution of z-directed magnetic field H_z at LC resonance frequency ($\nu_{LC} = 0.5$ THz) at the interface between the HR coat and HoFeO$_3$, plotted as the relative strength of the incident magnetic field. The *white dot* indicates the probe position where the Faraday rotation is measured. **d** Temporal waveforms of the magnetic field component H_z (*red line*) calculated by using the THz electric field measured (*blue line*) by an electro-optic sampling method shown in the inset

3 Results

HoFeO$_3$ is a typical weak ferromagnet and well investigated on spin dynamics in THz frequency region. As shown in Fig. 2a, the sample in the Γ_4 phase (T > 58 K) has two iron sublattice magnetizations m_i (i = 1, 2) which are almost antiferromagnetically aligned along the a-axis and slightly tilted toward the c-axis, giving rise to a spontaneous magnetization [4]. There are two spin-wave eigenmodes, called the anti-ferromagnetic-mode (AF-mode) and ferromagnetic-mode (F-mode). In our experimental setup, only the AF-mode can be excited by the THz magnetic near-field H_z, resulting in change of magnetization along the z-axis.

Figure 2b shows the temporal evolution of the Faraday rotation change induced by the generated THz magnetic near-field at several temperatures. The frequency of oscillation at each temperature is clearly assigned to that of the AF-mode [4]. To gain further insight into the spin motion and its driving force, we calculated the THz magnetic near-field induced magnetization dynamics and resultant polarization rotation of the probe pulse on the basis of the Landau-Lifshitz-Gilbert (LLG) equation [5]. We made numerical calculation with reported physical constants of HoFeO$_3$ and reproduced successfully the experimental results. One can see quite efficient build-up of AF-mode at 120 K, where the AF-mode is resonant with SRR mode, ν_{LC} = 0.5 THz.

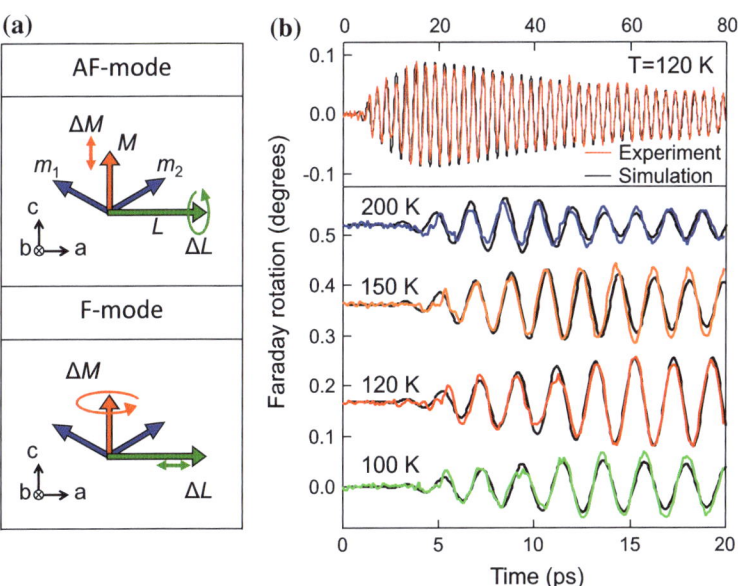

Fig. 2 **a** Schematics of static configuration of Fe sublattice magnetization. Magnetization motions for two modes are shown using ferromagnetic vector $M = m_1 + m_2$ and antiferromagnetic vector $L = m_1 - m_2$. **b** Measured temporal change of Faraday rotation angle after THz pump pulse excitation measured and simulation results (*black line*)

4 Summary

We studied the excitation of antiferromagnetic spin waves in $HoFeO_3$ crystal coupled to a SRR by using THz pulses. The magnetic field in the vicinity of the SSR induced by the incident THz-electric-field component excites and the Faraday rotation of the polarization of a near-infrared probe pulse directly measures the oscillations that correspond to the antiferromagnetic spin resonance mode. The good agreement of the observed signals with Landau-Lifshitz-Gilbert equation based on the two-lattice model confirms that the THz magnetic near-field is enhanced by the SRR by 20 times in amplitude compared to that of the incident THz magnetic field.

References

1. T. Kampfrath, A. Sell, G. Klatt, A. Pashkin, S. Mährlein, T. Dekorsy, M. Wolf, M. Fiebig, A. Leitenstorfer, and R. Huber, Nature Photon. **5**, 31 (2011).
2. N. Kumar, A. C. Strikwerda, K. Fan, X. Zhang, R. D. Averitt, P. C. M. Planken, and A. J. L. Adam, Opt. Express **20**, 11277 (2012).
3. H. Hirori, A. Doi, F. Blanchard, and K. Tanaka, Appl. Phys. Lett. **98**, 091106 (2011).
4. A. M. Balbashov, G.V. Kozlov, S. P. Lebedev, A. A. Mukhin, A. Yu. Pronin, and A. S. Prokhorov, Sov. Phys. JETP **68**, 629 (1989).
5. G. F. Herrmann, J. Phys. Chem. Solids **24**, 597 (1963).

Ultrafast Spin Dynamics in an Antiferromagnet NiO Observed in Pump-Probe and Terahertz Experiments

Takeshi Moriyasu, Suguru Wakabayashi, Hogyun Jinn
and Toshiro Kohmoto

Abstract We observed the ultrafast spin dynamics in an antiferromagnet NiO. The dynamics of the antiferromagnetic magnons and the magnetostriction was studied using the pump-probe technique and THz-TDS.

1 Introduction

The ultrafast spin dynamics and optical spin control in magnetic materials [1–4] are attractive topics because of the potential applications in the developments of ultrafast spin control, spintronics, quantum computing, and optical control of correlated spin systems. For the ultrafast spin control and device applications, solid-state materials are more desirable.

The magnon is one of elementary excitations in solid-state materials and plays important roles in magnetic materials. Usually, antiferromagnetic resonance frequencies, which are the precession frequencies of magnons, are higher than ferromagnetic resonance frequencies. Therefore, it is possible that antiferromagnets are used to realize the ultrafast spin control. In many cases, the antiferromagnetic resonance frequency lies in the terahertz region, and the spectroscopic investigation of antiferromagnets in this frequency region is very important.

In this work, we studied the ultrafast spin dynamics in an antiferromagnetic 3d transition metal monoxide NiO. To investigate the magnons and magnetostriction in NiO, we performed the two experiments of the pump-probe technique [5] and terahertz time-domain spectroscopy (THz-TDS). The two experiments are complementary, because the pump-probe technique is sensitive to Raman-active modes, and the THz-TDS is sensitive to IR-active modes. It is very useful to carry out the two experiments for the detailed investigation of the energy states.

T. Moriyasu (✉) · S. Wakabayashi · H. Jinn · T. Kohmoto
Graduate School of Science, Kobe University, Kobe 657-8501, Japan
e-mail: moriyaspin@gmail.com

© Springer International Publishing Switzerland 2015
K. Yamanouchi et al. (eds.), *Ultrafast Phenomena XIX*,
Springer Proceedings in Physics 162, DOI 10.1007/978-3-319-13242-6_161

2 Sample

NiO is a paramagnetic insulator with a cubic rock-salt structure above the Néel temperature 523 K. In the antiferromagnetic phase below the Néel temperature, the magnetic moments on Ni atoms align ferromagnetically within a (111) plane, and (111) planes are stacked antiferromagnetically in the direction normal to the (111) plane. In a simple antiferromagnet with easy-plane anisotropy such as MnO and NiO, two magnon modes, the high- and low-frequency modes, are expected theoretically [6]. In many cases, the high-frequency mode is observed in the THz region.

3 Experiment

In the pump-probe experiment, the pump pulse, whose wavelength and pulse width are 810 nm and 0.2 ps, is provided by a Ti:sapphire regenerative amplifier and the probe pulse, whose wavelength and pulse width are 900 nm and 0.2 ps, by an optical parametric amplifier. The population difference in the magnetic sublevels of the ground state or the fictitious magnetic field induced by the light-shift effect are instantaneously generated by the circularly polarized pump pulse, and then magnons are excited. The excited magnons are detected by a polarimeter as the change in the polarization of the linearly polarized probe pulse.

In the THz experiment, the terahertz wave excites magnons directly. The generation and detection of THz wave are carried out via the use of a large-area electro-optic (EO) THz emitter and a standard EO THz detector, respectively. The optical pulses for the THz-wave generation and detection are provided by the Ti:sapphire regenerative amplifier. Their wavelength, pulse width, pulse energy, and repetition rate are 810 nm, 0.15 ps, 700 μJ, and 1 kHz, respectively.

The sample, whose thickness is 0.1 mm for pump-probe technique and 5 mm for THz-TDS, is placed in a temperature-controlled cryostat.

4 Result and Discussion

The typical magnon signals in NiO observed by the pump-probe technique and THz-TDS are shown in Fig. 1a, c, and their Fourier transform are shown in Fig. 1b, d. The observed temperature dependence of the antiferromagnetic magnon modes up to the Néel temperature in NiO is shown in Fig. 1e. These modes are observed in a good accuracy by using the pump-probe and THz-TDS experiments than the result of Raman scattering and Brillouin scattering [7].

The magnetostrictive contribution to the refractive index was observed by using THz-TDS. Figure 2 shows the temperature dependence of the change Δn in refractive index obtained from the peak shift of the transmitted THz electric field.

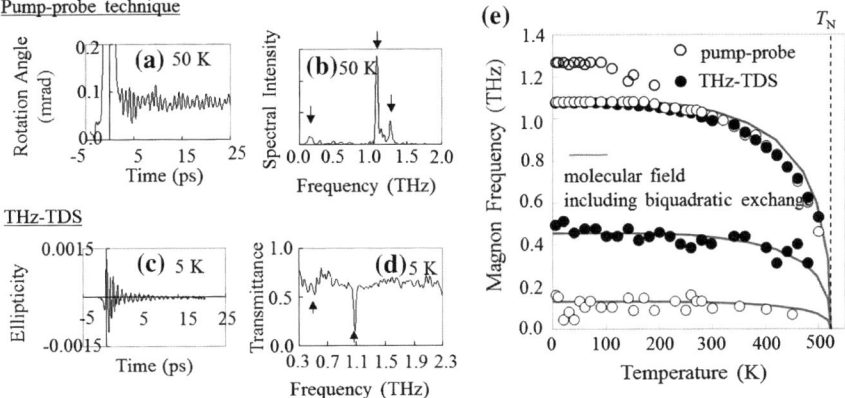

Fig. 1 Magnon signals in NiO observed by **a** the pump-probe technique and **c** THz-TDS. **b**, **d** Are the Fourier transform of the signals in (**a**, **c**). **e** Temperature dependence of the magnon frequencies observed in NiO

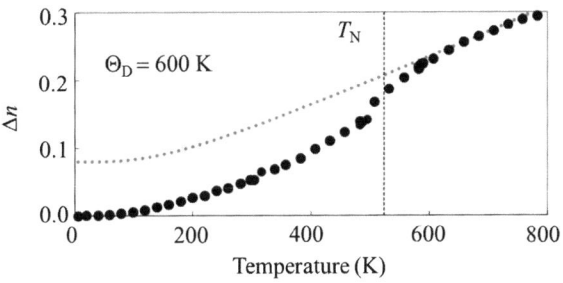

Fig. 2 Temperature dependence of the change Δn in refractive index obtained from the peak shift in the THz electric field

The dotted curve indicates the contribution from the thermal expansion calculated from the internal energy of phonons. The deviation from the calculated curve observed below the Néel temperature is considered to be a magnetostrictive contribution.

Our experimental results show that the time-domain spectroscopy have a large potential for the sensitive and accurate measurements in magnetic materials.

References

1. J. A. Gupta, R. Knobel, N. Samarth, and D. D. Awschalom, Science **292**, 2458-2461 (2001).
2. V. Kimel, A. Kirilyuk, and T. Rasing, Laser & Photon. Rev. **1**, 275-287 (2007).
3. R. C. Myers, M. H. Mikkelsen, J.-M. Tang, A. C. Gossard, M. E. Flatté, and D. D. Awschalom, Nat. Mater. 7, 203-208 (2008).

4. T. Moriyasu, D. Nomoto, Y. Koyama, Y. Fukuda, and T. Kohmoto, Phys. Rev. Lett. **103**, 213602 (2009).
5. M. Takahara, H. Jinn, S. Wakabayashi, T. Moriyasu, and T. Kohmoto, Phys. Rev. B **86**, 094301 (2012).
6. A. J. Sievers III, and M. Tinkham, Phys. Rev. **129**, 1566-1571 (1963).
7. J. Milano, L. B. Steren, and M. Grimsditch, Phys. Rev. Lett. **93**, 077601 (2004).

Part IX
Photoemitted Electron, Plasmon and Nanoplasmas

Controlling the Motion of Strong-Field, Few-Cycle Photoemitted Electrons in the Near-Field of a Sharp Metal Tip

Petra Groß, Björn Piglosiewicz, Slawa Schmidt, Doo Jae Park,
Jan Vogelsang, Jörg Robin, Cristian Manzoni, Paolo Farinello,
Giulio Cerullo and Christoph Lienau

Abstract The real-time probing of electron motion in solid nanostructures or the visualization of nanoplasmonic field dynamics may come into reach using electron pulses generated by strong-field tunneling from sharp gold tips irradiated by few-cycle laser pulses. The acceleration of the ultrashort electron wavepackets in the near field of the sharp gold tips introduces new possibilities of steering and control of electron wavepackets by light, which is expected to pave the way towards such ultrafast probing. Here we discuss the motion of these highly accelerated electrons in the near-field and demonstrate how the carrier-envelope phase admits a new control mechanism for their motion.

1 Introduction

Sharp nanometer-sized metallic tips recently emerged as a test bed for exploring strong-field phenomena such as high-harmonic generation and photoemission [1–5]. When such tips are illuminated with few-cycle laser pulses of sufficient field strength, optical field enhancement at the tip apex results in tunnelling of electrons out of the tip. The subsequent acceleration of the electrons within the local near-field gradient can be so powerful that their typical quiver motion in an oscillating laser field is fully suppressed. This results in sub-cycle electrons, which can traverse the near field within less than one half-cycle of the laser field [3–5].

Under these conditions, we enter a whole new regime of electron emission, which is characterized by the motion of the electrons in the oscillating, fast

P. Groß (✉) · B. Piglosiewicz · S. Schmidt · D.J. Park · J. Vogelsang · J. Robin · C. Lienau
Institut für Physik, Carl von Ossietzky Universität, 26129 Oldenburg, Germany
e-mail: petra.gross@uni-oldenburg.de

P. Groß · B. Piglosiewicz · S. Schmidt · D.J. Park · J. Vogelsang · J. Robin · C. Lienau
Center of Interface Science, Carl von Ossietzky Universität, 26129 Oldenburg, Germany

C. Manzoni · P. Farinello · G. Cerullo
IFN-CNR, Dipartimento di Fisica, Politecnico di Milano, 20133 Milan, Italy

© Springer International Publishing Switzerland 2015   659
K. Yamanouchi et al. (eds.), *Ultrafast Phenomena XIX*,
Springer Proceedings in Physics 162, DOI 10.1007/978-3-319-13242-6_162

decaying near-field with a decay length of only a few nm. This new sub-cycle regime is unique to solid state nanostructures with their short field decay length and steep field gradients. Recollisions, which are a key feature in atomic and molecular systems, are strongly suppressed, and instead, the electrons are accelerated along the field lines [4]. The shape of the nanostructure then determines the electrons trajectories.

2 Experimental Setup

We study photoemission from sharply etched, single-crystalline gold tips, which have an apex radius of down to 5 nm. Due to the small radius of curvature, the tips show a large field enhancement factor of ~9 and a short decay length of the local near field of ~2 nm [4, 5]. The gold tips are irradiated with 16-fs pulses (corresponding to 2.6 optical cycles) at a wavelength of 1.65 μm from a noncollinear optical parametric amplifier (NOPA) system followed by difference frequency generation (Fig. 1a). The combination of frequency conversion stages [6] ensures that the pulses have a highly stable CEP. Residual phase fluctuations were measured in a conventional *f*-to-2*f* interferometer to be as low as ~50 mrad over a time span of 20 min. The CEP is controlled via a pair of fused silica wedges, and the energy spectra of the emitted and accelerated electrons are measured as a function of CEP using a photo-electron spectrometer (PES).

Our experimental observations are corroborated by simulations of the energy spectra performed within a modified three-dimensional Simpleman model, taking

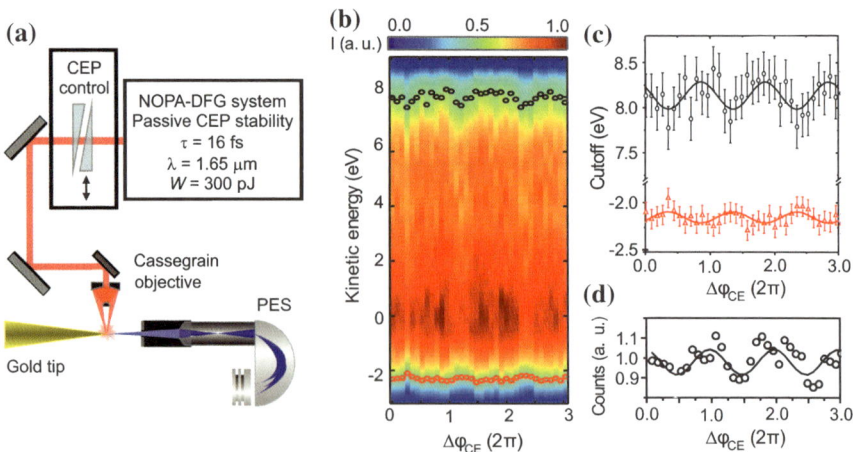

Fig. 1 **a** Experimental setup. **b** Electron kinetic energy spectra when the CEP is varied via the fused silica wedges. **c** High- and low energy cutoffs and **d** total electron yield extracted from (**a**), showing a clear modulation with the CEP

into account the repulsion between electrons [5, 7]. We take an analytical model for the optical near-field distribution around the apex. The local generation yield is deduced from a Fowler-Nordheim equation and the ejected electrons are accelerated as classical particles within the temporally oscillating and spatially varying near field. Electron kinetic energy spectra are generated by calculating the terminal kinetic energies of the electrons.

3 Results

Under these experimental conditions, we enter the sub-cycle regime, where electron emission predominantly takes place by strong-field tunneling, and where the electrons subsequently experience a strong acceleration allowing electrons to traverse the near-field within one optical half-cycle. The characteristic signature of sub-cycle electrons is the emergence of a high-kinetic-energy plateau in the measured kinetic energy spectra.

Kinetic energy spectra recorded while varying the CEP (Fig. 1b) show a clear modulation of the high- and low-energy cutoff. We see a high-energy cutoff variation by more than 1 eV, while the low-energy cutoff variation is less pronounced and amounts to 0.3 eV. Plotting both quantities together as a function of the CEP (as red and black circles in Fig. 1c), clearly inversely phased oscillations can be seen and thus a periodic broadening and narrowing of the spectrum. The same periodicity is found in the total electron yield (Fig. 1d).

From numerical simulation of the electron trajectories in the oscillating near-field we find that the acceleration of optical field-emitted electrons in the spatial near-field gradient is of key importance, and that this acceleration is governed by the CEP and the electron birth time with respect to the negative amplitude maximum of the driving laser field. The energetic width of the spectra is determined by the acceleration the sub-cycle electrons gain during the negative half-cycle of their generation. Their terminal kinetic energy depends on the field maximum of the respective half-cycle, and hence strongly on the CEP.

4 Conclusion

Taken together, our results present the first demonstration of CEP effects on the optical field emission of electrons from a single metallic nanostructure. The CEP controls the local amplitude of the near field during the generation cycle and therefore the electron acceleration within the near field gradient on a sub-cycle time scale. Previously we have shown that the fastest electrons are also spatially steered to follow the electric field lines, which was observed as a distinct narrowing of the emission angle [4]. This steering effect in principle allows influencing the electron trajectories via the nanostructure. Together, the spatial steering via the nanostructure and temporal steering via the laser pulse electric field constitute new control mechanisms on the spatial-temporal motion of electrons. We believe that such field-driven control of the electron motion in the near field of solid state nanostructures can be seen as a new form of quantum electronics, paving the way towards the generation, measurement, and application of attosecond electron pulses.

References

1. C. Ropers, D.R. Solli, C.P. Schulz, C. Lienau, and T. Elsässer, Phys. Rev. Lett. **98**, 43907 (2007)
2. M. Krüger, M. Schenk, and P. Hommelhoff, Nature **475**, 78 (2011)
3. G. Herink, D.R. Solli, M. Gulde, and C. Ropers, Nature **483**, 190 (2012)
4. D.J. Park, B. Piglosiewicz, S. Schmidt, H. Kollmann, M. Maschek, and C. Lienau, Phys. Rev. Lett. **109**, 244803 (2012)
5. B. Piglosiewicz, S. Schmidt, D.J. Park. J. Vogelsang, P. Groß, C. Manzoni, P. Farinello, G. Cerullo, C. Lienau, Nature Photon. **8**, 37 (2014)
6. C. Manzoni, G. Cerullo, and S. De Silvestri, Opt. Lett. **29**, 2668 (2004)
7. B. Piglosiewicz, J. Vogelsang, S. Schmidt, D.J. Park, P. Groß, and C. Lienau, Quantum Matter **3**, 297 (2014)

Velocity Map Imaging of Electrons Strong-Field Photoemitted from Si-Nanotip Arrays

Hong Ye, Jens S. Kienitz, Shaobo Fang, Sebastian Trippel,
Michael E. Swanwick, Phillip D. Keathley, Luis F. Velásquez-García,
Giovanni Cirmi, Giulio M. Rossi, Arya Fallahi, Oliver D. Mücke,
Jochen Küpper and Franz X. Kärtner

Abstract Ultrafast and ultra-bright electron sources with spatially structured emission are an enabling technology for free-electron lasers (FELs), electron diffractive imaging, compact coherent X-ray sources, and attosecond science.

Ultrafast and ultra-bright electron sources with spatially structured emission are an enabling technology for free-electron lasers (FELs), electron diffractive imaging, compact coherent X-ray sources [1] and attosecond science [2]. The interaction of sharp nanotips with intense few-cycle laser pulses are expected to potentially generate and provide precise control of confined coherent electron wave packets with attosecond duration. These wave packets can be used for the visualization of electron dynamics [3] and nanoplasmonic field dynamics [4]. In this work, we report on the characterization of the transverse electron momentum distribution upon photoemission from a surface when excited with 30 fs laser pulses at 800 nm, using a specially constructed velocity map imaging (VMI) electron spectrometer.

H. Ye (✉) · J.S. Kienitz · S. Fang · S. Trippel · G. Cirmi · G.M. Rossi ·
A. Fallahi · O.D. Mücke · J. Küpper · F.X. Kärtner
Center for Free-Electron Laser Science, Deutsches Elektronen-Synchrotron DESY,
Notkestraße 85, 22607 Hamburg, Germany
e-mail: hong.ye@desy.de

H. Ye · G.M. Rossi · J. Küpper · F.X. Kärtner
Department of Physics, University of Hamburg, Luruper Chaussee 149, 22761 Hamburg,
Germany

J.S. Kienitz · S. Fang · G. Cirmi · A. Fallahi · O.D. Mücke · J. Küpper · F.X. Kärtner
The Hamburg Center for Ultrafast Imaging, University of Hamburg, Luruper Chaussee 149,
22761 Hamburg, Germany

M.E. Swanwick · L.F. Velásquez-García
Microsystems Technology Laboratories, Massachusetts Institute of Technology,
77 Massachusetts Avenue, Cambridge, MA 02139, USA

P.D. Keathley · F.X. Kärtner
Department of Electrical Engineering and Computer Science and Research
Laboratory of Electronics, Massachusetts Institute of Technology, Cambridge,
MA 02139, USA

© Springer International Publishing Switzerland 2015   663
K. Yamanouchi et al. (eds.), *Ultrafast Phenomena XIX*,
Springer Proceedings in Physics 162, DOI 10.1007/978-3-319-13242-6_163

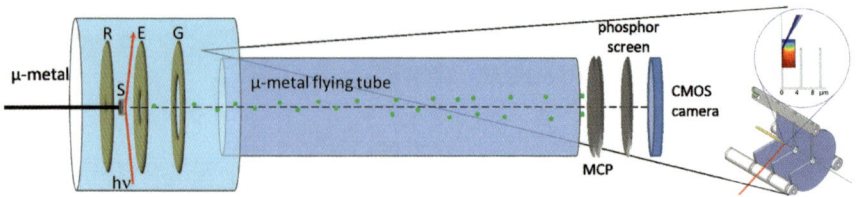

Fig. 1 Configuration of the VMI. The *red solid lines* represent the laser beam. *S* sample, *R* repeller, *E* extractor, *G* ground electrode. The *green dots* depict the emitted electrons. See text for details

The transverse momentum distribution determines the emittance of an electron gun. For calibration purposes, we first performed measurements from Au flat surfaces and then studied the emission from a novel ultrafast optical-field emission cathode comprised of a dense and highly uniform nanotip array. Such field emitters offer an attractive alternative to UV photocathodes, while providing a direct means of structuring the emitted electron beam.

The VMI technique has been developed to study dynamics of atoms and molecules through detection of the emitted electron or ion momentum distribution [5]. Figure 1 shows the schematic of our VMI for solid state samples, which consists of three electrodes (repeller, extractor, and ground electrode) [5], a 49 cm long drift tube, followed by the detector assembly of a multi-channel plate (MCP), phosphor screen and a CMOS camera. The full assembly is shielded by μ-metal tubing inside vacuum.

The Si-samples used in this experiment are uniform arrays of high-aspect-ratio doped silicon pillars, 38 μm high and 1 μm wide with a 5 μm pitch. Each pillar is topped by an ultra-sharp tip with an average radius of curvature of ~ 5 nm. The sample was treated with Piranha (H_2SO_4:H_2O_2 7:1) and BOE (NH_4F:HF 7:1) solutions to remove any organic residues and the thin native oxide. A 800 nm 30 fs Ti:sapphire laser amplifier with a 3-kHz repetition rate was used to illuminate the sample at an incident angle of $\sim 85°$, with a beam size of 30 μm FWHM, and p-polarization with respect to the nanotip direction. Single-shot electron images were read out at a 1 kHz rate, limited by the camera link, and averaged for 4 s and 60 s for spatial and velocity-map imaging, respectively.

A flat Au sample was tested first using ~ 13 nJ laser beam for the setup calibration, showing the results in Fig. 2a, b, where spatial images represent the sample surface morphology with a magnification ratio of ~ 23 and VMI images illustrate the transverse momentum distribution of the emitted electrons. In VMI mode, the x and y axes refer to the velocities V_x and V_y. Thus, higher transverse kinetic energies cause wider electron distributions in the images. In Fig. 2a, various surface contamination and defects were observed with sizes of a few μm and these artifacts reveal different momentum distributions. The mean transverse kinetic energy of emitted electrons is derived to be in the range of 0.44–0.55 eV, see Fig. 2c. Assuming that photoemission is dominated by multi-photon ionization with random direction, we use the Spicer's three-step model [6] to characterize the kinetic

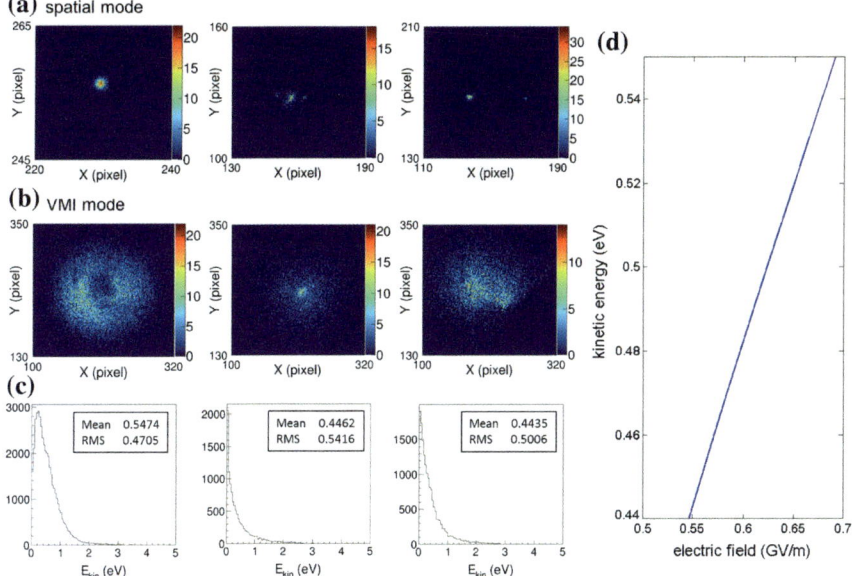

Fig. 2 Experimental results obtained from Au flat sample: **a** Spatial mode images showing defects on a flat Au-sample; **b** velocity-map images and **c** the extracted transverse kinetic energy distribution of emitted electrons from different defects, demonstrating the significant influence of surface morphology on the photoelectron momentum distributions; **d** electric field versus kinetic energy of emitted electrons simulated according to the three-step model [6]

energy. From Fig. 2d, for Au with the work function $\phi = 5.1$ eV and considering the Schottky effect, the electric field required to achieve a kinetic energy of 0.44–0.55 eV in a three-photon process (3×1.55 eV at 800 nm) is 0.54–0.68 GV/m, which is 1.6–2.2 times the initial electric field 0.32 GV/m in the laser beam itself. We believe this is caused both by the electric field enhancement resulting from the reflected laser beam and by defect edges.

Figure 3 shows the experimental results from the Si-nanotip arrays. The laser energy used is ~ 0.82 μJ resulting in an electric field of 2.6 GV/m (field enhancement not included). A higher intensity is observed on the left side of the VMI image (see Fig. 3a), which is consistent with numerical simulations of the electric field, indicating that the enhanced electric field is asymmetric and weaker on the side of the tip from which the laser beam comes in. We observe a peak in the kinetic energy of ~ 1 eV resulting from three-photon ionization, which is consistent with the work function for Si-nanotips of 3.6–4.05 eV depending on the oxidation layer thickness and Schottky effect [7]. From the center of the VMI image, an electron momentum distribution with a star-shaped structure of several sharp angles is observed with electron kinetic energies up to ~ 5 eV. These high electron energies demonstrate the strong-field emission from Si-nanotips. The angular structures are likely introduced by non-cylindrically-symmetric emitter tips.

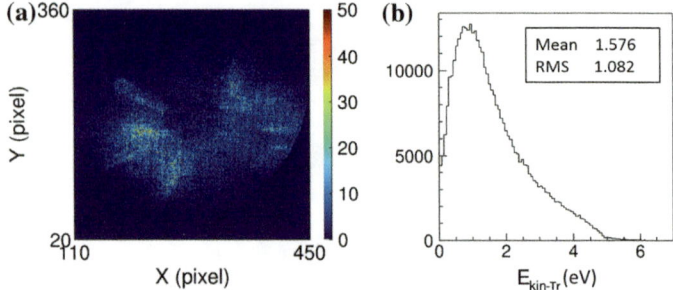

Fig. 3 Experimental results of a Si-nanotip array: **a** Measured transverse momentum distribution with VMI and **b** transverse kinetic energy distribution of emitted electrons

In conclusion, the constructed VMI enables us to visualize the transverse momentum distribution of electrons emitted from a surface. Measurements on Au surfaces gave consistent results relating excess electron energies upon multi-photon emission. In the strong-field photoemission regime from Si-nanotips we observe a strong emission peak corresponding to an excess energy of 1 eV and a star-shaped structure with transverse kinetic energies up to ~ 5 eV. By optimizing the VMI operation conditions and laser pulse parameters, we can further optimize the electron emission process under strong-field conditions to ultimately arrive at optimized electron guns for electron diffractive imaging and coherent X-ray sources.

Acknowledgments The authors like to thank Nele L.M. Müller for many helpful discussion on photocathode emittance measurements.

References

1. W. S. Graves *et al.*, Phys. Rev. Lett. **108**, 263904 (2012).
2. W. Putnam *et al.*, Ultrafast Phenomenon 2014, paper 08.Tue.B.3 (2014).
3. E. Goulielmakis *et al.*, Nature **466**, 739-743 (2010).
4. A. Kubo *et al.*, Nano Lett. **5**, 1123-1127 (2005).
5. A. T. J. B. Eppink and D. H. Parker, Rev. Sci. Instrum. **68**, 3477-3484 (1997).
6. D. H. Dowell and J. F. Schmerge, Phys. Rev. ST Accel. Beams **12**, 074201 (2009).
7. P. D. Keathley *et al.*, Ann. Phys. (Berlin) **525**, 144-150 (2013).

Visualization of Photocurrents in Nanoobjects by Ultrafast Low-Energy Electron Point-Projection Imaging

M. Müller, A. Paarmann and R. Ernstorfer

Abstract Ultrafast photocurrents in pin-type semiconductor InP nanowires are investigated with femtosecond time and nanometer spatial resolution. We demonstrate the capability of femtosecond point-projection imaging as a novel tool for spatiotemporal mapping of transient fields at nanostructure surfaces using low-energy electron pulses generated from a metal nanotip as a highly sensitive probe.

1 Introduction

Semiconductor nanowires (NWs) and, in particular, heterostructured NWs are promising candidates for future nanoscale electronic and optoelectronic devices, as well as ideal model systems for exploring fundamental semiconductor physics on nanometer length scales. Within the last years there has been vast progress in controlling the doping level in both radial and axial direction during NW growth [1]. In order to understand the physics of carrier transport in these structures, it is of major importance to study the dynamics of charge carriers upon photoexcitation. So far, typical studies such as time-resolved photoluminescence and photoemission electron microscopy provide either spatially- or time-averaged information, respectively. In this regard, it is most appealing to combine nanometer spatial with femtosecond temporal resolution to directly measure the spatio-temporal evolution of photoexcited charge distributions in such nanoobjects.

Here, we present the first femtosecond time- and nanometer-spatially resolved investigation of ultrafast photocurrents in heterostructured InP-NWs, employing our newly developed technique of femtosecond low-energy electron point-projection imaging (fsPPI). We show that low-energy electrons are ideally suited to detect transient fields in the near-surface region of nanoobjects generated by ultrafast photocurrents on nanometer dimensions. Due to the high sensitivity of this novel

M. Müller (✉) · A. Paarmann · R. Ernstorfer
Fritz-Haber-Institut der MPG, Faradayweg 4–6, 14195 Berlin, Germany
e-mail: m.mueller@fhi-berlin.mpg.de

© Springer International Publishing Switzerland 2015
K. Yamanouchi et al. (eds.), *Ultrafast Phenomena XIX*,
Springer Proceedings in Physics 162, DOI 10.1007/978-3-319-13242-6_164

approach, the detailed spatially resolved dynamics of weak fields can be studied in a wide range of low-dimensional systems, in particular semiconductor and plasmonic nanostructures.

2 Method

A schematic of our experimental apparatus is shown in Fig. 1a. We use polycrystalline tungsten nanotips as point source for femtosecond electron probe pulses with energies in the range of 20–200 eV generated by an ultrashort laser pulse [2]. The tip is placed ~ 20 μm in front of the sample and a point-projection image is recorded on the spatially resolving detector with a magnification of up to 10^4. Figure 1b shows projection images of gold NWs spanning across 2 μm holes in a thin carbon substrate in continuous field emission and laser-triggered photoemission mode, respectively. In the pulsed mode, an initial electron pulse duration of <10 fs can be achieved as shown by the interferometric autocorrelation of the photoemitted current in Fig. 1c, also demonstrating the three-photon character of the emission process. Time resolution is added to the experiment by varying the delay between the electron probe pulse and an optical pump pulse exciting the sample, see Fig. 1a.

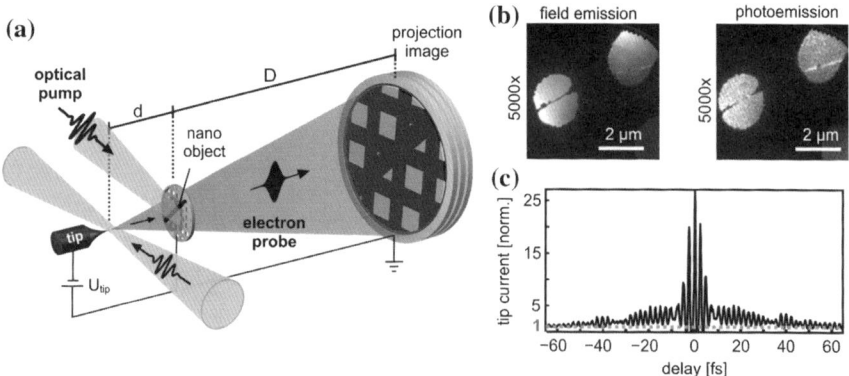

Fig. 1 **a** Scheme for femtosecond point projection imaging. **b** Projection images of a 50 nm gold NW spanning across 2 μm holes in a free-standing carbon thin film, for field emission (*left*) and photoemission (*right*) operation mode, respectively. **c** Interferometric autocorrelation of the nanotip current in the laser-triggered mode

3 Results and Discussion

First fsPPI data of axially doped p-i-n InP NWs are presented in Fig. 2. The projection image in Fig. 2a shows the NW before the optical pump pulse arrives. The projected wire diameter appears bright and much larger than its real space diameter (30 nm) due to static lensing effects such as the influence of the tip field and work function mismatches between the NW and the substrate. At temporal pump-probe overlap we observe a clear pump-induced change of the imaged wire profile, as shown in Fig. 2b, c. The intensity redistribution is most pronounced perpendicular to the wire, transiently changing the projected wire diameter. We also observe clear differences of these changes axially along the wire. The dynamics of these effects are analyzed in Fig. 2d, where we plot the fitted wire diameter as a function of time delay. At both axial positions of the NW the signal shows a rapid initial drop with a ~ 200 fs time constant, followed by the relaxation of the effect on a few ps time scale. While the time scales in both regions of the NW are similar, the magnitude is much larger in the NW segment of region 1 (squares in Fig. 2d).

We interpret the observations being a result of one-photon above-bandgap absorption from the 800 nm pump pulse and consecutive radial separation of the photo-generated charge carriers due to band bending at the semiconductor surface, which depends on the specific doping and surface condition of the respective NW segments. This would result in a transient reduction of the intrinsic surface band bending and could well explain the transient electric fields modifying the effective electron lensing along the NW. We also observe a slight axial redistribution of intensity indicating axial photocurrents, most probably induced by the axial p-i-n

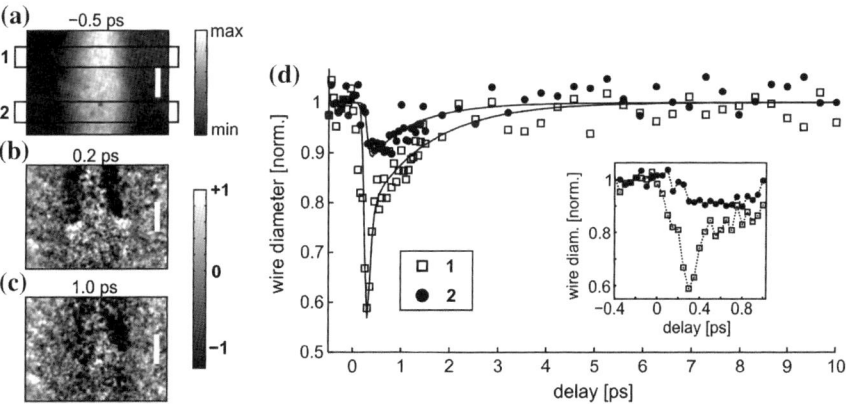

Fig. 2 **a** Point-projection image of a pin-type InP NW at negative delay times between pump and probe recorded at 70 eV electron energy. **b, c** Normalized difference images recorded at two different delays at 200 fs and 1 ps, respectively. *Scale bars* are all 500 nm. **d** Dynamics of the projected wire diameter for two segments in regions 1 and 2 along the nanowire as indicated by the *two boxes* in (**a**). The *inset* shows the fast initial dynamics of the diameter change on the sub-ps time scale

doping structure of the NW. A linear fluence dependence of the transient diameter changes supports the picture of one-photon absorption from the pump pulse. Noteworthy, the initial rise time of the effect is on the 200 fs time scale. Simulations studies [3] predict electron pulses with durations <50 fs in this operation mode, suggesting the initial dynamics are not limited by instrument resolution but instead are dominated by the dynamics of the photo-excited charge carriers.

4 Conclusions

Femtosecond low-energy electron point-projection investigations of axially doped pin-type InP nanowires are presented. We observe transient modifications of the projection images with a rise time of ~ 200 fs and decay times of a few ps, which we interpret being the result of separation of photo-carriers due to surface band bending, inducing a transient change of the electric fields surrounding the wire. Our data proves the capability of this novel experimental approach for detecting weak transient local electric fields on ~ 100 fs time scales and with nm spatial resolution, making it very attractive for investigations of various nanoscopic systems such as semiconducting and plasmonic nanostructures.

References

1. M. Hjort, J. Wallentin, R. Timm, A.A. Zakharov, U. Håkanson, J.N.Andersen, E. Lundgren, L. Samuelson, M.T. Borgström, and A. Mikkelsen, "Surface Chemistry, Structure and Electronic Properties from Microns to the Atomic Scale of Axially Doped Semiconductor Nanowires", ACS Nano **6**, 9679 (2012)
2. C. Ropers, D.R. Solli, C.P. Schulz, C. Lienau, and T. Elsaesser, "Localized Multiphoton Emission of Femtosecond Electron Pulses from Metal Nanotips", Phys. Rev. Lett. **98**, 043907 (2007)
3. A. Paarmann, M. Gulde, M. Müller, S. Schäfer, S. Schweda, M. Maiti, C. Xu, T. Hohage, F. Schenk, C. Ropers, and R. Ernstorfer, "Coherent Femtosecond Low-Energy Single-Electron Pulses for Time-Resolved Diffraction and Imaging: A Numerical Study" ,J. Appl. Phys. **112**, 113109 (2012).

Visualization of Charge Carrier Motion in Semiconductor Nanowires with Ultrafast Pump-Probe Microscopy

Michelle M. Gabriel, Erik M. Grumstrup, Justin R. Kirschbrown,
Christopher W. Pinion, Joseph D. Christesen, David F. Zigler,
Emma E.M. Cating, James F. Cahoon and John M. Papanikolas

Abstract Femtosecond pump-probe microscopy is used to directly visualize the diffusion of photogenerated charge carriers in undoped silicon nanowires, as well as charge separation in a nanowire encoded with an axial p-type/intrinsic/n-type (p-i-n) junction.

1 Introduction

A detailed understanding of the factors that govern the motion of mobile charge carriers through nanostructures is critical to many emerging nanotechnologies in electronics, optoelectronics and solar energy conversion. While the motion of charge carriers at low carrier densities is uncorrelated and easy to understand, many active electronic components operate at high carrier concentrations resulting from heavy doping or high injection. In this regime, carrier-carrier interactions and other many body effects (e.g. dopant/carrier interactions, electron screening, and electron-hole scattering) must be considered. We have combined ultrafast pump-probe spectroscopy with optical microscopy [1, 2] to directly image the charge carrier dynamics in individual Si nanowires (NWs) with both spatial and temporal resolution.

2 Pump-Probe Microscopy

The femtosecond pump-probe microscope is illustrated in Fig. 1. In these experiments, an individual NW is excited by a 425 nm femtosecond pump pulse that has been focused to a diffraction limited spot (350 nm) by a microscope objective,

M.M. Gabriel · E.M. Grumstrup · J.R. Kirschbrown · C.W. Pinion · J.D. Christesen ·
D.F. Zigler · E.E.M. Cating · J.F. Cahoon · J.M. Papanikolas (✉)
Department of Chemistry, University of North Carolina at Chapel Hill,
Chapel Hill, NC 27599-3290, USA
e-mail: john_papanikolas@unc.edu

© Springer International Publishing Switzerland 2015   671
K. Yamanouchi et al. (eds.), *Ultrafast Phenomena XIX*,
Springer Proceedings in Physics 162, DOI 10.1007/978-3-319-13242-6_165

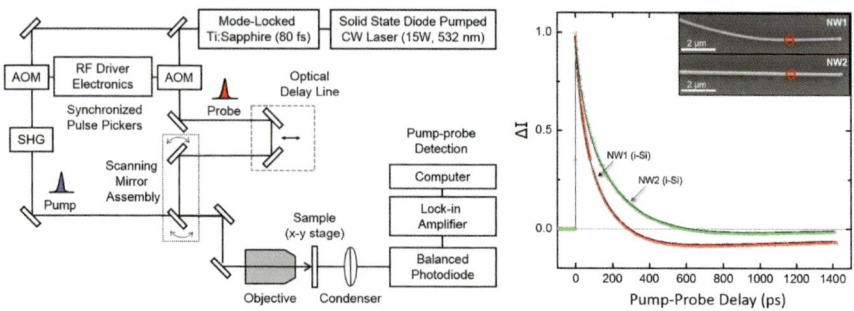

Fig. 1 *Left* Schematic diagram of the pump-probe microscope. Pump (425 nm) and probe (850 nm) pulses are focused to diffraction limited spots by a microscope objective. A sample stage allows the two pulses to be overlap with a specific structure. A scanning mirror assembly scans the angle of the probe beam as it enters the objective, enabling it to be spatially separated relative to the pump pulse. *Right* Pump-probe decay kinetics obtained with spatially overlapped pump and probe beams for two different Si nanowires; NW1 (*red*, 160 nm) and NW2 (*green*, 210 nm). Both were fit to a triexponential decay (*solid lines*). *Inset* SEM images showing the location of pump excitation as a *red circle*; *scale bars*, 2 μm

producing photogenerated carriers (electrons and holes) in a localized region of the structure. After a well-defined delay, pump-induced changes to the transmission of an 850 nm probe pulse are detected.

Figure 1 shows pump-probe decay kinetics when the probe beam is spatially overlapped with the pump pulse. Transients for two NWs are shown with the location of the excitation depicted by red circles in their corresponding SEM images. The photoinduced transparency observed at early times arises from band filling by the photogenerated free carriers. It decays over several hundred pico-seconds, and eventually changes sign to yield a low-amplitude absorptive component that persists beyond 1 ns and is attributed to a combination of trapped carriers and thermal effects. In this spatially-overlapped pump-probe (SOPP) configuration, the decay of the bleach reflects both electron-hole recombination and migration of carriers away from the excitation spot. The recombination time (∼200 ps) extracted from the bleach decay scales with NW diameter, consistent with surface mediated recombination.

Motion of the photogenerated carriers is observed using a spatially-separated pump-probe (SSPP) configuration, in which carriers are created in one location and detected in another, allowing direct imaging of charge carriers as move away from the excitation spot. In this SSPP configuration the pump beam is held fixed and the position of the probe beam is scanned by varying the angle of the probe beam as it enters the objective (Fig. 2a), resulting in a spatial map of the photoinduced transparency (and the free carriers) at a specified pump-probe delay. Images collected at a series of delays show the spatial-temporal evolution of the charge cloud following excitation, providing a direct visualization of carrier diffusion in undoped NWs and charge separation in Si p-i-n junctions.

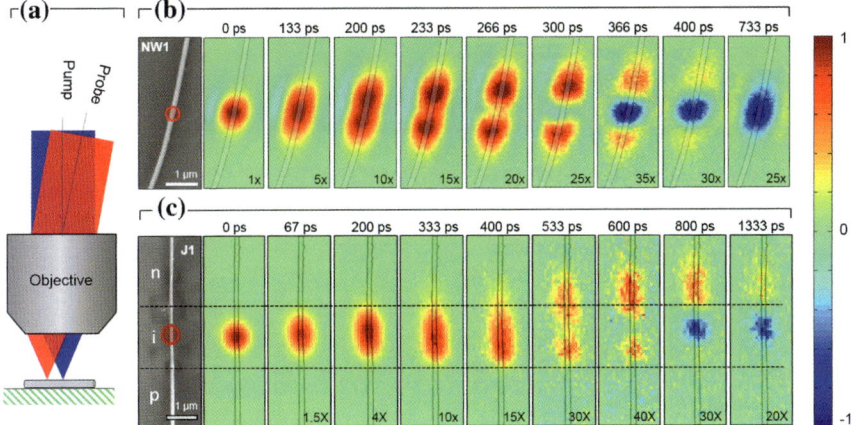

Fig. 2 Imaging of charge carrier motion in individual Si nanowires. **a** Schematic illustration of spatially separated scanning. **b** SSPP microscopy images depicting carrier diffusion in an undoped Si nanowire. *Left* SEM images of the NW with the location of the excitation spot depicted by the *red circle*. *Right* Series of SSPP images. Location of the nanowire is depicted by *faint lines*. Each image is depicted using a normalized *color scale* with relative amplitudes indicated by scaling factors. **c** SSPP images showing charge separation in a Si NW encoded with an axial p-i-n junction. Excitation is in the *middle* of the intrinsic segment, at the location of the *red circle*

Spatially-separated pump-probe images obtained on an undoped NWs are shown Fig. 2b [1]. The location of the pump pulse excitation is indicated by the red circle in the scanning electron microscopy (SEM) image at the left. The spatial extent of the photoinduced transparency created by the focused pump pulse is depicted by the red colors in the 0 ps frame. At later times, this transparency elongates, reflecting free carrier migration along the NW axis. The growth of the charge cloud is consistent with ambipolar diffusion in bulk Si. The blue spot that appears at longer times reflects the long-lived transient absorption (Fig. 1), which remains localized at the excitation spot.

Charge separation is apparent in SSPP images obtained from Si NWs encoded with axial p-type/intrinsic/n-type junctions (Fig. 2c) [2]. Here photoexcited carriers are produced at the center of the intrinsic region. Initially the carrier motion is dominated by carrier-carrier interactions, which opposes charge separation and causes the charge cloud to spreads in both directions. As it spreads, the carrier density decreases due to electron-hole recombination. By 300–400 ps it reaches the boundaries of the intrinsic segment, at which point its evolution becomes asymmetric. Over the next 200 ps, a bleach appears in the n-type region that then persists well beyond 1 ns, indicative of the formation of a long-lived charge separated state.

Acknowledgments Funding for this project was provided by the National Science Foundation under awards CHE-1213379 (J.M.P.) and DMR-1308695 (J.F.C.).

References

1. Gabriel, M. M.; Kirschbrown, J. R.; Christesen, J. D.; Pinion, C. W.; Zigler, D. F.; Grumstrup, E. M.; Mehl, B. P.; Cating, E. E. M.; Cahoon, J. F.; Papanikolas, J. M. Nano Lett., 13, 1336-1340 (2013).
2. Gabriel, M. M.; Grumstrup, E. M.; Kirschbrown, J. R.; Pinion, C. W.; Christesen, J. D.; Zigler, D. F.; Cating, E. E.; Cahoon, J. F.; Papanikolas, J. M. Nano Lett., 14, 3079 (2014).

Ultrafast Optical Control of Charge Dynamics in Organic and Hybrid Electronic Nanodevices

Artem A. Bakulin, Robert Lovrincic, Akshay Rao, Simon Gelinas,
Yu Xi, Oleg Selig, Zhuoying Chen, Richard H. Friend, Huib J. Bakker
and David Cahen

Abstract Using ultrafast visible/IR pulse-sequence spectroscopy combined with electric current detection, we engage vibronic and charge-delocalization phenomena to control the performance of optoelectronic devices base on organic semiconductors, colloidal quantum dots and conductive oxides.

OCIS Codes (320.7130) Ultrafast processes in condensed matter, semiconductors · (300.6340) Spectroscopy, infrared

Molecular electronics is a growing field with an ultimate goal of using organic and biological macromolecular systems such as conjugated polymers, molecular crystals, photochromic proteins or organic/inorganic hybrids in the development of nano-scale electrical circuits. Within the last decade it became clear that charge (de) localization, energetic disorder and vibronic coupling effects can be important roles for the conductivity of molecular-based materials. Until now, most efforts to control and use these effects have been confined to synthetic and material-processing approaches, aiming to develop materials of better quality by mostly trial-and-error exhaustive search.

In this work, we report a pioneering attempt to engage the charge-localization and vibronic-coupling effects optically, using ultrafast IR excitation of the vibrational or/and sub-gap electronic transitions. For this we combine ultrafast spec-

A.A. Bakulin (✉) · O. Selig · H.J. Bakker
FOM Institute AMOLF, Science Park 104, 1098 XG Amsterdam, The Netherlands
e-mail: A.Bakulin@AMOLF.nl

R. Lovrincic · Y. Xi · D. Cahen
Weizmann Institute of Science, 234 Herzl St., 76100 Rehovot, Israel

A. Rao · S. Gelinas · R.H. Friend
University of Cambridge, JJ Thomson Avenue, Cambridge CB30HE, UK

Z. Chen
ESPCI/CNRS/UPMC UMR 8213, 10 Rue Vauquelin, 75005 Paris, France

R. Lovrincic
Innovationlab, Braunschweig University, Speyerer Straße 4, 69115
Heidelberg, Germany

© Springer International Publishing Switzerland 2015
K. Yamanouchi et al. (eds.), *Ultrafast Phenomena XIX*,
Springer Proceedings in Physics 162, DOI 10.1007/978-3-319-13242-6_166

675

troscopy, photocurrent detection, [1] and ultrafast IR interferometry [2] providing the spectral resolution for selective vibrational excitation. We show that optical manipulation of vibrational and electronic states of the molecule gives control over charge separation and transport dynamics in the organic and hybrid materials.

Our main experimental tool is visible-pump—IR-push—photocurrent-probe spectroscopy (PPP), outlined in Fig. 1a. First, a 100-fs visible pump pulse illuminates the active layer of optoelectronic device introducing a population of charge carriers. The dynamics of carriers is then 'controlled' by an interferometer-controlled sequence of two 40-fs IR push pulses. Depending on the IR frequency, the push pulse can excite electronic transitions between localized and delocalized states, charges in low-energy 'trapping' states or molecular vibrations. The molecular-scale effect of IR excitation on the charge dynamics is observed through the corresponding change in the macroscopic photocurrent (δPC) as a function of the delay between 'pump' and push. A positive δPC value in such measurements is associated with the immobilized charge carriers, which can be made mobile again by the modification of the vibronic state of the molecular system.

In a broad range of organic systems we observed [1] that promotion of localized polaron states to delocalised band states is critical for long-range charge separation in OPV devices. The study of the material composition effect (Fig. 1b) on the charge separation dynamics indicates that it is the resonant coupling of photogenerated singlet excitons to a high-energy manifold of fullerene electronic states that enables efficient charge generation, bypassing localized charge-transfer states. These findings suggest that fullerene cluster size, concentration, and dimensionality are major factors in determining charge generation efficiency [3].

PPP experiments on colloidal quantum dots (Fig. 2a) and polymer-oxide hybrid materials reveal the presence of bound charges (Fig. 2b) or charge-transfer excitons (Fig. 2c) at the organic-inorganic interfaces and demonstrate their detrimental effect on the charge separation and transport [4, 5]. We show that the efficiency of exciton

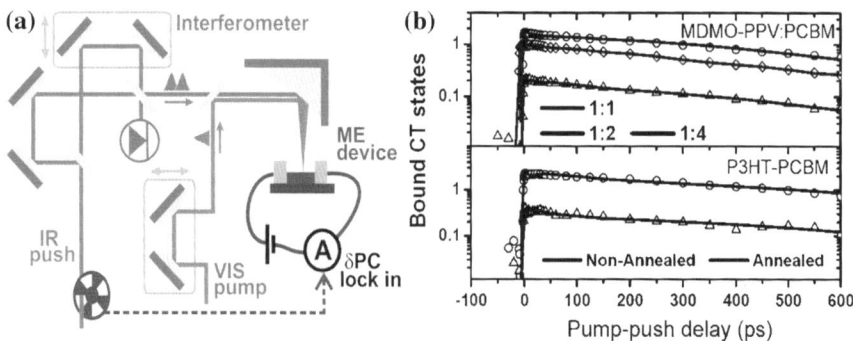

Fig. 1 **a** The layout of the experimental PPP setup; **b** PPP time-resolved data showing the effect of polymer-fullerene blend composition and morphology on charge dynamics in the organic solar cell devices

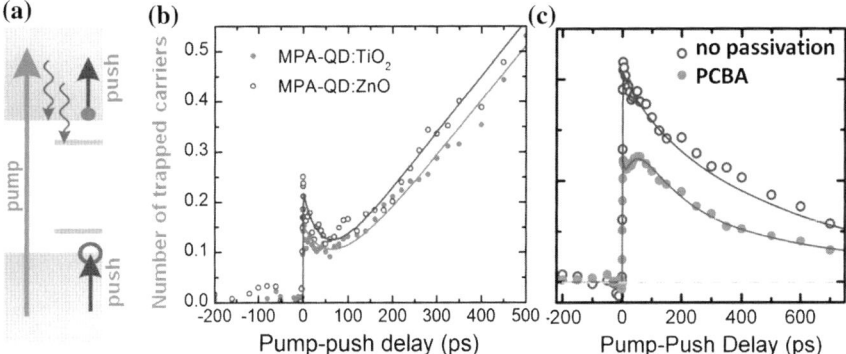

Fig. 2 **a** Energy diagram for the PPP experiment on the hybrid material; PPP time-resolved data showing the dynamics of immobile carriers in the *quantum-dot* **b** and polymer/oxide hybrid **c** solar cell devices

harvesting and dissociation to free charge carriers strongly depends on the properties of the interface, and can be substantially enhanced by the improved passivation of the inorganic surface.

Finally, we demonstrate that the effect of vibronic coupling [6] in pentacene/C60 bilayers can be used to optically control the conductivity in molecular crystal-based devices. The IR-push excitation of the aromatic C=C stretching mode at $1,440\ cm^{-1}$ induced the modulation of the nuclear coordinates, which appears to be favorable for the intermolecular charge transport (Fig. 3a, b). This observation opens the road for direct optical observation and utilization of nonadiabatic effects, which until now have been addressed mostly theoretically.

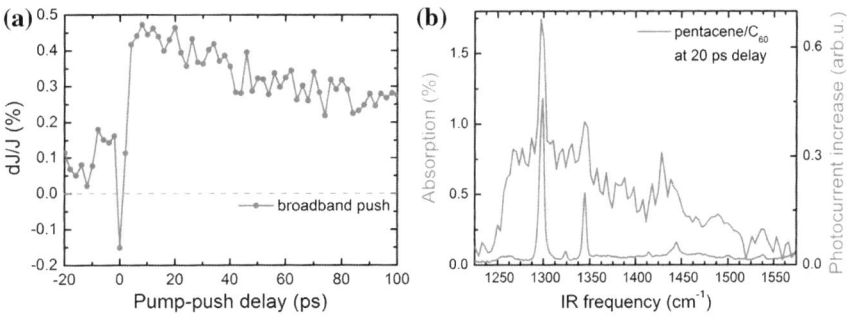

Fig. 3 PPP time-resolved (**a**) and push-frequency resolve (**b**) data showing the effect of vibrational and electronic excitation on the conductivity in the pentacene/C60 device

References

1. A. A. Bakulin, A. Rao, V. G. Pavelyev, P. H. M. v. Loosdrecht, M. S. Pshenichnikov, D. Niedzialek, J. Cornil, D. Beljonne and R. H. Friend, Science, **335**, 1340 (2012).
2. Helbing, J; Hamm, P, JOSA B, **28**, 171-178 (2011).
3. B.M. Savoie, A. Rao, A.A. Bakulin, S. Gelinas, B.Movaghar, R.H. Friend, T.J. Marks, M.A. Ratner, JACS, DOI: 10.1021/ja411859m (2014)
4. A.A. Bakulin, S. Neutzner, H.J. Bakker, L. Ottaviani, D. Barakel, Z. Chen, ACS Nano, **7**, 8771 (2013)
5. Y.Vaynzof, A.A. Bakulin, S. Gelinas, R.H. Friend, Phys. Rev. Lett., 108, 246605 (2012)
6. R.S. Sánchez-Carrera, P. Paramonov, G.M. Day, V. Coropceanu, and J.-L. Brédas, JACS, 132, 14437 (2010)

Ultrafast Non-thermal Response of Plasmonic Resonance in Gold Nanoantennas

Giancarlo Soavi, Giuseppe Della Valle, Paolo Biagioni,
Andrea Cattoni, Stefano Longhi, Giulio Cerullo
and Daniele Brida

Abstract Ultrafast thermalization of electrons in metal nanostructures is studied by means of pump-probe spectroscopy. We track in real-time the plasmon resonance evolution, providing a tool for understanding and controlling gold nanoantennas non-linear optical response.

In metal nanostructures, the collective oscillations of the conduction electrons, known as Localized Surface Plasmons (LSPs), dominate the light-matter interaction. LSPs give rise to intense optical resonances which are extremely sensitive to the dielectric environment and to the size and shape of the nanostructure as well. This feature makes plasmonic nanostructures particularly suitable for sensing applications in chemistry, biology and medicine [1]. Moreover, metal nanostructures are of great interest both from a fundamental perspective and for applications in novel devices combining electronics and optical generation/detection. For example, the capability of controlling light-matter interaction at the nanoscale leads to the confinement of large quantities of energy in sub-wavelength volumes allowing to access giant optical nonlinearities. Applications that aim to extreme light concentration, such as plasmonic antennas, lenses and resonators, are indeed the most successful up to date [2]. In this framework, manipulation and control of electron dynamics immediately after photo-excitation is of major importance in the development of active nanodevices. This clearly highlights the necessity of understanding, and experimentally accessing, the primary ultrafast electron relaxation processes of metals and confined metallic nanostructures.

G. Soavi · G.D. Valle · P. Biagioni · S. Longhi · G. Cerullo
Dipartimento di Fisica, Politecnico di Milano, P.zza L. Da Vinci 32,
20133 Milan, Italy

A. Cattoni
Laboratoire de Photonique et de Nanostructures, Route de Nozay,
91460 Marcoussis, France

D. Brida (✉)
Department of Physics and Center for Applied Photonics,
University of Konstanz, D-78457 Constance, Germany
e-mail: daniele.brida@uni-konstanz.de

© Springer International Publishing Switzerland 2015   679
K. Yamanouchi et al. (eds.), *Ultrafast Phenomena XIX*,
Springer Proceedings in Physics 162, DOI 10.1007/978-3-319-13242-6_167

Upon photoexcitation, the evolution of a plasmonic antenna starts with the radiative damping of the plasmon ($\approx$10 fs). In parallel, an out-of equilibrium electronic energy distribution is also created. Subsequent thermalization is driven by ultrafast electron-electron scattering, which leads to a thermal distribution within few hundreds of fs, and by electron-phonon scattering processes that become dominant on the ps timescale. Final cooling through coupling to the environment returns the system to the initial state over tens of ps. The plasmon dephasing time gives direct information about the quality of the nanostructure, its homogeneous broadening and the intrinsic loss (radiative or non-radiative) mechanisms [3]. Electron-electron and electron-phonon scattering mechanisms are instead crucial in defining the electrical and thermal conductivities of the sample and they offer a direct comparison between the bulk material and the nm-size properties, thus highlighting the non-trivial changes in the optical response induced by the presence of the LSP. While currently available ultrafast spectroscopy systems can easily study the electron-phonon scattering, the detection of the transition from an out-of-equilibrium to a thermalized distribution demands an extreme temporal resolution.

The aim of this study is to understand the effects of the early electron-electron thermalization process on plasmonic resonances. We perform ultrafast pump-probe measurements on an array of non-interacting gold nanoantennas obtained by electron beam lithography. A precise control of the nanoantenna size and shape allows a fine tuning of the plasmon resonance in the visible and near IR energy range (Fig. 1). Gold is a noble metal particularly suited to study the thermalization dynamics that takes place when an out-of equilibrium electron distribution is created in the conduction band by an impulsive optical excitation [4].

Here we focus on the ultrafast response of 40 × 150 nm gold nanoantennas. We excite the samples with a 13-fs near-IR pump centered at 900 nm and follow in real time, with a time-delayed probe, the evolution of the plasmonic resonance. Figure 2a reports the experimental differential transmission map spanning a broad spectral

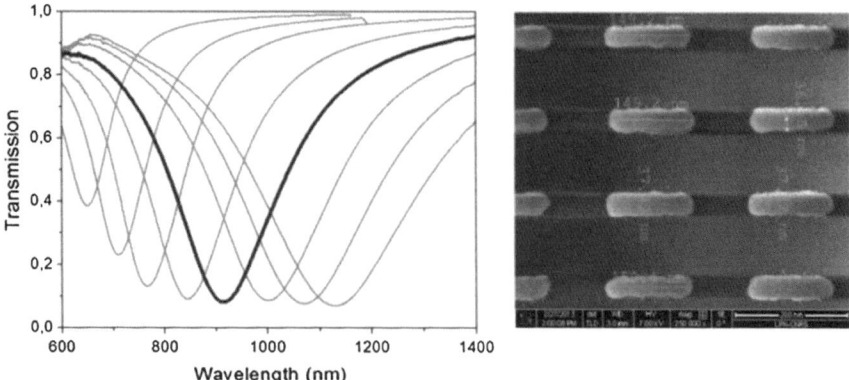

Fig. 1 *Left panel* Transmission spectra at the plasmonic resonance for the different gold nanoantennas. *Right panel* SEM image of the antenna used in the experiments

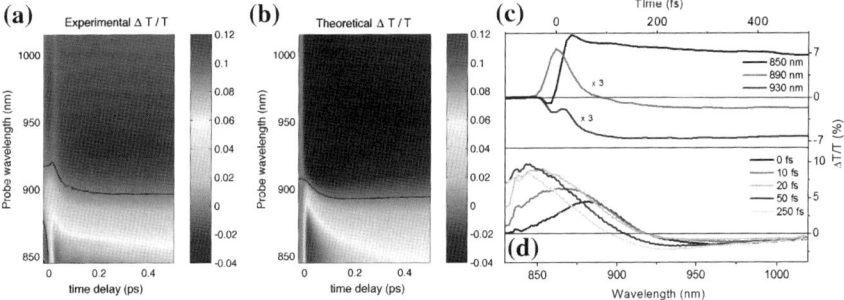

Fig. 2 **a** Pump-probe experimental map for the nanoantenna presented in Fig. 1 as a function of time and probe wavelength and **b** theoretical modeling based on a three temperature model. **c** Dynamics at 850 and 890 nm and **d** transient spectra at different pump-probe delays

range around the extinction peak of the nanoantenna (cf. left panel in Fig. 1). Interestingly, the dynamics below 920 nm wavelength (Fig. 2c) show a peculiar ultrafast (less than 100 fs) change in sign that can be observed also in the shift of the isosbestic point to higher energies in the transient spectra (Fig. 2d). The shape and temporal evolution of these transient spectra suggest that the overall electron dynamics in the nanoantenna is coupled to a strong modulation (broadening and shift) of the plasmonic resonance.

A theoretical modeling of the pump-probe map (Fig. 2b) shows very good agreement with the experimental results (Fig. 2a). The experimental data are simulated using a semiclassical model based on an extended version of the two-temperatures model (TTM), which has been recently demonstrated to quantitatively reproduce the ultrafast transient optical response of a thin gold film in the visible range [4]. In the present study the same model is applied to single out all the thermalization processes which contribute to the complex transient optical response exhibited by the nanoantenna in the near infrared.

According to these simulations, the peculiar dynamics detailed above is ascribable to the interplay between the two contributions arising from thermalized and non-thermalized carriers, giving rise to a modulation of the interband optical transitions of gold. The near infrared resonance of the antennas enhances the dynamics of non-thermal charges far from Fermi level. For this reason, the study of the plasmonic resonance allows investigations on fundamental electronic interactions.

Our experiments clarify the very early processes of electron thermalization in gold nanostructures and open a route to the possibility of controlling a plasmonic resonance by exploiting the transient evolution of a photoexcited electron distribution and its non-linear optical response.

References

1. K. A. Willets and R. P. Van Duyne, "Localized Surface Plasmon Resonance Spectroscopy and Sensing", Ann. Rev. Phys. Chem. **58**, 267-297 (2007).
2. J. A. Schuller, E. S. Barnard, W. Cai, Y. C. Jun, J. S. White and M. L. Brongersma, "Plasmonics for extreme light concentration and manipulation", Nat. Mater. **9**, 193-204 (2010).
3. A. Anderson, K. S. Deryckx, X. G. Xu, G. Steinmeyer and M. B. Raschke, "Few-Femtosecond Plasmon Dephasing of a Single Metallic Nanostructure from Optical Response Function Reconstruction by Interferometric Frequency Resolved Optical Gating", NanoLett. **10**, 2519-2524 (2010).
4. G. Della Valle, M. Conforti, S. Longhi, G. Cerullo and D. Brida, "Real-time optical mapping of the dynamics of nonthermal electrons in thin gold films", Phys. Rev. B **86**, 155139 (2012).

Control of Femtosecond Surface Plasmon Coupled onto a Gold Tapered Tip and Its Nonlinear Emission

Kazunori Toma, Yuta Masaki, Kenichi Hirosawa
and Fumihiko Kannari

Abstract Spatiotemporal nanofocusing of surface plasmon polaritons excited by femtosecond laser pulses on a tapered metallic tip with few tens nanometer radius is deterministically controlled using a measured plasmon response function.

1 Introduction

Metallic nanostructures supporting surface plasmon polaritons (SPPs) are recognized for demonstrating significantly enhanced near-fields, as well as energy transport with subwavelength lateral confinement. When combining ultrashort pulse laser and plasmonic nanophotonics, it is possible to confine laser excitation in both time and space and to realize space variant nonlinear excitation such as multiphoton excitation. Ropers et al. demonstrated efficient nonlocal optical excitation at the apex of a metal taper by grating-coupled SPPs excited by femtosecond laser pulses [1]. The temporal characteristics of SPPs coupled on an Au metal taper have been experimentally analyzed by Berweger et al. with interferometric frequency resolved optical gating (IFROG) using second-order harmonics generated at the apex of the tip [2]. They demonstrated that the spectral phase of SPP pulses can be controlled by shaping the excitation laser pulses.

In this paper, we employed cross-correlation dark-field image microscope to analyze SPP pulses coupled on an Au metal taper. We demonstrated deterministic arbitrary plasmon pulse control in both amplitude and phase by shaping excitation laser pulse. In addition, we controlled nonlinear radiation at the apex.

K. Toma (✉) · Y. Masaki · K. Hirosawa · F. Kannari
Department of Electronics and Electrical Engineering, Keio University,
3-14-1, Kohoku-Ku, Yokohama, Kanagawa, Japan
e-mail: kannari@elec.keio.ac.jp

© Springer International Publishing Switzerland 2015
K. Yamanouchi et al. (eds.), *Ultrafast Phenomena XIX*,
Springer Proceedings in Physics 162, DOI 10.1007/978-3-319-13242-6_168

683

2 Experimental Setup

The schematic of cross-correlation dark-field image measurement is shown in Fig. 1. We used an Au tapered tip with a tip edge radius of ∼20 nm and an opening angle of 15°. SPP can be launched on tapered metallic tip via grating coupling. A linear grating with a 1,730 nm period consists of 8 grooves with a width of 860 nm and a depth of 200 nm.

The femtosecond laser source is Ti:sapphire oscillator (8 fs (FWHM), $\Delta\lambda = 600$–1,000 nm). One of laser pulse beams split at a beam-splitter was focused onto the grating with a spot size of ∼10 μm using a microscope objective lens (×20, $NA = 0.35$ and a working distance of 20 mm) at an angle of incidence of 30°. Scattered light at the apex of the tip was collected by a microscope objective lens (×40, $NA = 0.55$ and a working distance of 7.5 mm) and imaged on a CCD camera. We measured fringe-resolved correlation images by varying an optical delay between two beams. This scheme will be a powerful tool to measure ultrafast SPP time histories both in phase and amplitude without any nonlinear optics. After the acquisition of a series of cross-correlation images, a cross-correlation waveform was obtained at the apex. Then, the spectral amplitude and phase of the SPP pulse were obtained by deconvolution using the reference pulse waveform which was fully characterized in advance. We obtained the plasmon response function for SPP coupling and propagation on the taper tip and control SPP pulse based on the response functions.

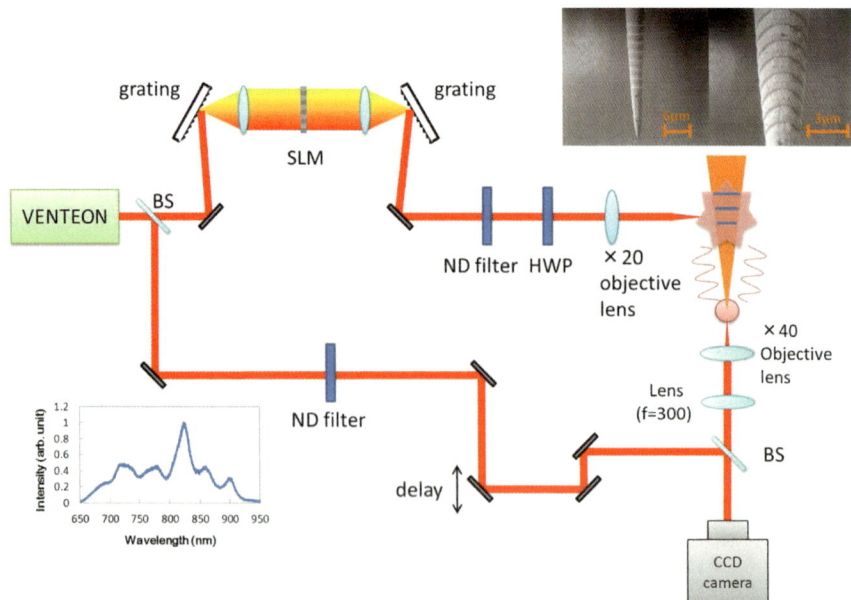

Fig. 1 Experimental setup of cross-correlation dark-field image measurement. *Insets* are excitation laser spectrum and a SEM pictures of Au tapered tip

3 Results and Discussion

Figure 2a shows measured response function of SPP at the apex. Once the response function is obtained, temporal plasmon characteristics can be deterministically designed. We shaped the excitation laser pulse by a femtosecond pulse shaper (Fig. 1) so that the dispersion of the response function is fully compensated and Fourier transform limited (FTL) plasmon pulse is generated at the apex. As a result, the SPP pulse width was reduced from 36.0 to 12.3 fs as shown in Fig. 2b.

In addition to the fundamental wavelength of SPP pulses, second harmonic (SH) can be generated because the axial symmetry at the apex is broken. We shaped the spectral phase of the femtosecond excitation pulse based on the response function. When third-order dispersions were designed for the SPP pulse at the apex, the peak of SH spectrum peak was exactly shifted in 370–420 nm to the point of inflection of the each third-order dispersion curve as shown in Fig. 2c. Therefore, we can selectively perform two-photon excitation at the apex with this control scheme. Similar control with spectral phase shaping can be applied for difference frequency mixing and four-wave mixing at the apex. Thus, nonlocal optical excitation of the apex of a nanostructured metal taper will be a versatile nano nonlinear light source.

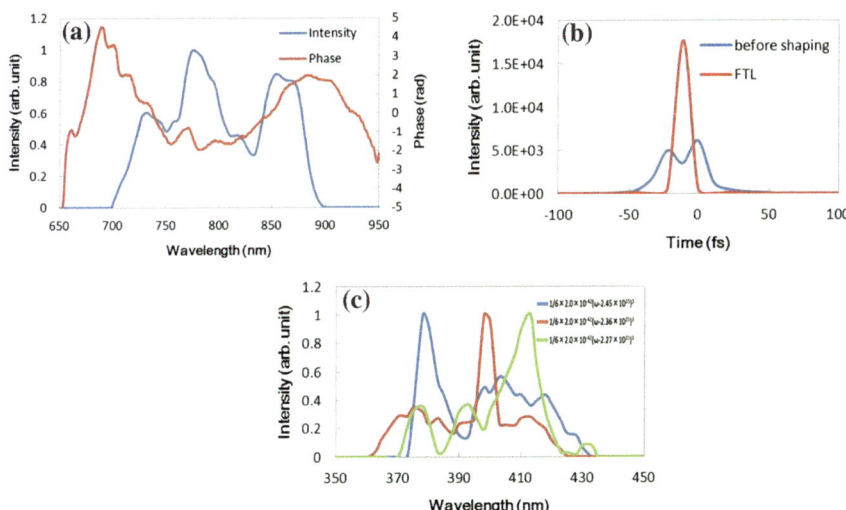

Fig. 2 **a** Measured plasmon response function, **b** measured plasmon pulses excited by FTL laser pulse and shaped laser pulse to generate a FTL plasmon pulse, and **c** control of SH spectrum at the apex by 3rd—order dispersion

References

1. C. Ropers et al. "Grating-coupling of surface plasmons onto metallic Tips: a nanoconfined light source," Nano Lett. **7**, 2784-2788 (2007).
2. S. Berweger et al. "Femtosecond nanofocusing with full optical waveform control," Nano Lett. **11**, 4309-4313 (2011).

Ultrafast Optical-Field Controlled Photoemission from Plasmonic Nanoparticle Arrays

W.P. Putnam, R.G. Hobbs, Y. Yang, K.K. Berggren and F.X. Kärtner

Abstract Exciting plasmonic nanoparticles with few-cycle optical pulses, we observe photoemission across few-micron gaps under ambient conditions. The photoemission current is modulated by the carrier-envelope phase of the excitation pulse.

Attosecond science fundamentally relies on the manipulation and control of ultrafast, attosecond bursts of electrons with optical waveforms. Extending this control from gaseous media to solid-state nanostructures would provide avenues to exciting scientific and technological developments. In the past few years, signatures of strong-field emission and electron re-scattering from nano-tip emitters [1–4] and plasmonic nanoparticles [5] have been observed. Additionally, a hallmark of optical waveform control, carrier-envelope phase (CEP) sensitivity, has been observed in the photoelectron energy spectra [2] and total emission current from single nano-tip emitters [4].

In this work, we explore photoemission from plasmonic nanoparticle arrays under ambient conditions, i.e. out of vacuum, and on the surface of a chip. We show that this photoemission current can be controlled by the CEP of the exciting laser pulse. Considering that the excitation pulse has less than half an octave of optical bandwidth at full width at half maximum, our devices might find applications as compact, solid-state CEP detectors.

Our experimental configuration is illustrated in Fig. 1a. Our laser source is an Er: fiber seeded supercontinuum (SC) femtosecond source [6, 7]. The SC pulses are

W.P. Putnam (✉) · R.G. Hobbs · Y. Yang · K.K. Berggren · F.X. Kärtner
Department of EECS and Research Laboratory of Electronics, Massachusetts Institute
of Technology, 77 Massachusetts Ave, Cambridge, MA 02139, USA
e-mail: bputnam@mit.edu

F.X. Kärtner
The Hamburg Center for Ultrafast Imaging, Luruper Chaussee 149, 22761 Hamburg,
Germany

F.X. Kärtner
Center for Free-Electron Laser Science, DESY and Department of Physics, University of
Hamburg, Notkestraße 85, 22607 Hamburg, Germany

© Springer International Publishing Switzerland 2015
K. Yamanouchi et al. (eds.), *Ultrafast Phenomena XIX*,
Springer Proceedings in Physics 162, DOI 10.1007/978-3-319-13242-6_169

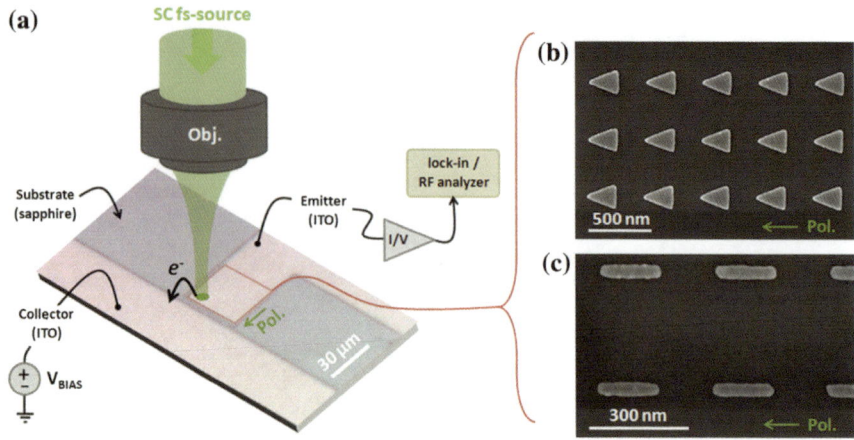

Fig. 1 Experimental configuration. **a** SC fs-pulses excite a plasmonic nanoparticle array resulting in electron emission. Electrons jump across a small gap from the emitter electrode to the positively biased collector electrode. The photoemission current is amplified in a transimpedance amplifier (I/V) and sent to a lock-in amplifier or a radio-frequency (RF) spectrum analyzer. **b** SEM image of Au nano-triangle array. **c** SEM image of Au nano-rod array (polarization direction labeled by Pol.)

about two and a half cycles ($\approx$10 fs) in duration at a central wavelength near 1.2 µm (SC spectrum ranges from 1 to 1.4 µm). The SC source has a repetition rate of 78.4 MHz and a pulse energy of up to 0.4 nJ. The carrier-envelope offset frequency (f_{CEO}) is stabilized to a local oscillator at 2 kHz. The SC pulses are focused by a reflecting objective to a 4.1 µm diameter spot on our devices.

Our devices consist of arrays of plasmonic nanoparticles. The arrays lie on patterned, indium-doped tin oxide (ITO) electrodes on a sapphire substrate. In the image in Fig. 1a, the array of nanoparticles is not visible, however, a light box outlines the array location for clarity. The darker regions of the image correspond to the insulating sapphire substrate and the lighter regions to the ITO electrodes. The emitter electrode connects to the nanoparticle array and is separated from the collector electrode by a 2.5 µm gap; this small gap allows for large bias fields with only modest bias voltages. When the SC pulses excite the device array, photoemitted electrons jump from the emitter to the collector (as illustrated in Fig. 1a). We use two nanoparticle array-types in our experiments—Au nano-triangles (SEM image in Fig. 1b) and Au nano-rods (SEM image in Fig. 1c). Both device-types are designed to have their resonance within the SC bandwidth (near 1.1 µm).

The results from the CEP-control measurements are shown in Fig. 2a, b. Figure 2a shows the radio-frequency spectra of the emitter currents from the nano-triangle and nano-rod devices. A large peak at $f_{CEO} = 2$ kHz is visible in the emitter current (I_E) from the nano-triangle devices, while no peak is visible for the nano-rods. The highly non-linear, strong-field photoemission process occurs only when the optical electric field points into the Au surface. Therefore, the asymmetric nano-triangles only emit electrons during every other half-cycle of the excitation pulse and, thereby, show

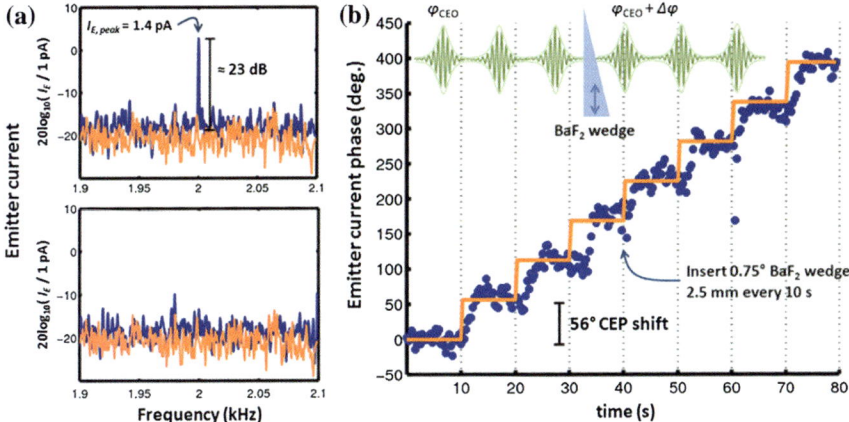

Fig. 2 CEP-control measurements. **a** RF spectra (1 Hz resolution-bandwidth) of emitter current (I_E) from nano-triangle devices (*top*) and nano-rods (*bottom*). The *orange* traces show the emitter current (I_E) without laser excitation. **b** Emitter current (I_E) phase shifts for the nano-triangle devices

a strong CEP sensitivity. The symmetric nano-rods, on the other hand, emit electrons every half-cycle, and accordingly, their CEP sensitivity is greatly reduced.

To confirm the CEP sensitivity, the phase of the emitter current at $f_{CEO} = 2$ kHz (measured on a lock-in amplifier) is measured as a BaF$_2$ wedge is moved across the SC pulse train (illustrated in Fig. 2b). The wedge is moved in 2.5 mm increments every 10 s. With a carrier wavelength of 1.2 µm, each increment should result in a ~56° phase-shift to the CEP of the SC pulses. This shift should then appear as a phase-shift on the lock-in amplifier. The experimental results for the nano-triangle devices very closely resemble this prediction, as denoted by the orange line in Fig. 2b.

In summary, we have shown photoemission from plasmonic nanoparticles across few-micron gaps under ambient conditions. We have demonstrated that the photoemission current consists of a strong carrier-envelope phase-sensitive component. These results are encouraging for the prospects of steering and manipulating attosecond electrons on a chip as well as for future compact, solid-state CEP detectors.

References

1. R. Bormann, M. Gulde, A. Weismann, S. V. Yalunin, C. Ropers, Phys. Rev. Lett., **105**, 147601 (2010).
2. M. Krüger, M. Schenk, P. Hommelhoff, Nature, **475**, 78 (2011).
3. G. Herink, D. R. Solli, M. Gulde, C. Ropers, Nature, **483**, 190 (2012).
4. B. Piglosiewicz, S. Schmidt, D. J. Park, J. Vogelsang, P. Groß, C. Manzoni, P. Farinello, G. Cerullo, C. Lienau, Nat. Photonics, **8**, 37 (2014).

5. P. Dombi, A. Hörl, P. Rácz, I. Márton, A. Trügler, J. R. Krenn, U. Hohenester, Nano Lett., **13**, 674 (2013).
6. J. A. Cox, W. P. Putnam, A. Sell, A. Leitenstorfer, F. X. Kärtner, Opt. Lett., **37**, 3579 (2012).
7. A. Sell, G. Krauss, R. Scheu, R. Huber, A. Leitenstorfer, Opt. Exp., **17**, 1070 (2009).

Real Space and Real Time Observation of Plasmon Wavepacket Dynamics in Single Gold Nanorod

Y. Nishiyama, T. Narushima, K. Imura and H. Okamoto

Abstract Plasmon dynamics in a single gold nanorod was visualized by ultrafast time-resolved near-field optical microscopy. On excitation of multiple plasmon modes, transient near-field images exhibit prominent temporal change that reflects a characteristic behavior of coherent superposition of the excited plasmons.

1 Introduction

Noble metal nanostructures attract much attention in various research fields because of their unique optical properties, such as confinement of the optical field in nano-scale, which arise from plasmon resonances. For understanding of such plasmonic properties of metal nanostructures, direct observation of spatio-temporal character-istics of plasmons is of fundamental importance. Experimentally, this is quite challenging because we need nanometric spatial and sub-20-fs time resolutions, considering the spatial scale and the dephasing time scale of plasmons [1]. Combination of scanning near-field optical microscopy (SNOM) with ultrafast spectroscopy satisfies these requirements [2], and we applied the ultrafast SNOM to spatio-temporal observation of plasmon wavepacket dynamics in a single gold nanorod. Spectrally broad ultrashort pulse can coherently excite two or more plas-mon modes that have different resonance frequencies and spatial structures. In the present study, two longitudinal plasmon modes of a gold nanorod were excited at the same time with spectrally-broad (near-field) femtosecond pulses. We found,

Y. Nishiyama · T. Narushima · H. Okamoto (✉)
Institute for Molecular Science and The Graduate University for Advanced Studies,
Myodaiji, Okazaki 444-8585, Japan
e-mail: aho@ims.ac.jp

K. Imura
School of Advanced Science and Engineering, Waseda University, Okubo, Shinjuku
Tokyo 169-8555, Japan

K. Yamanouchi et al. (eds.), *Ultrafast Phenomena XIX*,
Springer Proceedings in Physics 162, DOI 10.1007/978-3-319-13242-6_170

691

in the transient near-field images, two different features of plasmon modes appearing alternately with time at ~ 20 fs after the excitation. This is a characteristic behavior of coherent superposition of two (or more) states.

2 Experiment

The experimental setup for ultrafast SNOM is basically the same as that reported previously [2]. The system consists of a Ti:sapphire laser (~ 800 nm, 12 fs), a Michelson interferometer, dispersion compensation devices and a SNOM system. In the SNOM, a near-field probe with an aperture of ~ 100 nm diameter was installed as a source of near-field illumination. To obtain a near-field image, the distance between the probe tip and the sample surface was regulated to be constant (~ 10 nm). The optical fiber part of the probe broadens the pulse duration from original 12 fs to several ps. Therefore, precise dispersion compensation is indispensable to achieve high time resolution (<20 fs). We combined three devices, a grating pair, a chirped mirror pair, and a pulse shaping system equipped with a deformable mirror, which compensate for dispersions of different orders. In particular, the pulse shaping system enables flexible compensation of higher-order dispersion. After adaptive control of the pulse shape, we achieved 15-fs pulse duration at the probe tip. To obtain transient near-field image of the sample, we detected two-photon-induced photoluminescence (TPI-PL) from gold in the wavelength region of 500–675 nm as the optical signal. Since plasmon resonance is an intermediate state in the two-photon excitation process of the TPI-PL, the pump-probe TPI-PL signal reflects the plasmon dynamics after excitation by the pump pulse. We performed time-resolved measurements with an equal-pulse pump-probe scheme at every point of the scanning area of the sample. We then constructed transient near-field images by collecting TPI-PL signals at a fixed time delay between the pump and probe pulses. To clarify static nature of the plasmon modes (resonant wavelengths and spatial structures), we also performed two types of static near-field measurements. For the static transmission measurement, we used a Xe discharge lamp as a light source and analyzed the transmitted light by a spectrometer. Static two-photon excitation images were obtained by detecting TPI-PL under irradiation of narrow spectral-band pulses from a wavelength-tunable Ti:sapphire laser ($\lambda = 680$–1,600 nm, ~ 150 fs). The sample gold nanorod was fabricated by an electron-beam lithography lift-off technique. The actual sizes of the rod were 1,050 nm in length, 70 nm in width, and 20 nm in height.

3 Results and Discussion

Near-field extinction spectrum of the nanorod is shown in Fig. 1a. Two plasmon bands were found at 805 and 870 nm. Static two-photon excitation images at excitation wavelengths of 800 and 870 nm (Fig. 1b, c) exhibited spatial oscillation

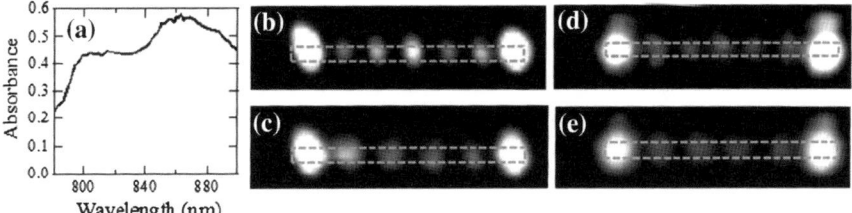

Fig. 1 **a** Near-field extinction spectrum of the gold nanorod. **b**, **c** Near-field two-photon excitation images of the nanorod excited at **b** 800 nm and **c** 870 nm. **d**, **e** Transient TPI-PL images of the same nanorod at the time delay of **d** 18.5 fs and **e** 19.9 fs. *Dashed lines* represent the outlines of the rod

features along the long axis of the nanorod, which reflected the respective standing wave functions of high-order longitudinal plasmon modes of the rod, as reported in the previous study [3]. From the spatial patterns of these images, plasmon resonances at 805 and 870 nm were identified as the 7th and 6th longitudinal modes, respectively. With the excitation pulses we used for the time-resolved measurement, these two modes can be excited simultaneously because they are spectrally overlapped with the pulses (780–890 nm). The transient near-field images show temporal changes. At zero time delay, the spatial oscillation was obscured particularly in the central part of the image due to simultaneous excitation of the two plasmon modes with different spatial patterns. At the time delay of ~ 20 fs, the spatial features of both modes (the 6th and 7th modes) were found to appear alternately with an interval of ~ 1.4 fs in the transient images (Fig. 1d, e). The alternating interval corresponds approximately to half of the oscillation period of the resonant plasmon modes. This result indicates that the two plasmon modes were out of phase to each other at the time delay of ~ 20 fs. From the oscillation frequencies of the modes, the two modes are predicted to be mutually out-of-phase at 19 fs after the simultaneous excitation, which explains well our experimental finding. From these results, we conclude that the wavepacket dynamics of the plasmons was visualized as the time-dependent near-field images.

References

1. C. Sönnichsen, T. Franzl, T. Wilk, G. von Plessen, J. Feldmann, O. Wilson, and P. Mulvaney, "Drastic reduction of plasmon damping in gold nanorods," *Phys. Rev. Lett.* **88**, 077402 (2002).
2. H. J. Wu, Y. Nishiyama, T. Narushima, K. Imura, and H. Okamoto, "Sub-20-fs time-resolved measurements in an apertured near-field optical microscope combined with a pulse-shaping technique," *Appl. Phys. Express* **5**, 062002 (2012).
3. H. Okamoto and K. Imura, "Near-field optical imaging of enhanced electric field and plasmon waves in metal nanostructures," *Prog. Surf. Sci.* **84**, 199 (2009).

Vector Pulse Shaped Ultrafast Plasmon Based on Response Functions Measured for Orthogonally Polarized Excitation

Yuta Masaki, Miyuki Kusaba, Kazunori Toma
and Fumihiko Kannari

Abstract For spatiotemporal vector pulse control of local plasmon at gold nanostructures, plasmon response functions at orthogonally polarized ultrafast laser excitation are measured by an electrical-field cross-correlation imaging method using a dark-field microscope. By shaping the vector pulse of the excitation laser, arbitrarily shaped plasmon vector pulses can be generated.

1 Introduction

Localized plasmon resonance in noble metal nanostructures has been receiving much attention to prepare a sub-wavelength-scale interaction platform between light and matter. Spatiotemporal control over the localized plasmon or surface plasmon polariton in nanostructures can be achieved using a vector pulse shaping technique for femtosecond excitation laser pulses [1]. So far, an adaptive control algorithm was used to shape an incident femtosecond laser pulse for generating a desired plasmon pulse [2]. In this study, we constructed a new scheme combining dark field microscopy with cross-correlation interferometer to measure plasmon response functions of gold nanostructures excited by 800-nm femtosecond laser pulses. The plasmon response functions can deterministically design the vector pulse shape of femtosecond excitation laser pulses to generate a desired plasmon vector pulse on the gold nanostructure.

Y. Masaki · M. Kusaba · K. Toma · F. Kannari (✉)
Department of Electronics and Electrical Engineering, Keio University, 3-14-1,
Kohoku-ku, Yokohama, Kanagawa, Japan
e-mail: kannari@elec.keio.ac.jp

© Springer International Publishing Switzerland 2015
K. Yamanouchi et al. (eds.), *Ultrafast Phenomena XIX*,
Springer Proceedings in Physics 162, DOI 10.1007/978-3-319-13242-6_171

694

2 Experimental Setup

The experimental setup of our cross-correlation measurement with femtosecond laser dark-field microscope is shown in Fig. 1a. We used an ultra-broadband femtosecond laser (VENTEON, 650–1,050 nm, rep. rate of 150 MHz). One of the split beam irradiates a gold nanostructure, and the scattered light is collected by the objective mirror, while the other laser beam directly reaches the CCD camera with a variable optical delay. We measured a series of image shots by varying the time delay and analyzed the fringe-resolved cross-correlation functions at a specific point in the dark-field image. Nanostructures we measured are cross-shaped Au with various aspect ratios R fabricated on SiO_2 substrate (Fig. 1b).

The spectral plasmon response functions are obtainable from the Fourier transform of a temporal cross-correlation function, when both the excitation and the reference pulses are known in their amplitude and phase [3].

3 Results and Discussion

First, the polarizations of the excitation and the measured light were set in parallel to the one of the nanocross arms. Figure 2a, b shows the plasmon spectra and the plasmon response functions, respectively, obtained at an Au nanorod with $R = 2.5$ and 3.5. We observed a clear plasmon resonance enhancement and a dependence on R.

Next, the polarizations of the excitation and the measured light were tilted at 45-degree to the Au nanocross arms. Figure 3a, b shows the measured plasmon spectrum and the plasmon spectrum predicted from the orthogonal response functions in Fig. 2b for the nanocross consisting of the nanorods with $R = 2.5$ and 3.5. The measured plasmon spectrum shows a strong resemblance to the calculated plasmon spectrum. Therefore, a linear description of the plasmon pulse based on the

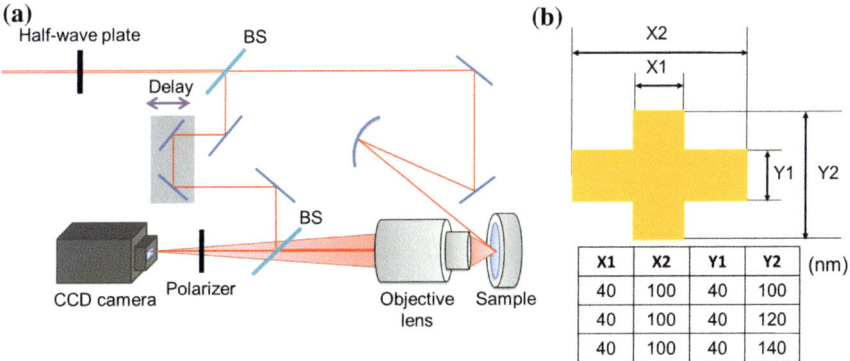

Fig. 1 **a** Experimental setup of dark-field microscopy. **b** Shape of Au cross-shaped nanostructure

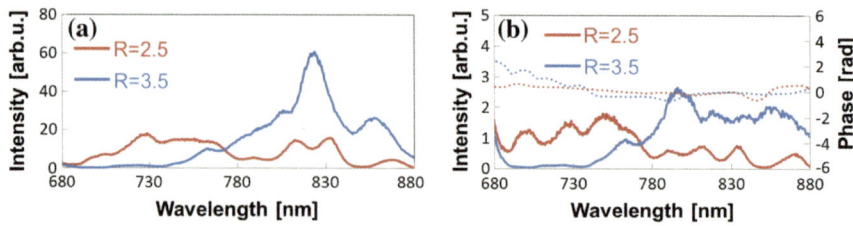

Fig. 2 **a** Plasmon spectra and **b** response functions of nanocross with R = 2.5 and 3.5

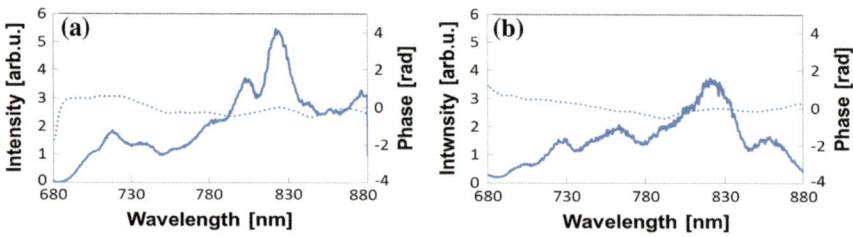

Fig. 3 **a** Measured plasmon spectrum when the excitation polarization was tilted at 45-degree from one of arms of nanocross consisting of nanorods of R = 2.5 and R = 3.5. **b** Calculated plasmon spectrum with the orthogonal response functions shown in Fig. 2b

orthogonal plasmon response functions was validated. Based on this linear description, one can deterministically shape the plasmon vector pulse by shaping the excitation laser vector pulses.

Figure 4 shows examples of vector shaped plasmon pulses designed on the nanocross of $R = 2.5$ and 3.5: a linearly polarized Fourier transform limited pulse at

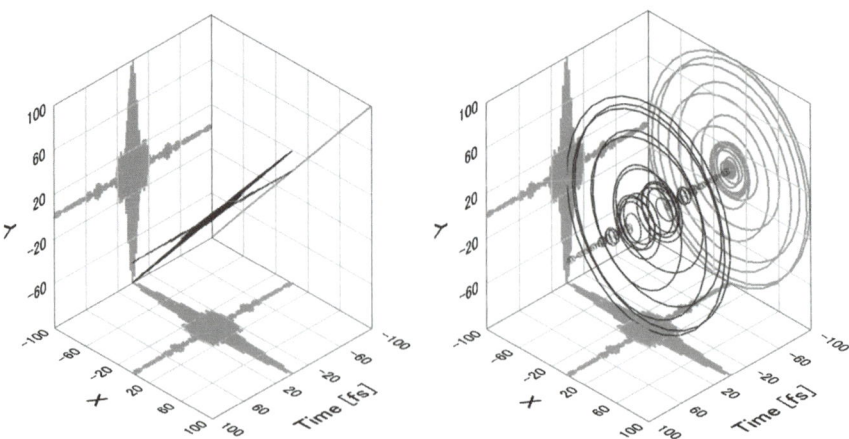

Fig. 4 Vector shaped plasmon pulse based on the orthogonal plasmon response functions

45-degree (left) and a circularly polarized Fourier transform limited pulse (right). To generate the linearly polarized FTL pulse at 45-degree, we applied the inverse spectral phase of the response function for each the orthogonally polarized excitation pulses and shaped the spectral amplitude identical for the both.

References

1. T. Brixner et al. "Ultrafast adaptive optical near-field control," Phys. Rev. B 73(12), 125437 (2006). K. Gallo and G. Assanto, "All-optical diode based on second-harmonic generation in an asymmetric waveguide," J. Opt. Soc. B 16, 267-269 (1999).
2. M. Aeschlimann et al. "Adaptive subwavelength control of nano-optical fields," Nature 446, 301-304 (2007)
3. S. Onishi et al. "Spatiotemporal control of femtosecond plasmon using plasmon response functions measured by near-field scanning optical microscopy (NSOM)," Opt. Express. 21, 26631-26641 (2013).

Few-Cycle Laser Pulse Induced Plasmon Assisted Thermionic Injection in Metal-Insulator-Metal Junctions

Matthias Hensen, Dominik Differt, Ingo Heesemann,
Christian Strüber, Adelheid Godt, Detlef Diesing and Walter Pfeiffer

Abstract Gold nanoparticles on a metal-insulator-metal junction locally enhance the absorption of few-cycle laser pulses. The locally heated electron gas leads to thermionic emission exceeding multiphoton emission and allows detection of single nanoparticles.

1 Introduction

In the interaction of intense laser pulses with nanostructures or optical antennas the electron emission is commonly discussed in the context of multiphoton processes [1] and strong field phenomena [2–4]. However, for sufficiently strong excitation and fast carrier thermalisation the local electronic system is heated to high temperatures leading to dominant thermionic emission depending on the work function of the metal. Qualitatively, thermionic emission is well described by the Richardson-Dushman equation. For a work function of 2–3 eV an electron gas temperature of several thousand Kelvin is required to obtain sufficiently high emission currents above 1 pA. This excitation regime has not yet been demonstrated for few-cycle laser pulse excitation since for homogeneous excitation of a surface such high electron gas temperatures also lead to substantial lattice heating and corresponding thermal surface degradation. However, local resonant excitation and the related

M. Hensen · D. Differt · C. Strüber · W. Pfeiffer (✉)
Fakultät für Physik, Universität Bielefeld, Universitätsstr. 25, 33615 Bielefeld, Germany
e-mail: pfeiffer@physik.uni-bielefeld.de

I. Heesemann · A. Godt
Fakultät für Chemie and Center for Molecular Materials, Universität Bielefeld,
Universitätsstr. 25, 33615 Bielefeld, Germany

D. Diesing
Fakultät für Chemie, Universität Duisburg-Essen, Universitätsstr. 5, 45141 Essen, Germany

© Springer International Publishing Switzerland 2015
K. Yamanouchi et al. (eds.), *Ultrafast Phenomena XIX*,
Springer Proceedings in Physics 162, DOI 10.1007/978-3-319-13242-6_172

field enhancement also significantly increase the locally absorbed energy density. The spatially inhomogeneous excitation reduces thermal degradation because of fast heat diffusion and thus opens the possibility to detect thermionic emission.

Here we demonstrate that few-cycle laser pulse excitation of a localized surface plasmon in a single metal nanoparticle leads to a strong spatially localized electron gas heating in the supporting layer, i.e. the top electrode of a metal-insulator-metal (MIM) junction, and induces thermionic injection currents in the MIM junction.

2 Experimental Setup

Few-cycle laser pulses are focused onto the MIM surface using a reflective objective (Fig. 1a). The sample is mounted on an xyz-stage and laser induced currents are measured using a lock-in amplifier. Citrate stabilized Au nanoparticles with diameters of 100 nm up to 200 nm were grown in aqueous solution and then dispersed on the surface by deposition of a few μl of solution on the junction and subsequent evaporation of the solvent. Comparison of scanning electron micros-copy (SEM) images and corresponding reflected light maps allows identifying individual nanoparticles (inset Fig. 1b) and performing spatial scans of the laser induced injection current for an individual nanoparticle (Fig. 1b).

3 Results and Discussion

For a bare MIM junction with a 40 nm thick Au top electrode an injection current signal is only measured at the rim of the junction, since in the thinning top electrode a higher excitation is achieved close to the metal-insulator interface. The intensity dependence (diamonds in Fig. 1c) yields a nonlinear exponent of 2.5 indicating a mixture of two-photon and three-photon processes as it is expected for the given MIM junction having an internal barrier height of 1.8 eV [5]. For a top electrode thickness of 40 nm the injection current drops below the detection limit (≈ 1 pA). Hence, these junctions are perfectly suited to investigate localized excitations. In the laser induced injection current map individual nanoparticles light up as highly localized spots with peak currents in the order of 0.2 nA (Fig. 1b).

Interestingly, the local injection current varies highly nonlinearly with the incident laser power. The nonlinear exponent of 9.0(1.5) (Fig. 1c) cannot be attributed to above threshold processes since for them also the lowest possible nonlinear order dominates the emission current since the contribution from higher order processes decreases exponentially [4]. The strong field emission regime leads to a reduced nonlinearity compared to the multiphoton regime and thus can also not account for the highly nonlinear intensity dependence. In contrast, according to the Richardson-Dushman equation, thermionic emission generates highly nonlinear

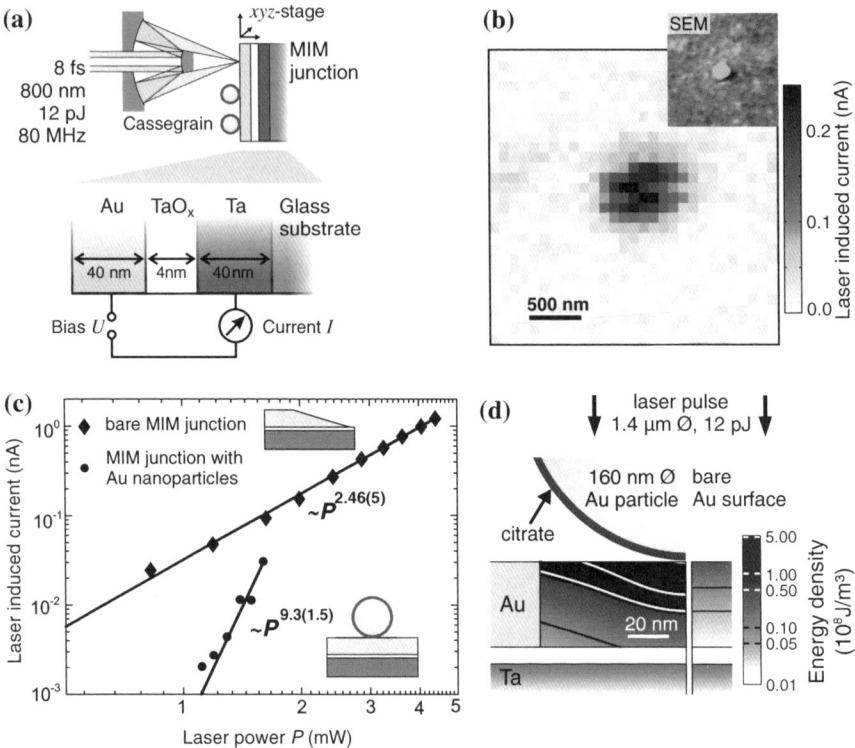

Fig. 1 a Scheme of the experimental setup. **b** Laser induced injection current in the MIM junction as the laser spot is scanned across the surface. The *inset* in the *upper right corner* shows an equally scaled SEM image of the Au nanoparticle that is positioned in the *centre* of the *scan area*. **c** Power dependence of the laser induced current excited at the rim of a MIM junction with a Au top electrode (*diamonds*) and due to excitation of a single Au nanoparticle with 160 nm diameter on *top* of a junction with a 40 nm Au top electrode (*circles*). The *solid lines* represent power law fits and the exponents are indicated. **d** Estimation of the locally deposited energy density in the Au *top* electrode for excitation of a Au nanoparticle positioned on *top* of the layer (*left part*) and for excitation of the bare junction (*right part*) as derived from FDTD calculations

intensity dependencies. As roughly estimated from FDTD calculations a gold nanoparticle deposited on a metal-insulator-metal junction enhances the excitation of the top metal layer by a factor of 50 compared to the bare junction (Fig. 1d) leading to locally deposited energy densities in the order of 10^8–10^9 J m^{-3}. Based on the electronic heat capacity of Au this corresponds to local electron gas temperature of about 3,000 K. Assuming an electron cooling time constant of about 1 ps and an effective excitation spot radius of 50 nm yields a local thermionic emission current of about 0.2 nA and a local nonlinear slope very close to 9. Thus, the simple thermionic emission model quantitatively explains both the measured laser induced current magnitude and the observed nonlinearity.

References

1. J. Lehmann, M. Merschdorf, W. Pfeiffer, A. Thon, S. Voll, G. Gerber, Phys. Rev. Lett. **85**, 2921–2924 (2000).
2. R. Bormann, M. Gulde, A. Weismann, S. V. Yalunin, C. Ropers, Phys. Rev. Lett. **105**, 147601 (2010).
3. P. Dombi, S. E. Irvine, P. Racz, M. Lenner, N. Kroó, G. Farkas, A. Mitrofanov, A. Baltuska, T. Fuji, F. Krausz, A. Y. Elezzabi, Opt. Express **18**, 24206–24212 (2010).
4. M. Krüger, M. Schenk, P. Hommelhoff, Nature **475**, 78–81 (2011).
5. D. Differt, W. Pfeiffer, D. Diesing, Appl. Phys. Lett. **101**, 111608 (2012).

Single Nanoparticles and Nanoplasmas in Femtosecond Laser Fields

Daniel D. Hickstein, Franklin Dollar, Jennifer L. Ellis,
Jim A. Gaffney, Mark E. Foord, George M. Petrov, Brett B. Palm,
Chengyuan Ding, K. Ellen Keister, Stephen B. Libby, Jose L. Jimenez,
Henry C. Kapteyn, Margaret M. Murnane and Wei Xiong

Abstract We combine an aerodynamic lens with a velocity-map-imaging spectrometer to make the first measurements of ultrafast dynamics in individual nanoplasmas. By using two laser pulses (800 and 400 nm) delayed by several picoseconds, we find that we can generate and control shock wave propagation in nanoplasmas, confirming a decade of theoretical predictions. Additionally, we observe pronounced asymmetries in the photoion angular distributions resulting from nanoparticles of different structure and composition, demonstrating the ability to observe nanoscale light absorption at laser intensities near the damage threshold.

OCIS Codes (160.4236) Nanomaterials · (190.4180) Multiphoton processes · (260.7120) Ultrafast phenomena

1 Introduction and Summary

The interaction of intense ($>1 \times 10^{14}$ W/cm^2) laser pulses with nanomaterials is of great interest across a wide range of fields, from physics to medicine. For example, nano-scale plasmas (nanoplasmas) can be used to create compact sources of

D.D. Hickstein (✉) · F. Dollar · J.L. Ellis · C. Ding · K. Ellen Keister · H.C. Kapteyn · M.M. Murnane · W. Xiong
JILA and Department of Physics, University of Colorado, Boulder, CO 80309, USA
e-mail: daniel.hickstein@colorado.edu

J.A. Gaffney · M.E. Foord · S.B. Libby
Physics Division, Physical and Life Sciences, Lawrence Livermore National Laboratory, Livermore, CA 94550, USA

G.M. Petrov
Plasma Physics Division, Naval Research Lab, Washington, DC 20375, USA

B.B. Palm · J.L. Jimenez
Department of Chemistry, University of Colorado, and CIRES, Boulder, CO 80309, USA

© Springer International Publishing Switzerland 2015
K. Yamanouchi et al. (eds.), *Ultrafast Phenomena XIX*,
Springer Proceedings in Physics 162, DOI 10.1007/978-3-319-13242-6_173

high-energy electrons/ions/photons, while laser-irradiated metal nanoparticles can be used to destroy tumors. However, studies of nanoparticles at laser intensities above the damage threshold are complicated by the need to use fresh sample for each laser shot, the fact that nanoparticles typically exist only in liquid solutions or on surfaces, and the inherent size and composition inhomogeneity of most nanoparticle samples.

In this study, we overcome these limitations by irradiating single ~ 100 nm nanoparticles in a vacuum environment. By coupling a velocity-map imaging (VMI) spectrometer to a nanoparticle aerosol source, we image plasma formation in individual, isolated nanoparticles for the first time, which allows us to observe several new phenomena. We find that at laser intensities $>1 \times 10^{14}$ W/cm^2, the entire nanoparticle is converted into a nanoplasma which expands on fs and ps timescales. If a second laser pulse arrives at an optimal time during the plasma expansion, a strong shockwave is formed, and this shock wave can be controlled by adjusting the time delay between the two laser pulses. Numerical hydrodynamic simulations confirm that the energy and amplitude of the shock waves depends on the intensity and time-delay of the laser pulses. This observation confirms a decade of theoretical predictions, which suggested that shock waves occur in laser-irradiated nanoplasmas and that these shock waves produce bursts of quasi-monoenergetic ions [1, 2].

2 Experiment

Nanocrystals of NaCl, KCl, KI, or NH_4NO_3 with diameters of ~ 100 nm are created using a compressed-gas atomizer and introduced into the vacuum chamber using an aerodynamic lens [3]. A nanoplasma is formed by irradiating the nanoparticle with a tightly focused 40 fs laser pulse (400 or 800 nm) with a variable intensity between 3×10^{13} and 4×10^{14} W/cm^2 (Fig. 1). The angle-resolved energy distribution of the ions created by an individual expanding nanoplasma is then recorded using a velocity-map-imaging (VMI) spectrometer [4, 5] resulting in a two-dimensional projection of the photoion/photoelectron angular distribution (PAD). Because the laser focal spot is small compared to the spacing between the nanoparticles, we probe approximately one nanoplasma every ~ 100 laser-shots. Most importantly, each PAD corresponds to a single particle (or no particle), thereby avoiding any laser-intensity or particle-size averaging effects. This allows us to capture previously unseen shock wave dynamics as a function laser intensity, particle size, and particle composition.

3 Shock Waves in Nanoplasmas

By recording the angular distribution of sudden bursts of more than 10^5 photoions per particle, we make the first observation of the formation of a nanoplasma in a single, isolated nanoparticle. We find that the laser intensity threshold for plasma

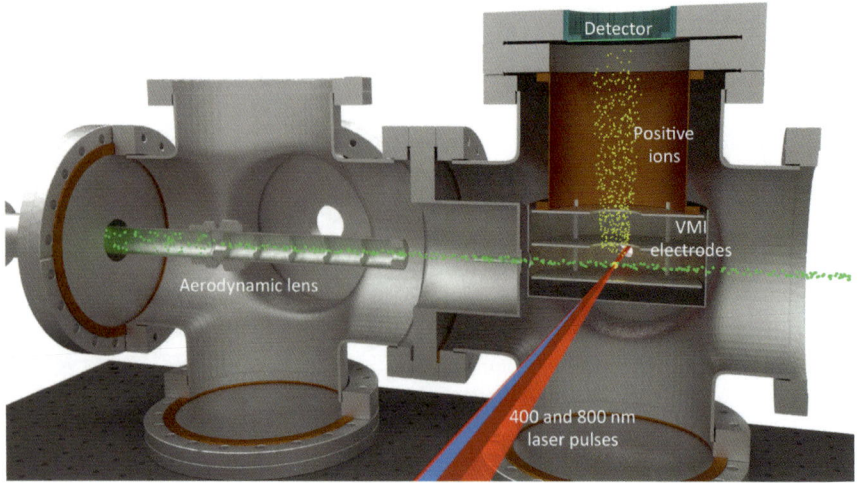

Fig. 1 Setup for imaging ultrafast dynamics in individual nanoparticles. An aerodynamic lens creates a collimated beam of nanoparticles, and two time-delayed laser pulses are used to excite and probe the sample. Photoions are collected by a microchannel-plate in a VMI geometry [6]

formation depends on the ionization potential of the nanocrystal, but remains in the range of 5×10^{13} to 1×10^{14} W/cm^2 for all materials in this study. Figure 2 demonstrates how two laser pulses can be used to control the ultrafast dynamics of the formation and propagation of shock waves in nanoplasmas formed from

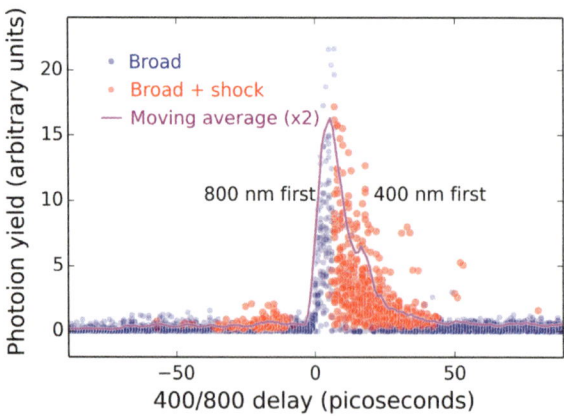

Fig. 2 Control of shock wave formation using two laser pulses. Each point indicates the total ion yield from a single nanoplasma explosion of an individual NH$_4$NO$_3$ nanoparticle as a function of the delay between 400 and 800 nm laser pulses. The *first pulse* forms an expanding nanoplasma. The *second pulse* causes a rapid pressure increase inside the nanoplasma, at a radius that depends on the delay between the pulses. When the delay between the two pulses is greater than 7 ps, shock waves are formed. The ion yield is higher when the 400 nm pulse precedes the 800 nm pulse because the 800 nm pulse is more effective at heating the expanded nanoplasma

ammonium nitrate (NH$_4$NO$_3$) nanocrystals. Changing the time delay not only controls the formation of the shock waves, but can also be adjusted to increase the yield of quasi-monochromatic ions.

4 Mapping Nanoscale Light Absorption with Plasma Explosion Imaging

When the laser intensity is set near the threshold to observe ion ejection from the nanoparticles, distinct asymmetries can be observed in the photoion angular distributions (Fig. 3). The asymmetric ion absorption contrasts with the more symmetric ion absorption that takes place at higher laser intensities and is a result from the formation of a localized plasma in just a small region of the nanoparticle. The plasma forms in a localized region because the laser intensity is not high enough to cause breakdown of the material without additional enhancement. The nanoparticle interacts with the light field, enhancing the laser field at specific locations that depend on the composition and geometry of the particle, thus providing the laser intensity needed to form a plasma.

We compare four different 100 nm diameter samples with different morphologies: single crystals of NaCl, aggregates of many 5 nm TiO$_2$ particles, composite particles

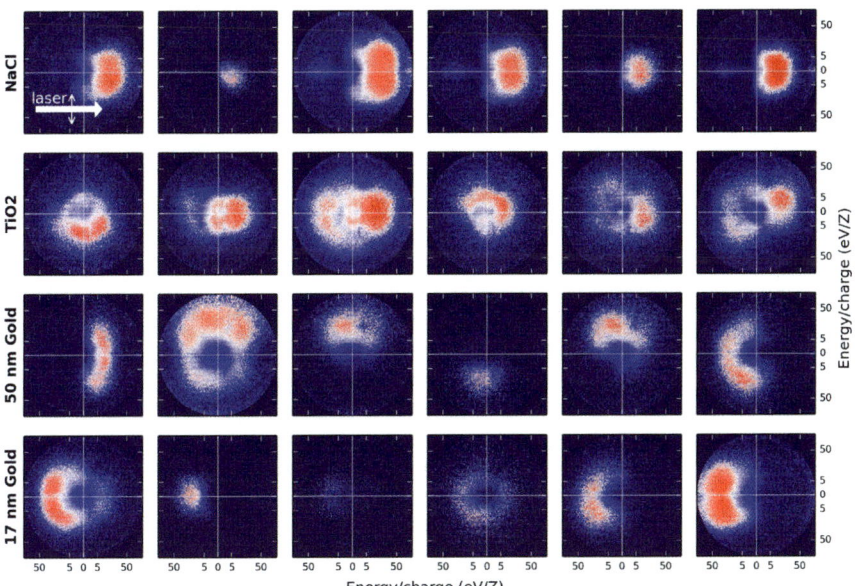

Fig. 3 Asymmetries in the photoion angular distributions reveal nanoscale light absorption in nanoparticles. The various nanoparticles feature different nanoscale structure, and therefore exhibit distinct nanoscale light absorption

consisting of a single 50 nm gold nanosphere imbedded in a ~ 100 nm dielectric sphere, and composite nanoparticles consisting of many 17 nm gold nanospheres imbedded in a ~ 100 nm dielectric sphere. For the NaCl, we observe ions ejected in the laser propagation direction, corresponding to a plasma formed on the backside of the NaCl particle due to the focusing of the light. For the TiO_2 particle, we observe a symmetric ion ejection corresponding to the enhancement of the light field between the TiO_2 nanoparticles, but with no directional preference. In the case of the 50 nm gold nanoparticles, the light field is enhanced near the surface of the gold nanoparticle and a plasma is formed that depends only on the orientation of the gold nanosphere within the dielectric sphere. For the 17 nm gold nanoparticle clusters, the light field is enhanced between adjacent gold nanoparticles, but the enhancement is greater on the illuminated side of the cluster, causing ejections of ions back towards the light source.

5 Conclusion

In summary, we present the first measurements of individual nanoplasmas, demonstrating a new method for studying laser-plasma and laser-nanoparticle interactions, which can be implemented using a tabletop apparatus and at a high repetition rate. We show that two laser pulses control the propagation of the shock waves and use numerical hydrodynamic simulations to unveil a mechanism for shock production in nanoplasmas. Furthermore, since these shocks are produced in plasmas with temperatures of just ~ 10 eV, this experiment enables a compact, inexpensive method for studying a relatively unexplored regime of dense, low-temperature nanoplasmas. Finally, we present a new method for characterizing nanoscale light absorption in single nanoparticles irradiated with intense femtosecond lasers.

References

1. A. E. Kaplan, B. Y. Dubetsky, and P. L. Shkolnikov, Phys. Rev. Lett. **91**, 143401 (2003).
2. F. Peano, R. A. Fonseca, J. L. Martins, and L. O. Silva, Phys. Rev. A **73**, 053202 (2006).
3. J. T. Jayne, D. C. Leard, X. Zhang, P. Davidovits, K. A. Smith, C. E. Kolb, and D. R. Worsnop, Aerosol Sci. Tech. **33**, 37 (2000).
4. W. Xiong, D. D. Hickstein, K. J. Schnitzenbaumer, J. L. Ellis, B. B. Palm, K. E. Keister, C. Ding, L. Miaja-Avila, G. Dukovic, J. L. Jimenez, M. M. Murnane, and H. C. Kapteyn, Nano Lett. **13**, 2924 (2013).
5. D. Hickstein, P. Ranitovic, S. Witte, X.-M. Tong, Y. Huismans, P. Arpin, X. Zhou, K. Keister, C. Hogle, B. Zhang, C. Ding, P. Johnsson, N. Toshima, M. Vrakking, M. Murnane, and H. Kapteyn, Phys. Rev. Lett. **109**, 073004 (2012).
6. A. T. J. B. Eppink and D. H. Parker, Rev. Sci. Instrum. **68**, 3477 (1997).

Part X
Novel Pulsed Sources
and Application

Passively CEP-Stable Front End for Frequency Synthesis

Hüseyin Çankaya, Anne-Laure Calendron and Franz X. Kärtner

Strong field physics requires high-energy pulses with multi-octave bandwidth to trigger efficiently sub-cycle events, most prominently isolated attosecond pulse generation. A laser driver based on frequency synthesis is a well-suited source for its flexibility in spectral shaping and scalability in spectrum and energy, as demonstrated in [1–3]. Energy scaling by amplification through optical parametric amplifiers (OPA) requires a broadband seed source and a high-energy pump. Due to the direct electric field driven processes, carrier envelope phase (CEP) stability of the seed is required. The front-end has thus to be CEP stable before the amplification with follow-on OPA's, typically at kHz repetition rate. The main methods to achieve CEP stability are either active stabilization [2], requiring several feedback loops increasing the system complexity, or passive CEP stabilization [3, 4]. Previously, passively CEP stable white-light super-continuum generation was shown by employing either difference frequency generation or idler of an OPA [3–5] based on Ti:Sapphire based pump source. The new scheme is based on white-light generation amplified by 2 OPA stages, and difference frequency generation by using the idler of one of the OPA's pumped with Yb based pump source which will

H. Çankaya (✉) · A.-L. Calendron · F.X. Kärtner
Center for Free-Electron Laser Science, Deutsches Elektronen Synchrotron,
and Department of Physics, University of Hamburg, Notkestrasse 85,
22607 Hamburg, Germany
e-mail: hcankaya@gmail.com

H. Çankaya · A.-L. Calendron · F.X. Kärtner
The Hamburg Centre for Ultrafast Imaging, Universität Hamburg,
Luruper Chaussee 149, 22761 Hamburg, Germany

F.X. Kärtner
Department of Electrical Engineering and Computer Science
and Research Laboratory of Electronics, Massachusetts Institute of Technology,
Cambridge, Massachusetts 02139, USA

© Springer International Publishing Switzerland 2015
K. Yamanouchi et al. (eds.), *Ultrafast Phenomena XIX*,
Springer Proceedings in Physics 162, DOI 10.1007/978-3-319-13242-6_174

enable higher amplification with respect to Ti:Sapphire technology based pump sources in further amplification stages.

The particular concept realized here uses slightly sub-picosecond pulses [6] from a multi-mJ regenerative amplifier, which greatly simplifies timing stabilization of seed and pump pulses in the follow on OPAs or OPCPAs. We demonstrate the CEP stability and two-octave coverage of the white light continuum generated by long-driver pulses.

The experimental setup, shown in Fig. 1, is driven by a high-energy Yb:KYW regenerative-amplifier, delivering 700 fs pulses after the first amplification stage [6]. The main part of this output will later be used to seed high-energy amplifiers to pump the follow-on OPCPA stages, and a minor portion, here 0.65 mJ, was used in the front-end. First, white-light was generated in a 10-mm long YAG crystal X1, then a spectral bandwidth of 200 nm centered at 2.18 µm was amplified to 7 µJ through two nearly degenerate OPA stages in BBO (X2 and X3). The passively CEP stable idler of the second OPA centered at 1.96 µm was then used to generate white-light super-continuum in a 3 mm YAG crystal (X4). CEP stability was confirmed by interfering the fundamental spectrum at 960 nm with the SHG of the remaining idler in an f-2f setup.

The generation of the white-light super-continuum with sub-picosecond pulses has been carefully studied for different materials such as YAG and Sapphire, with different lengths, focusing conditions and pulse durations. Optimized parameters to achieve a two-octave continuum as well as excellent pulse-to-pulse stability were determined. We also studied the influence of chirp on the long driver pulses on the WLG. The supercontinua obtained for compressed, negatively and positively chirped pulses, are shown on Fig. 2a. The pulse-to-pulse energy stability of the supercontinua reached 3.4 and 2.8 % for spectra below and above the driver wavelength, respectively: the chirp on the driving pulses did not influence the spectral bandwidth. Figure 2b shows signal and idler spectra at the end of the

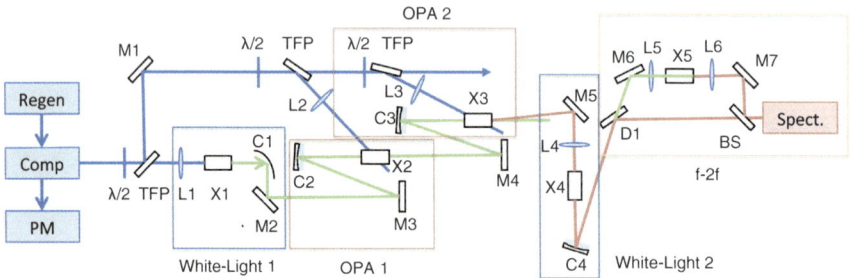

Fig. 1 Layout of the setup consisting of two white-light super-continua, OPA and f-2f. $\lambda/2$ stands for half-waveplate, TFP for thin-film polarizer, *M1–M7* flat mirrors (dielectric high reflectors for the pump and silver mirrors for the broadband pulses), *X1–X5* non-linear crystals and PM for power-meter blocking the main output of the compressor. *L1–L3* and *L4–L6* are focussing lenses for the 1 and 2 µm beams, respectively, whereas the broadband beam is collimated and focused through the parabolic mirror *C1* and curved mirrors *C2–C4*. D1 and BS stand for dichroic mirror and beam combiner, respectively

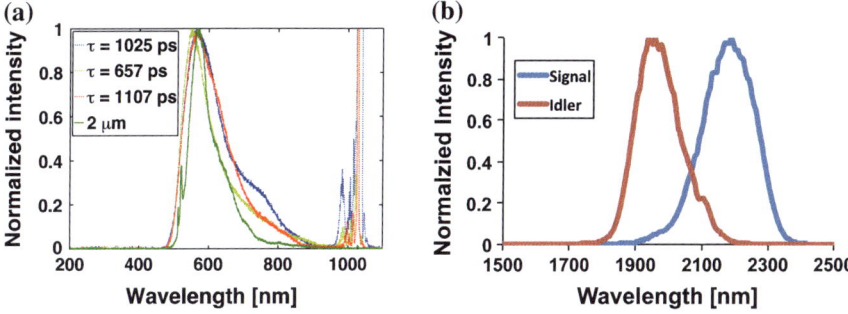

Fig. 2 a White-light continuua generated in YAG; the spectrum shown as *green solid curve* is generated by CEP stable 1.95 μm pulses shown as *red* in (**b**); the other, *dashed, curves* correspond to the white-light driven by compressed, negatively and positively chirped 1 μm pulses. **b** Spectra out of the 2nd OPA before the CEP stable white-light generation

second amplification stage at 2.18 μm. The signal spectrum was centered at 2178 nm with a spectral bandwidth of 210 nm (FWHM), while the idler spectrum had 153-nm bandwidth at 1956 nm, corresponding to 26 fs transform limited pulses.

The pulse amplitude fluctuations of the signal at the end of the second OPA was measured to be less than 2 % (rms), which enabled stable white-light continuum generation.

Figure 3a shows the beating of the white-light in the range of 940–1000 nm with the second harmonic of the remaining idler after super-continuum generation. The spectrum of the beat signal was averaged over 5,000 pulses. The fringes show clearly the CEP stability of the white-light continuum between 940 and 1000 nm over 4 h. Figure 3c shows calculated CEP drift. The rms CEP fluctuation was

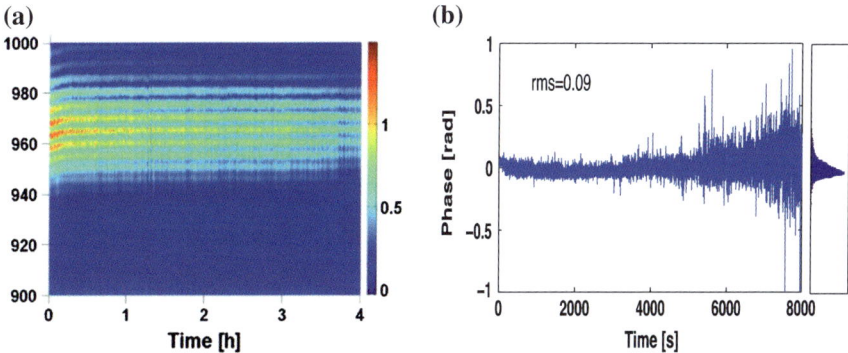

Fig. 3 a Fringes confirming the CEP stability of the white-light super-continuum in the range 940–1000 nm averaged over 5,000. **b** Calculated CEP over 4 h by using beat signal

calculated to be less than 150 mrad over 4 h. Previously, phase stability of white-light continua in bulk has only been shown for driver pulses with durations below 550 fs [4, 7, 8]. The measured spectral interferences confirmed that the CEP of the continuum was conserved for both short and long driver pulses.

In summary, we demonstrated passive CEP stable two-octave wide white-light continuum generation from 500 to 2 μm using slightly sub-picosecond pulses; the optimal parameters for white-light generation with long driver pulses have been identified. This source of CEP stable continuum is ideal for a two-octave wide high energy optical waveform synthesizer since it can be driven by the intermediate regenerative pre-amplifier for a high energy OPCPA pump line providing ultimately 1 J driver pulses at kHz repetition rate.

References

1. A. Wirth et al., "Synthesized Light Transients," Science 34, 195, (2011).
2. S-W. Huang et al., "High-energy pulse synthesis with sub-cycle waveform control for strong-field physics," Nature photonics, 5, 475 (2011).
3. O.D. Muecke, et al., "Millijoule-Level Parametric Synthesizer Generating Two-Octave-Wide Optical Waveforms for Strong-Field Experiments," San Jose, CLEO (2013).
4. G. Cerullo, et al., "Few-optical-cycle light pulses with passive carrier-envelope phase stabilization," Laser & Photonics Reviews. 5, 323-351, (2011).
5. A. Baltuska, et al. "Phase-Controlled Amplification of Few-Cycle Laser Pulses," IEEE JSTQE, 9, 972 (2003).
6. A-L Calendron et al. "High energy, high repetition rate pump laser system for OPCPAs," Europhoton, Neuchatel (2014).
7. M. Bradler, P. Baum, E. Riedle, "Femtosecond continuum generation in bulk laser host materials with sub-μJ pump pulses," Applied Physics B. 97 561–574 (2009).
8. P. Malevich. et al., "High-Repetition-Rate Multi-Millijoule Femtosecond 2.1-μm Ho:YAG Laser," Paris, ASSL (2013).

Tunable Few-Cycle Mid-IR Pulses Towards Single-Cycle Duration by Adiabatic Frequency Conversion

Peter R. Krogen, Haim Suchowski, Gregory J. Stein, Franz X. Kärtner and Jeffrey Moses

Abstract Using adiabatic difference frequency generation, we generate Fourier-limited, 4-cycle, tunable 2–4 μm mid-IR pulses at μJ-level, and 1.5-cycle, 3 μm pulses with controllable amplitude and phase by shaping before conversion from the near-IR.

The fields of nonlinear infrared spectroscopy and strong-field laser-matter interaction science call for intense and ever-shorter sources of mid-IR light [1, 2], from few-cycle down to single-cycle duration. In addition, experiments in these fields often demand spectral amplitude and phase control. Recently [3], we demonstrated the generation of an energetic, coherent, and octave-spanning mid-IR spectrum covering 2–5 μm by downconverting a broadband, chirped near-IR pulse in a quasi-phase-matching structure with an aperiodic poling period satisfying the conditions for adiabatic frequency conversion [4]. We achieved >80 % photon number conversion and one-to-one spectral amplitude transfer from the near-IR to the mid-IR, highlighting the capability of adiabatic frequency conversion to overcome the usual efficiency-bandwidth trade-off in wave mixing. However, as of yet there has been no experimental verification of the spectral phase transfer also expected in a chirped-pulse adiabatic conversion process, or compression of the coherent mid-IR spectrum to a near transform-limited pulse.

In this proof-of-principle study, we demonstrate compression of the mid-IR idler generated via adiabatic difference frequency generation (ADFG) to transform-limited duration by exploiting the near-IR to mid-IR spectral phase transfer inherent

P.R. Krogen (✉) · G.J. Stein · F.X. Kärtner · J. Moses
Department of Electrical Engineering and Computer Science and Research Laboratory of Electronics, Massachusetts Institute of Technology, Cambridge, MA 02139, USA
e-mail: pkrogen@mit.edu

H. Suchowski
NSF Nanoscale Science and Engineering Center, University of California, Berkeley, CA 94720, USA

F.X. Kärtner
DESY and Physics Department University of Hamburg, Center for Free-Electron Laser Science, Notkestraße 85, D-22607 Hamburg, Germany

© Springer International Publishing Switzerland 2015
K. Yamanouchi et al. (eds.), *Ultrafast Phenomena XIX*,
Springer Proceedings in Physics 162, DOI 10.1007/978-3-319-13242-6_175

to the process with a narrowband pump. Dechirping of the full mid-IR bandwidth will produce a single-cycle pulse. Limited by the acceptance bandwidth of our pulse characterization device, we have so far demonstrated an amplitude- and phase-shaped ~1.5-cycle source centered at 3.5 μm, or alternatively 4-cycle pulses tunable from 2.2 to 3.7 μm. In all cases, we obtained ~1-μJ energy. This result was made possible by adding only an ADFG stage and a bulk Si pulse compressor to the chirped output of a near-IR optical parametric chirped pulse amplifier (OPCPA). Since the spectral phase of the mid-IR idler is simply the combination of the phase on the input near-IR signal, the material dispersion in the ADFG crystal, and a constant conversion phase, control of the mid-IR spectral phase was achieved using a near-IR pulse shaper in the OPCPA.

Our experimental setup (Fig. 1) is the same modified OPCPA used in [3], with the addition of a bulk Si compressor. A Nd:YLF chirped pulse amplifier pumped both a 2-stage noncollinear optical parametric amplifier (NOPA) and the ADFG crystal, with a Ti:sapphire oscillator front end. Initially, numerical propagation simulations of the DFG process were performed to characterize the phase imparted by the ADFG crystal, from which it was concluded that dispersion is dominated by the material dispersion in LiNbO3. This is complicated by the fact that in ADFG different wavelengths are converted at different longitudinal locations in the crystal, resulting in a significant 'effective dispersion' that depends sensitively on the longitudinal variation of the poling period, $\Lambda(z)$. Our dispersion management scheme was designed based on the numerical simulations, which showed that the spectral phase could be well approximated by $\phi(\lambda) = L_{NIR}(\lambda)k_{NIR}(\lambda) + L_{MIR}(\lambda)k_{MIR}(\lambda)$, a combination of the material dispersion accumulated before and after frequency conversion, and approximating instantaneous conversion for each wavelength where it is exactly phase matched. For the generated mid-IR pulses in this experiment, this spectral phase was a smooth function that imparted ~1 ps group delay difference across the spectrum, which could be nearly fully removed by pre-compensation with traditional near-IR dispersion management techniques.

Our mid-IR source is a modification of a near-IR OPCPA, with the compressor replaced by the combination of an ADFG crystal and a bulk Si compressor. We adjusted the dispersive properties of the Dazzler to pre-chirp the near-IR pulse such that after conversion to the mid-IR and propagation through the Si compressor the pulse would be fully compressed. Care was taken while choosing compressor length and Dazzler settings to maintain a good temporal overlap between interacting pulses

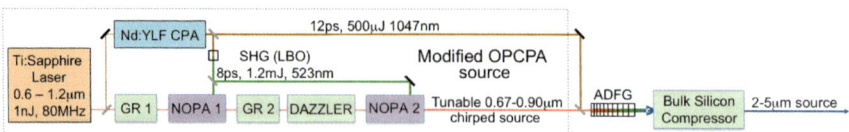

Fig. 1 Experimental setup. *Nd:YLF CPA* 12-ps, 1-kHz, 4-mJ chirped pulse amplifier; *GR1* grism pair GVD = −6,300 fs²; *NOPA1* 5-mm BBO; *GR2* grism pair GVD = −15,000 fs²; *Dazzler* (*Fastlite*) GVD = ~12,000 fs²; *NOPA2* 3-mm BBO; *ADFG* aperiodically poled, 20-mm MgO-doped congruent LiNbO3 grating; *Bulk Silicon Compressor* 25-mm, single pass

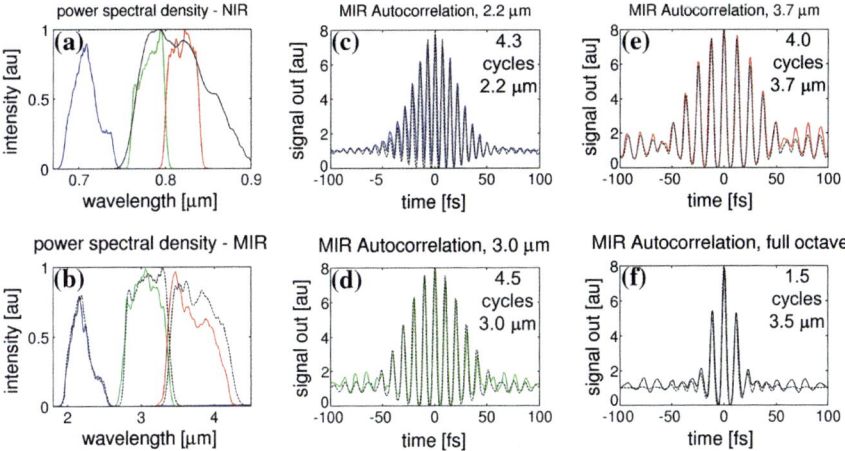

Fig. 2 Measured near-IR spectrum (**a**), expected mid-IR spectrum (**b**, *dashed*), measured mid-IR spectrum (**b**, *solid*), and autocorrelation traces (**c–f**), measured (*solid*) and expected as computed from the measured spectrum, assuming perfect compression of the mid-IR pulse (*dashed*), for 4 different input spectra (centered at 2.2, 3.0, 3.7 μm, and a broadband pulse at 3.5 μm)

in both the ADFG and second NOPA stages. The resulting compressed pulses were characterized in a homebuilt second-order interferometric autocorrelator, which uses either a 0.1-mm thick BBO crystal (for 2-μm operation) or a 0.16-mm thick AGS crystal (for 2.5–4.5-μm operation) for second harmonic generation (SHG), and an ext-InGaAs detector. The generated mid-IR spectrum can span from nearly 2–5 μm at −10 dB from peak, which supports a single-cycle pulse, however, the autocorrelator limited the detectable simultaneous bandwidth to one octave (2.4–4.8 μm, determined by available filters). Thus, we selected four narrower mid-IR spectral bands using the Dazzler to amplitude-shape the near-IR spectrum before conversion (Fig. 2a, b). The first three measured autocorrelation traces are plotted in Fig. 2c–e, along with the expected autocorrelations assuming perfect SHG phase matching and a perfectly compressed (Fourier-limited) pulse. The transform limits of the measured spectra are 31 fs at 2.2 μm, 46 fs at 3.0 μm, and 49 fs at 3.7 μm, each ∼4 optical cycles. The good match between the measured and expected autocorrelation traces shows there is an insignificant amount of spectral phase variation on the compressed mid-IR pulses. Thus, the pre-chirp imparted on the near-IR signal prior to conversion (experimentally optimized using the Dazzler by applying a quartic polynomial phase) compensated the spectral phase on the mid-IR pulse imparted by the adiabatic converter crystal and Si compressor. Note, an arbitrary phase could be added to the mid-IR pulse, if desired, by tuning the Dazzler. Finally, an autocorrelation trace of the pulses with a 2.4–4.8-μm mid-IR spectrum (selected to employ the full acceptance bandwidth of the autocorrelator) matches closely to the predicted autocorrelation trace of the transform-limited pulse (Fig. 2f), which corresponds to a 1.5-cycle pulse.

In conclusion, we have demonstrated a mid-IR source that delivers $\sim$ μJ energy pulses as short as 1.5 cycles (17 fs at 3.5 μm), with spectral phase and amplitude profiles that can be electronically tuned using a Dazzler. This demonstration of compression of an ADFG source also confirms that the spectral phase of the near-IR signal is transferred to the mid-IR idler, and the additional phase imparted by the adiabatic conversion process is smooth and easily removed using traditional dispersion management techniques. After improvement of the acceptance bandwidth of our characterization device, we anticipate the demonstration of single-cycle duration, with amplitude- and phase-tunability and 1-μJ energy.

References

1. P. Hamm and M. T. Zanni, *Concepts and Methods of 2d infrared Spectroscopy* (Cambridge University Press, 2011).
2. P. Colosimo, G. Doumy, C. I. Blaga, J. Wheeler, C. Hauri, F. Catoire, J. Tate, R. Chirla, A. M. March, G. G. Paulus, H. G. Muller, P. Agostini, and L. F. DiMauro, Nat. Physics **4**, 386–389 (2008).
3. H. Suchowski, P. R. Krogen, S.-W. Huang, F. X. Kärtner, and J. Moses, Opt. Express **21**, 28892-28901 (2013).
4. H. Suchowski, G. Porat, and A. Arie, Laser Photonics Rev. **8**, 333-367 (2014).

Carrier-Envelope Phase of Single-Cycle Pulses Generated Through Two-Color Laser Filamentation

Takao Fuji, Yutaka Nomura, Yu-Ting Wang, Atsushi Yabushita and Chih-Wei Luo

Abstract Carrier-envelope phase (CEP) of the single-cycle pulses generated through two-color filamentation has been investigated. It has been found that the phase of high frequency components of the generated pulses changes continuously and linearly with the relative phase between the two-color input pulses, whereas the phase of the low frequency components takes only two discrete values. To our knowledge, such behavior of the phase has never been discussed before. The transition of the phase behavior has been clearly observed by using frequency-resolved optical gating capable of CEP determination.

Passive stabilization of the carrier-envelope phase (CEP) of few-cycle pulses by the use of difference frequency generation (DFG) is a key technology for frequency comb and attosecond science [1, 2]. The CEP control at the passive stabilization scheme can be done by manipulating the relative phase between the two input pulses for the DFG. Similarly, the CEP of the ultrashort terahertz (THz) or mid-infrared (MIR) pulses generated through two-color laser filamentation is also passively stabilized. In this contribution, we report how the CEP of the single-cycle pulses generated through two-color laser filamentation is determined.

There are two basic physical models to explain the THz or MIR generation through two-color laser filamentation. The one is based on four-wave difference frequency generation (FWDFG) of the two-color input pulses and the other is a photocurrent model that accounts for electron motion in the two-color input field [3]. Here we use an one-dimensional model of the FWDFG. Assuming that the two input electric fields are $E_1(t) = \mathscr{E}_1(t) \exp(i\omega_1 t + i\phi_1) + \text{c.c.}$ and $E_2(t) = \mathscr{E}_2(t) \exp(i\omega_2 t + i\phi_2) + \text{c.c.}$, the nonlinear polarization for the process, $P_{\text{NL}}(t)$, can be written as

T. Fuji (✉) · Y. Nomura
Institute for Molecular Science, 38 Nishigonaka Myodaiji, Okazaki 444–8585, Japan
e-mail: fuji@ims.ac.jp

Y.-T. Wang · A. Yabushita · C.-W. Luo
National Chiao Tung University, Hsinchu 30010, Taiwan, Republic of China

© Springer International Publishing Switzerland 2015
K. Yamanouchi et al. (eds.), *Ultrafast Phenomena XIX*,
Springer Proceedings in Physics 162, DOI 10.1007/978-3-319-13242-6_176

$$P_{NL}(t) \propto E_1^2(t)E_2^*(t) = \mathscr{E}_1^2(t)\mathscr{E}_2^*(t)\exp(i(2\omega_1 - \omega_2)t + i(2\phi_1 - \phi_2)) + \text{c.c.}$$
$$= \mathscr{P}_{NL}(t)\exp(i\omega_0 t + i\Delta\phi) + \text{c.c.} \tag{1}$$

where $\mathscr{P}_{NL}(t) = \mathscr{E}_1^2(t)\mathscr{E}_2^*(t)$, $\omega_0 = 2\omega_1 - \omega_2$, $\Delta\phi = 2\phi_1 - \phi_2$. Assuming that $\ddot{P}_{NL}(t)$ is the far-field source, the generated field, $E_0(t)$, can be written as

$$E_0(t) \propto \ddot{P}_{NL}(t)$$
$$= \left(\ddot{\mathscr{P}}_{NL}(t) + i2\omega_0\dot{\mathscr{P}}_{NL}(t) - \omega_0^2\mathscr{P}_{NL}(t)\right)\exp(i\omega_0 t + i\Delta\phi) + \text{c.c.} \tag{2}$$

When the variation of the envelope of the FWDFG signal is faster than the difference of the two input frequencies, namely, $\ddot{\mathscr{P}}_{NL}(t) \gg \omega_0\mathscr{P}_{NL}(t) \gg \omega_0^2\mathscr{P}_{NL}(t)$, the field of the FWDFG reduces to $\ddot{\mathscr{P}}_{NL}(t)\exp(i\Delta\phi) + \text{c.c.}$ As a result, $\Delta\phi$ does not contribute to the phase but to the amplitude of the output field. On the other hand, when the variation of the envelope of the FWDFG signal is slower than the difference of the two input frequencies, namely, $\ddot{\mathscr{P}}_{NL}(t) \ll \omega_0\dot{\mathscr{P}}_{NL}(t) \ll \omega_0^2\mathscr{P}_{NL}(t)$, the real output field of the FWDFG can be written as $\omega_0^2\mathscr{P}(t)\exp(i\omega_0 t + i\Delta\phi) + \text{c.c.}$ In this case, the CEP of the field is $\Delta\phi$, which means that the relative phase between the two input pulses directly affect the CEP of the output pulse.

To experimentally investigate the CEP variation, we generated phase-stable MIR pulses through two-color laser filamentation in nitrogen [4] and characterized the pulse including its CEP information by using FROG capable of CEP determination (FROG-CEP) [5]. Figure 1a, b shows waveforms and spectral phases of the MIR pulses at different relative phases of the input pulses, respectively. Figure 1c shows phase change for each frequency components of the MIR pulses. The phase of the high frequency components ($\omega_0 > 3,000$ cm^{-1}) changes continuously and linearly with respect to the relative phase. On the other hand, the phase of the low frequency components ($\omega_0 < 3,000$ cm^{-1}) changes by 0 or π like a step function, which means that $\Delta\phi$ basically affect only the amplitude. The π phase jump means that

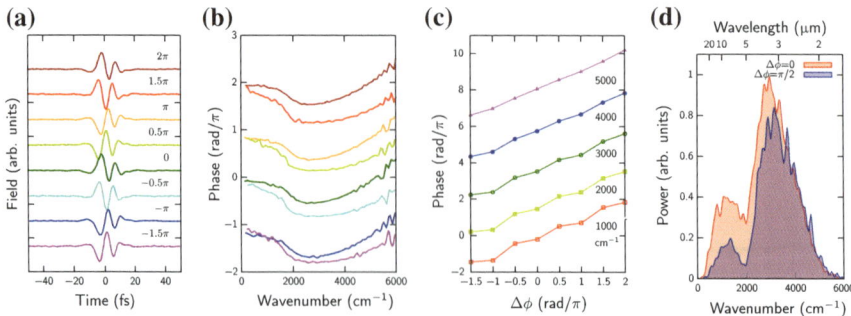

Fig. 1 **a** Waveforms and **b** spectral phases of the generated MIR pulses at each relative phase of the input pulses, respectively. **c** The relative phase dependence of the phases of the MIR pulses for each frequency component. **d** The power spectra of the generated MIR pulses

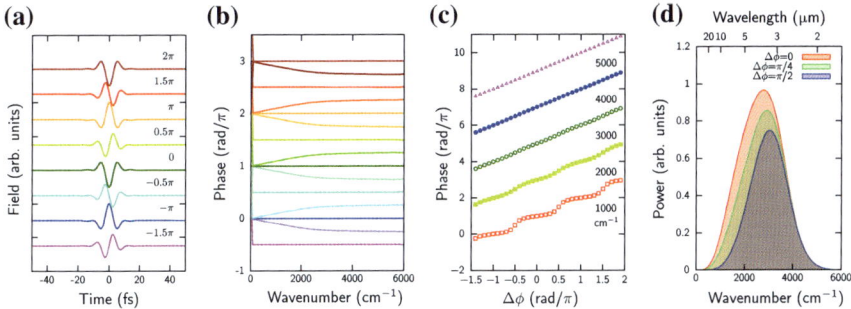

Fig. 2 Results of the numerical simulations. **a** Waveforms and **b** spectral phases of the generated MIR pulses at each relative phase of the input pulses, respectively. **c** The relative phase dependence of the phases of the MIR pulses for each frequency component. **d** The power spectra of the MIR pulses

only the sign of the amplitude (the sign of $\cos \Delta\phi$) changes. Figure 1d shows the power spectrum of the MIR pulse. The $\Delta\phi$ dependence is stronger for low frequency components ($<2{,}500$ cm^{-1}) whereas the high frequency components ($<2{,}500$ cm^{-1}) are much less sensitive to $\Delta\phi$.

We have also performed numerical simulations for the MIR pulses generated through two-color laser filamentation. The simulations are based on the above mentioned simple theory. The input electric fields were calculated from the fourier-transform of the spectra of the two-color pulses after the filamentation. We assumed no chirp for the both pulses in the simulation. Figure 2 shows the results of the numerical simulations. The stepwise variation of the phase and the phase dependence of the power spectrum is clearly reproduced with the simulations (see Fig. 2c, d).

In conclusion, the CEP of single-cycle pulses generated through two-color filament has been investigated by using FROG-CEP. It has been found that the relative phase dependence of the CEP is different for the generated frequency. We believe that these findings are useful for development of phase-stable single-cycle pulse generation. For example, the CEP at optical rectification would be even more robust than that at DFG since it does not depend on the delay between the two input pulses. Some artificial chirp might appear under some particular phase relationship between the two input pulses.

References

1. A. Baltuška, T. Fuji, T. Kobayashi, Phys. Rev. Lett. **88**, 133901 (2002)
2. G. Cerullo, A. Baltuška, O.D. Muecke, C. Vozzi, Laser Photon. Rev. **5**(3), 323 (2011). DOI 10.1002/lpor.201000013
3. M.D. Thomson, M. Kress, T. Loeffler, H.G. Roskos, Laser Photon. Rev. **1**(4), 349 (2007). DOI 10.1002/lpor.200710025

4. T. Fuji, Y. Nomura, Appl. Sci. **3**(1), 122 (2013). DOI 10.3390/app3010122 . URL http://www.mdpi.com/2076-3417/3/1/122

5. Y. Nomura, H. Shirai, T. Fuji, Nat. Commun. **4**, 2820 (2013). DOI 10.1038/ncomms3820

Phase-Locked Multi-THz High-Harmonic Generation by Dynamical Bloch Oscillations in Bulk Semiconductors

M. Hohenleutner, O. Schubert, F. Langer, B. Urbanek, C. Lange,
U. Huttner, D. Golde, T. Meier, M. Kira, S.W. Koch and R. Huber

Abstract Bloch oscillations are among the most spectacular quantum manifestations of electrons in crystalline solids. When an electric field accelerates an electron, its wavelength shortens. Once the latter equals twice the lattice constant, the wave should undergo Bragg reflection, causing electrons to oscillate in reciprocal and real space [F. Bloch, Z. Phys. 52, 555–600 (1928); C. Zener, Proc. R. Soc. A 137, 696–702 (1932)]. Ultrafast carrier scattering and dielectric breakdown under constant-field biasing have hampered the experimental observation of this long-standing prediction in bulk crystals [C. Zener, Proc. R. Soc. A 137, 696–702 (1932)]. Recently, high-harmonic generation has been attributed to a dynamical version of Bloch oscillations [S. Ghimire et al., Nature Phys. 7, 138–141 (2010)]. Controlling the precise shape of the optical fields, however, has been out of reach due to the fluctuating carrier-envelope phase (CEP) of the laser pulses. Novel developments in field-resolved multi-terahertz optics provide low-frequency CEP-stable electromagnetic wave-forms, which serve as a sub-cycle bias for high-field experiments [A. Sell et al., Opt. Lett. 33, 2767–2769 (2008); F. Junginger et al., Phys. Rev. Lett. 109, 147403 (2012)]. Here, we employ atomically strong and phase-locked multi-THz fields to control all-coherent charge transport in gallium selenide (GaSe) on femtosecond timescales. Off-resonantly driven coherent interband polarization and dynamical Bloch oscillations result in the emission of phase-stable high-order harmonics (HH) covering the frequency range from 0.1 to 675 THz [O. Schubert et al., Nature Photon. 8, 119-123 (2014)].

Bloch oscillations are among the most spectacular quantum manifestations of electrons in crystalline solids. When an electric field accelerates an electron, its wavelength shortens. Once the latter equals twice the lattice constant, the wave

M. Hohenleutner (✉) · O. Schubert · F. Langer · B. Urbanek · C. Lange · R. Huber
Department of Physics, University of Regensburg, 93040 Regensburg, Germany
e-mail: Matthias.Hohenleutner@physik.uni-regensburg.de

U. Huttner · D. Golde · M. Kira · S.W. Koch
Department of Physics, University of Marburg, 35032 Marburg, Germany

T. Meier
Department of Physics, University of Paderborn, 33098 Paderborn, Germany

© Springer International Publishing Switzerland 2015   721
K. Yamanouchi et al. (eds.), *Ultrafast Phenomena XIX*,
Springer Proceedings in Physics 162, DOI 10.1007/978-3-319-13242-6_177

should undergo Bragg reflection, causing electrons to oscillate in reciprocal and real space [1, 2]. Ultrafast carrier scattering and dielectric breakdown under constant-field biasing have hampered the experimental observation of this long-standing prediction in bulk crystals [2]. Recently, high-harmonic generation has been attributed to a dynamical version of Bloch oscillations [3]. Controlling the precise shape of the optical fields, however, has been out of reach due to the fluctuating carrier-envelope phase (CEP) of the laser pulses. Novel developments in field-resolved multi-terahertz optics provide low-frequency CEP-stable electromagnetic waveforms, which serve as a sub-cycle bias for high-field experiments [4, 5]. Here, we employ atomically strong and phase-locked multi-THz fields to control all-coherent charge transport in gallium selenide (GaSe) on femtosecond timescales. Off-resonantly driven coherent interband polarization and dynamical Bloch oscillations result in the emission of phase-stable high-order harmonics (HH) covering the frequency range from 0.1 to 675 THz [6].

We generate few-cycle multi-THz transients with adjustable CEP via difference frequency mixing [4]. Waveforms featuring a center frequency of 30 THz and peak electric fields of 72 MV/cm (Fig. 1a) are incident on a 220 μm thick sample of undoped bulk GaSe. Electro-optic detection of the transmitted waveform with 8-fs near-infrared gate pulses reveals low frequency components between 0.1 and 10 THz, stemming from optical rectification, as well as the second harmonic of the fundamental wave at a frequency of 60 THz (Fig. 1b). Intriguingly, ultrabroadband traces also exhibit spectral signatures at the third and fourth harmonics (Fig. 2a, blue shaded curve). Using an InGaAs and a Si spectrometer, a plateau-like region of

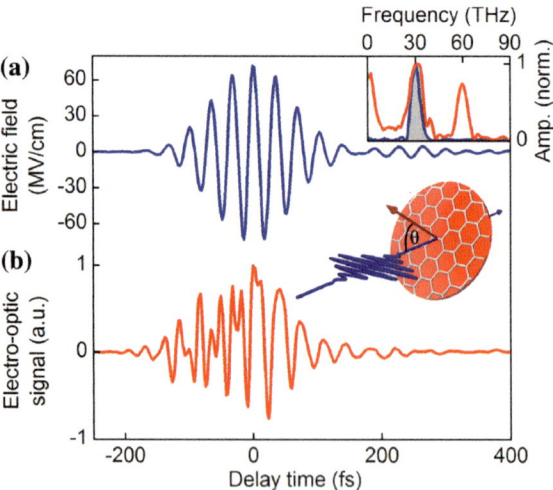

Fig. 1 a Electro-optically detected waveform of the multi-THz driving field featuring an amplitude of 72 MV/cm in air and a central frequency of 30 THz. **b** Electro-magnetic transient generated in a 220 μm thick GaSe sample (θ = 70°) by the pulse shown in a, revealing a superposition of fundamental, optical rectification and second harmonic components (*insets* corresponding spectra and experimental geometry defining the angle of incidence θ)

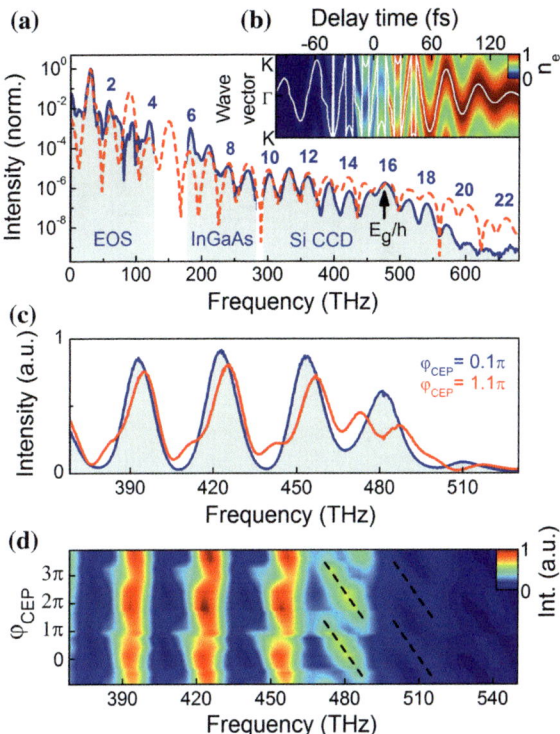

Fig. 2 a HH spectrum generated in GaSe by the waveform shown in Fig. 1a (*blue*). *Red dashed line* Calculated HH intensity obtained by the quantum-mechanical model. E_g: band gap energy of GaSe. **b** Computed dynamics of the electron distribution n_e in the first conduction band for internal peak fields of 14 MV/cm (*color-coded*). *White line* Center of the electron wave packet. **c** Measured HH intensity spectra generated by driving waveforms with carrier-envelope phases of $\varphi_{CEP} = 0.1\pi$ (*blue shaded*) and 1.1π (*red*). **d** Systematic dependence of HH spectra on φ_{CEP}: HH peaks of order $n \geq 15$ shift in frequency with a slope of -2.5 THz rad^{-1} (*dashed lines*)

harmonics of both even and odd order is observed up to the bandgap of GaSe at a frequency of 476 THz. Beyond this frequency, the harmonic intensity decreases due to interband absorption and is detected up to the 22nd order. Remarkably, all spectral components are absolutely phase-stable, as verified by electro-optic sampling of the lowest orders (Fig. 1b) and *f-2f*-interferometry for higher orders [6], indicating an all-coherent HH generation mechanism. The scaling of HH intensities with the peak driving field E reveals the non-perturbative character of the dynamics: While the nth order intensity I_n follows $I_n \sim E^{2n}$ for low fields, it approaches $I_n \sim E$, asymptotically, for large fields [6].

High-harmonic generation in atoms has been explained by a semiclassical three-step model including ionization, acceleration and recollision [7]. In solids, however, the quantum mechanical wave nature of electrons dominates the physics of high-field transport. We extend the many-body theory of [8] to a five band model, including inter- and intraband excitation as well as THz-induced band mixing. Figure 2b shows the calculated dynamics of the electron density n_e in the first conduction band for the experimental driving waveform and internal peak fields of 14 MV/cm. Carriers are dominantly injected at every second field maximum and simultaneously accelerated along the direction of the electric field. The electron wave packet starts to oscillate following the external field and, once the Brillouin zone boundary at the K-point is reached, enters the Brillouin zone from the opposite side again. For highest fields, multiple such Bragg reflections occur during one half cycle of the driving field, giving rise to HH emission. Our calculations reproduce the measured HH intensities very well (red dashed curve in Fig. 2a) and identify dynamical Bloch oscillations along with coherent interband excitation as the origin of HH emission. Due to quantum interference of multiple excitation pathways involving different electronic bands, HH generation is expected to depend sensitively on the amplitude and phase of the THz waveform. In our experiment, we vary the CEP of the driving field (φ_{CEP}) and simultaneously record HH spectra. A pure sign flip of the transient already radically alters the shape and intensity of the emitted spectra (Fig. 2c). For $\varphi_{CEP} \approx 0$, a sinusoidal modulation of the spectrum is observed, while additional peaks spaced by 15 THz appear for $\varphi_{CEP} \approx \pi$. For a comprehensive picture, we record intensity spectra while varying φ_{CEP} continuously from $-\pi$ to 4π (Fig. 2d). High-harmonic generation below 400 THz is only slightly affected, whereas the intensity maximum of the 16th harmonic shifts about a frequency of 480 THz, with a slope of -2.5 THz rad^{-1} (dashed black lines in Fig. 2d). These features are well reproduced by our full quantum mechanical theory [6].

In conclusion, coherent interband excitation and intraband acceleration in a bulk semiconductor are off-resonantly driven by intense, phase-stable multi-THz waveforms. Precise tuning of the CEP of the driving fields allows for sensitive control of the carrier dynamics, which leads to the emission of an all-coherent HH spectrum covering 12.7 optical octaves from THz to visible spectral ranges. The results highlight quantum phenomena relevant for future semiconductor devices at terahertz clock rates and open the door to a novel regime of high-field transport on timescales shorter than a single oscillation cycle of light.

References

1. F. Bloch, Z. Phys. **52**, 555–600 (1928).
2. C. Zener, Proc. R. Soc. A **137**, 696–702 (1932)
3. S. Ghimire et al., Nature Phys. **7**, 138–141 (2010)
4. A. Sell et al., Opt. Lett. **33**, 2767–2769 (2008)
5. F. Junginger et al., Phys. Rev. Lett. 109, 147403 (2012)
6. O. Schubert et al., Nature Photon. **8**, 119-123 (2014)
7. P.B. Corkum, F. Krausz, Nature Phys. **3**, 381–387 (2007)
8. D. Golde et al., Phys. Status Solidi B **248**, 863-866 (2011)

Direct Generation of 7 fs Whitelight Pulses from Bulk Sapphire

Emanuel Wittmann, Maximilian Bradler and Eberhard Riedle

Abstract Generation of sub-10 fs continuum pulses without external compression is demonstrated. We investigate the propagation of the newly generated wavelengths and find that a short crystal in combination with an achromatic and anastigmatic telescope (Schiefspiegler) leads to nearly chirp free continua.

1 New Insights into Bulk Filamentation

Supercontinuum generation (SCG) in bulk material is a generally applicable method to broaden the spectrum of femtosecond laser pulses at various wavelengths. The Fourier limit for a possible compression of, e.g., an 800 nm pumped continuum from sapphire, amounts to about 4 fs. Yet, no results have been published which show that bulk continua have intrinsically such short pulse durations. This is in striking contrast to the situation in continua generated in gas-filled hollow core fibers or in gas filamentation. There compression to below 4 fs has been shown. In precise investigations of the continuum generation and propagation we now find that the inability to compress the continuum stems from the highly wavelength dependent effective generation locus and propagation. This knowledge gives us the chance for ideal control of the process and therefore the ability to generate sub 10 fs pulses without the use of any external compression scheme. This validates that the new frequencies generated during filamentation develop highly coherently.

E. Wittmann (✉) · M. Bradler · E. Riedle
LS für BioMolekulare Optik, Ludwig-Maximilians-Universität München,
Oettingenstr. 67, 80538 Munich, Germany
e-mail: e.wittmann@physik.uni-muenchen.de

© Springer International Publishing Switzerland 2015   725
K. Yamanouchi et al. (eds.), *Ultrafast Phenomena XIX*,
Springer Proceedings in Physics 162, DOI 10.1007/978-3-319-13242-6_178

2 Propagation Properties of New Frequencies and Generation of Sub-10-fs Pulses During SCG

Two processes should be differentiated when the over-all appearance of continuum generation in a bulk material is considered. First, the spatial area or depth into the material where the new colors are developing has to be considered. Second, the propagation in the remaining material before exiting into free space has to be understood. That these issues are far from trivial is proven by the fact that a full collimation of a bulk continuum has not been reported and consequently the full temporal compression has not yet been achieved. In preliminary experiments we imaged the continuum from a 3 mm sapphire with a singlet lens and found that the blue part of the spectrum focuses earlier than the red part. In a semi-quantitative interpretation this could be attributed to the chromatic error. To circumvent this issue, we used a Schiefspiegler [1] that images all spectral components without chromatic error. We still found the blue spectral components focused earlier. An explanation would be that during filamentation all colors are generated at once, but short wavelengths fall behind the filament channel because of their lower group velocity. Without guiding by the filament, they start to diverge. The wavelengths close to the pump follow the channel longer and diverge later, leading to the observed color dependent propagation.

To verify this concept, a continuum is generated in ethanol and monitored from the side with a high resolution camera. To visualize the propagation of selected wavelength ranges we use the solvent (instead of sapphire) and add small amounts of Rhodamine 6G (absorbing around 530 nm) or alternatively Oxazine 1 (627 nm). These laser dyes absorb a small fraction of the newly generated light and fluoresce so that the beam propagation can be monitored from the side. Figure 1b shows the corresponding fluorescence signal and the side view from filamentation in ethanol with Rhodamine 6G (top). A strong signal is only found when the light has already broadened from the 8 μm filament and the dye absorption is not saturated as inside the channel. We find that different colors start to diverge at different positions in the solvent. Our proposed model is confirmed and we can explain the difficulty of properly imaging a continuum.

To avoid the wavelength selective propagation, the continuum has to be generated at the very end of the crystal. The crystal should be terminated at the dashed line in Fig. 1a so that no spectral and local separation occurs and all colors start to diverge simultaneously as shown in Fig. 1c. We find a short crystal length on the order of 1 mm suitable. This should lead to a chirp free continuum as the newly generated colors pass through no extra material, and all colors having the same spatial properties. With a 1 mm sapphire plate, we optimize the continuum generation onto the output face. A careful alignment still renders a continuum with little fluctuations. The pump source is a small fraction of the output of a Ti:sapphire amplifier (CPA 2001; Clark MXR) with a pulse duration of 170 fs and a central wavelength of 778 nm, focused with a f = 50 mm plano-convex lens. We obtain a continuum spanning down to 430 nm. For an anastigmatic and achromatic

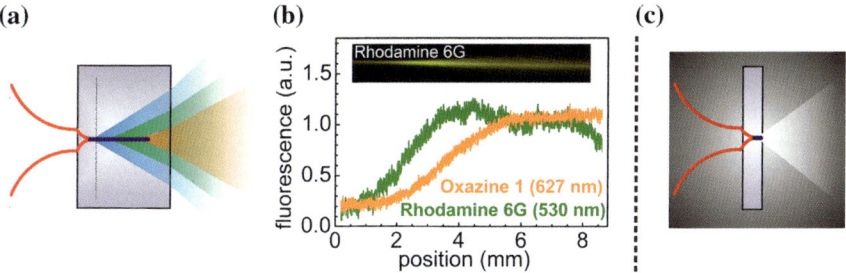

Fig. 1 a Scheme of bulk continuum generation and wavelength dependent beam propagation. **b** Fluorescence of dilute Rhodamine 6G (*green*) and Oxazine 1 (*orange*) solutions to monitor the beam propagation during filamentation in ethanol. **c** All colors diverge equally in a short crystal

collimation or imaging of the generated continuum we use a reflective Schiefs-piegler telescope [1], consisting of a suitable combination of a convex (R = −150 mm) and a concave mirror, R = 200 mm for imaging, 500 mm for collimation. This readily allows us to measure the pulse duration of the newly generated frequencies with a SHG intensity autocorrelator without the loss of any frequencies due to aberrations. The main peak of the autocorrelation signal corresponds to a 7 fs continuum pulse (Fig. 2a).

For a better insight into the spectrotemporal distribution of the continuum and to show that the autocorrelation signal does not correspond to a coherence spike [2] due to the complexity of the continuum pulse, the chirp of the sapphire continuum is determined with our transient spectrometer [3]. Figure 2b shows the transient signal of the continuum generated in the 1 mm sapphire plate. With a 200 nJ, 25 fs pump pulse at 470 nm and the sapphire continuum as probe we measured the cross correlation via two-photon absorption in a 130 μm GG400 (Schott AG) substrate. Two-photon absorption of pump and probe only occurs for temporal overlap and allows determining the group delay for every wavelength. Since only reflective optics are

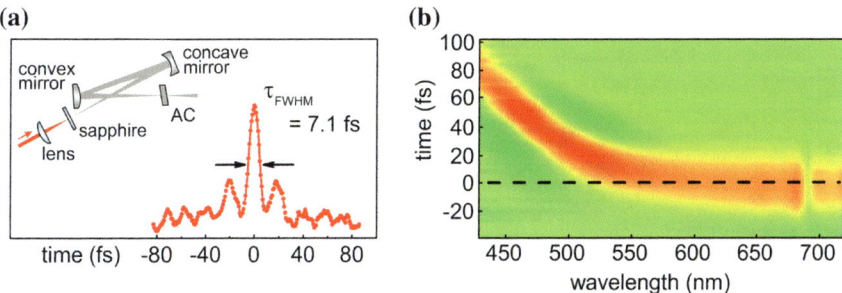

Fig. 2 a Schiefspiegler geometry for imaging the newly generated colors of a 1 mm sapphire continuum and autocorrelation trace demonstrating a 7 fs pulse. **b** Transient absorption measurement in 130 μm GG400 to determine the chirp of the continuum generated in a 1 mm sapphire plate

used in the probe beam path, the signal represents the intrinsic spectral chirp of the continuum after propagation through just a short length of air. Figure 2b shows that all spectral components from 500 to 700 nm coincide in time. The chirp for the short wavelength range originates both from the propagation in air and the small but finite region traveled at the end of the sapphire crystal. These chirped components of the pulse appear as broad and structured wings in the autocorrelation trace.

3 Conclusion and Outlook

It is possible to obtain sub-10 fs white light pulses directly from bulk filamentation. In the literature only a few examples for white light compression can be found, but until now the appropriate setup always is accompanied by huge complexity. By studying the propagation properties of the continuum we found a straightforward way to simplify the effort for the generation of short continuum pulses. An elaborate apparatus can be replaced by a lens, a 1 mm sapphire plate and the adequate adjustment for the incident pulse energy. With the imaging by a Schiefspiegler also any chromatic aberrations as well as astigmatism can be avoided. Such pulses are highly interesting for broadband amplification, ultrafast 2D spectroscopy, or spectroscopic experiments.

References

1. A. Kutter, Der Schiefspiegler: Ein Spiegelteleskop für hohe Bilddefinition (Weichhardt, 1953)
2. Rick Trebino, Frequency-Resolved Optical Gating: The measurement of Ultrashort Laser Pulses, (Kluwer Academic Publishers, Boston / Dordrecht / London, 2000), Chap. 4
3. U. Megerle, I. Pugliesi, C. Schriever, C.F. Sailer, E. Riedle, "Sub-50 fs broadband absorption spectroscopy with tunable excitation: putting the analysis of ultrafast molecular dynamics on solid ground", Appl. Phys. B **96**, 215-231 (2009).

Ultrafast 2 μm Laser Oscillators Based on Thulium-Doped ZBLAN Fibers

Yutaka Nomura, Masatoshi Nishio, Sakae Kawato and Takao Fuji

Abstract Mode-locked fiber laser oscillators are demonstrated by using thulium-doped ZBLAN fibers. Thanks to very low dispersion of ZBLAN glass fibers, pulses as short as 45 fs are generated at 1,900 nm.

In recent years, passively mode-locked fiber lasers operating around 1 and 1.5 μm have been studied extensively. Thulium-doped fiber lasers extend the operation wavelength to 2 μm region, which will be useful in many fields such as medical applications, eye-safe LIDAR, remote sensing, and mid-IR generation. Furthermore, broad emission spectra of thulium-doped fiber lasers make them promising light sources for ultrashort pulse generation in this wavelength region, which can be used for nonlinear processes such as micro-machining of transparent materials or efficient mid-infrared generation.

To obtain short pulses, the dispersion within the laser cavity must be compensated, which is not a trivial task for fiber lasers. The first mode-locked thulium-doped fiber laser is realized without any dispersion compensation and the duration of the generated pulses was ~ 500 fs [1]. Dispersion compensation was attempted with various methods such as inserting a stretcher in the cavity [2] or using normal-dispersion fibers that compensate the dispersion of ordinary fibers [3] and pulses as short as 119 fs are reported.

Y. Nomura (✉) · T. Fuji
Institute for Molecular Science, 38 Nishigo-Naka, Myodaiji, Okazaki 444-8585, Japan
e-mail: nomura@ims.ac.jp

T. Fuji
e-mail: fuji@ims.ac.jp

M. Nishio · S. Kawato
Graduate School of Engineering, University of Fukui, 3-9-1 Bunkyo, Fukui 910-8507, Japan
e-mail: mas2cee@gmail.com

S. Kawato
e-mail: kawato@u-fukui.ac.jp

Research and Education Program for Life Science, University of Fukui, 3-9-1 Bunkyo, Fukui 910-8507, Japan

K. Yamanouchi et al. (eds.), *Ultrafast Phenomena XIX*,
Springer Proceedings in Physics 162, DOI 10.1007/978-3-319-13242-6_179

729

Fig. 1 Schematic of the experimental setup

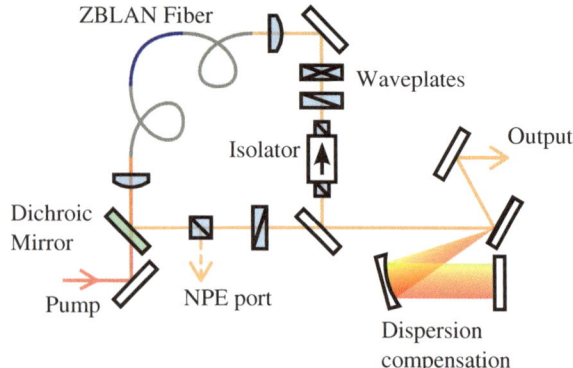

Here we demonstrate another approach, where we can significantly reduce the dispersion from the fiber itself. We used a thulium-doped fiber made of fluoride glass known as ZBLAN (ZrF_4-BaF_2-LaF_3-AlF_3-NaF). ZBLAN is known to have excellent optical properties in the mid-infrared region such as high transparency and low dispersion, which makes it a promising laser material. Using thulium-doped ZBLAN fibers, we have built a laser oscillator with a broad output spectra extending over 300 nm. The output pulses could be compressed down to 45 fs, which are the shortest pulses generated directly from laser oscillators around this wavelength region to our knowledge.

The laser setup is shown in Fig. 1. The cavity consisted of 0.2 m of single-mode, thulium-doped ZBLAN fiber (TDF) with the core diameter of 6 μm and NA of 0.2. The concentration of thulium ion is as high as 4 mol%. The absorption at 793 nm is estimated to be ~ 100 dBm^{-1}. On each end of the TDF, a passive single-mode ZBLAN fiber (SMF) with the same core diameter and NA as the TDF with the length of 1 m is attached to increase the cavity length and help mode-locking by nonlinear polarization evolution (NPE).

The active fiber is pumped by a Ti:sapphire laser (MaiTai, Spectra-Physics) operating at 793 nm. The pump beam is sent into the cavity through a dichroic mirror and coupled into a fiber with an aspherical lens ($f = 6$ mm). The coupling efficiency is estimated to be $\gtrsim 70$ %.

Unidirectional operation is enforced by an isolator placed in the free space section of the ring cavity. A half-wave plate (HWP) and a quarter-wave plate (QWP) are used to adjust the polarization state in the cavity to mode-lock the laser by NPE. A polarizing beam splitter (PBS) is used as a rejection port for the NPE. To adjust the dispersion within the cavity, a single-pass Martinez-type stretcher is placed in the cavity. Although the beam transmitted through this stretcher setup has spatial dispersion, the effect is rather small and did not deteriorate the laser operation. The zeroth-order reflection from the grating is used as the main output beam instead of the beam from the NPE rejection port. The HWP in front of the grating is used to optimize the output coupling ratio.

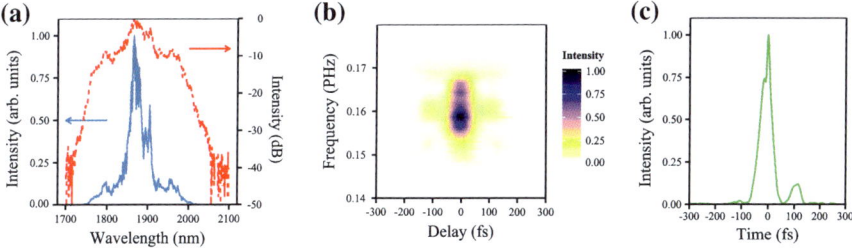

Fig. 2 Output from the oscillator. **a** Spectrum plotted in linear (*blue solid curve*) and log (*red dashed curve*) scale. **b** Measured FROG trace of compressed pulses. **c** Temporal profile retrieved from the FROG trace shown in **b**

Figure 2a shows the output spectrum measured with a monochromator. The spectrum extends from 1,730 to 2,050 nm at 30 dB below the peak. Stable pulses with the output power of ~ 13 mW are obtained from the main output at the pump power of 140 mW while pulses with ~ 10 mW power are obtained from the NPE rejection port. The pulse-to-pulse stability is ~ 1 % rms. When the pump power was increased to a higher level, the laser started to operate either in a multi-pulse regime or with a CW peak.

The pulse duration was measured with a home-built second-harmonic generation (SHG) frequency-resolved optical gating (FROG) device using a 30 μm-thick BBO crystal as the nonlinear medium. Figure 2b shows a typical FROG trace measured with the device. The pulse shape retrieved from this trace is shown on Fig. 2c. The pulse duration of 45 fs is obtained with a FROG error of 0.4 %, which is the shortest pulse generated from laser oscillators operating around the 2 μm region to the best of our knowledge.

Although our system uses a Ti:sapphire laser as the pump source, it would be beneficial if we could replace it with an inexpensive laser diode.

References

1. L.E. Nelson, E.P. Ippen, H.A. Haus, Appl. Phys. Lett. **67**(1), 19 (1995).
2. M. Engelbrecht, F. Haxsen, A. Ruehl, D. Wandt, D. Kracht, Opt. Lett. **33**(7), 690 (2008).
3. A. Wienke, F. Haxsen, D. Wandt, U. Morgner, J. Neumann, D. Kracht, Opt. Lett. **37**(13), 2466 (2012).

Characterizing Phase Fluctuations of Fiber Oscillators by Using External Optical Cavities

D. Schimpf, R. Schmeissner, J. Schulte, W. Liu, F. Kärtner
and N. Treps

Abstract We experimentally characterize amplitude and phase fluctuations of a femtosecond fiber oscillator close to the standard quantum limit (SQL). A passive optical cavity is employed to convert frequency noise to relative intensity noise (RIN) with close to quantum-limited sensitivity. The noise properties of the investigated oscillator reach the SQL (by 3 dB) at 0.1 μs timescales.

1 Introduction

To apply femtosecond oscillators to high precision metrology and quantum optics, their noise properties are required to be quantum limited at a given detection frequency [1, 2]. Typically, Titanium sapphire oscillators are employed. However, they are usually limited to laboratory setups due to cost and instabilities. In contrast, fiber oscillators are low cost and robust but little is known about their noise properties at high sideband frequencies. Direct phase noise measurements of fiber lasers require complex methods, such as the f-to-2f method succeeding super-continuum generation, which is based on nonlinear effects. A passive phase noise measurement can be realized by locking an external optical cavity to a laser and analysis of the error signal [3]. Even if this approach is relatively easy to implement; in practice, it is highly perturbed by electronic noise sources.

We propose and experimentally demonstrate a method that detects phase noise of ultrafast fiber lasers close to the standard quantum limit. The technique is readily applicable to any laser source and is based on analyzing the reflection from a passive cavity, a system known to contain all noise information of a light state [4].

D. Schimpf (✉) · J. Schulte · W. Liu · F. Kärtner
DESY and Department of Physics, Center for Free-Electron Laser Science,
University of Hamburg, Notkestraße 85, 22607 Hamburg, Germany
e-mail: damian.schimpf@desy.de

R. Schmeissner · N. Treps
Laboratoire Kastler Brossel, Université Pierre et Marie Curie, CNRS, ENS,
4 Place Jussieu, 75252 Paris Cedex, France

© Springer International Publishing Switzerland 2015
K. Yamanouchi et al. (eds.), *Ultrafast Phenomena XIX*,
Springer Proceedings in Physics 162, DOI 10.1007/978-3-319-13242-6_180

In the reflection from passive cavities phase noise translates to RIN—which is a well established characterization method for CW lasers [5]. We extend the technique to ultrafast lasers by noting that CEO phase noise is dominating timing jitter by several orders of magnitude. The method overcomes the drawback of the previous error signal method and achieves sensitivity close to the quantum limit.

2 Experimental Setup and Results

Figure 1 shows a schematic of the experimental setup. The Yb-doped fiber oscillator to be characterized is of stretched-pulse type and is mode-locked by a saturable-absorber mirror. The center wavelength of the spectrum is about 1,035 nm. The laser possesses a measured total net GDD of 0.012 ps^2. The repetition rate is 75 MHz. The high stability of the fiber laser cavity manifests itself in low repetition rate fluctuations, which is measured to be -80 dBc at 100 Hz sideband frequency. An external resonator cavity is locked to the fiber laser by using the Pound-Drever-Hall (PDH) technique. An electro-optical modulator imprints a 5 MHz phase modulation on the laser output. From the cavity reflection only the central part of the spectrum is used for the derivation of the PDH error signal. By controlling the position of a piezo-actuated mirror, stable locking is achieved over a time-span >10 min. To keep the design of the external resonator as simple as possible, we chose a semi confocal resonator with 300 MHz free-spectral-range. The cavity's resonance linewidth δv is about 200 kHz and can be regarded as a lower frequency

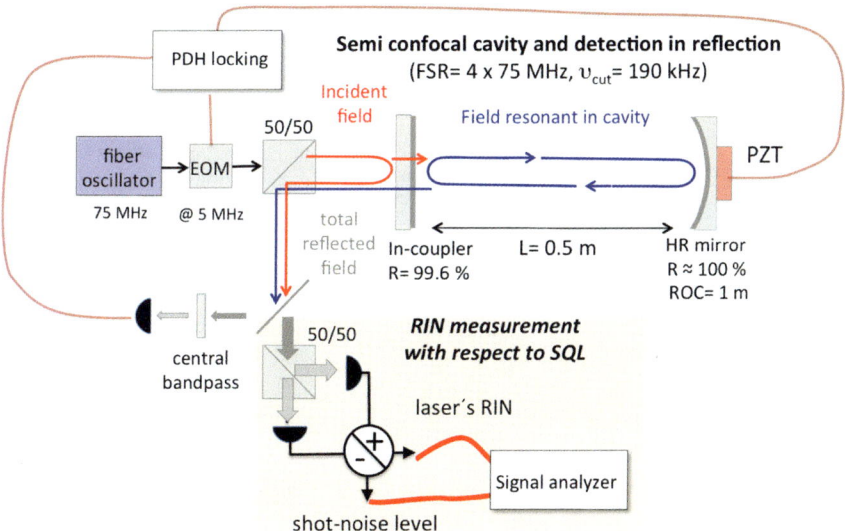

Fig. 1 Schematic of the experimental setup of locking an external optical resonator to the fiber laser

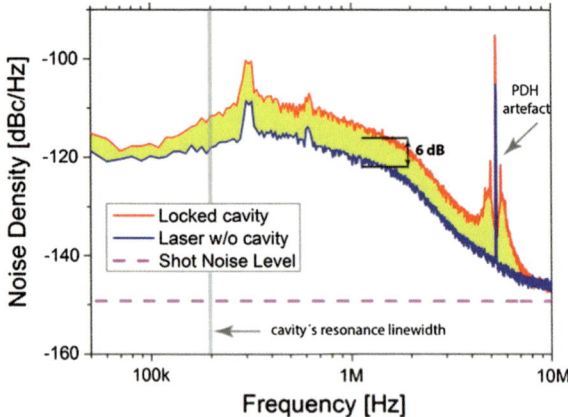

Fig. 2 Noise density of the laser without the effect of the cavity filter (*blue curve*) and for the case of the locked cavity (*red curve*). The additional noise is given by the difference (*yellow area*). The *grey vertical line* highlights the external resonator's line width

detection boundary for the herein proposed characterization technique. The beam that is reflected from the locked cavity is analyzed in terms of RIN by using a signal analyzer. Two cases are considered: First, the RIN measurement with the beam-path inside the cavity blocked, and second, a measurement with a locked cavity.

Figure 2 shows the effect of the cavity on the RIN measurement. When the beam path inside the cavity is blocked, the in-coupling mirror acts as a normal mirror and we perform a conventional RIN measurement of the fiber oscillator. The high levels of RIN, which are up to 30 dB above the SQL, are due to the RIN of the pump diodes. An independent balanced detector measurement allows a presentation of the data with respect to the SQL. If the cavity is locked, the laser phase noise is partially converted to amplitude fluctuations resulting in an increased RIN by 6 dB. This effect decreases for frequencies below the cavity's line width of 200 kHz and also depends on the electronic locking point. Above 8 MHz the two measurements collapse and are close to the standard quantum limit. Consequently, both amplitude and phase noise reach close to quantum limited properties. From the slope of the RIN's increase, estimation of the absolute level of CEO phase noise is possible [6].

3 Conclusion

The reflection of a laser output from a locked external cavity can be used to characterize phase (i.e. also frequency) noise of ultrafast lasers with high sensitivity. The advantages of the method are as follows: it is a passive technique and there is no need for generation of octave spanning spectra as compared to the f-to-2f method. Also, by converting phase noise to RIN, a standardized and quantitative comparison among different laser sources (including laser amplifier chains) is possible. The method can also be extended to a spectrally resolved analysis. And by designing a fiber oscillator with a pump diode exhibiting low RIN, quantum limited properties can be reached well below 8 MHz. This will render fiber oscillators to be robust and inexpensive sources suitable for quantum metrology.

References

1. B. Lamine, C. Fabre, and N. Treps, Phys, Rev. Lett., 101, 123601 (2008).
2. J. Roslund, R. Medeiros de Araújo, S. Jiang, C. Fabre, N. Treps, Nature photonics (2013).
3. Y. J. Cheng, P.L. Mussche, and A.E. Siegman, IEEE JQE 30, 1498-1504 (1994).
4. F. A. S. Barbosa, et al Phys. Rev. Lett. 111, 200402 (2013).
5. A. S. Villar, et al American J. of Physics 76, 922 (2008).
6. R Schmeissner, V. Thiel, C. Jacquard, C. Fabre, and N. Treps, Opt. Lett. 39, 3603 (2014)

Two Novel Schemes for Photon-Number Squeezed Pulse Generation in Ultrafast Nonlinear Fiber Optics

Aruto Hosaka, Shota Sawai, Kenichi Hirosawa and Fumihiko Kannari

Abstract We experimentally prove two novel techniques which solve issues in photon-number squeezed state generation with nonlinear Kerr effect: one is with an Er-doped fiber laser source, and the other is with a normal dispersion fiber at 800 nm. In the former case a collinear balanced detection technique is applied to eliminate spontaneous amplified emission noise. In the latter, negatively chirped laser pulses are used to achieve better matching with a local oscillator pulse.

1 Introduction

Current studies in squeezed laser pulse generation are motivated by the possibility to create and distribute continuous-variable entanglement for quantum communication. Squeezed light pulses at the telecommunication wavelength are important for optical fiber based quantum communication technology. However, erbium-doped fiber amplifiers (EDFA) used in present communication systems contain substantially large amplified spontaneous emission (ASE) noise which is much higher than the shot noise level (SNL). Therefore, alternatively, optical parametric oscillators of Ti:sapphire lasers or mode-locked Cr:YAG lasers have been employed to generate squeezed pulses. In this report, we applied a collinear balanced detection (CBD) technique [1] to eliminate ASE noise and achieved photon number squeezing at 1.55 μm using a noisy EDFA laser as a light source.

On the other hands, more experiments have been reported at ∼ 800 nm with a parametric down-conversion scheme to generate entangled state. However, it is difficult to generate fiber-optic squeezed pulses in the normal dispersion regime since pulses increase their pulse width by the chromatic dispersion and thus effective optical nonlinearity cannot take place, which is a large difference from

A. Hosaka · S. Sawai · K. Hirosawa · F. Kannari (✉)
Department of Electronics and Electrical Engineering, Keio University,
3-14-1 Hiyoshi, Kohoku-ku, Yokohama 223-8522, Japan
e-mail: kannari@elec.keio.ac.jp

© Springer International Publishing Switzerland 2015
K. Yamanouchi et al. (eds.), *Ultrafast Phenomena XIX*,
Springer Proceedings in Physics 162, DOI 10.1007/978-3-319-13242-6_181

soliton pulse propagation in the anomalous dispersion regime. Moreover, in a nonlinear polarization interferometer (NOPI) [1] which is commonly employed in experiments of photon-number squeezed pulse generation, mismatch of spectrum and temporal waveform of two pulses causes a deterioration of the squeezing level. In this work, we utilized spectral narrowing caused upon negatively chirped ultrashort pulses in the normal dispersion regime and achieved photon-number squeezing at 800 nm.

2 Photon Number Squeezing by Noise Reduced 1.5 μm EDFA with the CBD Technique

This technique compensates intensity noise at a specific radio-frequency by means of pulse splitting and recombination with a relative time delay τ [2]. Figure 1a shows the experimental setup. We used a femtosecond fiber laser (Femtolite, IMRA) as a light source. The center wavelength was 1,560 nm, the pulse width was ~160 fs (FWHM), and the repetition rate was 47.5 MHz. The intensity noise of a pulse train generated from the light source is higher than the SNL by 10 dB at an average optical power of 3 mW. Therefore, it is impossible to obtain squeezing without removing this excess noise. In our experiment, we used a NOPI to generate photon-number squeezing. Both orthogonally polarized pulses were divided into two pulse trains with a beam splitter for the CBD technique. We combined two pulse trains at PBS5 and detected them by one photodiode. As a result, a maximum photon-number squeezing of 2.6 dB was obtained when average power was 4 mW (see Fig. 1b) with the CBD technic at 1.5 μm.

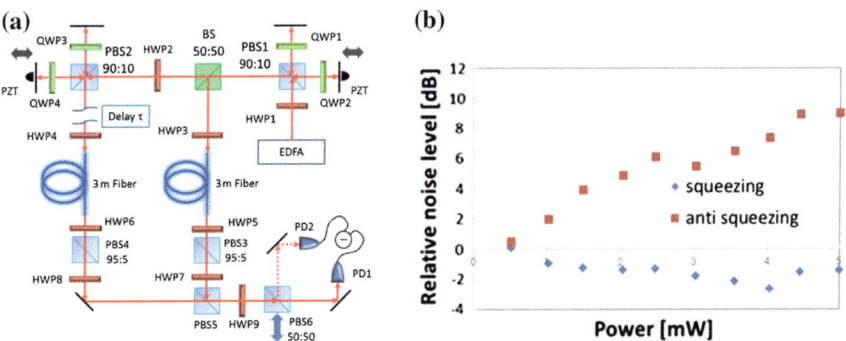

Fig. 1 a Experimental setup of CBD-photon number squeezing with EDFA laser using the nonlinear polarization interferometer. *HWP* half wave plate, *PBS* polarization beam splitter, *QWP* quarter wave plate, *BS* 50:50 beam splitter, *PZT* piezoelectric transducer. **b** Plots of intensity noise relative to the SNL as a function of laser power coupled into the fiber

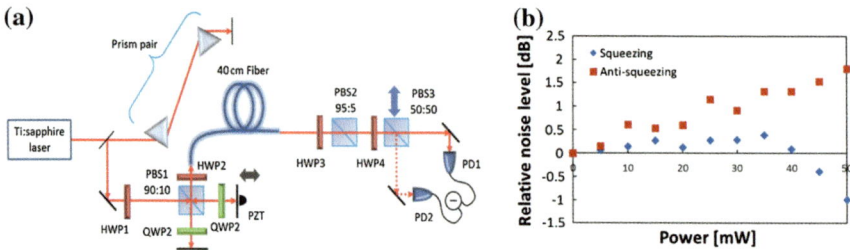

Fig. 2 **a** Experimental setup for generation of photon-number squeezed state by Ti:sapphire laser. *HWP* half wave plate, *PBS* polarization beam splitter, *QWP* quarter wave plate, *BS* 50:50 beam splitter, *PZT* piezoelectric transducer. **b** Plots of intensity noise relative to the SNL as a function of laser power coupled into the fiber

3 Photon Number Squeezing by 800 nm Pulses with Spectral Narrowing

In Fig. 2a a schematic view of the experimental setup for photon number squeezed pulse generation at 800 nm is shown. A mode-locked Ti:sapphire laser (Spectra-Physics, MaiTai-HP) was used as a light source. The center wavelength was 808 nm, the pulse width was ∼ 100 fs (FWHM), and the repetition rate was 80 MHz. We employed an NOPI using a conventional 40 cm long fiber.

When we added negative dispersion so that an Fourier transform limited pulse can be formed at the end of the fiber, significant spectral narrowing took place upon the intense laser pulse. Then, even with strong self-phase modulation, a better spectrum matching was obtained with the local oscillator pulse. Consequently, we obtained a maximum photon-number squeezing of 1.0 dB at an incident laser power of 50 mW (see Fig. 2b).

4 Conclusion

We achieved photon number squeezing at 1.55 μm using a noisy EDFA laser as a light source by use of the CBD technique. This experimental evidence indicates that our scheme makes it possible to observe phase-modulated vacuum noise entered at a beam splitter which separates two pulse trains by electrically cancelling substantial intensity noise of a light source at a specific RF frequency. We also achieved photon number squeezing at 800 nm using spectral narrowing of negatively chirped pulses through a normal dispersion fiber.

References

1. K. Nose, *et al.,* "Sensitivity enhancement of fiber-laser-based stimulated Raman scattering microscopy by collinear balanced detection technique." Opt. Express 20, 13958 (2012).
2. J. Higuchi, *et al.*, "Nonlinear Polarization Interferometer for Photon-Number Squeezed Light Generation." Jpn. J. Appl. Phys. 40, L1220 (2001).

Towards a Compact Fiber Laser for Multimodal Imaging

Bai Nie, Ilyas Saytashev and Marcos Dantus

Abstract We report on multimodal depth-resolved imaging of unstained living *Drosophila Melanogaster* larva using sub-50 fs pulses centered at 1060 nm wavelength. Both second harmonic and third harmonic generation imaging modalities are demonstrated.

1 Introduction

Due to the benefits of high contrast ratio, sub-micrometer resolution and depth resolved imaging multiphoton microscopy has been proven to be a powerful tool for studying living tissues [1, 2]. Especially for second harmonic generation (SHG) or third harmonic generation (THG) microscopy, no sample labeling is needed, which makes those methods preferable for non-invasive in vivo tissue imaging. In addition, SHG and THG provide complementary information due to their different optical-response mechanism. For both SHG and THG imaging, ultrashort laser pulses are preferred to achieve good multiphoton efficiency. It is found that SHG or THG efficiency is inversely proportional to the pulse duration or pulse duration square, respectively [3–6]. For clinical use, a compact and environmentally stable laser is need. In the past decade, fiber lasers have emerged as ideal ultrafast light sources [7]. Here an Yb fiber oscillator [8], capable of delivering pulses as short as ~50 fs at 1,060 nm central wavelength, is tested for multiphoton microscopy imaging. The capability of this laser for multiphoton microscopy is evaluated with different samples including prepared slides with stained mouse kidney and mouse intestine sections and unstained living whole *Drosophila Melanogaster* larva.

B. Nie · M. Dantus (✉)
Department of Physics and Astronomy, Michigan State University, East Lansing, MI 48824, USA
e-mail: dantus@msu.edu

I. Saytashev · M. Dantus
Department of Chemistry, Michigan State University, East Lansing, MI 48824, USA

© Springer International Publishing Switzerland 2015
K. Yamanouchi et al. (eds.), *Ultrafast Phenomena XIX*,
Springer Proceedings in Physics 162, DOI 10.1007/978-3-319-13242-6_182

Images generated by different modalities such as two-photon excited fluorescence (TPEF), SHG and THG are compared. Depth scan of SHG and THG is conducted and reconstructed 3D images are shown.

2 Experimental

An Yb fiber oscillator is operated at 43 MHz with average power up to 400 mW. This laser is based on an all-normal dispersion cavity and is similar to the design of the laser described in [8]. The output laser beam is guided through a 4-f folded pulse shaper (MIIPS Box 640, Biophotonic Solutions), which is used to compensate second order and higher phase distortions to deliver transform limited pulses at the focal plane. Output from the pulse shaper is directed to a laser-scanning multi-photon microscope. The laser beam is scanned by a galvanometer mirrors (QuantumDrive-1500, Nutfield Technology, Inc.) and coupled into a water-immersed objective (Zeiss LD C-APOCHROMAT 40x/1.1). The generated SHG and TPEF emissions from samples are collected in the Epi direction, being filtered out using a dichroic mirror (700DCSPXR, Chroma Technology Corp.) and a short-pass emission filter (ET680-SP-2P8, Chroma Technology Corp.). A photomultiplier (PMT, HC20-05MOD, Hamamatsu) is used to collect the SHG/TPEF signal. THG, which is primarily generated in the forward direction, is collected by a UV compatible objective (ReflX NT59-886, NA 0.28, Edmund Optics). The THG signal is also separated from the excitation light by a 400 nm short pass filter and detected by a PMT (H10720-210, Hamamatsu) whose signal is amplified (SRS445, Stanford Research Systems). The focal plane is moved to different layers using a step motor capable of making precisely controlled 2 μm height steps. All the SHG or THG images are then incorporated into a 3-D image.

3 Results and Discussion

Excitation laser pulses are compressed to about 50 fs at the focal plane of objective using the MIIPS enabled pulse shaper. To calibrate the microscope, two stained commercial samples (mouse kidney and mouse intestine, Molecular Probes) that have uniform thickness are imaged. For these two samples, the signal detected in Epi direction is mainly from two- or three-photon excited fluorescence. On the forward direction, mainly THG/three-photon excited fluorescence signal is detected. By combining the signal from Epi and forward directions, it is clearly seen that they provide complementary information for each other (see Fig. 1).

Beyond imaging pre-labeled samples, depth-resolved imaging of unstained live tissue is of greater importance. In a previous report [3], we demonstrated a fiber laser delivering 30 fs pulses used for multiphoton imaging of living tissues. However, the low pulse energy (about 1 nJ) limited the imaging depth capability.

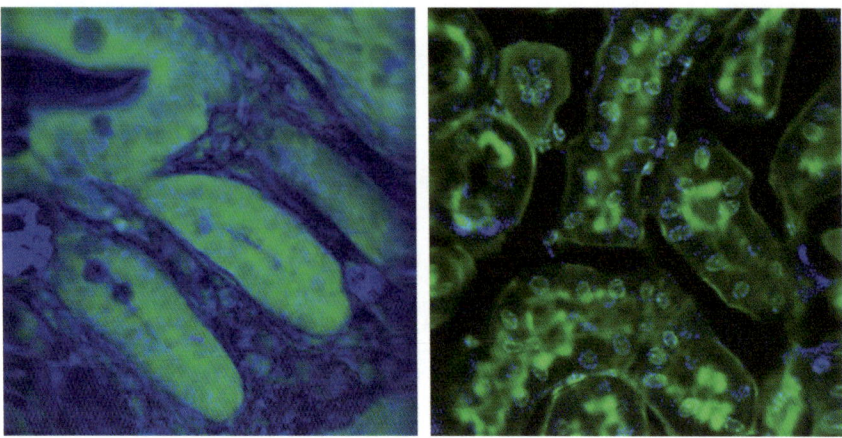

Fig. 1 Composition of TPEF (*green*, false color) and THG (*blue*, false color) imaging of mouse intestine (*left*) and mouse kidney (*right*), 150 μm × 150 μm area represented

The laser used in this work provides 10 times greater pulse energy and only slightly longer pulse duration. Depth resolved images of third instar *Drosophila* larva are shown in Fig. 2. The THG 3D image shows many more structures, for example the adipose tissue in the lower left corner, and less scattering than the SHG. The total scanned depth is about 90 μm.

The shorter pulse durations achieved by the laser greatly enhance two- and three-photon induced modalities in both stained and unstained living tissues.

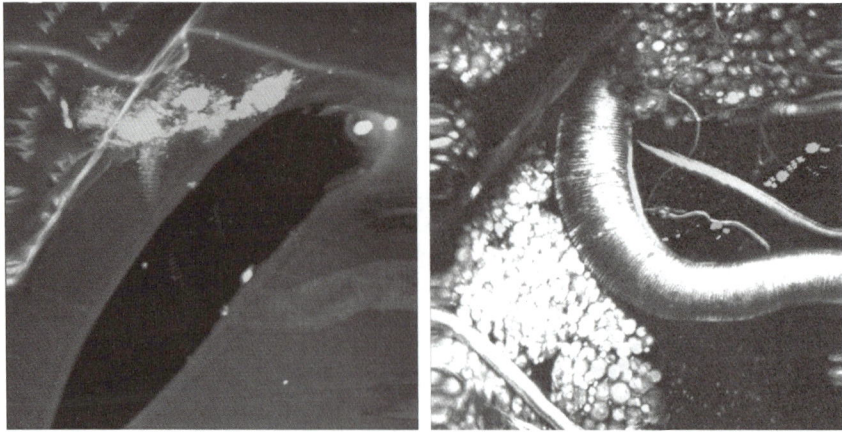

Fig. 2 Projection of 3-D images at 0° angle for SHG (*left*) and THG (*right*) microscopy of the third instar *D. Melanogaster* larva. Images are of the same 150 μm × 150 μm region centered at the trachea, but different contrast mechanisms highlight different organs

References

1. W. Denk et. al, "2-Photon Laser Scanning Fluorescence Microscopy," Science **248**, 73-76 (1990).
2. W. R. Zipfel et. al, "Nonlinear magic: multiphoton microscopy in the biosciences," Nature biotechnology **21**, 1369-1377 (2003).
3. B. Nie et. al, "Multimodal microscopy with sub-30 fs Yb fiber laser oscillator," Biomed Opt Express **3**, 1750-1756 (2012).
4. P. Xi et. al, "Two-photon imaging using adaptive phase compensated ultrashort laser pulses," J Biomed Opt **14** (2009).
5. P. Xi et. al, "Greater signal, increased depth, and less photobleaching in two-photon microscopy with 10 fs pulses," Opt Commun **281**, 1841-1849 (2008).
6. A. C. Millard et. al, "Third-harmonic generation microscopy by use of a compact, femtosecond fiber laser source," Appl Optics **38**, 7393-7397 (1999).
7. C. Xu, and F. W. Wise, "Recent advances in fibre lasers for nonlinear microscopy," Nature Photonics **7**, 875-882 (2013).
8. B. Nie et. al, "Generation of 42-fs and 10-nJ pulses from a fiber laser with self-similar evolution in the gain segment," Opt Express **19**, 12074-12080 (2011).

Measurement and Characterization of Sub-5 fs Broadband UV Pulses in the 230–350 nm Range

Valentyn I. Prokhorenko, Alessandra Picchiotti, Samansa Maneshi and R.J. Dwayne Miller

Abstract We report a new design of an all-reflective 3rd-order frequency resolved optical gating setup (FROG) for measurement and characterization of ultrashort UV-pulses in the 230–350 nm spectral range and tested it using 7.3 fs pulses generated in the 250–300 nm range. This setup allows also heterodyne detection which significantly increases its sensitivity.

1 Introduction

Time-resolved spectroscopy in the UV-range, in particular 2D-spectroscopy of DNA and proteins, requires development of broadband femtosecond UV-sources and corresponding diagnostic devices for direct measurement and characterization of light pulses. Among existing different methods the most appropriate for the 230–350 nm range is 3rd-order FROG [1] allowing "standalone" measurement and characterization of femtosecond pulses, without involving additional external pulses (as, e.g., in X-FROG and ZPA-SPIDER techniques [2, 3]). While the design of 3rd-order FROG is more complicated as compared to the 2nd-order FROG, registration and managing of second harmonic of 230–350 nm is much more difficult (115–175 nm belongs to the "vacuum" UV). In addition, there are no crystals available for generation of second harmonic in this spectral region.

Traditional design of 3rd-order FROG-apparatus is based on at least two conventional beam splitters made using thin substrates (0.5–1 mm); however, for the UV-range they are not appropriate due to strong stretching of pulses: for example, by passing a 7 fs pulse centered at 275 nm through 1 mm UV-grade fused silica plate it will be stretched to 110 fs. In order to avoid parasite phase distortions in measured pulses in refractive media, we developed an all-reflective FROG design based on a diffractive optic beam splitter (DO). This design is optimized for the UV

V.I. Prokhorenko (✉) · A. Picchiotti · S. Maneshi · R.J. Dwayne Miller
Max Planck Institute for Structure and Dynamics of Matter, Luruper Chaussee, 149, 22761 Hamburg, Germany
e-mail: valentyn.prokhorenko@mpsd.mpg.de

© Springer International Publishing Switzerland 2015
K. Yamanouchi et al. (eds.), *Ultrafast Phenomena XIX*,
Springer Proceedings in Physics 162, DOI 10.1007/978-3-319-13242-6_183

744

spectral range of 230–350 nm (spectral window $\sim 15,000$ cm^{-1}). The present results are representative for this wavelength range. The temporal resolution is currently limited by the spectral window of the spectrometer used and spectral width of DO. The estimated temporal resolution of the current setup is 3–4 fs; the shortest UV-pulse measured was 6 fs FWHM (spectral width of 40 nm FWHM, centered at 275 nm).

2 Experimental Layout

Figure 1 shows the design of developed all-reflective UV-FROG setup. The incoming beam, reflected by an auxiliary mirror towards an off-axis parabolic mirror OAPM$_1$ (Newport), is focused on the DO which is placed in the focal plane of OAPM (f = 150 mm).

An aluminum-coated reflective DO (crossed grating, Holoeye, custom design) diffracts the incoming beam into 4 first orders with efficiency of 60 %, then they are collimated and form 4 collinear beams separated by 12.5 × 12.5 mm. Upper beams 1 and 3 and lower beams (2) and (4) are passed through retroreflectors RR1,2 (PLX) and reflected from aluminum-coated right-angle prisms RP1,2 (custom design) then focused onto a 150 μm thick UV-grade fused silica plate by OAPM2 (Newport, f = 150 mm) for generating 3rd-order signal.

The measured signal is generated by four-wave mixing of beams 1, (2) and 3; the fourth beam (4) is blocked (B/F). Due to the symmetry of the optical design, beams 1 and (2) automatically coincide in time and have zero delay between them; acquiring of FROG-trace is done by moving RR$_1$ placed on a motorized translation

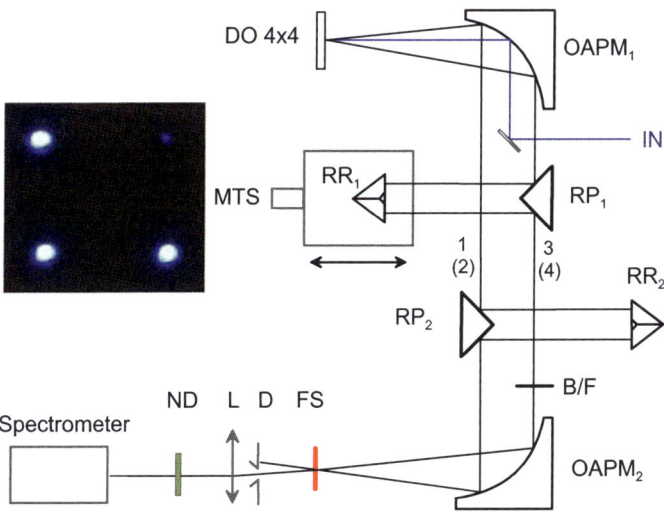

Fig. 1 Experimental layout of FROG setup. *Inset* shows a photograph of the incoming beams and the FROG-signal (*left-upper corner*) visible by eye

stage MTS equipped with an encoder (VP-25XL, Newport). Generated 3rd-order signal is passed through the block-diaphragm D and directed to a spectrometer (Avantes). For achieving high sensitivity the generated beam is directly focused onto a spectrometer input slit (25 μm). A neutral density filter ND adjusts the magnitude of measured signal. In the current configuration, this FROG setup allows measurement of sub-10 fs UV-pulses with an incoming energy of less than 30 nJ. This setup is very robust, doesn't need any "tweaking" for several months, and is insensitive to the pointing variations of incoming beam.

3 Results and Discussion

The sub-8 fs UV-pulses have been generated using achromatic frequency doubling (ASHG; the design is similar to the one published in [4]) of broadband VIS-pulses, generated by two-cascaded home-built non-collinear optical parametric amplifiers (NOPAs) pumped by a Coherent Elite USP laser system. The NOPA delivers light pulses with a spectral width of 490–600 nm and energy of 8 μJ. The generated UV-pulses, with the energy up to 750 nJ, were passed to the FROG setup for their characterization. For compression of UV-pulses, we used a deformable–mirror based compressor in a 4-f geometry, where for dispersion of the UV-light, a CaF_2 prism was used instead of a grating [5].

Figure 2 shows the FROG trace of uncompressed UV light pulses and the corresponding retrieved phase profile. Its shape contains basically only 3rd order

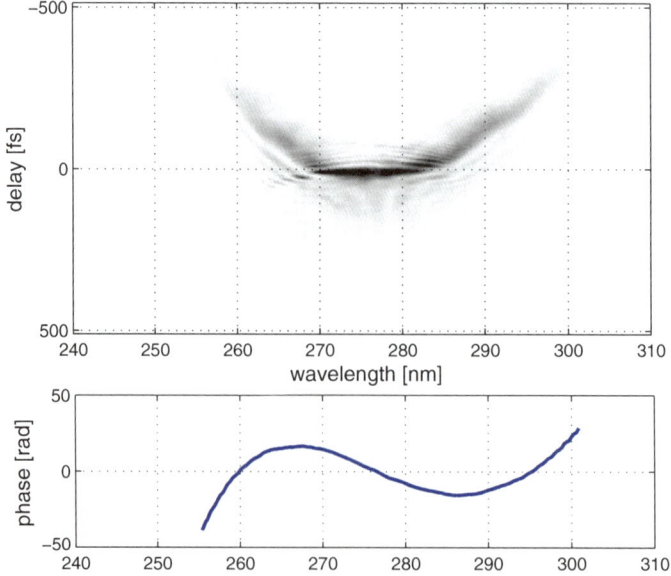

Fig. 2 FROG trace of uncompressed UV pulse and (*top*) and retrieved phase profile (*bottom*)

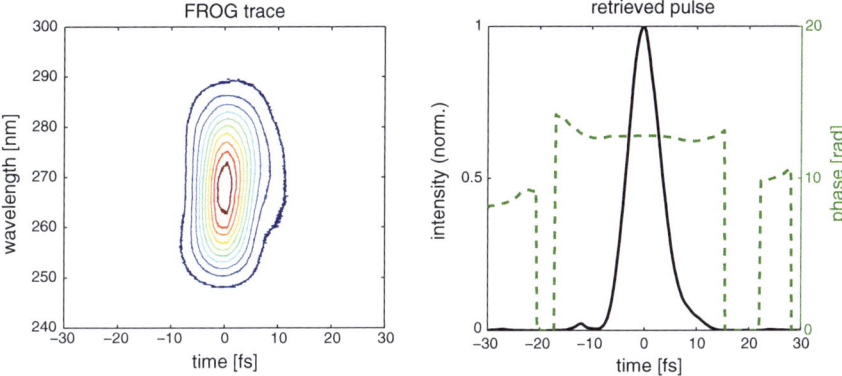

Fig. 3 FROG trace and retrieved temporal profile of compressed pulse (*solid*) together with it phase (*dashed*)

chirp since the second-order chirp is fully compensated by the prism chirp compressors, included into the architecture of the ASHG [4]. Figure 3 shows frequency-resolved FROG traces and the retrieved pulse with its phase after compression. As can be seen, the deformable-mirror based compressor reduces the FROG pulse duration from a couple of 100 fs to ~8 fs. For retrieving the pulse shape, we used a commercial program (Femtosoft Technologies). The measured UV-pulse with the spectral FWHM of 32 nm (center at 272 nm) has a duration of 7.3 fs, with an almost flat phase profile (Fig. 3 right), which is closed to a transform-limited pulse ($\Delta\nu\Delta\tau \sim 0.6$). Some distortions in the temporal profile (at the beginning of the pulse) are due to the residuals in phase that were not compensated by the deformable mirror used (Flexible Optical BV). To our knowledge, this is the shortest directly measured UV-pulse in the 250–300 nm spectral range.

It should be noted that replacing the beam blocker by a thin neutral density filter (position B/F) allows heterodyne measurement of FROG signals which significantly increases the sensitivity of the setup. Using a 200 µm thick filter with OD = 2, we were able to detect and measure FROG-signals with incoming pulse energies of ~3 nJ, i.e. an increase in sensitivity by one order of magnitude better than for homodyne-detected FROGs. However, currently there are no appropriate algorithms for retrieving the pulse shape and phase from heterodyne-measured FROG traces.

References

1. R. Trebino, K.W. DeLong, D.N. Fittinghoff, J.N. Sweester, M.A. Krumbugel, B.A. Richman, Rev. Sci. Instrum. **68**, 3277 (1997).
2. S. Linden, H. Giessen, J. Kuhl, Phys. Stat. Sol. **206**, 119 (1998).
3. P. Baum, E. Riedle, J. Opt. Soc. **B 22**, 1875 (2005).

4. P. Baum, S. Lochbrunner, E. Riedle, Opt. Lett. **29**, 1686 (2004).
5. E. Zeek, R. Bartels, M. M. Murnane, H. C. Kapteyn, S. Backus, G. Vdovin, Opt. Lett. **25,** 587 (2000).

Generation and Characterization of Tunable μJ-Level, Sub-10 fs UV Pulses

Rocìo Borrego-Varillas, Alessia Candeo, Sandro De Silvestri, Giulio Cerullo and Cristian Manzoni

Abstract We present a scheme for the generation of tunable ultrashort UV pulses based on sum-frequency generation between a broadband visible pulse and a narrowband tunable pulse. It allows for broadband UV pulses tunable from 0.3 to 0.4 μm, with energy up to 1.5 μJ. Full characterization of the UV pulse is obtained by 2D spectral shearing interferometry. We demonstrate 8.4 fs UV pulses.

Femtosecond light pulses in the ultraviolet (UV) spectral range ($\lambda < 400$ nm) are required to study fundamental chemical and biological processes in biomolecules. High time resolution and two-dimensional spectroscopy experiments [1], in particular, call for ultrabroadband, few-optical-cycle pulses. The generation of sub-10 fs UV pulses is however not straightforward, since this spectral region poses serious challenges in terms of: (i) broad pulse bandwidth; (ii) pulse energy; (iii) spectral phase handling; (iv) pulse characterization. Most approaches rely on nonlinear frequency conversion to shift ultrashort visible and infrared pulses to the UV range. However, frequency conversion of broadband pulses to short wavelengths needs to compromise between the competing requirements of phase-matching and high efficiency. In addition, linear dispersion by propagation in transparent media is particularly severe in the UV spectral range, and careful control of the spectral phase to achieve transform-limited (TL) pulse duration is difficult due to higher order dispersion introduced by pulse compressors. Finally, pulse characterization in the UV is complicated by the short-wavelength absorption of nonlinear materials and the unfavourable phase-matching conditions.

In this work, we introduce a simple scheme for the generation of ultra-broadband UV pulses which overcomes these challenges. Our system is based on collinear sum-frequency generation (SFG) between broadband visible pulses and narrowband pulses (NP) tunable from 0.6 to 1 μm; the phase-matching configuration used in our scheme balances the trade-off between bandwidth and efficiency. This process also allows us to efficiently manipulate the spectral phase of the UV light

R. Borrego-Varillas (✉) · A. Candeo · S. De Silvestri · G. Cerullo · C. Manzoni
IFN-CNR, Dipartimento Di Fisica, Politecnico Di Milano,
Piazza Leonardo Da Vinci, 32, 20133 Milan, Italy
e-mail: rocio.borrego@polimi.it

© Springer International Publishing Switzerland 2015
K. Yamanouchi et al. (eds.), *Ultrafast Phenomena XIX*,
Springer Proceedings in Physics 162, DOI 10.1007/978-3-319-13242-6_184

749

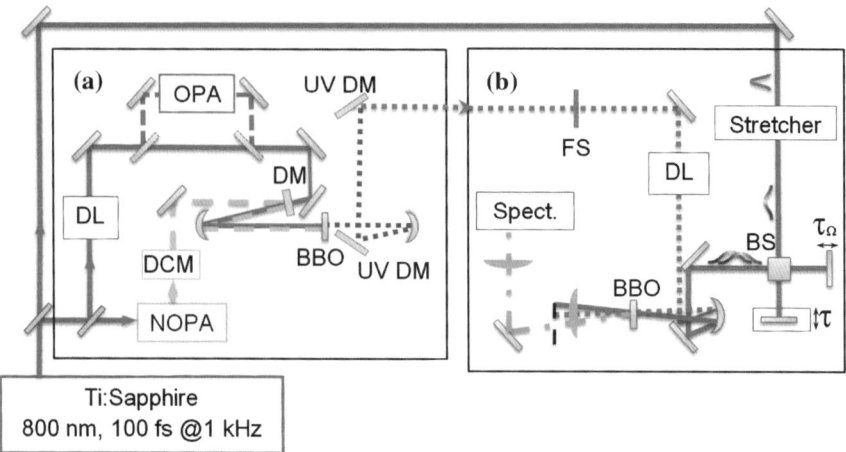

Fig. 1 a Experimental setup for the generation of the UV pulses. *DL* delay line; *DM* dichroic mirror; **b** 2DSI setup based on DFG between UV and stretched replicas of the FF. *BS* beamsplitter; *FS* fused silica plate

through parametric transfer [2]. The SFG bandwidth depends on the group velocity mismatch (GVM) between the broadband UV and visible pulses. The configuration with the smallest GVM is Type II ($e_{NP} + o_{vis} \rightarrow e_{UV}$), which allows for an efficient transfer of the broad visible bandwidth to the UV.

The experimental setup for ultra-broadband UV pulse generation is shown in Fig. 1a. The system starts with a regeneratively amplified Ti:Sapphire laser (Libra, Coherent), which delivers 100 fs, 1 kHz, 800 nm pulses. A fraction of the laser light drives a single-pass visible non-collinear optical parametric amplifier (NOPA), pumped by the second harmonic (SH) and seeded by a white-light continuum. This NOPA produces broadband pulses with spectrum extending from 525 to 740 nm, corresponding to TL pulse duration of 6.2 fs, and with energy higher than 10 μJ. The spectral phase of the NOPA pulses is manipulated by reflections onto a pair of dielectric Double Chirped Mirrors (DCMs).

The visible pulses are mixed with narrowband pulses, either derived from the fundamental (FF) beam or generated by a second narrowband OPA. SFG occurs in a 50 μm-thick BBO crystal cut for Type II interaction. This crystal thickness is a good trade-off, balancing efficiency and bandwidth of the SFG process and allowing to up-convert the whole NOPA spectrum. A sequence of SFG spectra, for different narrowband up-converting pulses, is shown in Fig. 2; the energy of the UV pulses is higher than 1 μJ. The spectral phase of the UV light is controlled by the mechanism of indirect phase transfer: the SFG process transfers the spectral phase of the NOPA pulse to the up-converted light. To compensate for the positive chirp acquired by propagation to the measurement point, we impart a slight negative chirp to the visible pulses by additional reflections on the DCMs, resulting in negatively chirped UV pulses. Fine tuning of the spectral phase of the UV is then achieved by adding a small amount of material dispersion.

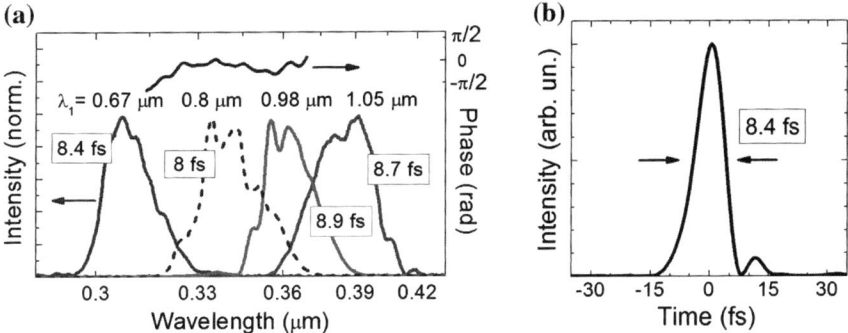

Fig. 2 **a** Tunability of the UV pulses from SFG between the 600 nm broadband NOPA pulse and a narrowband pulse with wavelength indicated on *top*. We also provide the spectral phase of the pulse with *dashed* spectrum, measured by 2DSI. **b** Retrieved time intensity of the pulse corresponding to the *dashed* spectrum of panel (**a**)

The measurement of the duration of the UV pulses is challenging since, due to the lack of transparency and of a suitable phase-matching configuration, no traditional techniques based on SH or SFG can be applied. We developed a system based on the 2D spectral shearing interferometry (2DSI) technique [3], which allows characterizing the spectral phase of ultrabroadband pulses. The 2DSI method relies on spectral shearing interferometry with zero delay. In traditional 2DSI, two highly chirped replicas of a broadband gate pulse undergo SFG with the test pulse in a nonlinear crystal, to generate two spectrally-sheared signals. The delay of one of the chirped auxiliary pulses is scanned over a few optical cycles and the spectrum of the up-converted signal is recorded as a function of the phase delay, yielding a 2D map which encodes the group delay (GD) of the test pulse. In our approach, the two replicas of the gate, provided by splitting a chirped portion of the FF light in a Michelson interferometer (Fig. 1b), are mixed with the UV pulse in a 10-μm thick Type-I BBO crystal, to generate two spectrally down-converted signals by difference frequency generation (DFG). One of the arms of the interferometer is equipped with a high precision translation stage, which scans the delay between the two gate pulses by −/+10 fs.

The GD is extracted from the 2DSI map by Fourier analysis along the scanning delay. By integrating the GD, we could estimate the spectral phase (Fig. 2a, top solid line) and the temporal profile of the corresponding pulse (see panel (b)). The resulting pulse has 8.4-fs width, very close to the 8-fs TL value [4].

In conclusion, we have demonstrated a simple and robust scheme for the generation of tunable UV pulses with energy in excess of 1 μJ and sub-10 fs duration. In addition we demonstrated a technique for the characterization of the UV spectral phase and anticipate the use of these pulses for high time resolution spectroscopy of biomolecules.

This work was supported by the ERC Advanced Grant STRATUS (ERC-2011-AdG No. 291198), the MIUR FIRB Grant No. RBFR12SW0J, and the European Commission (Marie Curie actions, FP7-PEOPLE-263 IEF-2012).

References

1. Jiang J. and Mukamel S. (2010) Angew. Chem. Int. Ed. 49:9666.
2. Tan H.-S., Schreiber E., and Warren W. S. (2002) Opt. Lett. 27:439.
3. Birge J. R., Ell R., and Kaertner F.X. (2006) Opt. Lett. 31:2063.
4. Borrego-Varillas R., Candeo A., Viola D., De Silvestri S., Garavelli M., Cerullo G. and Manzoni C. (2014) Opt. Lett. 39:3849.

Femtosecond Pulses in 375–405 nm Region by Chirped Sum Frequency

Prem B. Bisht and S. Akbar Ali

Abstract Tunable femtosecond pulses in UV region from 375–405 nm have been generated directly by chirped sum frequency mixing by simple experimental geometry at a single pump wavelength with nJ energy. To estimate the UV pulse characteristics, the fundamental pulse of the Ti: Sapphire oscillator has been characterized by using with a small modification to the chirped sum frequency setup.

1 Introduction

Ultrafast tunable pulses in near UV region have applications in time-resolved spectroscopy besides in several other areas viz., waveguide fabrication, bio-photonics, bio-medicine and in nanotechnology. Second harmonic generation of the fundamental wavelength of the Ti: Sapphire laser can generate pulses in near UV. Sum frequency generation process has also been used to generate pulses in UV region by changing the chirp of the input pulses [1]. However, the obtained wavelength range in these is limited by the tunability of the fundamental beam. Earlier, broadly tunable parametric line emission from β-barium borate on pumping with picosecond pulses was reported by us [2]. In this report, we present the generation of tunable fs pulses in near UV range by using the sum frequency of the chirped and the main pulse of Ti: Sapphire oscillator by using a simple experimental geometry. Further, with a small modification of this set up into the spectral phase interferrometry for direct electric filed reconstruction (SPIDER) technique [3], we have characterized the fundamental pulses to estimate the properties of the obtained UV pulses.

P.B. Bisht (✉) · S. Akbar Ali
Physics Department, Indian Institute of Technology Madras, Chennai 600036, India
e-mail: bisht@iitm.ac.in

© Springer International Publishing Switzerland 2015
K. Yamanouchi et al. (eds.), *Ultrafast Phenomena XIX*,
Springer Proceedings in Physics 162, DOI 10.1007/978-3-319-13242-6_185

2 Experimental

The schematic setup is shown in Fig. 1. A Ti: Sapphire oscillator (TISSA 100) with ~ 100 fs duration pulses (2 nJ, 775 nm, 82 MHz) was used in the experiments. The pulses from the oscillator were divided into two replica pulses. One part of the beam was used to generate chirped pulses by passing through the SF 57 glass slab of 60 mm length. The chirped pulses along with the other replica of the pulses were focused with two focusing mirrors of 200 mm focal length. The sum frequency was generated by overlapping spatially and temporally the two portions on a 2 mm β-barium borate (BBO) crystal (Castech) cut at an angle of 29°. The spectra of the generated sum frequency signals were measured by using the spectrometer (Spectra Pro-2150i) and air-cooled CCD (PIXIS: 100).

3 Results and Discussion

As shown in Fig. 1, for generation of the tunable pulses, it is important to have a frequency chirped portion of the pulse. A portion of the main pulse is passed through the delay line for temporal overlap with the chirped pulse on to the BBO crystal. The sum frequency between the main fundamental pulse and various frequency components of the chirped pulses contribute to the observed tunability of the near UV pulses. With a suitable phase matching angle, tunable pulses from 375–405 nm can be generated. Figure 2 shows the experimentally recorded spectra

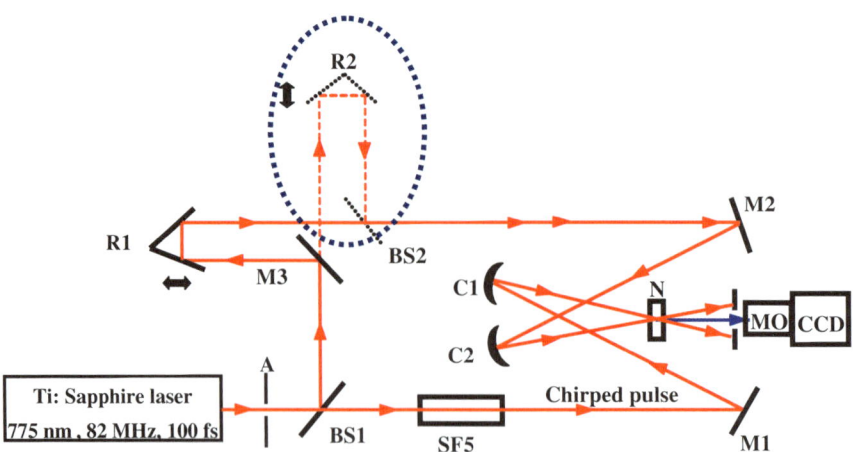

Fig. 1 Experimental set up for chirped sum-frequency generation scheme: *BS*1, *BS*2 Beam splitters, *A* Aperture, *SF*57 Glass block, *M*1, *M*2, *M*3 Plane mirrors, *R*1, *R*2 Retro reflectors, *C*1, *C*2 Concave mirrors, *N* Nonlinear crystal (BBO), *MO* Monochromator, *CCD* Charge-coupled device. The *encircled part* of the optical set up is used in for characterization of the pulse

of the generated pulses (Panel A). Panel B shows the simulations for the observed signal wavelengths versus the crystal phase matching angles [4]. It can be seen that the phase matching angle varies from 31.2° to 28.8° for the observed tuning range.

In order to characterize the pulses, an additional delay line was inserted in Fig. 1 as shown by the dotted path. The obtained UV pulses with the sum frequency of the dotted path create a Michelson-type arrangement generating another replica pulse with a fixed time delay with respect to the existing one. These two replica pulses along with the chirped pulse are overlapped temporally and spatially inside the BBO crystal to generate SPIDER interferogram (see Fig. 3). The distribution of the spectral phase can be obtained by applying Fourier transform to the Fig. 3 and is found to be flat throughout spectrum. The estimated pulse of the tunable pulses is about 550 fs.

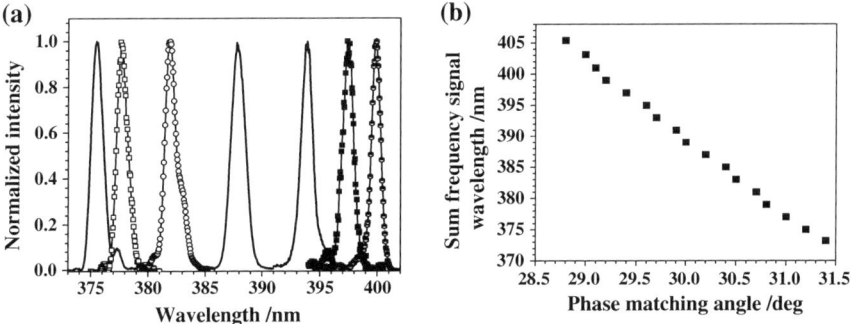

Fig. 2 Experimentally recorded sum frequency signals (*Panel A*). *Panel B* gives the plot obtained for sum frequency signal wavelength versus the phase matching angle

Fig. 3 Measured SPIDER interferogram by using the combination of monochromator and CCD

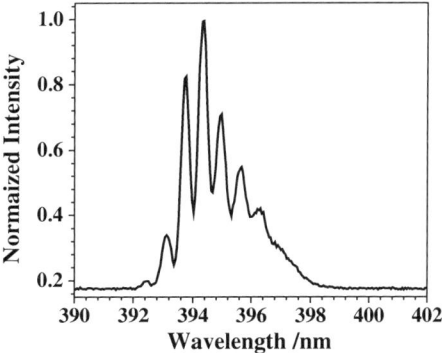

4 Conclusions

Generation of tunable fs pulses in near UV region has been demonstrated directly from nJ pulse energies of Ti: Sapphire oscillator at a fixed pump wavelength. The fs pulses can be tuned continuously with the steps of about 1 nm. The characterization of the output in UV has been done based on the estimates from the measurements of the fundamental pulse. We believe that this technique will be useful with higher energy pulses from the amplifiers.

References

1. Osvay K, Ross IN (1999) Efficient tunable bandwidth frequency mixing using chirped pulses. Opt Commun 166: 113-119.
2. Nautiyal A, Bisht PB (2008) Broadly tunable parametric line emission from β-barium borate on pumping with picosecond pulses. Opt Commun 281: 3351-3355.
3. Iaconis C, Walmsley IA (1998) Spectral phase interferometry for direct electric-field reconstruction of ultrashort optical pulses. Opt Lett 23: 792-794.
4. Smith AV (2000) How to select nonlinear crystals and model their performance using SNLO software in Nonlinear Materials, Devices and Applications, Vol. 3928, SPIE proceedings, pp. 62-69.

Pushing the NOPA to New Frontiers: Output to Below 400 nm, MHz Operation and ps Pump Duration

Maximilian Bradler, Lamia Kasmi, Peter Baum and Eberhard Riedle

Abstract Two sub-ps MHz range Yb-based lasers are used to pump NOPAs at 343 nm. A SHG driven supercontinuum allows tuning down to 395 nm. For a 1 ps pump, supercontinuum seeding is still applicable, the pulses are compressed to the 20 fs regime with a potential for sub-10 fs duration.

1 Optical Parametric Amplifiers Pumped by Yb-Based Lasers

For many years the generation of powerful ultrashort pulses was dominated by Ti: sapphire oscillators and amplifiers. Optical parametric amplifiers (OPA) were developed to provide full spectral tunability. The introduction of noncollinear phase matching (NOPA) and chirped parametric amplification (OPCPA) enables to generate some of the most intense, shortest and most widely tunable pulses available today. NOPAs were originally pumped by the SHG of kHz Ti:sapphire systems with a pulse duration around 100 fs. This allows efficient continuum generation to be used as seed light and provides a high intensity in the amplifier. Recently, the development of femtosecond lasers has turned to Yb-based active media, because higher average powers are achievable with diode laser pumping. The consequences are a central wavelength around 1,030 nm and pulse durations not significantly below one picosecond. Pumping a NOPA with the SHG at 515 nm [1] and BBO as active material resulted in a shortest output wavelength of 600 nm. With THG pumping [2] 387 nm should be reachable, but was not yet demonstrated due to the lack of suited seed light. We now show in two setups employing newly available Yb-based systems at up to 1 MHz as pump that the sub-400 nm range can be

M. Bradler · E. Riedle (✉)
LS Für BioMolekulare Optik, LMU München, Oettingenstr. 67, 80538 Munich, Germany
e-mail: Riedle@physik.uni-muenchen.de

L. Kasmi · P. Baum
Ludwig-Maximilians-Universität München, Am Coulombwall 1, 85748 Garching, Germany

© Springer International Publishing Switzerland 2015
K. Yamanouchi et al. (eds.), *Ultrafast Phenomena XIX*,
Springer Proceedings in Physics 162, DOI 10.1007/978-3-319-13242-6_186

reached if a SHG-pumped continuum is used as seed. We generate tunable blue pulses with Fourier limits in the 10 fs range. With a 1 ps, 300 kHz pump system we also succeed to generate a stable continuum and subsequently amplify major spectral parts into powerful pulses with the potential of sub-10 fs duration.

2 Fully Utilizing the Tuning Possibility of a 343 nm Pumped NOPA

In our first series of experiments we use a 20 W fiber-based system (Tangerine fs, Amplitude Systems) delivering 1,030 nm pulses with 20 μJ pulse energy and 300 fs pulse duration. The system is operated at 200 kHz for development and 1 MHz for full output. To generate 343 nm pump pulses (see Fig. 1a) we frequency-triple the major part of the pulses in a sequential arrangement of a type I doubling BBO and a type II BBO with intrinsic group velocity compensation [2] to obtain 3 μJ UV pulses.

For the seed light generation, part of the 1,030 nm pump is frequency doubled in a second 0.8 mm type I BBO crystal. The SHG pulses with an energy of 750 nJ are spectrally filtered with dielectric mirrors and focused onto a 4 mm thick YAG crystal [3]. The observed continuum (Fig. 1b) contains spectral components well below 400 nm and with a thicker sapphire crystal down to 350 nm. The high-wavelength cutoff is above 650 nm. Significant parts of this extremely wide spectrum are amplified in a NOPA employing a 3 mm type I BBO crystal cut at 32.5°. Typical output spectra are shown in Fig. 1b with Fourier limits between 10 and 20 fs for the entire range. The shortest central wavelength reached is 395 nm, limited by idler absorption at around 3 μm. When optimized for output power, the pulse energy exceeds 1 μJ, corresponding to a 40 % quantum efficiency in the nonlinear conversion of the pump light.

The pulses are compressed with fused silica prisms. A typical autocorrelation at 425 nm is shown in Fig. 1c, showing a sub-20 fs duration that is limited by higher-order chirp, but still highly competitive with alternative methods for generating

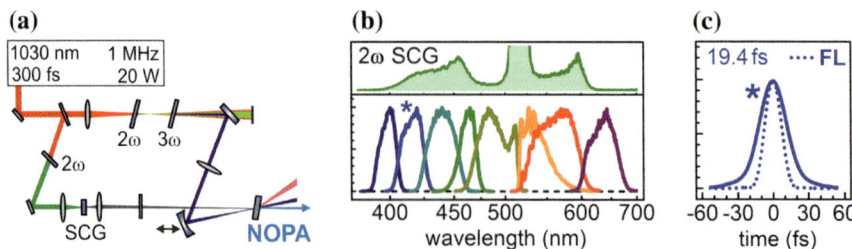

Fig. 1 a NOPA pumped by 343 nm pulses. **b** Supercontinuum (SCG) generated by pumping with 515 nm pulses. Spectra of pulses tunable from 395 to 630 nm with Fourier limits below and close to 10 fs. **c** Autocorrelation of 425 nm pulses

tunable blue pulses at high repetition rate. We also frequency double the NOPA output and generate pulses down to 210 nm in a single additional conversion stage utilizing a 300 μm BBO. The observed spectral width is only limited by the acceptance bandwidth of the BBO crystal, providing possible pulse durations of below 20 fs. Altogether, the system now provides fully tunable 10–20 fs pulses from the deep UV all the way up to 630 nm without any wavelength gaps. The additional range up to 950 nm can be reached with a NOPA pumped by the residual 515 nm light. Such a stage is already integrated in the actual amplifier layout.

3 Continuum-Seeded NOPA with 1 ps 130 W Pump Pulses

The second pump system is a diode-pumped regenerative Yb:YAG disk amplifier based on a previous design [4], but operating at 50–300 kHz. The output power is up to 130 W at a central wavelength of 1,032 nm and the pulses have 1 ps duration. The main challenge with these long pulses is the seed light generation. Here, we report on a stable supercontinuum generated with 7 μJ at 1,032 nm in a 4 mm YAG crystal. This is the basis for the subsequent amplification in the NOPA, without resorting to external seed sources such as a synchronized Ti:sapphire oscillator. The continuum extends from below 500 nm up to the pump wavelength (Fig. 2a). Using frequency-tripled 6.7 μJ pulses for NOPA pumping, we are able to generate tunable pulses in the green spectral range with up to 1 μJ output energy and Fourier limits close to 10 fs (Fig. 2c). At specific phase matching conditions, pulses with a Fourier limit of 7 fs can be obtained as a consequence of the long pump pulse length. With a compressor made of two SF10 prisms at less than 1 m of separation, the pulses were compressed to the 20 fs regime (Fig. 2b). This already represents a pulse shortening by a factor of 40.

In a first application, we produced narrow-band tunable pulses with a slit in the prism compressor and studied the performance of a femtosecond electron gun in dependence on the photoemission wavelength. We found that the two-photon photoemission regime with visible pulses reproduces earlier results with ultraviolet

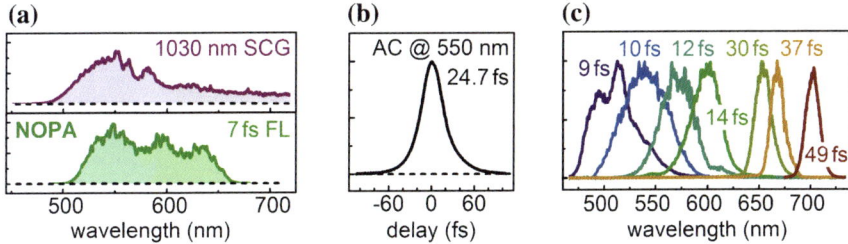

Fig. 2 a Continuum generated from 1 ps 1,030 nm pulses and broadband NOPA output. **b** Autocorrelation of 550 nm pulses. **c** Tuning range of the 344 nm-pumped NOPA with Fourier limits

pulses [5]. The duration of the electron pulses decreases and the coherence increases as the photon energy approaches half of the work function. This result proves the applicability of a powerful picosecond laser at 50–300 kHz, which is powerful enough for providing sufficiently strong tunable pump pulses for femtosecond electron diffraction and microscopy.

4 Towards New Applications of Wavelength-Tunable MHz 10 fs Pulses

The presented experimental results show that the presently evolving Yb-based ultrashort pulse sources operating at up to MHz repetition rate and with pulse durations close to 1 ps can be efficiently used to generate pulses tunable from 210 to 950 nm. Pulse durations below 20 fs are demonstrated and sub-10-fs capability is found. With both pump lasers, a supercontinuum generated in a YAG crystal is used as seed and avoids the need for elaborate temporal synchronization. Due to the high repetition rates, many new applications that need extensive scanning or a high degree of averaging together with a resonant excitation now become feasible.

References

1. C. Schriever, S. Lochbrunner, P. Krok, and E. Riedle, Opt. Lett. **33**, 192-194 (2008).
2. C. Homann, C. Schriever, P. Baum, and E. Riedle, Opt. Express **16**, 5746-5756 (2008).
3. M. Bradler, P. Baum, and E. Riedle, Appl. Phys. B **97**, 561-574 (2009).
4. T. Metzger, A. Schwarz, C. Y. Teisset, D. Sutter, A. Killi, R. Kienberger, and F. Krausz, Opt. Lett. **34**, 2123-2125 (2009).
5. P. Baum, Chem. Phys. **423**, 55–61 (2013).

2 MHz Tunable Non Collinear Optical Parametric Amplifiers with Pulse Durations Down to 6 fs

Julien Nillon, Olivier Crégut, Christian Bressler and Stefan Haacke

Abstract We present a 2 MHz non collinear optical parametric amplifier for high repetition rate time resolved X-ray or optical spectroscopy, with pulse durations down to 6.0 fs and energies in the 30–800 nJ range.

OCIS Codes (060.2320) Fiber optics amplifiers and oscillators · (190.4970) Parametric oscillators and amplifiers · (190.7110) Ultrafast nonlinear optics · (320.5520) Pulse compression

1 Introduction

Ultrafast spectroscopy requires femtosecond pulses with excitation wavelengths matching the absorption bands of the studied material system. Providing continuously tunable emission throughout the visible and infrared, Optical Parametric Amplification (OPA) has proven to be a very attractive solution to fill the gaps between discrete laser wavelengths, with tunable broadband amplification being possible in non-collinear OPA's [1]. With additional Second Harmonic Generation (SHG) or Difference Frequency Generation (DFG) of the amplified pulses, the tunability could be extended from the UV to the mid-IR [2, 3]. With the commercial availability of ultrafast fiber lasers delivering energies of several tens of microjoules, such devices can be designed to operate at unprecedented repetition rates of a few MHz [4, 5], enabling shorter acquisition times as well as better signal-to-noise ratio. This is of key importance for optical pump/X-ray probe time resolved studies performed at synchrotron facilities like Petra III, ESRF or PSI. Depending on the

J. Nillon (✉) · O. Crégut · S. Haacke
Institut de Physique et Chimie Des Matériaux de Strasbourg, Université de Strasbourg - CNRS, 67034 Strasbourg, France
e-mail: nillon@ipcms.unistra.fr

J. Nillon · C. Bressler
European XFEL, Albert Einstein Ring 19, 22761 Hamburg, Germany

© Springer International Publishing Switzerland 2015
K. Yamanouchi et al. (eds.), *Ultrafast Phenomena XIX*,
Springer Proceedings in Physics 162, DOI 10.1007/978-3-319-13242-6_187

chosen filling pattern, the repetition rate of synchrotrons lies in the MHz to GHz range, which means that most X-ray photons are lost when using conventional kHz OPA pumps.

We report on the development of a new non-collinear OPA (NOPA) pumped by a 2 MHz commercial fiber laser [6]. Our versatile device was designed to achieve the highest flexibility on every laser parameters, with repetition rates from single shot to 2 MHz, tuning from UV to NIR, narrowband or broadband amplification allowing for instance the generation of 6.0 fs pulses at a wavelength of 850 nm. While all these parameters can be adjusted within seconds or a couple of minutes, special attention was paid to beam stability, with shot-to-shot power fluctuations of 1.3 % RMS thanks to a compact and robust mechanical design.

2 Results

The Tangerine Fiber Chirped Pulse Amplifier (FCPA) from Amplitude Systèmes (20 W, 300 fs), offering a repetition rate tunable from 200 to 2,000 kHz, pumps our NOPA (Fig. 1). If needed, an external AOM is used to adjust the repetition rate from single shot to 2 MHz depending on the experiments. The device can be alternatively pumped by the second or third harmonic of the 1,030 nm fundamental (SH- or TH-NOPA). 20 % of the fundamental is used for white-light (WL) generation in a 4 mm thick YAG plate. The horizontally polarized pumps and vertically polarized WL are overlapped in two type I, AR-coated BBO crystals. One is cut at $\theta = 33°$ with a thickness of 2 mm (TH-NOPA), the other is 4 mm thick and cut at $\theta = 23, 5°$ (SH-NOPA). The pumps are focused inside the crystals to a spot size of 120 μm FWHM (TH-NOPA) and 150 μm FWHM (SH-NOPA), respectively.

The NOPAs can either run in narrowband operation if wavelength-selective excitation of the material system under study is required (Fig. 2), or in broadband amplification followed by convenient pulse compression, if the applications call for sub-10 fs pulses (Fig. 3). Switching between the two modes is easily achieved by inserting a suitable glass rod on the continuum path, for example SF10 (visible

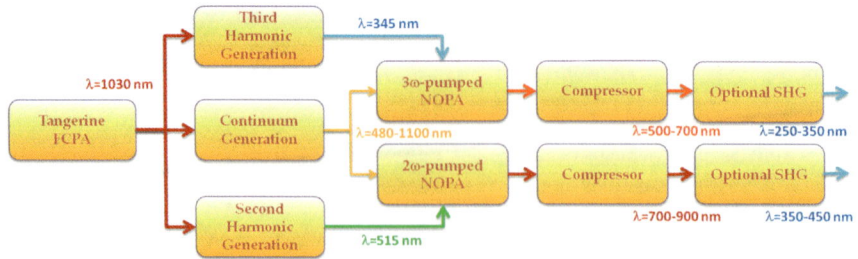

Fig. 1 2 MHz non-collinear optical parametric amplifier setup

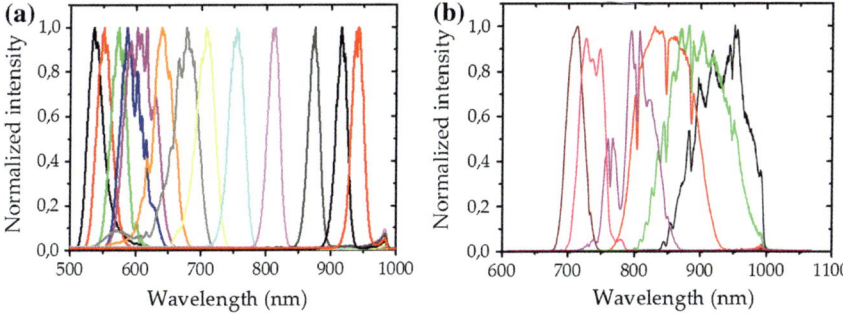

Fig. 2 Spectra of the amplified white-light pulses produced by the TH-NOPA (**a**) and by the SH-NOPA (**b**) in tunable narrow-band operation

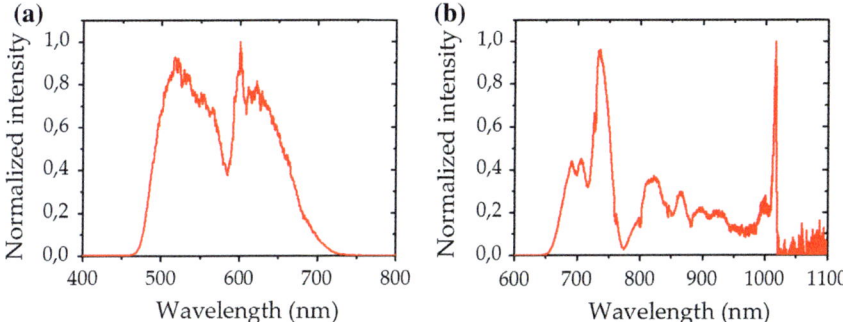

Fig. 3 Spectra of the amplified white-light pulses produced by the TH-NOPA (**a**) and by the SH-NOPA (**b**) in ultra-broadband mode, with a spectral width could support 4.2 fs FT pulses (TH-NOPA) or 5.4 fs FT pulses (SH-NOPA)

range) or ZnSe (IR range), and to compensate for the delay. Indeed, in magic angle configuration, the temporal overlap is the only parameter to adjust in order to tune the wavelength on a broad spectral range.

The output energy of the two NOPAs at 2 MHz operation is higher than 30 nJ from 500 to 900 nm for the TH-NOPA with a maximum of 170 nJ at 580 nm (6.8 % conversion efficiency). In contrast, the SH-NOPA provides energies higher than 100 nJ from 710 to 1,000 nm with a maximum of 750 nJ at 920 nm (18.7 % conversion efficiency). The efficiency of the TH-NOPA is rather low due to the lower intensity inside the crystal. Increasing the intensity eventually results in damage of the BBO crystal, the origin of which is presently studied. The output power of both NOPAs is remarkably stable. Long-term fluctuations are 0.5 % RMS over a recording period of 90 min with pulse-to-pulse fluctuations of 1.3 % RMS over 5 min for the SH-NOPA.

Using a standard fused silica prism compressor, we achieve sub-25 fs pulse width in narrowband operation. For the amplified broadband spectra a double

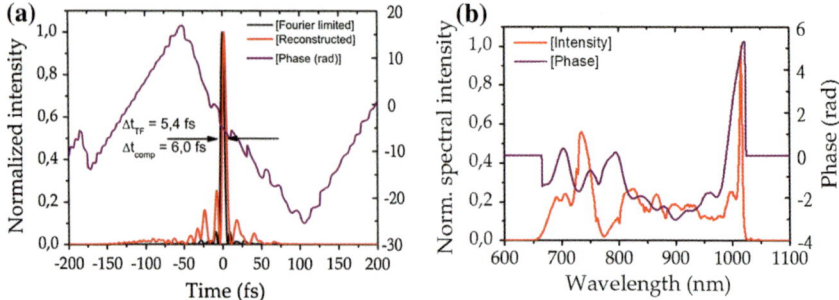

Fig. 4 Intensity and phase in the time (*left*) and spectral (*right*) domain of the ultrashort 6.0 fs pulses at 840 nm, as determined by the FC-SPIDER (APE GmbH, Berlin). The intensity dip at 770 nm is due to parasitic SHG

chirped mirror compressor is needed. We use commercial DCMs (Femtolasers) for the near-IR pulses. Only 40 bounces are needed for optimum compression, maintaining the throughput of the compressor to 73 %. Work with custom-made DCMs for compressing the broadband VIS pulses (Fig. 3a) is in progress. Fine tuning of the dispersion was done by inserting fused silica wedges on the beam path. We succeeded to compress the pulses down to 6.0 fs (2.1 optical cycles) with a TF duration of 5.4 fs (Fig. 4). This is to our knowledge the shortest pulse duration obtained from a compact white light seeded high repetition NOPA.

3 Conclusions

We have demonstrated the generation of tunable pulses between 520 and 1,000 nm with pulse durations in the range 14–30 fs and energies from 30 to 750 nJ at a high repetition rate of 2 MHz. With minor adjustments, our NOPA can also be operated in a broadband configuration able to generate 6.0 fs pulses at 850 nm as well as a spectrum supporting 4.2 fs FT pulses at 570 nm. Thanks to the 2 MHz repetition rate, ultrafast X-ray studies will now benefit from an improved data acquisition speed as well as a better signal-to-noise ratio. Time-resolved optical pump/X-Ray probe experiments are planned at ESRF Grenoble and Petra III in Hamburg.

References

1. E. Riedle, M. Beutter, S. Lochbrunner, J. Piel, S. Schenkl, S. Spörlein, W. Zinth, "Generation of 10 to 50 fs pulses tunable through all of the visible and the NIR", Appl. Phys. B 71, 457–465 (2000).
2. M. Beutler, M. Ghotbi, F. Noack, D. Brida, C. Manzoni, and G. Cerullo, "Generation of high-energy sub-20 fs pulses tunable in the 250-310 nm region by frequency doubling of a high-power noncollinear optical parametric amplifier," Optics Letters, vol. 34, no. 6, pp. 710–712, 2009.

3. C. Ventalon, J. M. Fraser, J.-P. Likforman, D. M. Villeneuve, P. B. Corkum, and M. Joffre, "Generation and complete characterization of intense mid-infrared ultrashort pulses," Journal of the Optical Society of America B, vol. 23, no. 2, pp. 332–340, 2006.
4. S. Hädrich, S. Demmler, J. Rothhardt, C. Jocher, J. Limpert, and A. Tünnermann, "High-repetition-rate sub-5-fs pulses with 12GW peak power from fiber-amplifier-pumped optical parametric chirped-pulse amplification," Optics Letters, vol. 36, no. 3, pp. 313–315, 2011.
5. C. Homann, C. Schriever, P. Baum, and E. Riedle, "Octave wide tunable uv-pumped NOPA : pulses down to 20 fs at 0.5 MHz repetition rate," Optics Express, vol. 16, no. 8, pp. 5746–5756, 2008.
6. J. Nillon, O. Crégut, C. Bressler, and S. Haacke, "2 MHz tunable non collinear optical parametric amplifiers with pulse durations down to 6 fs," Opt. Express, 22, 14964–14974, 2014.

Fiber-Slab-Pumped OPCPA for XUV-Based Time-Resolved Photoelectron Spectroscopy at 500 kHz Repetition Rate

Michele Puppin, Yunpei Deng, Oliver Prochnow, Jan Matyschok, Thomas Binhammer, Uwe Morgner, Martin Wolf and Ralph Ernstorfer

Abstract A passive optically-synchronized OPCPA based on a combination of fiber and slab pump lasers is presented. We demonstrate 30 µJ, sub-20 fs, 790 nm pulses at 500 kHz repetition rate, suitable for high harmonic generation.

1 Introduction

Time- and angle-resolved photoelectron spectroscopy (trARPES) is a powerful tool to probe photo-induced processes and excited state dynamics at solid surfaces in the femtosecond range. Up to now, there have been two complementary approaches distinguished by the employed probe laser pulses: UV sources with 100 s of kHz repetition rate provide sufficient counting statistics to allow for the investigation of transiently populated states [1], whereas high harmonic generation (HHG) based

M. Puppin (✉) · Y. Deng · M. Wolf · R. Ernstorfer
Fritz-Haber Institut der Max Planck Gesellschaft, Faradayweg 4-6, 14195 Berlin, Germany
e-mail: puppin@fhi-berlin.mpg.de

Y. Deng
e-mail: deng@fhi-berlin.mpg.de

M. Wolf
e-mail: wolf@fhi-berlin.mpg.de

R. Ernstorfer
e-mail: ernstorfer@fhi-berlin.mpg.de

O. Prochnow · J. Matyschok
VENTEON Laser Technologies GmbH, Hertzstr. 1B, 30827 Garbsen, Germany

T. Binhammer
VENTEON Laser Technologies GmbH, Hollerithallee 17, 30419 Hannover, Germany

J. Matyschok · U. Morgner
Institute of Quantum Optics, Leibniz Universität Hannover Welfengarten 1, 30167 Hannover, Germany

© Springer International Publishing Switzerland 2015
K. Yamanouchi et al. (eds.), *Ultrafast Phenomena XIX*,
Springer Proceedings in Physics 162, DOI 10.1007/978-3-319-13242-6_188

XUV sources provide spectroscopic access to the electronic structure in the full Brillouin zone but at limited repetition rate of up to 10 kHz [2]. We aim at bridging this technology gap by developing an efficient femtosecond XUV source with 500 kHz repetition rate.

The key component of this approach is an optical parametric chirped-pulse amplifier (OPCPA) providing tunable visible to the near-infrared ultra-short pulses. We report on hybrid fiber-slab Ytterbium-based laser system and a white-light seeded single stage OPCPA providing 30 µJ pulses with sub-20 fs duration suitable for HHG at 500 kHz repetition rate.

2 Fiber Oscillator and Hybrid Fiber-Slab Amplifier

A combination of fiber master oscillator and fiber-slab power amplifier system provides both pump and seed for the OPCPA (Fig. 1). A 25 MHz mode-locked Ytterbium fiber oscillator working in all normal dispersive regime [3] provides a spectral bandwidth of 10 nm full width half maximum (FWHM) at a central wavelength of 1,030 nm.

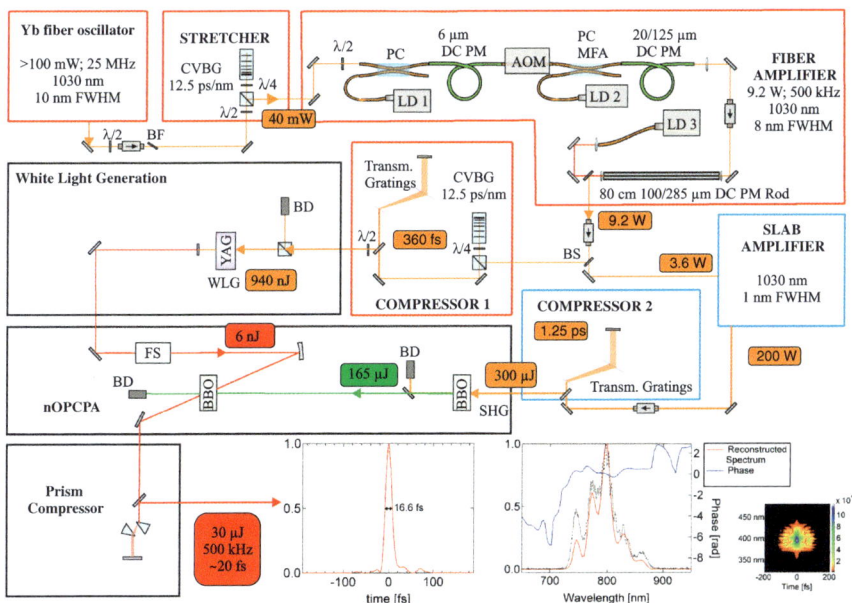

Fig. 1 Scheme of the OPCPA. List of abbreviations: *CVBG* Chirped Volume Bragg Grating; *LD* laser diode; *PC* pump combiner; *MFA* mode field adapter; *DC PM*: Double clad polarization maintaining; *SHG* second harmonic generation; *WLG* white light generation; *FS* fused silica; *FROG* frequency resolved optical gating. *Insets:* SHG-FROG results. *Bottom left* graph: retrieved temporal intensity showing a 16.6 fs full width at half maximum. *Bottom center* graph: Retrieved (*red line*) and measured (*black line*) spectral intensity; blue line: retrieved spectral phase. *Bottom right* graph: measured SHG FROG trace

The stretched output of the oscillator is coupled into a dual-stage fiber preamplifier including a fiber-coupled acousto-optical modulator for reducing the repetition rate to the range between 300 kHz and 1 MHz. These pulses are further amplified in an 80 cm long rod-type photonic crystal fiber (NKT, DC 285/100 PM-Yb-ROD) to more than 9 W average power. The system is typically operated at a repetition rate of 500 kHz. A portion of the output directly seeds a commercial Yb:YAG SLAB amplifier (Amphos 200). 3.6 W of seed power are sufficient to reach a saturated output power exceeding 200 W. The gain-narrowed spectrum centered at 1,030 nm output is compressed to ∼1.25 ps pulse duration by a transmission grating compressor with 75 % efficiency.

3 OPCPA Results and Discussion

The remaining portion of the fiber amplifier output is compressed to ∼360 fs pulse duration and used for the generating the OPCPA seed. 940 nJ of the compressed output of the fiber amplifier are used to generate a white light (WL) continuum in a 4 mm thick YAG crystal: the resulting 6 nJ pulses in the spectral range 610–940 nm (−10 dBc) seed the OPCPA.

An 82.5 W average power pump for the OPCPA is generated by frequency doubling the slab amplifier output in a 2 mm thick BBO crystal. The WL seed is stretched by additional 65 mm fused silica and its mode size is matched to the 975 μm diameter pump spot. The two beams are overlapped non-collinearly in a 4 mm thick BBO crystal (θ = 24.3°, internal angle 2.4°, Type I, walk-off compensated configuration); an amplification to more than 18 W average power is obtained, at a central frequency of 790 nm with a bandwidth of 134 nm (−10 dBc); the parasitic second harmonic generation for the signal is below 260 mW.

The pulses are de-chirped in a Brewster cut fused silica prism compressor and characterized by SHG FROG: a 16.6 fs FWHM temporal envelope is reconstructed and we obtain a compressed power of 15 W. An upper boundary for amplified parametric superfluorescence of 60 mW is measured after the compressor by blocking the seed beam.

In our WLG based approach the pump and seed arms are split after the fiber amplifier: this is comparatively a shorter path difference than direct seeding of the OPCPA with the output of a broadband oscillator seeding a similar fiber/slab amplifier chain. The pump pulses amplified in the slab follow an optical path length of less than 10 m before interacting with the white light pulses in the OPCPA crystal: the drift of the central wavelength is below 0.5 % over 1.5 h without any active phase and path length stabilization.

This novel light source is well suited for HHG in noble gases [3]. The OPCPA concept provides tunability in both central wavelength and spectral width of the output. The energy resolution in trARPES is dictated by the spectral width of the XUV probe pulses, which depends on the bandwidth of the driver light. Our approach will allow for optimization of the experimental energy and time resolution by bandwidth tuning and represents a very attractive light source for HHG-based spectroscopies at high repetition rates.

References

1. Schmitt, F et al.,"Transient electronic structure and melting of a charge density wave in TbTe3". Science 321(5896), 1649–52 (2008).
2. Rohwer, T. et al., "Collapse of long-range charge order tracked by time-resolved photoemission at high momenta", Nature, 471(7339), 490–493 (2011).
3. Chong, A. et al., "Properties of normal-dispersion femtosecond fiber lasers," J. Opt. Soc. Am. B 25, 140-148 (2008).
4. Krebs, M. et al., "Towards isolated attosecond pulses at megahertz repetition rates", *Nature Photonics*, 7(7), 555–559 (2013).

Sub-100 fs Mid-Infrared Pulses as Driver for a Table-Top Hard X-Ray Source

Jannick Weisshaupt, Vincent Juvé, Shian Ku, Marcel Holtz,
Michael Woerner, Thomas Elsaesser, Skirmantas Ališauskas,
Audrius Pugžlys and Andrius Baltuška

Abstract Midinfrared powerful 90 fs pulses at a wavelength of $\lambda = 3.9$ μm drive a femtosecond hard x-ray source (Cu Kα: $\hbar\omega = 8.05$ keV). Up to 10^8 X-ray photons/pulse are generated which is twice as many as with 800 nm drivers of a 100 times higher peak intensity.

1 Introduction

Ultrashort hard x-ray pulses are sensitive probes of structural dynamics on the picometer length and femtosecond time scales of electronic and atomic motions [1, 2]. Recent progress in generating such pulses has initiated new directions of condensed matter research [3, 4]. When applying laser-driven x-ray sources in such studies, the signal-to-noise ratio is eventually determined by the shot noise in detecting x-ray photons. Thus, new concepts for generating a higher hard x-ray flux are requested and of high current interest. Here, we present the first table-top femtosecond hard x-ray source driven by intense mid-infrared pulses ($\lambda = 3.9$ μm).

Table-top femtosecond x-ray sources are based on the interaction between an intense optical field and a metallic target. The key steps for producing hard x-ray photons are depicted by Fig. 1a: (1) field-induced extraction of electrons from the metal target, (2) electron acceleration in vacuum by the very strong laser field and (3) electron re-entrance into the target, which leads to x-ray generation via collisional inner-shell ionization followed by a radiative transition of an outer-shell

J. Weisshaupt · V. Juvé (✉) · S. Ku · M. Holtz · M. Woerner · T. Elsaesser
Max-Born-Institut Für Nichtlineare Optik Und Kurzzeitspektroskopie,
12489 Berlin, Germany
e-mail: juve@mbi-berlin.de

S. Ališauskas · A. Pugžlys · A. Baltuška
Photonics Institute, Vienna University of Technology,
Gusshausstrasse 27-387, 1040 Vienna, Austria
e-mail: baltuska@tuwien.ac.at

© Springer International Publishing Switzerland 2015
K. Yamanouchi et al. (eds.), *Ultrafast Phenomena XIX*,
Springer Proceedings in Physics 162, DOI 10.1007/978-3-319-13242-6_189

electron into the unoccupied inner shell. For laser intensities above a few 10^{15} W/cm^2, the tunneling barrier is nearly suppressed and electrons close to the Fermi level are extracted with a probability close to one [5]. The two key parameters influencing the generated X-ray flux are the wavelength λ and intensity I of the optical driver. The maximum kinetic energy gained by the electrons during the acceleration process is proportional to $I\lambda^2$. One expects that the electron acceleration efficiency is much better at longer wavelengths since the optical period is longer.

2 Results

In our experiments we use powerful mid-infrared pulses generated in an Optical Parametric Chirped Pulse Amplification (OPCPA) scheme [6] which delivers up to $W = 6$ mJ per pulse (pulse duration $\Delta t = 90$ fs) at a wavelength of $\lambda = 3.9$ μm and a repetition rate of 20 Hz. The p-polarized light was steered into a vacuum chamber and focused down to a diameter of $d_{FWHM} = 21$ μm onto the copper target (thickness of 20 μm) under an angle of incidence of $\theta = 41°$. Of the entire x-ray flux emitted in a solid angle of 4π, a small part corresponding to 2.4×10^{-8} sr, is detected in a transmission geometry using a calibrated energy-resolving CdTe detector. The typical spectrum ($W = 5.2$ mJ, Fig. 1b) consists of the characteristic line emission of Cu $\hbar\omega_{K\alpha} \simeq 8.0$ keV and $\hbar\omega_{K\beta} \simeq 8.9$ keV and the Bremsstrahlung radiation which extends up to 70 keV.

In Fig. 1c, we plot the Kα photons per shot emitted into the full solid angle 4π (red circles: $\lambda = 3.9$ μm) as a function of the driver intensity. For comparison we show results measured with a $\lambda = 800$ nm driver (blue circles) [7, 8]. The $\lambda = 3.9$ μ m data show a steep increase of the Kα flux of about 3 orders of magnitude (maximum: 8×10^7 Kα photons per shot) when increasing the intensity on the target from $I = 10^{15}$ to 10^{16} W/cm^2. For the $\lambda = 800$ nm driver we observe a maximum of 4×10^7 Kα photons per shot and a similar increase for the intensity range from $I = 10^{17}$ to 10^{18} W/cm^2, i.e., for 100 times higher intensities. The experimental data (symbols in Fig. 1c) are in excellent quantitative agreement with theoretical calculations based on the three steps model described above (solid lines in Fig. 1c). Our model predicts a further increase of Kα-flux with intensity for the $\lambda = 3.9$ μm driver whereas the $\lambda = 800$ nm drivers are close to their saturation limits. Thus, the experiments presented here demonstrate the expected $I\lambda^2$ scaling for laser-driven hard X-ray sources paving the way for future table-top systems with up to 10^{13} Kα photons per second using mid-infrared drivers in the kHz regime.

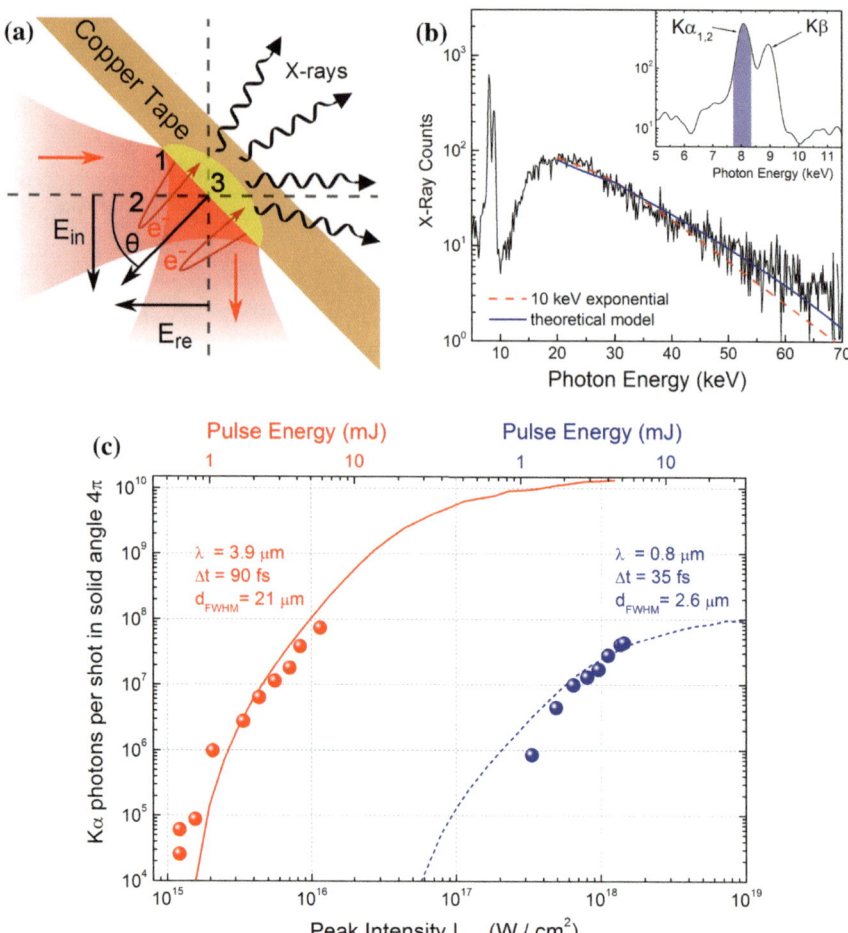

Fig. 1 **a** Schematic of the laser-target interaction geometry and the x-ray generation process. **b** Measured X-ray spectrum using the $\lambda = 3.9\,\mu m$ driver. The two *peaks* are the Cu characteristic *lines* $K\alpha_{1,2} \simeq 8.0\,keV$ and $K\beta = 8.9\,keV$. The *dashed line* is an exponential fit to the Bremsstrahlung with a photon temperature of 10 keV. *Inset* Characteristic X-ray emission *lines*. The *blue-shaded* area denotes the energy interval over which the $K\alpha$ flux is integrated in the intensity dependent data. **c** Comparison of experiment (*symbols*) with theory (*solid* and *dashed lines*) for a 20 μm thick Cu band illuminated under $\theta = 41°$ angle of incidence. *Plotted* is the number of $K\alpha$ photons per shot emitted in the solid angle 4π as a function of the peak intensity for the mid-infrared (*red symbols*) and 800 nm pulses [7, 8] (*blue symbols*). *Solid lines* in (**b**) and (**c**) are results of model calculations

References

1. C. Bostedt et al., *Ultra-fast and ultra-intense x-ray sciences: first results from the Linac Coherent Light Source free-electron laser*, Journal of Physics B 46, 164003 (2013).
2. T. Elsaesser, M. Woerner *Perspective: Structural dynamics in condensed matter mapped by femtosecond x-ray diffraction*, Journal of Chemical Physics **140**, 020901 (2014)
3. B. Freyer, F. Zamponi, V. Juvé, J. Stingl, M. Woerner,T. Elsaesser and M. Chergui, *Ultrafast inter-ionic charge transfer of transition-metal complexes mapped by femtosecond X-ray powder diffraction*, The Journal of Chemical Physics **138**, 14 (2013)
4. V. Juvé, M. Holtz, F. Zamponi, M. Woerner, T. Elsaesser, and A. Borgschulte, *Field-driven dynamics of correlated electrons in LiH and NaBH$_4$ revealed by femtosecond x-ray diffraction*, Phys. Rev. Lett. **111**, 217401 (2013)
5. S. V. Yalunin, M. Gulde, C. Ropers, *Strong-field photoemission from surfaces: Theoretical approaches*, Phys. Rev. B **84**, 195426 (2011)
6. G. Andriukaitis, T. Balčiūnas, S. Ališauskas, A. Pugžlys, A. Baltuška, T. Popmintchev, M. Chen, M. Murnane and H. Kapteyn, *90 GW peak power few-cycle mid-infrared pulses from an optical parametric amplifier*, Opt. Lett. **36**, 2755–2757 (2011)
7. N. Zhavoronkov, Y. Gritsai, M. Bargheer, M. Woerner, T., Elsaesser, F. Zamponi, I. Uschmann and E. Förster, *Microfocus CuK$_\alpha$ source for femtosecond x-ray science*, Opt. Lett. **30**, 1737–1739 (2005)
8. F. Zamponi, Z. Ansari, C.v.K. Schmising, P. Rothhardt, N. Zhavoronkov, M. Woerner, T. Elsaesser, M. Bargheer, T. Trobitzsch-Ryll and M. Haschke, *Femtosecond hard x-ray plasma sources with a kilohertz repetition rate*, Appl. Phys. A **96**, 51–58 (2009)

Generation of Stationary On-Axis Optical Filaments by Means of Dammann Lenses

J. Pérez-Vizcaíno, O. Mendoza-Yero, R. Borrego-Varillas,
G. Mínguez-Vega, J.R. Vázquez de Aldana and J. Láncis

Abstract We demonstrate the utilization of Dammann lenses encoded onto a spatial light modulator (SLM) for triggering non-linear effects. For continuous illumination Dammann lenses generate a multifocal pattern characterized by a set of N foci diffraction orders, all with the same intensity. We theoretically show that for pulses shorter than 100 fs the effects of chromatic aberrations influence the uniformity of the generated pattern. Multifocal second harmonic generation (SHG) and on-axis multiple filamentation are produced and actively controlled in β-BaB$_2$O$_4$ (BBO) and fused silica samples, respectively, with an amplified Ti:Sapphire femtosecond laser (30 fs at FWHM). Our proposal allows us to dynamically control both the quantity of foci and the distance among them. The output diffraction pattern is in good agreement with theoretical calculations. The measured spectra at the rear face of the supercontinuum sample for different separation among foci are also provided. The potential of this technique is very promising in different fields of non-linear optics or in applications of in-depth materials microprocessing.

1 Introduction

The high peak power achieved in amplified femtosecond pulses [1] has motivated the development of applications benefited from the non-linear processes [2] that take place in their interaction with matter. It is of great interest the effects that such non-linear interactions have in the laser pulse itself, provided that they allow the modification of certain properties like the spectral content. In general, second-order parametric processes [3] are efficiently used to get tunability of the central wavelength

J. Pérez-Vizcaíno (✉) · O. Mendoza-Yero · R. Borrego-Varillas ·
G. Mínguez-Vega · J.R.V. de Aldana · J. Láncis
Instituto de Nuevas Tecnologías de la Imagen (INIT), Universidad Jaume I,
Castellón E 12080, Spain
e-mail: jvizcain@uji.es

R. Borrego-Varillas
Departamento de Física Aplicada, Universidad de Salamanca, Salamanca 37008, Spain

through SHG or sum/difference-frequency generation: they all expand the accessible wavelength range of standard Ti:Sapphire laser systems. Third-order processes are, for instance, very often exploited to increase the spectral width of femtosecond pulses through optical-Kerr effect (self-phase modulation), or in combination with strong-field ionization and higher order processes (supercontinuum generation [4] and filamentation [5]), thus allowing, for instance, the subsequent compression of the pulse to a much shorter temporal duration [6].

On the other hand, Dammann lenses are binary phase distributions of alternative $0, \pi$ zones for well defined transient points [7]. The resultant binary lens function will create a series of N focal planes having focal lengths of f/n with equal intensities, where $n = (..., -3, -1, 1, 3,...)$, f or an even number of orders, and $n = (..., -2, -1, 0, 2,...)$, for an odd number of orders [8].

In this contribution, we introduce the use of Dammann lenses implemented in SLM to trigger non-linear effects. Specifically, we demonstrate the application of Dammann lenses to SHG and filamentation in transparent dielectrics [9].

2 Experimental Set-up

In Fig. 1 a schematic diagram of the optical set-up is shown. For the experiment we use a Ti:Sapphire femtosecond laser that emits linearly polarized pulses of about 30 fs at FWHM, centered at $\lambda_0 = 800$ nm with 50 nm at FWHM of spectral bandwidth and a repetition rate of 1 kHz.

The reflected beam from the SLM passes through lenses L_1 and L_2, with focal lengths $f_1 = f_2 = 100$ mm, that form a 1x telescope where the zero order is filtered and the first order is selected. The third lens (L_3), with focal length F, must be placed at a distance F from the SLM screen image plane (SIP) formed by the telescope. This is necessary to submit the requirements of equal spacing between all diffracted orders. Effectively, note that for negative values of n, we obtain diverging beams that create virtual spots, as well as distances between the spots are not equal.

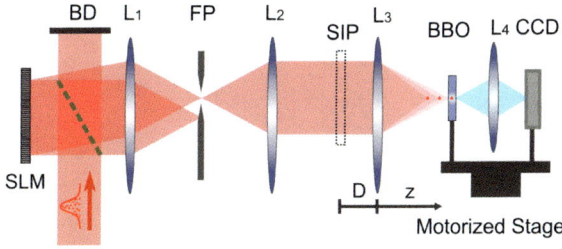

Fig. 1 Experimental set-up. The input beam is split in two rays by a beam splitter (*green dotted line*) to work with the SLM at normal incidence. A beam dumper (*BD*) is used to block the unused part of the beam

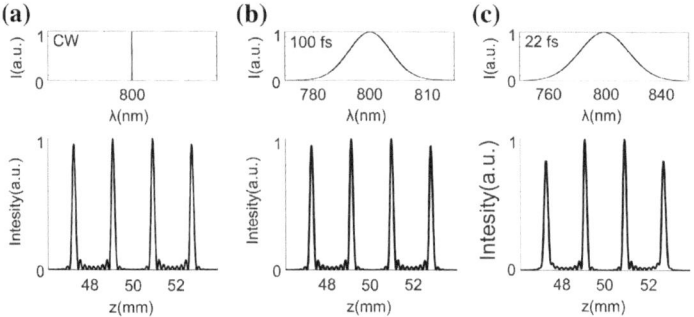

Fig. 2 Simulated irrandiance pattern obtained with a Dammann lens which is illuminated with **a** monochromatic beam, **b** 100 fs laser pulse, and **c** 22 fs laser pulses

This configuration allows us to obtain a number N of all real (non virtual) equally spaced focused spots at focal planes located at a distance $z = f_n$ from L_3, defined as

$$1/f_n = 1/F - n/(nD - f).$$ (1)

by simply applying the Gauss equation for image formation.

At this point we analyze the influence of broadband light spectra on the irradiance given by Dammann lenses under ultrashort pulsed illumination. In the bottom part of Fig. 2 the Fresnel diffraction integral is used to simulate the irradiance $I(z)$. Here, the on-axis irradiance obtained with a Dammann lens of 4 equal foci separated by 1.92 mm is shown. As it can be seen in the top part of Fig. 2, the simulation was carried out for two different Fourier transform limited Gaussian laser pulses, and an ideal CW source.

The theoretic irradiance uniformity in the case of the CW source is approximately 10^{-5}. However, small discrepancies in the height of the peaks are caused mainly due to deviations of the 0-π transition locations from their ideal values when encoding Dammann lenses into a SLM [10, 11]. For an ultrashort laser pulse of 100 fs falls of 0.2 and 1.4 % (for n = ±1 and n = ±3 orders, respectively) are registered. The equivalent values for 22 fs pulses were 5 and 16 %. Therefore, chromatic aberrations will increase the spatial width and decrease the peak irradiance of the foci, reducing the uniformity of the generated irradiance patterns.

3 Experimental Results

In Fig. 3 our experimental results are shown. A fused silica sample (20 × 10 × 5 mm³, all faces polished) was placed along the multifocal axial distribution. The number and separation of the focal spots can be controlled through the parameters of the Dammann lenses. The plasma emission of the filaments was registered in a

(a)   **(b)**   **(c)**

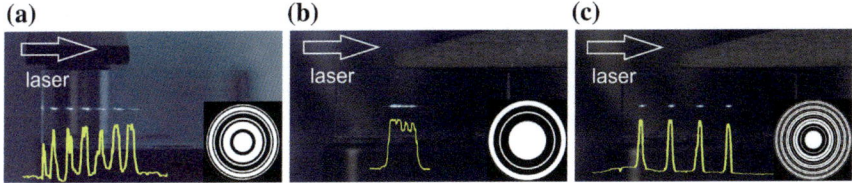

Fig. 3 Experimental on-axis plasma filaments generated in a fused silica glass by means of programmable Dammann lenses. The laser beam propagates from *left* to *right*

CCD from one of the sides of the sample. In Fig. 2a six foci having a distance of 0.85 mm among them, are generated. Figure 3b, c correspond to four equal foci with spatial period of 0.4 and 2.1 mm, respectively. At the bottom of Fig. 1 both the binary phase masks and the central irradiance profile of the filaments are included as insets. As it can be seen in Fig. 3, the shape and peak intensity of the filaments slowly change with the sample depth. This might be expected from the spherical aberration induced by the refraction at the air-glass interface.

4 Conclusions Results

In summary, by using a phase-only SLM programmable, on-axis multiple MFs have been generated in a fused silica glass with femtosecond laser pulses. By changing the Dammann lenses parameters we demonstrated a complete control over the position, width and peak intensity of the MFs. Several applications are expected to benefit from these results, such as in-depth parallel processing of transparent dielectrics or the creation of long filaments by the concatenation of shorter ones.

References

1. D. Strickland and G. Mourou, "Compression of amplified chirped optical puses" Opt. Comm. 56, 219 (1985).
2. R. W. Boyd, *Nonlinear Optics* (Academic, 2008).
3. P. Franken, A. Hill, C. Peters, and G. Weinreich, "Generation of optical harmonics," Phys. Rev. Lett. 7, 118 (1961).
4. R. Alfano, *The supercontinuum laser source* (Springer 2006).
5. A. Couairon, and A. Mysyrowicz, "Filamentation of ultrashort laser pulses in transparent media", Phys. Rep. 441, 47 (2007).
6. Nisoli, S. De Silvestri, and O. Svelto, "Generation of high energy 10 fs pulses by a new pulse compression technique" Appl. Phys. Lett. 68, 2793 (1996).
7. Moreno, J. A. Davis, D. M. Cottrell, N. Zhang and X.-C. Yuan, "Encoding generalized phase functions on Dammann gratings" Opt. Lett. **35**, 1536 (2010).

8. Davis, I. Moreno, J. L. Martínez, T. J. Hernández, and D. M. Cottrell, "Creating three-dimensional lattice patterns using programmable Dammann gratings", App. Opt. 50, 3653 (2011).
9. J. Pérez-Vizcaíno, O. Mendoza-Yero, R. Borrego-Varillas, G. Mínguez-Vega, J. R. Vázquez de Aldana, and J. Lancis, "On-axis non-linear effects with programmable Dammann lenses under femtosecond illumination," Optics Letters 38, 1621-1623 (2013).
10. J. Yu, C. Zhou, W. Jia, W. Cao, S. Wang, J. Ma and H. Cao, "Three-dimensional Dammann array", App. Opt. **51**, 1619 (2012).
11. D. C. O´Shea, "Reduction of the zero-order intensity in binary Dammann gratings" App. Opt. **34**, 6533 (1995).

Wavefront Analysis of High-Efficiency, Large-Scale, Thin Transmission Gratings

Chun Zhou, Takashi Seki, Tsuyoshi Kitamura, Yoshiyuki Kuramoto, Takashi Sukegawa, Nobuhisa Ishii, Teruto Kanai, Jiro Itatani, Yohei Kobayashi and Shuntaro Watanabe

Abstract Large-scale transmission gratings with groove densities of 1,250 and 1,740 lines/mm have been developed with diffraction efficiencies above 95 %. The minimized bending of the grating results in a negligible wavefront distortion of a pulse compressor.

1 Introduction

High-peak-power, ultrashort-pulse lasers with a high repetition rate are required in the generation of high-order harmonics and attosecond pulses [1]. Large-scale, high efficiency transmission gratings (TGs) are attractive for a compressor of terawatt (TW)-class Ti:sapphire lasers because of high damage threshold and long lifetime.

Here, we developed new TGs with a size of $180 \times 60 \times 1$ mm^3 and groove densities of 1,740 and 1,250 lines/mm by using optical lithography [2, 3]. Both gratings showed diffraction efficiencies above 95 % and a compressor throughput up to 80 %. Thin grating reduces the threshold of white-light continuum generation due to self-phase modulation, but is easily bended by anti-reflection (AR) coating. We characterized the bending of a grating due to the AR coating by measuring the

C. Zhou (✉) · S. Watanabe
Research Institute for Science and Technology, Tokyo University of Science,
Noda, Chiba 278-8510, Japan
e-mail: czhou@rs.noda.tus.ac.jp

T. Seki · T. Kitamura · Y. Kuramoto · T. Sukegawa
Corporate R&D Headquarters, CANON Inc., Utsunomiya 321-3292, Japan

N. Ishii · T. Kanai · J. Itatani · Y. Kobayashi
Institute for Solid State Physics, University of Tokyo, Kashiwa, Chiba 277-8581, Japan

C. Zhou · S. Watanabe
CREST, Japan Science and Technology Agency (JST), Chiyoda, Tokyo 102-0075, Japan

© Springer International Publishing Switzerland 2015   779
K. Yamanouchi et al. (eds.), *Ultrafast Phenomena XIX*,
Springer Proceedings in Physics 162, DOI 10.1007/978-3-319-13242-6_191

wavefronts of the reflected beams from the substrate surface. By improving the evaporation process, the bending due to AR coating was minimized to 2.9 λ at 633 nm. The wavefront distortion of a pulse compressor is analyzed numerically.

2 Design and Analysis of Transmission Gratings

We used a conventional projection system for semiconductor lithography to fabricate the TGs. The original pattern was exposed to a resist on a 1 mm-thick fused quartz substrate with a diameter of 200 mm by a 4× reduction lens system, and the size of one grating unit was 10 mm × 10 mm. The 180 mm × 60 mm grating is made by continuously connecting the 18 × 6 matrices of one unit to form a single grating with an accuracy of 5 nm. The grooves are notched by etching the pattern on a fused quartz substrate. The back side of the grooves is AR coated.

Figure 1a [3] shows the calculated contour map of the efficiency at an incident angle of 44° with a 1,740 lines/mm grating. The wavelength dependences of the diffraction efficiencies are shown in Fig. 1b [3] for 1,740 and 1,250 lines/mm at incident angles of 44° and 30°, respectively. The maximum efficiency was over 95 % at 800 nm. In the case of the 1,250 lines/mm grating, efficiency reaches up to 97 % at 800 nm and a compressor throughput is above 80 % for 20 fs pulses.

3 Spatial Distortion and Wavefront Analysis

The grating is held at the four corners by pressing to polished flat planes (1 × 1 mm²) by elastic plates. To determine the deformation of the TGs, we measured the wavefront distortion of the beam reflected from a TG by a Zygo

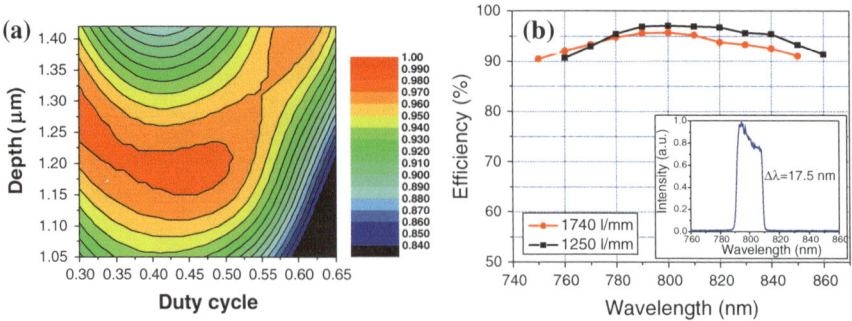

Fig. 1 a Calculated contour map of the efficiencies at 800 nm for a 1,740 lines/mm grating. **b** Dependences of the diffraction efficiencies on the wavelength for the TE polarization in 1,740 (*red*) and 1,250 (*black*) lines/mm gratings. The *inset* shows a typical spectrum of the probe [3], [Fig. 1a, b was reprinted with permission from Optical Society of America (OSA)]

interferometer equipped with a He–Ne laser (633 nm). The two-dimensional (2-D) wavefront distortion $W(x, y)$ is related to the deformation of the grating $H(x, y)$ by $W(x, y) = 2H(x, y)$, where x and y are the axes across and along the groove direction, respectively. For simplicity, we define $W(x)$ as the average of $W(x, y)$ from $y = 27.5\text{–}32.5$ mm and $H(x)$ represents the one-dimensional (1-D) distortion.

We observed a clear parabolic bending of the grating with AR coating by the ion-assisted deposition method. The bending direction was from the groove to AR surface (positive). The peak to valley (PV) values of the bending over the total area are distributed from 44 to 52 λ at 633 nm among 5 samples with 1,250 and 1,740 lines/mm groove densities. The bending was reduced by one order of magnitude by adding a negative bending with a SiO_2 under coat and by AR coating with e-beam deposition. Figure 2a [3] shows a wavefront profile in the central region $W(x)$. The bending is considerably reduced to PV $= 2.9$ λ in $H(x)$ [PV $= 5.8$ λ in $W(x)$].

By a simply analysis, $W(x) = 2(\cos\beta_0 - \cos\alpha_0)H(x)$ for a double-passed TG, while $W(x) = 2(\cos\beta_0 + \cos\alpha_0)H(x)$ for a reflective grating (RG), where β_0 and α_0 are the diffracted and incident angles, respectively. This relation shows that the spatial bending does not induce a wavefront distortion through the TG in the Littrow condition of a monochromatic beam, while the wavefront distortion is approximately two times the spatial bending in the case of a RG. The wavefronts are almost flat in TG but largely parabolic in RG respectively as in Fig. 2b near the Littrow condition. Although femtosecond pulses contain a broad spectrum, this fact explains why a pulse compressor with TGs is considerably less sensitive to the bending than that of RGs.

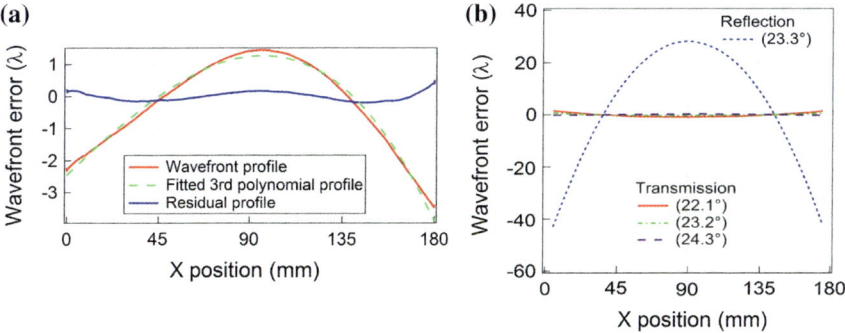

Fig. 2 **a** Wavefront profile (*red solid line*) averaged from $y = 27.5\text{–}32.5$ mm and fitted to a 3rd-order polynomial curve (*green dash line*). **b** Wavefront errors in λ at 633 nm versus grating position along the dispersive direction for TG and RG, respectively. PV $= 44$ λ in $H(x)$ [3], [Fig. 2a was reprinted with permission from Optical Society of America (OSA)]

4 Ray Tracing and Summary

Ray tracing calculation based on the measured bending shows that the group delay between 750 and 850 nm can be compensated within the spatial variation of <0.3 fs in a folded compressor and <0.003 fs in a four-grating compressor using 1,250 lines/mm gratings for a 60 mm-diameter beam with a bandwidth of 100 nm. We measured the spatial pulse front distortion in a TW-class Ti:sapphire laser in which the beam size in the dispersive direction was 85 mm on a grating. However, we did not observe an apparent distortion across a 20 mm-diameter beam within a detection limit (<1.5 fs). This is consistent with the above ray tracing analysis.

References

1. F. Krausz, and M. Ivanov, Rev. Mod. Phys. **81**, 163 (2009).
2. C. Zhou, T. Seki, T. Sukegawa, T. Kanai, J. Itatani, Y. Kobayashi, and S. Watanabe, Appl. Phys. Express **4**, 072701 (2011).
3. C. Zhou, T.Seki, T. Kitamura, Y. Kuramoto, T. Sukegawa, N. Ishii, T. Kanai, J. Itatani, Y. Kobayashi, and S. Watanabe, Opt. Express **22**, 5995 (2014).

Part XI
Pulse Shaping and Manipulations

High-Energy Sub-Optical-Cycle Parametric Waveform Synthesizer

Giovanni Cirmi, Giulio M. Rossi, Shaobo Fang, Shih-Hsuan Chia,
Oliver D. Mücke, Cristian Manzoni, Paolo Farinello, Giulio Cerullo
and Franz X. Kärtner

Abstract We compressed and characterized using FROG a 3-channel, multi-mJ, sub-optical-cycle parametric waveform synthesizer in the visible and infrared (0.5–2.2 µm). The synthesizer will become a versatile tool for strong-field physics experiments.

Waveform Nonlinear Optics aims to study and control the nonlinear interaction of matter with extremely short optical waveforms custom-tailored within an optical cycle of light. Several new applications arise in attoscience and strong-field physics, e.g., the generation of intense isolated attosecond XUV pulses [1], relativistic laser-plasma interactions and laser-driven electron acceleration [2], launching valence-electron wavepacket dynamics in atoms and molecules [3], and control of sub-cycle electron transport in solids [4].

Different technological approaches [5] have recently been pursued to generate such waveforms. Waveform synthesis based on supercontinuum generation in hollow-core fiber compressors represents a rather mature technology and has led to the generation of sub-cycle optical pulses (covering 260–1,100 nm) [3], however,

G. Cirmi (✉) · G.M. Rossi · S. Fang · S.-H. Chia · O.D. Mücke · F.X. Kärtner
Center for Free-Electron Laser Science, Deutsches Elektronen-Synchrotron DESY,
Notkestraße 85, 22607 Hamburg, Germany
e-mail: giovanni.cirmi@cfel.de

F.X. Kärtner
Physics Department, University of Hamburg, Luruper Chaussee 149, 22761 Hamburg,
Germany

G. Cirmi · G.M. Rossi · S. Fang · S.-H. Chia · O.D. Mücke · F.X. Kärtner
The Hamburg Center for Ultrafast Imaging, Luruper Chaussee 149, 22761 Hamburg,
Germany

F.X. Kärtner
Department of Electrical Engineering and Computer Science and Research Laboratory
of Electronics, Massachusetts Institute of Technology, Cambridge, MA 02139, USA

C. Manzoni · P. Farinello · G. Cerullo
IFN-CNR Dipartimento di Fisica, Politecnico di Milano, Piazza Leonardo da Vinci 32,
20133 Milan, Italy

© Springer International Publishing Switzerland 2015
K. Yamanouchi et al. (eds.), *Ultrafast Phenomena XIX*,
Springer Proceedings in Physics 162, DOI 10.1007/978-3-319-13242-6_192

the energy throughput is typically limited to several tens or hundreds of µJs (depending on bandwidth) due to ionization in the gas medium. Alternatively, parametric waveform synthesizers [6–8] feature straightforward energy scalability [1, 2] and can also be extended into the important mid-IR region for the realization of bright coherent tabletop high-harmonic sources in the water-window and keV X-ray region [9]. Parametric waveform synthesizers can be classified into (i) sequential and (ii) parallel schemes. In the sequential approach [2, 8], different spectral regions are amplified in subsequent amplification stages employing different phase-matching conditions without ever splitting the beam. The advantage of this sequential approach is the avoidance of synchronization issues, which comes at the expense of the grand challenge to compress the entire bandwidth with the same dispersive elements at once. In contrast, parallel synthesis schemes [3, 6, 7] offer much greater flexibility by splitting the bandwidth into separate channels, in which the different spectral regions can individually be amplified and/or compressed. Importantly, the coherent synthesis of reproducible waveforms in parallel schemes requires sub-cycle timing synchronization as well as tight stabilization of the relative phases among the different channels, and carrier-envelope phase (CEP) stability.

In 2013, we proposed and already reported first amplification results (see Fig. 1) for a 3-channel multi-mJ sub-cycle parametric waveform synthesizer [7]. Our synthesizer is driven by a non-CEP-stable cryogenically cooled Ti:sapphire amplifier (150 fs, 18 mJ, 1 kHz). We generate a CEP-stable seed continuum (0.5–2.5 µm) by white-light generation in a YAG crystal pumped by the second harmonic (1.06 µm) of the CEP-stable idler of a NIR OPA. The continuum is then split with dichroic mirrors and seeds three BBO-based OPA channels: a VIS non-collinear OPA (NOPA), a NIR and an IR degenerate OPA (DOPA), pumped by the pulses at 0.8 µm (IR DOPA) and by its second harmonic at 0.4 µm (VIS NOPA, NIR DOPA).

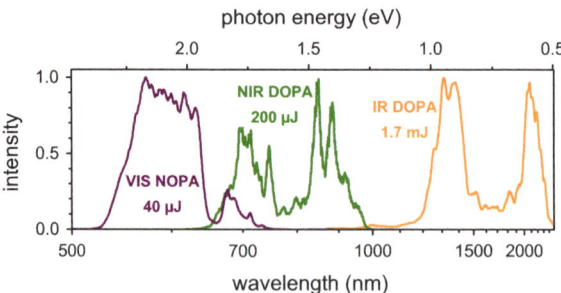

Fig. 1 Output spectra so far obtained from the 2-octave-wide multi-mJ parametric synthesizer: second-stage output spectrum of the VIS NOPA (*violet*), third-stage spectra of the NIR DOPA (*green*) and IR DOPA (*orange*). None of these spectra is contaminated by superfluorescence backgrounds. The spectra (VIS NOPA/NIR DOPA/IR DOPA) support 6.7/5.2/5.2 fs pulses corresponding to 3.4/2.1/1.1 optical cycles at 595/750/1390 nm center wavelengths. The transform-limited (TL) pulse duration from the synthesis of these three spectra (assuming 1:1.3:1 intensity weighting and all CEPs to be 0) is 1.9 fs FWHM (not shown), corresponding to 0.7 optical cycles at 785 nm center wavelength

Figure 1 shows the output spectra and pulse energies obtained so far from the three OPA channels. After two amplification stages only, all channels yield ∼40–80 μJ each.

Here, we present first FROG pulse-characterization results of all three spectral channels, up to the second stage for the VIS NOPA and for the NIR DOPA, and up to the third stage for the IR DOPA. The FROG measurement of the last IR DOPA stage, not shown here, yields a 9.9-fs pulse duration. These results demonstrate the feasibility to recompress all three channels simultaneously close to the Fourier limit and show the flexibility of our dispersion compensation scheme for different experimental situations. The compressed third stage of the IR DOPA is currently delivering 0.3 mJ energy per pulse, but we have already demonstrated the energy scalability up to 1.7 mJ (superfluorescence free) as shown in Fig. 1. We emphasize that the FROG results shown in Fig. 2 were measured at the actual waveform synthesis point, i.e., each of the three pulses propagates through its complete synthesizer channel and the beam combiner and thus experiences the correct dispersion and nonlinearity on its way. In other words, the synthesis can later directly be performed from these pulses after locking their relative timing using feedback loops with balanced optical cross-correlators, that can achieve sub-cycle synchronization with <30-as RMS timing jitter [6]. The dispersion compensation scheme [10] includes custom-designed double-chirped mirror (DCM) pairs and dichroic beam splitters/combiners, plates and wedges (SiO$_2$, ZnSe) for dispersion fine-tuning, and the CaF$_2$ window of our future experiment's vacuum chamber.

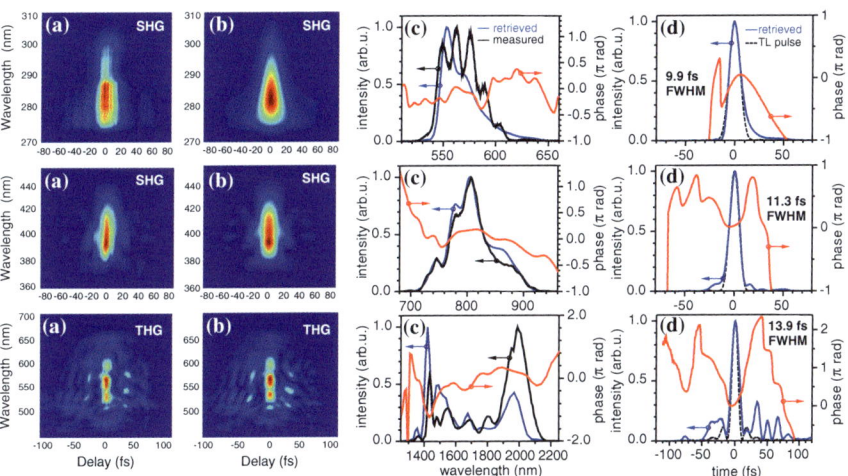

Fig. 2 FROG characterization of the second-stage OPA outputs from the three channels at the synthesis point. The VIS NOPA (*top row*) and NIR DOPA (*middle row*) are characterized by means of SHG-FROG, the IR DOPA (*bottom row*) using surface THG-FROG. **a** Measured and **b** retrieved FROG traces. **c** Measured spectrum, retrieved spectral intensity and phase. **d** Retrieved temporal intensity and phase profiles as well as TL intensity profile. The retrieved FWHM pulse durations are indicated. No marginal correction was applied to the THG-FROG trace in the *bottom row*

The next step will be the waveform synthesis from all three channels at the mJ level. Synthesized waveforms (not shown here) as short as 2.7 fs FWHM can be obtained from the second OPA stages for different relative phases, computed from the measured spectral amplitudes and phases in Fig. 2c.

Our parametric waveform synthesizer spanning more than two octaves already provides unprecedented opportunities for strong-field experiments on solids [4] and nanostructured solid-state devices [11]. After compressing the full 3-channel 3-stage system [7], it will become a versatile tool for controlling strong-field interactions in atoms and molecules and for attosecond pump-probe spectroscopy using VIS/IR and XUV/soft-X-ray pulses.

References

1. E. J. Takahashi et al., Nat. Commun. **4**, 2691 (2013); E. J. Takahashi et al., Phys. Rev. Lett. **104**, 233901 (2010).
2. L. Veisz et al., CLEO Pacific Rim 2013, TuD2-3.
3. A. Wirth et al., Science **334**, 195 (2011). Also see: T. T. Luu et al., QELS 2013, QF1C.6.
4. O. D. Mücke, Phys. Rev. B **84**, 081202(R) (2011); S. Ghimire et al., Nature Phys. **7**, 138 (2011); A. Schiffrin et al., Nature **493**, 70 (2013); M. Schultze et al., Nature **493**, 75 (2013).
5. C. Manzoni et al., Laser Photonics Rev., DOI 10.1002/lpor.201400181 (2014).
6. S.-W. Huang et al., Nature Photonics **5**, 475 (2011); C. Manzoni et al., Opt. Lett. **37**, 1880 (2012).
7. G. Cirmi et al., UFO IX, 2013, We3.3; O. D. Mücke et al., CLEO 2013, CTh3H.3; S. Fang et al., CLEO Pacific Rim 2013, WB3-1; G. M. Rossi et al., CLEO 2014, SF1E.3.
8. A. Harth et al., Opt. Express **20**, 3076 (2012).
9. E. J. Takahashi et al., Phys. Rev. Lett. **101**, 253901 (2008); T. Popmintchev et al., Science **336**, 1287 (2012).
10. S.-H. Chia et al., Optica **1**, 315 (2014); S.-H. Chia et al., patent EP14155053, "Chirped dichroic mirror and a source for broadband light pulses."
11. P. D. Keathley et al., Ann. Phys. **525**, 144 (2013); H. Ye et al., Ultrafast Phenomena 2014, Chapter no. 163.

Above-Millijoule Optical Waveforms Compressible to Sub-fs Using Induced-Phase Modulation in a Neon-Filled Hollow-Core Fiber

Shaobo Fang, Hong Ye, Giovanni Cirmi, Giulio M. Rossi, Shih-Hsuan Chia, Oliver D. Mücke and Franz X. Kärtner

Abstract We demonstrate 1.7-mJ optical waveforms based on induced-phase modulation for generating sub-femtosecond optical pulses. Using custom-designed double-chirped mirrors and a spatial light modulator, such optical waveforms will become a versatile tool for strong-field attoscience.

Coherent optical waveform synthesis aims to generate intense, custom-tailored, sub-cycle optical waveforms, which is currently one of the most intriguing and promising frontiers of attoscience and strong-field physics. Up to now, coherent waveform synthesis based on self-phase modulation (SPM) in a neon-filled hollow-core fiber (HCF) compressor allowed the generation of sub-cycle ~ 300-μJ optical pulses [1, 2]. For pursuing the synthesis of the shortest possible pulses within this scheme, the total output energy of this multi-channel synthesizer is limited to a few tens of μJs mainly by the UV channel containing the smallest pulse energy [1, 2], which might not be sufficient for many interesting applications in attoscience. A potential solution out of this dilemma is the application of induced-phase modulation (IPM) [3, 4] based on the interaction between two (or more) co-propagating optical pulses of different color and relatively long pulse duration in a gas-filled HCF. The IPM technique offers control over the spectral shape by adjusting the relative intensity ratio and the relative delay between the input pulses and allows

S. Fang (✉) · H. Ye · G. Cirmi · G.M. Rossi · S.-H. Chia · O.D. Mücke · F.X. Kärtner
Center for Free-Electron Laser Science, Deutsches Elektronen-Synchrotron DESY,
Notkestraße 85, 22607 Hamburg, Germany
e-mail: shaobo.fang@cfel.de

S. Fang · G. Cirmi · O.D. Mücke · F.X. Kärtner
The Hamburg Center for Ultrafast Imaging, Luruper Chaussee 149,
22761 Hamburg, Germany

H. Ye · G.M. Rossi · S.-H. Chia · F.X. Kärtner
Department of Physics, University of Hamburg, Luruper Chaussee 149,
22761 Hamburg, Germany

F.X. Kärtner
Department of Electrical Engineering and Computer Science and Research Laboratory
of Electronics, Massachusetts Institute of Technology, Cambridge, MA 02139, USA

© Springer International Publishing Switzerland 2015   789
K. Yamanouchi et al. (eds.), *Ultrafast Phenomena XIX*,
Springer Proceedings in Physics 162, DOI 10.1007/978-3-319-13242-6_193

a more efficient generation of ultrabroadband optical pulses than those produced solely by SPM. Such an IPM-based synthesizer is expected to greatly relieve the energy-scaling bottleneck in the UV region [5–7], and the enhanced spectral broadening of the UV region is particularly appealing for the realization of ultrahigh HHG conversion efficiencies in bright tabletop high-harmonic sources.

By employing the advantages of IPM, we have demonstrated an isolated 1.3-cycle pulse with 2.6-fs duration and 3.6-µJ energy centered at 600 nm [3]. Later, much broader spectra (270–1,000 nm) with several hundred µJs have been generated, supporting 1.5-fs transform-limited (TL) pulses [4]. Most importantly, attosecond optical waveforms with above-mJ energy would have a tremendous impact for applications in attoscience and strong-field physics. In this work, we demonstrate for the first time an above-mJ IPM waveform synthesizer driven by a carrier-envelope phase (CEP) stabilized chirped-pulse amplification (CPA) system. Using a neon-filled fused-silica HCF, we achieved a 1.72-mJ CEP-locked super-continuum spanning the range 340–950 nm, which is straightforwardly compressible to terawatt attosecond optical waveforms.

The experimental setup is shown in Fig. 1. The output beam of an 800-nm, 30-fs, 5.5-mJ CEP-stabilized Ti:sapphire CPA system with a 3 kHz repetition rate was divided into two beams by a beam splitter (BS) with a splitting ratio of 55:45 (reflectance/transmittance). The reflected pulses were used as fundamental pulses. The transmitted pulses passed through a half-wave plate (HWP), followed by a 0.5-mm-thick type-I β-barium borate (BBO) crystal to generate the second-harmonic (2ω) pulses at 400 nm with the same polarization as the reflected pulses at 800 nm. Two pairs of harmonic separators (HSs) were used to filter out the residual fundamental pulses from the second-harmonic pulses, as well as to adjust the time delay Δt for optimum IPM. The ω and 2ω pulses were recombined using a dichroic mirror and reflected by a concave silver mirror with a focal length f = 1,900 mm, which focused the combined beam into a fused-silica HCF (length L = 1,150 mm, diameter d = 500 µm). The fiber was placed in the middle of a 400-cm-long glass tube sealed at the two ends with 3-mm-thick CaF_2 windows at Brewster angle. To obtain the broadest IPM output spectrum, a fundamental pulse with 2.3-mJ energy and a second-harmonic pulse with 420-µJ energy were focused and injected into the

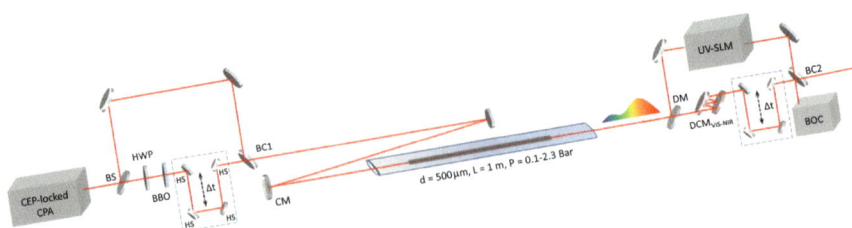

Fig. 1 Sub-femtosecond IPM-based optical waveform synthesizer: *BS* beam splitter; *BBO* β-barium borate crystal; *HWP* half-wave plate; *HS* harmonic separator; *BC* beam combiner; *CM* concave mirror; *DM* dichroic mirror; *DCM* double-chirped mirror; *SLM* spatial light modulator; *BOC* balanced optical cross-correlator

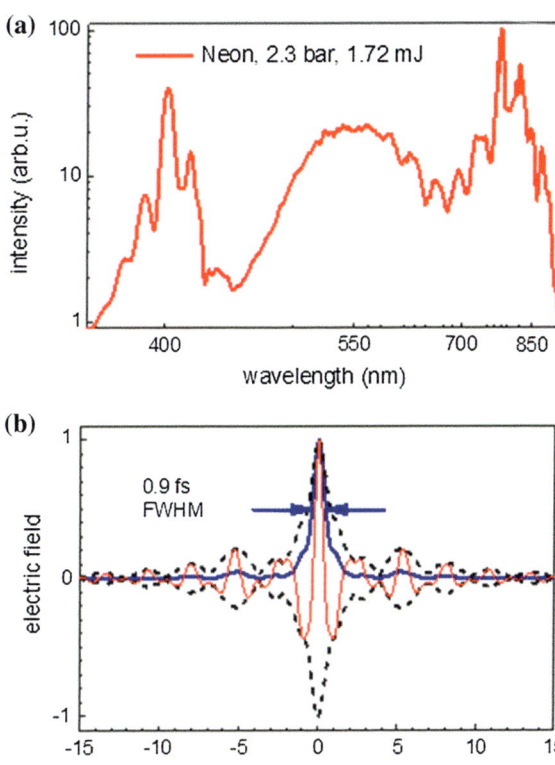

Fig. 2 a Supercontinuum created by IPM, **b** the corresponding TL pulse (*thin line* electric field $E(t)$; *dashed* field envelope; *thick line* intensity $I(t)$)

HCF filled with neon at the pressure of 2.3 bar. Figure 2a shows a 1.72-mJ IPM output spectrum (covering 365–930 nm, measured at 1 % of the peak intensity) supporting 0.9-fs pulses (Fig. 2). After splitting this broadband light source into two different wavelength channels (UV, VIS-NIR) using a broadband dichroic mirror (DM), we plan to compress the different spectral regions using custom-designed double-chirped mirrors ($DCM_{VIS-NIR}$) and a UV spatial light modulator (SLM) [8, 9]. Finally, we will recombine the two channels by another broadband dichroic mirror. We also need to tightly lock the relative timing of the two pulses using a feedback loop employing a balanced optical cross-correlator (BOC), that can achieve sub-cycle synchronization with <30-as (rms) timing jitter [10–12]. To demonstrate the feasibility of compressing this broadband spectrum, we present for the first time SHG-FROG pulse-characterization results (Fig. 3) of the NIR channel. The dispersion compensation scheme includes $DCM_{VIS-NIR}$ pairs (covering 660–1,000 nm), plates and wedges (SiO_2) for dispersion fine-tuning, and the CaF_2 window of our future experiment's vacuum chamber. The SHG-FROG measured and retrieved temporal intensity and phase profiles as well as the TL intensity profile are presented in Fig. 3. The compressed 9.3-fs pulses are very close to the 8-fs Fourier limit of the NIR channel.

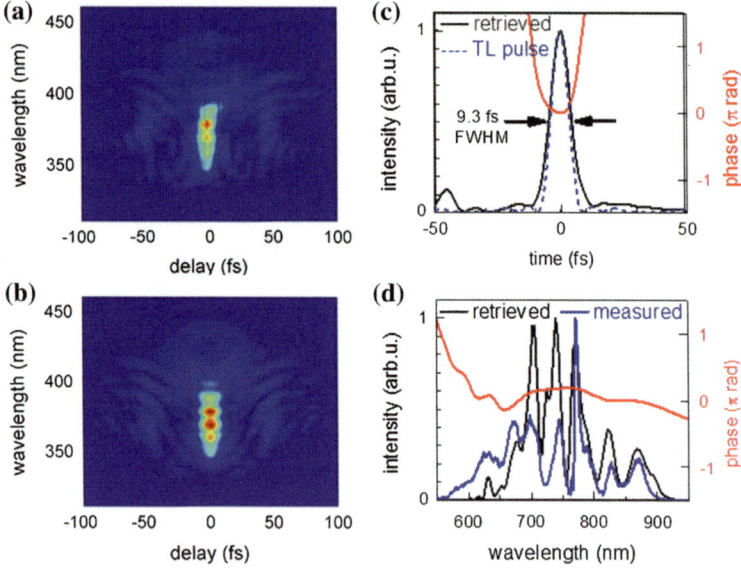

Fig. 3 SHG-FROG characterization of the NIR channel. **a** Measured and **b** retrieved FROG traces. **c** FROG retrieved temporal intensity and phase profiles as well as TL intensity profile. **d** Measured spectrum, FROG retrieved spectral intensity and phase

In conclusion, we have discussed the technological challenges of an above-mJ IPM-based optical waveform synthesizer and presented first spectro-temporal characterization results of the NIR channel. By employing custom-designed DCMs and a UV-SLM in the near future, we foresee that this high-energy sub-fs synthesizer will become a versatile tool for nonlinear attosecond optics experiments [13, 14].

References

1. A. Wirth *et al.*, Science **334**, 195-200 (2011).
2. T. T. Luu *et al.*, QELS 2013, paper QF1C.6 (2013).
3. E. Matsubara *et al.*, JOSA B **24**, 985–989 (2007).
4. S. Fang *et al.*, IEEE Phot. Tech. Lett. **23**, 688-690 (2011).
5. S. Fang *et al.*, CLEO Pacific Rim 2013, paper WB3-1 (2013).
6. S. Fang *et al.*, HILAS 2014, paper HW1C.2 (2014).
7. G. M. Rossi *et al.*, CLEO 2014, paper SF1E.3 (2014).
8. T. Tanigawa *et al.*, Opt. Lett. **34**, 1696–1698 (2009).
9. J. Zhu *et al.*, Applied Optics. **49**, 350-357 (2010).
10. S.-W. Huang *et al.*, Nat. Photonics **5**, 475 (2011).
11. S.-W. Huang *et al.*, J. Phys. B **45**, 074009 (2012).
12. C. Manzoni *et al.*, Opt. Lett. **37**, 1880 (2012).
13. E. J. Takahashi *et al.*, Nat. Commun. **4**: 2691 (2013).
14. E. J. Takahashi *et al.*, Phys. Rev. Lett. **104**, 233901 (2010).

Isolating Quantum Coherence Using Coherent Multi-dimensional Spectroscopy with Spectrally Shaped Pulses

Jonathan O. Tollerud, Christopher R. Hall and Jeffrey A. Davis

Abstract We demonstrate how spectral shaping in coherent multidimensional spectroscopy can isolate specific signal pathways and directly access quantitative details. We identify, isolate and analyse weak coherent coupling between spatially separated excitons in asymmetric double quantum-wells.

1 Introduction

Coherent multidimensional spectroscopy (CMDS) of electronic transitions has become an increasingly useful and versatile tool for understanding complex systems, from semiconductor nanostructures [1] to light-harvesting complexes that exist in photosynthetic organisms [2, 3]. Much of the analysis of the results from these experiments, however, is qualitative and relies on fitting complex models to reproduce the experimental data. Here we describe pathway-selective coherent multidimensional spectroscopy based on spectral shaping of the excitation pulses as a means to isolate specific signal pathways and directly access quantitative details. We apply this approach to selectively excite coherent superpositions of excitons that are spatially separated and localised to different semiconductor quantum wells.

Coherent coupling between spatially separated systems has long been explored as a necessary requirement for quantum information and cryptography [4]. Recent discoveries suggest such phenomena appear in a much wider range of processes, including light-harvesting in photosynthesis [2, 3]. With the approach we describe here, identification and quantitative analysis of this type of weak coherent coupling between distant quantum systems becomes simpler and more powerful.

J.O. Tollerud · C.R. Hall · J.A. Davis (✉)
Centre for Quantum and Optical Science, Swinburne University of Technology,
John St, Hawthorn, Melbourne 3122, Australia
e-mail: JDavis@swin.edu.au

© Springer International Publishing Switzerland 2015 793
K. Yamanouchi et al. (eds.), *Ultrafast Phenomena XIX*,
Springer Proceedings in Physics 162, DOI 10.1007/978-3-319-13242-6_194

2 Experimental and Results

We utilize a CMDS setup based on spatial light modulators, similar to the one developed by Nelson et al. [5]. With this setup we not only have precise control of the pulse delays, phase and spectral phase, but we also have control over the spectral amplitudes, which allows us to spectrally shape the individual pulses to excite specific quantum pathways.

In the asymmetric double quantum wells studied here there are two GaAs quantum wells of different width separated by an AlGaAs barrier. In general, there are four exciton states that can be observed (Fig. 1a), two localized to the wide well and two localized to the narrow well. In this system, it is expected that coherent coupling between excitons localized in the same well is strong, while coherent

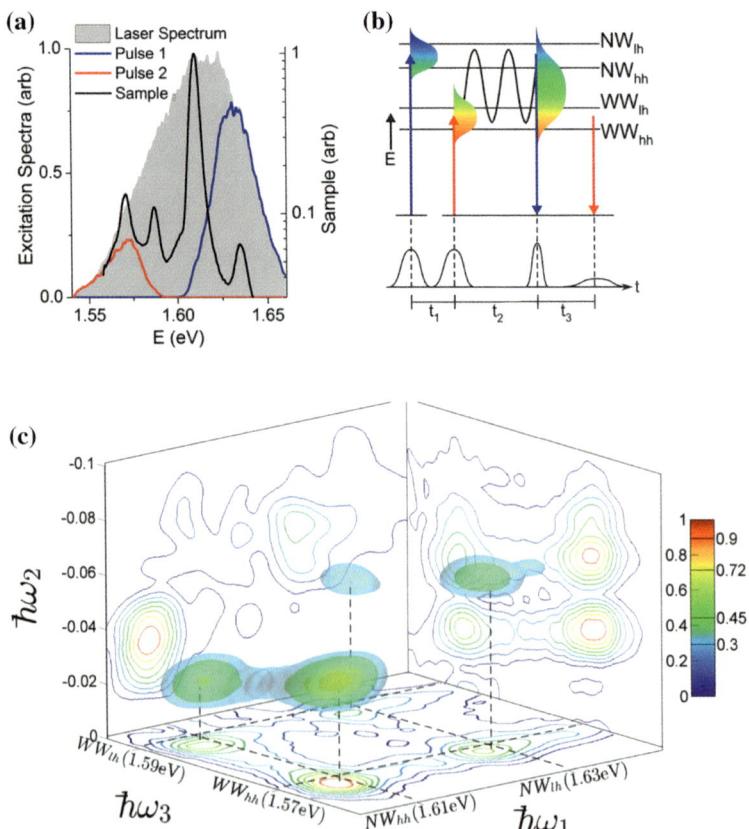

Fig. 1 a Shows the spectrum of the asymmetric double quantum well sample (*black line*), with the four spectral peaks, together with the laser spectrum for the broadband (*shaded*) and the two shaped pulses (*red* and *blue*). **b** Shows the coherence specific excitation scheme utilized and **c** shows the 3D plot under these excitation conditions. Adapted from [6]

coupling between excitons localized in different wells is weak. In broadband CMDS experiments clear signatures of coherence between excitons in the same well can be identified and even analyzed due to them being well-separated in a 3D spectrum. Signatures of coherence between two pairs of excitons localized in different wells can also be identified, but for these peaks little other analysis is possible because the weak coherence signal is swamped by other signal contributions.

By spectrally shaping the first two pulses so that the first is resonant only with the excitons in the narrow well and the second pulse is resonant only with excitons in the wide well, as shown in Fig. 1b, we are able to selectively excite the coherences between excitons localized in different quantum wells. The resulting 3D spectrum, Fig. 1c, shows not only peaks due to the two inter-well coherences identified in the broadband experiment, but also two due to additional inter-well coherences. All four peaks are 'cross-peaks' in the projection onto the (ω_1, ω_3) plane (the standard 2D spectrum) and are shifted along ω_3 by the expected energy differences for these coherent superpositions.

The identification of all four of these inter-well coherences, where the excitons are well-separated with no spatial overlap, is a significant result in itself, however, the enhanced signal to noise and ability to isolate in three spectral dimensions allows further unprecedented analysis [6]. We have developed new and enhanced existing techniques to analyze such three-dimensional peak shapes and reveal details of the interactions between the different spatially separate states and the extent of any correlated broadening [6, 7]. Finally, with a dynamic range $>10^4$ in electric field amplitude this approach allows us to quantitatively compare different signal pathways and in the present case precisely determine coupling strengths and their dependence on a range of parameters.

3 Conclusions

We have devised a pathway specific CMDS experiment that combines an ability to selectively excite specific quantum pathways by spectrally shaping the excitation pulses with the many of the benefits of CMDS. The ensuing ability to identify, isolate and analyse coherences, and indeed any specific signal pathway, can provide significant insight into the interactions and dynamics in a range of complex systems. In photosynthetic light harvesting complexes, for example, this type of approach has the potential to resolve important questions regarding the nature and role of quantum effects in efficient energy transfer.

References

1. X. Q. Li, T. H. Zhang, C. N. Borca, and S. T. Cundiff, "Many-body interactions in semiconductors probed by optical two-dimensional fourier transform spectroscopy," Phys. Rev. Lett. 96 (2006).
2. G. S. Engel, T. R. Calhoun, E. L. Read, T. K. Ahn, T. Mancal, Y. C. Cheng, R. E. Blankenship, and G. R. Fleming, "Evidence for wavelike energy transfer through quantum coherence in photosynthetic systems," Nature 446, 782–786 (2007).
3. D. B. Turner, R. Dinshaw, K.-K. Lee, M. S. Belsley, K. E. Wilk, P. M. G. Curmi, and G. D. Scholes, "Quantitative investigations of quantum coherence for a light-harvesting protein at conditions simulating photosynthesis," Phys. Chem. Chem. Phys 14, 48 (2012).
4. M. A. Nielsen and I. L. Chuang, Quantum Information and Quantum Computation (Cambridge University Press, 2000).
5. D. B. Turner, K. W. Stone, K. Gundogdu, and K. A. Nelson, "Invited article: The coherent optical laser beam recombination technique (colbert) spectrometer: Coherent multidimensional spectroscopy made easier," Review of Scientific Instruments 82 (2011).
6. C. R. Hall, J. O. Tollerud and J. A. Davis, "Isolating quantum coherence using coherent multi-dimensional spectroscopy with spectrally shaped pulses,"Optics Express 22, 6719 (2014).
7. C. R. Hall, J. O. Tollerud, H. M. Quiney, and J. A. Davis, "Three-dimensional electronic spectroscopy of excitons in asymmetric double quantum wells," New Journal of Physics 15, 045028 (2013).

Spatiotemporal Dynamics of Femtosecond Pulses Shaped by Diffractive Optical Elements

Rocío Borrego-Varillas, Benjamín Alonso, Jorge Pérez-Vizcaíno,
Isabel Gallardo-González, Gladys Mínguez-Vega,
Omel Mendoza-Yero, Jesús Lancis, Andrew Forbes and Íñigo J. Sola

Abstract We present a complete experimental characterization and simulation of the spatiotemporal and spatio-spectral effects taking place when a femtosecond pulse is shaped by a diffractive optical element.

Ultrafast femtosecond lasers have found many applications in fields like spectroscopy, high-field physics, medicine or laser processing. Some of these applications have been demonstrated to benefit from diffractive shaped beams [1]. For example, flat-top beams (FTBs) are preferred to Gaussian profiles in micromachining since they yield minimal edge roughness.

A common approach to obtain a shaped profile is by means of a phase-only diffractive optical element (DOE) and a Fourier lens. However, due to the broadband nature of femtosecond pulses, space-time coupling effects can happen since the shaped profile is formed for a given wavelength λ at a plane [2]

$$z = f \frac{\lambda_{DOE}}{\lambda} \tag{1}$$

being f the focal length of the Fourier transforming lens and λ_{DOE} the wavelength the DOE was designed for.

Hence, a complete characterization is mandatory to assess the usefulness of diffractive shaped pulses for further applications. In this contribution, we report the complete spatiotemporal dynamics and propagation of femtosecond shaped pulses by Gaussian-to-flat-top DOEs and diffractive lenses (DLs).

R. Borrego-Varillas (✉) · B. Alonso · Í.J. Sola
Universidad de Salamanca, Plz. de la Merced s/n, 37008 Salamanca, Spain
e-mail: rocio.borrego@polimi.it

R. Borrego-Varillas · J. Pérez-Vizcaíno · G. Mínguez-Vega · O. Mendoza-Yero · J. Lancis
Universitat Jaume I, Av. Sos Baynat s/n, 12080 Castelló de la Plana, Spain

I. Gallardo-González
Centro de Láseres Pulsados, C/Adaja s/n, 37185 Salamanca, Spain

A. Forbes
Council for Scientific and Industrial Research, Pretoria 0001, South Africa

© Springer International Publishing Switzerland 2015
K. Yamanouchi et al. (eds.), *Ultrafast Phenomena XIX*,
Springer Proceedings in Physics 162, DOI 10.1007/978-3-319-13242-6_195

For the experimental characterization we employed the STARFISH technique, which consists on measuring the spatially-resolved spectral interferences by using a fiber coupler as interferometer [3]. The experimental measurements were corroborated by simulations. For this purpose, the Fresnel integral was numerically solved:

$$U(r,z,\lambda) = \frac{2\pi}{iz\lambda}\exp\left(\frac{i\pi r^2}{z\lambda}\right)\int_0^\infty U_0(\rho,\lambda)\exp\left(-i\frac{\pi\rho^2}{f\lambda_{DOE}}+i\frac{\pi\rho^2}{z\lambda}\right)J_0\left(\frac{2\pi r\rho}{\lambda z}\right)\rho\,d\rho$$

(2)

being U_0 the electric field at the plane of the DOE.

The experimental setup is shown in Fig. 1. The laser delivers 100 fs pulses at 1 kHz repetition rate and a central wavelength of 795 nm. A spatial filter was used to obtain a clean Gaussian beam; the profile and wavefront were checked with a CCD camera and a wavefront sensor. The beam was then split in two replicas: the first one acted as a reference and was temporally characterized by means of the FROG technique. The DOE was placed in the second arm, which has a motorized fiber spatially scanning the test pulse along a transverse axis. A CCD was put at the Fourier plane in order to monitor the spatial profile of the shaped beams.

As DOEs we employed a DL and a Gaussian-to-flat-top shaper. The focal length of the DL depended on the wavelength according to (1) and was designed with a focal length of $f = 106.6$ mm for a wavelength of 795 nm. The FTB shaper was designed to work at 800 nm with a 200 mm focal length lens, and provided a 4 mm-FTB from a 6 mm Gaussian beam.

The spatiotemporal propagation of the pulses focused with a DL is shown in Fig. 2a. The central panel shows the dynamics at the position of the focus for the central wavelength. In the spatio-spectral domain the beam waist is minimal at 795 nm and increases gradually for shorter and longer wavelengths, as predicted by (1) (redder frequencies are diverging, whereas bluer ones are still converging).

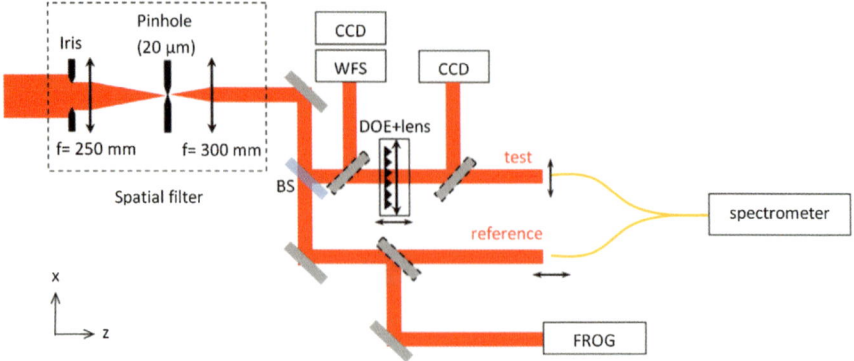

Fig. 1 Experimental setup for the spatiotemporal characterization of diffractive shaped beams (see details in the text)

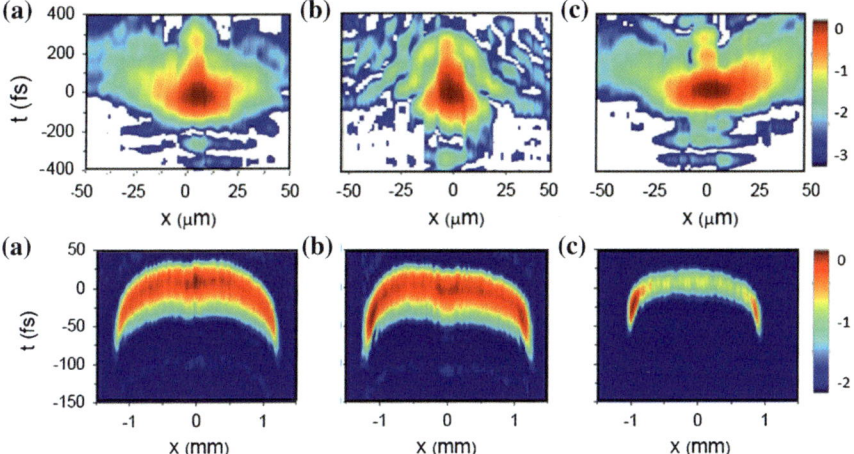

Fig. 2 Experimental spatiotemporal traces for 100-fs pulses (*top*) shaped by a DL at **a** 105.5 mm, **b** 106.6 mm and **c** 107.6 mm propagation distances and 30-fs FTB pulses (*bottom*) at **a** 180 mm, **b** 200 mm and **c** 220 mm propagation distances

The spatiotemporal trace corresponds to the far-field structure with a main broadened central peak, and a train of pulses in the wings coming from the ring structure [4]. Before and after the focus (left and right panels in Fig. 2a), the pulse front curvature is similar.

Figure 2b shows the results of 30-fs FTBs. The central panel corresponds to the position where the FTB is formed for the central wavelength of the laser. In the spatio-spectral domain, the beam size changed monotonically with the wavelength. This wavelength dependence also manifests on the form of side tails in the temporal domain, being the pulse lengthened at the edges of the flat-top. The pulse front presented a curvature in such a way that the pulse is delayed at the borders of the FTB by approximately 50 fs with respect to the central part.

In conclusion, we have demonstrated the potential of the STARFISH technique for the spatiotemporal characterization of DOEs. Space-time coupling effects have been observed for both, DLs and top-hat shapers. The results of this study will be helpful, for example, to understand the filamentation dynamics with DLs [5] and to assess the validity of flat-top shapers for femtosecond laser processing.

References

1. Dickey F.M., Holswade S. C. (2000) Laser Beam Shaping: theory and techniques.
2. Forbes A., Dickey F. M., DeGama M. and du Plessis A. (2012) Opt. Lett. 37:49.
3. Alonso B., Sola I. J., Varela O., Hernandez-Toro J., Mendez C., San Roman J., Zair A., and Roso L. (2010) JOSA B 27:933.

4. Alonso B., Borrego-Varillas R., Mendoza-Yero O., Sola I. J., San Roman J., Minguez-Vega G. and Roso L. (2012) JOSA B 29:1993.
5. Borrego-Varillas R., Romero C., Mendoza-Yero O., Minguez-Vega G., Gallardo I. and Vazquez de Aldana J. R. (2013) JOSA B 30:2059.

Spectral Shaping and Continuous Tuning of Multi-color Carrier-Envelope Phase Locked Pulse

A. Yabushita, C.-H. Kao and T. Kobayashi

Abstract We have proposed and demonstrated two schemes to generate multi-color CEP-locked beams by using the NOPA. In the first scheme, a spatial filter in the spectrally dispersed seed light made the seed spectrum have three peaks, and the generated idler beam also had three peaks. The spectrum of the CEP-locked pulse can be shaped into arbitral spectral shape. In the second scheme, parabolic chirp was introduced on the seed pulse to generate two-color CEP-locked pulse whose colors are tunable.

1 Introduction

Carrier envelope phase (CEP) of the laser pulses are generally not locked in conventional laser systems. Meanwhile, CEP-locked laser pulse has various applications such as soft X-ray generation [1], frequency metrology [2], and control of diatomic molecule dissociation [3]. There are two kinds of methods to lock the

A. Yabushita (✉) · C.-H. Kao · T. Kobayashi
Department of Electrophysics, National Chiao-Tung University, Hsinchu 300, Taiwan
e-mail: yabushita@mail.nctu.edu.tw

C.-H. Kao
e-mail: vb9999v@yahoo.com.tw

T. Kobayashi
e-mail: kobayashi@ils.uec.ac.jp

A. Yabushita · T. Kobayashi
Core Research for Evolutional Science and Technology, Japan Science and Technology Agency, Chiyoda-ku, Tokyo 102-0076, Japan

T. Kobayashi
Department of Applied Physics and Chemistry and Institute for Laser Science, University of Electro-Communications, Chofu, Tokyo 182-8585, Japan

T. Kobayashi
Institute of Laser Engineering, Osaka University, Suita, Osaka 565-0971, Japan

© Springer International Publishing Switzerland 2015   801
K. Yamanouchi et al. (eds.), *Ultrafast Phenomena XIX*,
Springer Proceedings in Physics 162, DOI 10.1007/978-3-319-13242-6_196

CEP. In active CEP-lock method [4], the laser spectrum is broadened over one octave using a fiber. The high frequency component of the broadened spectrum is mixed with the double frequency of the low-frequency component and their interference is observed (f-2f interferometer). The observed beat reflecting the change of CEP was used as feed-back signal to stabilize the CEP. Following that, we have developed another type of CEP-lock method called as passive CEP-lock method [5] using a non-collinear optical parametric amplifier (NOPA) [6]. The NOPA system was pumped with second harmonic (SH) of the Ti:sapphire laser pulse and seeded with the white-light continuum generated by the same SH. The CEP of the pump pulse randomly shifts its value pulse-to-pulse, but the amount of shift is equal to that of the seed pulse. Therefore, the idler pulse emerging from the NOPA is locked to a constant value not experiencing any frequency shift in pulse-to-pulse. In the present work, we have developed and demonstrated two schemes to generate multi-color CEP-locked pulses. The CEPs of the pulses are self-stabilized by the NOPA. A seed pulse of the NOPA was spectral or temporally modified to generate multi-color CEP-locked pulse.

2 Experimental

The CEP-locked pulse with broadband spectrum was generated by a NOPA as follows. The SH of the regenerative amplifier was separated into two pulses by a beam sampler. A small portion of the SH pulse reflected on the beam sampler was focused into a calcium fluoride (CaF_2) plate to generate white light (WL) by self-phase modulation. The CaF_2 plate was pasted on a mechanical rotation stage and the plate was shaken continuously minimizing damage on the plate under irradiation of the SH beam. The WL pulse was used as a seed pulse of the NOPA. In the light path of the WL pulse, we have inserted a prism compressor, which consists of a pair of equilateral dispersive prism. The prism compressor had two functions in the present work. One is to adjust the chirp of the NOPA seed beam, resulting in wavelength tunability of the NOPA idler beam. Another function of the prism compressor is for arbitrary modulation in the NOPA seed spectrum, which could be performed inserting a spatial filter in the beam path of the seed beam spatially dispersed in the prism compressor. A large portion of the SH pulse transmitted through the beam sampler. The intense SH beam stretched its pulse duration passing through a fused silica glass block with length of 40 mm, and used as a pump pulse of the NOPA.

3 Results and Discussion

The duration of the SH pulse was shorter than that of the WL pulse. When the SH pulse overlaps with the center wavelength component of the WL pulse, the NOPA output has the broadest band. We have inserted the spatial filter in the light path of

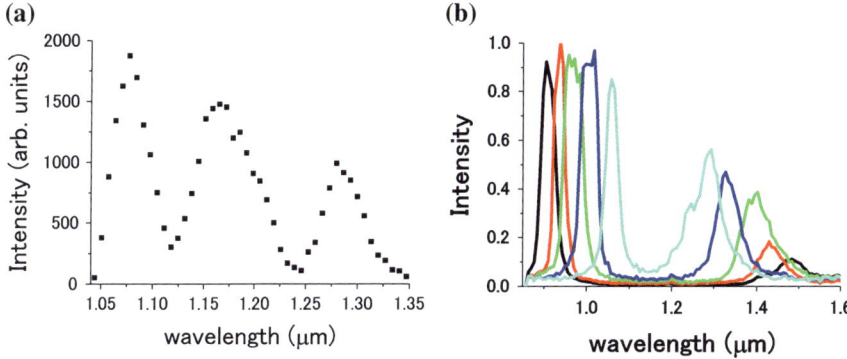

Fig. 1 CEP-locked idler spectrum with **a** three colors and **b** tunable double color

the seed beam, which modified the seed spectrum to have three peaks. Amplifying the three-peak seed spectrum in NOPA, the CEP-locked idler pulse also has three peaks in its spectrum (see Fig. 1a). The simulation results of group delay showed that the seed beam has the remaining parabolic chirp. When the SH pulse does not overlap with the center wavelength component of the WL pulse, the NOPA output will have two peaks in its spectrum. The wavelengths of the two peaks can be tuned by adjusting the delay between the SH pulse and the WL pulse. To adjust the delay between the SH pulse and the WL pulse, a corner reflector was inserted in the light path of the SH pulse. The position of the corner reflector was scanned by manipulating a micrometer of a manual positioning stage set under the corner reflector. Figure 1b shows the idler spectrum observed by changing the optical delay between the SH pulse and the WL pulse at 40 μm step. The result shows that we could obtain two-color CEP-locked pulse with wavelength tunability.

4 Summary

We have demonstrated two schemes to generate multi-color CEP-locked pulse. In the first scheme, we have modified the seed spectrum inserting a prism compressor and a spatial filter in the light path of the seed beam. The three color seed beam was amplified in the NOPA to obtain the three color idler beam whose CEP is self-stabilized. In the second scheme, we have adjusted the optical delay between the SH pump pulse and the WL seed pulse to obtain two color CEP-locked idler pulse whose wavelength is tunable. The two color amplification of the seed (and idler) beam was caused by the high order chirp introduced by the prism compressor in the light path of the WL seed beam.

Acknowledgments This work was supported by the National Science Council of Taiwan and partially by the Japan Science and Technology Agency (JST). Authors thank Prof. Baltuska for his kind discussion and suggestions.

References

1. A. Blatuska et al., "Attosecond control of electronic processes by intense light fields", Nature **421**, 611-615 (2003).
2. D. J. Jones, et al., "Carrier-Envelope Phase Control of Femtosecond Mode-Locked Lasers and Direct Optical Frequency Synthesis", Science **288**, 635-639 (2000).
3. V. Roudnev, B.D. Esry, I. Ben-Itzhak, "Controlling HD^+ and H_2^+ Dissociation with the Carrier-Envelope Phase Difference of an Intense Ultrashort Laser Pulse", Phys. Rev. Lett. **93**, 163601 (2004).
4. H.R. Telle et al., "Carrier-envelope offset phase control: a novel concept for absolute optical frequency measurement and ultrashort pulse generation", Appl. Phys. B **69**, 327-332 (1999).
5. A. Baltuska, T. Fuji, T. Kobayashi, "Controlling the Carrier-Envelope Phase of Ultrashort Light Pulses with Optical Parametric Amplifier", Phys. Rev. Lett. **88**, 133901 (2002).
6. T. Wilhelm, J. Piel, E. Riedle, "Sub-20-fs pulses tunable across the visible from a blue-pumped single-pass non-collinear parametric converter", Opt. Lett. **22**, 1494-1496 (1997).

Tailoring of High-Field Multi-THz Waveforms with Sub-cycle Precision

B. Mayer, C. Schmidt, J. Bühler, J. Fischer, D.V. Seletskiy, D. Brida, A. Pashkin and Alfred Leitenstorfer

Abstract Shaping of extremely intense mid-infrared transients by means of time-domain slicing and frequency-domain synthesis is demonstrated. We achieve phase-stable transients at multiple MV/cm peak fields with single-cycle duration and strong polar asymmetry.

We present two alternative approaches to control duration and optical cycles of multi-THz waveforms. The first approach relies on waveform slicing directly in the time domain by means of an ultrafast semiconductor switch [1]. Such amplitude modulation inherently increases the spectral bandwidth by generation of new spectral components. We demonstrate for the first time sub-cycle slicing for the generation of high-field and phase-locked multi-THz transients [2] containing only few cycles of light. The frequency domain approach is based on coherent $\omega - 2\omega$ synthesis resulting in high-field multi-THz waveforms with strong polar asymmetry. To this end, we generate broadband second harmonic of the few-cycle fundamental waveform leading to an octave-spanning phase-locked THz spectrum. By controlling the phase difference between the first and second harmonics, we demonstrate fully tunable polar asymmetry of the synthesized THz transients.

The ultrafast time-domain slicing is achieved by a controlled reflection of intense multi-THz pulses from an intrinsic germanium (i-Ge) wafer placed at Brewster's angle. The reflected THz field is detected as a function of electro-optic sampling (EOS) delay time t_1. The ultrafast switching is triggered by 18-fs-short pulses with a center wavelength of 1,080 nm and energies up to 15 μJ. We employ a two-stage non-collinear optical parametric amplifier [3] in order to achieve intense broadband control pulses necessary for an efficient switching process. The temporal offset t_2 of the control pulse with respect to the THz transient is adjusted with a delay stage. By adjusting the delay time t_2, we are able to precisely control the temporal position of the leading edge and hence the pulse duration of the THz transient. This is illus-

B. Mayer · C. Schmidt · J. Bühler · J. Fischer · D.V. Seletskiy · D. Brida · A. Pashkin · A. Leitenstorfer (✉)
Department of Physics and Center for Applied Photonics, University of Konstanz, 78457 Konstanz, Germany
e-mail: aleitens@uni-konstanz.de

© Springer International Publishing Switzerland 2015
K. Yamanouchi et al. (eds.), *Ultrafast Phenomena XIX*,
Springer Proceedings in Physics 162, DOI 10.1007/978-3-319-13242-6_197

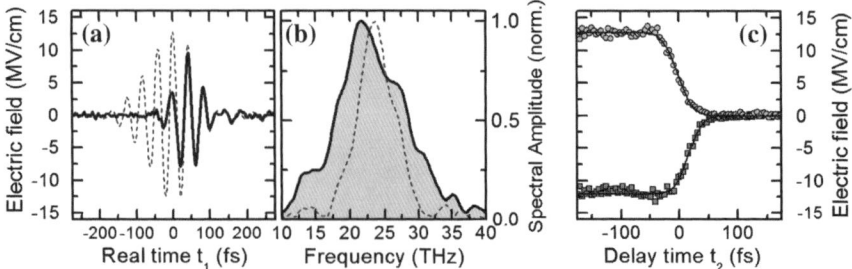

Fig. 1 a Time-domain profiles of multi-THz transients. The *dashed line* corresponds to the reference transient fully reflected by the photoexcited Ge wafer. The *solid line* shows the sliced single-cycle THz waveform. **b** Corresponding normalized amplitude spectra of the reference (*dashed line*) and the sliced (*solid line*) waveforms indicate a notable spectral broadening produced by the sub-cycle switching. **c** Dynamics of the slicing process illustrated by electric fields measured at a fixed EOS time t_1 as a function of the delay time t_2. The evolution of the reflected field maximum (*circles*) and the adjacent minimum (*squares*) is shown together with fitting functions (*solid lines*)

trated in Fig. 1a which shows the reference high-field multi-THz waveform containing about 2.5 oscillation cycles and the waveform as sliced by the control pulse with the fluence of 13.7 mJ/cm^2. The resulting THz transient contains only a single optical cycle. It has a remarkably high peak electric field of 10 MV/cm. The generation of new spectral components results in nearly doubling of the bandwidth of the sliced transient as clearly seen in Fig. 1b. Figure 1c shows the reflected THz field at the maximum and the minimum of the reference waveform as a function of the delay time t_2. The obtained reflectivity onset time of 45 fs is smaller than the period of the multi-THz oscillation indicating slicing operation in the sub-cycle range [4].

In the second approach, we combine a phase-locked fundamental and collinearly-generated second harmonic of the THz transient to synthesize a novel functional waveform with controlled polar asymmetry of the temporal envelope. Such asymmetry is directly achieved by the change of the relative phase of the injected second harmonic with respect to the fundamental wave. The asymmetry allows for coherent control of ultrafast transport in condensed matter [5], as well as sub-cycle control of XUV and THz generation mechanisms in plasmas [6, 7]. For generation of the synthetic field we obtain the second harmonic (2ω) by placing an $AgGaSe_2$ nonlinear crystal ($\theta = 56°$, $\varphi = 45°$, thickness of 350 μm) in the intermediate focus (60 μm FWHM) of the fundamental (ω) phase-locked transient. The emerging $\omega - 2\omega$ field is re-collimated and transmitted through a dispersive element (500-μm-thick InAs wafer), which serves as phase plate for control of the relative phase of the fundamental and second harmonic, and hence the asymmetry of the synthesized waveform. The resulting THz field is resolved by means of broadband EOS in a 10-μm-thick ZnTe crystal, ensuring a detection bandwidth spanning the synthesized spectrum. Figure 2 shows the waveform for the condition of π relative phase shift, resulting in maximum (positive) asymmetry in the envelope. Strong envelope

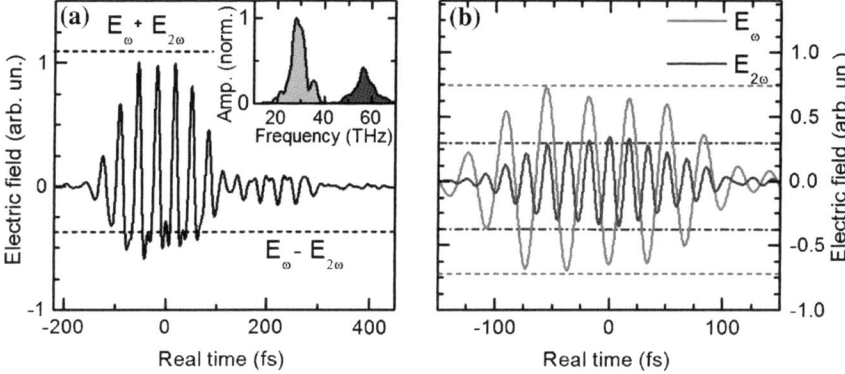

Fig. 2 **a** Time-domain profile of synthetic two-color THz waveform. Strong envelope asymmetry is clearly visible by comparison of the *shaded areas*; Fourier components of the transient in (**a**) are depicted in the *inset*; **b** Decomposition of the spectrum into the fundamental (*light gray*) and second harmonic (*dark gray*) transients allows determination of the maximal field amplitude of each component (E_ω, $E_{2\omega}$, respectively). Comparison of the maximum ($E_\omega + E_{2\omega}$) and minimum ($E_\omega - E_{2\omega}$) possible field values (*dotted lines* in panel a) are in very good agreement with the observed asymmetry in the synthesized waveform

asymmetry is clearly visible in the time domain, due to the efficient second harmonic generation process (49 % yield in electric field amplitude).

In conclusion, we have demonstrated time-domain slicing based on an ultrafast semiconductor mirror which enables the control of ultrashort optical waveforms on a sub-cycle level of precision. Using this technique, we succeeded in generation of single-cycle multi-THz transients with peak electric fields of 10 MV/cm. This approach can be scaled up for even higher THz fields and extended by implementing an additional "switch-off" device through gated transmission in Ge for arbitrary tailoring of multi-THz waveforms. In another approach, we generated and detected synthetic THz transients with strong and controlled polar asymmetry. Such transients may be used for novel experiments on field-controlled transport and light-matter interaction, particularly accessing non-perturbative regimes.

References

1. C. Rolland and P. B. Corkum, "Generation of 130-fsec midinfrared pulses," J. Opt. Soc. Am. B **12**, 1625 (1986).
2. A. Sell, A. Leitenstorfer and R. Huber, "Phase-locked generation and field-resolved detection of widely tunable terahertz pulses with amplitudes exceeding 100 MV/cm," Opt. Lett. **33**, 2767 (2008).
3. C. Manzoni, D. Polli and G. Cerullo, "Two-color pump-probe system broadly tunable over the visible and the near infrared with sub-30 fs temporal resolution," Rev. Sci. Instrum. **77**, 023103 (2006).

4. B. Mayer, C. Schmidt, J. Bühler, D. V. Seletskiy, D. Brida, A. Pashkin and A. Leitenstorfer, "Sub-cycle slicing of phase-locked and intense mid-infrared transients," New J. Phys. **16**, 063033 (2014).
5. R. Atanasov, A. Haché, J. L. P. Hughes, H. M. van Driel, and J. E. Sipe, "Coherent control of photocurrent generation in bulk semiconductors," Phys. Rev. Lett. **76**, 1703-1706 (1996).
6. J. Kim, C. M. Kim, H. T. Kim, G. H. Lee, Y. S. Lee, J. Y. Park, D. J. Cho, and C. H. Nam, "Highly Efficient High-Harmonic Generation in an Orthogonally Polarized Two-Color Laser Field," Phys. Rev. Lett. **94**, 243901 (2005).
7. K. Y. Kim, A. J. Taylor, J. H. Glownia and G. Rodriguez, "Coherent control of terahertz supercontinuum generation in ultrafast laser–gas interactions", Nature Photon. **2**, 605-609 (2008).

Over 1-mJ Intense Ultrashort Optical-Vortex Pulse Generation with Programmable Topological-Charge Control by Chirped-Pulse Amplification

Keisaku Yamane, Asami Honda, Yasunori Toda and Ryuji Morita

Abstract We demonstrated the generation of over 1-mJ intense optical-vortex pulses of which topological charges were programmable controlled by computer-generated holograms. The pulse duration was characterized to be 27 fs by two-dimensional spectral shearing interferometry.

1 Introduction

Optical vortices attract growing attention during last decade because of their various applications—quantum information processing, laser ablation, optical tweezers, etc. The essential property of optical vortex is its spatially-varying phase distribution. Combination of recent phase control techniques in time or frequency domain with spatial phase control to generate optical vortices is a reasonable approach. It enables us to carry out ultrafast nonlinear spectroscopy and high-field interaction with matter.

We have already demonstrated the generation of 56-μJ, 5.9-fs ultrashort optical-vortex pulses by using optical parametric amplification (OPA) [1]. The key device in our previous work was an optical-vortex converter with a spatial-variant wave-plate. This device has advantages of ultrabroadband applicability and of high

K. Yamane (✉) · A. Honda · Y. Toda · R. Morita
Department of Applied Physics, Hokkaido University, Kita-13, Nishi-8, Kita-ku,
Sapporo 060-8628, Japan
e-mail: k-yamane@eng.hokudai.ac.jp

A. Honda
e-mail: asami-h@eng.hokudai.ac.jp

Y. Toda
e-mail: toda@eng.hokudai.ac.jp

R. Morita
e-mail: morita@eng.hokudai.ac.jp

K. Yamane
CREST, JST, Chiyoda-ku, Japan

© Springer International Publishing Switzerland 2015
K. Yamanouchi et al. (eds.), *Ultrafast Phenomena XIX*,
Springer Proceedings in Physics 162, DOI 10.1007/978-3-319-13242-6_198

throughput. On the other hand, the topological charge of optical vortices generated by it is uniquely determined by the structure or design of the spatial-variant waveplate. Therefore, in order to change topological-charge value, replacement of the waveplate and realignment of the amplifier are required, which implies the low flexibility of the experimental setup.

The computer-generated hologram (CGH) [2] is widely used for the generation of optical vortices, enabling us to generate light with arbitrary topological charges, even as a charge-mixed state. However, it is not suitable for ultrashort (or ultrabroadband) input pulses because its essential effect of diffraction causes angular dispersion. In earlier study [3], although a 4-f setup was proposed to compensate for the angular dispersion, the total throughput is quite low (less than several percents) owing to the low diffraction efficiency (~ 10–~ 20 %) of the CGH.

In the present paper, we report the generation of mJ-class intense ultrashort optical-vortex pulses by Ti:sapphire-based chirped-pulse amplification of the seeding optical vortices generated by the 4-f vortex converter with a CGH. Low throughput of the 4-f setup can be easily recovered in the amplification process, thus our configuration enables us to obtain intense ultrashort optical-vortex pulses with programmable topological-charge control.

2 Experiments and Results

Our system consists of two Ti:sapphire-based amplifiers and a 4-f optical-vortex converter located between them. The output from a Ti:sapphire laser oscillator (repetition rate: 80 MHz, 730–880 nm, averaged power: ~ 500 mW) was temporally-stretched by a grating-based pulse stretcher and amplified by a home-built 1st-stage Ti:sapphire regenerative laser amplifier (repetition rate: 1 kHz). A 2-μm-thick etalon filter [4] was utilized to compensate for the so-called gain narrowing effect due to bandwidth limitations of a Pockels cell, polarizers and other components. The output from the regenerative amplifier was attenuated to ~ 70 μJ/pulse, which is below the damage threshold of a spatial light modulator (SLM, Hamamatsu photonics X10468, 792 × 600 pixels, design wavelength range: 700–900 nm) in our 4-f optical-vortex converter.

The primary reason why optical-vortex conversion was done after regenerative amplification is that the amplification process is greatly affected by the dominant transverse eigenmodes of the amplifier cavity itself, thus it is difficult to well amplify the seeding vortices. Moreover, if we amplify very weak nJ-level optical-vortex pulses directly in the amplifier, non-negligible amplified spontaneous emission (ASE) might arise in the beam center by the spatially-Gaussian-shaped pump beam, which is quite different from the doughnut-like beam shape of the seeding optical vortex. This undesirable ASE can be greatly decreased by optical-vortex conversion after pre-amplification. After passing through the optical-vortex converter, the generated optical vortices were amplified again by a 2nd-stage 4-pass amplifier, and the frequency-chirp was compensated for by a grating-based pulse

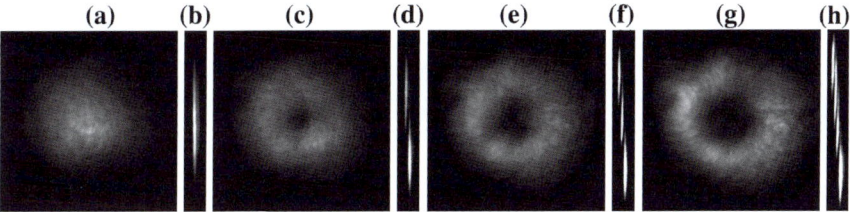

Fig. 1 Measured beam profiles of the amplified optical-vortex pulses with topological charges of **a** $m = 0$, **c** $m = 1$, **e** $m = 2$ and **g** $m = 3$, together with the corresponding interferograms by astigmatic transformation in the cases of **b** $m = 0$, **d** $m = 1$, **f** $m = 2$ and **h** $m = 3$, respectively

compressor. This 2-stage amplification enables us to generate intense broadband optical-vortex pulses with pulse energy more than damage threshold of the optical-vortex converter with the SLM.

The beam profiles of the amplified optical vortices and the corresponding interferograms by the astigmatic transformation [5] are shown in Fig. 1. Seen clearly, the topological charges of the output were well-controlled with topological-charge flexibility by changing the CGH patterns, and the amplified vortices have the common center point in all cases. The pulse energies after frequency-chirp compensation were 1.1, 1.3, 1.2 and 1.0 mJ for topological charges of $m = 0$, 1, 2, and 3, respectively. The typical temporal profile of the amplified optical-vortex pulses ($m = 0$–3) were characterized by a two-dimensional shearing interferometry apparatus (Fig. 2a). As shown in Fig. 2b, the pulse duration was evaluated to be 27 fs (the corresponding Fourier-transform limited pulses: 24 fs), which is close to that of one of the typical shortest pulses directly from Ti:sapphire amplifiers.

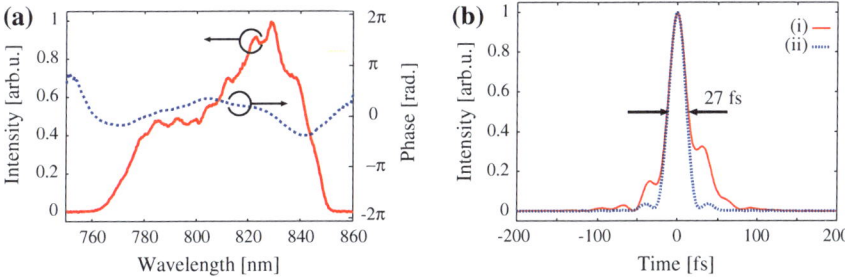

Fig. 2 **a** Measured spectrum and reconstructed spectral phase of the amplified optical-vortex pulses ($m = 0$). **b** Temporal profiles of (i) the reconstructed (27 fs, $m = 0$) and (ii) the corresponding Fourier-transform-limited optical pulses (24 fs)

References

1. K. Yamane, T. Toda and R. Morita, "Ultrashort optical-vortex pulse generation in few-cycle regime," Opt. Express **20**, 18986–18993 (2012).
2. A. Mair, A. Vaziri, G. Weihs and A. Zeilinger, "Entanglement of the orbital angular momentum states of photons," Nature **412**, 313–316 (2001).
3. I. Zeylikovich, H. I. Sztul, V. Kartazaev, T. Le and R. R. Alfano, "Ultrashort Laguerre-Gaussian pulses with angular and group velocity dispersion compensation," Opt. Lett. **32**, 2025–2027 (2007).
4. C. P. J. Barty, G. Korn, F. Raksi, C. Rose-Petruck, J. Squier, A.-C. Tien, K. R. Wilson, V. V. Yakovlev and K. Yamakawa, "Regenerative pulse shaping and amplification of ultrabroadband optical pulses," Opt. Lett. **21**, 219–221 (1996).
5. V. Denisenko, V. Shvedov, A. S. Desyatnikov, D. N. Neshev, W. Krolikowski, A. Volyar, M. Soskin and Y. S. Kivshar, "Determination of topological charges of polychromatic optical vortices," Opt. Express **17**, 23374–23379 (2009).

The Influence of Generalized Focusing on Polarization-Shaped Few-Cycle Pulsed Beams

Balázs Major, Miguel A. Porras, Attila P. Kovács
and Zoltán L. Horváth

Abstract In this work we show how generalized focusing affects the polarization-state of polarization-shaped pulsed beams being only a few cycles long. We present simulations to exemplify how this phenomenon can cause the rotation of the so-called 'polarization gating' pulse commonly used for isolated attosecond pulse generation. The general validity of the recently published simple rules for these diffraction-caused changes is also demonstrated.

1 Introduction

It is well-known that the polarization state of the statistically-stationary radiation of partially coherent or partially polarized electromagnetic sources can change on free-space propagation [1, 2]. On the other hand, for deterministic, fully coherent and fully polarized pulsed beams used in experiments nowadays [3–5], such changes are only expected for beams having inhomogeneous polarization [6], or being tightly focused [6, 7]. However, it has been recently shown that focusing affects the polarization state of such pulsed beams, if they are only few cycles long [8, 9]. Here we summarize some properties of this phenomenon, and present an example to show how these findings can affect the polarization state of few-cycle pulses used in the various fields of science.

B. Major (✉) · A.P. Kovács · Z.L. Horváth
Department of Optics and Quantum Electronics, University of Szeged, Dóm tér 9,
Szeged 6720, Hungary
e-mail: bmajor@titan.physx.u-szeged.hu

M.A. Porras
Departamento de Física Aplicada a los Recursos Naturales, Grupo de Sistemas Complejos,
Universidad Politécnica de Madrid, Rios Rosas 21, 28003 Madrid, Spain

© Springer International Publishing Switzerland 2015 813
K. Yamanouchi et al. (eds.), *Ultrafast Phenomena XIX*,
Springer Proceedings in Physics 162, DOI 10.1007/978-3-319-13242-6_199

2 Focusing-Induced Polarization-State Changes

Similarly to [8, 9], we describe the polarization state of a polarization-shaped few-cycle pulse by its instantaneous polarization ellipse. In this description, at every moment of time a polarization ellipse is associated with the electric field in the same way as in the case of monochromatic waves, and the polarization state is primarily defined by the orientation $\psi(t)$ and ellipticity $\chi(t)$ of the instantaneous ellipse. Following the calculation given in [9], based on a general model of ideal focusing with only two perpendicular components of the electric field propagating independently, it can be shown that the instantaneous properties of the polarization state change according to the following approximate rules while the pulse having ω_0 carrier angular frequency propagates from the focusing element to the focal point (in this model the same as propagating to the far field):

$$\psi^{(f)}(\tau) - \psi(\tau) \simeq \frac{1}{\omega_0} \frac{1}{1 - \tan^2 \chi(\tau)} \frac{d \tan \chi(\tau)}{d\tau}, \tag{1}$$

$$\tan \chi^{(f)}(\tau) - \tan \chi(\tau) \simeq \frac{1}{\omega_0} [1 - \tan^2 \chi(\tau)] \frac{d \psi(\tau)}{d\tau}, \tag{2}$$

where τ is the local time, which takes into account the time needed for the pulse to propagate from the source to the point of interest [9]. The expressions above show that the time-variation of the instantaneous ellipticity (orientation) induces a change in the instantaneous orientation (ellipticity) during focusing. This effect is demonstrated to be caused by diffraction, that is by the finite size of the source [9], or in other words by the Wolf effect [10]. As the temporal variation of $\psi(t)$ or $\chi(t)$ can only originate from the fast change of the phase difference or the rapid variation of the envelope ratio of the orthogonal components [8], this effect is more pronounced for few-cycle pulses, and disappears for monochromatic waves.

3 An Example: The Polarization Gating Pulse

In [9] the polarization gating (PG) pulse was an example to demonstrate (1) independently from (2). In this paper we expand this simple model of the PG pulse by taking into account the pulse chirp. Figure 1a depicts the temporal evolution of the electric field in the case of the chirped PG pulse. For better visualization of the polarization state, the Stokes parameters of the instantaneous polarization ellipse are plotted in Fig. 1b in the case of the initial and focused pulses. In this figure it is easy to notice that the primary effect of focusing is a constant rotation of the instantaneous ellipse, due to its time varying ellipticity (see (1)). However, it is important to note that due to the temporal chirp, the initial orientation is also time-dependent (linearly), but the effect in this case for this parameter is almost negligible. Also note that in

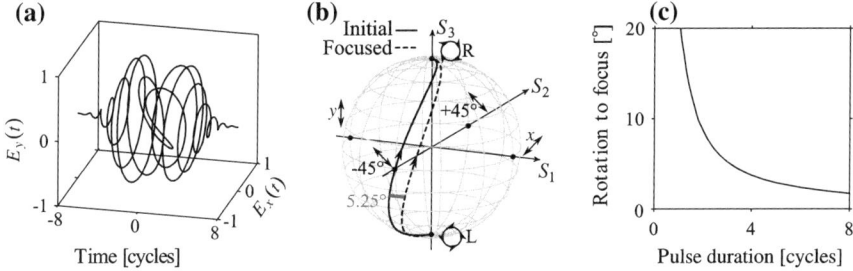

Fig. 1 a The 800 nm chirped polarization gating pulse acquired by sending a three-cycle pulse (in intensity-FWHM ΔT) through two quarter-wave plates (as described in [4]). The quadratic term of temporal phase is: $\exp(i0.3t^2/\Delta T^2)$. Based on the simple model of the PG pulse in [9]. **b** Representation of the instantaneous polarization ellipse of the polarization gating pulse on the Poincaré sphere. The *continuous curve* represents the initial, the *dashed curve* the focused pulse. The *arrows* indicate the time evolution. **c** Scaling of the rotation effect with pulse duration

case of zero chirp, the orientation would be time-independent, and in Fig. 1b the curves would go along the lines of longitude of the Poincaré sphere. In Fig. 1c it is shown how this effect, which is independent of focal length or beam size, disappears for longer and is enhanced for close to single-cycle pulses.

4 Conclusions

We have shown that focusing in general affects the polarization state of polarization-shaped few-cycle electromagnetic pulsed beams, and these changes are caused by diffraction. We demonstrated our findings with the chirped polarization gating pulse.

Acknowledgement The project was partially funded by TÁMOP-4.2.2.A-11/1/KONV-2012-0060 "Impulse lasers for use in materials science and biophotonics", supported by the European Union and co-financed by the European Social Fund. M. A. P. acknowledges financial support from Project MTM2012-39101-C02-01 of Ministerio de Economía y Competitividad of Spain.

References

1. D.F.V. James, J. Opt. Soc. Am. A **11**(5), 1641 (1994).
2. O. Korotkova, E. Wolf, Opt. Commun. **246**(1–3), 35 (2005).
3. F. Weise, S. Weber, M. Plewicki, A. Lindinger, Chem. Phys. **332**(2-3), 313 (2007).
4. I.J. Sola, A. Zaïr, R. López-Martens, P. Johnsson, K. Varjú, E. Cormier, J. Mauritsson, A. L'Huillier, V. Strelkov, E. Mével, E. Constant, Phys. Rev. A: At., Mol., Opt. Phys. **74**(1), 013810 (2006).

5. C. Ruchert, C. Vicario, C.P. Hauri, Phys. Rev. Lett. **110**(12), 123902 (2013).
6. L.E. Helseth, Phys. Rev. E: Stat., Nonlinear, Soft Matter Phys. **72**(4), 047602 (2005).
7. C. Spindler, W. Pfeiffer, T. Brixner, Appl. Phys. B: Lasers Opt. **89**(4), 553 (2007).
8. M.A. Porras, J. Opt. Soc. Am. B **30**(6), 1652 (2013).
9. B. Major, M.A. Porras, Z.L. Horváth, J. Opt. Soc. Am. A **31**(6), 1200 (2014).
10. Z. Dačić, E. Wolf, J. Opt. Soc. Am. A **5**(7), 1118 (1988).

Simultaneous Selective Two-Photon Microscopy Using MHz Rate Pulse Shaping and Quadrature Detection of the Time-Multiplexed Signal

Ilyas Saytashev, Bingwei Xu, Marshall T. Bremer and Marcos Dantus

Abstract We demonstrate a method for simultaneous fast selective two-photon excited fluorescence (TPEF) microscopy imaging of two different fluorophores using quadrature detection of the signal from a single PMT detector.

1 Introduction

The broad bandwidth and high peak intensity of femtosecond lasers enable the two-photon excitation of many different fluorophores with high efficiency using a single laser setup. Selective excitation of particular fluorophores can be achieved by a computer controlled pulse shaper taking advantage of multiphoton intrapulse interference (MII) [1–4]. Phase shaping, in particular, can maintain the high efficiency, but excite a particular frequency within the two-photon bandwidth with high spectral resolution. Unfortunately, pulse shapers appropriate for microscopy have slow refresh rates which limit the speed with which multiple fluorophores can be measured with a single detector. Here we achieve 162 MHz rate pulse shaping on a near octave spanning laser using passive components. With a single photo-multiplier tube (PMT) detector, we are able to selectively image two fluorophores with excitation separated by 50 nm.

I. Saytashev · M. Dantus (✉)
Department of Chemistry, Michigan State University, East Lansing, MI 48824, USA
e-mail: dantus@msu.edu

B. Xu · M. Dantus
Biophotonic Solutions Inc, East Lansing, MI 48823, USA

M.T. Bremer · M. Dantus
Department of Physics and Astronomy, Michigan State University, East Lansing,
MI 48824, USA

© Springer International Publishing Switzerland 2015   817
K. Yamanouchi et al. (eds.), *Ultrafast Phenomena XIX*,
Springer Proceedings in Physics 162, DOI 10.1007/978-3-319-13242-6_200

2 Experimental

An ultra-broad bandwidth oscillator and pulse shaper (femtoAdaptiv, BioPhotonic Solutions Inc.) with central wavelength at 812 nm (Fig. 1) is capable of exciting two-photon transitions in the 380–500 nm range. Selective excitation in the blue and red portions of the spectrum is achieved by a combination of second order dispersion (SOD) and third order dispersion (TOD) [2, 3]. As it is shown in Fig. 2a, the output from the laser was split into two arms with different second order dispersion (SOD). The recombined beams create a train of pulses with spectral phase switching at 162 MHz rate. Each pulse induces selective TPEF on the sample at wavelengths determined by the amount of SOD and TOD in the beam [6].

Fluorescence is detected by a single photomultiplier tube (PMT) detector. Signal from the PMT detector contains fluorescence signals from two different selectively-excited fluorophores. The two separate signals are isolated by quadrature detection using a lock-in amplifier as shown in Fig. 2b. Images are obtained from two different fluorophores simultaneously. This allows fast frame-rate imaging.

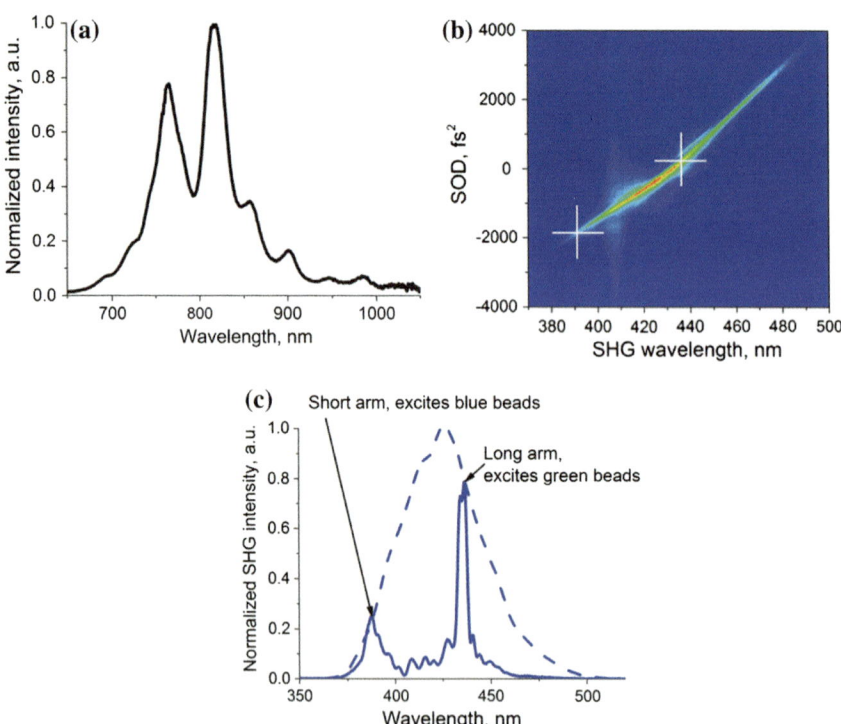

Fig. 1 **a** Emission spectrum of laser oscillator; **b** MIIPS scan [5] of the laser pulse without compensation mask applied. **c** SHG spectrum of laser pulse with applied phase compensation (*dashed*), pre-chirped and delayed in different arms (*solid*). All measurements made at the focal plane of a Zeiss LD C-APOCHROMAT 40×/1.1 objective

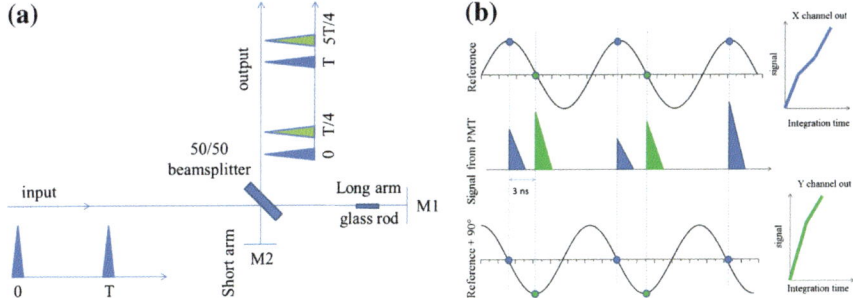

Fig. 2 **a** Schematic representation of excitation beam quarter-period delay line with different amount of SOD in arms. **b** Simultaneous acquisition of signals from two different fluorophores at single photo-detector by quadrature detection using lock-in amplifier

3 Results and Discussion

Excitation laser pulses with different shapes (typical energy ~ 110 pJ per pulse at the focus of the objective) are temporally delayed by ~ 3 ns. The microscopy image (shown in Fig. 3) provides selective signal from 10 μm blue microspheres when detecting signal from the X channel of the lock-in amplifier. The image from the 6 μm beads is obtained when detecting signal from the Y channel of the lock-in amplifier (with 90° phase difference). The combination of both signals provides the full image.

Wide tunability of the two-photon excitation wavelength, fast switching rate between the selective excitation, and low photodamage (due to the higher peak intensity of shorter pulses [7]) enables application of this method for in vivo dynamic imaging in biological samples.

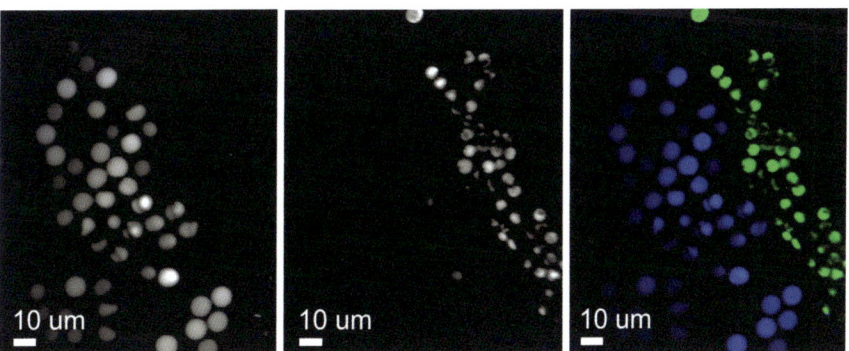

Fig. 3 Microscopy images of fluorescent beads, using the signal from X channel (*left image*) and Y channel (*center image*) of the lock-in amplifier. False colored image obtained by adding the X and Y channel signals (*right image*)

References

1. K. A. Walowicz et. al, "Multiphoton intrapulse interference 1; Control of multiphoton processes in condensed phases." J. Phys. Chem. A **106**, 9369-9373, (2002).
2. V. V. Lozovoy et. al, "Multiphoton intrapulse interference II. Control of two- and three-photon laser induced fluorescence with shaped pulses." J. Chem. Phys. **118**, 3187-3196 (2003).
3. I. Pastirk et. al, "Selective two-photon microscopy with shaped femtosecond pulses." Optics Express **11**, 1695-1701 (2003).
4. M. Dantus and V.V. Lozovoy, "Experimental Coherent Laser Control of Physicochemical Processes." Chem. Reviews **104**, 1813-1860 (2004).
5. Y. Coello et. al, "Interference without an interferometer: a different approach to measuring, compressing, and shaping ultrashort laser pulses," J. Opt. Soc. Am. B **25**, A140-A150 (2008).
6. G. Labroille et. al, "Dispersion-based pulse shaping for multiplexed two-photon fluorescence microscopy", Opt. Letters **35**, 3444 (2010).
7. P. Xi et. al, "Greater Signal and Less Photobleaching in Two-Photon Microscopy with Ultrabroad Bandwidth Femtosecond Pulses," Opt. Commun. **281**, 1841-1849 (2008).

Simultaneous Spatial and Temporal Focusing of Femtosecond Laser Pulses for Directly Writing Optical Waveguides in Pr^{3+} Doped ZBLAN Glass

Yusuke Yamanaka, Kenichi Hirosawa and Fumihiko Kannari

Abstract Characteristics of clad-drawn optical waveguides in a Pr^{3+} doped ZrF_4-BaF_2-LaF_3-AlF_3-NaF (ZBLAN) glass using simultaneous spatial and temporal focusing (SSTF) of femtosecond laser pulses are described. Thanks to the SSTF, the axial two-photon excitation length is confined to $\sim 20\ \mu m$. At a laser repetition rate of 1 kHz, chirped longer laser pulses always induce larger refractive index changes than those by a 40-fs laser pulse. An optical waveguide exhibiting an NA of ~ 0.04 is fabricated.

1 Introduction

Direct writing of optical waveguides into bulk materials using femtosecond laser pulses has been attracting much attention in various applications [1, 2]. The type of modification depends on the material composition and laser pulse parameters. Using positive change of refractive index, one can fabricate a circular-cross-section core of a waveguide. However, the direct laser drawing of a core may cause degradation of its optical property. On the other hand, using negative change of refractive index, the modified volume can be used as a clad of a waveguide. ZrF_4-BaF_2-LaF_3-AlF_3-NaF (ZBLAN) glass shows negative change in refractive index when low repetition rates ($< \sim 10$ kHz) femtosecond laser pulses are focused [3]. The ZBLAN glass is one of promising hosts of trivalent praseodymium (Pr^{3+}) that exhibits efficient stimulated emission in the visible region.

In this paper, we present fabrication of a Pr^{3+} doped ZBLAN (Pr:ZBLAN) waveguide using low repetition rate (1 kHz) femtosecond laser toward a diode-pumped visible waveguide laser. For achieve minute modification sizes, we employed a simultaneous spatial and temporal focusing (SSTF) scheme. SSTF has

Y. Yamanaka · K. Hirosawa · F. Kannari (✉)
Department of Electronics and Electrical Engineering, Keio University,
3-14-1, Hiyoshi, Kohoku-ku, Yokohama 223-8522, Japan
e-mail: kannari@elec.keio.ac.jp

© Springer International Publishing Switzerland 2015
K. Yamanouchi et al. (eds.), *Ultrafast Phenomena XIX*,
Springer Proceedings in Physics 162, DOI 10.1007/978-3-319-13242-6_201

been used to form a frequency-distributed array of low numerical aperture (NA) beam-lets. Only at a temporal focus, all of the frequencies overlap spatially and temporally, and the pulse is compressed to a Fourier transform-limit in time as well as diffraction limit in space. One of significant advantages of the SSTF scheme is that the intensity of input pulse is suppressed out of the temporal focus. Therefore, two-photon excitation can be limited in a small volume near the temporal focus. The SSTF scheme brings better quality and higher flexibility in waveguide drawing.

2 Experiments and Results

Our experimental setup for processing waveguide with SSTF is shown in Fig. 1a. Femtosecond laser pulses are angularly dispersed by the 600 line/mm diffraction grating, and are collimated by the $f = 500$ mm cylindrical lens. The dispersed light is temporally compressed only at the focal plane of the $f = 30$ mm focusing lens. In the monochromatic perspective, the beam passes through two cylindrical lenses of which separation is set as $2f$ (=1,000 mm), therefore the beam diameter does not change at the focusing lens. The femtosecond laser pulses (800 nm, 50 fs (FWHM)) generated from a Ti:Sapphire regenerative amplifier were used. To fabricate a clad structure, we mounted a $6 \times 10 \times 30$ mm^3 Pr:ZBLAN bulk glass on a 3-axis motorized stage and moved the stage in double track pattern as illustrated in Fig. 1b. We succeeded to suppress the axial processing length down to ~ 20 μm by the SSTF. However, the refractive index change was too small to quantitatively measure.

For inducing larger refractive index change, we stretched the pulse duration from 50 fs to 3.5 ps by adding a second-order dispersion. In the following experiment, we used only spatial focusing with 20X objective lens ($f = 9$ mm, NA $= 0.4$) because of just simplicity in the optical layout. Processing with the chirped longer

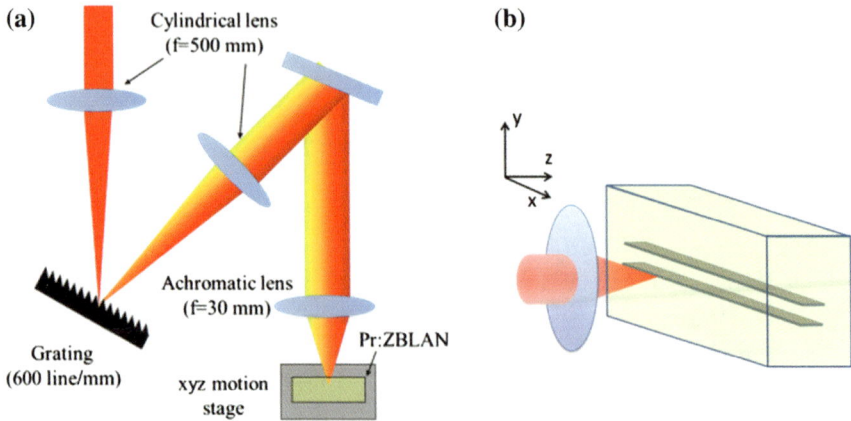

Fig. 1 **a** Experimental setup of waveguide fabrication. **b** Fabrication scheme of waveguide clad

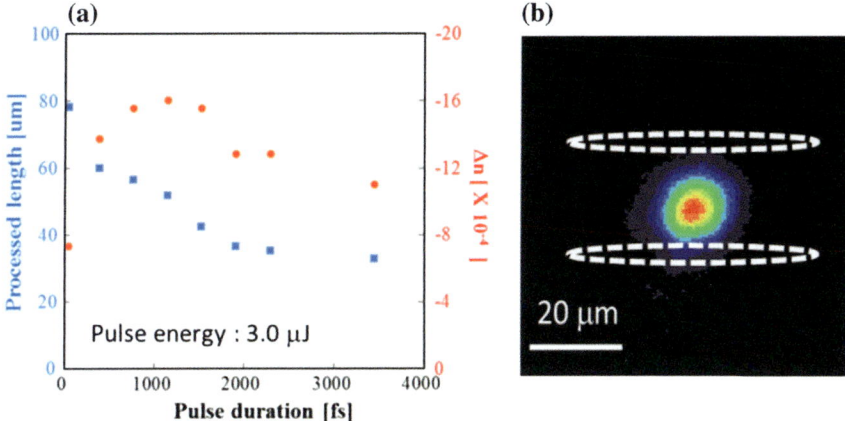

Fig. 2 **a** Dependence of refractive index change and processed length on the pulse duration of input pulse. **b** Near-field image of a fabricated waveguide at 630 nm

pulse induces shorter processed depth lengths and larger refractive index change than those achieved by short pulse processing (Fig. 2a). In our experiment, maximum refractive index change was -1.6×10^{-3}, when the pulse duration was 1.15 ps, whereas the refractive index change was less than half of that with a 50 fs laser pulses. With an incident laser pulse energy of 3.0 μJ and a pulse duration of 1.15 ps, we fabricate two parallel single modified tracks separated by 20 μm at a velocity of 100 μm/s. The fabricated waveguide length was 29.5 mm. Figure 2b shows the near-field image of the guided mode of the fabricated waveguide at 630 nm. The modified region is illustrated as a white dashed line in Fig. 2b. Good waveguiding was observed in the area between two tracks and the NA of waveguide is calculated as ∼0.04 from a far-field image.

3 Conclusion

We fabricated optical waveguides in a Pr:ZBLAN glass by direct clad drawing using femtosecond laser pulses. The SSTF enables to shorten the axial processing length, and chirped longer laser pulses cause larger refractive index change and shorter processed depth lengths. The fabricated waveguide shows NA of ∼0.04. We believe that this technology supports to create complex optical waveguide circuits in functional materials.

Acknowledgments This research was supported by a Grant-in-aid from the Ministry of Education, Culture, Sports, Science, and Technology, Japan for the Photon Frontier Network Program and by JSPS KAKENHI Grant Number 22656020

References

1. R. R. Gattass and E. Mazur, "Femtosecond laser micromachining in transparent materials," Nature Photon. **2**, 219-225 (2008).
2. G. Della Valle, R. Osellame, and P. Laporta, "Micromachining of photonic devices by femtosecond laser pulses," J. Opt. A 11, 013001 (2009).
3. J. P. Bérubé, M. Bernier, and R. Vallée, "Femtosecond laser-induced refractive index modifications in fluoride glass," Opt. Mater. Express **3**, 598-611 (2013).

Printed by Printforce, the Netherlands